AF249108

Advances in the
Applications of Nonstandard
Finite Difference Schemes

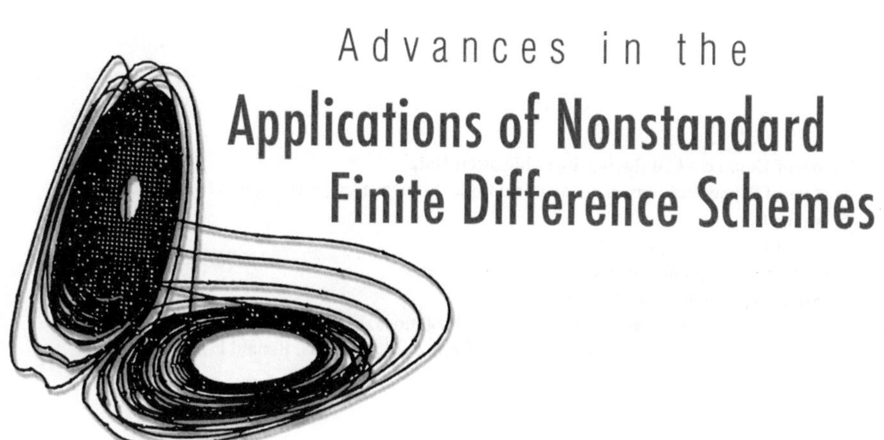

editor

Ronald E. Mickens

Clark Atlanta University, USA

World Scientific

NEW JERSEY · LONDON · SINGAPORE · BEIJING · SHANGHAI · HONG KONG · TAIPEI · CHENNAI

Published by

World Scientific Publishing Co. Pte. Ltd.

5 Toh Tuck Link, Singapore 596224

USA office: 27 Warren Street, Suite 401-402, Hackensack, NJ 07601

UK office: 57 Shelton Street, Covent Garden, London WC2H 9HE

Library of Congress Cataloging-in-Publication Data
Advances in the applications of nonstandard finite difference schemes / edited by Ronald
  E. Mickens.
    p. cm.
  Includes bibliographical references.
  ISBN 981-256-404-7 (alk. paper)
    1. Differential equations, Partial--Numerical solutions. 2. Differential
equations--Numerical solutions. 3. Finite differences. I. Mickens, Ronald E., 1943–

QA374.A256 2005
518'.64--dc22

               2005051429

British Library Cataloguing-in-Publication Data
A catalogue record for this book is available from the British Library.

Printed by FuIsland Offset Printing (S) Pte Ltd, Singapore

PREFACE

Since the publication of my edited volume[a], on nonstandard finite difference (NSFD) methods for the numerical integration of differential equations, a number of new results and applications of these techniques have been obtained. The field of constructing NSFD schemes had its genesis in a paper published in 1989.[b] In recent years, progress has occurred in both the general foundational basis and range of physical/engineering phenomena for which such schemes have been applied. The development of this new edited volume indicates the interest and value of this topic by an increasing number of researchers. An additional, but significant, aspect of this work is that the mathematical basis for NSFD methods is beginning to be studied. These efforts give insight into why such techniques "work" and allow the possibility for their generalization.

This book may be considered an update and extension of the previous edited volume. It consists of fourteen chapters arranged by alphabetical ordering based on the names of the first author for each individual chapter. Our editorial work consisted of having each chapter reviewed for technical correctness and checking for minor errors resulting from typos, misspellings, etc. We have not attempted to change any of the written text into standard American English, but, have pragmatically allowed the use of "international scientific English."

This book contains fourteen chapters, each written by researchers who have applied NSFD methods to investigate particular systems in the natural and/or engineering sciences. The first chapter provides a very brief overview of the fundamental rules for constructing NSFD schemes and a definition

---

[a]R. E. Mickens (editor), *Application of Nonstandard Finite Difference Schemes* (World Scientific, Singapore, 2000).

[b]R. E. Mickens, "Exact solutions to a finite difference model of a reaction-advection equation: Implications for numerical analysis," *Numerical Methods for Partial Differential Equations* **5** (1989), 313–325.

of dynamic consistency (DC). This principle, along with the concept of positivity preserving schemes, plays an extremely important role in the formulation of NSFD schemes for several classes of ordinary and partial differential equations. The chapter concludes with a partial listing of several outstanding problems that need to be resolved within the framework of NSFD methods. The other thirteen chapters consider a wide range of topics:

- simulation of robotic systems
- boundary value problems for Bratu-Gel'Fand and related problems
- computational electromagnetics
- nonlinear micro heat transport
- single and multi-species interacting populations
- non-smooth mechanical systems
- asymptotic consistency in discrete models arising in population biology
- SI, SIS, and multi-population competition models
- robust discretizations and time step-size behavior for chaotic systems
- contributions to the theory of NSFD methods
- singular perturbation problems
- frequency accurate finite difference methods
- Lotka-Volterra systems

The preface to the previous (2000) edited volume contains the following paragraph: "As a contributor to and editor of this volume, I look forward to both personally extending the current knowledge of nonstandard schemes and for advances that will come from the efforts of others. While these schemes may not presently resolve all of the difficulties involved with finite difference models of differential equations, their use clearly gives in many cases much better discrete models than ones obtained using standard methods. My general view is that nonstandard schemes have an exciting future and will provide exciting opportunities for new results in pure mathematics and improved numerical solutions of differential equations." The contributions in this current volume show these sentiments to be true.

Finally, I want to express appreciation to my many colleagues for both their interest in NSFD methods and for the many collaborations that have arisen. As always, I am particularly grateful to Annette Rohrs for her editorial work and related activities that led to the smooth integration of the various manuscripts into a document from which this volume was con-

structed. Clearly, without her efforts, this publishing project could not have
been completed. For support of my research on NSFD methods, during the
period 1988–2005, I wish to thank the following agencies and programs for
funds: Army Research Office, Department of Energy, the MBRS-SCORE
Program at Clark Atlanta University, and NASA.

Ronald E. Mickens
Atlanta, Georgia USA
June 2005

CONTENTS

6 Reliable Finite Difference Schemes with Applications in Mathematical Ecology 249
D. T. Dimitrov, H. V. Kojouharov and B. M. Chen-Charpentier

7 Applications of the Non-Standard Finite Difference Method in Non-Smooth Mechanics 287
Y. Dumont

8 Finite Difference Schemes on Unbounded Domains 343
M. Ehrhardt

CHAPTER 1

NONSTANDARD FINITE DIFFERENCE METHODS

Ronald E. Mickens

Clark Atlanta University
Box 172 – Physics Department
Atlanta, GA 30314, USA
rohrs@math.gatech.edu

A brief history of nonstandard finite difference (NSFD) methods is given along with clarifying remarks related to the views of some others on these techniques. We also present the basic rules governing the formulation of NSFD schemes and discuss the concept of dynamic consistency and its role in the construction of such schemes. A list of some outstanding problems is given to indicate possible future research directions for the continued investigation of NSFD for the purpose of numerical integration of differential equations.

1. General Comments

Nonstandard finite difference (NSFD) methods for the numerical integration of differential equations had their genesis in a paper published in 1989 [1]. The basic rules to construct such schemes [2] and their application to specific nonlinear equations appear in a variety of publications [3,4]. In recent years, NSFD discrete models have been constructed and/or tested for a wide range of nonlinear dynamical systems:

- singular boundary value problems expressed in cylindrical or spherical coordinates [5],
- a generalized Nagumo reaction-diffusion model [6],
- equations modeling stellar structure [7],
- the dynamics of HIV transmission [8],
- modified linear heat/diffusion transport problems [9].

1

An essentially complete listing and summary of publications using NSFD methods, up to 2004, is presented in the paper by Patidar [10]. This paper [10] and other published works [3]–[9] provide ample evidence that NSFD schemes are enjoying a growing applicability as the practical users of numerical techniques for differential equations become aware of the advantages and power of these methods. However, there is the view, held by some individuals [11,12], that the NSFD method depends on "using the known solution of the differential equation or by 'ad hoc' experimentation ..." [12]. A detailed study and examination of the results produced to date easily show this interpretation of what has been accomplished using NSFD methods to be false.

An essential issue, coming from both my work and of others on NSFD methods, is the realization that each differential equation has to be considered a "unique" mathematical structure and, consequently, must be discretely modeled in a unique manner. This is a very important aspect of NSFD methods and the contributors to this book illustrate this point in their presented works.

In the next section, we give a brief outline of the basic rules defining the fundamental techniques that generate NSFD schemes for differential equations. This is followed, in Section 3, by a similar brief discussion of the principle of dynamic consistency. Since excellent and extensive published presentations already exist on these two topics [2,3,13,14], there is no need to provide full information here on this topic.

2. NSFD Methods: Basic Principles

Detailed studies of so-called exact finite difference schemes [2] form the foundation of NSFD methods [1,2,15]. The extension and generalization of these results to special groups of differential equations for which exact schemes are not available has also provided additional insight into the required structural properties of NSFD methods [16]. Based on this work, the following rules for constructing nonstandard schemes follow:

Rule 1. The orders of the discrete derivatives should be equal to the orders of the corresponding derivatives appearing in the differential equations.

Comment 1. If the orders of the discrete derivatives are larger than those occurring in the differential equations, then numerical instabilities will in general occur [2,17].

Rule 2. Discrete representations for derivatives must, in general, have nontrivial denominator functions.

Comment 2. Consider the first-order derivative of $x(t)$ and its discrete analog; its most general form is

$$\frac{dx}{dt} \rightarrow \frac{x_{k+1} - \psi(h)x_k}{\phi(h)}, \tag{1}$$

where ψ and ϕ are functions of the step-size, $h = \Delta t$, and are called, respectively, the "numerator" and "denominator" functions; $t_k = hk$, and $x(t) \rightarrow x_k$; and where the (ψ, ϕ) have the properties

$$\psi(h) = 1 + O(h), \qquad \phi(h) = h + O(h^2). \tag{2}$$

Note that conventional discrete representation for the first derivative take $\psi(h) = 1$ and $\phi(h) = h$ [17]. For systems of coupled, first-order, ordinary differential equations there exists a systematic method for constructing denominator functions [2,16]. Also, unless the "system" has dissipation, the numerator function is usually equal to one [2].

Rule 3. Nonlinear terms should, in general, be replaced by nonlocal discrete representations.

Comment 3. The simplest illustration of this requirement is the logistic equation, i.e.,

$$\frac{dx}{dt} = x(1 - x). \tag{3}$$

A straightforward standard discretization gives

$$\frac{x_{k+1} - x_k}{h} = x_k(1 - x_k). \tag{4}$$

This equation can be transformed into the logistic difference equation [2]

$$z_{k+1} = \lambda z_k(1 - z_k), \qquad \lambda = 1 + h \tag{5}$$

which has periodic and chaotic solutions, and thus cannot be a valid discrete model for Eq. (3). However, the use of a nonlocal representation for x^2, i.e.,

$$x^2 \rightarrow x_{k+1}x_k, \tag{6}$$

gives

$$\frac{x_{k+1} - x_k}{h} = x_k(1 - x_{k+1}), \tag{7}$$

or

$$x_{k+1} = \left(\frac{1 + h}{1 + hx_k}\right) x_k. \tag{8}$$

This discrete model gives numerical solutions for Eq. (3) that are qualitatively correct for $x_0 > 0$ and all $h > 0$; see Mickens [2].

Rule 4. Special conditions that hold for either the differential equation and/or its solutions should also hold for the difference equation model and/or its solutions.

Comment 4. An example is an ordinary differential equation for which the substitution, $t \to -t$, leaves the equation invariant. If the discrete model does not also have this property, then numerical instabilities may occur.

Comment 5. In general, numerical instabilities are solutions to the discrete equations that do not correspond to any solution of the corresponding differential equations.

Definition. A nonstandard finite difference scheme is any discrete representation of a system of differential equations that is constructed based on the above rules.

3. Dynamic Consistency

Dynamic consistency (DC) is an important concept that can be used to guide the construction of valid discrete models for differential equations. DC is always formulated in terms of particular properties of a system [13,18]. Below, we define DC in terms of ordinary differential equations (DOE); however, it should be clear that the definition can be easily extended to partial differential equations (PDE).

Definition

Consider the differential equation

$$\frac{dx}{dt} = f(x, t, \lambda), \tag{9}$$

where λ represents the parameters defining the system modeled by Eq. (9). Let a finite difference scheme for Eq. (9) be

$$x_{k+1} = F(x_k, t_k, h, \lambda). \tag{10}$$

Let the differential equation and/or its solutions have property P. The discrete model, Eq. (10), is dynamically consistent with Eq. (9), if it and/or its solutions also has property P.

Comment 6. For many systems in the natural and engineering sciences, properties of particular importance include [19,20,21]:

- positivity
- boundedness
- monotonicity
- fixed-points and their stability properties
- integer valued dependent variables

- existence of special solutions (traveling waves, solitons, rational, etc.)
- limit cycles and other periodic solutions.

A given system might include one or more of these properties or others not listed. The critical issue is that a valid finite difference model will not exist unless all of the essential properties of the original differential equations are incorporated into it. From this viewpoint, it follows that the existence of numerical instabilities is an indication that some "physical principle" underlying the dynamics of the original system is not included in the discrete model.

Comment 7. Let, for the same initial conditions, $x(t)$ and x_k be the solutions, respectively, to Eqs. (9) and (10). If for any time step-size, $h > 0$, we have

$$x(t_k) = x_k, \tag{11}$$

then Eq. (10) is an exact difference scheme for Eq. (9) [2].

Comment 8. Note that NSFD rule 4, given in the previous section, incorporates the principle of DC.

To illustrate the use of DC in the construction of NSFD schemes, consider the decay equation

$$\frac{dx}{dt} = -\lambda x, \qquad x(0) = x_0, \tag{12}$$

where λ is a positive parameter. Applying the qualitative theory of differential equations [22], it can be directly shown, even without knowledge of the explicit solution, that solutions to Eq. (12) have the following properties:

P_1 – $x(t) = 0$ is a fixed-point;

P_2 – given $x_0 \neq 0$, then $x_0 x(t) > 0$ for $t > 0$;

P_3 – $x(t)$ monotonically decreases in magnitude to zero for any $x_0 \neq 0$.

The standard forward-Euler scheme

$$\frac{x_{k+1} - x_k}{h} = -\lambda x_k \quad \text{or} \quad x_{k+1} = (1 - \lambda h)x_k, \tag{13}$$

violates P_2 and P_3 if λh is sufficiently large. Consequently, this scheme is not DC with Eq. (12). However, the NSFD scheme

$$\frac{x_{k+1} - x_k}{\phi(h)} = -\lambda x_k, \qquad \phi(h) = \frac{1 - e^{-\lambda h}}{\lambda}, \tag{14}$$

can be easily demonstrated to satisfy P_1, P_2, and P_3 for all step-sizes, $h > 0$. (This scheme follows directly from the work given in Mickens [16]; see also

Mickens [2].) Thus, the NSFD scheme of Eq. (14) is DC with Eq. (12). Another interesting feature of this representation is that it is also an exact finite difference scheme for Eq. (12); see Mickens [2].

The publications [6,7,8,9,13,14,18] provide a wide range of the application of the principle of DC to differential equations for which the property of positivity holds.

4. Some Outstanding Problems

Progress in the creation/construction/understanding of the NSFD set of procedures requires the investigation of a set of related issues. The following is a listing of some of the problems for which resolution is needed.

1) Coupled ODE's having fixed-points that are either linear or nonlinear centers are very difficult to deal with in an efficient manner [23,24,25]. Such systems may have several time scales. Also the numerical computed periods of oscillations can depend on the time step-size [24]. One consequence is that the integration step-size must be small relative to these scales if the computed numerical solutions are to give meaningful solutions. Violation of this restriction will always give "physically meaningless" solutions.

2) To date, essentially all of the NSFD schemes have been studied for one-space dimension systems. Thus, it is of great importance to see how the current techniques can be extended to higher space dimension equations. Preliminary results show that the algebraic work increase rapidly with an increase in the number of space variables [26,27].

3) To date, little effort has gone into the investigation of implicit schemes, especially for PDE's [2]. One way to proceed would be to reconsider some of the model linear and nonlinear PDE's already studied using explicit methods. A restriction to be placed on these implicit schemes is to have all the variables, evaluated at the advanced discrete-time level, appear linearly in the discrete finite difference equations.

4) Cross-diffusion occurs when, in a system of coupled PDE's the diffusion coefficients depend on variables other than those appearing in the evolutionary term of a particular equation. Such terms regularly occur in the dynamics of cancer [28] and the spatial interactions of several populations [29,30]. An example of such a set of equations is [28]

$$\frac{\partial u}{\partial t} = u(1-u) - \frac{\partial}{\partial x}\left(u\frac{\partial c}{\partial x}\right), \qquad \frac{\partial c}{\partial t} = -uc^2, \tag{15}$$

where $u(x,t)$ and $c(x,t)$ satisfy a positivity condition, i.e.,

$$u(x,0) \geq 0, \quad c(x,0) \geq 0 \Rightarrow u(x,t) \geq 0, \quad c(x,t) \geq 0 \qquad \text{for } t > 0. \tag{16}$$

The second term on the right-side of the "u"-evolution equation is the cross-diffusion term. The issue to be studied is how to construct NSFD schemes such that the positivity condition, Eq. (16), holds for the solutions to the discrete equations.

5) The NSFD schemes to date have been of rather low order accuracy in the step-sizes. Procedures exist to calculate second-order accurate NSFD models [27]. However, such efforts so far give schemes that may violate the positivity condition. This is a fundamental problem whose solution is still unknown.

6) Finally, it should be mentioned that a firm theoretical basis is needed to fully understand NSFD methods. Some progress in this direction is being made by Jean M.-S. Lubuma and his collaborators [31].

Acknowledgements

The results given in this chapter have been supported over the previous several decades by research grants from ARO, DOE, NASA, and NIH-MBRS. I wish to thank a number of collaborators for their stimulating scientific discussions and research productivity in the area of NSFD methods: Abba B. Gumel (University of Manitoba), Pedro M. Jordan (NRL, Stennis Space Center), Kale Oyedeji (Morehouse College), and Sandra A. Rucker (Clark Atlanta University). I also would like to acknowledge several individuals who through their work has influenced and helped shape my own current views of NSFD methods and related issues: Ron Buckmire (Occidental College) and Jean M.-S. Lubuma (University of Pretoria).

References

1. R. E. Mickens, "Exact solutions to a finite-difference model of a nonlinear reaction-advection equation: Implications for numerical analysis," *Numerical Methods for Partial Differential Equations* **5** (1989), 313–325.
2. R. E. Mickens, *Nonstandard Finite Difference Models of Differential Equations* (World Scientific, Singapore, 1994).
3. R. E. Mickens (editor), *Applications of Nonstandard Finite Difference Schemes* (World Scientific, Singapore, 2000).
4. Special Issues Dedicated to Professor Ronald E. Mickens on the occasion of his 60th Birthday; Guest Editor, A. B. Gumel, *Journal of Difference Equations and Applications* **9** (#11), 989–1056 (2003); **9** (#12), 1059–1112 (2003).
5. R. Buckmire, "Investigations of nonstandard, Mickens-type, finite-difference schemes for singular boundary value problems in cylindrical or spherical co-

ordinates, *Numerical Methods for Partial Differential Equations* **19** (2003), 380–398.

6. Z. Chen, A. B. Gumel, and R. E. Mickens, "Nonstandard discretizations of the generalized Nagumo reaction-diffusion equation," *Numerical Methods for Partial Differential Equations* **19** (2003), 363–369.

7. R. E. Mickens, "A non-standard finite difference scheme for the equations modelling stellar structure," *Journal of Sound and Vibration* **265** (2003), 1116–1120.

8. A. B. Gumel, S. M. Maghadas, and R. E. Mickens, "Effect of a preventive vaccine on the dynamics of the HIV transmission," *Communications in Nonlinear Science and Numerical Simulation* **9** (2004), 649–659.

9. R. E. Mickens and P. M. Jordan, "A positivity-preserving nonstandard finite difference scheme for the damped wave equation," *Numerical Methods for Partial Differential Equations* **20** (2004), 639–649.

10. K. C. Patidar, "On the use of nonstandard finite difference methods," *Journal of Difference Equations and Applications* (reviewed and accepted for publication).

11. R. P. Agarwal, Book review: R. E. Mickens, Nonstandard Finite Difference Models of Differential Equations, *SIAM Review* **37** (1995), 459.

12. C. M. García-López, "Piecewise-linearized and linearized θ-methods for ordinary and partial differential equations," *Computers and Mathematics with Applications* **45** (2003), 351–381.

13. R. E. Mickens, "Dynamic consistency: A fundamental principle for constructing nonstandard finite difference for differential equations," *Journal of Difference Equations and Applications* (reviewed and accepted for publication).

14. R. E. Mickens, "The role of positivity in the construction of NSFD schemes for PDE's," in D. Schultz et al. (editors), *Proceedings of International Conference on Scientific Computing and Mathematical Modeling* (University of Wisconsin-Milwaukee, May 25–27, 2000); pps. 294–307.

15. R. E. Mickens and A. Smith, "Finite difference models of ordinary differential equations: Influence of denominator functions," *Journal of the Franklin Institute* **327** (1990), 143–145.

16. R. E. Mickens, "Finite-difference schemes having the correct linear stability properties for all finite step-sizes II," *Dynamic Systems and Applications* **1** (1992), 329–340.

17. F. B. Hildebrand, *Finite Difference Equations and Simulations* (Prentice-Hall; Englewood Cliffs, NJ; 1968).

18. R. E. Mickens, "Discrete models of differential equations: The roles of dynamic consistency and positivity," *Journal of Difference Equations and Applications* (reviewed and accepted for publication).

19. E. S. Oran and J. P. Boris, *Numerical Simulation of Reactive Flow* (Elsevier, New York, 1987).

20. E. Hairer, C. Lubich, and G. Wanner, *Geometric Numerical Integration: Structure-Preserving Algorithms for Ordinary Differential Equations* (Springer, Berlin, 2002).

21. J. Crank, *The Mathematics of Diffusion*, 2nd Edition (Clarendon Press, Oxford, 1975).

22. R. E. Mickens, *Mathematical Methods for the Natural and Engineering Sciences* (World Scientific, Singapore, 2004); see sections 4.2 and 4.3.

23. R. E. Mickens, "A nonstandard finite-difference scheme for the Lotka-Volterra system," *Journal of Applied Numerical Mathematics* **45** (2003), 309–314.

24. A. B. Gumel and R. E. Mickens, "Numerical study of a nonstandard finite-difference scheme for the van der Pol equation," *Journal of Sound and Vibration* **250** (2002), 955–963.

25. R. E. Mickens, "Step-size dependence of the period for a forward-Euler scheme for the van der Pol equation," *Journal of Sound and Vibration* **258** (2002), 199–202.

26. R. E. Mickens, "Exact finite-difference schemes for two-dimensional advection equations," *Journal of Sound and Vibration* **207** (1997), 426–428.

27. E. H. Twizell, A. B. Gumel, and Q. Cao, "A second-order scheme for the 'Brusselator' reaction-diffusion system," *Journal of Mathematical Chemistry* **26** (1999), 297–316.

28. B. P. Marchant, J. Norbury, and A. J. Perumpani, "Traveling shock waves arising in a model of malignant invasion," *SIAM Journal of Applied Mathematics* **60** (2000), 463–476.

29. W.-M. Ni, "Diffusion, cross-diffusion, and their spike-like steady states," *Notices of the American Mathematical Society* **45** (1998), 9–25.

30. J. D. Murray, *Mathematical Biology* (Springer-Verlag, Berlin, 1989); see section 9.4

31. R. Anguleov and J. M.-S. Lubuma, "Contributions to the mathematics of the nonstandard finite difference method and applications," *Numerical Methods for Partial Differential Equations* **17** (2001), 518–543.

32. R. Anguelov and J.M.-S. Lubuma, "Nonstandard finite difference methods by nonlocal approximation," *Mathematics and Computer in Simulation* **61** (2003), 465–475.

33. R. Anguelov, P. Kama, and J. M.-S. Lubuma, "On nonstandard finite difference models of reaction-diffusion equations, *Journal of Computational and Applied Mathematics* **175** (2005), 11–29.

CHAPTER 2

APPLICATION OF NONSTANDARD FINITE DIFFERENCE SCHEMES TO THE SIMULATION STUDIES OF ROBOTIC SYSTEMS

R. F. Abo-Shanab

Department of Mechanical Engineering
Assiut University
Assiut, Egypt
roshdy_foaad@hotmail.com

N. Sepehri and C. Q. Wu

Department of Mechanical and Manufacturing Engineering
The University of Manitoba
Winnipeg, Manitoba, Canada
nariman@cc.umanitoba.ca; cwu@cc.umanitoba.ca

The application of the nonstandard finite difference schemes to the challenging task of obtaining stable numerical solutions for highly nonlinear and coupled differential equations that describe the dynamics of robotic manipulators is investigated in this chapter. It is shown that despite its simplicity, an appropriate form of discrete derivatives of the original differential equations could greatly reduce numerical instabilities while expediting computation time. Here, two nonstandard schemes are employed to construct the discrete derivatives. In the first scheme, the orders of the discrete derivatives are equal to the orders of the corresponding derivatives of the differential equations and, the denominators of the discrete derivatives take on a more complicated function of the step-size to ensure that the fixed (equilibrium) points of the resulting discrete system has the same stability properties as those of the original system. The second scheme has the same characteristics as the first one, with the addition of having the nonlinear terms replaced by nonlocal discrete representations. Both schemes are evaluated and compared with the popular fourth-order Runge-Kutta method, through simulating the motion of a two degree-of-freedom planar manipulator. It is demonstrated that firstly, using nonstandard schemes, the possibility of hav-

11

ing spurious solutions is eliminated. Secondly, nonstandard finite difference derivatives are numerically stable given any large step-sizes. This is significant since using nonstandard schemes, numerical simulation of the complex dynamics of robotic systems can potentially be expedited, thereby allowing close to real-time simulations of robotic systems with reliable results.

1. Introduction

Numerical integration of ordinary differential equations (ODEs) using conventional methods could result in numerical instabilities, i.e., solutions of the discrete derivatives could qualitatively become different from those of the original ODEs. This is due to the fact that first, discretization of continuous systems introduces an extra parameter (step-size) to the original system of equations. Using a value for this parameter larger than some relevant time scale, which is often not readily known for conventional finite difference schemes, lead to solutions that may not reflect the actual dynamics of the system. Secondly, if the orders of the discrete derivatives are higher than those of the original ODEs, the discrete systems will have more fixed points than those of the original systems, known as spurious solutions. Chen and Solis [1] discussed the appearance of the spurious solutions when first-order ODEs are discretized by the popular Runge-Kutta integration method. They concluded that the reliability of the numerical solutions to a particular ODE could be verified only by constructing several discrete schemes and comparing their results with some known properties of the exact solution.

Recently, Mickens [2,3] introduced the principle of exact finite difference scheme as the one for which the difference equations have the general solutions as those of the original differential equations. A major advantage of using this principle for a differential equation is that issues related to the usual considerations of consistency, stability, and convergence do not arise. However, it is impossible to construct an exact discrete model for an arbitrary ODE. Mickens [4], then, introduced certain rules to follow in order to obtain the best difference equations. In general, these rules, also called "nonstandard finite difference schemes" offer the prospect of obtaining finite difference models that do not possess the standard numerical instabilities caused by discretizing the differential equations using conventional methods. Anguelov and Lubuma [5] defined the nonstandard finite difference scheme as the one that meets at least one of the following two conditions suggested by Mickens: (i) denominator functions for the discrete derivatives are expressed in terms of more complicated functions of the step-sizes than those conventionally used, and (ii) nonlinear terms are modeled non-locally on the computation grid. Recently, Mickens [6] showed that NSFD schemes, eliminate all the numerical instabilities found by Chen and Solis [1]. Also, Sekhavat et al. [7] used a NSFD scheme to calculate Lyapunov exponents for control stability analysis and showed that the scheme could greatly reduce the computational time.

In this chapter, we explore the application of the NSFD schemes to the area of simulation studies of the motion of robotic systems, in terms of producing more reliable results and reducing the computational time towards real-time simulations. To the best of the authors' knowledge, the applications of NSFD schemes to the simulation of the motion of robotic manipulators have not yet been addressed in any of the previous literature.

2. Dynamic Model of the Robotic Systems

The dynamic equations for a general serial n-link manipulator can be described, using the Lagrange-Euler method, as follows [8,9]:

$$\boldsymbol{\tau} = \mathbf{M}(\mathbf{q})\ddot{\mathbf{q}} + \mathbf{H}(\mathbf{q}, \dot{\mathbf{q}}) + \mathbf{G}(\mathbf{q}) \tag{1}$$

Vector $\mathbf{q} = \{q_1, q_2, \ldots, q_n\}^T$ denotes the generalized joint angular/linear displacement coordinates and $\boldsymbol{\tau} = \{\tau_1, \tau_2, \ldots, \tau_n\}^T$ is the generalized torque/force vector including frictional forces. $\mathbf{M}(\mathbf{q})$ is a $n \times n$ symmetric, positive definite inertial matrix whose elements are

$$\tilde{M}_{ij} = \text{Trace}\left\{ \Delta_i \left[\sum_{p=j}^{n} \mathbf{T}_p \mathbf{I}_p \mathbf{T}_p^T \right] \Delta_j^T \right\} \qquad j \geq i \tag{2a}$$

$$\tilde{M}_{ji} = \tilde{M}_{ij} \tag{2b}$$

In (2), $\mathbf{T}_p$ is the homogeneous transformation matrix from coordinate system p to the inertial coordinate system. $\mathbf{I}_p$ is defined as

$$\mathbf{I}_p = \begin{bmatrix} \mathbf{I}_p^p + m_p \mathbf{r}_p^p {\mathbf{r}_p^p}^T & m_p \mathbf{r}_p^p \\ m_p {\mathbf{r}_p^p}^T & m_p \end{bmatrix} \tag{3}$$

where $\mathbf{I}_p^p$ is the 3×3 inertial matrix of link p about its center of mass with respect to its own coordinate frame. $\mathbf{r}_p^p$ is the position vector of mass center of link p expressed in the coordinate system p, and m_p is the mass of link p. Term Δ_i in (2) is a differential operator:

$$\Delta_i = \begin{bmatrix} \lambda_i \hat{\mathbf{z}}_{i-1} & (\lambda_i \hat{\mathbf{p}}_{i-1} + \varphi_i \mathbf{E}) \mathbf{z}_{i-1} \\ \mathbf{0} & 0 \end{bmatrix} \tag{4}$$

where $\mathbf{z}_i$ is the *z-axis* of the coordinate frame i. $\mathbf{p}_i$ is the position vector of the origin of the coordinate frame i. $\lambda_i = 1$ for revolute (rotating) joints and $\lambda_i = 0$ for prismatic (sliding) joints. $\varphi_i = 1 - \lambda_i$ and $\mathbf{E}$ is a 3×3 unity matrix. $\hat{\mathbf{z}}$ and $\hat{\mathbf{p}}$ are 3×3 skew symmetric matrices constructed based on the 3×1 vectors $\mathbf{z} = \{z_x, z_y, z_z\}^T$ and $\mathbf{p} = \{p_x, p_y, p_z\}^T$,, respectively. For example,

$$\hat{\mathbf{z}} = \begin{bmatrix} 0 & -z_z & z_y \\ z_z & 0 & -z_x \\ -z_y & z_x & 0 \end{bmatrix} \tag{5}$$

$\mathbf{H}(\mathbf{q}, \dot{\mathbf{q}})$s in (1) is the $n \times 1$ nonlinear centripetal and Coriolis torque vector whose elements are

$$\mathbf{H}(\mathbf{q}, \dot{\mathbf{q}}) = \{h_1, h_2, \ldots, h_n\}^T \tag{6}$$

where

$$h_i(\mathbf{q}, \dot{\mathbf{q}}) = \sum_{j=1}^{n} \sum_{k=1}^{n} h_{ijk} \dot{q}_j \dot{q}_k \tag{7}$$

and

$$h_{ijk} = \text{Trace} \left\{ \Delta_i \left[\sum_{p=k}^{n} \mathbf{T}_p \mathbf{I}_p \mathbf{T}_p^T \right] \Delta_k^T \Delta_j \right\} \qquad i < k, \quad j \leq k \tag{8}$$

Finally, $\mathbf{G}(\mathbf{q})$ in (1) is the $n \times 1$ gravity loading torque vector whose elements are

$$\mathbf{G}(\mathbf{q}) = \{G_1, G_2, \ldots, G_n\}^T \tag{9}$$

where

$$G_i(\mathbf{q}) = -\mathbf{g}^T \Delta_i \left[\sum_{p=i}^{n} m_p \mathbf{T}_p (\mathbf{r}_p^p; 1) \right] \tag{10}$$

where $\mathbf{r}_i^i$ is a 3×1 position vector of mass center of link i with respect to its own coordinate frame, and $\mathbf{g}$ is the gravity vector.

Equation (1) can be re-written in the following form:

$$\ddot{\mathbf{q}} = \mathbf{D}[\tau - \mathbf{H}(\mathbf{q}, \dot{\mathbf{q}}) - \mathbf{G}(\mathbf{q})] \tag{11}$$

where $\mathbf{D} = \mathbf{M}^{-1}(\mathbf{q})$. The following states are now used:

$$\dot{x}_1 = x_2 = \dot{q}_1$$
$$\dot{x}_2 = \ddot{q}_1$$
$$\dot{x}_3 = x_4 = \dot{q}_2$$
$$\dot{x}_4 = \ddot{q}_2$$
$$\vdots$$
$$\dot{x}_{2n-1} = x_{2n} = \dot{q}_n$$
$$\dot{x}_{2n} = \ddot{q}_n$$

In a general form, one can write

$$f_{i(i=1,2,\ldots,2n)}(x) = \begin{cases} x_{i+1} & i = 1,3,5,\ldots,2n-1 \\ \{\mathbf{D}[\tau - \mathbf{H}(\mathbf{x}) - \mathbf{G}(\mathbf{x})]\}_i & i = 2,4,6,\ldots,2n \end{cases} \tag{12}$$

where n is the number of the manipulator's degrees of freedom, and $\{\ldots\}_i$ represents the ith element of the vector $\{\ldots\}$.

3. Constructing the Discrete Derivatives

Two nonstandard finite difference schemes are used to construct the discrete derivatives of the above system. In scheme I, the discrete derivatives are expressed locally similar to the standard forward-Euler representation:

$$x_i^{k+1} = x_i^k + \varphi f_i = F_i(\mathbf{x}^k) \qquad i = 1,2,\ldots,2n \tag{13}$$

where f_i is given by (12). The effective step-size, φ (also termed "denominator function") takes on a more complicated function as will be described shortly.

In scheme II, the nonlinear terms are further replaced by nonlocal discrete representation:

$$x_i^{k+1} = \frac{x_i^k + \varphi v_i}{1 - \varphi u_i} = F_i(\mathbf{x}^k) \qquad i = 1,2,\ldots,2n \tag{14}$$

where v_i and u_i are defined, for the robotics systems under consideration, as follows:

$$v_i = \begin{cases} x_{i+1}^k & i = 1,3,\ldots,2n-1 \\ \{\mathbf{D}[\tau - \mathbf{G}(\mathbf{x}^k)]\}_{i/2-w_i} & i = 2,4,\ldots,2n \end{cases} \tag{15}$$

$$u_i = \begin{cases} 0 & i = 1,3,\ldots,2n-1 \\ b_i & i = 2,4,\ldots,2n \end{cases} \tag{16}$$

Terms w_i and b_i, in Equations (15) and (16), are defined based on the way the nonlinear terms are expressed nonlocally. Here, the nonlinear terms are expressed as $x_i^2 = x_i^{k+1} x_i^k$ and $x_i x_j = x_i^{k+1} x_j^k$. From Equations (1) and (14), we have:

$$w_i = \sum_{m=1}^{n} D_{\hat{i},m} \left[\sum_{\substack{j=1 \\ j \neq \hat{i}}}^{n} \sum_{\substack{\ell=1 \\ \ell \neq \hat{i}}}^{n} h_{mj\ell} x_{2j}^k x_{2\ell}^k \right] \qquad i = 2\hat{i} = 2, 4, \ldots, 2n \qquad (17)$$

$$b_i = \sum_{m=1}^{n} D_{\hat{i},m} \left[h_{m\hat{i}\hat{i}} x_{2\hat{i}}^k + 2 \sum_{\substack{j=1 \\ j \neq \hat{i}}}^{n} h_{m\hat{i}j} x_{2j}^k \right] \qquad i = 2\hat{i} = 2, 4, \ldots, 2n \qquad (18)$$

The effective step-size, φ, for both schemes, is determined by the requirement of having the discrete system preserves the stability properties of the equilibrium points of the original system. It is, therefore, selected such that its value is never larger than the smallest time scale of the system. Here, φ is defined as following:

$$\varphi = \frac{1 - e^{-\lambda^* h}}{\lambda^*} \qquad (19)$$

where $h > 0$ is the selected (also called "actual") step-size, and $\lambda^* = \max\{\lambda_i, i = 1, 2, \ldots, 2n\}$ is the largest eigenvalue of the Jacobian matrix $R = \frac{\partial \mathbf{F}(\mathbf{x}^k)}{\partial \mathbf{x}^k}\big|_{\bar{\mathbf{x}}}$ (where $\mathbf{F}(\mathbf{x}^k) = \{F_1(\mathbf{x}^k), F_2(\mathbf{x}^k), \ldots, F_n(\mathbf{x}^k)\}^T$) pertaining to fixed (equilibrium) point, $\bar{\mathbf{x}}$, of the original differential system. The use of function (19) allows the value of h to be selected much larger than one normally choose, because it is the effective step-size φ that determines the stability and not the actual step-size h. An acceptable value for φ is the one for which the maximum eigenvalue of the Jacobian matrix R is less than 1. In such a case, the fixed point is numerically stable [3], i.e., the fixed point of the discrete derivatives has the same stability properties as those of the critical (fixed) point of the original system. This can be easily demonstrated for scheme I. Rewriting Equation (13), we have

$$\frac{\mathbf{x}^{k+1} - \mathbf{x}^k}{\frac{1 - e^{-\lambda^* h}}{\lambda^*}} = \mathbf{f}(\mathbf{x}^k) \qquad (20)$$

where $\mathbf{x} = \{x_1, \ldots, x_2, \ldots, x_n\}^T$ and $\mathbf{f} = \{f_1, f_2, \ldots, f_n\}^T$. Knowing that at fixed (equilibrium) point, $\bar{\mathbf{x}} = \{\bar{x}_1, \bar{x}_2, \ldots, \bar{x}_n\}^T$, $\mathbf{f}(\bar{\mathbf{x}}) = \mathbf{0}$, one can write

$$\mathbf{f}(\mathbf{x}^k) = \frac{\partial \mathbf{f}}{\partial \mathbf{x}^k}\bigg|_{\bar{\mathbf{x}}} \varepsilon^k \qquad (21)$$

where $\epsilon^k = (\bar{\mathbf{x}}^k - \bar{\mathbf{x}}) = \{\varepsilon_1^k, \varepsilon_2^k, \ldots, \varepsilon_n^k\}^T$ represents any perturbation about the fixed point.

Now, consider the value of the parameter λ^* that makes the scheme numerically stable given any value of the parameter h. From Equations (20) and (21), we have,

$$\frac{\epsilon^{k+1} - \epsilon^k}{\frac{1-e^{-\lambda^* h}}{\lambda^*}} = \left.\frac{\partial \mathbf{f}}{\partial \mathbf{x}^k}\right|_{\bar{\mathbf{x}}} \epsilon^k \tag{22}$$

Assume that function $\mathbf{f}$ is sufficiently smooth, then the Jacobian matrix $R = \left.\frac{\partial \mathbf{f}}{\partial \mathbf{x}^k}\right|_{\bar{\mathbf{x}}}$. Thus,

$$\epsilon^{k+1} - \epsilon^k = R\epsilon^k \frac{1 - e^{-\lambda^* h}}{\lambda^*} \tag{23}$$

$$\epsilon^{k+1} = \left[I + R\frac{1 - e^{-\lambda^* h}}{\lambda^*}\right] \epsilon^k \tag{24}$$

$$\epsilon^k = \left[I + R\frac{1 - e^{-\lambda^* h}}{\lambda^*}\right] \epsilon^0 \tag{25}$$

Assume that the eigenvalues $(\lambda_1, \lambda_2, \ldots, \lambda_n)$ of the matrix R have nonzero real parts, then from [5], we have

$$\varepsilon_i^k = \left[I + \frac{\lambda_i}{\lambda^*}(1 - e^{-\lambda^* h})\right]^k \varepsilon_i^0 \qquad i = 1, 2, \ldots, n \tag{26}$$

Now, if we choose $\lambda^* = \max\{|\lambda_i|\}$, then
(i) if the fixed point is stable, i.e., $\mathrm{Re}(\lambda_i) < 0$, then $\varepsilon_i^k \to 0$ as $k \to \infty$ and,
(ii) if the fixed point is unstable, at least the real part of one of the eigenvalues is greater than 0, then $\varepsilon_i^k \to \infty$ as $k \to \infty$.

From (i) and (ii), one concludes that the critical (fixed) point of the discrete system has the same stability properties as those of the fixed point of the original system.

In summary, using the NSFD schemes described by Equations (13) and (14), the orders of the discrete derivatives are equal to the orders of the differential equations, the denominator function (19) is expressed, in terms of the step-size, in a more complicated manner than those used conventionally and, for scheme II, described by Equation (14), the nonlinear terms in $f_i(x)$ are approximated in a nonlocal manner, by a suitable function of several points of the mesh. These properties, as will be demonstrated next, eliminate the possibility of having spurious solutions and ensure that the fixed (equilibrium) points of the discrete model have the same stability properties as those of the continuous one given any step-size.

4. Simulation Studies

The application of the two schemes as outlined above, is now exemplified with simulation of a two-link planar manipulator shown in Figure 1.

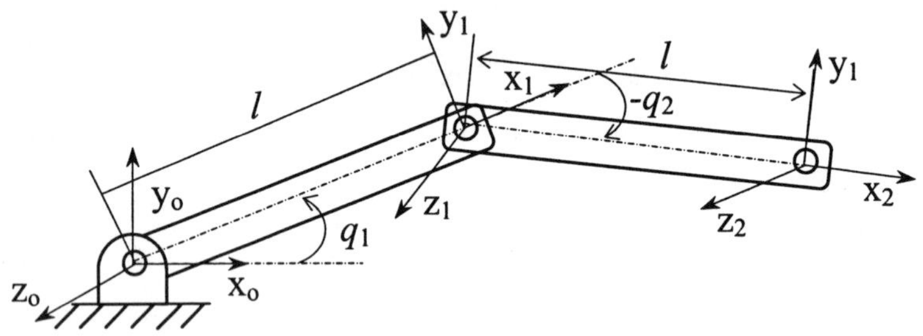

Fig. 1.   Two-link manipulator.

The dynamic equations for this system are derived from the general Equation (1). The elements of the inertial matrix $\mathbf{M}$ are:

$$M_{11} = \frac{1}{3} m_1 \ell^2 + \frac{4}{3} m_2 \ell^2 + m_2 C_2 \ell^2 \tag{27a}$$

$$M_{12} = M_{21} = \frac{1}{3} m_2 \ell^2 + \frac{1}{2} m_2 \ell^2 C_2 \tag{27b}$$

$$M_{22} = \frac{1}{3} m_2 \ell^2 \tag{27c}$$

The elements of the nonlinear Coriolis and centrifugal torque vector $\mathbf{H}$ are:

$$H_1 = -\frac{1}{2} m_2 S_2 \ell^2 \dot{q}_2^2 - m_2 S_2 \ell^2 \dot{q}_1 \dot{q}_2 \tag{28a}$$

$$H_2 = \frac{1}{2} m_2 S_2 \ell^2 \dot{q}_1^2 \tag{28b}$$

The elements of the gravity loading torque vector $\mathbf{G}$ are:

$$G_1 = \frac{1}{2} m_1 g \ell C_1 + \frac{1}{2} m_2 g \ell C_{12} + m_2 g \ell C_1 \tag{29a}$$

$$G_2 = \frac{1}{2} m_2 g \ell C_{12} \tag{29b}$$

Note in the above equations $C_1 = \cos(q_i)$, $S_i = \sin(q_i)$ and $C_{ij} = \cos(q_i + q_j)$.

4.1. Case Study I

4.1.1. Description of the Case

This case study simulates the free motion of the manipulator under the gravitational effect. The initial states are $q_1 = -60°$, $\dot{q}_1 = 0$ deg/s, $q_2 = 25°$ and $\dot{q}_2 = 0$ deg/s. The frictions at the manipulator joints are considered to be viscous. Thus, $\tau_1 = -k_1\dot{q}_1$ and $\tau_2 = -k_2\dot{q}_2$, where k_1 and k_2 are the viscous friction coefficients. It is clear that the manipulator finally settles to the stable position $q_1 = -90°$ and $q_2 = 0°$.

The dynamic equations are written in the state space form as follows:

$$f_1 = \dot{x}_1 = x_2$$

$$f_2 = \dot{x}_2 = D_{11}\left(-k_1 x_2 + \frac{1}{2} m_2 S_3 \ell^2 x_4^2 + m_2 S_3 \ell^2 x_2 x_4 \right.$$

$$\left. - \frac{1}{2} m_1 g \ell C_1 - \frac{1}{2} m_2 g \ell C_{13} - m_2 g \ell C_1 \right)$$

$$+ D_{12}\left(-k_2 x_4 - \frac{1}{2} m_2 S_3 \ell^2 x_2^2 - \frac{1}{2} m - 2 g \ell C_{13}\right)$$

$$f_3 = \dot{x}_3 = x_4$$

$$f_4 = \dot{x}_4 = D_{21}\left(-k_1 x_2 + \frac{1}{2} m_2 S_3 \ell^2 x_4^2 + m_2 S_3 \ell^2 x_2 x_4 \right.$$

$$\left. - \frac{1}{2} m_1 g \ell C_1 - \frac{1}{2} m_2 g \ell C_{13} - m_2 g \ell C_1 \right)$$

$$+ D_{22}\left(-k_2 x_4 - \frac{1}{2} m_2 S_3 \ell^2 x_2^2 - \frac{1}{2} m_2 g \ell C_{13}\right) \tag{30}$$

where $x_1 = q_1$, $x_2 = \dot{q}_1$, $x_3 = q_2$, $x_4 = \dot{q}_2$, $C_i = \cos(x_i)$, $S_i = \sin(x_i)$ and $C_{ij} = \cos(x_i + x_j)$. D_{11}, D_{12}, D_{21} and D_{22} are the elements of the matrix $\mathbf{D} = \mathbf{M}^{-1}$. Both nonstandard schemes (13) and (14), are used for numerical simulations of this system:

$$x_i^{k+1} = x_i^k + \varphi f_i \qquad i = 1, 2, 3, 4 \tag{31}$$

$$x_i^{k+1} = \frac{x_i^k + \varphi v_i}{1 - \varphi u_i} \qquad i = 1, 2, 3, 4 \tag{32}$$

The elements of vectors v_i and u_i are determined from general Equations (15) and (16) as follows:

$$v_1 = x_2^k \tag{33}$$

$$v_2 = -D_{11}G_1 + D_{12}(\tau_2 - G_2) + \frac{1}{2} D_{11} m_2 S_3 \ell^2 (x_4^k)^2$$

$$v_3 = x_4^k$$

$$v_4 = D_{21}(\tau_1 - G_1) - D_{22} G_2 - \frac{1}{2} D_{22} m_2 S_3 \ell^2 (x_2^k)^2$$

$$u_1 = 0 \tag{34}$$

$$u_2 = -D_{11} m_2 S_3 \ell^2 x_4^k + \frac{1}{2} D_{12} m_2 S_3 \ell^2 x_2^k + D_{11} k_1$$

$$u_3 = 0$$

$$u_4 = -\left(\frac{1}{2} x_4^k + x_2^k \right) D_{21} m_2 S_3 \ell^2 + D_{22} k_2$$

Note that

$$G_1 = \frac{1}{2} m_1 g \ell C_1 + \frac{1}{2} m_2 g \ell C_{13} + m_2 g \ell C_1 \tag{35}$$

and

$$G_2 = \frac{1}{2} m_2 g \ell C_{13} \tag{36}$$

For each scheme, the elements of the Jacobian matrix $R = \left. \frac{\partial \mathbf{F}(\mathbf{x}^k)}{\partial \mathbf{x}^k} \right|_{\bar{\mathbf{x}}}$ are calculated for the stable fixed point $\mathbf{x} = \{-90°, 0, 0, 0, \}^T$. The maximum eigenvalues of these matrices are then calculated to determine the bound on effective step-size for stable numerical integration.

4.1.2. *Numerical Results*

First, the dynamic equations of the manipulator are integrated using fourth-order Runge-Kutta (RK4) numerical scheme. Figure 2 shows the joint angle trajectories using step-size 0.001s, which shows correct convergence. Figure 3 shows the simulation results using RK4 method but with larger step-sizes. For step-sizes greater than 0.054s, the integration routine noticeably converges to wrong solutions. Such solutions are known as "spurious asymptotic" solutions. Further increasing the effective step-size leads to non-convergent numerical results as seen in Figure 4.

The same system is now simulated with the discrete derivatives obtained using the nonstandard scheme I. Figure 5 shows the relation between the effective step-size, φ, and the maximum eigenvalue, $\lambda_{\max}$, of the Jacobian matrix $R = \frac{\partial \mathbf{F}(\mathbf{x}^k)}{\partial \mathbf{x}^k}\big|_{\bar{\mathbf{x}}}$, of the discrete system at the fixed point $\bar{\mathbf{x}}$. As is seen, for $0 < \varphi < 0.0396s$, the maximum eigenvalue, $\lambda_{\max}$, stays below 1, meaning that the scheme is numerically stable for this range of step-sizes. The simulation results shown in Figure 6 demonstrate this fact that the numerical results converge to the actual solution as long as $0 < \varphi < 0.0396s$.

We now employ the discrete derivatives obtained using the nonstandard scheme II for simulations. Figure 7 shows the relation between the effective step-size, φ, and the maximum eigenvalue, $\lambda_{\max}$, for the discrete system at the fixed point, $\bar{\mathbf{x}}$. As is seen for $0 < \varphi < 0.1413s$, $\lambda_{\max} < 1$. Thus, scheme II is numerically stable for values of φ larger than those allowable using the scheme I. Numerical simulation results shown in Figures 8 and 9 show that the solutions give right indication about the stability of the system even with a large effective step-size of $\varphi = 0.08s$ (providing simulation time to continue long enough). Therefore, the stability property of the fixed point can be preserved.

R. F. Abo-Shanab, N. Sepehri and C. Q. Wu

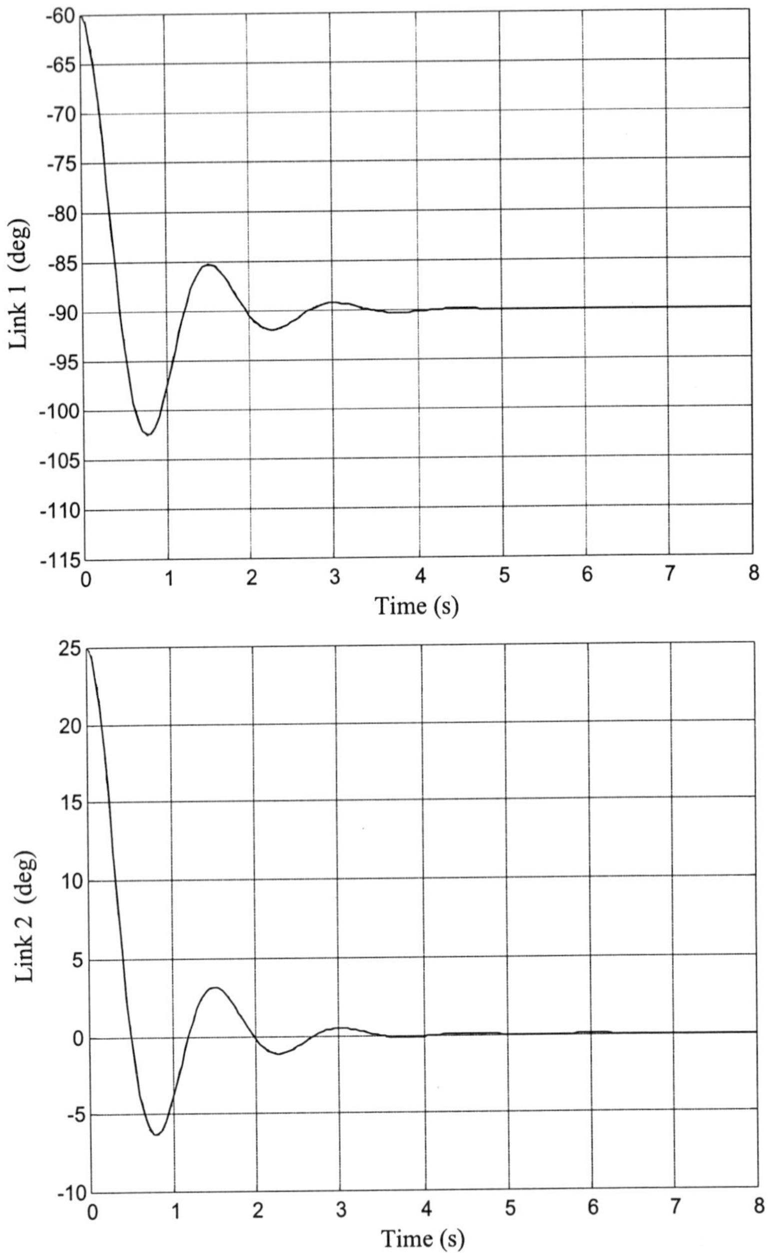

Fig. 2. Simulation results for case study I using RK4 method with step-size 0.0001s.

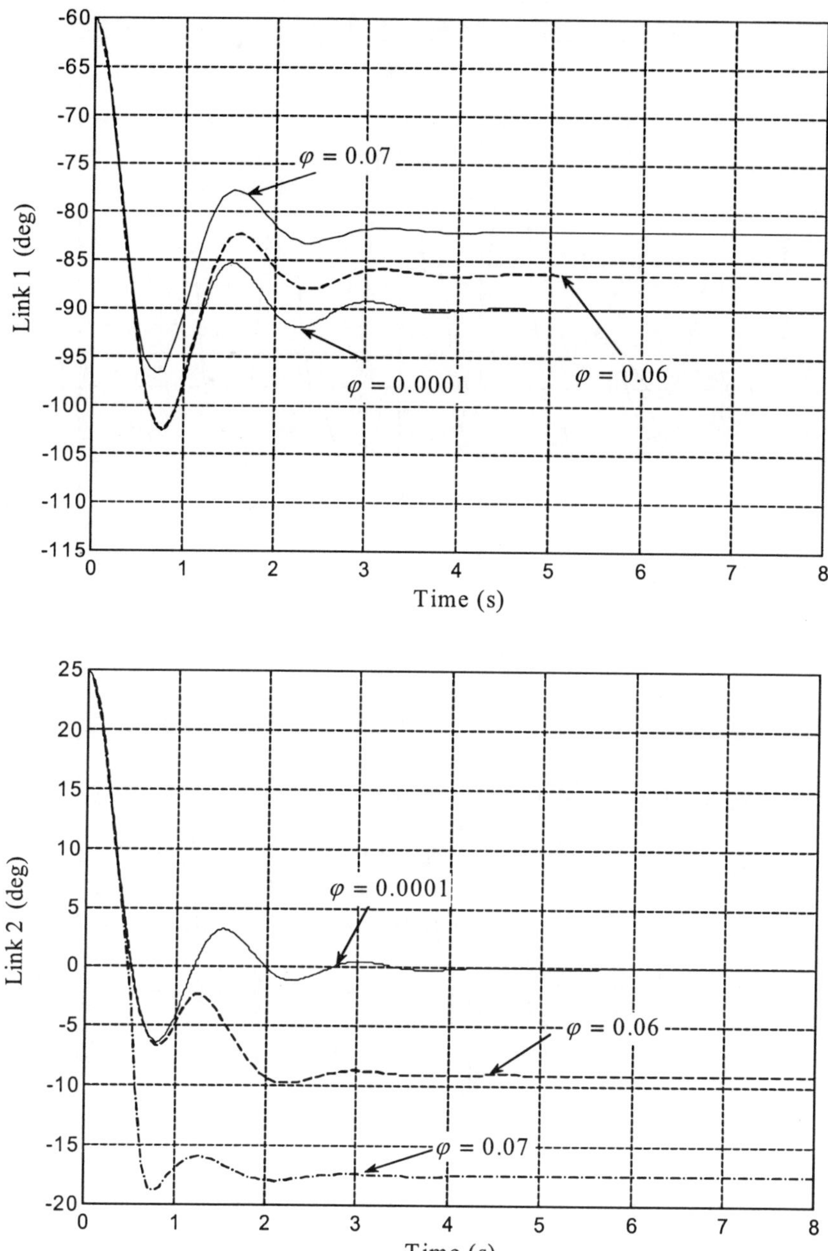

Fig. 3.   Spurious solutions obtained by RK4 method (case study I).

R. F. Abo-Shanab, N. Sepehri and C. Q. Wu

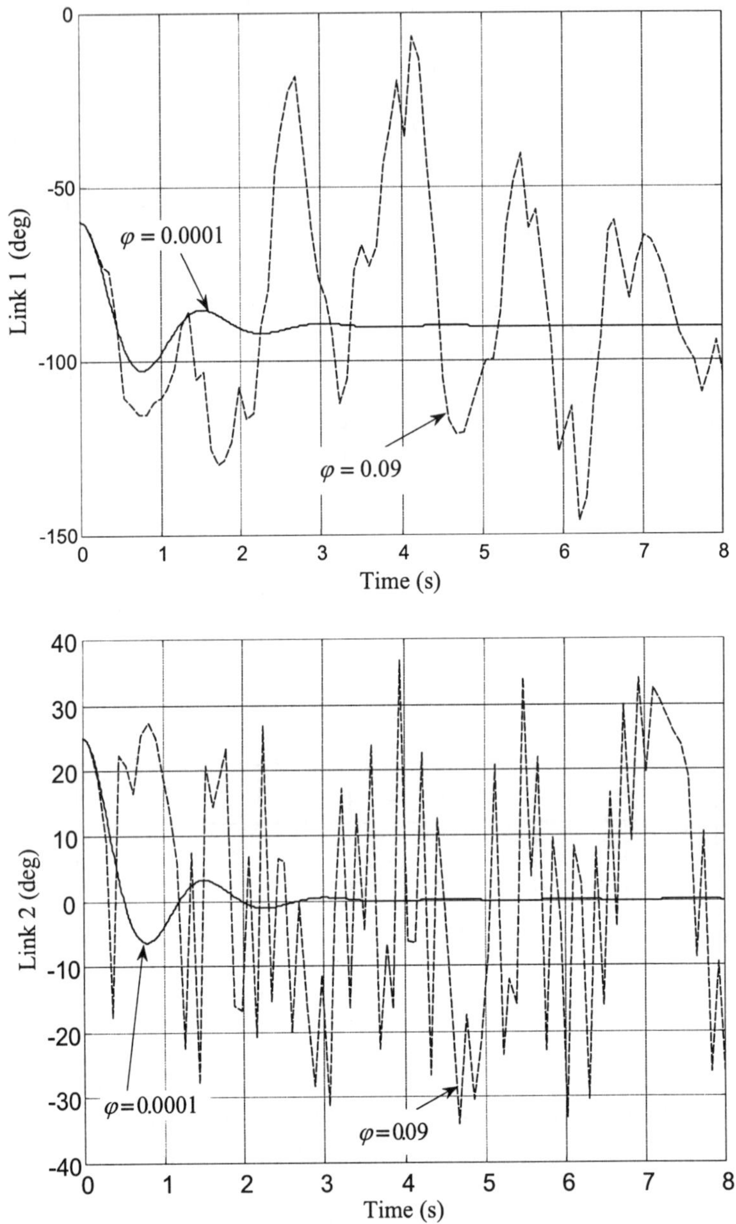

Fig. 4.   Nonconvergent numerical results obtained by RK4 method.

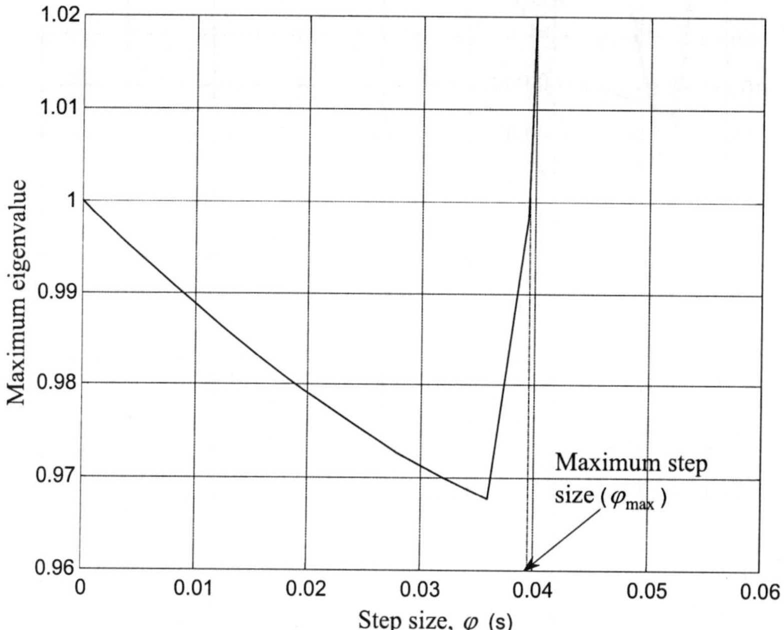

Fig. 5. Relation between effective step-size, φ, and the maximum eigenvalue of the discrete system obtained by nonstandard scheme I (case study I).

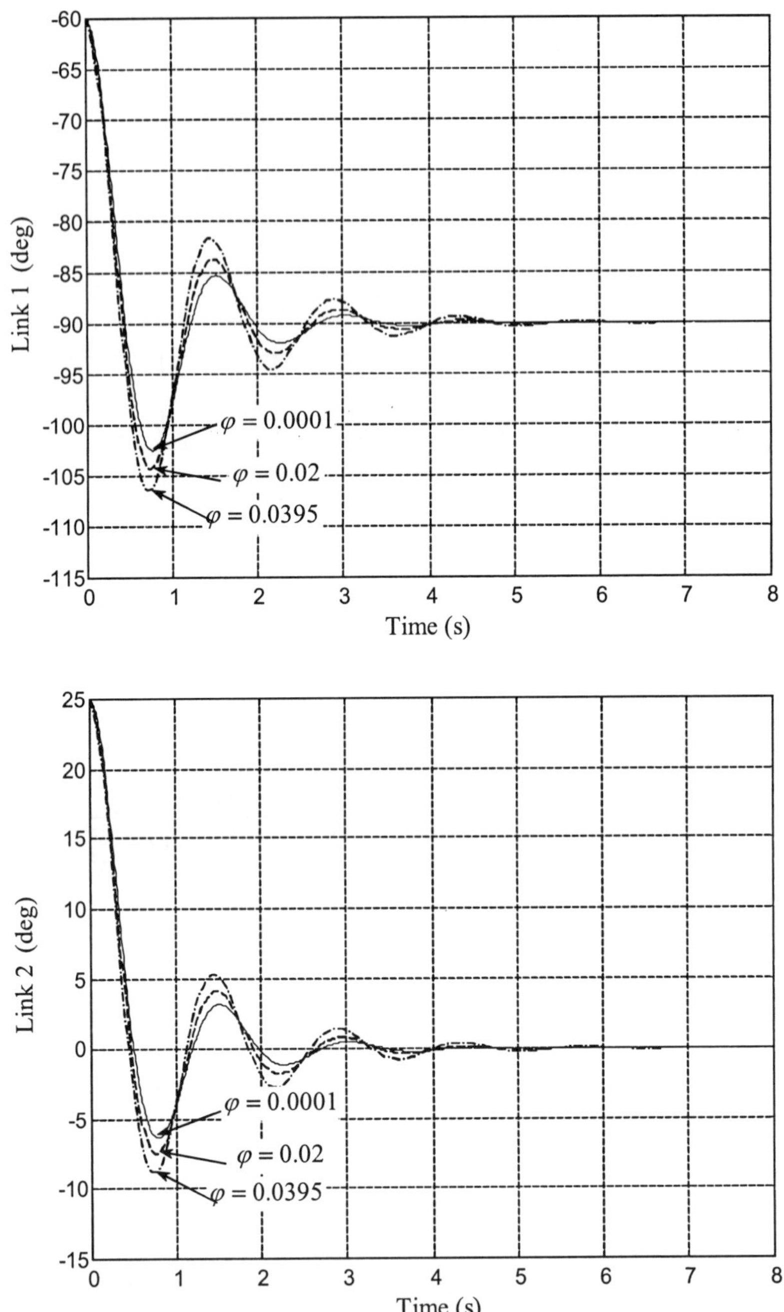

Fig. 6.   Simulation results using nonstandard scheme I with effective step-sizes less than the critical step-size.

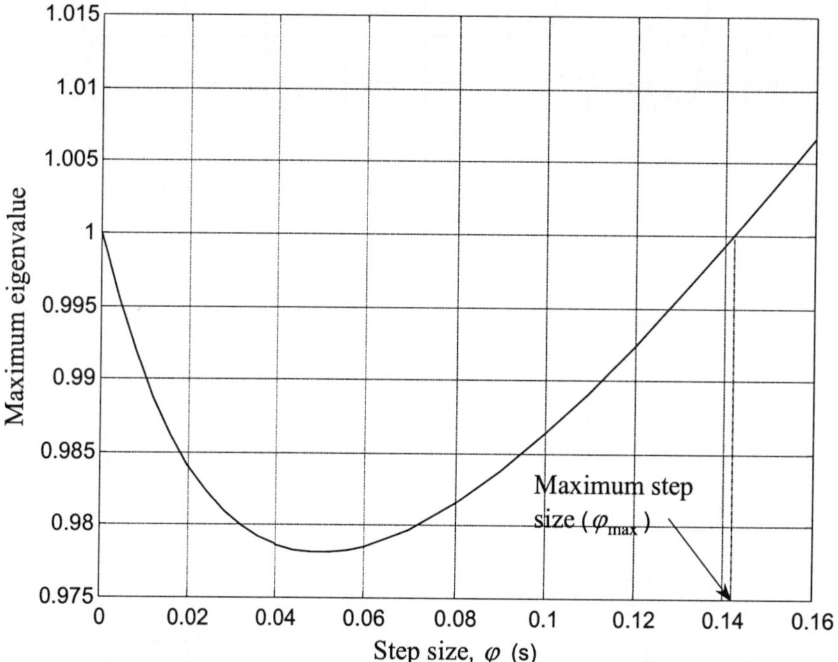

Fig. 7. Relation between the effective step-size, φ, and the maximum eigenvalue of the discrete system obtained by nonstandard scheme II (case study I).

   R. F. Abo-Shanab, N. Sepehri and C. Q. Wu

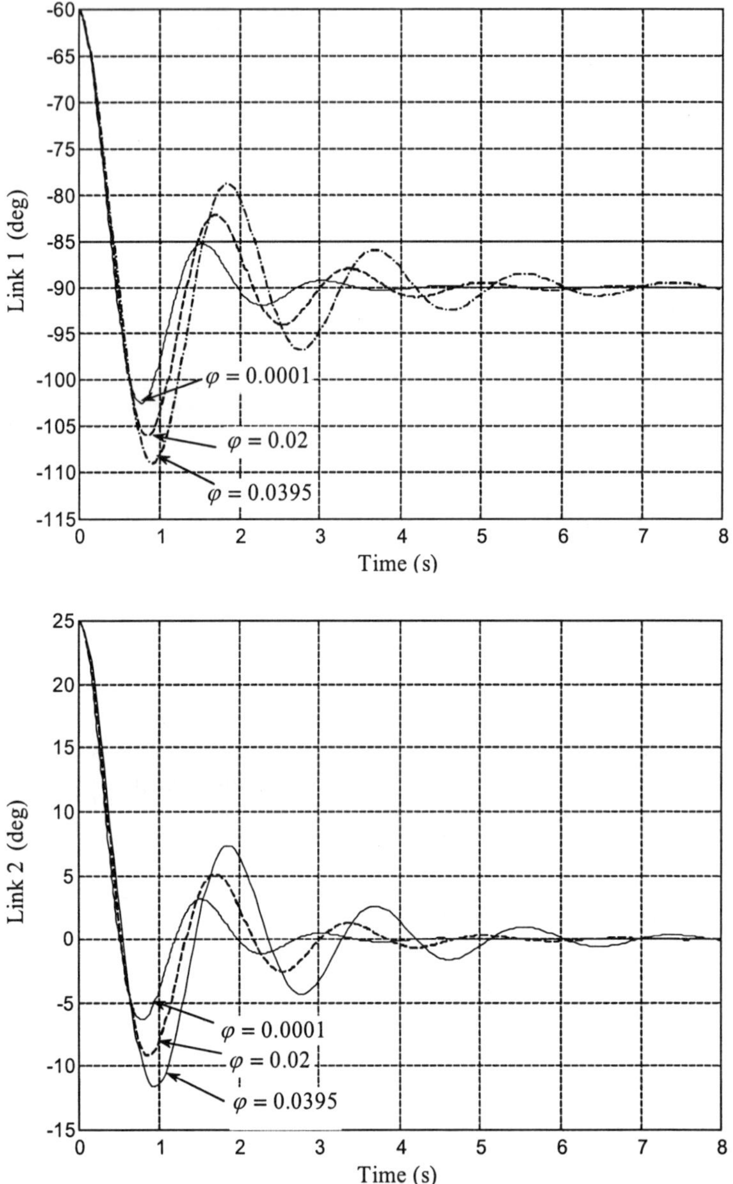

Fig. 8.   Simulation results using nonstandard scheme II with different effective step-sizes.

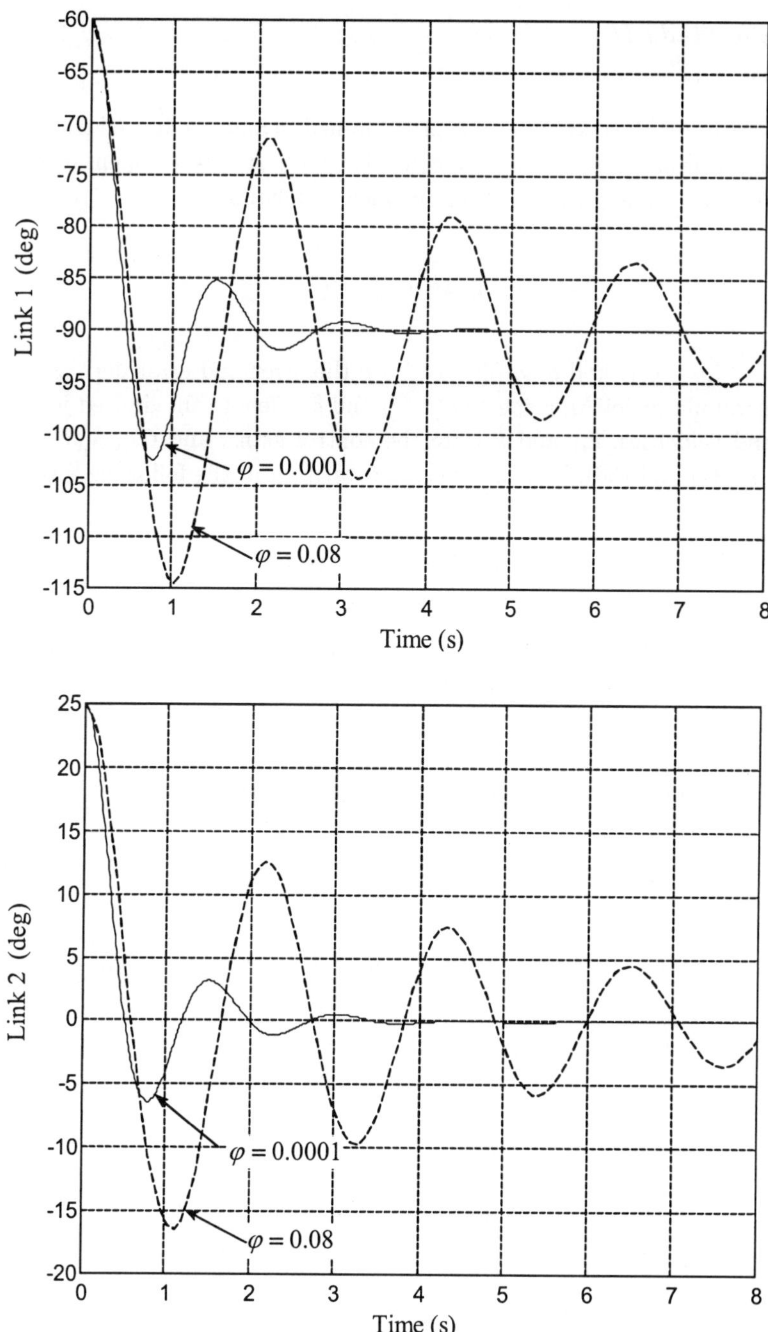

Fig. 9. Simulation results using nonstandard scheme II with large effective step-sizes.

4.2. Case Study II

4.2.1. Description of the Case

In this case study, the manipulator is programmed to follow a desired trajectory using a proportional-derivative controller with gravity compensation. Thus, control torques τ_i $(i = 1, 2)$ are defined as follows:

$$\tau_i = -k_{i1}e_i - k_{i2}\dot{e}_i + G_i \tag{37}$$

where $e_i = (q_i - q_{id})$ and $\dot{e}_i = (\dot{q}_i - \dot{q}_{id})$ are the error and error derivative at the manipulator joints, respectively. q_{id} and $\dot{q}_{id}$ denote the desired joint angles and velocities. k_{i1} and k_{i2} are the control gains, and $G_{i(i=1,2)}$ are defined in (35) and (36). Equation (11) is re-written in the following form:

$$\ddot{\mathbf{e}} = \mathbf{D}[\boldsymbol{\tau} - \mathbf{H}(\mathbf{e} + \mathbf{q}_d, \dot{\mathbf{e}} + \dot{\mathbf{q}}_d) - \mathbf{G}(\mathbf{e} + \mathbf{q}_d)] - \ddot{\mathbf{q}}_d \tag{38}$$

where $\mathbf{e} = \mathbf{q} - \mathbf{q}_d$ is the vector of joint coordinate errors. Alternatively, the above equation can be given as below:

$$\ddot{e}_1 = D_{11}\left(-k_{11}e_1 - k_{12}\dot{e}_1 - \frac{1}{2}m_2 S_2 \ell^2 (\dot{e}_2 + \dot{q}_{2d})^2\right.$$
$$\left. - m_2 S_2 \ell^2 (\dot{e}_1 + \dot{q}_{1d})(\dot{e}_2 + \dot{q}_{2d})\right) \tag{39a}$$
$$+ D_{12}\left(-k_{21}e_2 - k_{22}\dot{e}_2 + \frac{1}{2}m_2 S_2 \ell^2 (\dot{e}_1 + \dot{q}_{1d})^2\right) - \ddot{q}_{1d}$$

$$\ddot{e}_2 = D_{21}\left(-k_{11}e_1 - k_{12}\dot{e}_1 - \frac{1}{2}m_2 S_2 \ell^2 (\dot{e}_2 + \dot{q}_{2d})^2\right.$$
$$\left. - m_2 S_2 \ell^2 (\dot{e}_1 + \dot{q}_{1d})(\dot{e}_2 + \dot{q}_{2d})\right) \tag{39b}$$
$$+ D_{22}\left(-k_{21}e_2 - k_{22}\dot{e}_2 + \frac{1}{2}m_2 S_2 \ell^2 (\dot{e}_1 + \dot{q}_{1d})^2\right) - \ddot{q}_{2d}$$

where $C_i = \cos(e_i + q_{id})$, $S_i = \sin(e_i + q_{id})$ and $C_{ij} = \cos(e_i + q_{id} + e_j + q_{jd})$. In the state space form, Equations (39) can be written as follows (to be used by scheme I):

$$\dot{x}_1 = x_2 \tag{40}$$

$$\dot{x}_2 = D_{11}\left(-k_{11}x_1 - k_{12}x_2 - \frac{1}{2}m_2S_2\ell^2(x_4 + \dot{q}_{2d})^2\right.$$
$$\left. - m_2S_2\ell^2(x_2 + \dot{q}_{1d})(x_4 + \dot{q}_{2d})\right)$$
$$+ D_{12}\left(-k_{21}x_3 - k_{22}x_4 + \frac{1}{2}m_2S_2\ell^2(x_2 + \dot{q}_{1d})^2\right) - \ddot{q}_{1d}$$

$$\dot{x}_3 = x_4$$

$$\dot{x}_4 = D_{21}\left(-k_{11}x_1 - k_{12}x_2 - \frac{1}{2}m_2S_2\ell^2(x_4 + \dot{q}_{2d})^2\right.$$
$$\left. - m_2S_2\ell^2(x_2 + \dot{q}_{1d})(x_4 + \dot{q}_{2d})\right)$$
$$+ D_{22}\left(-k_{21}x_3 - k_{22}x_4 + \frac{1}{2}m_2S_2\ell^2(x_2 + \dot{q}_{1d})^2\right) - \ddot{q}_{2d}$$

where $S_2 = \sin(x_3 + q_{2d})$, $x_1 = e_1$, $x_2 = \dot{e}_1$, $x_3 = e_2$ and $x_4 = \dot{e}_2$. Terms of the nonstandard scheme II are obtained from the general Equations (15) and (16):

$$u_1 = 0 \tag{41}$$

$$u_2 = D_{11}\left[k_{12} - m_2S_2\ell^2(x_4^k + \dot{q}_{2d})\right] + \frac{1}{2}D_{12}m_2S_2\ell^2(x_2^k + 2\dot{q}_{1d})$$

$$u_3 = 0$$

$$u_4 = D_{21}\left[-\frac{1}{2}m_2S_2\ell^2(x_4^k + 2\dot{q}_{2d}) - m_2S_2\ell^2(x_2^k + \dot{q}_{1d})\right] + D_{22}k_{22}$$

and

$$v_1 = x_2^k \tag{42}$$

$$v_2 = D_{11}\left[-k_{11}x_1^k - \frac{1}{2}m_2S_2\ell^2(x_4^k + \dot{q}_{2d})^2 - m_2S_2\ell^2\dot{q}_{1d}(x_4^k + \dot{q}_{2d})\right]$$
$$+ D_{12}\left(-k_{21}x_3^k + k_{22}x_4^k + \frac{1}{2}m_2S_2\ell^2\dot{q}_{1d}^2\right) - \ddot{q}_{1d}$$

$$v_3 = x_4^k$$

$$v_4 = D_{21}\left(-k_{11}x_1^k + k_{12}x_2^k - \frac{1}{2}m_2S_2\ell^2\dot{q}_{2d}^2 - m_2S_2\ell^2\dot{q}_{2d}(x_2^k + \dot{q}_{1d})\right)$$

$$+ D_{22}\left(-k_{21}x_3^k + \frac{1}{2}m_2 S_2 \ell^2 (x_2^k + \dot{q}_{1d})^2\right) - \ddot{q}_{2d}$$

4.2.2. *Numerical Results*

The desired trajectories, q_{id} ($i = 1, 2$), are defined in the joint coordinate space as follows:

$$q_{id} = q_{iI} + (q_{iF} - q_{iI})[10t_n^3 - 15t_n^4 + 6t_n^t] \tag{43a}$$

$$t_n = \frac{t - t_I}{t_F - t_I} \tag{43b}$$

q_{iI} and q_{iF} are the initial and final positions and, t_I and t_F are the initial and final time instants. Table 1 shows the values of these parameters. Figure 10 shows the desired trajectory of joint 1 and joint 2 during the task.

Table 1: Initial and final values of the joints coordinate.

Joint	t_I	t_F	q_{iI}	q_{iF}
1	3 s	5 s	$-90°$	$-70°$
2	4 s	6 s	$0°$	$10°$

The simulation study is first conducted using RK4 method. Figure 11 shows the tracking errors (e_1 and e_2) using a small step-size of 0.001 s. The results with larger step-sizes of 0.007 s and 0.009 s, are shown in Figures 12 and 13. As is seen, the solution converges to right values for only small step-sizes. As the step-size becomes greater than 0.005 s, the numerical results converge to either spurious solutions or (by further increasing the step-size) do not converge.

Figure 14 shows the tracking errors using scheme I and with step-size $h = 0.001$s. The effective step-size, φ, during numerical simulation is shown in Figure 15. An acceptable value for φ is the one for which the maximum eigenvalue of the Jacobian matrix R is less than 1 to ensure the discrete system preserves the stability properties of the equilibrium points of the original system. Note that in this case study, the elements of the Jacobian matrix R changes with trajectory. Consequently, the maximum eigenvalue and the effective step-size change with time. Figure 16 shows the simulation results for $h = 10$s and the effective step-size is shown in Figure 17.

Figure 18 shows the tracking errors using scheme II and with step-size $h = 0.001$s. The effective step size is shown in Figure 19. Figure 20 shows the simulation of the tracking errors given the step-size $h = 10$s. As is seen, the scheme still indicates correct indication regarding the stability of the fixed point even with this large step-size. Figure 21 shows that the effective step-size, φ, has increased to ≈ 0.025s when the actual step-size was changed to 10s.

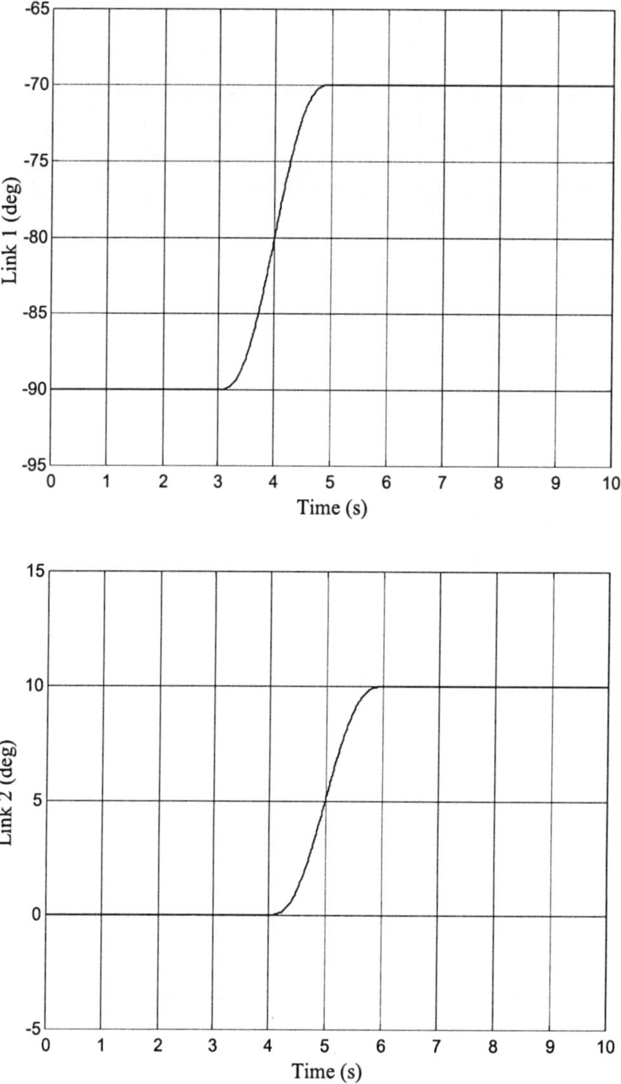

Fig. 10.   Desired joint trajectories (case study II).

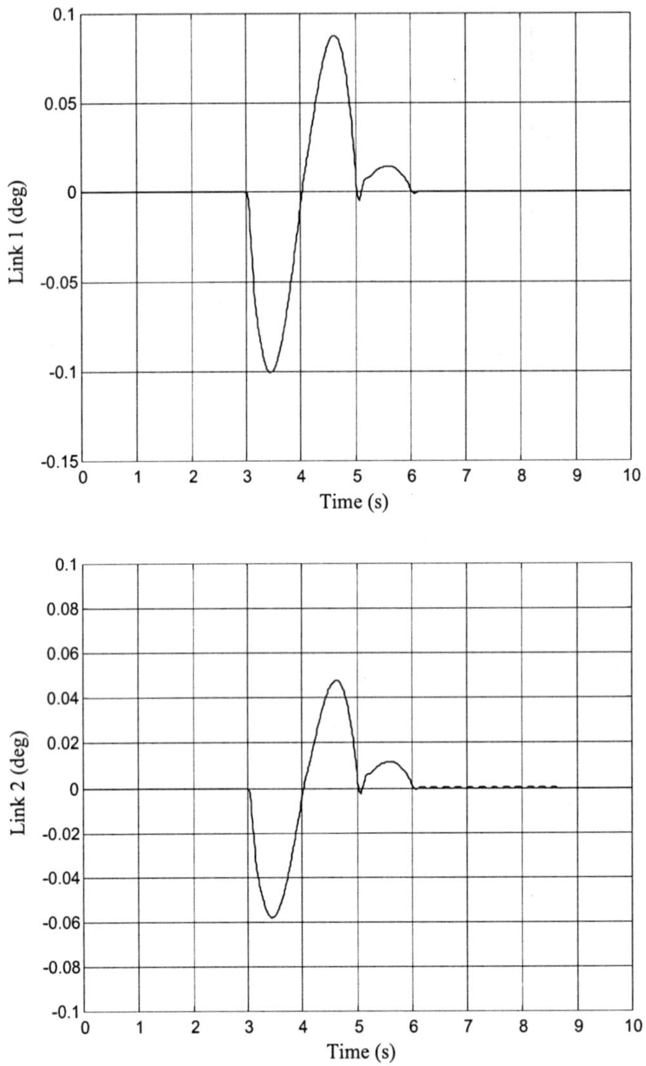

Fig. 11.   Tracking errors using RK4 method with step-size 0.001s.

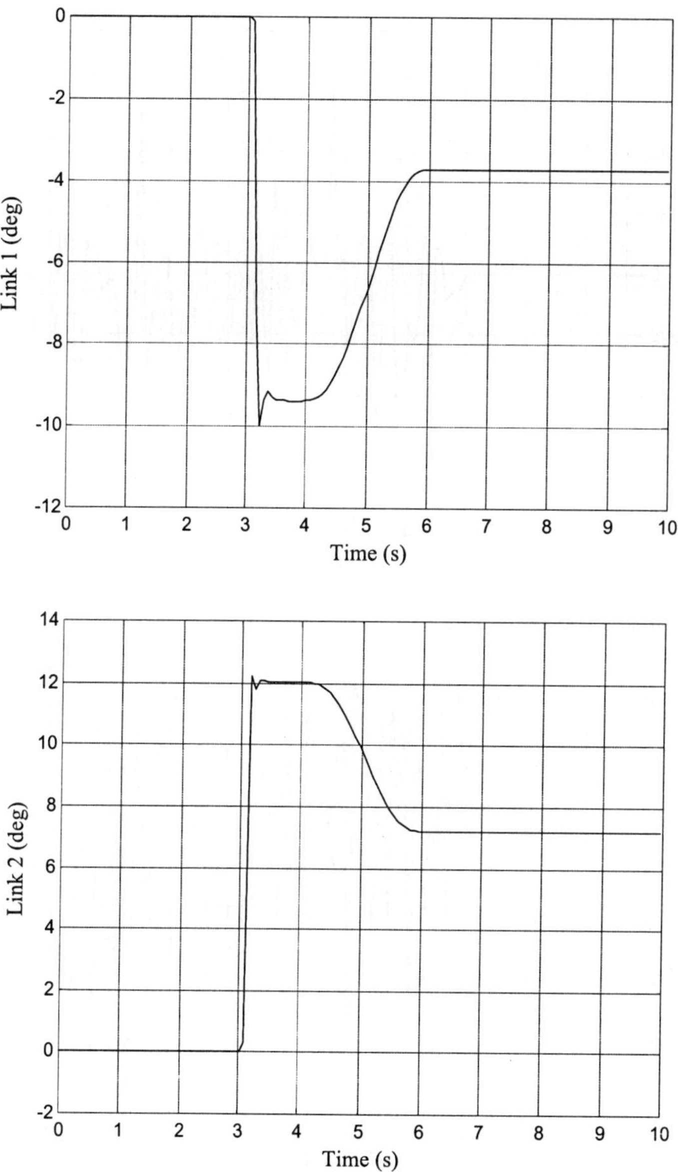

Fig. 12.   Tracking errors using RK4 method with step-size 0.007s.

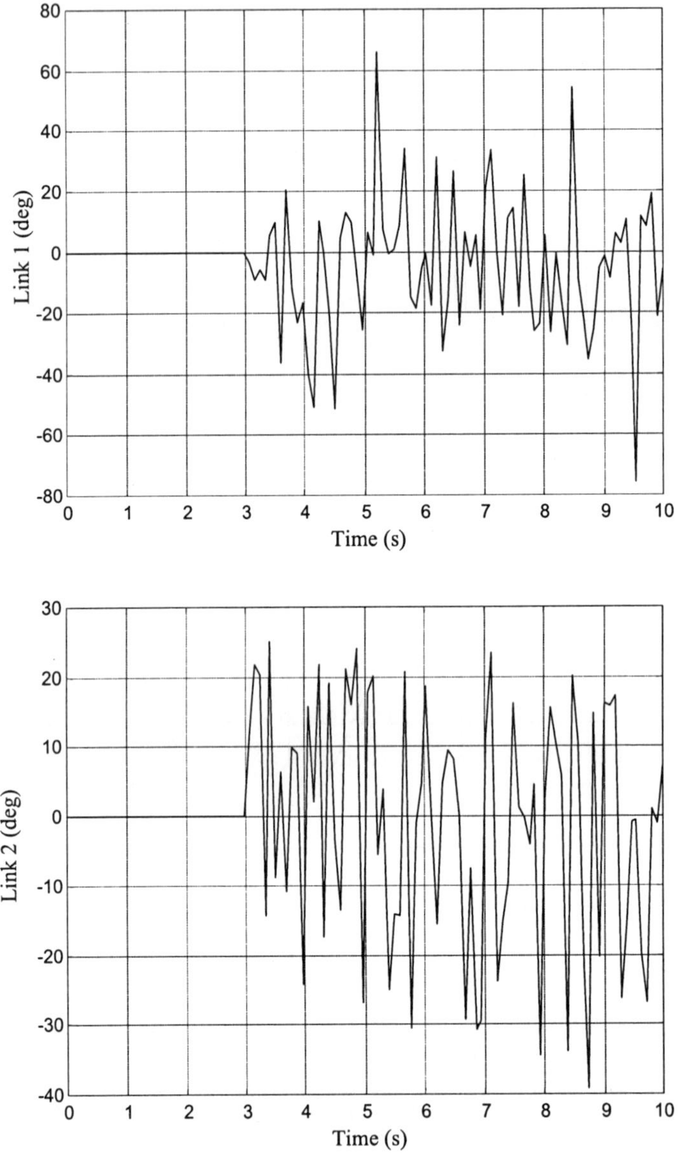

Fig. 13.   Tracking errors using RK4 method with step-size 0.009s.

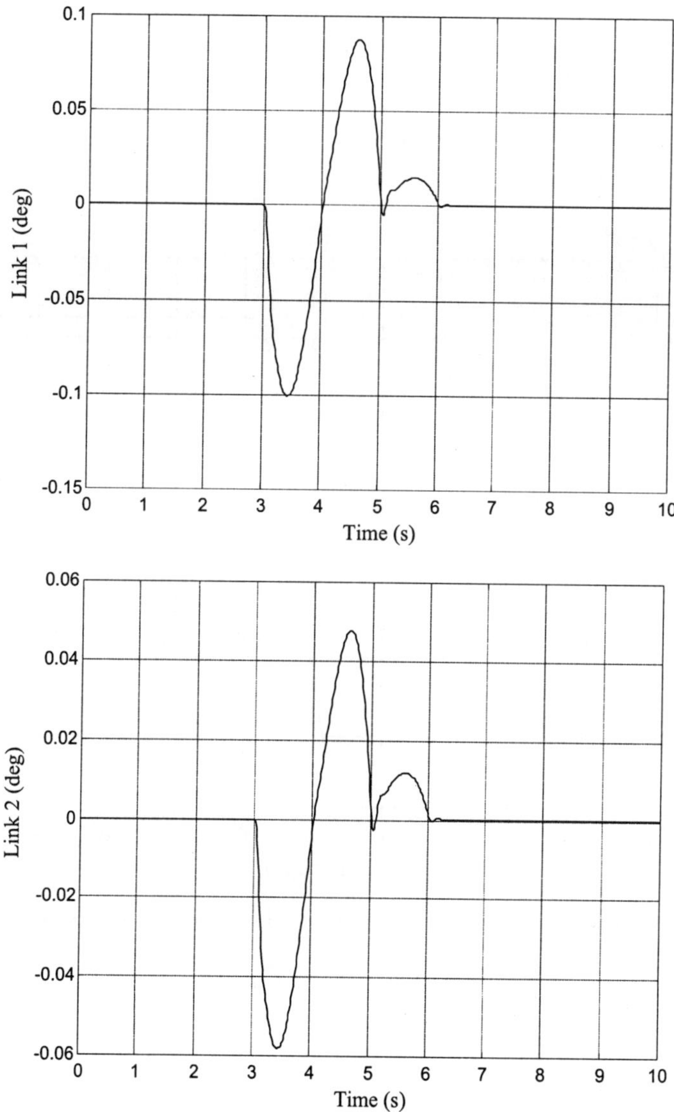

Fig. 14.   Tracking errors using nonstandard scheme I with step-size $h = 0.001$s.

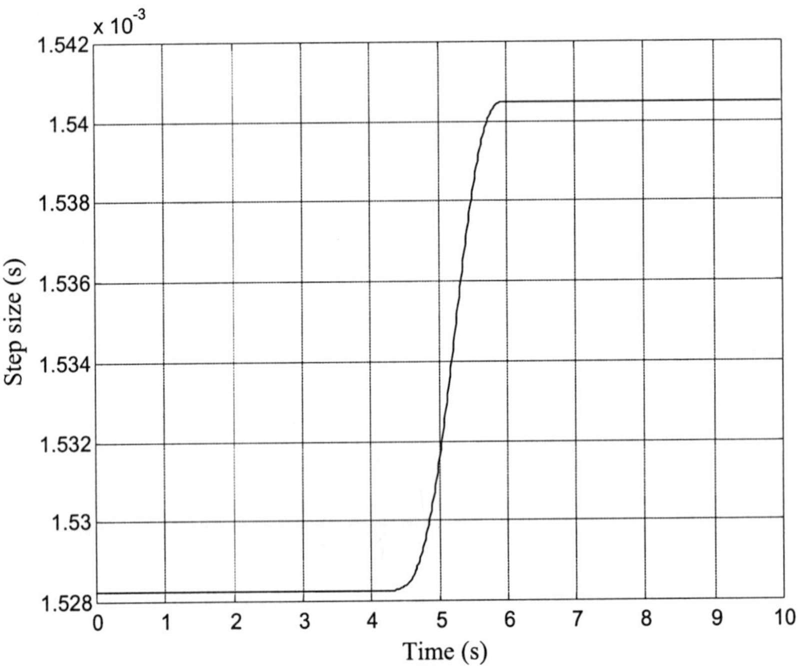

Fig. 15.   Changes in the effective step-size, φ, using nonstandard scheme I (step-size $h = 0.001$s).

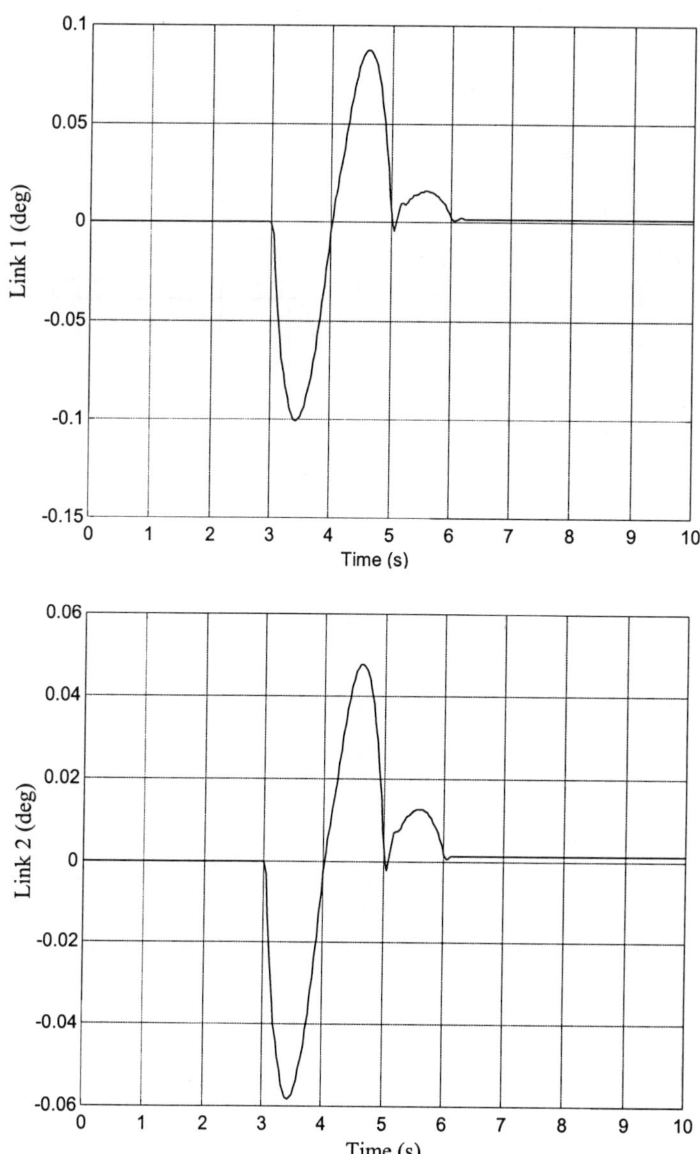

Fig. 16. Tracking errors using nonstandard scheme I with step-size $h = 10$s.

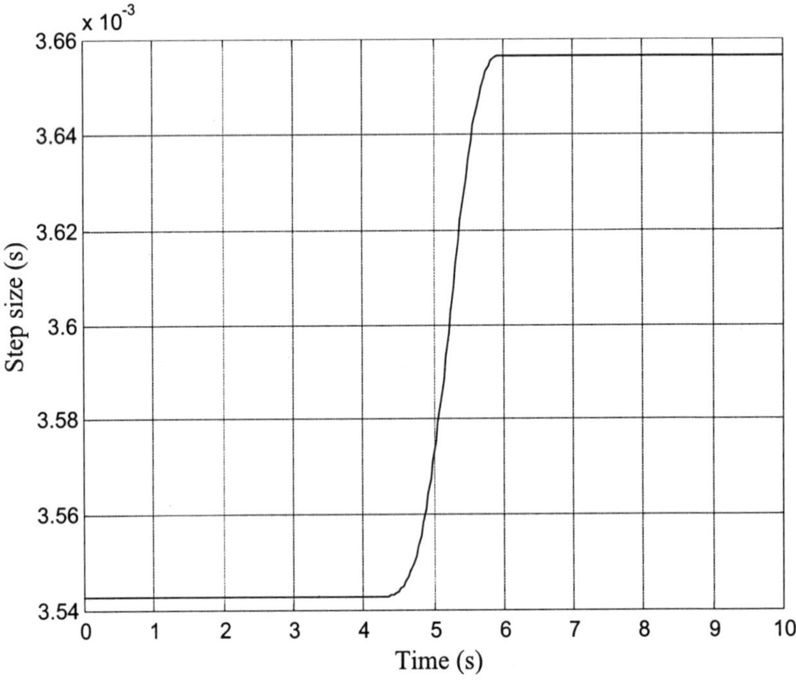

Fig. 17. Changes in the effective step-size, φ, using nonstandard scheme I (step-size $h = 10$s).

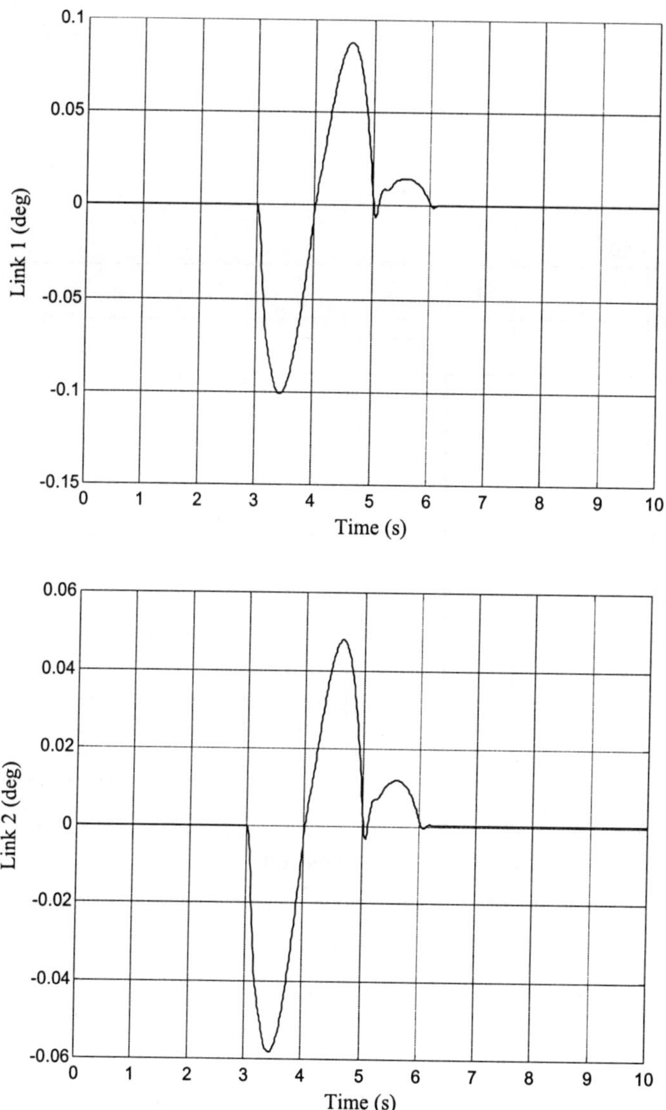

Fig. 18. Tracking errors using nonstandard scheme II with step-size $h = 0.001$s.

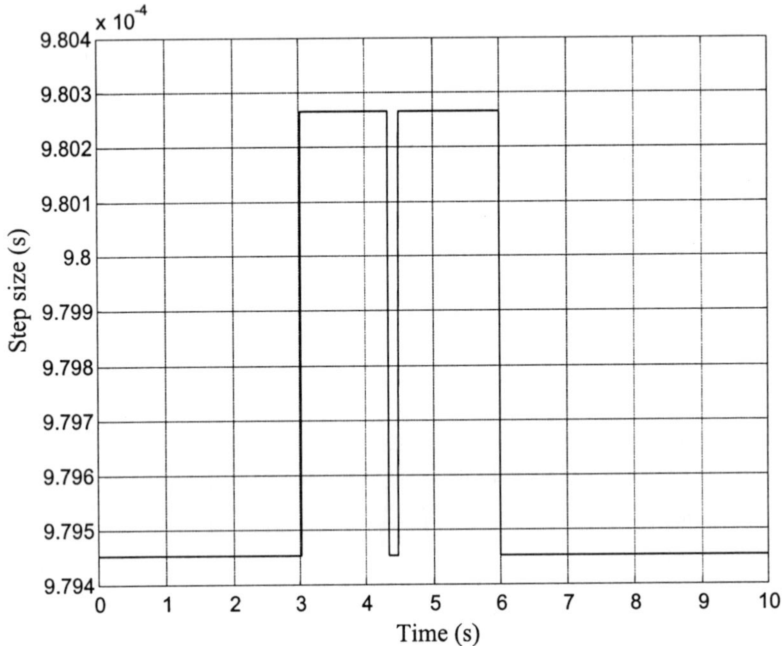

Fig. 19.   Changes in the effective step-size, φ, using nonstandard scheme II (step-size $h = 0.001$s).

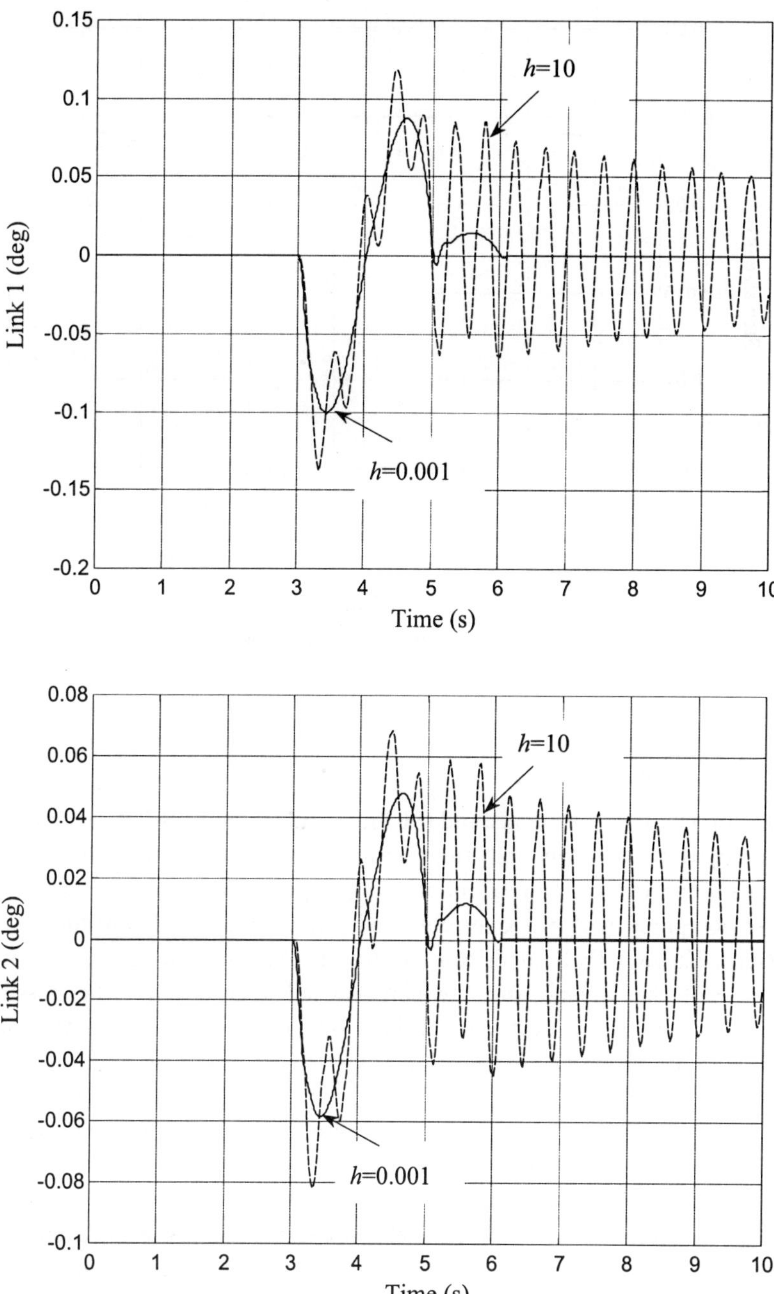

Fig. 20. Comparison between simulation results of tracking errors using small and large values for the step-size h (scheme II).

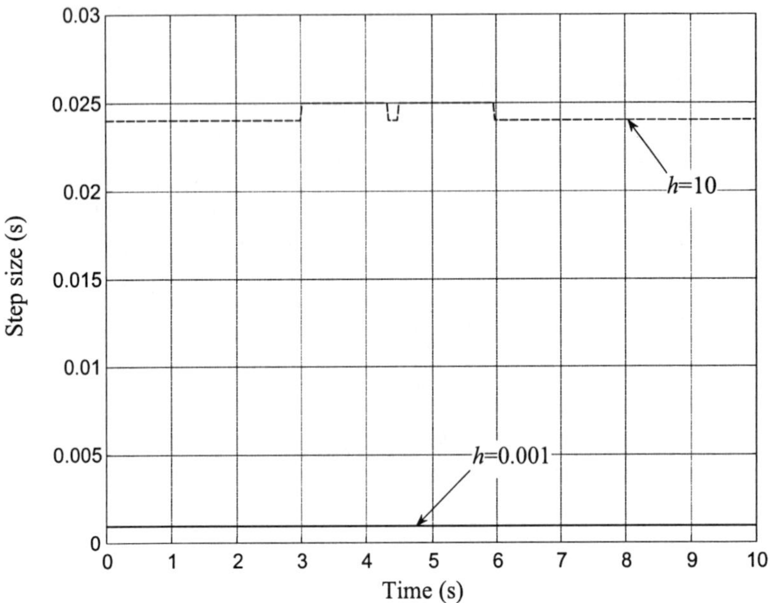

Fig. 21. Changes in the effective step-size, φ, using nonstandard scheme II (step-sizes $h = 0.001$s and $h = 10$s).

5. Conclusions

In this chapter, two nonstandard finite difference (NSFD) schemes were employed to, for the first time, study numerical solutions of the equations of motion of a typical robot manipulator. The simulation results were also compared with the ones obtained using the conventional fourth-order Runge-Kutta (RK4) algorithm. It was shown that by using the proper choice of the scheme and the nonlocal substitutions, (i) the possibility of having spurious solutions obtained by the RK4 method is eliminated, (ii) stability properties of the original system are preserved, and (iii) simulation results are numerically stable given any selected value of the step-sizes. This is significant since using the presented schemes, numerical simulation of the complex dynamics of robotic systems can be expedited, without facing numerical instabilities.

A difficulty of using NSFD schemes is the on-line calculation of the effective step-size for solving discrete derivatives. As the number of states increases, the computational load increases especially for the case in which the effective step-size changes as a function of the manipulator trajectory.

The relation between the step-size and the transient responses obtained by numerical simulations needs to be also investigated in detail. Future research in this area should also focus on obtaining "the best discretization" (i.e., the one that really reflects the dynamics of the differential equations) by other nonlocal approximations.

References

1. B. Chen and F. Solis, Discretizations of nonlinear differential equations using explicit finite order methods, *Journal of Computational and Applied Mathematics* **90** (1998), 171–183.
2. R. E. Mickens, *Nonstandard Finite Difference Models of Differential Equations*, (World Scientific, Singapore, 1994).
3. A. Serfaty de Markus and R. E. Mickens, Suppression of numerically induced chaos with nonstandard finite difference scheme, *Journal of Computational and Applied Mathematics* **106** (1999), 317–324.
4. R. E. Mickens, *Applications of Nonstandard Finite Difference Schemes*, (World Scientific, Singapore, 2000).
5. R. Anguelov and J. Lubuma, Contributions to the mathematics of the nonstandard finite difference method and applications, *Journal of Numerical Methods for Partial Differential Equations* **17** (2001), 518–543.
6. R. E. Mickens, Discretizations of nonlinear differential equations using explicit nonstandard methods, *Journal of Computational and Applied Mathematics* **110** (1999), 181–185.
7. P. Sekhavat, N. Sepehri and Q. Wu, Calculation of Lyapunov exponent using nonstandard finite difference discretizations scheme: a case study, *Journal of Differential Equations and Applications* **10** (2004), 369–378.
8. M. M. Sallam, R. F. Abo-Shanab and A.A. Nasser, Modified methods for dynamic modeling of robot manipulators, *Proceedings ASME Design Engineering Technical Conference* (1998), Paper DETC98/MECH5860.
9. K. S. Fu, R. C. Gonzalez and C.S.G. Lee, *Robotics: Control, Sensing, Vision, and Intelligence*, (McGraw-Hill, Singapore, 1987).

CHAPTER 3

APPLICATIONS OF MICKENS FINITE DIFFERENCES TO SEVERAL RELATED BOUNDARY VALUE PROBLEMS

Ron Buckmire

Mathematics Department
Occidental College
1600 Campus Road
Los Angeles, CA 90041-3338, U.S.A.
buckmire@oxy.edu

The initial discovery and implementation by the author of a particular kind of nonstandard finite difference (NSFD) scheme called a "Mickens finite difference" (MFD) for approximating the radial derivatives of the Laplacian in cylindrical coordinates is reviewed. The development of a similar scheme for the spherical coordinates case is also recounted. Examples of application of the schemes to several related (singular and nonsingular, linear and nonlinear) boundary value problems are given. Examples of applying Buckmire's MFD scheme to the bifurcatory, nonlinear eigenvalue problems of Bratu and Gel'fand are also presented. The results support the utility and versatility of MFD schemes for boundary value problems with singularities or bifurcations.

Keywords: NSFD, MFD, Bratu problem, Gelfand problem, nonstandard finite differences, Mickens finite difference, bifurcation, singular, nonlinear, eigenvalue

1. Introduction

In this paper a review of the discovery, development and various implementations by the author of a specific Mickens finite difference (MFD) shall be presented. Application of this MFD leads to nonstandard finite difference (NSFD) schemes which can be used to efficiently approximate solutions to various boundary value problems. Boundary value problems associated

---

The author is Associate Professor and Chairperson of Mathematics at Occidental College.

with Bratu, Gel'fand and others are considered. The MFD schemes given here produce new and different ways to discretize the Laplacian operator $r^p \dfrac{d}{dr}$, where $p = 1$ is the cylindrical case and $p = 2$ is the spherical case. They are examples of the kinds of numerical methods Professor Ronald E. Mickens of Clark Atlanta University has analyzed and popularized for years ([31], [32], [33], et cetera).

In Section 2 of this paper the discovery and development of the Mickens finite difference by the author is recounted. The initial application was to a mixed-type, nonlinear elliptic-hyperbolic partial differential equation with singular boundary conditions found in theoretical aerodynamics. These results have previously appeared in [6], [7] and [8] and are reviewed here.

In Section 3, the application of MFD schemes to two linear singular boundary value problems with known exact solutions which are related to the transonic aerodynamics problem discussed in Section 2 are recounted. These results have appeared in a more detailed fashion in [9].

In Section 4, an MFD scheme is applied to a different nonlinear, singular boundary value problem which is related to the linear, singular boundary value problems of Section 3 and also has known exact multi-valued solutions. These results have appeared in a more detailed fashion in [10].

In Section 5, an MFD scheme is applied to a nonsingular, nonlinear boundary value problem related to the singular, nonlinear boundary value problem of Section 4 which also has known exact multi-valued solutions related to the problem in Section 4. These results have not previously appeared in print.

The paper concludes with discussion about the versatility of Mickens finite differences for application with diverse kinds of boundary value problems and some suggestions for future work.

2. The Buckmire MFD Scheme

The goal of this section of the paper is to provide the background history for the development of the Mickens finite difference which is the main topic of this paper and is discused further in Section 3 and Section 4. In Buckmire's 1994 thesis [6] this MFD was introduced in order to find particular slender bodies of revolution that possess shock-free flows as specific numerical solutions of a mixed-type, singular boundary value problem. The problem is formulated using transonic small disturbance theory found in [12], [13] and [14], among other sources. Cole & Schwendeman announced the first computation of a fore-aft, symmetric, shock-free transonic slender body in

[16]. This work was expanded in [6], which led to the first computation of shock-free, transonic, slender bodies with axisymmetry but without fore-aft symmetry. Basically, the problem involves numerically solving a boundary value problem with an elliptic-hyperbolic partial differential equation (the Kármán-Guderley equation) in cylindrical coordinates, with a singular inner Neumann boundary condition at $r = 0$ and a non-singular outer Dirichlet boundary condition far away from $r = 0$. Namely,

$$(K - (\gamma + 1)\phi_x)\phi_{xx} + \phi_{\tilde{r}\tilde{r}} + \frac{1}{\tilde{r}}\phi_{\tilde{r}} = 0. \tag{1}$$

$$\begin{aligned}
\phi(x,\tilde{r}) &\rightarrow S(x)\log\tilde{r} + G(x), \quad &&\text{as } \tilde{r} \rightarrow 0, \ |x| \leq 1 \\
\phi(x,\tilde{r}) &\text{ bounded}, \quad &&\text{for } \tilde{r} = 0, \ |x| > 1.
\end{aligned} \tag{2}$$

$$\phi(x,\tilde{r}) \rightarrow \frac{\mathcal{D}}{4\pi} \frac{x}{(x^2 + K\tilde{r}^2)^{3/2}}, \qquad \text{as } (x^2 + \tilde{r}^2)^{1/2} \rightarrow \infty. \tag{3}$$

In (1), (2) and (3) the variable $\tilde{r}$ is a scaled cylindrical coordinate, K is the transonic similarity parameter, $\mathcal{D}$ is a dipole strength and $\phi(x,\tilde{r})$ is a velocity disturbance potential. Both $S(x)$ and $G(x)$ are bounded functions. The main point of sketching the boundary value problem here is to emphasize that the function $G(x)$ which occurs in (2) needs to be computed very accurately, because the pressure coefficient on the body depends directly on $G'(x)$. It is the pressure coefficient which allows the determinination of whether the body possesses a shock-free flow. Computing it is complicated by the fact that $\phi(x,\tilde{r})$ and $S(x)\log\tilde{r}$ are both becoming singular as $\tilde{r} \rightarrow 0$, which is where the boundary condition must be evaluated, and the quantity $G'(x)$ we require is the *derivative* of the difference between these two large quantities. Thus a numerical method was needed to compute the solution $\phi(x,\tilde{r})$ particularly accurately as $\tilde{r} \rightarrow 0$. It was discovered that an exact, nonstandard finite-difference scheme existed for a simpler, related boundary value problem. This discovery was the motivation for adoption of the scheme introduced in [6] and analyzed and discussed in more detail in [7] and [8]. Upon further analysis the author found other nonstandard finite difference schemes which could be derived for slightly different boundary value problems, and then extended this concept to boundary value problems in spherical coordinates. It is these results which were presented in detail in [9] that are summarized below.

2.1. *Derivation of the Buckmire MFD Scheme*

This subsection shall explain the formal derivation of the Buckmire MFD scheme. It is a nonstandard discretization of the Laplacian operator $\mathcal{R}_p = r^p \dfrac{d}{dr}$, where $p = 1$ or $p = 2$. Laplace's equation in cylindrical coordinates is given by

$$\frac{1}{r}(ru_r)_r + u_{\theta\theta} = 0$$

and clearly contains the $\mathcal{R}_1$ operator. Laplace's equation in spherical coordinates is given by

$$\frac{1}{r^2}(r^2 u_r)_r + u_{\theta\theta} = 0$$

and clearly contains the $\mathcal{R}_2$ operator. The Kármán-Guderley equation (1) which was the subject of [6] also contains $\mathcal{R}_1$, the radial derivatives of the Laplacian in cylindrical coordinates. The first step in the discretization of the $\mathcal{R}_p$ operator is to choose a grid $\{r_j\}_{j=0}^N$ on the interval $0 \leq r \leq 1$ where

$$0 \leftarrow r_0 < r_1 < r_2 < \ldots < r_j < \ldots < r_N = 1. \tag{4}$$

In the subsequent subsections the derivation of the Buckmire MFD scheme for the cylindrical and spherical cases will be given.

2.1.1. *The cylindrical case*

Consider $B(r)$ which is defined as

$$B(r) = \mathcal{R}_1 u = \frac{r\,du}{dr},$$

where $u = u(r)$ is an unknown function (the solution) the operator $\mathcal{R}_1$ acts on. The problem at hand requires determining a numerical discretization or approximation for $\mathcal{R}_1$.

There are several choices for discretizing $B(r)$, the radial derivatives in cylindrical coordinates, but the standard forward-difference approximation method and the new nonstandard scheme were selected and will be compared with each other. Note that the discrete quantity B is actually defined in between grid points, not on them.

$$B_{j+1/2}^{(SFD)} = r_{j+1/2} \frac{u_{j+1} - u_j}{r_{j+1} - r_j} \tag{5}$$

$$B_{j+1/2}^{(MFD)} = \frac{u_{j+1} - u_j}{\log(r_{j+1}) - \log(r_j)} = \frac{u_{j+1} - u_j}{\log(r_{j+1}/r_j)} \tag{6}$$

The scheme in (5) shall be referred to as the SFD scheme and the new scheme in (6) shall be referred to as the MFD scheme. The MFD scheme can be obtained by assuming that $B(r)$ should be constant on each subinterval (r_j, r_{j+1}) of the grid. If one relates $B(r)$ back to the physical fluid mechanics problem we want to solve, it corresponds to a mass flux. Therefore the condition is being imposed that the mass flux be constant, which is physically appropriate. The relationship between $B_{j+1/2}$ and u_j and u_{j+1} solves the simple boundary value problem

$$ru' = B_{j+1/2} = \text{constant} \tag{7}$$

$$u(r_j) = u_j \tag{8}$$

$$u(r_{j+1}) = u_{j+1}. \tag{9}$$

The solution to this is $u(r) = B_{j+1/2} \log r + C$, which, when one applies the boundary conditions (8) and (9) leads to the formula

$$B_{j+1/2} = \frac{u_{j+1} - u_j}{\log(r_{j+1}/r_j)}.$$

Thus the MFD is a nonstandard, *exact* finite-difference scheme for the ordinary differential equation

$$r\frac{du}{dr} = B, \text{where } B \text{ is a known constant.} \tag{10}$$

2.1.2. *The spherical case*

In a similar fashion to the procedure outlined above, a nonstandard, exact finite-difference scheme can be obtained for the spherical analogue to (10). The spherical version of the differential equation comes from setting $A(r) = \mathcal{R}_2 u$ equal to a constant, producing

$$r^2\frac{du}{dr} = A, \text{where } A \text{ is a known constant.} \tag{11}$$

Even though it does not have the same physical significance of a mass flux as it did in cylindrical co-ordinates, we can still obtain a relationship between $A_{j+1/2}$, u_j and u_{j+1} by solving (11) using the conditions (8) and (9). The solution in this case is $u(r) = -A/r + C$, which when one applies the boundary condition leads to the difference equation

$$A_{j+1/2} = \frac{u_{j+1} - u_j}{\left(-\dfrac{1}{r_{j+1}}\right) - \left(-\dfrac{1}{r_j}\right)}.$$

This can be re-arranged to produce

$$A^{(MFD)}_{j+1/2} = r_j r_{j+1} \frac{u_{j+1} - u_j}{r_{j+1} - r_j}. \tag{12}$$

This formula is again an *exact*, nonstandard finite-difference scheme for (11). The SFD scheme for this differential equation would be

$$A^{(SFD)}_{j+1/2} = r^2_{j+1/2} \frac{u_{j+1} - u_j}{r_{j+1} - r_j}. \tag{13}$$

Notice that in the spherical MFD scheme (12) it is the non-local discretization of r^2 which makes it nonstandard or "Mickens finite difference." In the cylindrical MFD scheme (6) it is the presence of nonlinear functions (logarithms) and the non-local discretization which make it nonstandard. Regardless, both schemes have zero local truncation error; they are exact.

2.2. *Informal Derivation of the Nonstandard Schemes*

One can also derive the form of the nonstandard schemes given in the previous section by using a more intuitive but less rigorous approach involving differentials. The differential equation to be approximated is re-arranged through the use of differentials and then the differentials are approximated by finite Δs.

$$B = r \frac{du}{dr} = \frac{du}{\dfrac{dr}{r}} = \frac{du}{d(\log(r))} \approx \frac{\Delta u}{\Delta(\log(r))}$$

This approximate version of the rearranged ODE is actually the cylindrical MFD (6).

$$\frac{\Delta u}{\Delta(\log(r))} = \frac{u_{j+1} - u_j}{\log(r_{j+1}) - \log(r_j)} = B^{(2)}_{j+1/2}$$

Similarly, one can derive the spherical MFD found in (12) by rearranging the ODE in (11)

$$A = r^2 \frac{du}{dr} = \frac{du}{\dfrac{dr}{r^2}} = \frac{du}{d\left(-\dfrac{1}{r}\right)} \approx \frac{\Delta u}{\Delta\left(-\dfrac{1}{r}\right)}$$

$$= \frac{u_{j+1} - u_j}{\left(-\dfrac{1}{r_{j+1}}\right) - \left(-\dfrac{1}{r_j}\right)} = r_j r_{j+1} \frac{u_{j+1} - u_j}{r_{j+1} - r_j} = A^{(2)}_{j+1/2}$$

Conclusion

In this section two very different derivations of MFD schemes which discretize the $\mathcal{R}_p$ operator in cylindrical ($p = 1$) or spherical ($p = 2$) coordinates were given. The new schemes were presented adjacent to the SFD methods for the same Laplacian operator to highlight the unusual features of the nonstandard schemes. In the subsequent sections, both types of schemes will be applied to particular singular boundary value problems and the numerical results will demonstrate the superior utility and versatility of the MFD schemes.

3. MFD Application to Two Singular Linear Boundary Value Problems

Introduction

To illustrate the efficacy and accuracy of the MFD schemes derived in (6) and (12) in Section 2, they are applied to a number of *linear* singular boundary value problems related to the original problem discussed in Buckmire's thesis [6]. These other problems possess the same essential singular nature near the origin due to the nature of the Laplacian operator in cylindrical and spherical coordinates. The reason for the choice of these particular linear singular boundary value problems is that they have easily found exact solutions which involve logarithms or Bessel functions. Thus these problems are highly suitable for the benchmarking of the Buckmire MFD scheme introduced in [6] and discussed in [7], [8] and Section 2. The goal of this section of this paper is to present a brief summary of the results reported in [9] which indicate the utility of MFD schemes for the numerical solution of singular boundary value problems.

3.1. *The First Model Problem*

The Kármán-Guderley equation (1) and the associated boundary conditions of (2) and (3) can be directly related to the simple boundary value problem given below

$$\frac{1}{r}\frac{d}{dr}\left(r\frac{du}{dr}\right) - m^2 u = 0, \qquad m \text{ constant} \qquad (14)$$

$$r\frac{du}{dr}\bigg|_{r=0} = S, \qquad (15)$$

$$u(1) = G. \qquad (16)$$

The singular boundary value problem has a known exact solution involving logarithms or Bessel functions, depending on the value of m, where m is a natural number. The exact solutions to the model boundary value problem in cylindrical coordinates can be written as

$$m = 0, \qquad u(r) = S \log r + G \tag{17}$$

$$m > 0, \qquad u(r) = -SK_0(rm) + (G + SK_0(m))\frac{I_0(rm)}{I_0(m)}. \tag{18}$$

Note that these solutions to the cylindrical model problem have the required singular behavior $(\mathcal{O}(\log r))$ as $r \to 0$.

3.1.1. *The $m \neq 0$ cylindrical case*

When $m \neq 0$ the model differential equation in (14) becomes

$$r^2 u''(r) + r u'(r) - m^2 r^2 u(r) = 0, \tag{19}$$

which after the scaling $s = mr$ can be seen to be the zeroth-order Bessel's equation

$$s^2 u''(s) + s u'(s) - s^2 u(s) = 0.$$

Using a standard discretization, the difference equation for the $m \neq 0$ form of (14) is

$$\frac{1}{r_j}\left(r_{j+1/2}\frac{u_{j+1} - u_j}{r_{j+1} - r_j} - r_{j-1/2}\frac{u_j - u_{j-1}}{r_j - r_{j-1}}\right) - m^2 u_j = 0. \tag{20}$$

Using the Buckmire MFD, the nonstandard discretization will produce

$$\frac{1}{r_j}\left(\frac{u_{j+1} - u_j}{\log(r_{j+1}/r_j)} - \frac{u_j - u_{j-1}}{\log(r_j/r_{j-1})}\right) - m^2 u_j = 0. \tag{21}$$

In the $m \neq 0$ cases, solutions are obtained by solving a tri-diagonal system of equations for $\{u_j\}_{j=0}^{N}$. In the $m \neq 0$ case the nonstandard scheme is not exact, but it can be clearly seen from the numerical results given later in this section that the MFD scheme does a better job of approximating the exact solution than the standard scheme does, especially near the $r = 0$ singularity. Recall, that in the original problem, it is near $r = 0$ that accuracy is most required. Before examining the numerical results of solving the difference equations in (20) and (21), the application of Buckmire's MFD to another singular boundary value problem will be presented in the next subsection.

3.2. *The Second Model Problem*

The second model problem (in spherical coordinates) is not directly motivated from the Kármán-Guderley boundary value problem as the cylindrical version is. It is simply an analogous extrapolation from the cylindrical model problem given in (14),

$$\frac{1}{r^2}\frac{d}{dr}\left(r^2\frac{du}{dr}\right) - n^2 u = 0, \qquad n \text{ constant} \tag{22}$$

$$r^2\frac{du}{dr}\bigg|_{r=0} = S, \tag{23}$$

$$u(1) = G. \tag{24}$$

The exact solution of the model spherical differential equation can be found to consist of hyperbolic sines and hyperbolic cosines after noticing that (22) can be rewritten (when $n \neq 0$) as

$$r^2 u''(r) + 2r u'(r) - n^2 r^2 u(r) = 0. \tag{25}$$

This looks very similar to the Bessel's equation from the cylindrical coordinates problem (19), but the solutions are very different. The derivatives can be grouped so that if $v = ru$ the equation becomes

$$r^2 u'' + 2r u' - n^2 r^2 u = (ru)'' - n^2(ru)$$
$$= v''(r) - n^2 v(r)$$
$$= 0.$$

The general solution to (25) is $u(r) = C_1 \dfrac{\sinh(nr)}{r} + C_2 \dfrac{\cosh(nr)}{r}$. The exact solutions to the model boundary value problem in spherical coordinates given in (22), (23) and (24) can be written as

$$n = 0, \quad u(r) = -\frac{S}{r} + S + G \tag{26}$$

$$n > 0, \quad u(r) = \frac{-S\cosh(nr)\sinh(n) + (G + S\cosh(n))\sinh(nr)}{r\sinh(n)}. \tag{27}$$

These solutions above also exhibit singular behavior as $r \to 0$, albeit much more strongly than their cylindrical coordinate counterparts. The solutions in spherical coordinates have singular behavior $(\mathcal{O}(\frac{1}{r}))$ as $r \to 0$.

3.2.1. *The $n \neq 0$ spherical case*

Using a standard discretization for the $n \neq 0$ form of (22) is

$$\frac{1}{r_j^2}\left(r_{j+1/2}^2 \frac{u_{j+1} - u_j}{r_{j+1} - r_j} - r_{j-1/2}^2 \frac{u_j - u_{j-1}}{r_j - r_{j-1}}\right) - n^2 u_j = 0. \tag{28}$$

The Buckmire MFD produces a nonstandard discretization which is

$$\frac{1}{r_j^2}\left(r_j r_{j+1}\frac{u_{j+1}-u_j}{r_{j+1}-r_j}-r_j r_{j-1}\frac{u_j-u_{j-1}}{r_j-r_{j-1}}\right)-n^2 u_j=0. \qquad (29)$$

Like the $m\neq 0$ cylindrical case, the solutions to the $n\neq 0$ spherical case are obtained by solving a tri-diagonal system of equations for $\{u_j\}_{j=0}^N$. Fortunately, exact solutions can be found for all values of m (cylindrical cases) and n (spherical cases). In the $n\neq 0$ cases the MFD scheme (29) is not exact, but the numerical results given here will demonstrate that it does a much better job of approximating the exact solution than the standard scheme (29) does.

3.3. *Numerical Results*

In this subsection the numerical results will be given which indicate the effectiveness of the MFD schemes in approximating the solution to the singular boundary value problems in cylindrical and spherical coordinates. This is done by comparing the solutions to the cylindrical and spherical model problems generated by the numerical schemes given in (20) & (21) and (28) & (29) to the exact solutions given in (18) and (27).

Numerically one cannot actually evaluate the Neumann boundary conditions (15) and (23) at $r=0$ exactly. Instead one chooses a small parameter ϵ and evaluates the boundary condition at $r=\epsilon$ repeatedly with values of ϵ that approach zero. For the results displayed in Figure 1, ϵ=.1, .01, .001, .0001 and .00000001. The filled dots in the figure represent the error due to the SFD scheme while the empty dots represent the error due to the MFD scheme. Note that in both the cylindrical and spherical case the error due to the MFD scheme near the boundary (i.e. as $\epsilon\to 0$) is consistently smaller than the error due to the SFD scheme. It is these results which demonstrate the ability of the nonstandard scheme to better handle the singular nature of the pertinent boundary value problems.

Since the MFD schemes in cylindrical coordinates and spherical coordinates are exact (have zero error) in the $m=0$ and $n=0$ cases, respectively they have not been included here but are available in [9].

For the $m\neq 0$ and $n\neq 0$ cases we need to compare the difference between the nonstandard scheme's solution and the exact solution with the difference between the standard scheme's solution and the exact solution. In addition, since the motivation for the scheme was the ability to evaluate the solution near $r=0$, the comparison of the numerical solutions with the exact solution at ever smaller values of ϵ is important. Figure 1 depicts

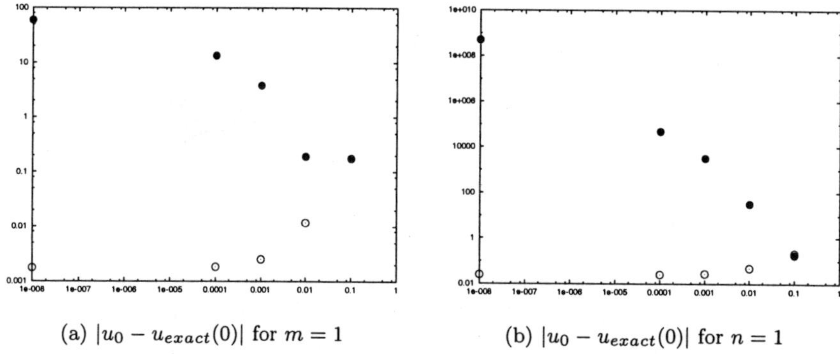

(a) $|u_0 - u_{exact}(0)|$ for $m = 1$ (b) $|u_0 - u_{exact}(0)|$ for $n = 1$

Fig. 1.   Numerical error at $r = \epsilon \to 0$ for cylindrical ($m = 1$) and spherical ($n = 1$) data

the error between the exact solution and the numerical solution that each numerical method makes as the Neumann boundary conditions (15) and (23) are approximated at ever smaller values of r ($\epsilon \to 0$) for both the cylindrical and spherical model boundary value problems.

The two types of finite-difference schemes (standard and nonstandard) approximate the solutions to these problems with wildly varying accuracy, with the MFD scheme being more successful by *orders of magnitude*. At $r = \epsilon = .0001$ the standard scheme produces an error of about 10^2 while the nonstandard scheme produces an error of about 10^{-1}. In Figure 2 the graphs show the error on a log-log scale with each curve representing a solution computed at a different value of ϵ. Notice in Figure 2(b) that the nonstandard scheme's error actually *decreases* as the boundary condition is evaluated at a more singular value closer to the origin, while the reverse is true for the standard scheme in Figure 2(a).

The corresponding graphs of the error made by the two competing schemes in solving the spherical model problem are given below in Figure 3. In Figure 3(a) one can notice that the order of magnitude of the error made by the standard scheme is gigantic ($\approx 10^4$) while in Figure 3(b) it is clear that the nonstandard scheme has only a modest error ($\approx 10^{-1}$), even when the inner boundary condition is being evaluated at the relatively small value of $\epsilon = .0001$.

All the calculations performed in this section used a uniform discrete grid with $N = 101$ grid points, with a grid separation which varied depending on ϵ. The known constants in the boundary conditions were taken to be $G = 2$ and $S = 5$ for no particular reason.

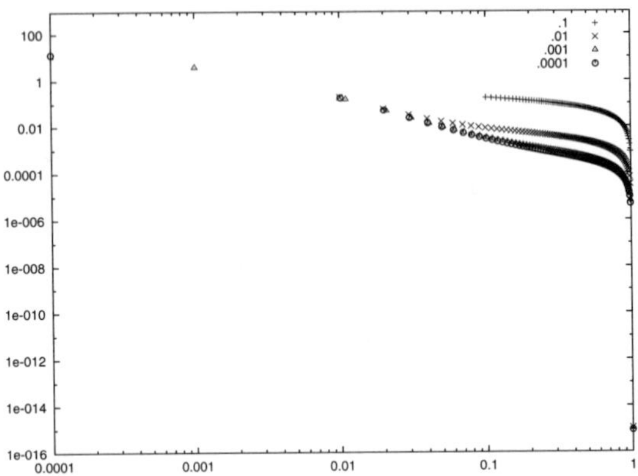

(a) $m = 1$ error using standard finite differences

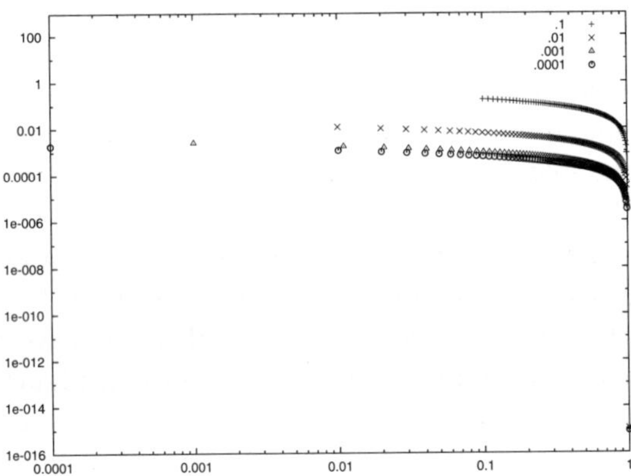

(b) $m = 1$ error using nonstandard finite differences

Fig. 2.   Numerical error comparison as $\epsilon \to 0$ for cylindrical solutions of (14)

Conclusions

In this section of the paper, MFD schemes have been applied to solve singular boundary value problems with differential equations in cylindrical or spherical coordinates. These model boundary value problems were simplified versions of the original boundary value problem the Buckmire MFD scheme was first invented for in [6]. The numerical results presented here

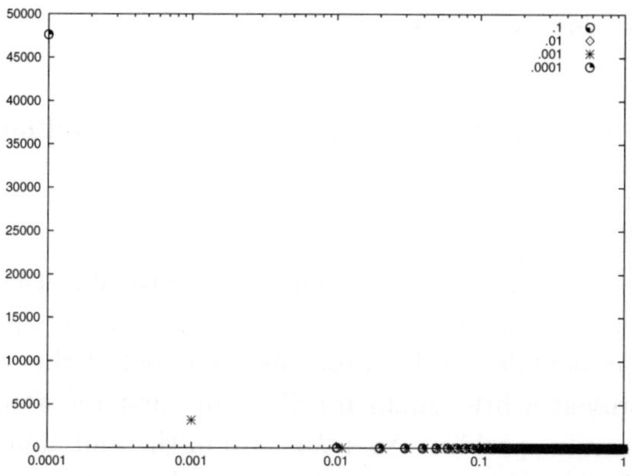

(a) $n = 1$ error using standard finite difference

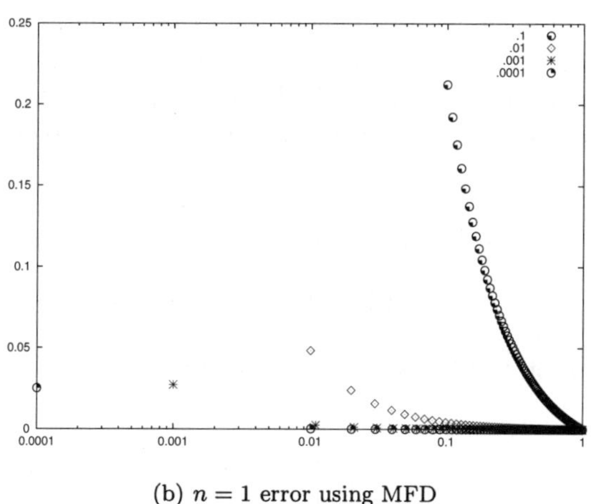

(b) $n = 1$ error using MFD

Fig. 3.   Numerical error comparison as $\epsilon \to 0$ for spherical solutions of (22)

show that the MFD schemes appear to tackle singular boundary value problems more accurately and efficiently than standard finite-difference schemes. In particular, the nonstandard schemes easily approximate the solution near the singularity at the origin where the standard schemes generally fail and where accuracy of the solution was most desired. In the next section, MFD schemes will be applied to some more singular boundary value problems, related to the ones considered in this section. However, the singular bound-

ary value problem in the next section are nonlinear and possess bifurcations
with multiple-valued solutions.

4. MFD Application to the Cylindrical Bratu-Gel'fand Problem

Introduction

Following application of MFD to a couple of singular linear boundary value
problems in the previous section, in this section MFD are applied to a singular nonlinear boundary value problem. The goal of this section of this
paper is to present a brief summary of results first reported in [10]. The
nonlinear eigenvalue problem $\Delta u + \lambda e^u = 0$ in the unit square with $u = 0$
on the boundary is often referred to as "the classical Bratu problem" or
"Bratu's problem." By changing the geometry to a unit circle the classical
Bratu problem is known as the Bratu-Gelfand problem [20]. It is a nonlinear eigenvalue problem with two known bifurcated solutions for $\lambda < \lambda_c$, no
solutions for $\lambda > \lambda_c$ and a unique solution when $\lambda = \lambda_c$. Due to the nature of the Laplacian operator in cylindrical coordinates, the Bratu-Gelfand
problem is also a singular nonlinear boundary value problem.

The Bratu-Gelfand problem can also be written as

$$u''(r) + \frac{1}{r}u'(r) + \lambda e^{u(r)} = 0 \qquad 0 \leq r \leq R,$$

$$\text{with } u(0) < \infty \quad \text{and} \quad u(R) = 0.$$

The exact solution to (30) is given in [36] and is

$$u(r; \lambda) = \ln \left[\frac{b}{\left(1 + \frac{\lambda b}{8} r^2\right)^2} \right], \tag{30}$$

where b is given by

$$b = \frac{32}{\lambda^2 R^4}\left(1 - \frac{\lambda R^2}{4} \pm \sqrt{1 - \frac{\lambda R^2}{2}}\right). \tag{31}$$

Clearly there are only solutions when $\lambda \leq \dfrac{2}{R^2}$. When $R = 1$ and more
specifity about the inner boundary condition is given (i.e. $u'(0) = 0$) equations (30) and (31) can be combined to write down the solution to (30)

as

$$u(r;\lambda) = \ln \left[\frac{\frac{32}{\lambda^2}\left\{1 - \frac{\lambda}{4} \pm \sqrt{1 - \frac{\lambda}{2}}\right\}}{\left(1 + \frac{4r^2}{\lambda}\left\{1 - \frac{\lambda}{4} \pm \sqrt{1 - \frac{\lambda}{2}}\right\}\right)^2} \right].$$ (32)

The above expression in (32) has two values for every value of $0 < \lambda < 2$. For example, Figure 4 depicts the bifurcated behavior of the solution by depicting the two solutions for $\lambda = 1$ in relation to the unique solution obtained when $\lambda = 2$. When $\lambda = 1$ the solution obtained from taking the positive square root in (32) when $\lambda = 1$ shall be denoted as $u_+(r;1)$ and $u_-(r;1)$ as the solution obtained when taking the negative square root in (32).

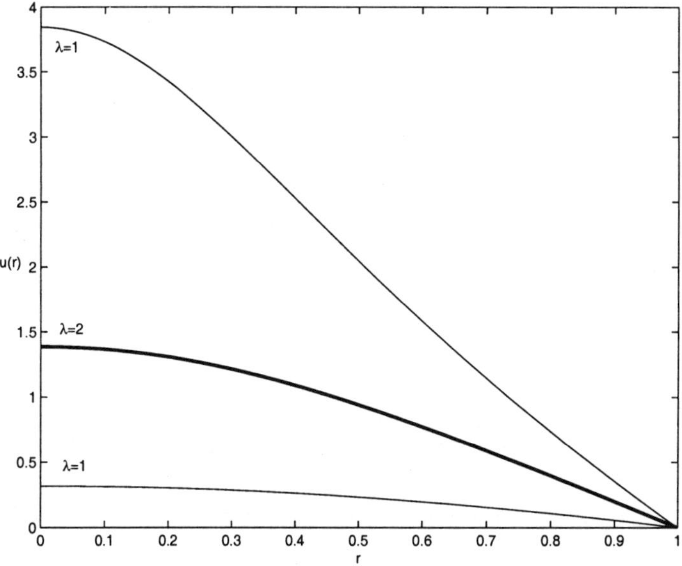

Fig. 4. Exact solutions to the Bratu-Gelfand problem when $\lambda = 1$ (bifurcated) and $\lambda = 2$ (unique)

The exact form of the upper curve in Figure 4 is given by

$$u_+(r;1) = \ln \left[\frac{24 + 16\sqrt{2}}{\left(1 + r^2(3 + 2\sqrt{2})\right)^2} \right]$$ (33)

and the exact form of the lower curve is given by

$$u_-(r;1) = \ln \left[\frac{24 - 16\sqrt{2}}{\left(1 + r^2(3 - 2\sqrt{2})\right)^2} \right]. \tag{34}$$

The maximum value $\|u\|_\infty$ of both curves occurs at $r = 0$, and $u_+(0;1) = \ln(4) + \ln(6 + 4\sqrt{2}) = 3.84218871$ and $u_-(0;1) = \ln(4) + \ln(6 - 4\sqrt{2}) = 0.31669436$.

Another way to illustrate the bifurcated nature of the solution is to graph the maximum value of $u(r)$ on $0 \le r \le 1$ versus λ, as shown in Figure 5. This also clearly shows the "turning point" in the solution at the critical value of $\lambda_c = 2$.

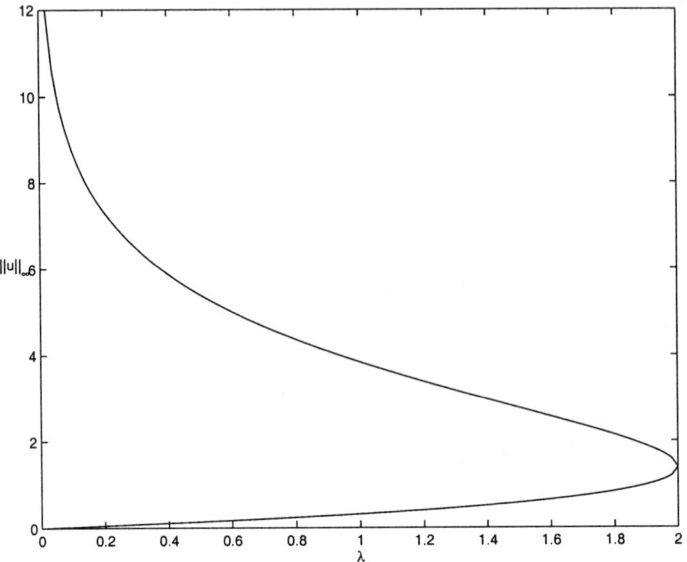

Fig. 5.   Maximum value of $u(r)$ versus λ depicting the turning point at $\lambda = 2$

The single-valued version of (32) that occurs when $\lambda = 2$ is astonishingly simple:

$$u(r) = \ln \left[\frac{4}{(1 + r^2)^2} \right] = \ln(4) - 2\ln(1 + r^2). \tag{35}$$

The graph of this function (35) is depicted in Figure 6. It is the exact solution to (30) and clearly obeys the boundary conditions $u(1) = 0$ and

$u'(0) = 0$. Note also that its maximum value occurs at $r = 0$ and is exactly $\ln(4) = 1.38629436....$

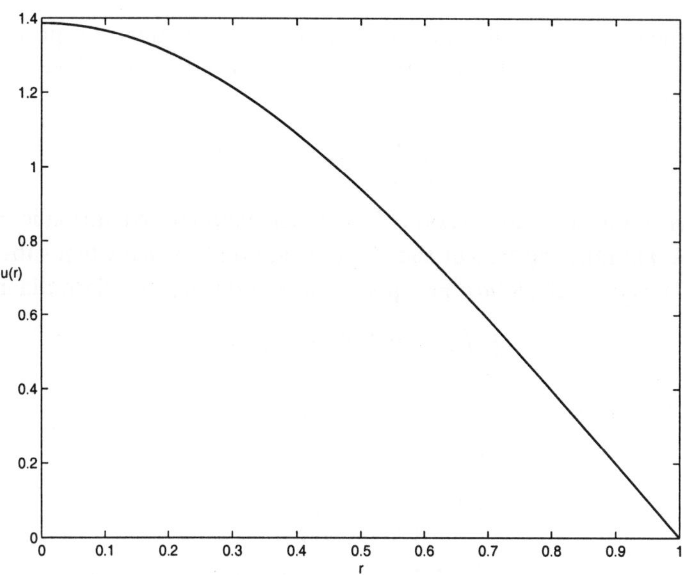

Fig. 6. Exact solution of the Bratu-Gelfand problem when $\lambda = 2$

4.1. *Numerical Solutions of the Bratu-Gelfand Problem*

Standard finite differences and Buckmire's MFD were used to compute numerical solutions to the Bratu-Gelfand problem (30) in order to compare them. Both methods involve forming discrete versions of the boundary value problem by approximating the derivatives and boundary conditions and solving the resulting system of nonlinear difference equations using Newton's Method. The first step in the numerical solution is to discretize the domain of the problem. The grid chosen was $\{r_j\}_{j=0}^N$ on the interval $0 \leq r \leq 1$ where

$$0 = r_0 < r_1 < r_2 < \ldots < r_j < \ldots < r_N = 1. \tag{36}$$

For a uniform grid, the grid separation parameter h is constant and $h = 1/N$ with $r_k = 0 + kh$ for $k = 0$ to N. Using the standard finite-difference scheme the discrete version of the Bratu-Gelfand problem (30) will be

$$\frac{1}{r_j} \left(r_{j+1/2} \frac{u_{j+1} - u_j}{r_{j+1} - r_j} - r_{j-1/2} \frac{u_j - u_{j-1}}{r_j - r_{j-1}} \right) + \lambda e^{u_j} = 0. \tag{37}$$

The nonstandard finite-difference scheme for (30) will be

$$\frac{1}{r_j}\left(\frac{u_{j+1}-u_j}{\log(r_{j+1}/r_j)}-\frac{u_j-u_{j-1}}{\log(r_j/r_{j-1})}\right)+\lambda e^{u_j}=0. \tag{38}$$

Note: since this is a singular problem at $r = 0$, r_0 must be positive, i.e. $0 < r_0 \ll 1$. A simple discrete version of the inner boundary condition $u'(0) = 0$ is

$$\frac{u_1-u_0}{r_1-r_0}=0 \Rightarrow u_1 = u_0. \tag{39}$$

Another, more accurate, version of the inner boundary condition is that the flux (i.e. ru' must be zero at the "first" grid point, which when substituted into (37) leads to the following equation at $j = 0$ using standard differencing

$$\frac{1}{r_0}\left(r_{1/2}\frac{u_1-u_0}{r_1-r_0}\right)+\lambda e^{u_0}=0. \tag{40}$$

Using the nonstandard difference method (38) the discrete version of the inner boundary condition is

$$\frac{1}{r_0}\left(\frac{u_1-u_0}{\log(r_1/r_0)}\right)+\lambda e^{u_0}=0. \tag{41}$$

The discrete version of the outer boundary condition $u(1) = 0$ is

$$u_N = 0. \tag{42}$$

When $0 < \lambda < 2$ the system of nonlinear equations due to the standard discretization $((37),(40),(42))$ and the system due to the Mickens discretization $((38),(41),(42))$ are each solved very easily using Newton's Method. Computations are conducted using the exact solution $u(r;2)$ (35) as an initial guess, with a tolerance of 10^{-8}. The numerical errors generated by the two competing methods for $\lambda = 1$ and for various values of increasing N are given in Figure 7 and Figure 8. Notice the completely different quantitative and qualitative nature of the graphs. The graph in Figure 8 illuminates the error behavior of the nonstandard method by using a semilog scale. The smallest maximum error in Figure 7 (the N=1000 curve) is greater than the largest maximum error in Figure 8 (the N=100 curve). Clearly the solution produced by the Mickens scheme has superior accuracy over the one generated using standard finite differences when $\lambda = 1$.

At the turning point $\lambda = 2$ the system of equations generated by using the standard finite difference scheme refuses to converge. This is not unexpected since it is widely known that numerically computing solutions at or near the turning point is difficult using standard methods [26]. However, the Mickens finite-difference method has no problem generating numerical solution at this critical value of the parameter λ.

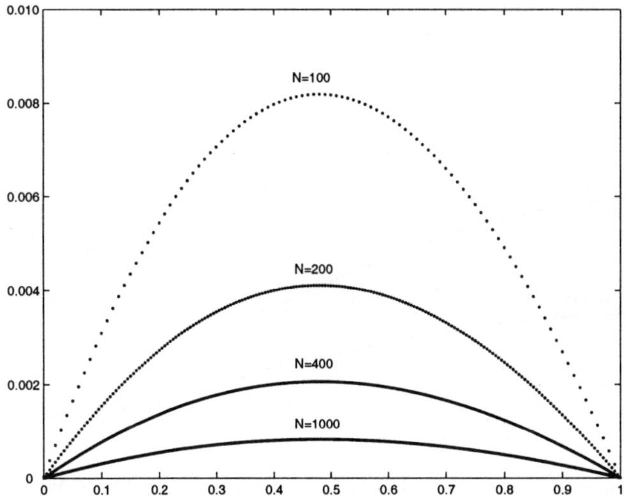

Fig. 7.   Numerical error versus r for standard method when $\lambda = 1$

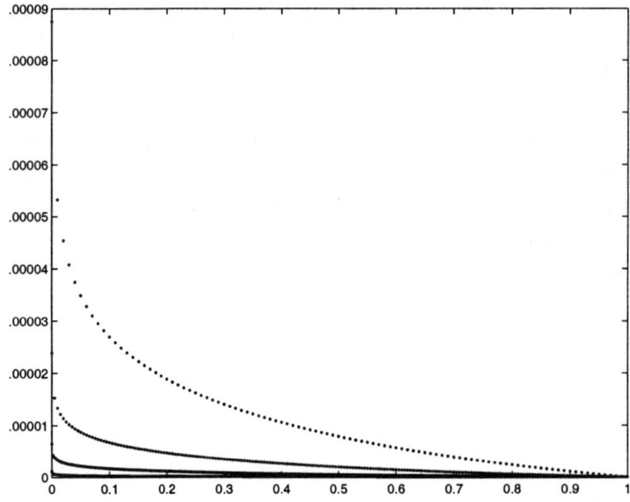

Fig. 8.   Numerical error versus r for Mickens method when $\lambda = 1$

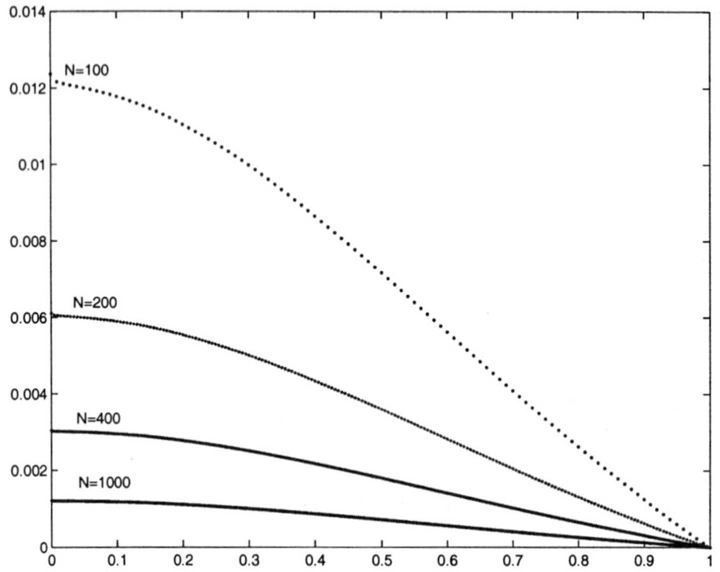

Fig. 9.   Numerical error using Mickens finite differences with $N = 100, 200, 400$ and 1000 for $\lambda = 2$

The numerical results of solving the Bratu-Gelfand problem at the critical value of $\lambda = 2$ are depicted in Figure 9. This shows that the error is greatest at $r = 0$ (as expected) but that the error over the entire domain $0 \leq r \leq 1$ clearly goes to zero as the number of grid points N increases. These results show the efficacy and versatility of using MFD methods to solve this singular nonlinear boundary value problem which also happens to possess a double-valued solution. In the next section, the application of MFD to a boundary value problem which is related to the Bratu-Gel'fand problem but is a *non-singular* nonlinear boundary value problem which also happens to have a double-valued solution for values less than a given parameter will be examined.

5. MFD Application to the One-Dimensional Bratu Problem

Introduction

In this section of the paper the results of applying a MFD scheme to a nonsingular boundary value problem related to the Bratu-Gelfand problem discussed in Section 4 shall be presented. In this section, a Mickens finite difference is applied to the 1-dimensional planar Bratu problem. This version of the Bratu problem, $u'' + \lambda e^u = 0$ with $u(0) = u(1) = 0$, has two known, bifurcated, exact solutions for values of $\lambda < \lambda_c$ and no solutions for $\lambda > \lambda_c$. The value of λ_c is simply $8(\alpha^2 - 1)$ where α is the fixed point of the hyperbolic cotangent function. The Bratu problem is an elliptic partial differential equation which comes from a simplification of the solid fuel ignition model in thermal combustion theory [19]. It is also a nonlinear eigenvalue problem that is often used as a benchmarking tool for numerical methods ([3], [4], [17]). In [22], Jacobsen and Schmitt provide an excellent summary of the significance and history of the Bratu problem. The goal of this section will be to compare the numerical solutions to the planar one-dimensional Bratu problem produced by MFD to solutions produced by other numerical techniques. The work in this section has not been previously published.

The classical Bratu problem is

$$\Delta u + \lambda e^u = 0 \quad \text{on } \Omega : \{(x,y) \in 0 \le x \le 1, 0 \le y \le 1\} \tag{43}$$
$$\text{with } u = 0 \quad \text{on } \partial\Omega. \tag{44}$$

The 1-dimensional (planar) version of this problem is

$$u''(x) + \lambda e^{u(x)} = 0 \quad 0 \le x \le 1, \tag{45}$$
$$\text{with } u(0) = 0 \text{ and } u(1) = 0. \tag{46}$$

In Section 5.1 of this paper the exact solution of the one-dimensional planar Bratu problem will be presented. Details of the bifurcated nature of the solution are given. In Section 5.2 brief explanations of the various methods chosen to solve the will be presented. In Section 5.3 numerical solutions generated using Mickens finite differences will be compared to solutions produced using different numerical methods: standard finite differences, a pseudospectral method due to Boyd [4] and the Adomian polynomial decomposition algorithm [17]. All of the approximate solutions are compared to the exact solution. This section shall conclude with some overall comments and observations based on the numerical result.

5.1. *The 1-dimensional Planar Bratu Problem*

The exact solution to (45) is given in [3] and [17] and presented here as

$$u(x) = -2\ln\left[\frac{\cosh((x - \tfrac{1}{2})\tfrac{\theta}{2})}{\cosh(\tfrac{\theta}{4})}\right] \tag{47}$$

                    R. *Buckmire*

where θ solves

$$\theta = \sqrt{2\lambda} \cosh\left(\frac{\theta}{4}\right). \tag{48}$$

There are two solutions to (48) for values of $0 < \lambda < \lambda_c$. For $\lambda > \lambda_c$ there are no solutions. Note that this property is similar to the Bratu-Gel'fand problem discussed in Section 4. The solution (47) is only unique for a critical value of $\lambda = \lambda_c$ which solves

$$1 = \sqrt{2\lambda_c} \sinh\left(\frac{\theta_c}{4}\right)\frac{1}{4}. \tag{49}$$

By graphing the line $y = \theta$ and the curve $y = \sqrt{2\lambda} \cosh(\frac{\theta}{4})$ for fixed values of $\lambda = 1, 2, 3, 4$ and 5 the solutions of (48) can be seen as the two points of intersections of the curve and the line in Figure 10. Clearly, there is only one solution when the $y = \theta$ line is exactly tangential to the $y = \sqrt{2\lambda} \cosh(\frac{\theta}{4})$ curve, which leads to the condition given in (49).

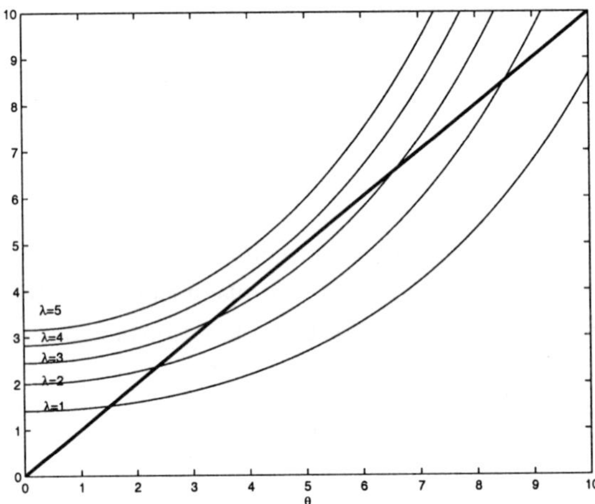

Fig. 10.   Graphical depiction of dependence of solutions of (48) upon λ

Dividing (49) by (48) produces:

$$\frac{4}{\theta_c} = \tanh\left(\frac{\theta_c}{4}\right)$$

$$\Rightarrow \frac{\theta_c}{4} = \coth\left(\frac{\theta_c}{4}\right)$$

$$\Rightarrow \alpha = \coth\left(\alpha\right)$$

The critical value θ_c is four times α, which is the positive fixed point of the hyperbolic cotangent function, 1.19967864.

$$\theta_c = 4.79871456 \,. \tag{50}$$

The exact value of θ_c can therefore be used in (49) to obtain the exact value of λ_c:

$$\lambda_c = \frac{8}{\sinh^2\left(\frac{\theta_c}{4}\right)} = 8(\alpha^2 - 1) \Rightarrow \lambda_c = 3.513830719 \,. \tag{51}$$

The relationship between λ and θ for some values of λ less than λ_c are given in Table 1. Obviously, when $\lambda = \lambda_c$ then $\theta_1 = \theta_2 = \theta_c$ and when $\lambda > \lambda_c$ there are no solutions to (48).

Table 1.   Corresponding values of
θ for various $\lambda \leq \lambda_c$

λ	θ_1	θ_2
0.5	1.0356946	13.0382393
1.0	1.5171645	10.9387028
1.5	1.9397652	9.5816998
2.0	2.3575510	8.5071995
2.5	2.8115549	7.5480981
3.0	3.3735077	6.5765692
3.5	4.5518536	5.0543427
λ_c	4.7987146	4.7987146

Figure 11 shows how the maximum value of the solution function (47) depends on the nonlinear eigenvalue λ with the critical value of λ_c highlighted at the "turning point." Notice how similar the bifurcation diagram in Figure 11 resembles the bifurcation diagram in Figure 5. Table 1 and Figure 11 are two different ways of depicting the property of the solution that it is double-valued for $\lambda < \lambda_c$. In the next subsection, numerical methods to compute these solutions to (45) are presented.

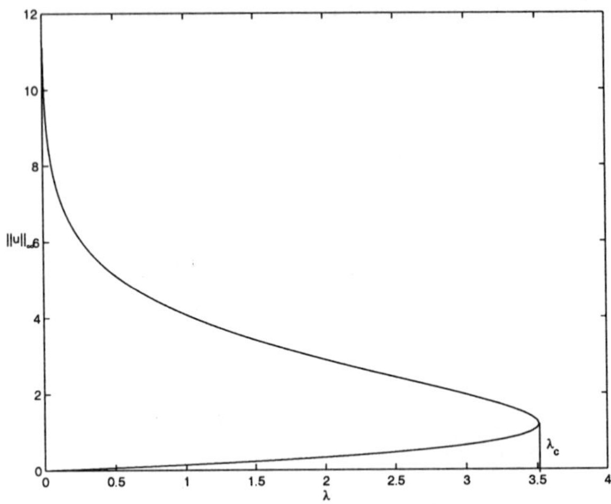

Fig. 11. Bifurcated nature of the exact solution to the 1-D Bratu problem

5.2. *Numerical Methods*

In this section the details of various numerical methods used to compute solutions to (45) shall be given. The first method involves approximating the differential equation with finite differences. Both standard and non-standard (Mickens) finite-difference schemes are used. In addition to the methods which use finite-differences, two pseduospectral methods are used. The first, due to Boyd [4], uses Gegenbauer polynomials as basis functions. The second, due to Adomian [2] assumes the solution can be represented as an infinite series of polynomials. Lastly the problem was also solved using a shooting method easily available in `Matlab`.

5.2.1. *Finite difference methods*

To solve a boundary value problem using finite differences involves discretizing the differential equation and boundary conditions. This method transforms the problem into a system of simultaneous nonlinear equations which are then usually easily solved using Newton's method. There are many choices for how to approximate the derivatives which appear in a differential equation. The first step in the computation of the numerical solution of (45) using a finite-difference method is to approximate the continuous domain of the problem with a discrete grid. The grid chosen was

$\{x_j\}_{j=0}^{N}$ on the interval $0 \leq x \leq 1$ where

$$0 = x_0 < x_1 < x_2 < \ldots < x_j < \ldots < x_N = 1. \tag{52}$$

For a uniform grid, the grid separation parameter h is constant and $h = 1/N$ with $x_k = 0 + kh$ for $k = 0$ to N. Using a standard finite-difference scheme, the discrete version of the planar Bratu problem (45) will be

$$\frac{u_{j+1} - 2u_j + u_{j-1}}{h^2} + \lambda e^{u_j} = 0, \qquad j = 1, 2, \ldots, N - 1. \tag{53}$$

A nonstandard finite-difference scheme for (45) is

$$\frac{u_{j+1} - 2u_j + u_{j-1}}{2 \ln[\cosh(h)]} + \lambda e^{u_j} = 0, \qquad j = 1, 2, \ldots, N - 1. \tag{54}$$

The boundary conditions given in (46) become

$$u_0 = u_N = 0. \tag{55}$$

The discretization given in (54) is another example of a MFD. The nonstandard finite difference scheme given in (54) is a MFD for the second derivative

$$u'' \approx \frac{u_{j+1} - 2u_j + u_{j-1}}{\phi(h)}$$

where the denominator function $\phi(h) = 2 \ln[\cosh(h)] = h^2 + o(h^2)$. Thus, in the limit as $h \to 0$ the standard finite-difference scheme (53) and the Mickens finite-difference scheme (54) will be identical. However, for the finite values of h at which numerical computations are conducted the hypothesis is that the nonstandard form of the denominator function $\phi(h)$ will lead to improved accuracy for the Micken finite difference over the standard finite difference.

5.2.2. *Boyd collocation*

Boyd [4] developed a pseudospectral method to produce approximate solutions to the classical two-dimensional planar Bratu problem

$$\frac{\partial^2 u}{\partial x^2} + \frac{\partial^2 u}{\partial y^2} + \lambda e^u = 0 \text{ on } \{(x, y) \in -1 \leq x \leq 1, -1 \leq y \leq 1\} \tag{56}$$

with $u = 0$ on the boundary of the square. The basic idea is that the unknown solution $u(x, y)$ can be completely represented as an infinite series of spectral basis functions

$$u(x, y) = \sum_{k=1}^{N} a_k \phi_k(x, y). \tag{57}$$

The basis functions $\phi_k(x, y)$ are chosen so that they obey the boundary conditions and have the property that the more terms of the series that are kept, the more accurate the representation of the solution $u(x, y)$ is. In other words, as $N \to \infty$ the error diminishes to zero. For finite N the series expansion in (57) is substituted into (56) to produce the residual R. The residual function will depend on the spatial variables (x, y), the unknown coefficients a_k and the parameter λ. The goal of Boyd's pseudospectral method is to find a_k so that the residual function R is zero at N "collocation points." The collocation points are usually chosen to be the roots of orthogonal polynomials that fall into the same family as the basis functions $\phi_k(x, y)$. Boyd [4] uses the Gegenbauer polynomials to define the collocation points. The Gegenbauer polynomials [18] are orthogonal on the interval $[-1, 1]$ with respect to the weight function $w(x) = (1 - x^2)^b$ where $b = -1/2$ corresponds to the Chebyshev polynomials and $b = 1$ is the choice Boyd uses. The second-degree Gegenbauer polynomial is

$$G_2(x) = \frac{3}{2}(5x^2 - 1), \qquad -1 \le x \le 1. \tag{58}$$

In other words, using a 1-point collocation method at the point $x_1 = \left(\frac{1}{\sqrt{5}}, \frac{1}{\sqrt{5}}\right)$ and the choice of $\phi_1(x, y) = (1 - x^2)(1 - y^2)$ Boyd is able to obtain an approximation to the value of λ_c with a relative error of 8% [4]. Note that this choice for $\phi_1(x, y)$ satisfies the boundary conditions (44) since $\phi(1, y) = \phi(-1, y) = \phi(x, -1) = \phi(x, 1) = 0$.

In the rest of this section Boyd's collocation method described above for Bratu's problem in planar two-dimensional coordinates (56) shall be extrapolated to solve the planar one-dimensional Bratu probem (45). The most obvious difference is the change in the domain from a square $[-1, 1] \times [-1, 1]$ to an interval $[0, 1]$. Using a linear transformation of $z = \frac{x + 1}{2}$ the Gegenbauer polynomial $G_2(x)$ defined on $[-1, 1]$ found in (58) becomes

$$G_2(2z - 1) = 6(1 - 5z + 5z^2), \qquad 0 \le z \le 1. \tag{59}$$

The corresponding collocation point to x_1 becomes $z_1 = \frac{1}{10}(5 + \sqrt{5})$ with $\phi_1(z) = z(1 - z)$ and assuming $u(z) = A\phi_1(z)$. (Note this form of $\phi(z)$ satisfies the boundary conditions that $\phi(0) = \phi(1) = 0$.) Substituting z_1 and $\phi_1(z)$ into the one-dimensional planar Bratu problem (replace x by z) produces an equation for the residual which is constrained to be zero.

$$R[z_1; \lambda, A] = R[\frac{1}{10}(5 + \sqrt{5}); \lambda, A] = -2A + \lambda e^{\frac{1}{5}A} = 0. \tag{60}$$

Solving (60) for λ produces

$$\lambda_1(A) = 2Ae^{-0.2A}. \tag{61}$$

The expression (61) attains its maximum value of λ_c at $A = 5$. In other words, using 1-point Boyd collocation produces an estimate of $\lambda_c = 10e^{-1} = 3.67879441$ which is 4.7% greater than the exact value of $\lambda_c = 3.513830719$.

To increase accuracy the number of collocation points is increased. However, the number of residual equations (and their complexity) will simultaneously also increase. Using 2-point collocation the form of $u(x)$ is assumed to be

$$u(z) = A\phi_1(z) + B\phi_2(z) = Az(1-z) + Bz(1-z)(2z-1)^2, \quad 0 \le z \le 1. \tag{62}$$

The above two-point collocation expansion corresponds to the expansion $u(x) = A(1 - x^2) + Bx^2(1 - x^2)$ which would be valid on $[-1, 1]$. The fourth-degree Gegenbauer polynomial, defined on $[-1, 1]$ is

$$G_4(x) = \frac{15}{8}(1 - 14x^2 + 21x^4), \qquad -1 \le x \le 1 \tag{63}$$

which on transformation to $[0, 1]$ becomes

$$G_4(2z - 1) = 15(1 - 14z + 56z^2 - 84z^3 + 42z^4), \qquad 0 \le z \le 1 \tag{64}$$

$G_4(x)$ has four roots on the interval $[-1, 1]$ symmetrically distributed around the origin. The two largest roots are selected as the collocation points for the 2-point Boyd collocation method. The two residual equations are formed by substituting (62) into the planar Bratu equation at the collocation points.

$$R\left[\frac{1}{42}\left(21 + \sqrt{21(7 - 2\sqrt{7})}\right); \lambda, A, B\right] = -8A - 8B + \frac{32B}{\sqrt{7}}$$
$$+ \lambda e^{\frac{2}{63}(21A + 3\sqrt{7}A + 5B - \sqrt{7}B)} = 0$$

$$R\left[\frac{1}{42}\left(21 + \sqrt{21(7 + 2\sqrt{7})}\right); \lambda, A, B\right] = -8A - 8B - \frac{32B}{\sqrt{7}}$$
$$+ \lambda e^{\frac{2}{63}(21A - 3\sqrt{7}A + 5B + \sqrt{7}B)} = 0.$$

The method of solution is to find closed-form expressions for λ and B in terms of either A. This is not easy to do with the system as currently constituted. However by eliminating terms in the exponentials which are significantly smaller than the other terms it turns out that a closed form expression for $\lambda(A)$ obtained from the 2-point Boyd collocation method can be found.

$$\lambda_2(A) = \frac{64\sqrt{7}Ae^{\frac{2}{21}(-7+\sqrt{7})A}}{-7+4\sqrt{7}+(7+4\sqrt{7})e^{\frac{4A}{3\sqrt{7}}}}. \tag{65}$$

The maximum value of the expression (65) is $\lambda_c = 3.45611039$, which is only 1.64% smaller than the exact value (51).

5.2.3. *Adomian polynomial decomposition*

Adomian [2] developed a "polynomial decomposition" method of representing solutions to boundary value problems of the form

$$u'' = -F(u) \tag{66}$$
$$u(0) = \alpha \quad \text{and} \quad u(1) = \beta.$$

The exact solution to (66) can be represented by a Green's Function

$$u(x) = \lambda \int_0^1 g(x,s)F(u(s))ds + (1-x)\alpha + \beta x \tag{67}$$

where

$$g(x,s) = \begin{cases} s(1-x), & 0 \le s \le x \\ x(1-s), & x \le s \le 1. \end{cases} \tag{68}$$

Adomian's decomposition method assumes that the unknown solution $u(x)$ and the given nonlinear functional $F(u)$ can each be represented as infinite series.

$$u = \sum_{i=0}^{\infty} u_i = u_0 + u_1 + u_2 + \dots \tag{69}$$

and

$$F(u) = \sum_{i=0}^{\infty} A_i = A_0 + A_1 + A_2 + \dots. \tag{70}$$

In the case of $F(u)$ the infinite series is a Taylor Expansion about u_0. In other words

$$F(u) = F(u_0) + F'(u_0)(u - u_0) + F''(u_0)\frac{(u - u_0)^2}{2}$$
$$+ F^{(3)}(u_0)\frac{(u - u_0)^3}{3} + \dots. \tag{71}$$

By re-writing (69) as $u - u_0 = u_1 + u_2 + u_3 + \ldots$, substituting it into (71) and then equating the two expressions for $F(u)$ found in (71) and (70) defines formulas for the "Adomian polynomials."

$$F(u(s)) = A_0 + A_1 + A_2 + \cdots = F(u_0) + F'(u_0)(u_1 + u_2 + u_3 + \ldots)$$
$$+ F''(u_0)\frac{(u_1 + u_2 + u_3 + \ldots)^2}{2!} + \cdots . \tag{72}$$

By equating terms in (72) the first few Adomian polynomials A_0, A_1, A_2 are given...

$$A_0 = F(u_0)$$
$$A_1 = u_1 F'(u_0)$$
$$A_2 = u_1^2 F''(u_0)/2! + u_2 F'(u_0)$$
$$A_3 = u_1^3 F^{(3)}(u_0)/3! + 2u_1 u_2 F''(u_0)/2! + u_3 F'(u_0)$$
$$A_4 = u_1^4 F^{(4)}(u_0)/4! + 3u_1^2 u_2 F^{(3)}(u_0)/3! + (2u_1 u_3 + u_2^2)F''(u_0)/2!$$
$$+ u_4 F'(u_0) .$$

Now that the $\{A_k\}_{k=0}^{\infty}$ are known, (70) can be substituted in (67) to specify the terms in the expansion for the solution (69).

$$u(x) = \lambda \int_0^1 g(x, s) \sum_{i=0}^{\infty} A_i \, ds + (1 - x)\alpha + \beta x \,,$$

$$\sum_{i=0}^{\infty} u_i = \alpha(1 - x) + \beta x + \lambda \sum_{i=0}^{\infty} \int_0^1 g(x, s) A_i \, ds \,.$$

Equating the terms yields

$$u_0 = \alpha(1 - x) + \beta x$$
$$u_1 = \lambda \int_0^1 g(x, s) A_0(s) ds$$
$$u_2 = \lambda \int_0^1 g(x, s) A_1(s) ds$$
$$\vdots \quad \vdots$$
$$u_k = \lambda \int_0^1 g(x, s) A_{k-1}(s) ds \,.$$

Now the $\{u_k\}_{k=0}^{\infty}$ are known, so the solution is given by $u = u_0 + u_1 + u_2 + u_3 + \ldots$.

To apply the Adomian polynomial decomposition method to solve the one-dimensional planar Bratu problem (45) involves setting $\alpha = \beta = 0$ and

$F(u) = e^u$. A happy accident is that the kth derivative of $F(u)$, $F^{(k)}(u) = e^u$ so that choosing $u_0 = 0$ greatly simplifies the formulas for the Adomian polynomials $\{A_k\}$ since $e^{u_0} = 1$.

$$A_0 = 1$$
$$A_1 = u_1$$
$$A_2 = u_1^2/2! + u_2$$
$$A_3 = u_1^3/3! + u_1 u_2 + u_3$$
$$A_4 = u_1^4/4! + u_1^2 u_2/2 + u_1 u_3 + u_2^2/2! + u_4$$
$$\vdots \quad \vdots$$

Knowing the $\{A_k\}$ terms leads to the calculation of the $\{u_k\}$ terms

$$u_0 = 0$$
$$u_1 = \lambda \int_0^1 g(x,s) \cdot 1 \ ds = \lambda \int_0^x s(1-x) \ ds + \lambda \int_x^1 x(1-s) \ ds$$
$$= \frac{1}{2}(1-x)x\lambda$$
$$u_2 = \lambda \int_0^1 g(x,s) \cdot \frac{1}{2}(1-s)s\lambda \ ds$$
$$= \lambda^2(1-x) \int_0^x \frac{1}{2}(1-s)s^2 \ ds + \lambda^2 x \int_x^1 \frac{1}{2}(1-s)^2 s \ ds$$
$$= \frac{1}{24}(1 - 2x^2 + x^3)x\lambda^2$$
$$u_3 = \lambda \int_0^1 g(x,s) \cdot A_2(s) \ ds$$
$$= \frac{1}{1440}(9 - 10x^2 - 15x^3 + 24x^4 - 8x^5)x\lambda^3$$
$$u_4 = \lambda \int_0^1 g(x,s) \cdot A_3(s) ds$$
$$\vdots \quad \vdots$$

5.2.4. *Shooting method*

The last and probably the most obvious method used to obtain a numerical solution of the planar Bratu problem is the nonlinear shooting method. This involves converting the nonlinear boundary value problem (45) into a system of nonlinear initial value problems which look like

$$\frac{d}{dt}\vec{y} = \vec{f}(\vec{y}), \qquad \vec{y}(0) = \vec{y}_0,$$

with the choice of $\vec{y} = \begin{bmatrix} y_1 \\ y_2 \\ y_3 \\ y_4 \end{bmatrix}$, $\vec{f} = \begin{bmatrix} f_1(\vec{y}) \\ f_2(\vec{y}) \\ f_3(\vec{y}) \\ f_4(\vec{y}) \end{bmatrix} = \begin{bmatrix} y_2 \\ -e^{y_1} \\ y_4 \\ -e^{y_3} \end{bmatrix}$ and $\vec{y}_0 = \begin{bmatrix} 0 \\ s_0 \\ 0 \\ 1 \end{bmatrix}$.

The shooting method works by choosing a value s_0 for $u'(0)$, solving the initial value problem (using a standard ODE solver like Runge-Kutta) and then comparing the value of $y_1(b)$ with the expected value of $u(b) = 0$. A new value of s_k is chosen by using Newton's Method, where

$$s_{k+1} = s_k - \frac{y_1(b) - u(b)}{y_3(b)}. \tag{73}$$

The method is said to converge when the difference between subsequent values of s_k fall below a given tolerance, in other words $y_1(b)$ is very close to $u(b)$.

In the next subsection, the results of using the numerical methods detailed in this section are given.

5.3. *Numerical Results*

The results of applying various numerical methods to produce solutions of the planar one-dimensional Bratu problem (45) are given in this section. We shall begin with considering the results obtained using finite differences. A comparison of the errors generated using Mickens finite differences and standard finite differences are illustrated in Figure 12. By examining Figure 12 it can be observed that the error due to each finite difference method decreases proportionally to with $h^2 = \frac{1}{N^2}$. Also note that the Mickens finite difference error (solid line) is consistently smaller than the standard discretization error (dashed line). The value of the parameter λ shall be taken to be one.

Interestingly, despite the fact that there are two solutions to (45) at $\lambda = 1$ as shown in Figure 13, the standard finite difference scheme will only converge to one of them, the "lower" solution, i.e. the one below the $\lambda = \lambda_c$ solution. The MFD scheme will converge to either solution, depending on the initial guess chosen for all values $0 < \lambda < \lambda_c$. Neither discretization method will converge to the unique solution at $\lambda = \lambda_c$. Note that this is different behavior than what happened when standard finite differences and

    R. Buckmire

MFD were used to solve the Bratu-Gel'fand problem in Section 4. There, the MFD scheme converged at $\lambda = \lambda_c$ but the standard finite difference scheme did not.

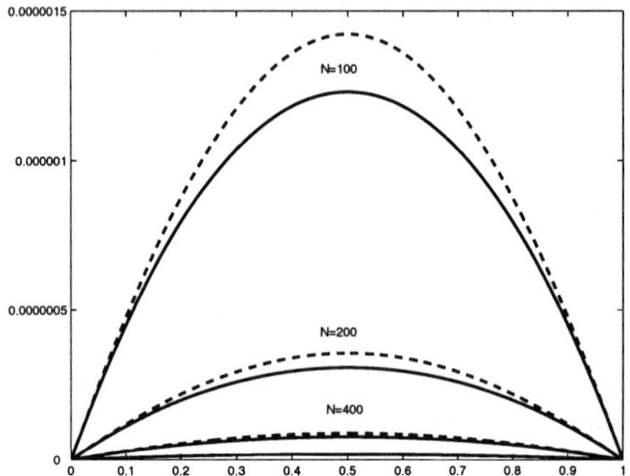

Fig. 12.   Comparison of standard error and Mickens error for $N = 100, 200, 400$ and 800 when $\lambda = 1$

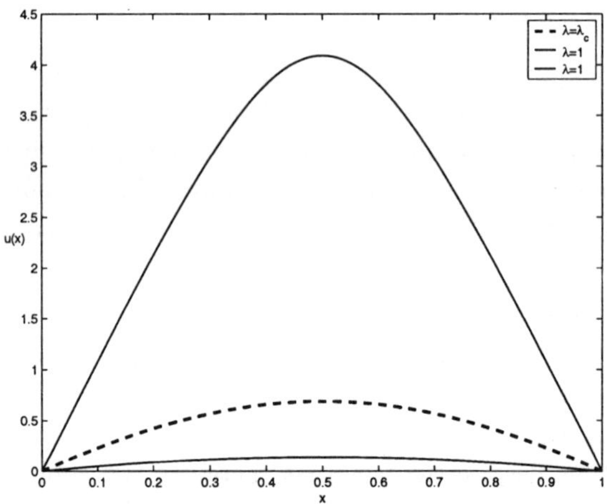

Fig. 13.   The two solution curves for $\lambda = 1$ and the unique solution curve for $\lambda = \lambda_c$

The solution produced by Boyd's pseudospectral method does not have the deficiency of being unable to converge to both solutions of the Bratu problem for $\lambda < \lambda_c$ which the standard finite-difference method and Adomian decomposition method both have. Boyd's method is able to produce continuous expressions for λ versus the maximum value of the solution. In Figure 14 the behavior of Boyd solutions produced using 1-point and 2-point collocation is compared with the exact solution's bifurcated behavior (as depicted in Figure 11), which indicates the multivalued nature of the Boyd solutions.

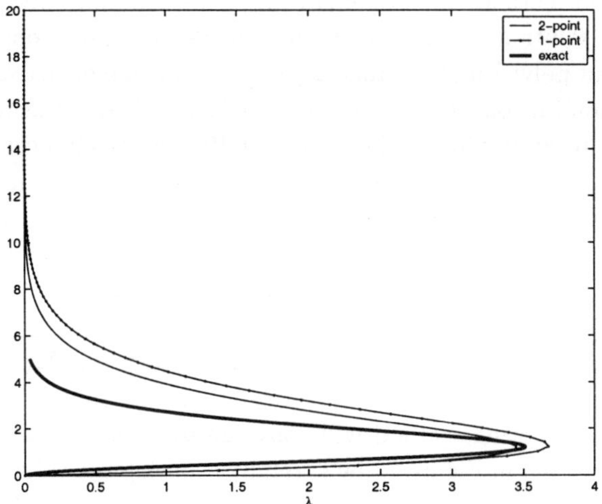

Fig. 14.   Dependence of Boyd pseudospectral solutions on λ

When $\lambda = 1$ there are two solutions to the Bratu problem (45), which are depicted in Figure 13 and called the "upper" solution and the "lower" solution. In Figure 15 the results of producing solutions using Boyd's pseudospectral method to the Bratu problem when $\lambda = 1$ are depicted. The exact solution is the dark solid line, with the solution from the 1-point collocation depicted using a continuous dotted line and the solution from the 2-point collocation depicted using a continuous solid line. Interestingly, Boyd's method does very well with just 1-point collocation to approximate the lower solution. The 1-point Boyd collocation method doesn't do a very good job of approximating the solution to the "upper" Bratu solution, though the 2-point Boyd collocation does much better, as seen in Figure 15. This is not a surprise, since the expectation is that using more collocation points will decrease the error. By looking at Figure 14 it is clear that at

$\lambda = 1$ the three curves (exact solution, 1-point and 2-point) are close together at the lower arc of the bifurcation curve corresponding to the "lower solution" and they are not close together at the upper arc of the bifurcation curve corresponding to the "upper solution." The proximity of the curves is indicative of the numerical error, and the error in approximating the lower solution is smaller than the error in approximating the upper solution.

The solutions generated by the Adomian polynomial decomposition only approach the exact solution for small values of $\lambda \leq 1$. Like the standard discretization method, the Adomian method's solution only converges to the "lower" solution at $\lambda = 1$. In Figure 16 the first three nonzero terms of the Adomian polynomial expansion (solid curves) are depicted next to the exact solution (unconnected dotted line). Clearly, these terms $(u_1 + u_2 + u_3)$ are enough to approximate the exact solution relatively accurately when $\lambda = 1$. However, if $\lambda = \lambda_c$ is selected one needs far more than three terms of the series $\{u_k\}_{k=0}^{\infty}$ to converge to the exact solution, as can be seen in Figure 16. Deeba et. al. [17] obtained identical results when they applied Adomian's polynomial decomposition method to the same boundary value problem (45).

The shooting method was implemented using the MATLAB routine `ode45` and produces accurate numerical solutions rapidly for values of $\lambda < \lambda_c$. The shooting method will converge to both the upper and lower solutions depicted in Figure 13 by carefully choosing the value of the initial slope s_0 in (73). However, when $\lambda = \lambda_c$, like the finite-difference methods in Section 5.2.1, the shooting method will not converge to a solution within the given tolerance. Figure 17 depicts the numerical error produced by the nonlinear shooting method as it approximates both Bratu solutions at $\lambda = 1$. Since the numerical error of the shooting method depends on the tolerance of the ODE solver, and not the grid separation, N was chosen to be 100 with a `RelTol` of 10^{-10}.

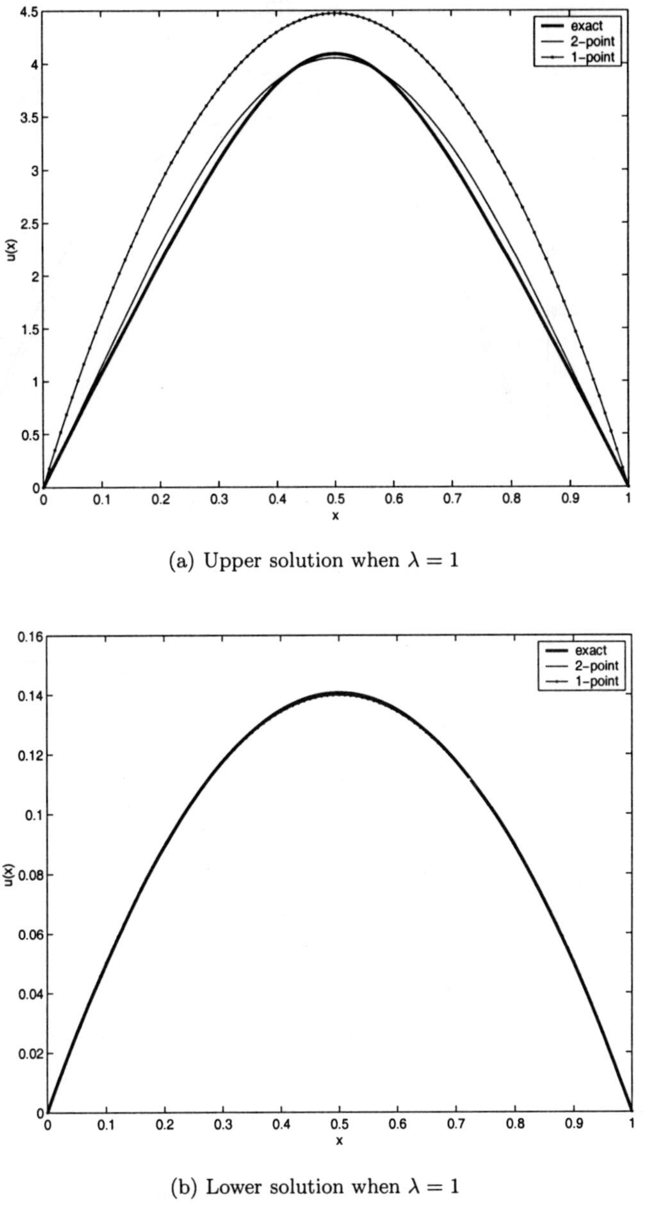

(a) Upper solution when $\lambda = 1$

(b) Lower solution when $\lambda = 1$

Fig. 15.   Comparison of Boyd solutions generated by one-point and two-point collocation

R. *Buckmire*

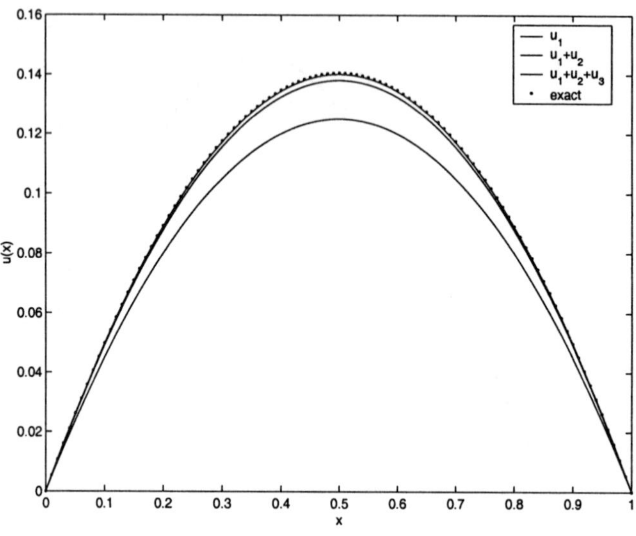

(a) Adomian polynomial solution for $\lambda = 1$

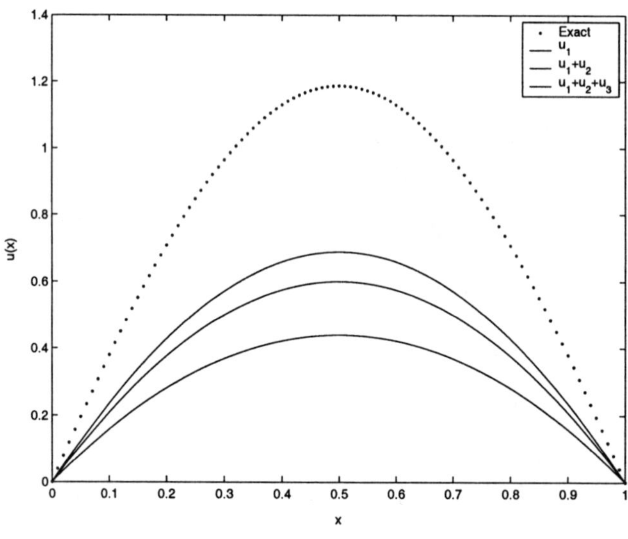

(b) Adomian polynomial solution for $\lambda = \lambda_c$

Fig. 16.   Comparison of using first three non-zero terms of Adomian polynomial solution for $\lambda = 1$ and $\lambda = \lambda_c$

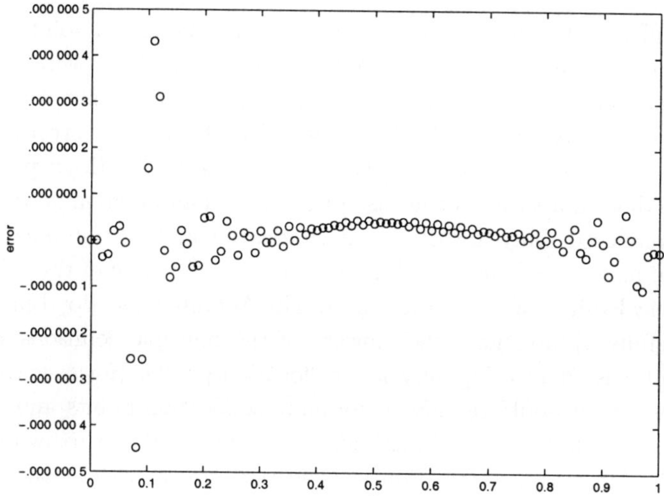

(a) Error due to nonlinear shooting for $\lambda = 1$ "upper" solution

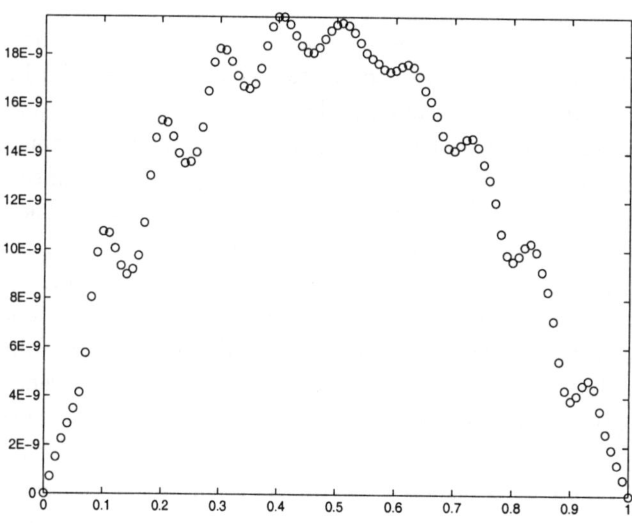

(b) Error due to nonlinear shooting for $\lambda = 1$ "lower" solution

Fig. 17. Errors generated by the nonlinear shooting method when $\lambda = 1$ using $N = 100$

Conclusion

Five different methods were used to generate numerical solutions of the planar one-dimensional Bratu problem. The five methods were, two finite-difference methods, two spectral methods and a nonlinear shooting method. The methods were chosen for their ease of use for relative error generated. This is why only a few terms (two in the case of the Boyd pseudospectral method and three in the case of the Adomian polynomial decomposition) were used. Only the Mickens discretization and the nonlinear shooting method had no difficulty handling the bifurcated nature of the solution for subcritical values of the parameter λ. The Adomian and Boyd methods do successfully approximate the "lower" of the multiple solutions when the value of λ is small using very few collocation points. However to increase their accuracy would require many more collocation points and would no longer make these "simple" methods to implement. It is worthwhile to note that the Mickens discretization method performs as well as the nonlinear shooting method, and is also very easy to implement.

6. Conclusion

The goal of this paper has been to demonstrate the versatility of Mickens finite differences through the application of these nonstandard methods to singular linear boundary value problems, singular nonlinear boundary value problems with bifurcations and a nonsingular, nonlinear boundary value problem with bifurcations. In Section 3 the singular nature of the solution is much better captured near r equals zero by the numerical solution generated by the MFD compared to standard finite differences. In Section 4 the MFD is able to generate a numerical solution at exactly the same critical value of the nonlinear eigenvalue where the standard finite difference solution fails to converge. In Section 5 the MFD produces numerical solutions more accurate than the solution produced by standard finite differences and with less work than solutions generated by two different pseudospectral methods. The overall conclusion of this paper is that Mickens finite differences in general should be considered for use as a solution technique for a wide variety of problems. In particular, when the problem involves approximating derivatives in cylindrical or spherical coordinates, Buckmire's MFD scheme should be used.

Future work shall involve application of MFD to the "other" Bratu-Gel'fand problem, i.e. the classic Bratu problem in spherical coordinates. This problem has no known exact solution and possesses a much more complicated bifurcation curve [23]. It will be interesting to see whether MFD can be used to better resolve the limiting details of the bifurcation curve. In addition, there are other important nonlinear boundary value problems with and without bifurcations in cylindrical coordinates and spherical coordinates to which Buckmire's MFD scheme should be applied.

Acknowledgements

The author would like to thank Tom Witelski for first mentioning the name "Mickens" to me. The numerical calculations for this paper were performed using `Matlab` and `Mathematica` and the figures were all generated using `Matlab`.

References

1. J. P. Abbott, "An efficient algorithm for the determination of certain bifurcation points" *Journal of Computational and Applied Mathematics* Volume **4** Number 1 (1978), 19–27.
2. G. Adomian, *Solving Frontier Problems of Physics: The Decomposition Method* (Kluwer, Dordrecht, 1994).
3. U. M. Ascher, R. M. M. Matheij and R. D. Russell, *Numerical Solution of Boundary Value Problems for Ordinary Differential Equations* (Society for Industrial and Applied Mathematics, Philadelphia, 1995).
4. J. P. Boyd, "An analytical and numerical study of the two-dimensional Bratu equation," *Journal of Scientific Computing* Volume **1** Number 2 (1986), 183–206.
5. G. Bratu, "Sur les équations intégrales non-linéaires," *Bulletins of the Mathematical Society of France* Volume **42** (1914), 113–142.
6. R. Buckmire, *The Design of Shock-free Transonic Slender Bodies of Revolution* (Ph.D. thesis, Rensselaer Polytechnic Institute, Troy, 1994).
7. R. Buckmire, "A new finite-difference scheme for singular boundary value problems in cylindrical or spherical coordinates" *Mathematics Is For Solving Problems* (Editors L. Pamela Cook and Victor Roytburd, Society for Industrial and Applied Mathematics, Philadelphia, 1996), 3–9.
8. R. Buckmire, "On the design of shock-free transonic slender bodies of revolution," **AIAA Paper 98-2686** *American Institute of Aeronautics and Astronautics 2nd Theoretical Fluid Mechanics Meeting*, Albuquerque, 1998).
9. R. Buckmire, "Investigations of nonstandard, Mickens-type, finite-difference schemes for singular boundary value problems in cylindrical or spherical coordinates," *Numerical Methods for Partial Differential Equations* Volume **19** Number 3 (2003), 380–398.
10. R. Buckmire, "Application of a Mickens finite-difference scheme to the cylindrical Bratu-Gelfand problem," *Numerical Methods for Partial Differential Equations* Volume **20** Number 3 (2004), 327–337.

11. S. Chandrasekhar, *An Introduction to the Study of Stellar Structure* (Dover Publications, New York, 1957).

12. J. D. Cole and L. P. Cook, *Transonic Aerodynamics* (North-Holland, 1986).

13. J. D. Cole and N. Malmuth, "Shock Wave Location on a Slender Transonic Body of Revolution," *Mechanics Research Communications* **16(6)** (1989), 353–357.

14. J. D. Cole and A. F. Messiter. "Expansion Procedures and Similarity Laws for Transonic Flow," *Z.A.M.P.* **8** (1957), 1–25.

15. J. D. Cole and E. M. Murman, "Calculation of Plane Steady Transonic Flows," *AIAA Journal* Volume **9** Number 1 (January 1971), 114–121.

16. J. D. Cole, and D. W. Schwendeman, "Hodograph Design of Shock-free Transonic Slender Bodies," (*3rd International Conference on Hyperbolic Problems* Editors Bjorn Engqvist and Bertil Gustafsson, Uppsala, Sweden, June 11-15 1990).

17. E. Deeba, S. A. Khuri and S. Xie, "An algorithm for solving boundary value problems," *Journal of Computational Physics* Volume **159** (2000), 125–138.

18. B. A. Finlayson, *The Method of Weighted Residuals,* (Academic Press, New York, 1972).

19. D. A. Frank-Kamenetski, *Diffusion and Heat Exchange in Chemical Kinetics,* (Princeton University Press, Princeton, 1955).

20. I.M. Gelfand, "Some problems in the theory of quasi-linear equations," *Amer. Math. Soc. Transl. Ser. 2* Volume **29** (1963), 295–381.

21. R. Haberman, *Elementary Applied Partial Differential Equations* (Prentice Hall, Englewood Cliffs, 1987).

22. J. Jacobsen, and K. Schmitt, "The Liouville-Bratu-Gelfand problem for radial operators," *Journal of Differential Equations* **184** (2002), 283–298.

23. D.D. Joseph, and T. S. Lundgren, "Quasilinear dirichlet problems driven by positive sources," *Archive for Rational Mechanics and Analysis* Volume **145** Number 10 (1998), 241–269.

24. J. A. Krupp and E. M. Murman, "Computation of Transonic Flows past Lifting Airfoils and Slender Bodies," *AIAA Journal* Volume **10** Number 7 (July 1972), 880–886.

25. J. Liouville, "Sur l'équation aux différences partielles $\dfrac{d^2 \log \lambda}{dudv} \pm \lambda 2a^2 = 0$, *J. Math. Pure Appl.*, **18** (1853), 71–72.

26. J. S. McGough, "Numerical continuation and the Gelfand problem," *Applied Mathematics and Computation* Volume **89** (1998), 225–239.

27. R. E. Mickens, "Difference equation models of differential equations having zero local truncation errors," *Differential Equations.*(Birmingham, Ala., 1983), 445-449, North-Holland Math. Stud., 92, (North-Holland, Amsterdam-New York, 1984).

28. R. E. Mickens, "Exact solutions to difference equation models of Burgers' equation," *Numerical Methods for Partial Differential Equations* Volume **2** Issue 2 (1986), 123–129.

29. R. E. Mickens, "Difference Equation Models of Differential Equations," *Mathematical and Computer Modelling* Volume **11** (1988), 528–530.

30. R. E. Mickens, and A. Smith. "Finite-difference Models of Ordinary Differential Equations: Influence of Denominator Functions," *Journal of the Franklin Institute* **327** (1990), 143–145.

31. R. E. Mickens, *Nonstandard Difference Models of Differential Equations* (World Scientific, Singapore, 1994).

32. R. E. Mickens (editor), *Applications of Nonstandard Finite Differences*, (World Scientific, Singapore, 2000).

33. R. E. Mickens, "Nonstandard Finite Difference Schemes for Differential Equations," *Journal of Difference Equations and Applications* Volume **8** Issue 9. (2002), 823–847.

34. J. Ockendon, S. Howison, A. Lacey and A. Movchan, *Applied Partial Differential Equations,* (Oxford University Press, October 1999).

35. D. L. Scharfetter, and H. K. Gummel, "Large-Signal Analysis of a Silicon Read Diode Oscillator," *IEEE Transactions on Electron Devices* Volume ED-16, Number 1, (January 1969), 64–84.

36. K. Wanelik, "On the iterative solutions of some nonlinear eigenvalue problems," *Journal of Mathematical Physics* **30** Number 8 (August 1989), 1707–12.

CHAPTER 4

HIGH ACCURACY NONSTANDARD FINITE-DIFFERENCE TIME-DOMAIN ALGORITHMS FOR COMPUTATIONAL ELECTROMAGNETICS: APPLICATIONS TO OPTICS AND PHOTONICS

James B. Cole

University of Tsukuba
Institute for Information Science and Electronic
Tsukuba, Ibaraki 305-8573, Japan
cole@is.tsukuba.ac.jp

We apply nonstandard finite difference (NSFD) models to develop new high accuracy finite-difference time-domain (FDTD) algorithms for computational electromagnetics and optics. The basic FDTD algorithm is simple and easy to program, but the accuracy of conventional FDTD algorithms is low. For step size h, the error is $\epsilon \sim h^2$ and the computational cost is $C \sim h^4$ in three dimensions. Thus halving h reduces the ϵ by a factor of four, but C rises by a factor of sixteen. Using NSFD models we constructed FDTD algorithms for which $\epsilon \sim h^6$ with about the same computational cost. We introduce NSFD versions of the FDTD algorithm to solve the wave equation, Maxwell's equations (Yee algorithm) and a new version of the Mur absorbing boundary condition. We tested our new algorithms by computing scattering off cylinders and spheres in the Mie regime and comparing with the analytic solutions. The accuracy of the NSFD algorithms is superior to the conventional FDTD ones. We illustrate our methods with some example applications from our current research. Among the topics we cover are propagation in dispersive media and surface plasmons, light propagation in structures with subwavelength features including conducting diffraction gratings and biological structures, and the improvement of light coupling through media interfaces using subwavelength structures.

1. Introduction

In this chapter we develop nonstandard finite difference models of the wave equation and Maxwell's equations, and derive new high accuracy finite-difference time-domain (FDTD) algorithms to solve problems in optics and

89

computational electromagnetics.

There is an old proverb, "the devil is in the details." Although the algorithms themselves are simple, using them to solve realistic problems requires a deep knowledge of physics and mathematics, and very careful attention to detail. We address these details in some example calculations, such as light propagation in structures with subwavelength features including photonic crystals, diffraction gratings, and biological structures. We also introduce algorithms to investigate and surface plasmons in thin metallic surfaces.

The physics of electromagnetic and acoustical waves is described by the wave equation, and by Maxwell's equations, but by themselves these equations are of little use for solving most practical problems with irregularly-shaped boundaries between different media. Except for a few simple shapes that are highly symmetric or infinitely periodic, analytic solutions do not exist, and numerical methods are a necessity.

Many numerical methods available, but the second-order finite-difference time-domain (FDTD) algorithm is esthetically appealing because it derives directly from the original differential equation, and can handle boundaries of arbitrary shape. The boundary conditions are a consequence of Maxwell's equations or of the wave equation and are thus implicitly enforced as the FDTD algorithm is iterated, so long as they were satisfied by the initial wave fields.

The FDTD algorithm is popular because it is simple and easy to program. Its main drawback is low accuracy. For grid spacing h the error is $\varepsilon \sim h^2$. Accuracy can be increased by reducing h, but as the grid fineness increases, the computational cost rises much faster than the accuracy. When h is decreased, the time step, Δt, must decrease in proportion to h in order to maintain algorithm stability (see Sect. 6). The computational cost, C, which is proportional to the number of space-time grid points, is thus $C \sim 1/h^{d+1}$ in d spatial dimensions. For example, in three dimensions $C \sim 1/h^4$. Halving h ($h \to h/2$) reduces the error by a factor of four ($\varepsilon \to \varepsilon/4$) but C rises by a factor of sixteen ($C \to 16C$). Even on a supercomputer, high accuracy solutions of large-scale problems are out of reach withthe conventional second order FDTD algorithms. It is possible to increase accuracy by using higher order finite-difference approximations, but this not only complicates the algorithm, and increases the computational cost, but the algorithm can become unstable. Moreover, at the boundaries between different media, the error of higher-order algorithms can actually increase.

In this chapter we derive second-order FDTD algorithms for which the solution error is $\varepsilon \sim h^6$, by using what are called nonstandard finite-difference (NSFD) models. The high accuracy NSFD versions of the FDTD algorithms are only slightly more complicated than the conventional ones, and existing computer programs can easily be modified to run the NSFD algorithms.

2. Basic Concepts

2.1. *Finite-Difference Models and the FDTD Algorithm*

There is an old Chinese proverb, "One example is worth 10,000 equations." In this spirit, let us forthwith introduce the FDTD algorithm. The wave equation in one dimension is

$$\left(\partial_t^2 - v^2 \partial_x^2\right) \psi(x, t) = 0, \tag{1}$$

where v is the phase velocity. The second-order central finite-difference (FD) approximation to first derivative is

$$f'(x) \cong \frac{d_x}{h} f(x), \tag{2}$$

where the difference operator d_x is defined by

$$d_x f(x) = f(x + h/2) - f(x - h/2). \tag{3}$$

For reasons that will become apparent later, we call (2) the standard finite difference (SFD) approximation. Replacing f with f' in (2) gives

$$f''(x) \cong \frac{d_x^2}{h^2} f(x), \tag{4}$$

where $d_x^2 = d_x d_x$. Evaluating $d_x^2 f$, we find

$$d_x^2 f(x) = f(x + h) + f(x - h) - 2f(x). \tag{5}$$

Substituting FD approximations for the derivatives in the wave equation, we have

$$\left(d_t^2 - \bar{v}^2 d_x^2\right) \psi(x, t) = 0, \tag{6}$$

where

$$\bar{v}^2 = \left(\frac{v \Delta t}{h}\right)^2. \tag{7}$$

From now on we shall call (6) the standard finite-difference (SFD) model of the wave equation. Rearranging (6), and expanding $d_t^2 \psi(x, t)$ with (5) we obtain

$$\psi(x, t + \Delta t) + \psi(x, t - \Delta t) - 2\psi(x, t) = \bar{v}^2 d_x^2 \psi(x, t). \tag{8}$$

Solving for $\psi(x, t + \Delta t)$ gives the basic FDTD algorithm,

$$\psi(x, t + \Delta t) = -\psi(x, t - \Delta t) + 2\psi(x, t) + \bar{v}^2 d_x^2 \psi(x, t). \qquad (9)$$

From now on we shall call (9) the standard finite-difference time-domain (S-FDTD) algorithm. Given the initial fields, $\psi(x, 0)$ and $\psi(x, \Delta t)$, ψ can be found at all future times by iterating (9) on the space-time grid,

$$t = 0, \Delta t, 2\Delta t \cdots, \qquad (10)$$

$$x = 0, h, 2h, \cdots, N_x h, \qquad (11)$$

where $N_x + 1$ is the number of grid points on the x-axis.

Algorithm (9) is a second-order algorithm because it derives from second-order FD approximations. Expanding $d_t^2 f(t)/\Delta t^2$ in a Taylor series we have

$$\frac{d_t^2}{\Delta t^2} f(t) = f''(t) + \frac{1}{12}\Delta t^2 f''''(t) + \cdots. \qquad (12)$$

The leading error term of (4) is proportional to Δt^2, whence the term "second order FD approximation."

One might try to construct a more accurate FD model of the wave equation by using higher-order FD approximations. For example, the fourth-order FD approximation to f'' is

$$f''(t) = \frac{1}{\Delta t^2} \left[\frac{4}{3}[f(t + \Delta t) + f(t - \Delta t)] \right. \qquad (13)$$

$$\left. - \frac{1}{12}[f(t + 2\Delta t) + f(t - 2\Delta t)] - \frac{5}{2}f(t) \right] - \frac{1}{90}\Delta t^4 f^{(6)}(t).$$

Replacing the derivatives of the wave equation with fourth-order FD approximations yields a fourth-order FDTD algorithm. To iterate the fourth-order algorithm, four initial values of ψ are needed, even though solutions of the wave equation itself are completely specified by just two initial values. The fourth-order FD model is thus fundamentally different from the wave equation [1], and some of its solutions are very different from those of the wave equation. The second-order FD model (6), even though it is less accurate, is actually a better model of the wave equation because its solution, like that of the wave equation, is specified by two initial values of ψ.

Let us now investigate the error of the SFD model (6). The most general solution of the one-dimensional wave equation (1) is a linear superposition of forward ($+x$-direction) and backward ($-x$-direction) moving waves,

$$\psi_\pm(x,t) = f(x \mp vt), \tag{14}$$

where f is an arbitrary function. Substituting (14) into the FD model yields

$$\left(d_t^2 - \bar{v}^2 d_x^2\right)\psi_\pm(x,t) = \varepsilon. \tag{15}$$

If (6) were an exact model of (1), ε would vanish. The extent to which it does not is a measure of the model error. Directly evaluating (15) we find

$$\varepsilon = f(x \mp vt + v\Delta t) + f(x - vt - v\Delta t) - 2f(x \mp vt) \tag{16}$$
$$- \bar{v}^2\left[f(x + h \mp vt) + f(x - h \mp vt) - 2f(x \mp vt)\right].$$

Examination of (16) shows that by setting $h = v\Delta t$, ε can be made to vanish. Thus the condition

$$\frac{v\Delta t}{h} = 1 \tag{17}$$

specifies an exact model of the wave equation.

Even though its constituent FD approximations are only second order, at $\bar{v} = 1$ the SFD model is exact because the errors in the FD approximations to the space and time derivatives mutually cancel. Although this is a special case, it does demonstrate that we can improve the accuracy of FD models by a clever choice of the model parameters.

2.2. *A Nonstandard Finite Difference Model*

Backward- and forward-moving monochromatic solutions of the wave equation with angular frequency ω and wavenumber k are given by

$$\varphi_\pm(x,t) = e^{i(kx \mp \omega t)}, \tag{18}$$

where

$$v = \frac{\omega}{k}, \tag{19}$$

$\lambda = 2\pi/k$ is the wavelength, and $T = 2\pi/\omega$ is the wave period. A general monochromatic solution is a linear superposition of φ_+ and φ_- of the form

$$\varphi = a_+\varphi_+ + a_-\varphi_-, \tag{20}$$

where the $a_\pm$ are constants. Computing $d_x^2\varphi$ and $d_t^2\varphi$ we find

$$d_x^2\varphi = -4\sin^2\left(kh/2\right)\varphi, \tag{21}$$
$$d_t^2\varphi = -4\sin^2\left(\omega\Delta t/2\right)\varphi. \tag{22}$$

We can generalize the SFD model (6) by replacing $\bar{v}^2$ with a free parameter, which we call u^2. The model then becomes,

$$\left(d_t^2 - u^2 d_x^2\right) \psi(x,t) = 0. \tag{23}$$

Substituting a monochromatic solution into (23), the model error is

$$\varepsilon = \left(d_t^2 - u^2 d_x^2\right) \varphi(x,t) \tag{24}$$
$$= 4\left[u^2 \sin^2\left(kh/2\right) - \sin^2\left(\omega\Delta t/2\right)\right] \varphi(x,t).$$

We see that ε can be made to vanish with the choice $u^2 = u_0^2$, where

$$u_0^2 = \frac{\sin^2\left(\omega\Delta t/2\right)}{\sin^2\left(kh/2\right)}. \tag{25}$$

We now define the nonstandard finite difference (NSFD) model [1] to be

$$\left(d_t^2 - u_0^2 d_x^2\right) \psi(x,t). \tag{26}$$

Replacing $\bar{v}^2$ with u_0^2 in (9) gives the nonstandard finite-difference time-domain (NS-FDTD) algorithm.

$$\psi(x, t + \Delta t) = -\psi(x, t - \Delta t) + 2\psi(x,t) + u_0^2 d_x^2 \psi(x,t). \tag{27}$$

The NSFD model and the NS-FDTD algorithms are exact with respect to monochromatic solutions of the wave equation. If v changes with position we must use $k = k(x)$ in (25), where

$$k(x) = \frac{\omega}{v(x)}. \tag{28}$$

For $h = v\Delta t$ we have $u_0^2 = 1$, and the NS-FDTD and S-FDTD algorithms are equivalent, and exact with respect to general solutions (14) of the wave equation.

The NSFD model can be deduced by a different line of reasoning. The SFD approximation (2) to the derivative can be generalized by replacing h by a function $s(h)$ so that

$$f'(x) \cong \frac{d_x}{s(h)} f(x). \tag{29}$$

Equation (29) is called a nonstandard finite difference (NSFD) approximation [1]. One might suppose that setting

$$s(h) = \frac{d_x f(x)}{f'(x)}, \tag{30}$$

would yield an exact FD expression for the derivative of any function, but this is not true. For (29) to be a valid approximation, we must have

$$\lim_{h \to 0} \frac{d_x f(x)}{s(h)} = f'(x).$$ (31)

Expanding $s(h)$ in a Taylor series,

$$s(h) = s(0) + hs'(0) + \cdots,$$ (32)

we obtain the conditions,

$$\lim_{h \to 0} s(h) = h,$$ (33)

$$s'(0) = 1.$$ (34)

In the FDTD algorithm f an unknown function so (30) may not be computable from the data if s is a function of x, even if it satisfies (33) and (34).

Monochromatic solutions of the wave equation belong to the set $S_1 = \{\sin(kx), \cos(kx), e^{\pm ikx}\}$, and (30) gives

$$s(h) = \frac{2}{k} \sin\left(kh/2\right),$$ (35)

which is independent of x, and satisfies (33) and (34). Thus an exact FD expression for $f \in S_1$ is

$$f'(x) = \frac{d_x}{s(h)} f(x).$$ (36)

For the more general class of functions $S_0 = \{e^{ax}, \sinh(ax), \cosh(ax)\}$, where a is complex, (30) gives

$$s(h) = \frac{2}{a} \sinh(ah/2),$$ (37)

and again (36) holds exactly. Furthermore for $f \in S_0$ or S_1, we have

$$f''(x) = \frac{d_x^2}{s(h)^2} f(x).$$ (38)

Replacing the derivatives in the wave equation with exact NSFD expressions of the form of (38) we arrive at the NSFD model of (1).

2.3. *Extension to Two and Three Dimensions*

The wave equation in three dimensions is

$$\left(\partial_t^2 - v^2 \nabla^2\right) \psi(\mathbf{x}, t) = 0, \tag{39}$$

where $\mathbf{x} = (x, y, z)$. We now proceed to construct a FD model of (39). The SFD approximation to $\nabla^2 \psi$ is

$$\nabla^2 \psi \cong \frac{\mathbf{D}_1^2 \psi}{h^2}, \tag{40}$$

where

$$\mathbf{D}_1^2 = d_x^2 + d_y^2 + d_z^2. \tag{41}$$

Using $\partial_t^2 \psi \cong d_t^2 \psi / \Delta t^2$ gives the SFD model of the wave equation,

$$\left(d_t^2 - \bar{v}^2 \mathbf{D}_1^2\right) \psi(\mathbf{x}, t) = 0, \tag{42}$$

where $\bar{v}^2 = v^2 \Delta t^2 / h^2$. Expanding $d_t^2 \psi(\mathbf{x}, t)$ and solving for $\psi(\mathbf{x}, t + \Delta t)$, we obtain the S-FDTD algorithm,

$$\psi(\mathbf{x}, t + \Delta t) = -\psi(\mathbf{x}, t - \Delta t) + 2\psi(\mathbf{x}, t) + \bar{v}^2 \mathbf{D}_1^2 \psi(\mathbf{x}, t). \tag{43}$$

On a uniform grid, $\Delta x = \Delta y = \Delta z = h$, we discretize space and time by

$$t = 0, \Delta t, 2\Delta t, \cdots \tag{44}$$
$$x = 0, h, 2h, \cdots, N_x h \tag{45}$$
$$y = 0, h, 2h, \cdots, N_y h \tag{46}$$
$$z = 0, h, 2h, \cdots, N_z h. \tag{47}$$

It is possible to use different grid spacings on each axis, or even variable grid spacing on the same axis, but this complicates the algorithm, and we omit this topic.

Replacing $\bar{v}^2$ with a free parameter u^2 in the SFD model, we now seek a value of u^2 for which the error vanishes. The SFD model becomes

$$\left(d_t^2 - u^2 \mathbf{D}_1^2\right) \psi(\mathbf{x}, t) = 0. \tag{48}$$

A monochromatic plane wave solution of (39) moving in direction $\mathbf{k}$ is

$$\varphi_{\mathbf{k}}(\mathbf{x}, t) = e^{i(\mathbf{k} \cdot \mathbf{x} - \omega t)}, \tag{49}$$

where $\mathbf{k} = (k_x, k_y, k_z)$ is the wave vector. A general monochromatic solution,

$$\varphi = \sum_{|\mathbf{k}|=k} \varphi_{\mathbf{k}}, \tag{50}$$

is a superposition of plane waves moving in different directions of common angular frequency, and wavenumber magnitude $k = |\mathbf{k}|$.

Substituting $\varphi_{\mathbf{k}}$ into (48), let us define the model error to be

$$\varepsilon = \frac{\left(d_t^2 - u^2 \mathbf{D}_1^2\right) \varphi_{\mathbf{k}}}{\varphi_{\mathbf{k}}}. \tag{51}$$

For simplicity, let us first evaluate ε in two dimensions. Here $\mathbf{D}_1^2 = d_x^2 + d_y^2$, $\mathbf{x} = (x, y)$, and $\mathbf{k} = (k_x, k_y) = k\,(\cos\theta, \sin\theta)$. We find

$$\frac{\mathbf{D}_1^2 \varphi_{\mathbf{k}}}{\varphi_{\mathbf{k}}} = -4 \left[\sin^2(k_x h/2) + \sin^2(k_y h/2)\right], \tag{52}$$

$$\frac{d_t^2 \varphi_{\mathbf{k}}}{\varphi_{\mathbf{k}}} = -4 \sin^2\left(\omega \Delta t/2\right), \tag{53}$$

thus the model error is

$$\varepsilon = 4 \left[u^2 \left(\sin^2(k_x h/2) + \sin^2(k_x h/2)\right) - \sin^2(\omega \Delta t/2)\right]. \tag{54}$$

We see that ε can be made to vanish with the choice

$$u^2(k_x, k_y) = \frac{\sin^2(\omega \Delta t/2)}{\sin^2(k_x h/2) + \sin^2(k_y h/2)}, \tag{55}$$

but unfortunately it vanishes for only one paricular direction of $\mathbf{k}$. What we really need is a parameter that is independent of θ for which ε vanishes over all plane wave directions. Expanding $\mathbf{D}_1^2 \varphi_{\mathbf{k}}/\varphi_{\mathbf{k}}$ in a Taylor series about $\sin\theta = 0$, we have

$$\frac{\mathbf{D}_1^2 \varphi_{\mathbf{k}}}{\varphi_{\mathbf{k}}} = -4 \sin^2\left(kh/2\right) - \frac{1}{24} \left(kh\right)^4 \sin^2\left(2\theta\right) + \cdots. \tag{56}$$

If we could construct an FD operator for $\nabla^2 \varphi_{\mathbf{k}}$ with smaller angular dependence, we could reduce the FD approximation error. There is no second-order FD approximation which is completely independent of θ, but as we shall see, we can construct one which is nearly so.

In two dimensions there is another FD approximation to $\nabla^2 \psi$ given by

$$\nabla^2 \psi \cong \frac{\mathbf{D}_2^2 \psi}{h^2}, \tag{57}$$

where $\mathbf{D}_2^2$ is defined by

$$2\mathbf{D}_2^2 f(x, y) = f(x + h, y + h) + f(x + h, y - h) \tag{58}$$

$$+ f(x - h, y + h) + f(x - h, y - h) - 4f(x, y).$$

Evaluating $\mathbf{D}_2^2 \varphi_{\mathbf{k}}/\varphi_{\mathbf{k}}$, and expanding in a Taylor series we find

$$\frac{\mathbf{D}_2^2 \varphi_{\mathbf{k}}}{\varphi_{\mathbf{k}}} = -4\sin^2(kh/2) + \frac{1}{12}(kh)^4 \sin^2(2\theta) + \cdots . \tag{59}$$

Taking γ to be a free parameter, let us now try to find a superposition of $\mathbf{D}_1^2$ and $\mathbf{D}_2^2$

$$\mathbf{D}_\gamma^2 = \gamma \mathbf{D}_1^2 + (1-\gamma)\mathbf{D}_2^2 \tag{60}$$

for which the angular dependence of $\mathbf{D}_\gamma^2 \varphi_{\mathbf{k}}/\varphi_{\mathbf{k}}$ is minimal. Comparing the Taylor expansions (56) and (59), we see that the choice $\gamma = 2/3$ cancels the $(kh)^4 \sin^2(2\theta)$-terms, leaving

$$\frac{\mathbf{D}_{2/3}^2 \varphi_{\mathbf{k}}}{\varphi_{\mathbf{k}}} = -4\sin^2(kh/2) - \frac{1}{720}(kh)^6 \sin^2(2\theta) + \cdots . \tag{61}$$

There is an even better choice [2,3], given by $\gamma = \gamma_0$, where,

$$\gamma_0 = 1 - \frac{\sin^2(k_x' h/2) + \sin^2(k_y' h/2) - \sin^2(kh/2)}{2\sin^2(k_x' h/2)\sin^2(k_y' h/2)}, \tag{62}$$

and $(k_x', k_y') = k\left(2^{-1/4}, \sqrt{1-2^{-1/2}}\right)$. Defining

$$\mathbf{D}_0^2 = \gamma_0 \mathbf{D}_1^2 + (1-\gamma_0)\mathbf{D}_2^2, \tag{63}$$

we have

$$\frac{\mathbf{D}_0^2 \varphi_{\mathbf{k}}}{\varphi_{\mathbf{k}}} = -4\sin^2(kh/2) \tag{64}$$

$$-\frac{1}{24192}(kh)^8 \left[\left(\sqrt{2}-1\right)\sin^2 2\theta - \frac{1}{2}\sin^4 2\theta\right] + \cdots .$$

A convenient expression for γ is

$$\gamma_0(k) = \frac{2}{3} - \frac{1}{90}(kh)^2 - \frac{1}{15120}(kh)^4\left(11 - 5\sqrt{2}\right) \tag{65}$$

$$-\frac{1}{907200}(kh)^6\left(29 - 17\sqrt{2}\right) - \cdots .$$

The angular dependence of $\mathbf{D}_0^2 \varphi_{\mathbf{k}}/\varphi_{\mathbf{k}}$ is very much smaller than that of either $\mathbf{D}_1^2 \varphi_{\mathbf{k}}/\varphi_{\mathbf{k}}$ or $\mathbf{D}_2^2 \varphi_{\mathbf{k}}/\varphi_{\mathbf{k}}$. Defining $s(h)$ by (35), a nearly exact FD expression for $\nabla^2 \varphi_{\mathbf{k}}$ is

$$\nabla^2 \varphi_{\mathbf{k}} \cong \frac{\mathbf{D}_0^2 \varphi_{\mathbf{k}}}{s(h)^2}. \tag{66}$$

Making the substitution $\mathbf{D}_1^2 \to \mathbf{D}_0^2$ in the FD model (48), the best choice of u^2 is u_0^2 given by (25). We now call

$$\left(d_t^2 - u_0^2 \mathbf{D}_0^2\right) \psi(\mathbf{x}, t) = 0 \tag{67}$$

the NSFD model of the wave equation. Defining $\varepsilon_{\text{nsfd}}$ in analogy to (51), and making a Taylor expansion we find

$$\varepsilon_{\text{nsfd}} = -\frac{1}{29192} k^2 (kh)^6 \left[\left(\sqrt{2} - 1\right) \sin^2 (2\theta) - \frac{1}{2} \sin^4 (2\theta)\right] + \cdots . \tag{68}$$

Similarly, the error of the SFD model is

$$\varepsilon_{\text{sfd}} = -\frac{1}{12} k^2 (kh)^2 \cos^2 \theta + \cdots . \tag{69}$$

The error of the SFD model is proportional to $(h/\lambda)^2$, whereas the error of the NSFD model is proportional to $(h/\lambda)^6$.

The nonstandard FDTD (NS-FDTD) algorithm, which follows simply from the NSFD model, is

$$\psi(\mathbf{x}, t + \Delta t) = -\psi(\mathbf{x}, t - \Delta t) + 2\psi(\mathbf{x}, t) + u_0^2 \mathbf{D}_0^2 \psi(\mathbf{x}, t). \tag{70}$$

In three dimensions $\mathbf{D}_0^2$ is a superposition of three different FD operators,

$$\mathbf{D}_0^2 = \gamma_1 \mathbf{D}_1^2 + \gamma_2 \mathbf{D}_2^2 + \gamma_3 \mathbf{D}_3^2, \tag{71}$$

where $\mathbf{D}_1^2$ is given by (41) and

$$
\begin{aligned}
4\mathbf{D}_2^2 f(x, y, z) = {}& f(x + h, y + h, z + h) + f(x - h, y - h, z - h) \\
& + f(x + h, y - h, z + h) + f(x - h, y + h, z - h) \\
& + f(x - h, y + h, z + h) + f(x + h, y - h, z - h) \\
& + f(x - h, y - h, z + h) + f(x + h, y + h, z - h) \\
& - 8f(x, y, z),
\end{aligned}
\tag{72}
$$

$$
\begin{aligned}
4\mathbf{D}_3^2 f(x, y, z) = {}& f(x + h, y + h, z) + f(x - h, y - h, z) \\
& + f(x + h, y - h, z) + f(x - h, y + h, z) \\
& + f(x + h, y, z + h) + f(x - h, y, z - h) \\
& + f(x + h, y, z - h) + f(x - h, y, z + h) \\
& + f(x, y + h, z - h) + f(x, y - h, z - h) \\
& - 12f(x, y, z).
\end{aligned}
\tag{73}
$$

It can be shown that the optimum superposition weights are

$$
\begin{aligned}
\gamma_1 &= \eta_0\left(1 - \gamma_0\right) + \left(2\gamma_0 - 1\right), \\
\gamma_2 &= \eta_0\left(1 - \gamma_0\right) \\
\gamma_3 &= 1 - \gamma_1 - \gamma_2,
\end{aligned}
\tag{74}
$$

where

$$
\eta_0 = \frac{1}{s_1^2} + \frac{2s_1^2 - s_2^2}{4\left(1 - \gamma_0\right)s_1^4 s_2^2} - \left(\frac{1 + \gamma_0}{1 - \gamma_0}\right)\frac{1}{2s_2^2},
\tag{75}
$$

$s_1 = \sin(kh/4)$, and $s_2 = \sin\left(kh\sqrt{2}/4\right)$. Expression (75) is equivalent to eq. (26) in ref. [2] with $\phi' = \pi/4$. In ref. [2] $\mathbf{D}_3^2$ is gven incorrectly in eq. (22c). The correct form is given here in (73). Expanding (75) in a Taylor series gives

$$
\begin{aligned}
\eta_0(k) = \frac{2}{5} &+ \left(\frac{1913}{50400} - \frac{5}{252}\sqrt{2}\right)(kh)^2 \\
&+ \left(\frac{1457}{151200} - \frac{17}{30240}\sqrt{2}\right)(kh)^4 + \cdots.
\end{aligned}
\tag{76}
$$

It can be shown by direct Taylor expansion that the error of the three-dimensional NSFD model is the same as the two-dimensional one with respect to monochromatic plane waves.

3. High Accuracy FDTD Algorithm for the Absorbing Wave Equation

3.1. *Exact Algorithm in One Dimension*

The absorbing wave equation is given by

$$
\left(\partial_t^2 - v^2\partial_x^2 + 2\alpha\partial_t\right)\psi(x,t) = 0
\tag{77}
$$

where $\alpha \geq 0$ is the absorption, and v is a velocity parameter. With the space-time discretizations of (10) and (11), $\partial_t\psi$ cannot be approximated by (2) because ψ is unknown at $t \pm \Delta t/2$. Defining the difference operator

$$
d_t' f(t) = f(t + \Delta t) - f(t - \Delta t),
\tag{78}
$$

we have

$$
f'(t) \cong \frac{d_t'}{2\Delta t}f(t).
\tag{79}
$$

Now replacing $\partial_t^2\psi$ and $\partial_x^2\psi$ with FD approximations (4) we obtain the SFD model of (77),

$$
\left(d_t^2 - \bar{v}^2 d_x^2 + \bar{a}d_t'\right)\psi(x,t) = 0,
\tag{80}
$$

where

$$\bar{v}^2 = \frac{v^2 \Delta t^2}{h^2}, \tag{81}$$

$$\bar{a} = \alpha \Delta t. \tag{82}$$

Rearranging (80), and using the definitions of $d_t^2 \psi$ and $d_t' \psi$, we solve for $\psi(x, t + \Delta t)$ to obtain the S-FDTD algorithm,

$$\psi(x, t + \Delta t) = -\left(\frac{1-\bar{a}}{1+\bar{a}}\right)\psi(x, t - \Delta t) + \left(\frac{2}{1+\bar{a}}\right)\psi(x, t) \tag{83}$$
$$+ \left(\frac{1}{1+\bar{a}}\right)\bar{v}^2 d_x^2 \psi(x, t).$$

Because the FD approximations are second order, the SFD model error is proportional to h^2.

We now seek FD model parameters for which the error vanishes. Defining the difference operator

$$W(u, a) = d_t^2 - u^2 d_x^2 + a d_t', \tag{84}$$

and regarding both a and u as free parameters, a generalized FD model of (77) is

$$W(u, a)\psi(x, t) = 0. \tag{85}$$

The choices $u^2 = \bar{v}^2$, and $a = \bar{a}$ define the SFD model (80). Let us now evaluate the error of the generalized FD model with respect to solutions of the absorbing wave equation. Forward- and backward- moving solutions of (77) are

$$\varphi_{\pm}(x, t) = e^{i\left(kx \mp \omega' t\right)} e^{-\alpha t}, \tag{86}$$

where

$$\omega'^2 = \omega^2 - \alpha^2, \tag{87}$$

$$v = \omega/k. \tag{88}$$

The condition

$$\alpha < \omega \tag{89}$$

ensures that (86) is an oscillatory solution. A general decaying harmonic solution of (77) is a superposition of forward- and backward-moving solutions given by

$$\varphi = a_+ \varphi_+ + a_- \varphi_-. \tag{90}$$

Rewriting (86) in the form $\varphi_\pm = e^{ikx}e^{-z_\pm t}$, where $z_\pm = -(\alpha \pm i\omega')$, and using

$$d_x^2 e^{ikx} = -4\sin^2{(kh/2)}\, e^{ikx}, \tag{91}$$

$$d_t' e^{zt} = 2\sinh{(z\Delta t)}\, e^{zt}, \tag{92}$$

$$d_t^2 e^{zt} = 4\sinh^2{(z\Delta t/2)}\, e^{zt}, \tag{93}$$

we seek values of u and a for which $W(u,a)\varphi_\pm = 0$. Comparing real and imaginary parts, we find that the imaginary part of $W(u,a)\varphi_\pm$ vanishes for the choice $a = a_0$, where

$$a_0 = \tanh(\alpha \Delta t), \tag{94}$$

while the real part vanishes for the choice $u^2 = u_0^2$, where

$$u_0^2 = \frac{\sin^2(\omega'\Delta t/2) + \sinh^2(\alpha\Delta t/2)}{\sin^2(kh/2)\cosh(\alpha\Delta t)}. \tag{95}$$

In the limit $\alpha \to 0$, u_0^2 reduces, as expected, to (25). We now define the NSFD model of the absorbing wave equation to be $W(u_0, a_0)\psi(x,t) = 0$, or more explicitly,

$$\left(d_t^2 - u_0^2 d_x^2 + a_0 d_t'\right)\psi(x,t) = 0. \tag{96}$$

The NSFD model is exact with respect to decaying harmonic solutions. Making the substitutions $\bar{v}^2 \to u_0^2$ and $\bar{a} \to a_0$ in (83) gives the NS-FDTD algorithm,

$$\psi(x, t+\Delta t) = -\left(\frac{1-a_0}{1+a_0}\right)\psi(x, t-\Delta t) + \left(\frac{2}{1+a_0}\right)\psi(x,t) \tag{97}$$

$$+ \left(\frac{1}{1+a_0}\right)u_0^2 d_x^2 \psi(x,t).$$

3.2. *Extension to Two and Three Dimensions*

In two and three dimensions the absorbing wave equation is

$$\left(\partial_t^2 - v^2\nabla^2 + 2\alpha\partial_t\right)\psi(\mathbf{x},t) = 0. \tag{98}$$

Following the developments of the previous sections, the SFD model of (98) is

$$\left(d_t^2 - \bar{v}^2\mathbf{D}_1^2 + \bar{a}d_t'\right)\psi(\mathbf{x},t) = 0, \tag{99}$$

where $\bar{v}^2$ and $\bar{a}$ are given by (3.1.5) and (3.1.6). Solving for $\psi(\mathbf{x}, t + \Delta t)$ yields the S-FDTD algorithm,

$$\psi(\mathbf{x}, t + \Delta t) = -\left(\frac{1 - \bar{a}}{1 + \bar{a}}\right)\psi(\mathbf{x}, t - \Delta t) + \left(\frac{2}{1 + \bar{a}}\right)\psi(\mathbf{x}, t) \qquad (100)$$
$$+ \left(\frac{1}{1 + \bar{a}}\right)\bar{v}^2 \mathbf{D}_1^2 \psi(\mathbf{x}, t).$$

The NSFD model of (98) is obtained with the replacements $\mathbf{D}_1^2 \to \mathbf{D}_0^2$ (63), $\bar{a} \to a_0$, and $\bar{v}^2 \to u_0^2$ in (99) giving,

$$\left(d_t^2 - u_0^2 \mathbf{D}_0^2 + a_0 d_t'\right)\psi(\mathbf{x}, t) = 0. \qquad (101)$$

The error of the NSFD model is given by (68).

Making the same replacements in the S-FDTD algorithm (100) gives the NS-FDTD algorithm,

$$\psi(\mathbf{x}, t + \Delta t) = -\left(\frac{1 - a_0}{1 + a_0}\right)\psi(\mathbf{x}, t - \Delta t) + \left(\frac{2}{1 + a_0}\right)\psi(\mathbf{x}, t) \qquad (102)$$
$$+ \left(\frac{1}{1 + a_0}\right)u_0^2 \mathbf{D}_0^2 \psi(\mathbf{x}, t).$$

4. Nonstandard FDTD Solution of the Conducting Maxwell Equations

4.1. *The Yee Algorithm*

Using the relationship between Maxwell's equations and the wave equation, we can extend the developments of the previous sections to construct a high accuracy FDTD algorithm to solve Maxwell's equations.

In a linear conducting medium, Maxwell's equations are

$$\mu \partial_t \mathbf{H}(\mathbf{x}, t) = -\nabla \times \mathbf{E}(\mathbf{x}, t), \qquad (103)$$
$$\varepsilon \partial_t \mathbf{E}(\mathbf{x}, t) = \nabla \times \mathbf{H}(\mathbf{x}, t) - \sigma \mathbf{E}(\mathbf{x}, t), \qquad (104)$$

where ε is the relative electric permittivity, μ the magnetic permeability, and σ is the conductivity. Replacing the derivatives with SFD approximations, the SFD model of Maxwell's equations is

$$d_t \mathbf{H}(\mathbf{x}, t) = -\frac{1}{\mu}\frac{\Delta t}{h}\mathbf{D}_1 \times \mathbf{E}(\mathbf{x}, t), \qquad (105)$$

$$d_t \mathbf{E}(\mathbf{x}, t + \Delta t/2) = \frac{1}{\varepsilon}\frac{\Delta t}{h}\mathbf{D}_1 \times \mathbf{H}(\mathbf{x}, t + \Delta t/2) \qquad (106)$$
$$- \frac{\sigma}{2\varepsilon}\Delta t \left[\mathbf{E}(\mathbf{x}, t + \Delta t) + \mathbf{E}(\mathbf{x}, t)\right].$$

J. B. Cole

Here $\mathbf{D}_1$ is a vector difference operator for the gradient (∇) defined by

$$\mathbf{D}_1 = \hat{\mathbf{x}}d_x + \hat{\mathbf{y}}d_y + \hat{\mathbf{z}}d_z, \tag{107}$$

where $\hat{\mathbf{x}}$, $\hat{\mathbf{y}}$, and $\hat{\mathbf{z}}$ are unit vectors along the principal coordinate axes.

Solving for $\mathbf{H}(\mathbf{x}, t + \Delta t/2)$, and $\mathbf{E}(\mathbf{x}, t + \Delta t)$ yields a FDTD algorithm, called the Yee algorithm [7],

$$\mathbf{H}(\mathbf{x}, t + \Delta t/2) = \mathbf{H}(\mathbf{x}, t - \Delta t/2) - \frac{1}{\mu(\mathbf{x})}\frac{\Delta t}{h}\mathbf{D}_1 \times \mathbf{E}(\mathbf{x}, t), \tag{108}$$

$$\mathbf{E}(\mathbf{x}, t + \Delta t) = \left(\frac{1 - \sigma\Delta t/2\varepsilon}{1 + \sigma\Delta t/2\varepsilon}\right)\mathbf{E}(\mathbf{x}, t) \tag{109}$$

$$+ \frac{1}{\varepsilon}\left(\frac{1}{1 + \sigma\Delta t/2\varepsilon}\right)\mathbf{D}_1 \times \mathbf{H}(\mathbf{x}, t + \Delta t/2).$$

From now on we call this the standard Yee (S-Yee) algorithm. In order to use central finite-difference approximations for the time derivatives, the $\mathbf{E}$- and $\mathbf{H}$-fields must be evaluated at different time steps. Although it is not evident from our notation, each electromagnetic field component lies at a different position on the numerical grid so that central FD approximations can be used for the spatial derivatives. In our notation, $\mathbf{H}(\mathbf{x}, t - \Delta t/2)$ and $\mathbf{E}(\mathbf{x}, t)$ are abbreviations for

$$H_x(\mathbf{x}, t - \Delta t/2) = H_x(x, y - h/2, z - h/2, t - \Delta t/2), \tag{110}$$
$$H_y(\mathbf{x}, t - \Delta t/2) = H_y(x - h/2, y, z - h/2, t - \Delta t/2), \tag{111}$$
$$H_z(\mathbf{x}, t - \Delta t/2) = H_z(x - h/2, y - h/2, z, t - \Delta t/2), \tag{112}$$
$$E_x(\mathbf{x}, t) = E_x(x - h/2, y, z, t), \tag{113}$$
$$E_y(\mathbf{x}, t) = E_y(x, y - h/2, z, t), \tag{114}$$
$$E_z(\mathbf{x}, t) = E_z(x, y, z - h/2, t). \tag{115}$$

Other arrangements of the electromagnetic fields on the numerical grid are also possible.

4.2. Maxwell's Equations and the Wave Equation

In a uniform medium in which ε, μ, and σ are constant, Maxwell's equations reduce to the wave equation. Applying $\nabla \times$ to both sides of (103), and using the vector identity

$$\nabla \times (\nabla \times \mathbf{V}) = \nabla(\nabla \bullet \mathbf{V}) - \nabla^2\mathbf{V}, \tag{116}$$

we obtain

$$\left(\partial_t^2 - \frac{1}{\varepsilon\mu}\nabla^2 + \frac{\sigma}{\varepsilon}\partial_t\right)\mathbf{E}(\mathbf{x}, t) = 0. \tag{117}$$

Here we assume that the net charge density vanishes, so that $\nabla \bullet \mathbf{E} = 0$ in accordance with Gauss' law. Similarly applying $\nabla\times$ to both sides of (104), and using the fact that $\nabla \bullet \mathbf{H} = 0$, we obtain

$$\left(\partial_t^2 - \frac{1}{\varepsilon\mu}\nabla^2 + \frac{\sigma}{\varepsilon}\partial_t\right)\mathbf{H}(\mathbf{x}, t) = 0. \tag{118}$$

Comparing these wave equations with (98), we can make the identifications

$$v^2 = \frac{1}{\varepsilon\mu}, \tag{119}$$

$$\alpha = \frac{\sigma}{2\varepsilon}. \tag{120}$$

At first sight, the FD model of Maxwell's equations, (105) and (106), may seem different from that of the wave equation (99), but in each region of a uniform medium they are equivalent. Analagously to (116), we have

$$\mathbf{D}_1 \times (\mathbf{D}_1 \times \mathbf{V}) = \mathbf{D}_1 (\mathbf{D}_1 \bullet \mathbf{V}) - \mathbf{D}_1^2\mathbf{V}. \tag{121}$$

Assuming that the charge density vanishes, $\mathbf{D}_1 \bullet \mathbf{E} = 0$. Making the substitution $t \to t - 1/2$ in (106) we obtain

$$\begin{aligned} d_t\mathbf{E}(\mathbf{x}, t) = {}& \frac{1}{\varepsilon}\frac{\Delta t}{h}\mathbf{D}_1 \times \mathbf{H}(\mathbf{x}, t) \\ & - \frac{\sigma}{2\varepsilon}\Delta t\left[\mathbf{E}(\mathbf{x}, t + \Delta t/2) + \mathbf{E}(\mathbf{x}, t - \Delta t/2)\right]. \end{aligned} \tag{122}$$

Applying d_t to both sides and using (105), we find

$$d_t^2\mathbf{E}(\mathbf{x}, t) = \left[\frac{1}{\varepsilon\mu}\frac{\Delta t^2}{h^2}\right]\mathbf{D}_1^2\mathbf{E}(\mathbf{x}, t) - \left[\frac{\sigma}{2\varepsilon}\Delta t\right]d_t'\mathbf{E}(\mathbf{x}, t), \tag{123}$$

where d_t' is given by (78). We have used the identity

$$d_t f(t + \Delta t/2) + d_t f(t - \Delta t/2) = d_t' f(t). \tag{124}$$

Comparing with the FD model of the wave equation (99) we obtain the same identifications of (119) and (120).

Because Maxwell's equations reduce to the wave equation in each electromagnetic field component, it might seem that we could solve the wave equation with the FDTD algorithm, instead of solving Maxwell's equations with more complicated Yee algorithm. In a uniform medium each field propagates independently according to the wave equation, but at boundaries between different media, this is no longer true. Where ε and μ change with position, the field components no longer propagate independently. Furthermore, the wave equation and Maxwell's equations have different boundary conditions.

Let B be a boundary between two non-conducting dielectric materials, and let $\hat{\mathbf{n}}$ be a unit vector normal to B, and $\hat{\mathbf{t}}$ a unit vector tangent to B. At the boundary the scalar field of the wave equation must satisfy

$$\psi \qquad \text{continuous,} \tag{125}$$

$$(\hat{\mathbf{n}} \bullet \nabla)\,\psi \; \text{continuous,} \tag{126}$$

where $(\hat{\mathbf{n}} \bullet \nabla)\,\psi$ is the derivative along the normal to B.

On the other hand, at an interface between non-conducting two dielectrics it is a consequence of Maxwell's equations that the electromagnetic fields satisfy,

$$\hat{\mathbf{t}} \bullet \mathbf{E} \; \text{continuous,} \tag{127}$$

$$\hat{\mathbf{t}} \bullet \mathbf{H} \; \text{continuous.} \tag{128}$$

In each uniform dielectric region, the electromagnetic field components independently satisfy the wave equation, but on the boundaries they satisfy (127) and (128), rather than the wave equation boundary conditions. There is, however, one special case for which the Maxwell equation boundary conditions and the wave equation boundary conditions are equivalent.

In two dimensions the electromagnetic fields can be decomposed into two independent modes, the transverse magnetic (TM) mode, and the transverse electric (TE) mode. The TE mode is defined by

$$E_x = E_y = H_z = 0, \tag{129}$$
$$H_x, H_y, E_z \neq 0,$$

and the TM mode by

$$H_x = H_y = E_z = 0, \tag{130}$$
$$E_x, E_y, H_z \neq 0.$$

Some authors interchange these definitions so one must always verify their meaning. Here we follow the conventions of the optics community. Maxwell's equations in the TE mode reduce to,

$$\mu \partial_t H_x = -\partial_y E_z, \tag{131}$$

$$\mu \partial_t H_y = \partial_x E_z, \tag{132}$$

$$\varepsilon \partial_t E_z = \partial_x H_y - \partial_y H_x - \sigma E_z. \tag{133}$$

In the TM mode Maxwell's equations are,

$$\mu \partial_t H_z = -\partial_x E_y + \partial_y E_x, \tag{134}$$

$$\varepsilon \partial_t E_x = \partial_y H_z - \sigma E_x - \sigma E_x \tag{135}$$

$$\varepsilon \partial_t E_y = -\partial_x H_z - \sigma E_y. \tag{136}$$

Setting $\sigma = 0$, and taking the time dependence to be $e^{-i\omega t}$, and writing $\mathbf{E}(\mathbf{x}, t) = \mathbf{E}(\mathbf{x})e^{-i\omega t}$, $\mathbf{H}(\mathbf{x}, t) = \mathbf{H}(\mathbf{x})e^{-i\omega t}$, Maxwell's equations become,

$$i\omega\mu\mathbf{H}(\mathbf{x}) = \nabla \times \mathbf{E}(\mathbf{x}), \tag{137}$$

$$i\omega\varepsilon\mathbf{E}(\mathbf{x}) = -\nabla \times \mathbf{H}(\mathbf{x}). \tag{138}$$

In the TE mode $\mathbf{E}(\mathbf{x}) = \hat{\mathbf{z}}E_z(\mathbf{x})$. Assuming that all dielectric boundaries are infinitely long and invariant in the z-direction, $\mathbf{E}$ is parallel (tangential) to B. Taking μ to be constant everywhere (but not ε), then as we have just seen E_z satisfies the wave equation in each region of constant ε. On the interface the boundary conditions (127) and (125) are equivalent because $\mathbf{E}$ is tangential to B

Now let us examine $\mathbf{H}$. Let $\hat{\mathbf{n}} = n_x\hat{\mathbf{x}} + n_y\hat{\mathbf{y}}$ be the normal to B. The unit tangent vector to B is $\hat{\mathbf{t}} = \hat{\mathbf{n}} \times \hat{\mathbf{z}} = \hat{\mathbf{x}}n_y - \hat{\mathbf{y}}n_x$. For the TE mode eq. (137) gives

$$i\omega\mu\mathbf{H} = \hat{\mathbf{x}}\partial_y E_z - \hat{\mathbf{y}}\partial_x E_z. \tag{139}$$

Taking the dot product of both sides with $\hat{\mathbf{t}}$ gives,

$$i\omega\mu\mathbf{H} \bullet \hat{\mathbf{t}} = \hat{\mathbf{x}}n_x\partial_x E_z + \hat{\mathbf{y}}n_y\partial_y E_z \tag{140}$$

$$= (\hat{\mathbf{n}} \bullet \nabla) E_z. \tag{141}$$

Thus the requirement that $\mathbf{H} \bullet \hat{\mathbf{t}}$ be continuous across B is equivalent to the requirement that $(\hat{\mathbf{n}} \bullet \nabla) E_z$ be continuous across B.

Hence for harmonically varying electromagnetic fields in the TE mode, the boundary condition (127) on E_z is equivalent to (126). Thus E_z can be computed from the wave equation across a boundary between two different

                                     J. B. Cole

dielectrics. It can be shown, however, that this is not the case for the **H**-fields, in either the TE or TM mode.

Sometimes the one-dimensional forms of Maxwell's equations are used. In the TE mode (129), let us impose the additional constraint that $H_x = 0$. This describes an infinite plane wave plane wave propagating in the x-direction. Maxwell's equations then reduce to

$$\mu \partial_t H_y = \partial_x E_z, \tag{142}$$

$$\varepsilon \partial_t E_z = \partial_x H_y - \sigma E_z. \tag{143}$$

Similarly, in the TM mode (130) adding the constraint $E_y = 0$ gives

$$\mu \partial_t H_z = -\partial_x E_y, \tag{144}$$

$$\varepsilon \partial_t E_y = -\partial_x H_z - \sigma E_y. \tag{145}$$

Actually the TM- and TE-modes are equivalent in one dimension. Rotating the coordinate system about the x-axis such that $\hat{\mathbf{y}} \to \hat{\mathbf{z}}$ and $\hat{\mathbf{z}} \to -\hat{\mathbf{y}}$, we have $E_z \to E_y$ and $H_y \to -H_z$ in the TE mode, and eqs. (142) and (143) transform to (144) and (145) of the TM mode. It can be shown that the boundary conditions are those of the one-dimensional wave equation.

4.3. *NSFD Version of the Yee Algorithm*

To construct an NSFD model of Maxwell's equations we would like to have a vector difference operator $\mathbf{D}_0$ for which $\mathbf{D}_0 \bullet \mathbf{D}_0 = \mathbf{D}_0^2$, but unfortunately such an operator does not exist. It is, however, possible to find one which has the property,

$$\mathbf{D}_1 \bullet \mathbf{D}_0 = \mathbf{D}_0^2, \tag{146}$$

even though $\mathbf{D}_0 \bullet \mathbf{D}_0 \neq \mathbf{D}_0^2$. Let us now proceed to determine $\mathbf{D}_0$.

In two dimensions $\mathbf{D}_0$ is a superposition of two difference operators for ∇

$$\mathbf{D}_0 = \gamma' \mathbf{D}_1 + (1 - \gamma') \mathbf{D}_2, \tag{147}$$

where γ' is a parameter to be determined. There is a second vector difference operator for the gradient defined by

$$\mathbf{D}_2 = \hat{\mathbf{x}} d_x^{(2)} + \hat{\mathbf{y}} d_y^{(2)}, \tag{148}$$

where $d_x^{(2)}$, and $d_y^{(2)}$ are alternate difference operators for ∂_x and ∂_y. These operators are defined by

$$2 d_x^{(2)} f(x,y) = f(x + h/2, y + h) + f(x + h/2, y - h) \tag{149}$$
$$- f(x - h/2, y + h) - f(x - h/2, y - h),$$

$$2d_y^{(2)}f(x,y) = f(x+h,y+h/2) + f(x-h,y+h/2) \qquad (150)$$
$$- f(x+h,y-h/2) - f(x-h/2,y-h/2).$$

We now determine γ' by requiring that $\mathbf{D}_0^2 = \mathbf{D}_1 \bullet \mathbf{D}_0$, or

$$\gamma_0\mathbf{D}_1^2 + (1-\gamma_0)\mathbf{D}_2^2 = \mathbf{D}_1 \bullet [\gamma'\mathbf{D}_1 + (1-\gamma')\mathbf{D}_2]. \qquad (151)$$

Directly evaluating $[\mathbf{D}_1 \bullet \mathbf{D}_2]f(x,y)$, we find that

$$\mathbf{D}_1 \bullet \mathbf{D}_2 = 2\mathbf{D}_2^2 - \mathbf{D}_1^2. \qquad (152)$$

Solving (151), we obtain $\gamma' = (1+\gamma_0)/2$. Thus,

$$\mathbf{D}_0 = \left(\frac{1+\gamma_0}{2}\right)\mathbf{D}_1 + \left(\frac{1-\gamma_0}{2}\right)\mathbf{D}_2 \qquad (153)$$

is the operator that satisfies (146).

In three dimensions $\mathbf{D}_0$ is a superposition of three independent difference operators for ∇, given by

$$\mathbf{D}_0 = \gamma_1'\mathbf{D}_1 + \gamma_2'\mathbf{D}_2 + (1-\gamma_1'-\gamma_2')\mathbf{D}_3, \qquad (154)$$

where again γ_1' and γ_2' are parameters that must be determined. In three dimensions

$$\mathbf{D}_2 = \hat{\mathbf{x}}d_x^{(2)} + \hat{\mathbf{y}}d_y^{(2)} + \hat{\mathbf{z}}d_z^{(2)}, \qquad (155)$$

where

$$4d_x^{(2)}f(x,y,z) = f(x+h/2,y+h,z+h) + f(x+h/2,y+h,z-h) \qquad (156)$$
$$+ f(x+h/2,y-h,z+h) + f(x+h/2,y-h,z-h)$$
$$- f(x-h/2,y+h,z+h) - f(x-h/2,y+h,z-h)$$
$$- f(x-h/2,y-h,z+h) - f(x-h/2,y-h,z-h),$$

$$4d_y^{(2)}f(x,y,z) = f(x+h,y+h/2,z+h) + f(x+h,y+h/2,z-h) \qquad (157)$$
$$+ f(x-h,y+h/2,z+h) + f(x-h,y+h/2,z-h)$$
$$- f(x+h,y-h/2,z+h) - f(x+h,y-h/2,z-h)$$
$$- f(x-h,y-h/2,z+h) - f(x-h,y-h/2,z-h),$$

$$4d_z^{(2)}f(x,y,z) = f(x+h,y+h,z+h/2) + f(x+h,y-h,z+h/2) \qquad (158)$$
$$+ f(x-h,y+h,z+h/2) + f(x-h,y-h,z+h/2)$$
$$- f(x+h,y+h,z-h/2) + f(x+h,y-h,z-h/2)$$
$$- f(x-h,y+h,z-h/2) + f(x-h,y-h,z-h/2).$$

J. B. Cole

The third difference operator for ∇ is

$$\mathbf{D}_3 = \hat{\mathbf{x}}d_x^{(3)} + \hat{\mathbf{y}}d_y^{(3)} + \hat{\mathbf{z}}d_z^{(3)}, \tag{159}$$

where

$$
\begin{aligned}
4d_x^{(3)} f(x,y,z) = {}& f(x+h/2, y+h, z) + f(x+h/2, y-h, z) \\
& + f(x+h/2, y, z+h) + f(x+h/2, y, z-h) \\
& - f(x-h/2, y+h, z) - f(x-h/2, y-h, z) \\
& - f(x-h/2, y, z+h) - f(x-h/2, y, z-h),
\end{aligned} \tag{160}
$$

$$
\begin{aligned}
4d_y^{(3)} f(x,y,z) = {}& f(x+h, y+h/2, z) + f(x-h, y+h/2, z) \\
& + f(x, y+h/2, z+h) + f(x, y+h/2, z-h) \\
& - f(x+h, y-h/2, z) - f(x-h, y-h/2, z) \\
& - f(x, y-h/2, z+h) - f(x, y-h/2, z-h),
\end{aligned} \tag{161}
$$

$$
\begin{aligned}
4d_z^{(2)} f(x,y,z) = {}& f(x+h, y, z+h/2) + f(x+h, y, z+h/2) \\
& + f(x, y+h, z+h/2) + f(x, y-h, z+h/2) \\
& - f(x+h, y, z-h/2) - f(x+h, y, z-h/2) \\
& - f(x, y+h, z-h/2) - f(x, y-h, z-h/2).
\end{aligned} \tag{162}
$$

Evaluating $[\mathbf{D}_1 \bullet \mathbf{D}_2] f(x,y,z)$, and $[\mathbf{D}_1 \bullet \mathbf{D}_3] f(x,y,z)$ we find

$$\mathbf{D}_1 \bullet \mathbf{D}_2 = 3\mathbf{D}_2^2 - \mathbf{D}_3^2, \tag{163}$$

$$\mathbf{D}_1 \bullet \mathbf{D}_3 = 2\mathbf{D}_3^2 - \mathbf{D}_1^2. \tag{164}$$

We can now find γ_1' and γ_2' by requiring that $\mathbf{D}_0^2 = \mathbf{D}_1 \bullet \mathbf{D}_0$, or

$$
\gamma_1 \mathbf{D}_1^2 + \gamma_2 \mathbf{D}_2^2 + (1 - \gamma_1 - \gamma_2)\mathbf{D}_3^2 =
$$
$$
\mathbf{D}_1 \bullet [\gamma_1' \mathbf{D}_1 + \gamma_2' \mathbf{D}_2 + (1 - \gamma_1' - \gamma_2')\mathbf{D}_3], \tag{165}
$$

where γ_1, and γ_2 are given by (74). Using (163) and (164) to solve (165), we obtain

$$\gamma_1' = \frac{1}{3}\eta_0 (1 - \gamma_0) + \gamma_0, \tag{166}$$

$$\gamma_2' = \frac{1}{3}\eta_0 (1 - \gamma_0), \tag{167}$$

$$\gamma_3' = 1 - \gamma_1' - \gamma_2', \tag{168}$$

where η_0 is given by (75). Further details about the derivation of $\mathbf{D}_0$ are given in [2,3].

We can now construct a high accuracy NSFD model of Maxwell's equations with the replacement $\mathbf{D}_1 \to \mathbf{D}_0$ in (105), while retaining $\mathbf{D}_1$ in (106). Rewriting (119) and (120) we have

$$\frac{1}{\mu} = v\sqrt{\frac{\varepsilon}{\mu}}, \tag{169}$$

$$\frac{1}{\varepsilon} = v\sqrt{\frac{\mu}{\varepsilon}}. \tag{170}$$

Following Sect. 3.1, and eqs. (95) and (94), we make the replacements

$$\frac{v\Delta t}{h} \to u_0, \tag{171}$$

$$\frac{\sigma\Delta t}{2\varepsilon} \to \tanh\left(\frac{\sigma\Delta t}{2\varepsilon}\right), \tag{172}$$

where u_0 is computed from ε and σ using (120). Inserting (169)-(172) in the SFD model of Maxwell's equations, (105) and (106), gives the NSFD model,

$$d_t\mathbf{H}(\mathbf{x}, t) = -u_0\sqrt{\frac{\varepsilon}{\mu}}\mathbf{D}_0 \times \mathbf{E}(\mathbf{x}, t) \tag{173}$$

$$d_t\mathbf{E}(\mathbf{x}, t + \Delta t/2) = u_0\sqrt{\frac{\mu}{\varepsilon}}\mathbf{D}_1 \times \mathbf{H}(\mathbf{x}, t + \Delta t/2) \tag{174}$$

$$- \tanh\left(\frac{\sigma\Delta t}{2\varepsilon}\right)[\mathbf{E}(\mathbf{x}, t + \Delta t) + \mathbf{E}(\mathbf{x}, t)].$$

Solving for $\mathbf{H}(\mathbf{x}, t+\Delta t/2)$ and $\mathbf{E}(\mathbf{x}, t+\Delta t)$, we obtain the NS (nonstandard)-Yee algorithm,

$$\mathbf{H}(\mathbf{x}, t + \Delta t/2) = \mathbf{H}(\mathbf{x}, t - \Delta t/2) - u_0\sqrt{\frac{\varepsilon}{\mu}}\mathbf{D}_0 \times \mathbf{E}(\mathbf{x}, t) \tag{175}$$

$$\mathbf{E}(\mathbf{x}, t + \Delta t) = \left(\frac{1 - \tanh\left(\sigma\Delta t/2\varepsilon\right)}{1 + \tanh\left(\sigma\Delta t/2\varepsilon\right)}\right)\mathbf{E}(\mathbf{x}, t) \tag{176}$$

$$+ \frac{u_0\sqrt{\mu/\varepsilon}}{1 + \tanh\left(\sigma\Delta t/2\varepsilon\right)}\mathbf{D}_1 \times \mathbf{H}(\mathbf{x}, t + \Delta t/2).$$

It should be noted that the NS-Yee algorithm given here superceeds the one given in [3].

The error of the NS-Yee algorithm is proportional to $(h/\lambda)^6$ in regions of constant ε, μ, and σ with respect to decaying harmonic plane waves. Although the error rises at the boundaries between different media, it is still quite small as we shall later show. The error of the S-Yee algorithm is proportional to $(h/\lambda)^2$ and also rises at the media boundaries, and is frequency-dependent.

4.4. *Comparison with Analytic Solutions*

Although in principle the NS-Yee algorithm is more accurate than the S-Yee one, its performance must be verified on actual problems. A severe test of any algorithm is Mie scattering. In the Mie regime, the feature size of the scatterer is comparable to the wavelength of the incident radiation. and analytic solutions exist only for a few simple shapes such as spheres and cylinders.

In Fig. 1 an infinite plane wave (not shown) is incident from the left upon a dielectric cylinder. The scattered intensity is calculated with the S- and NS-FDTD algorithms and is compared with the analytic solutions [6]. As we can see, the NS-FDTD algorithm gives much better results than the S-FDTD one. Outside the cylinder $\lambda/h = 8$, but inside λ/h is only 5. The scattered field due to an infinite plane wave can be computed in a finite computational domain by decomposing the total field into the sum of the scattered and incident field and solving for the scattered field alone as explained in Sect. 7.3.

5. Nonstandard Finite Difference Version of the Mur Absorbing Boundary

5.1. *Introduction*

All FDTD calculations are carried out in a finite computational domain, D. The fields at a given point, $\mathbf{x}$, are updated using field values of the neighboring points, thus if $\mathbf{x}$ lies on the boundary of D it has no neighbors on one or more sides, and there is not enough information to update the field.

For example in one dimension, to compute $\psi(0, t + \Delta t)$ on the domain

$$x = 0, h, 2h, \cdots N_x h, \tag{177}$$

we must evaluate $d_x^2 \psi(0, t) = \psi(h, t) + \psi(-h, t) - 2\psi(0, t)$, but $\psi(-h, t)$ is unknowable because $x = -h$ lies outside D. Similarly, to compute $\psi(N_x h, t)$, the unknowable value, $\psi(N_x h + h, t)$ is needed. There are three basic solutions to this problem.

(**1**) Impose the conditions $\psi(0, t) = 0$, and $\psi(N_x h, t) = 0$, thus avoiding the need to compute ψ on the boundary of D - it is already specified for all

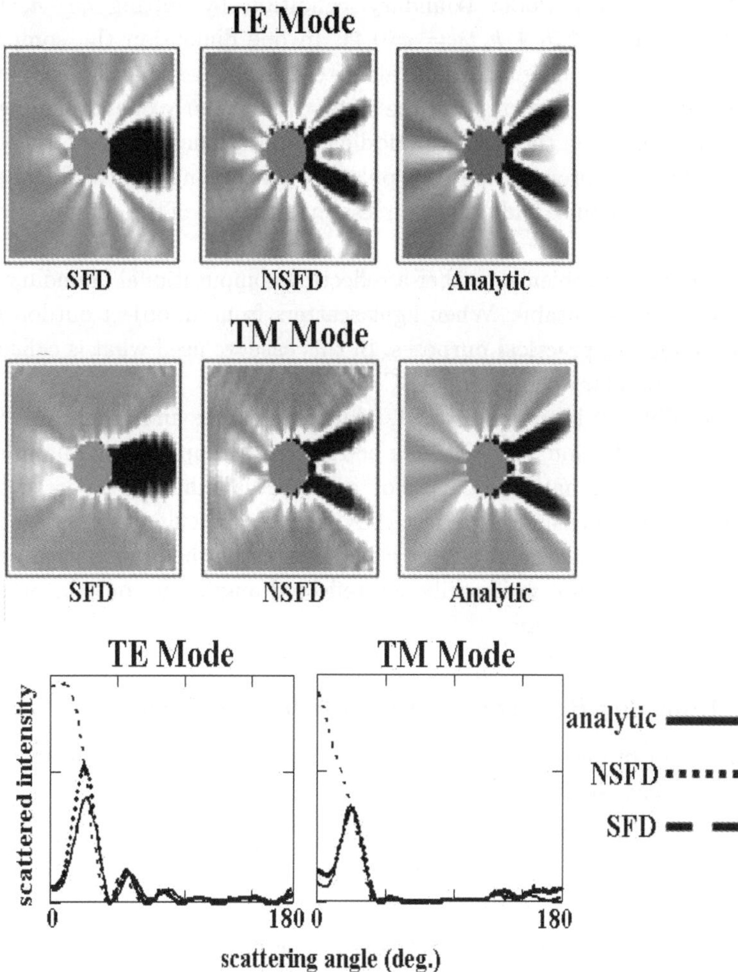

Fig. 1.   **Mie Scattering off an Infinite Dielectric Cylinder in Two Dimensions.**
Scattered intensity is computed using the SFD and NSFD algorithms, is compared with
the analytic solution. Grey scale visualization of $|E_y|^2$ for the TM case, and of $|E_z|^2$ for
the TE case. Interior fields are not shown. In the bottom panel, scattered intensity is
plotted as a function of angle from the incident field direction. Cylinder radius $= \lambda_0$,
index of refraction $= 1.6$, $\lambda_0/h = 8$, where λ_0 is the free space wavelength.

time. This is called a reflecting boundary condition because signals incident
on the computational boundary are reflected inwards. A reflecting boundary
condition is appropriate to model wave propagation in a reflecting cavity.

(2) Define a periodic boundary condition by setting $\psi(-h,t) = \psi(N_x h, t)$, and $\psi(N_x h + h, t) = \psi(0, t)$. In one dimension the computational domain is topologically equivalent to a circle. A signal that impinges on the left boundary emerges on the right, and similarly one impinging on the right emerges on the left. A periodic computational domain is suitable for modeling the propagation of a pulse in a very long optical fiber, provided the spatial pulse width is much less than the size of the computational domain.

(3) In many problems, neither a reflecting computational boundary nor a periodic one is suitable. When light scatters from an object outdoors, it disappears for all practical purposes. In this case we need what is called an absorbing boundary condition (ABC).

An annihilator is an operator, A, which annihilates an incident wave, ψ, without any reflection. An ABC can be realized by applying an annihilator on the computational boundary. For a perfect annihilator $A\psi = 0$, but except in one dimension, a perfect ABC is very difficult to construct. An approximate annihilator gives $A\psi = \varepsilon\psi$, where ε is the annihilation error. The quantity ε is not necessarily the reflected amplitude, but the smaller ε, the lower the reflection.

5.2. *Absorbing Boundary Conditions in One Dimension*

Let us begin in one dimension, where it is possible to construct an exact ABC. The wave equation,

$$\left(\partial_t^2 - v^2 \partial_x^2\right) \psi(x,t) = 0, \tag{178}$$

can be factored into

$$\left(\partial_t + v\partial_x\right)\left(\partial_t - v\partial_x\right) \psi(x,t) = 0. \tag{179}$$

This yields two wave equations,

$$\left(\partial_t + v\partial_x\right) \psi(x,t) = 0, \tag{180}$$

$$\left(\partial_t - v\partial_x\right) \psi(x,t) = 0. \tag{181}$$

General solutions of $\left(\partial_t \pm v\partial_x\right)\psi = 0$ are $f\left(x \mp vt\right)$, which are forward ($+x$-direction) and backward ($-x$-direction) moving waves. For this reason (180) is called the forward wave equation, and (181) is called the backward wave equation. Collectively they are known as the one-way wave equations. The differential operators, $A_\pm = \left(\partial_t \pm v\partial_x\right)$, thus annihilate forward- and backward-moving waves, respectively. We can now construct an ABC for

FDTD calculations from finite-difference realizations of these annihilation operators.

Simple Finite Difference Model for the One-Way Wave Equations

Let us first develop the simplest model. In (180) and (181) we approximate $\partial_t \psi$ using a forward finite difference (FFD) approximation of the form,

$$\frac{d}{dt} f(t) \cong \frac{f(t + \Delta t) - f(t)}{\Delta t}. \tag{182}$$

The FFD expression for $f(t)$ could also be regarded as a central finite difference (CFD) approximation to $f'(t+\Delta t/2)$. The space derivatives must be approximated using points inside the computational domain, so in the forward wave equation we approximate $\partial_x \psi(N_x h, t)$ with a backward finite difference (BFD) approximation,

$$\frac{d}{dx} f(x) \cong \frac{f(x) - f(x - h)}{h}. \tag{183}$$

Similarly in the backward wave equation we approximate $\partial_x \psi(0, t)$ with a FFD expression. The FFD and BFD approximations are called one-sided FD approximations. Substituting one-sided FD approximations into the one-way wave equations on the computational boundary gives the FD models,

$$\psi(N_x h, t + \Delta t) - \psi(N_x h, t) = -\frac{v \Delta t}{h} \left[\psi(N_x h, t) - \psi(N_x h - h, t) \right], \tag{184}$$

$$\psi(0, t + \Delta t) - \psi(0, t) = \frac{v \Delta t}{h} \left[\psi(h, t) - \psi(0, t) \right]. \tag{185}$$

Solving for ψ at $t + \Delta t$ gives the ABC

$$\psi(N_x h, t + \Delta t) = \psi(N_x h, t) - \frac{v \Delta t}{h} \left[\psi(N_x h, t) - \psi(N_x h - h, t) \right], \tag{186}$$

$$\psi(0, t + \Delta t) = \psi(0, t) - \frac{v \Delta t}{h} \left[\psi(0, t) - \psi(h, t) \right]. \tag{187}$$

Let us simplify the notation by writing $\psi(x, t) = \psi_x^t$, $\psi(x, t \pm \Delta t) = \psi_x^{t \pm 1}$, and $\bar{v} = v \Delta t / h$. Let b denote the x-coordinate of a boundary point, and i the coordinate one grid spacing inside the boundary. Define $b = 0$ and $i = b + h$ on the left computational boundary, while $b = N_x h$, $i = b - h$ on the right. The FD models of both one-way wave equations, can now be compactly expressed in the form

$$\psi_b^{t+1} - \psi_b^t + \bar{v} \left(\psi_b^t - \psi_i^t \right) = 0, \tag{188}$$

and the ABC becomes

$$\psi_b^{t+1} = \psi_b^t - \bar{v}\left(\psi_b^t - \psi_i^t\right). \tag{189}$$

Let us define the left side of (188) to be $A\psi(b,t)$, and evaluate the error $\varepsilon = A\varphi_\pm(b,t)/\varphi_\pm(b,t)$ of the FD model with respect to monochromatic waves moving in the $\pm x$-directions, $\varphi_\pm = e^{i(kx \mp \omega t)}$. We find

$$\varepsilon = -\left(1 - e^{\mp i\omega \Delta t}\right) + \bar{v}\left[1 - e^{\mp ikh}\right]. \tag{190}$$

For the special case $v\Delta t/h = 1$ we see that ε vanishes, and (189) reduces to

$$\psi_b^{t+1} = \psi_i^t. \tag{191}$$

In this case $Af(x \pm vt) = 0$, which means that A is a perfect annihilator. Although (191) is a special case, it is useful in certain problems.

For $v\Delta t/h < 1$, the error is $\varepsilon \sim h$, and the first order ABC of (189) is inadequate. Following the previous developments, we could replace $\bar{v}$ by a free parameter u, and seek a value of u for which ε vanishes. As we can see from (190), u would be complex. Unless we use an FDTD algorithm which computes both the real and complex parts of ψ this approach is not very useful, and we will not consider it further.

Centered Finite Difference Model for the One-Way Wave Equations

The error of the simple ABC (189) is first order because the FD model (188) uses one-sided FD approximations. If central FD approximations were used, the error could be reduced to $\varepsilon \sim h^2$.

In (188) the one-way wave equations were modeled at the space-time point (b,t), but they can also be modeled at the midpoint $m = (b+i)/2$, at time $t + \Delta t/2$. We now proceed to construct a CFD model of

$$\left(\partial_t \pm v\partial_x\right)\psi(m, t + \Delta t/2) = 0. \tag{192}$$

On the right $m = (N_x - 1/2)h$, and a CFD approximation for $\partial_x\psi$ in the forward wave equation is

$$\partial_x\psi(m,t) \cong \frac{\psi(b,t) - \psi(i,t)}{h}. \tag{193}$$

On the left side where $m = h/2$, we have

$$\partial_x\psi(m,t) \cong \frac{\psi(i,t) - \psi(b,t)}{h}. \tag{194}$$

The CFD approximation for the time derivative is

$$\partial_t\psi(m, t+\Delta t/2) \cong \frac{\psi(m, t+\Delta t) - \psi(m, t)}{\Delta t}. \tag{195}$$

Using

$$\psi_m^t \cong \frac{\psi_b^t + \psi_i^t}{2} \tag{196}$$

we obtain

$$\partial_t\psi(m, t+\Delta t/2) = \frac{1}{2\Delta t}\left[\left(\psi_b^{t+1} - \psi_b^t\right) + \left(\psi_i^{t+1} - \psi_i^t\right)\right]. \tag{197}$$

Similarly, using

$$\psi(m, t+\Delta t/2) \cong \frac{\psi(m, t) + \psi(m, t+\Delta t)}{2}, \tag{198}$$

we have

$$\partial_x\psi(m, t+\Delta t/2) \cong \frac{1}{2h}\left[\left(\psi_b^{t+1} - \psi_i^{t+1}\right) + \left(\psi_b^t - \psi_i^t\right)\right]. \tag{199}$$

The centered FD model then becomes,

$$\left(\psi_b^{t+1} - \psi_b^t\right) + \left(\psi_i^{t+1} - \psi_i^t\right) + \bar{v}\left[\left(\psi_b^{t+1} - \psi_i^{t+1}\right) + \left(\psi_b^t - \psi_i^t\right)\right] = 0. \tag{200}$$

We use the term "centered FD model," rather than "central FD model," because the derivatives at $(m, t+\Delta t/2)$ are approximated by averaging. We now call (200) the SFD model of (192) because it derives from standard finite difference approximations. Solving for ψ_b^{t+1}, we obtain the ABC

$$\psi_b^{t+1} = \psi_i^t + \left(\frac{1-\bar{v}}{1+\bar{v}}\right)\left(\psi_b^t - \psi_i^{t+1}\right), \tag{201}$$

which we now call the S-ABC. For the special case $\bar{v} = 1$, the S-ABC reduces to the exact ABC of (191). The field values at the interior points, ψ_i^t and ψ_i^{t+1} are computed from the FDTD algorithm for the wave equation.

Let us now evaluate the error of the SFD model. Defining the left side of (200) to be $A\psi(m, t+\Delta t/2)$, the error with respect to the monochromatic solutions is

$$\varepsilon = \frac{A\varphi_\pm(m, t+\Delta t/2)}{\varphi_\pm(m, \Delta t/2)}, \tag{202}$$

$$= \mp 4i\sin\left(\omega\Delta t/2\right)\cos\left(kh/2\right) \pm 4i\bar{v}\sin\left(kh/2\right)\cos\left(\omega\Delta t/2\right).$$

For $0 < \bar{v} < 1$, the error is now $\varepsilon \sim h^2$, while at $\bar{v} = 1$ it vanishes for any wave of the form $f(x \mp vt)$. In this case (201) reduces to the perfect ABC of (191).

Examination of (202) reveals that ε can be made to vanish by replacing $\bar{v}$ with

$$u_1 = \frac{\tan\left(\omega\Delta t/2\right)}{\tan\left(kh/2\right)}. \tag{203}$$

The NSFD model of (192) is thus

$$\left(\psi_b^{t+1} - \psi_b^t\right) + \left(\psi_i^{t+1} - \psi_i^t\right) + u_1\left[\left(\psi_b^{t+1} - \psi_i^{t+1}\right) + \left(\psi_b^t - \psi_i^t\right)\right] = 0, \tag{204}$$

and the nonstandard-ABC (NS-ABC) becomes,

$$\psi_b^{t+1} = \psi_i^t + \left(\frac{1-u_1}{1+u_1}\right)\left(\psi_b^t - \psi_i^{t+1}\right). \tag{205}$$

For $\bar{v} = 1$, the NS-ABC also reduces to the perfect ABC of (191).

5.3. *Nonstandard Finite Difference Model of Engquist-Majda One-Way Wave Equations*

In two and three dimensions there is no exact ABC, but many good one are now available. In general the better the ABC, the more complicated, costly, and difficult it is to implement. For example the perfectly matched layer (PML) is the probably the best ABC if high absorption is the only criterion, but its computational cost is high [8]. At the other end of the spectrum is the second-order Mur ABC [9]. The second-order Mur ABC is a FD algorithm to solve the Engquist-Majda (EM) one-way wave equations [10]. It is low-cost, and simple to implement, but it performs poorly on a coarse grid.

In this section we use NSFD methods to derive a simple modification of the Mur ABC that greatly increases its absorption without increasing its computational cost. This new ABC is very effective for the low values of $\bar{v}$ on the computational boundary in photonic crystal simulations.

Two Dimensions

For simplicity let us first consider two dimensions with the space-time grid

$$t = 0, \Delta t, 2\Delta t, \cdots \tag{206}$$

$$x = 0, h, 2h, \cdots, N_x h, \tag{207}$$

$$y = 0, h, 2h, \cdots, N_y h. \tag{208}$$

The two-dimensional wave equation can be expressed in the form

$$\left[\partial_t^2 - v^2\left(\partial_x^2 + \partial_y^2\right)\right]\psi(\mathbf{x}, t) = 0, \tag{209}$$

where $\mathbf{x} = (x, y)$. Defining

$$P = \sqrt{\partial_t^2 - v^2 \partial_y^2}, \tag{210}$$

the wave equation factors into,

$$(P + v\partial_x)(P - v\partial_x)\psi(\mathbf{x}, t) = 0, \tag{211}$$

and yields two one-way wave equations of the form,

$$(P \pm v\partial_x)\psi(\mathbf{x}, t) = 0. \tag{212}$$

These equations are called the Engquist-Majda (EM) one-way wave equations. As they stand, the EM equations are of little use, but they can be recast into a more useful form.

Squaring (210) we have

$$P^2 = \partial_t^2 \left(1 - v^2 \partial_y^2 / \partial_t^2\right), \tag{213}$$

and expanding $(P^2)^{1/2}$ via the binomial theorem, gives

$$P = \partial_t - \frac{1}{2} v^2 \frac{\partial_y^2}{\partial_t} + \cdots . \tag{214}$$

Substituting the first two terms of P into (212) and multiplying by ∂_t we arrive at the second-order EM one-way wave equations,

$$\left(\partial_t^2 \pm v\partial_{xt} - \frac{1}{2} v^2 \partial_y^2\right)\psi(\mathbf{x}, t) = 0. \tag{215}$$

The operator,

$$W_\pm = \partial_t^2 \pm v\partial_{xt} - \frac{1}{2} v^2 \partial_y^2, \tag{216}$$

is an approximate annihilation operator. Let us evaluate how effectively $W_\pm$ annihilates a monochromatic plane wave $\varphi_\mathbf{k} = e^{i(\mathbf{k}\bullet\mathbf{x} - \omega t)}$, where $\mathbf{k} = (k_x, k_y) = k(\cos\theta, \sin\theta)$. The annihilation error, $\varepsilon_{\mathrm{EM}} = W_\pm \varphi_\mathbf{k} / \varphi_\mathbf{k}$, is

$$\varepsilon_{\mathrm{EM}}(\theta) = -k^2 v^2 \left(1 - \cos\theta - \frac{1}{2}\sin^2\theta\right) \tag{217}$$

$$= -k^2 v^2 \left(\frac{1}{8}\sin^4\theta - \frac{1}{16}\sin^6\theta + \cdots\right),$$

where we have used the fact that $\omega = vk$.

To construct a usable ABC, we must formulate the second-order EM equations as difference equations. Following the previous section, let us write $\psi(x, y, t) = \psi_x^t$, and $\psi(x, y, t \pm \Delta t) = \psi_x^{t\pm1}$. Here we take $b = h$, and $b = (N_x - 1)h$ on the left, and right computational boundaries, respectively.

We have found that these choices give a better result than $b = 0$, and $b = N_x h$. As before $i = b + h$ on the left, $i = b - h$ on the right, and the midpoint is $m = (b + i)/2$.

We now proceed to construct a FD model of the second-order EM equations at (m, y, t)

$$\left(\partial_t^2 \pm v \partial_{xt} - \frac{1}{2} v^2 \partial_y^2 \right) \psi(m, y, t) = 0. \tag{218}$$

Using CFD approximations and averaging, we have

$$\psi(m, y, t) \cong \frac{1}{2} \left(\psi_b^t + \psi_i^t \right), \tag{219}$$

and thus

$$\partial_t^2 \psi(m, y, t) \cong \frac{1}{2\Delta t^2} d_t^2 \left(\psi_b^t + \psi_i^t \right), \tag{220}$$

$$\partial_y^2 \psi(m, y, t) \cong \frac{1}{2h^2} d_y^2 \left(\psi_b^t + \psi_i^t \right), \tag{221}$$

$$\partial_x \psi(m, y, t) \cong \frac{\psi_b^t - \psi_i^t}{h}, \tag{222}$$

$$\partial_t \partial_x \psi(m, y, t) \cong \frac{1}{2\Delta t} \left[\left(\frac{\psi_b^{t+1} - \psi_i^{t+1}}{h} \right) - \left(\frac{\psi_b^{t-1} - \psi_i^{t-1}}{h} \right) \right]. \tag{223}$$

Because ψ is unknown at $t + \Delta t/2$, we approximate $\partial_{xt} \psi(m, y, t)$ using (79). Substituting these approximations into (218) gives the Mur centered FD scheme,

$$d_t^2 \left(\psi_b^t + \psi_i^t \right) + \bar{v} \left[(\psi_b^{t+1} - \psi_i^{t+1}) - (\psi_b^{t-1} - \psi_i^{t-1}) \right] \tag{224}$$

$$- \frac{1}{2} \bar{v}^2 d_y^2 \left(\psi_b^t + \psi_i^t \right) = 0,$$

which we now call the SFD model. Because of the way we have defined b and i it applies to wave moving in both the $\pm x$-directions.

Expanding $d_t^2 \psi_b^t$ and solving for ψ_b^{t+1} gives the second-order Mur ABC, which we call the S-ABC.

$$\psi_b^{t+1} = \psi_b^t + (\psi_i^t - \psi_i^{t-1}) + \left(\frac{1 - \bar{v}}{1 + \bar{v}} \right) \left[(\psi_b^t - \psi_b^{t-1}) - (\psi_i^{t+1} - \psi_i^t) \right] \tag{225}$$

$$+ \frac{1}{2} \left(\frac{\bar{v}^2}{1 + \bar{v}} \right) d_y^2 \left(\psi_b^t + \psi_i^t \right).$$

As in one dimension, the interior fields ψ_i^t and ψ_i^{t+1} are evaluated using the wave equation FDTD algorithm. ABC (225) has been formulated for waves striking the left or right boundaries, but all the above developments can be extended to the top and bottom boundaries of the computational domain.

On the corners $\psi(b, b, t)$ is indeterminate because it can be calculated by either the x- or y-directional form of (225). One solution is the average the x- and y-directional ABCs, but relatively large reflections can still occur. We have found that these reflections can be suppressed by taking $b = h$ and $b = (N_x - 1)h$ instead of $b = 0$ and $b = N_x h$, on the left and right sides, respectively, and then setting $\psi(0, y, t) = \psi(N_x h, y, t) = \psi(x, 0, t) = \psi(x, N_y h, t) = 0$. The interior of the computational domain is thus

$$[2h \leq x \leq (N_x - 2)\, h] \times [2h \leq y \leq (N_y - 2)\, h]. \tag{226}$$

Let us now evaluate the performance of the S-ABC. Defining the left side of (224) to be $A_{\text{sfd}}\psi(m, y, t)$, the annihilation error of A_{sfd} is

$$\varepsilon_{\text{sfd}} = \frac{A_{\text{sfd}}\varphi_{\mathbf{k}}(m, y, t)}{\varphi_{\mathbf{k}}(m, y, t)}. \tag{227}$$

For notational convenience, let us define $\bar{\omega} = \omega\Delta t$, $\bar{k} = kh$, $\bar{k}_{x,y} =, k_{x,y}h$, and $\varepsilon'_{\text{sfd}} = \varepsilon_{\text{sfd}}/8\sin^2(\bar{\omega}/2)$. Evaluating $\varepsilon'_{\text{sfd}}$ we have

$$\varepsilon'_{\text{sfd}}(\theta) = -\cos(\bar{k}_x/2) + \bar{v}\frac{\sin(\bar{k}_x/2)}{\tan(\bar{\omega}/2)} + \frac{1}{2}\bar{v}^2\frac{\sin^2(\bar{k}_y/2)}{\sin^2(\bar{\omega}/2)}\cos(\bar{k}_x/2). \tag{228}$$

Expanding in a Taylor series about $\sin\theta = 0$ gives

$$\varepsilon'_{\text{sfd}}(\theta) = \left[-\cos(\bar{k}/2) + \bar{v}\frac{\sin(\bar{k}/2)}{\tan(\bar{\omega}/2)} \right]$$

$$+ \left[-\frac{1}{4}\bar{k}\sin(\bar{k}/2) - \frac{1}{4}\bar{v}\bar{k}\frac{\cos(\bar{k}/2)}{\tan(\bar{\omega}/2)} + \frac{1}{8}\bar{v}^2\bar{k}^2\frac{\cos(\bar{k}/2)}{\sin^2(\bar{\omega}/2)} \right]\sin^2(\theta) + \cdots. $$

$$\tag{229}$$

Taylor expanding the terms of (229) in powers of $\bar{k}$ gives

$$\varepsilon'_{\text{sfd}}(\theta) = \frac{1}{12}\left(1 - \bar{v}^2\right)\bar{k}^2 + \left[\frac{1}{12}\bar{v}^2 - \frac{1}{8}\right]\bar{k}^2\sin^2\theta + \cdots. \tag{230}$$

This analysis shows that $\varepsilon'_{\text{sfd}}(\theta) \sim h^2 + h^2\theta^2$, which means that $\varepsilon'_{\text{sfd}}(0)$ does not vanish. Comparing with (217), we see that A_{sfd} is a poor FD realization of the EM annihilator, $W_{\pm}$.

Using a NSFD model of the EM equations, we shall now show that it is possible to derive a much better FD annihilator. Let us form a generalized model of the EM one-way wave equations with the subsitutions $\bar{v} \rightarrow u_1$, and $\bar{v}^2 \rightarrow u_2^2$ in (224),

$$d_t^2\left(\psi_b^t + \psi_i^t\right) + u_1\left[(\psi_b^{t+1} - \psi_i^{t+1}) - (\psi_b^{t-1} - \psi_i^{t-1})\right] \tag{231}$$

$$-\frac{1}{2}u_2^2 d_y^2\left(\psi_b^t + \psi_i^t\right) = 0,$$

where u_1 and u_2 are parameters to be determined. Define the left side of (231) to be $A_{\mathrm{nsfd}}\psi(m, y, t)$, where A_{nsfd} is the NSFD annihilator, and evaluate

$$\varepsilon_{\mathrm{nsfd}} = \frac{A_{\mathrm{nsfd}}\varphi_{\mathbf{k}}(m, y, t)}{\varphi_{\mathbf{k}}(m, y, t)}. \tag{232}$$

For convenience we define $\varepsilon'_{\mathrm{nsfd}} = \varepsilon_{\mathrm{nsfd}}/8\sin^2(\bar{\omega}/2)$, and making the substitutions $\bar{v} \to u_1$, and $\bar{v}^2 \to u_2^2$ in (228), we obtain

$$\varepsilon'_{\mathrm{nsfd}}(\theta) = -\cos\left(\bar{k}_x/2\right) + u_1\frac{\sin\left(\bar{k}_x/2\right)}{\tan\left(\bar{\omega}/2\right)} + \frac{1}{2}u_2^2\frac{\sin^2\left(\bar{k}_y/2\right)}{\sin^2\left(\bar{\omega}/2\right)}\cos\left(\bar{k}_x/2\right). \tag{233}$$

At $\theta = 0$, where $k_x = k$ and $k_y = 0$, the choice in

$$u_1 = \frac{\tan\left(\omega\Delta t/2\right)}{\tan\left(kh/2\right)} \tag{234}$$

in (233) gives $\varepsilon'_{\mathrm{nsfd}}(0) = 0$. Notice that the u_1 obtained here is exactly the same as the u_1 of (203) for the one-dimensional NS-ABC. Substituting (234) into (233) gives

$$\varepsilon'_{\mathrm{nsfd}}(\theta) = -\cos\left(\bar{k}_x/2\right) + \cot\left(\bar{k}/2\right)\sin\left(\bar{k}_x/2\right) \tag{235}$$
$$+ \frac{1}{2}u_2^2\frac{\sin^2\left(\bar{k}_y/2\right)}{\sin^2\left(\bar{\omega}/2\right)}\cos\left(\bar{k}_x/2\right).$$

We now proceed to choose u_2^2. Writing

$$u_2^2 = w_2^2\sin^2\left(\bar{\omega}/2\right), \tag{236}$$

and Taylor expanding (235) about $\sin\theta = 0$, we obtain

$$\varepsilon'_{\mathrm{nsfd}}(\theta) = \left[-\frac{1}{4}\frac{\bar{k}}{\sin\left(\bar{k}/2\right)} + \frac{1}{8}w_2^2\bar{k}^2\cos\left(\bar{k}/2\right)\right]\sin^2\theta + \cdots. \tag{237}$$

The choice,

$$w_2^2 = \frac{4}{\bar{k}\sin\bar{k}} \tag{238}$$

in (237) cancels the $\sin^2\theta$ term leaving $\varepsilon'_{\mathrm{nsfd}}(\theta) \sim \sin^4\theta$. Instead of simply canceling the $\sin^2\theta$-term, however, we can choose w_2^2 in such a way that it partially cancels out the higher order terms numerically. For example, we can choose w_2^2 such that

$$\langle\varepsilon'_{\mathrm{nsfd}}\rangle = \frac{1}{\phi}\int_0^{\phi}\varepsilon'_{\mathrm{nsfd}}(\theta)d\theta = 0. \tag{239}$$

Over the range of incident angles, $0 \leq \theta \leq \phi$ the mean value of $\varepsilon'_{\text{nsfd}}$ vanishes. We might be tempted to choose $\phi = \pi/2$, but the larger ϕ, the greater the root mean square (rms) deviation of $\varepsilon'_{\text{nsfd}}$ from zero. We have found that that the choice $\phi = \pi/6$ ensures "small" rms values of $\varepsilon'_{\text{nsfd}}$.

Taking $\phi = \pi/6$, and using Taylor expansions about $\sin \theta = 0$ we can approximately evaluate $\langle \varepsilon'_{\text{nsfd}} \rangle$. Retaining only the most significant terms and simplifying, we find

$$w_2^2 = \frac{21}{5} \frac{1}{\bar{k} \sin \bar{k}}. \tag{240}$$

One could also seek a value of w_2^2 which minimizes the rms values $\varepsilon'_{\text{nsfd}}$, but the result is similar to (240). Using (236) we now take

$$u_2^2 = \frac{21}{5} \frac{\sin^2 (\omega \Delta t/2)}{(kh) \sin(kh)}. \tag{241}$$

The nonstandard version of the second-order Mur ABC (NS-ABC) is thus

$$\psi_b^{t+1} = \psi_i^t + (\psi_i^t - \psi_i^{t-1}) + \left(\frac{1 - u_1}{1 + u_1} \right) \left[(\psi_b^t - \psi_b^{t-1}) - (\psi_i^{t+1} - \psi_i^t) \right] \tag{242}$$

$$+ \frac{1}{2} \left(\frac{u_2^2}{1 + u_1} \right) d_y^2 \left(\psi_b^t + \psi_i^t \right).$$

For special applications in which most of the radiation impinges on the computational boundary at angle ϕ, $\varepsilon'_{\text{nsfd}}(\phi)$ can be made to vanish exactly with the choice $w_2^2 = w_2^2(\phi)$, where

$$w_2^2(\phi) = 2 \frac{\cos \left(\bar{k}'_x/2 \right) - \cot \left(\bar{k}/2 \right) \sin \left(\bar{k}'_x/2 \right)}{\sin^2 \left(\bar{k}'_y/2 \right) \cos \left(\bar{k}'_x/2 \right)}, \tag{243}$$

where $(k'_x, k'_y) = k(\cos \phi, \sin \phi)$. In this case

$$u_2^2 = w_2^2(\phi) \sin^2 (\omega \Delta t/2). \tag{244}$$

Although the NS-ABC optimally absorbs monochromatic plane waves, it also absorbs pulses more effectively than the S-ABC. The performance of both the S-ABC and NS-ABC, are frequency-dependent.

Three Dimensions

The three-dimensional wave equation is

$$\left[\partial_t^2 - v^2 \left(\partial_x^2 + \partial_y^2 + \partial_z^2 \right) \right] \psi(\mathbf{x}, t) = 0, \tag{245}$$

where $\mathbf{x} = (x, y, z)$. Defining

$$P = \sqrt{\partial_t^2 - v^2 \left(\partial_x^2 + \partial_y^2 \right)}, \tag{246}$$

the wave equation factors into

$$(P + v\partial_z)(P - v\partial_z)\,\psi(\mathbf{x}, t) = 0, \tag{247}$$

and the one-way wave equations become,

$$(P \pm v\partial_z)\,\psi(\mathbf{x}, t) = 0. \tag{248}$$

Defining the transverse Laplacian, $\nabla_t^2 = \partial_x^2 + \partial_y^2$, and replacing ∂_y^2 by ∇_t^2 in (213) - (215), the three-dimensional the second-order EM one-way wave equations become

$$\left(\partial_t^2 \pm v\partial_{zt} - \frac{1}{2}v^2\nabla_t^2\right)\psi(\mathbf{x}, t) = 0. \tag{249}$$

The three-dimensional second-order EM annihilation operator is

$$W_\pm = \partial_t^2 \pm v\partial_{zt} - \frac{1}{2}v^2\nabla_t^2. \tag{250}$$

We now proceed to construct a finite difference model of the three-dimensional EM equations. The transverse Laplacian is approximated by

$$\nabla_t^2 f(\mathbf{x}) \cong \frac{\mathbf{D}_t^2 f(\mathbf{x})}{h^2}, \tag{251}$$

where $\mathbf{D}_t^2 = d_x^2 + d_y^2$. The replacements $d_y^2 \to \mathbf{D}_t^2$ in (224) give the three-dimensional SFD model of the EM equations,

$$d_t^2\left(\psi_b^t + \psi_i^t\right) + \bar{v}\left[\left(\psi_b^{t+1} - \psi_i^{t+1}\right) - \left(\psi_b^{t-1} - \psi_i^{t-1}\right)\right] \tag{252}$$

$$-\frac{1}{2}\bar{v}^2\mathbf{D}_t^2\left(\psi_b^t + \psi_i^t\right) = 0.$$

Here b denotes the z-coordinate of an x-y boundary plane, with $b = h$ on one side, and $b = N_z h$ on the opposite side of the domain. In our notation $\psi_z^t = \psi(x, y, z, t)$. The second-order Mur ABC in three dimensions becomes

$$\psi_b^{t+1} = \psi_b^t + \left(\psi_i^t - \psi_i^{t-1}\right) + \left(\frac{1-\bar{v}}{1+\bar{v}}\right)\left[\left(\psi_b^t - \psi_b^{t-1}\right) - \left(\psi_i^{t+1} - \psi_i^t\right)\right] \tag{253}$$

$$+\frac{1}{2}\left(\frac{\bar{v}^2}{1+\bar{v}}\right)\mathbf{D}_t^2\left(\psi_b^t + \psi_i^t\right),$$

which we call the S-ABC.

Defining the left side of (252) to be $A_{\mathrm{sfd}}\psi(x, y, m, t)$, let us now compare how effectively $W_\pm$ and A_{sfd} annihilate a monochromatic plane wave, $\varphi_\mathbf{k} = e^{i(\mathbf{k}\bullet\mathbf{x}-\omega t)}$, where

$$\mathbf{k} = \left(k_t\cos\phi,\, k_t\sin\phi,\, k\cos\theta\right), \tag{254}$$

$k_t = k\sin\theta$. Computing $\varepsilon_{\mathrm{EM}} = W_{\pm}\varphi_{\mathbf{k}}/\varphi_{\mathbf{k}}$, we obtain the same result as (217) in two dimensions, and hence $\varepsilon_{\mathrm{EM}}$ is independent of ϕ.

On the other hand the annihilation error, $\varepsilon_{\mathrm{sfd}} = A_{\mathrm{sfd}}\varphi_{\mathbf{k}}/\varphi_{\mathbf{k}}$, is a function of ϕ because $\mathbf{D}_t^2\varphi_{\mathbf{k}}$ is given by (52). We could try to remove the azimuthal dependence by employing a two-dimensional NSFD approximation to $\nabla_t^2\varphi_{\mathbf{k}}$, but the superposition parameter, $\gamma_0 = \gamma_0(k_t)$, is a function of θ, so the ϕ dependence of $\varepsilon_{\mathrm{sfd}}$ can only be removed at one value of θ.

Actually the azimuthal dependence is much less of a concern than it seems. The ϕ dependence is is governed by the discretization of the transverse wavelength, $\lambda_t = 2\pi/k_t$. At small θ, h/λ_t is very small, and from (56) the ϕ dependence of $\mathbf{D}_t^2\varphi_{\mathbf{k}}$ is also small. As θ increases h/λ_t decreases, the ϕ-dependence of $\varepsilon_{\mathrm{sfd}}$ grows. At large θ, however, $\varepsilon_{\mathrm{sfd}}$ is large anyway, and there is little to be gained by removing its ϕ-dependence. At small θ, where the Mur ABC error is small, h/λ_t is also small and

$$\frac{\mathbf{D}_t^2\varphi_{\mathbf{k}}}{\varphi_{\mathbf{k}}} = -4\sin^2(k_t h/2) \tag{255}$$

is a good approximation.

Using (255) with the substitutions $\bar{v} \to u_1$ and $\bar{v}^2 \to u_2^2$ in (252) and (253), we can now construct the three-dimensional the NSFD model of the EM equations, and obtain the NS-ABC. We have

$$d_t^2\left(\psi_b^t + \psi_i^t\right) + u_1\left[\left(\psi_b^{t+1} - \psi_i^{t+1}\right) - \left(\psi_b^{t-1} - \psi_i^{t-1}\right)\right] \tag{256}$$
$$-\frac{1}{2}u_2^2\mathbf{D}_t^2\left(\psi_b^t + \psi_i^t\right) = 0,$$

and

$$\psi_b^{t+1} = \psi_b^t + \left(\psi_i^t - \psi_i^{t-1}\right) + \left(\frac{1-u_1}{1+u_1}\right)\left[\left(\psi_b^t - \psi_b^{t-1}\right) - \left(\psi_i^{t+1} - \psi_i^t\right)\right] \tag{257}$$
$$+\frac{1}{2}\left(\frac{u_2^2}{1+u_1}\right)\mathbf{D}_t^2\left(\psi_b^t + \psi_i^t\right).$$

5.4. *Comparisons and Practical Tests*

In Fig. 2 we compare the annihilation errors of $W_{\pm}$, A_{sfd}, and A_{nsfd} in two dimensions with respect to infinite monochromatic plane waves, by plotting $\varepsilon'_{\mathrm{EM}} = \varepsilon_{\mathrm{EM}}/8\sin^2(\bar{\omega}/2)$, $\varepsilon'_{\mathrm{sfd}}$, and $\varepsilon'_{\mathrm{nsfd}}$, as a functions of θ for two different space-time discretizations. Except near the zero-crossing of $\varepsilon'_{\mathrm{sfd}}$, we see that $|\varepsilon'_{\mathrm{nsfd}}| < |\varepsilon'_{\mathrm{sfd}}|$. In the left figure $\lambda/h = 8$, $T/\Delta t = 10$ is suitable to model a dielectric scatterer (refractive index $n \geq 1$) embedded in vacuum. Here $\bar{v}$ satisfies the NSFD stability requirement (see Sect. 6) that

$\bar{v} \leq 0.84$. In the right figure, the discretization $\lambda/h = 8$, $T/\Delta t = 34$ can be used to model a computational domain filled with dielectric containing vacuum-filled holes. The dielectric has refractive index $n = 3.4$, and taking $\bar{v} = 8/34$ in the dielectric guarantees that $\bar{v} = (8/34)n$ in the holes satisfies the NSFD stability requirement.

In many simulations the wave fields are pulses. In Fig. 3 we compare the performance of the S-ABC with the NS-ABC using gaussian pulses. The pulse is incident from left along the x-axis and impinges perpendicularly upon the right computational boundary, where it is partially reflected. The reflected intensity is visualized in a grey scale, and intensity plots along x-axis through the centers of the incident and reflected pulses are shown. Except at the leading trailing edges of the pulse, the NS-ABC gives nearly perfect absorption, while S-ABC gives a large reflection. We have used the discretization $\lambda/h = 8$, $T/\Delta t = 10$.

In Fig. 4 the pulse of Fig. 3 is incident on the left computational boundary at 30 degrees to the normal. Here we use the discretization $\lambda/h = 8$, $T/\Delta t = 34$. Let the reflected intensity of the NS-ABC be R_{nsfd}, and that of the S-ABC be R_{sfd}. We find that $R_{\mathrm{nsfd}}/R_{\mathrm{sfd}} = 2.7 \times 10^{-2} \cong 1/37$.

5.5. *Summary of the NS-ABC Method*

In the previous sections, we sought to improve the FD models of the wave equation, and Maxwell's equations. Here the goal is different. The second-order one-way EM equations are only approximate so we do not really need a faithful FD model. The actual goal is to find a FD operator that annihilates incident waves on the numerical grid; the EM equations merely guide its construction. At small incidence angles, where $W_\pm$ is a quite good annihilator, it is fruitful to accurately model the EM equations, but at higher angles it does not matter if the FD annihilation operator deviates from $W_\pm$ so long as it annihilates incident waves. Fixing u_1 with (234) ensures that $\varepsilon_{\mathrm{nsfd}}(0) = \varepsilon_{\mathrm{EM}}(0) = 0$. The fact that $\lim_{k \to 0} u_1 = \bar{v}$ indicates that A_{nsfd} is a good approximation to $W_\pm$ at small θ. Had we fixed u_2^2 with (238) we would have had $\lim_{k \to 0} u_2^2 = \bar{v}^2$, and the NSFD model (231) would be faithul at larger angles. But the larger θ, the larger $\varepsilon_{\mathrm{EM}}$, so it is better to deviate from modeling the EM equations. Choosing u_2^2 to be given by (241), we find that $\lim_{h \to 0} u_2^2 = 1.2\bar{v}^2$, but this choice improves the ABC at high incidence angles.

The goal, therefore, of further research on ABCs for FDTD simulation ought to be a better FD annihilator, with less attention paid to FD models of differential annihilation operators.

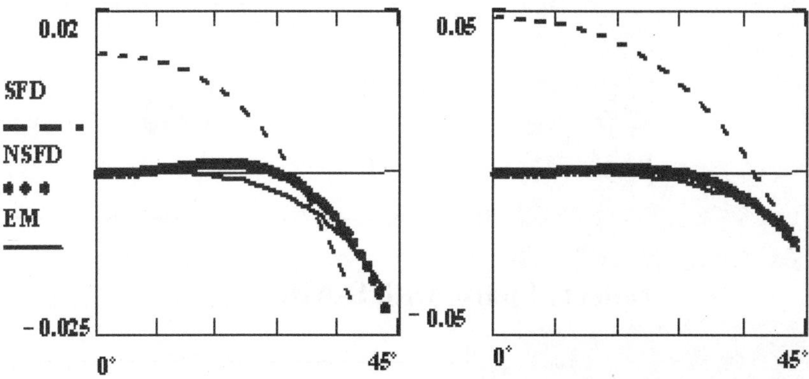

Fig. 2.   **Annihilation Error**. Annihilation error of the S-ABC, NS-ABC, and second-order EM operator, versus incidence angle for an infinite plane wave. Left figure: $\lambda/h = 8$, $T/\Delta t = 10$. Right figure: $\lambda/h = 8$, $T/\Delta t = 34$.

6. Discretization and Numerical Stability

6.1. *Introduction*

We have discussed finite difference models of several differential equations, and have shown how changing the parameters of these models can yield high accuracy FDTD solutions. We have neglected, however, to mention the two most important parameters of all: the spatial step, h, and the time step, Δt. Their sizes are not arbitrary, but are constrained both by the size scales of the problem to be solved.

Sampling Density

In FDTD calculations the wave field is sampled on a space-time grid

$$t = 0, \Delta t, 2\Delta t \cdots , \tag{258}$$

$$x = 0, h, 2h, \cdots , \tag{259}$$

and the solution is represented as a set of space-time samples. In general the denser the sampling, the more accurate the representation, but when the solution is known to be periodic, the sampling can be quite sparse. Consider a signal of wavelength λ, and period T in one dimension,

$$\psi(x,t) = A \sin\left(\frac{2\pi}{\lambda}x - \frac{2\pi}{T}t + \phi\right), \tag{260}$$

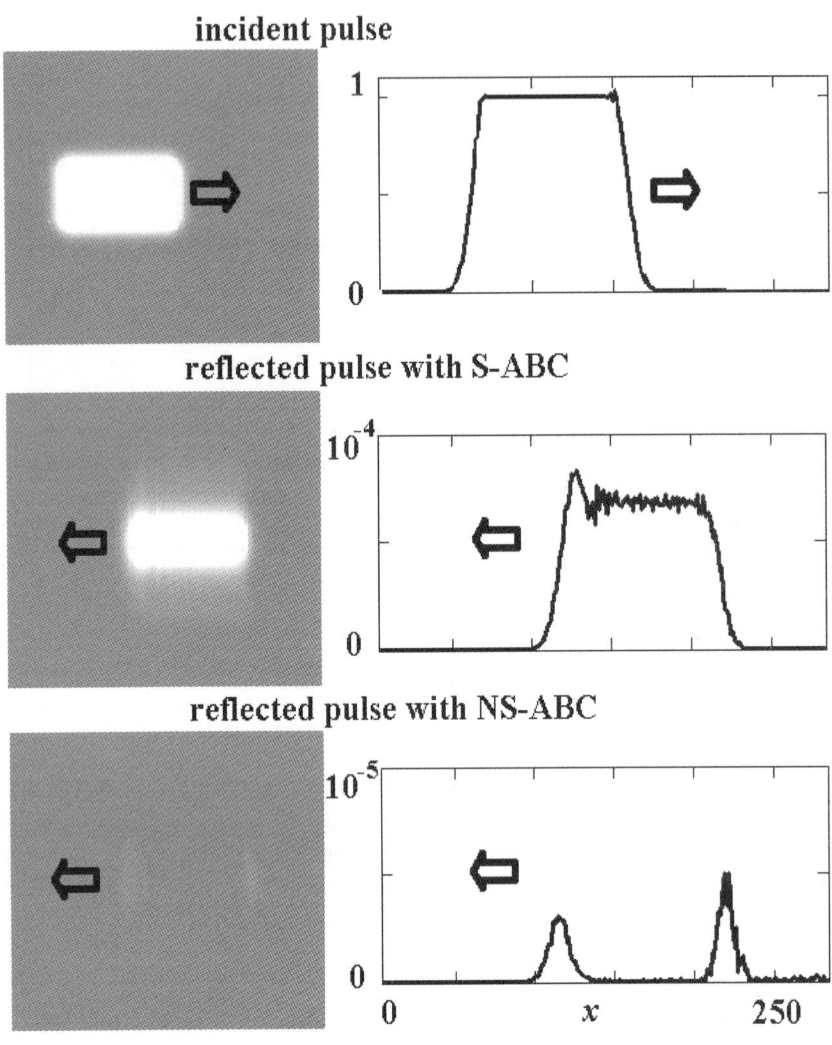

Fig. 3. NS-ABC vs. S-ABC at Normal Incidence. The standard second order Mur absorbing boundary condition (S-ABC) is compared with the NS-ABC at normal incidence. $\lambda/h = 8$, $T/\Delta t = 10$.

where A is the amplitude and ϕ the phase. If A, ϕ, λ, and T can be determined from the sample data, $\psi(x,t)$ is completely specified. It is intuitively obvious that Δt cannot be larger than T. According to the Nyquist Sam-

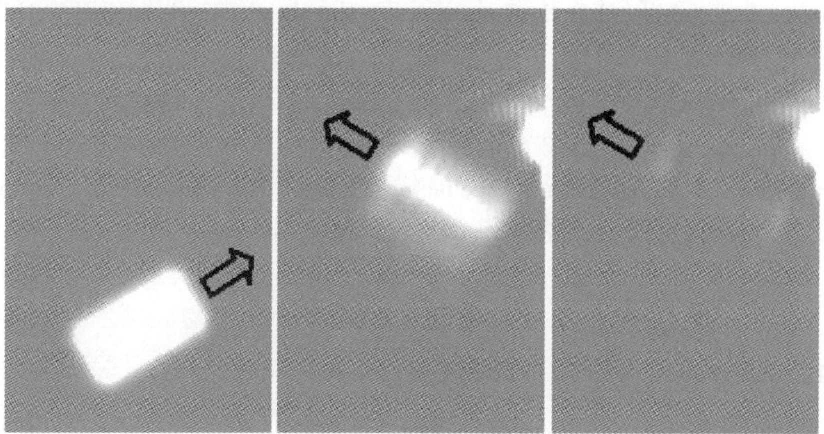

Fig. 4. **NS-ABC vs. S-ABC at 30 deg. Incidence** We compare the performance of the S-ABC with the NS-ABC at 30 deg. incidence. $\lambda/h = 8$, $T/\Delta t = 34$.

pling Theorem, the condition

$$\frac{T}{\Delta t} > 2 \qquad (261)$$

ensures that both T and ϕ can be determined at a fixed position. More than two time samples per period are needed to specify both the amplitude and phase, although two are sufficient to determine only amplitude. On a d-dimensional grid dimensional grid, the largest space step is $h\sqrt{d}$ (along the diagonals), so the Nyquist limit becomes

$$\frac{\lambda}{h} > 2\sqrt{d}. \qquad (262)$$

The Nyquist limits determine the maximum frequency, and the minimum wavelength that can be accurately represented on the space-time numerical grid, but the Nyquist limits are not the only constraints on h and Δt.

Numerical Stability

Whenever an algorithm is iterated, numerical instabilities can arise. Let $\mathbf{x}_0$ be either a single number or a wave field, and let $\mathbf{x}_n = A^n \mathbf{x}_0$ be the result of acting n times on $\mathbf{x}_0$ with algorithm A. If $\lim_{n \to \infty} |\mathbf{x}_n| \to \infty$, then the algorithm is said to be unstable. We do not attempt a general analysis of numerical stability, but limit ourselves to the algorithms developed in this chapter.

Consider a finite-difference algorithm of the form

$$\psi(t + \Delta t) = p\psi(t - \Delta t) + 2q\psi(t), \tag{263}$$

where p and q are constants - the factor of 2 is for later convenience. Taking $\tau = 0, 1, 2, \cdots$, let us define

$$t = \tau\Delta t, \tag{264}$$

$$\psi_\tau = \psi(\tau\Delta t). \tag{265}$$

In this notation, algorithm (263) becomes,

$$\psi_{\tau+1} = p\psi_{\tau-1} + 2q\psi_\tau. \tag{266}$$

Combining ψ_τ and $\psi_{\tau+1}$ into a vector,

$$|\tau\rangle = \begin{bmatrix} \psi_\tau \\ \psi_{\tau+1} \end{bmatrix}, \tag{267}$$

we can express (266) as a matrix equation,

$$|\tau\rangle = A\,|\tau - 1\rangle, \tag{268}$$

where

$$A = \begin{bmatrix} 0 & 1 \\ p & 2q \end{bmatrix}. \tag{269}$$

The initial state is

$$|0\rangle = \begin{bmatrix} \psi_0 \\ \psi_1 \end{bmatrix}, \tag{270}$$

and all subsequent states are given by

$$|\tau\rangle = A^\tau\,|0\rangle. \tag{271}$$

The algorithm is therefore stable if $\lim_{\tau\to\infty} \|A^\tau\|^2$ is finite.

Letting I be the 2×2 unit matrix, the eigenvalues of A are found by solving

$$\det(A - \lambda I) = 0, \tag{272}$$

which is equivalent to the quadratic equation,

$$\lambda^2 - 2q\lambda - p = 0. \tag{273}$$

Solving it we find the eigenvalues,

$$\lambda_\pm = q \pm \sqrt{q^2 + p}. \tag{274}$$

The corresponding normalized eigenvectors are,

$$|\lambda_\pm\rangle = \frac{1}{\sqrt{1 + |\lambda_\pm|^2}} \begin{bmatrix} 1 \\ \lambda_\pm \end{bmatrix}. \tag{275}$$

In this notation, the dual of

$$|v\rangle = \begin{bmatrix} v_1 \\ v_2 \end{bmatrix}, \tag{276}$$

is

$$\langle v| = \begin{pmatrix} v_1^* & v_2^* \end{pmatrix}, \tag{277}$$

and the magnitude squared of $|v\rangle$ is $\|v\|^2 = \langle v \mid v\rangle$, where z^* is the complex conjugate of z.

Expanding $|0\rangle$ in terms of the eigenvectors we have

$$|0\rangle = c_+ |\lambda_+\rangle + c_- |\lambda_-\rangle, \tag{278}$$

where the expansion coefficients are $c_\pm = \langle \lambda_\pm \mid 0\rangle$. Eq. (271) then becomes

$$|\tau\rangle = c_+ \lambda_+^\tau |\lambda_+\rangle + c_- \lambda_-^\tau |\lambda_-\rangle. \tag{279}$$

Algorithm (263) is thus numerically stable if

$$|\lambda_\pm| \leq 1. \tag{280}$$

Condition (280) is a form of the CFL (Courant, Friedrich, Levy) [12] stability condition.

Suppose that $|\lambda_+| \leq 1$, $|\lambda_-| > 1$, and $c_- = 0$, then

$$|0\rangle = c_+ |\lambda_+\rangle + 0 |\lambda_-\rangle. \tag{281}$$

It might appear that $|\tau\rangle$ would remain finite. In the absence of computational round off error it would, but since any digital calculation is eventually rounded off, after many (N) iterations, the $|\lambda_-\rangle$ eigenvector appears, and we have

$$|N\rangle = c_+ \lambda_+^N |\lambda_+\rangle + \varepsilon |\lambda_-\rangle. \tag{282}$$

Further iteration gives

$$|N + \tau\rangle = c_+ \lambda_+^{N+\tau} |\lambda_+\rangle + \varepsilon \lambda_-^\tau |\lambda_-\rangle, \tag{283}$$

and eventually $|\tau\rangle$ diverges.

The numerical stability of all the algorithms given in this chapter can now be analyzed by expressing them in form (266).

6.2. *Stability of the Wave Equation FDTD Algorithm*

The wave equation is

$$\left(\partial_t^2 - v^2 \nabla^2\right) \psi(\mathbf{x}, t) = 0. \tag{284}$$

Its generalized FD model is

$$\left(d_t^2 - u^2 \mathbf{D}^2\right) \psi(\mathbf{x}, t) = 0, \tag{285}$$

and the FDTD algorithm is

$$\psi(\mathbf{x}, t + \Delta t) = -\psi(\mathbf{x}, t - \Delta t) + \left(2 + u^2 \mathbf{D}^2\right) \psi(\mathbf{x}, t). \tag{286}$$

In the S-FDTD algorithm (43) $\mathbf{D}^2 = \mathbf{D}_1^2$ and $u^2 = v^2 \Delta t^2 / h^2$, while in the NS-FDTD one (70) $\mathbf{D}^2 = \mathbf{D}_0^2$ and $u^2 = \sin^2\left(\omega \Delta t / 2\right) / \sin^2\left(kh/2\right)$.

To cast the FDTD algorithm into the form of (263) we need to evaluate $\mathbf{D}^2 \psi$. Replacing ψ with a monochromatic plane wave $\varphi_{\mathbf{k}} = e^{i(\mathbf{k} \bullet \mathbf{x} - \omega t)}$, we can express $\mathbf{D}^2 \varphi_{\mathbf{k}}$ in the form

$$\mathbf{D}^2 \varphi_{\mathbf{k}} = -2D^2 \varphi_{\mathbf{k}}(\mathbf{x}, t), \tag{287}$$

where D^2 is given in Sect 6.6.

Following the notation of Sect. 5, we write $\psi(\mathbf{x}, t) = \psi_{\mathbf{x}}^t$ and the FDTD algorithm becomes

$$\varphi_{\mathbf{x}}^{t+1} = -\varphi_{\mathbf{x}}^{t-1} + 2\left(1 - D^2 u^2\right) \varphi_{\mathbf{x}}^t. \tag{288}$$

It is now in the form of (266) with

$$p = -1 \tag{289}$$
$$q = 1 - D^2 u^2. \tag{290}$$

Using (274), the eigenvalues of A are

$$\lambda_{\pm} = 1 - u^2 D^2 \pm \sqrt{\left(1 - u^2 D^2\right)^2 - 1}. \tag{291}$$

For $z_{\pm} = x \pm \sqrt{x^2 - 1}$, it can be shown that both $|z_{\pm}| \leq 1$ only if x is both real and $|x| \leq 1$. Thus to fulfill the stability condition $|\lambda_{\pm}| \leq 1$, we must have

$$\left(1 - u^2 D^2\right)^2 \leq 1. \tag{292}$$

If both $D^2 > 0$ and $u^2 > 0$, as is the case for non-vanishing monochromatic plane waves, the stability condition becomes

$$u^2 \leq \frac{2}{D^2}. \tag{293}$$

Since D^2 is a function of both the magnitude and direction of $\mathbf{k}$, stability can assured by putting the maximum value of D^2, $\max\left(D^2\right)$, in the place of D^2 in (293). The stability condition is thus,

$$u^2 \leq \frac{2}{\max\left(D^2\right)}. \tag{294}$$

It is convenient to express (294) as a relationship between the space step, h, and the time step, Δt, in the form

$$\frac{v\Delta t}{h} \leq c. \tag{295}$$

Clearly, at fixed h, the larger c, the greater the allowed value of Δt. The larger Δt, the fewer the number of iterations needed to solve a problem. At wavelength λ, and period T, (295) can be written

$$\frac{\lambda/h}{T/\Delta t} \leq c. \tag{296}$$

In d dimensions, the Nyquist limit, (262) requires that $c < \sqrt{d}$. Except in one dimension, as subsequent analysis will show, the upper limit on c is actually lower. In all dimensions we must have $c \leq 1 \Rightarrow T/\Delta t \geq \lambda/h$.

SFD Stability

In the SFD algorithm, $\mathbf{D}^2 = \mathbf{D}_1^2$, and from Sect. 6.6, we find $\max\left(D_1^2\right) = 2d$ in $d = 1$, 2 and 3 dimensions. Using the fact that $u^2 = v^2\Delta t^2/h^2$, the stability condition (294) becomes,

$$\frac{v\Delta t}{h} \leq c_d^{\mathrm{sfd}} = \frac{1}{\sqrt{d}}. \tag{297}$$

Evaluating the values of c_d^{sfd}, and rounding downwards we have,

$$c_1^{\mathrm{sfd}} = 1 \tag{298}$$

$$c_2^{\mathrm{sfd}} = \frac{\sqrt{2}}{2} \cong 0.70 \tag{299}$$

$$c_3^{\mathrm{sfd}} = \frac{\sqrt{3}}{3} \cong 0.57. \tag{300}$$

NSFD Stability

In the NSFD algorithm, $\mathbf{D}^2 = \mathbf{D}_0^2$, and from Sect. 6.6, we have $\max\left(D_0^2\right) = 2$, $8/3$, and 3.158 for $d = 1$, 2, and 3, respectively. Using the fact that $u^2 = \sin^2\left(\omega\Delta t/2\right)/\sin^2\left(kh/2\right)$, the stability condition (294) becomes,

$$\frac{\sin\left(\omega\Delta t/2\right)}{\sin\left(kh/2\right)} \leq \sqrt{\frac{2}{\max\left(D_0^2\right)}}. \tag{301}$$

We would now like to express this in the form of (295). Writing $\bar{\omega} = \omega \Delta t$ and $\bar{k} = kh$, let $\bar{\omega} = c\bar{k}$, where c is a value to be determined such that (301) holds. In d dimensions write $\max\left(D_0^2\right) = m_d^{\text{nsfd}}$, and (301) can be expressed as

$$\sin\left(c\bar{k}/2\right) \leq \sqrt{\frac{2}{m_d^{\text{nsfd}}}} \, \sin\left(\bar{k}/2\right). \tag{302}$$

In the range $0 \leq x \leq \pi/2$, $\sin x$ monotonically rises with increasing x, so $\sin x \leq \sin y \Rightarrow x \leq y$. The Nyquist limit (262) can be expressed in the form $\bar{k} < \pi/\sqrt{d}$. Since we will never formulate a FDTD algorithm which violates (262), and assuming that $c \leq 1$, we conclude from (302) that

$$c < \frac{2\sqrt{d}}{\pi} \arcsin\left[\sqrt{\frac{2}{m_d^{\text{nsfd}}}} \, \sin\left(\frac{\pi}{2\sqrt{d}}\right)\right]. \tag{303}$$

Inserting the numerical values of m_d^{nsfd} from Sect. 6.6 into (302), we now have

$$c_1^{\text{nsfd}} = 1, \tag{304}$$

$$c_2^{\text{nsfd}} = \frac{2\sqrt{2}}{\pi} \arcsin\left[\frac{\sqrt{3}}{2} \sin\left(\frac{\pi}{2\sqrt{2}}\right)\right] \cong 0.84, \tag{305}$$

$$c_3^{\text{nsfd}} = \frac{2\sqrt{3}}{\pi} \arcsin\left[\sqrt{\frac{2}{m_3^{\text{nsfd}}}} \, \sin\left(\frac{\pi}{2\sqrt{3}}\right)\right] \cong 0.78. \tag{306}$$

Comparing with the SFD results, we find that in one dimension $c_1^{\text{nsfd}} = c_1^{\text{sfd}}$, but in two and three dimensions $c_{2,3}^{\text{nsfd}} > c_{2,3}^{\text{sfd}}$, which implies that a longer time step can be used in the NS-FDTD algorithms.

6.3. *Stability of the Absorbing Wave Equation FDTD Algorithm*

The absorbing wave equation is,

$$\left(\partial_t^2 - v^2\nabla^2 + 2\alpha\partial_t\right)\psi(\mathbf{x}, t) = 0, \tag{307}$$

where $\alpha \geq 0$ is the absorption. Its generalized finite-difference model is

$$\left(d_t^2 - u^2\mathbf{D}^2 + ad_t'\right)\psi(\mathbf{x}, t) = 0, \tag{308}$$

and the corresponding FDTD algorithm is

$$\psi(\mathbf{x}, t + \Delta t) = \frac{1}{1+a}\left[-\left(1-a\right)\psi(\mathbf{x}, t - \Delta t) + \left(2 + u^2\mathbf{D}^2\right)\psi(\mathbf{x}, t)\right]. \tag{309}$$

In the S-FDTD algorithm (100) $\mathbf{D}^2 = \mathbf{D}_1^2$, $a = \alpha\Delta t$, and $u^2 = v^2\Delta t^2/h^2$, while in the NS-FDTD one (102) $\mathbf{D}^2 = \mathbf{D}_0^2$, $a = \tanh(\alpha\Delta t)$, and $u^2 = u_0^2$, where

$$u_0^2 = \frac{\sin^2(\omega'\Delta t/2) + \sinh^2(\alpha\Delta t/2)}{\sin^2(kh/2)\cosh(\alpha\Delta t)}, \tag{310}$$

$\omega'^2 = \omega^2 - \alpha^2$, $\alpha < \omega$, and $v = \omega/k$.

Replacing ψ by a decaying harmonic plane wave $\varphi_{\mathbf{k},\alpha} = e^{i(\mathbf{k}\bullet\mathbf{x}-\omega't)}e^{-\alpha t}$, and using (287) we cast the FDTD algorithm into the form of (266),

$$\varphi_{\mathbf{x}}^{t+1} = -\left(\frac{1-a}{1+a}\right)\varphi_{\mathbf{x}}^{t-1} + \frac{2}{1+a}\left(1 - D^2 u^2\right)\varphi_{\mathbf{x}}^t, \tag{311}$$

with

$$p = -\left(\frac{1-a}{1+a}\right), \tag{312}$$

$$q = \frac{1}{1+a}\left(1 - D^2 u^2\right). \tag{313}$$

From (274) the eigenvalues of A are,

$$\lambda_{\pm} = \frac{1}{1+a}\left[1 - u^2 D^2 \pm \sqrt{(1 - u^2 D^2)^2 - (1 - a^2)}\right]. \tag{314}$$

Consider the functions,

$$\lambda_{\pm}(x) = \frac{x \pm \sqrt{x^2 + a^2 - 1}}{1+a}. \tag{315}$$

For $-1 \leq x \leq 1$ it is easily shown that $|\lambda_{\pm}(x)| \leq 1$. Setting $\delta > 0$ and Taylor expanding $\lambda_+(1 + \delta)$ about one, we see that $|\lambda_+(1 + \delta)| > 1$, and similarly that $|\lambda_-(-1 - \delta)| > 1$. The stability condition for (311) is thus

$$\left(1 - u^2 D^2\right)^2 \leq 1, \tag{316}$$

which is the same as (292) for both the SFD, and NSFD algorithms. The stability condition for the absorbing wave equation S-FDTD algorithm is exactly the same as that of the non-absorbing wave equation. In the NS-FDTD algorithm u^2 is given by (310). Defining $\bar{\omega} = \omega\Delta t$, $\bar{k} = kh$, and $\bar{\alpha} = \alpha\Delta t$, let us write $\bar{\omega} = c\bar{k}$, and $\bar{\alpha} = fc\bar{k}$, where $0 \leq f < 1$. We can now determine the maximum value of c for which (294) is satisfied by solving

$$\frac{\sin^2\left(c\bar{k}\sqrt{1 - f^2}/2\right) + \sinh^2\left(fc\bar{k}/2\right)}{\cosh\left(fc\bar{k}\right)\sin^2\left(\bar{k}/2\right)} \leq \frac{2}{m_d^{\text{nsfd}}}. \tag{317}$$

It is difficult to derive an analytic formula for c, so this condition should be tested in advance of each computation.

136　　　　　　　　　　　　　　　*J. B. Cole*

6.4. *Stability of the Yee Algorithm*

As we have seen in Sect. 4, the Yee algorithm, and the wave equation FDTD algorithm are closely related. From (119) we have

$$v^2 = \frac{1}{\varepsilon\mu}, \tag{318}$$

so if the ε and μ parameters are both positive, the stability conditions for the Yee algorithm are exactly the same as those of the wave equation FDTD algorithm.

On the other hand if one of these parameters is negative and the other positive, $v^2 < 0$, and in (292) u^2 also becomes negative. For negative u^2 the stabilty the Yee algorithm is unstable. Materials with negative ε and μ actually exist. At some frequencies, metals have negative values of ε, while μ is positive, and "metamaterials" with negative ε and μ have recently been fabricated. Algorithms to simulate electromagnetic propagation in metals with negative ε are introduced in Sect. 9.

Certain kinds of diffracting structures with subwavelength features, such as photonic crystals, have effectively negative v^2, but so long as the actual values of ε and μ, are both positive positive in the Yee algorithm (SFD or NSFD version), it remains numerically stable.

6.5. *Stability of the Second-Order Mur Absorbing Boundary Condition*

One Dimension

The centered absorbing boundary conditions (201) and (205) can be expressed in the form

$$\psi_b^{t+1} = \psi_i^t + \left(\frac{1-u}{1+u}\right)\left(\psi_b^t - \psi_i^{t+1}\right). \tag{319}$$

For the S-ABC, $u = v\Delta t/h$, and for the NS-ABC $u = \tan(\omega\Delta t/2)/\tan(kh/2)$.

On the right side of the computational domain, $b = N_x h$ and $i = (N_x - 1)h$, and with respect to forward-moving waves, $\varphi(x,t) = e^{i(kx-\omega t)}$, and we have $\varphi_i^t = \varphi_b^t e^{-ikh}$. Substituting this expression for φ_i^t into (319), and solving for ψ_b^{t+1} gives a simple difference equation of the form

$$\psi_b^{t+1} = d\varphi_b^t, \tag{320}$$

where

$$d = \frac{\left(\frac{1-u}{1+u}\right) + e^{-ikh}}{1 + \left(\frac{1-u}{1+u}\right) e^{-ikh}}. \tag{321}$$

With the time discretization $t = 0,\ \Delta t,\ 2\Delta t, \cdots$, the solution of (320) is

$$\varphi_b^t = d^{t/\Delta t} \varphi_b^0. \tag{322}$$

Algorithm (319) is stable if $|d| \leq 1$. Multiplying both the numerator and denominator of d by $e^{ikh/2}$, we see that they are complex conjugates of each other. For any complex number $z = \rho e^{i\theta}$, it is true that $|z^*/z| = 1$. Thus $|d| = 1$ and algorithm (319) is stable for any value of u at $b = N_x h$. Such an algorithm is said to be unconditionally stable.

On the right side, where $b = 0$ and $i = h$, with respect to a backward-moving wave, $\varphi(x,t) = c^{i(kx+\omega t)}$, and we have $\varphi_i^t = \varphi_b^t e^{ikh}$. Substituting into (319), and solving for ψ_b^{t+1} gives

$$\varphi_b^{t+1} = \frac{1}{d}\varphi_b^t. \tag{323}$$

Since $|d| = 1 \Rightarrow |1/d| = 1$, algorithm (319) is also unconditionally stable at $b = 0$.

The centered ABC is thus unconditionally stable.

Two Dimensions

The centered ABC is

$$\psi_b^{t+1} = \psi_b^t + (\psi_i^t - \psi_i^{t-1}) + \left(\frac{1 - u_1}{1 + u_1}\right)\left[(\psi_b^t - \psi_b^{t-1}) - (\psi_i^{t+1} - \psi_i^t)\right] \tag{324}$$

$$+ \frac{1}{2}\left(\frac{u_2^2}{1 + u_1}\right) d_y^2 \left(\psi_b^t + \psi_i^t\right).$$

Here we temporarily regard u_1 and u_2 as free parameters. In the S-ABC, $u_1 = u_2 = v\Delta t/h$, while in the NS-ABC, u_1 and u_2 are given by (234), and (240), respectively.

Let us return to the FD model of the EM one-way wave equations at $m = (b + i)/2$ in (218),

$$d_t^2\left(\psi_b^t + \psi_i^t\right) + u_1\left[(\psi_b^{t+1} - \psi_i^{t+1}) - (\psi_b^{t-1} - \psi_i^{t-1})\right] \tag{325}$$

$$-\frac{1}{2}u_2^2 d_y^2\left(\psi_b^t + \psi_i^t\right) = 0.$$

On the right boundary $b = (N_x - 1)h$ and $i = b - h$, and for a right-moving monochromatic plane wave, $\varphi_{\mathbf{k}} = e^{i(\mathbf{k} \bullet \mathbf{x} - \omega t)}$, we have

$$\varphi_b^t = \varphi_m^t e^{ik_x h/2}, \tag{326}$$

$$\varphi_i^t = \varphi_m^t e^{-ik_x h/2}, \tag{327}$$

where $\varphi_m^t = \varphi_{\mathbf{k}}(m, y, t)$. From these expressions we have

$$\varphi_b^t + \varphi_i^t = 2\varphi_m^t \cos(k_x h/2), \tag{328}$$

$$\varphi_b^t - \varphi_i^t = 2i\varphi_m^t \sin(k_x h/2). \tag{329}$$

Using the fact that $d_y^2 \varphi_{\mathbf{k}} / \varphi_{\mathbf{k}} = -4\sin^2(k_y h/2)$ and substituting into (325) we obtain

$$c_x \left(\varphi_m^{t+1} + \varphi_m^{t-1} - 2\varphi_m^t \right) + iu_1 s_x \left(\varphi_m^{t+1} - \varphi_m^{t-1} \right) + 2c_x s_y^2 u_2^2 \varphi_m^t = 0. \tag{330}$$

For brevity we have written $s_x = \sin(k_x h/2)$, $s_y = \sin(k_y h/2)$, and $c_x = \cos(k_x h/2)$. Solving for φ_m^{t+1} we have

$$\varphi_m^{t+1} = -\left(\frac{c_x - iu_1 s_x}{c_x + iu_1 s_x} \right) \varphi_m^{t-1} + 2c_x \left(\frac{1 - s_y^2 u_2^2}{c_x + iu_1 s_x} \right) \varphi_m^t, \tag{331}$$

which is in the form of (266). Writing $\alpha = c_x + iu_1 s_x$ and $\beta = 1 - s_y^2 u_2^2$, we have $p = -\alpha^*/\alpha$ and $q = c_x \beta / \alpha$. The eigenvalues corresponding to (331) are given by (274),

$$\lambda_{\pm}^{\text{right}} = \frac{1}{\alpha} \left[\beta c_x \pm \sqrt{\beta^2 c_x^2 - |\alpha|^2} \right]. \tag{332}$$

On the left boundary, where $b = h$ and $i = 2h$, a left-moving monochromatic plane wave is $\varphi_{\mathbf{k}} = e^{i(\mathbf{k} \bullet \mathbf{x} + \omega t)}$. In terms of φ_m^t, we have

$$\varphi_b^t = \varphi_m^t e^{-ik_x h/2}, \tag{333}$$

$$\varphi_i^t = \varphi_m^t e^{ik_x h/2}. \tag{334}$$

Equation (328) is unchanged, but (329) becomes

$$\varphi_b^t - \varphi_i^t = -2i\varphi_m^t \sin(k_x h/2). \tag{335}$$

In place of (331) we have

$$\varphi_m^{t+1} = -\left(\frac{c_x + iu_1 s_x}{c_x - iu_1 s_x} \right) \varphi_m^{t-1} + 2c_x \left(\frac{1 - s_y^2 u_2^2}{c_x - iu_1 s_x} \right) \varphi_m^t, \tag{336}$$

which is just the complex conjugate of (331). Interchanging the roles of α and α^* in (332), the eigenvalues on the left side are

$$\lambda_{\pm}^{\text{left}} = \frac{1}{\alpha^*} \left[\beta c_x \pm \sqrt{\beta^2 c_x^2 - |\alpha|^2} \right]. \tag{337}$$

If

$$\beta^2 c_x^2 \leq |\alpha|^2 , \tag{338}$$

both $\left|\lambda_\pm^{\text{left}}\right| \leq 1$ and $\left|\lambda_\pm^{\text{right}}\right| \leq 1$, which satisfies the stability condition of (280). On the other hand, if $\beta^2 c_x^2 > |\alpha|^2$, it can be shown that one of the eigenvalues is greater than one, making the ABC unstable. Rewriting (338), the stability condition is

$$\frac{\left(1 - s_y^2 u_2^2\right)^2}{1 + u_1^2 \left(s_x^2/c_x^2\right)} \leq 1. \tag{339}$$

Since $1 + u_1^2(s_x^2/c_x^2) \geq 1$, (339) is satisfied if

$$\left(1 - s_y^2 u_2^2\right)^2 \leq 1. \tag{340}$$

Since $s_y^2 \leq 1$, (340) gives

$$u_2^2 \leq 2, \tag{341}$$

which guarantees the stability of the ABC, regardless of the value of u_1.

For the S-ABC (225), $u_1 = u_2 = v\Delta t/h$ and

$$\frac{v\Delta t}{h} \leq \sqrt{2}, \tag{342}$$

satisfies (341) and guarantees stabilty. For the NS-ABC, u_2^2 is given by (241), and (341) becomes

$$\frac{21}{5} \frac{\sin^2(\omega\Delta t/2)}{(kh)\sin(kh)} \leq 2. \tag{343}$$

Three Dimensions

The analysis is similar to the two-dimensional case. The FD model is given by (256), and we consider monochomatic plane waves that propagate at angle θ to z-axis and impinge on an x-y plane at $z = b$. On one side $b = h$, and on the opposite side $b = (N_z - 1)h$. In three dimensions $d_y^2 \rightarrow D_t^2 = d_x^2 + d_y^2$, and $s_y^2 \rightarrow s_t^2$ in (340). In analogy with the definition of s_y^2, we have

$$s_t^2 = \sin^2[k_t h \cos(\phi/2)] + \sin^2[k_t h \sin(\phi/2)], \tag{344}$$

where ϕ is the azimuthal angle of $\mathbf{k}$ on the x-y plane. The stability condition (340) now becomes

$$\left(1 - s_t^2 u_2^2\right)^2 \leq 1. \tag{345}$$

Since $s_t^2 \leq 2$, we have

$$u_2^2 \leq 1. \tag{346}$$

The stability condition for the S-ABC now becomes

$$\frac{v\Delta t}{h} \leq 1, \tag{347}$$

and the stability condition for the NS-ABC becomes

$$\frac{21}{5}\frac{\sin^2(\omega\Delta t/2)}{(kh)\sin(kh)} \leq 1. \tag{348}$$

For most values of λ/h and $T/\Delta t$ that are used in simulations, u_2^2 easily satisfies the stability condition.

6.6. *Supplementary Derivations*

In the stability analysis we have used $\max(D^2)$, where D is defined by $\mathbf{D}^2 e^{i\mathbf{k}\bullet\mathbf{x}} = -2D^2 e^{i\mathbf{k}\bullet\mathbf{x}}$. We now proceed to derive the values of $\max(D^2)$.

SFD Algorithms

In the SFD algorithms $\mathbf{D}^2 = \mathbf{D}_1^2$, where

$$\begin{aligned}
\mathbf{D}_1^2 &= d_x^2, & d &= 1 \\
\mathbf{D}_1^2 &= d_x^2 + d_y^2, & d &= 2 \\
\mathbf{D}_1^2 &= d_x^2 + d_y^2 + d_z^2, & d &= 3
\end{aligned} \tag{349}$$

in d dimensions. Defining

$$\mathbf{D}_1^2 e^{i\mathbf{k}\bullet\mathbf{x}} = -2D_1^2 e^{i\mathbf{k}\bullet\mathbf{x}}, \tag{350}$$

we have

$$D_1^2 = 2\sin^2(k_x h/2), \qquad\qquad\qquad d = 1$$

$$D_1^2 = 2\left[\sin^2(k_x h/2) + \sin^2(k_y h/2)\right], \qquad\qquad d = 2 \tag{351}$$

$$D_1^2 = 2\left[\sin^2(k_x h/2) + \sin^2(k_y h/2) + \sin^2(k_z h/2)\right], \; d = 3$$

where $\mathbf{k} = k_x$, (k_x, k_y), (k_x, k_y, k_z) for $d = 1, 2, 3$, respectively. Regarding D_1^2 as a function of $\mathbf{k}$, $D_1^2 = D_1^2(\mathbf{k})$, the maximum of D_1^2, is easily see to be

$$\max\left(D_1^2\right) = 2d. \tag{352}$$

NSFD Algorithms

In one dimension $\mathbf{D}_0^2 = \mathbf{D}_1^2 = d_x^2$, and $\max\left(D_0^2\right) = 1$.

For $d = 2$, there is a second FD operator for the laplacian given by (58)

$$2\mathbf{D}_2^2 f(x, y) = f(x + h, y + h) + f(x + h, y - h) \tag{353}$$
$$+ f(x - h, y + h) + f(x - h, y - h) - 4f(x, y).$$

Defining

$$\mathbf{D}_2^2 e^{i\mathbf{k}\bullet\mathbf{x}} = -2D_2^2 e^{i\mathbf{k}\bullet\mathbf{x}}, \tag{354}$$

we find

$$D_2^2 = 1 - \cos(k_x h)\cos(k_y h), \tag{355}$$

which defines $D_1^2 = D_1^2(\mathbf{k})$.

For the weighted superposition,

$$\mathbf{D}_\gamma^2 = \gamma \mathbf{D}_1^2 + (1 - \gamma)\,\mathbf{D}_1^2, \tag{356}$$

where $0 \leq \gamma \leq 1$ is the superposition weight, D_γ^2 is given by

$$\mathbf{D}_\gamma^2 e^{i\mathbf{k}\bullet\mathbf{x}} = -2D_\gamma^2 e^{i\mathbf{k}\bullet\mathbf{x}}, \tag{357}$$

and

$$D_\gamma^2(\mathbf{k}) = \gamma D_1^2(\mathbf{k}) + (1 - \gamma)\,D_2^2(\mathbf{k}). \tag{358}$$

We can determine $\max\left(D_\gamma^2\right)$ by solving for (k_x, k_y) such that

$$\frac{\partial}{\partial k_x} D_\gamma^2 = 0, \tag{359}$$

$$\frac{\partial}{\partial k_y} D_\gamma^2 = 0. \tag{360}$$

We find that

$$\max\left(D_\gamma^2\right) = 2, \qquad 0 \ \leq \gamma \leq 1/2$$
$$\tag{361}$$
$$\max\left(D_\gamma^2\right) = 4\gamma, \qquad 1/2 \leq \gamma \leq 1.$$

Up to now we have not specified the value of γ. For the NSFD algorithm we choose $\gamma = \gamma_0$, where γ_0 is given by (62). From its definition it is easy to show that $\max(\gamma_0) = 2/3$. Thus for

$$\mathbf{D}_0 = \gamma_0 \mathbf{D}_1^2 + (1 - \gamma_0)\,\mathbf{D}_1^2, \tag{362}$$

$$D_0^2(\mathbf{k}) = \gamma_0 D_1^2(\mathbf{k}) + (1 - \gamma_0)\,D_2^2(\mathbf{k}), \tag{363}$$

we have

$$\max\left(D_0^2\right) = \frac{8}{3}, \tag{364}$$

in two dimensions.

In three dimensions there are three FD operators for the lapalacian: $\mathbf{D}_1^2$ is given in (349), while $\mathbf{D}_2$ and $\mathbf{D}_3^2$ are defined in (72) and (73). Evaluating $D_2^2 = -2D_2^2 e^{i\mathbf{k}\bullet\mathbf{x}}$, and $D_3^2 = -2D_3^2 e^{i\mathbf{k}\bullet\mathbf{x}}$, we have

$$D_2^2 = 1 - \cos(k_x h)\cos(k_y h)\cos(k_z h), \tag{365}$$

$$2D_3^2 = 3 - \cos(k_x h)\cos(k_y h) - \cos(k_x h)\cos(k_z h) \tag{366}$$
$$- \cos(k_y h)\cos(k_z h).$$

Letting $0 \le \gamma \le 1$ and $0 \le \eta \le 1$ be, as yet, unspecified parameters, in analogy to (74) let us define

$$\begin{aligned} \gamma_1 &= \eta(1-\gamma) + (2\gamma_0 - 1), \\ \gamma_2 &= \eta(1-\gamma) \\ \gamma_3 &= 1 - \gamma_1 - \gamma_2, \end{aligned} \tag{367}$$

and

$$\mathbf{D}_{\gamma,\eta}^2 = \gamma_1 \mathbf{D}_1^2 + \gamma_2 \mathbf{D}_2^2 + (1 - \gamma_1 - \gamma_2)\,\mathbf{D}_3^2. \tag{368}$$

We define

$$\mathbf{D}_{\gamma,\eta}^2 e^{i\mathbf{k}\bullet\mathbf{x}} = -2D_{\gamma,\eta}^2 e^{i\mathbf{k}\bullet\mathbf{x}}, \tag{369}$$

and

$$D_{\gamma,\eta}^2(\mathbf{k}) = \gamma_1(\gamma,\eta)D_1^2(\mathbf{k}) + \gamma_2(\gamma,\eta)D_2^2(\mathbf{k}) + \gamma_3(\gamma,\eta)D_3^2(\mathbf{k}). \tag{370}$$

We can determine $\max\left(D_{\gamma,\eta}^2\right)$ by solving for (k_x, k_y, k_z) such that

$$\frac{\partial}{\partial k_x}D_{\gamma,\eta}^2 = 0, \tag{371}$$

$$\frac{\partial}{\partial k_y}D_{\gamma,\eta}^2 = 0, \tag{372}$$

$$\frac{\partial}{\partial k_z}D_{\gamma,\eta}^2 = 0. \tag{373}$$

We find that $\max\left(D_{\gamma,\eta}^2\right) = m(\gamma,\eta)$, where

$$m(\gamma,\eta) = 2\eta(2\gamma + 1) + 6(1 - \eta)(2\gamma - 1). \tag{374}$$

In accord with the Nyquist sampling theorem (262), we must always take $\lambda/h \ge 2\sqrt{3} \Rightarrow 0 \le kh \le \pi/\sqrt{3}$. Over this range of k, γ_0, and η_0 lie in the ranges R_γ, and R_η, respectively, where

$$R_\gamma : 0.6270 \cong \gamma_0(\pi/\sqrt{3}) \le \gamma \le 2/3 = \gamma_0(0), \tag{375}$$

$$R_\eta : 0.4345 \cong \eta_0(\pi/\sqrt{3}) \le \eta \le 2/5 = \eta_0(0). \tag{376}$$

On $R_\gamma \times R_\eta$, we find that

$$\max\left(m(\gamma, \eta)\right) = m\left(2/3, \eta(\pi/\sqrt{3})\right), \tag{377}$$
$$\cong 3.158.$$

Thus in the stabilty analyses we take

$$\max\left(D_0^2\right) = m\left(2/3, \eta(\pi/\sqrt{3})\right), \tag{378}$$
$$\cong 3.158.$$

7. Numerical Models

7.1. *Introduction*

The high accuracy algorithms that we have introduced are of little use by themselves. As we have already seen in Sect. 5, good results cannot be obtained with the NS-FDTD or NS-Yee algorithms unless the computational boundary is properly terminated. In many cases the accuracy of the solution is not limited by the that of NS-FDTD algorithms, but by the accuracy of the numerical model of the problem.

Let us try to compute the reflection coefficient for a plane wave incident on a dielectric slab of width $w = 2h$, and refractive index $n = 2$ in one dimension. Let the incident wavelength be $\lambda_0 = 8h$, and assume that the slab is immersed in vacuum. The wavelength inside the slab is $\lambda_1 = \lambda_0/n = 4h$. Several practical problems now arise.

7.2. *Scatterer Model*

Shape Representation

The slab must be represented on the numerical grid. Two possible models are shown in Fig. 5. Which is correct? Let us define a "scatterer function," S for which

$$S(x) = 1, \quad x \in \text{scatterer}, \tag{379}$$
$$S(x) = 0, \quad x \notin \text{scatterer}.$$

Integrating S, over the scatterer interior and the immediate exterior region, we have

$$I_S = \frac{1}{h} \int S(x)dx. \tag{380}$$

It is logical to expect that $I_S = \bar{w}$, where $\bar{w} = w/h$. Approximating the integral by a sum, Σ_S, we have

$$\int S(x)\,dx \cong \sum_{\bar{x}} S_{\bar{x}} \qquad (381)$$

where $\bar{x} = x/h$, $S_{\bar{x}} = S(x)$, and the summation is carried out over all a region that includes all interior grid points and the neighboring points immediately outside the scatterer. As 5 shows, $\Sigma_S = 3$ in model A, and $\Sigma_S = 2$ in model B.

We can directly verify that model B is correct because the solution to this particular problem is known. The reflected amplitude is zero. The analytic solution, found by solving the one-dimensional wave equation subject to the boundary conditions of (125) and (126), is given in [6]. Intuitively this result can be understood as destructive interference between the wave which passes the first face of the slab, and the wave reflected in the slab interior at the second face. In Fig. 6 we compare the reflected amplitude from models A and B using the NS-FDTD algorithm. Comparing Fig. 7 with model B in Fig. 6, we see that the NS-FDTD algorithm is much more accurate than the S-FDTD one.

We have taken the slab width to be an integral multiple of h. What if $\bar{w}$ is not an integer, and the edge of the slab lies between grid points? Let us make make an imaginary subgrid including the slab and its immediate exterior, as shown in as shown in Fig. 8. Let x_1 be some point on the grid. Using the subgrid we can estimate what fraction of the neighborhood, $x_1 - h/2 \le x \le x_1 + h/2$, lies inside the slab. If fraction f lies inside, we set $S(x_1) = f$. We call this the "fuzzy" representation in analogy with the concept of a fuzzy logical variable which takes values between 0 and 1. The model in $S = 0$ and 1, only, is called the "staircase" representation after its appearance in Fig. 9. The fuzzy representation can be extended to two and three dimensions to model objects of arbitrary size, shape and position on the numerical grid, as shown in Fig. 9.

Size and Scale

In FDTD calculations, as we have seen, the error of the solution depends on the ratio of the wavelength to the space step, λ/h. The ratio of the size, s, of the scatterer to λ is equally important. The geometrical optics regime is defined by $s \gg \lambda$. In the geometrical optics regime, the wave properties of electromagnetic radiation can be neglected, and there is no need to use the FDTD method. The Mie regime is defined by $s \sim \lambda$. In the Mie regime diffraction and interference effects are important, and a reasonable representation of the scatterer shape is essential. On the other hand in the Rayleigh regime, defined by $s \ll \lambda$, the shape representation of the scatterer is less critical, so long as the integral of refractive index over the scatterer volume is given correctly.

Thus in the numerical model it is best to express all sizes and distances in terms of the as ratios of the wavelength. It is not the absolute size that is important, rather it is s/r that determines the nature of the solutions.

Refractive Index

Suppose that a scatterer of refractive n_1 is immersed in a medium of refractive index n_0. This is exactly equivalent to a scatterer of refractive index of $n = n_1/n_0$ immersed in a medium of refractive index 1. It does not matter which refractive index is larger, and $n_1 < n_0 \Rightarrow 0 < n < 1$. There are thus two classes of problem, $n > 1$, and $0 < n < 1$.

For $n > 1$, the values of n that can be represented on the numerical grid is constrained by the Nyquist limit (262). In d dimensions, let the the wavelength outside of the scatterer be λ_0, while inside it is $\lambda_1 = \lambda_0/n < \lambda_0$. The Nyquist limit implies the constraint,

$$n \le \frac{\lambda_0/h}{2\sqrt{d}}. \tag{382}$$

Thus at a given spatial discretization, there is a maximum value of n that can be represented on the grid. Conversely, for a given value of n, we must set

$$\frac{\lambda_0}{h} \ge 2n\sqrt{d}. \tag{383}$$

For the case $n < 1 \Rightarrow \lambda_0 < \lambda_1$, if the Nyquist limit is satisfied inside the scatterer, it always satisfied outside. On the other hand the stability constraint (296) must be satisfied both inside and outside, which implies the constraint,

$$\frac{T}{\Delta t} \ge \frac{\lambda_0/h}{c}, \tag{384}$$

where c is given by eqs. (298)-(300) or eqs. (304)-(306). To represent vacuum holes embedded in a dielectric of refractive index n, the larger n, the greater $T/\Delta t$ must be.

Magnetic Permeability and Electric Permittivity

For most materials, it is an excellent approximation to assume that the magnetic permeability μ is equal to the magnetic permeability, μ_0, of the vacuum. From (119) we have

$$\frac{v_0^2}{n^2} = \frac{1}{\mu_0 \varepsilon}, \tag{385}$$

where v_0^2 is the vacuum propagation velocity. Thus we have

$$\varepsilon = n^2, \qquad \text{for} \quad \mu = \mu_0. \tag{386}$$

In this case the remarks about the representation of refractive index can be extended to ε.

Taking $v_0 = c$, equation (385) implies that thus $\mu_0 = 1/c^2$. If the computational domain is filled with material of refractive index $n_0 > 1$, we have

$$\mu_0 = \frac{n_0^2}{c^2}. \tag{387}$$

Conductivity

In the Yee algorithm (109) or (176) the important quantity is $\alpha \Delta t$, where $\alpha = \sigma/2\varepsilon$. For the absorbing wave equation to have oscillatory solutions, we must have $\alpha \leq \omega$, thus it is sensible to express $\alpha \Delta t$ as a fraction of $\omega \Delta t$.

7.3. *Sources and Scattered Fields*

Sources in the Wave Equation and Maxwell's Equations

Given two initial wavefields, $\psi(\mathbf{x}, 0)$, and $\psi(\mathbf{x}, \Delta t)$, ψ can be found at all subsequent times by iterating the FDTD algorithm. Sometimes initial values of $\psi(\mathbf{x}, 0)$ and $\psi(\mathbf{x}, \Delta t)$ are not known, or it is inconvenient to insert them into the computational domain. For example, in Figs. 3 and 4 initial pulses have been inserted into the computational domain. The longer the pulse the more space that is required.

Initial fields can be generated by a source, $s(\mathbf{x}, t)$, which specifies the location of sources and their time dependence. The wave equation (98) with a source is

$$\left(\partial_t^2 - v^2 \nabla^2 + 2\alpha \partial_t\right) \psi(\mathbf{x}, t) = s(\mathbf{x}, t). \tag{388}$$

The wave equation FDTD algorithm becomes

$$\psi(\mathbf{x}, t + \Delta t) = \frac{1}{1+a} \left[-\left(1-a\right)\psi(\mathbf{x}, t - \Delta t) + \left(2 + u^2 \mathbf{D}^2\right)\psi(\mathbf{x}, t)\right] \tag{389}$$

$$+ \Delta t s(\mathbf{x}, t),$$

where following (309) we have left $\mathbf{D}^2$ and u^2 unspecified (SFD or NSFD version). The source turns on at $t = 0$, thus $s(\mathbf{x}, t \leq 0) = 0$, and we take the initial fields to be $\psi(\mathbf{x}, -\Delta t) = 0$, and $\psi(\mathbf{x}, 0) = 0$. These initial fields satisfy the boundary conditions, and so do all subsequent fields.

Electromagnetic fields are generated by current sources, $\mathbf{J}(\mathbf{x}, t)$. Maxwell's equations (103) and (104) with a source current become

$$\mu \partial_t \mathbf{H}(\mathbf{x}, t) = -\nabla \times \mathbf{E}(\mathbf{x}, t), \tag{390}$$

$$\varepsilon \partial_t \mathbf{E}(\mathbf{x}, t) = \nabla \times \mathbf{H}(\mathbf{x}, t) - \sigma \mathbf{E}(\mathbf{x}, t) - \mathbf{J}(\mathbf{x}, t). \tag{391}$$

The Yee algorithm, (175) and (176), becomes

$$\mathbf{H}(\mathbf{x}, t + \Delta t/2) = \mathbf{H}(\mathbf{x}, t - \Delta t/2) - u\sqrt{\frac{\varepsilon}{\mu}} \mathbf{D} \times \mathbf{E}(\mathbf{x}, t) \tag{392}$$

$$\mathbf{E}(\mathbf{x}, t + \Delta t) = \left(\frac{1 - \sigma'}{1 + \sigma'}\right) \mathbf{E}(\mathbf{x}, t) \tag{393}$$

$$+ u\sqrt{\frac{\mu}{\varepsilon}} \left(\frac{1}{1 + \sigma'}\right) \mathbf{D}_1 \times \mathbf{H}(\mathbf{x}, t + \Delta t/2) - \left(\frac{\Delta t}{\varepsilon}\right) \mathbf{J}(\mathbf{x}, t).$$

In the S-Yee algorithm $\mathbf{D} = \mathbf{D}_1$, $u = \Delta t/h\sqrt{\varepsilon\mu}$, and $\sigma' = \sigma\Delta t/2\varepsilon$. In the NS-Yee algorithm $\mathbf{D} = \mathbf{D}_0$ in (392), $\mathbf{D} = \mathbf{D}_1$ in (393), $u = u_0$ as given by (95), and $\sigma' = \tanh(\sigma\Delta t/2\varepsilon)$. Again we specify that $\mathbf{J}(\mathbf{x}, t \leq 0) = 0$, and take $\mathbf{H}(\mathbf{x}, -\Delta t/2) = \mathbf{E}(\mathbf{x}, 0) = 0$. The initial values of the electromagnetic fields vanish and satisfy the boundary conditions as do all the subsequently computed fields.

Computing Scattered Fields

At the beginning of this section we discussed computing the reflection coefficient of a slab. The incident wave field could be generated with a source positioned in front of the slab, but the reflected and the incident fields interfere and it is impossible to distinguish the reflected field alone.

Let an incident field ψ_0 scatter off an object of refractive index n, and absorption α. In the medium surrounding the scatterer, $n = 1$, $\alpha = 0$, and $v = v_0$. Within the scatterer, $v = v_0/n$. The total field, ψ, can be decomposed into the sum of the incident field, ψ_0, and the scattered field, ψ_s, thus

$$\psi = \psi_0 + \psi_s. \tag{394}$$

Setting the source term to zero in (388), ψ and ψ_0 are given by

$$\left(\partial_t^2 - \frac{v_0^2}{n(\mathbf{x})^2}\nabla^2 + 2\alpha(\mathbf{x})\partial_t\right)\psi(\mathbf{x}, t) = 0, \tag{395}$$

$$\left(\partial_t^2 - v_0^2\nabla^2\right)\psi(\mathbf{x}, t) = 0, \tag{396}$$

where outside of the scatterer $n = 1$ and $\alpha = 0$. By definition the incident field "sees" only vacuum, so $n = 1$ and $\alpha = 0$ in (396). Subtracting (396) from (395), we have

$$\left(\partial_t^2 - \frac{v_0^2}{n(\mathbf{x})^2} \nabla^2 + 2\alpha(\mathbf{x})\partial_t \right) \psi_s(\mathbf{x}, t) = s_0(\mathbf{x}, t), \tag{397}$$

where

$$s_0(\mathbf{x}, t) = - \left[2\alpha(\mathbf{x})\partial_t + \left(\frac{n(\mathbf{x})^2 - 1}{n(\mathbf{x})^2} \right) \partial_t^2 \right] \psi_0(\mathbf{x}, t), \tag{398}$$

is a source term that arises from the action of the incident field on the scatterer. Here we have used the fact that $\nabla^2 \psi_0 = \partial_t^2 \psi_0$. Outside the scatterer, $n = 1$ and the source term vanishes. The scattered field thus obeys the wave equation with a source (388), where the source is given by (398). The initial fields are taken to be $\psi_s(\mathbf{x}, -\Delta t) = \psi_s(\mathbf{x}, 0) = 0$, and $s_0(\mathbf{x}, t \leq 0) = 0$.

Next consider a scatterer of conductivity σ and electric permittivity ε, immersed in a non-conducting medium where $\varepsilon = 1$. The electromagnetic fields can be decomposed into the sum of the incident and scattered fields, in the form

$$\mathbf{H} = \mathbf{H}_0 + \mathbf{H}_s, \tag{399}$$

$$\mathbf{E} = \mathbf{H}_0 + \mathbf{E}_s. \tag{400}$$

Assuming μ to be everywhere constant, the Maxwell equation for the **H**-fields are

$$\mu\partial_t \mathbf{H}(\mathbf{x}, t) = -\nabla \times \mathbf{E}(\mathbf{x}, t), \tag{401}$$

$$\mu\partial_t \mathbf{H}_0(\mathbf{x}, t) = -\nabla \times \mathbf{E}_0(\mathbf{x}, t). \tag{402}$$

Subtracting of the second equation from the first yields the Maxwell equation for the scattered **H**-field,

$$\mu\partial_t \mathbf{H}_s(\mathbf{x}, t) = -\nabla \times \mathbf{E}_s(\mathbf{x}, t), \tag{403}$$

which is exactly the same as the equation obeyed by the total **H**-field. For the **E**-field we have,

$$\varepsilon(\mathbf{x})\partial_t \mathbf{E}(\mathbf{x}, t) = \nabla \times \mathbf{H}(\mathbf{x}, t) - \sigma(\mathbf{x})\mathbf{E}(\mathbf{x}, t) \tag{404}$$

$$\partial_t \mathbf{E}_0(\mathbf{x}, t) = \nabla \times \mathbf{H}_0(\mathbf{x}, t). \tag{405}$$

Subtracting the second equation from the first yields the Maxwell equation for the scattered **E**-field,

$$\varepsilon(\mathbf{x})\partial_t \mathbf{E}_s(\mathbf{x}, t) = \nabla \times \mathbf{H}_s(\mathbf{x}, t) - \sigma(\mathbf{x})\mathbf{E}_s(\mathbf{x}, t) - \mathbf{J}_0(\mathbf{x}, t), \tag{406}$$

where

$$\mathbf{J}_0(\mathbf{x}, t) = \left[(\varepsilon(\mathbf{x}) - 1)\partial_t + \sigma(\mathbf{x})\right]\mathbf{E}_0(\mathbf{x}, t), \qquad (407)$$

is a current source term that arises from the interaction of the incident electromagnetic field on the scatterer. Outside the scatterer $\mathbf{J}_0$ vanishes. The scattered electromagnetic fields can now be computed from the Yee algorithm (392) and (393) with the source current given by $\mathbf{J}_0(\mathbf{x}, t)$. Again we take $\mathbf{J}_0(\mathbf{x}, t \leq 0) = 0$, and $\mathbf{H}(\mathbf{x}, -\Delta t/2) = \mathbf{E}(\mathbf{x}, 0) = 0$.

Time Dependence of Sources

Whenever a source turns on suddenly, it produces a broadband of transient signals that can intefere with wavefield we want to compute. For example,

$$s(t) = \sin(\omega t), \qquad t > 0, \qquad (408)$$
$$s(t) = 0, \qquad t < 0, \qquad (409)$$

goes from zero to one in just $1/4$ of a wave period. To suppress the production of transients, sources should be turned on slowly over several waveperiods. For example,

$$\bar{s}(t) = s(t)\left(1 - e^{-(t/\tau)^2}\right), \qquad t > 0, \qquad (410)$$
$$\bar{s}(t) = 0, \qquad t \leq 0. \qquad (411)$$

Similarly pulses should be turned on and off slowly. An example of such a pulse is shown in Fig. 3. In our algorithms we use pairs of sources of opposite phase, with a slow turn-on such as that of (410).

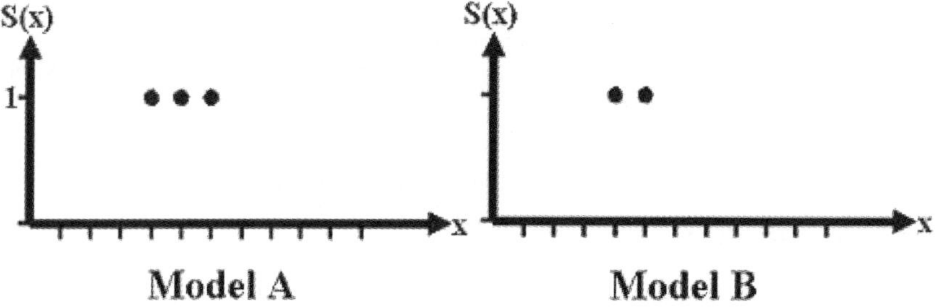

Fig. 5.   Models of a Slab on the Numerical Grid.
The grid spacing is h, and the slab width is $w = 2h$. Model B is the best one because $\Sigma_S = w/h$, whereas, $\Sigma_S = (w + 1)/h$ for A.

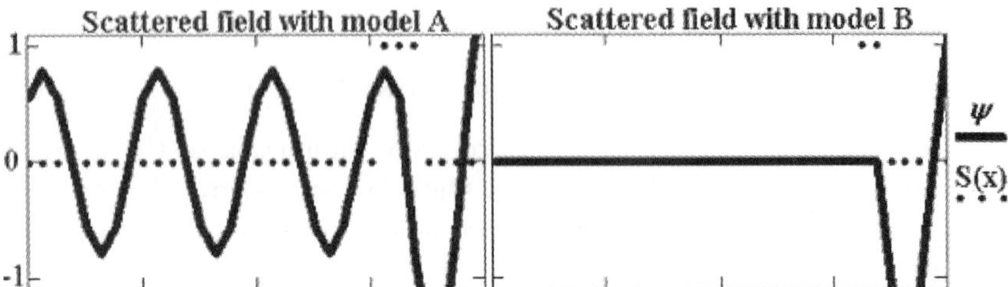

Fig. 6. **Scattering off a One-Dimensional Slab.**
Scattered field, ψ, due to an infinite plane wave incident from the left onto a slab of width $w = 2h$, and refractive index $n = 2$, wavelength $\lambda_0 = 8h$ using models A and B in 5 to represent the slab. NS-FDTD algorithm is used in both calculations. NS-FDTD agrees well with the analytic result (reflected field = 0) in model B.

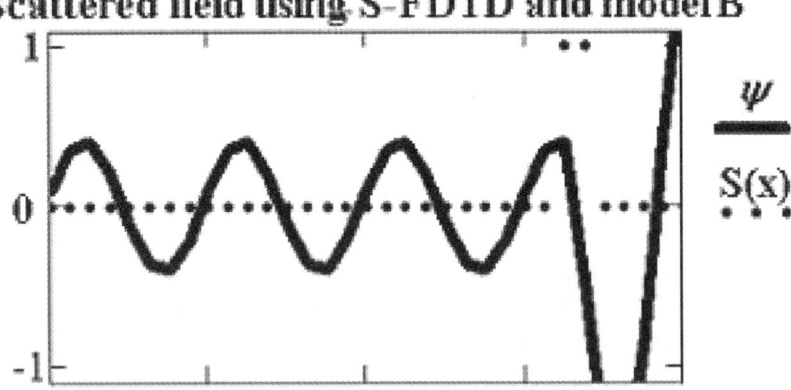

Fig. 7. **Scattering off a One-Dimensional Slab Using S-FDTD.**
Same slab used in 6, with model A of 5 using the S-FDTD algorithm. The error is very large compared to the NS-FDTD result.

8. Light Propagation in Subwavelength Structures*

8.1. *Introduction*

The interaction of light with structures having periodically modulated refractive index is an important subject of study in optics. In two dimensions planar optical gratings are fabricated by inducing periodic modulation of the electrical properties (dielectric constant or conductivity (absorption

---

*Contributed by Saswatee Banerjee, University of Tsukuba, Japan.

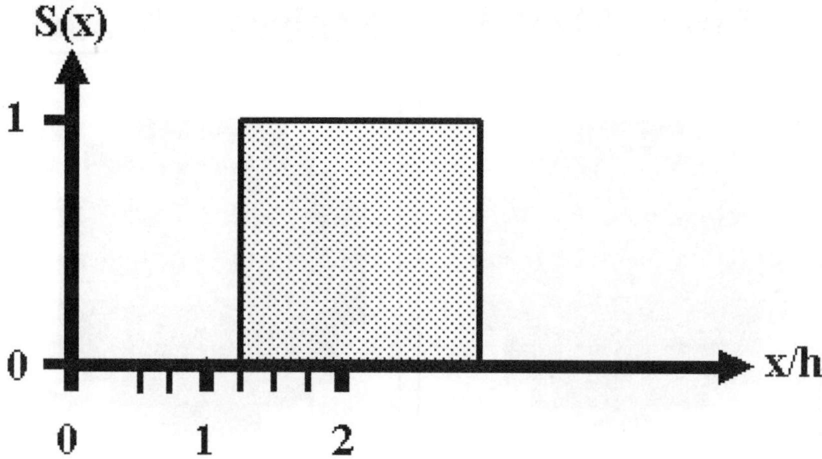

Fig. 8. **Fuzzy Model of a Slab.**
The grid spacing is h, but in neighborhood of the slab boundary, we make a subgrid of spacing $h/4$. The point $x_1/h = 1$ is outside the slab. But $1/4$ of the neighborhood $x_1 - h/2 \leq x \leq x_1 + h/2$ lies within it, so we set $S(x_1) = 1/4$. On the other hand $S(0) = 0$

coefficient) or combination of both). Gratings may also be surface-relief (corrugated) type with periodic variations in the surface of a dielectric or conducting material. Three dimensional grating structures are fabricated by periodically arranging or embedding macroscopic dielectrics within surrounding media. The periodic nature of the structure imparts photonic bandgaps, within which photons are forbidden to propagate in certain directions. Such grating structures display strong diffraction and dispersion effects.

Electromagnetic modeling and simulation play an important role in design and development of artificial optical materials and components. To compute light propagation in artificial media with wavelength-scale modulation (less than 5λ, λ being the wavelength of the incident radiation in vacuum) it is necessary to solve the full form of Maxwell's equation with suitable boundary conditions. Analytical solution for finite diffractive optical structures with arbitrary and sub-wavelength features do not exist.

In this section, we use NS-FDTD methods to solve some example problems from our research projects: (1) biological optical structures which produce strong color effects (structural color), (2) subwavelength structured interfaces to couple light from a high- to a low-refractive index medium,

Fuzzy Model   Staircase Model

Fig. 9. **Fuzzy Model of a Circle.**
Using a subgridding of $h/16$ we compute the fuzzy values of $S(x,y)$ $(0 \leq S \leq 1)$ for a circle of radius $6.48h$, and centered at $(31.63, 31.18)h$ on the grid. The fuzzy model is compared with the staircase model $(S = 0 \text{ or } 1)$.

and (3) conducting diffraction gratings.

We begin by casting the conducting Maxwell's equations (103) and (104) into a form more suitable for our purposes. Assuming the time dependence of the electromagnetic fields to be $e^{-i\omega t}$, Maxwell's equations become

$$-i\omega\mu_0\mathbf{H} = -\nabla \times \mathbf{E}, \tag{412}$$

$$-i\omega\hat{\varepsilon}\mathbf{E} = -\nabla \times \mathbf{H}, \tag{413}$$

where $\hat{\varepsilon}$, the complex electric permittivity defined by

$$\hat{\varepsilon} = \varepsilon + i\frac{\sigma}{\omega}. \tag{414}$$

Eliminating $\mathbf{H}$ from equations (412) into (413), and assuming that $\nabla \bullet \mathbf{E} = 0$, one obtains the Helmholtz equation,

$$\left(\nabla^2 + \hat{k}^2\right)\mathbf{E} = 0, \tag{415}$$

where

$$\hat{k}^2 = \omega^2\mu_0\varepsilon + i\omega\mu_0\sigma. \tag{416}$$

The quantity $\hat{k}$ is the magnitude of the complex propagation vector. μ is set to be μ_0, the vacuum magnetic permeability. Taking analogy from the dielectric case (386) a complex refractive index can be defined as

$$\hat{n}^2 = \hat{\varepsilon}. \tag{417}$$

Comparing the real and imaginary parts of $\hat{n} = n_{\rm r} + in_{\rm i}$ with $\hat{\varepsilon}$ one finds

$$\varepsilon = n_{\rm r}^2 - n_{\rm i}^2, \tag{418}$$

$$\sigma = 2n_{\rm r}n_{\rm i}^2\omega. \tag{419}$$

These relationships hold only if ε and σ are real constants. In reality they are functions of frequency, but if the frequency of the incident light is not close to a plasma frequency of the material, ε and σ can be well approximated by real constants [23].

Although actual structures are three-dimensional, signicant observations can be made from two-dimensional simulations if the spatial variation of the electromagnetic fields in the third dimension is small compared to that in the other two. Light in any arbitrary polarization state can be decomposed into TE (129) and TM-polarizations (130). Thus the three-dimensional problem can be reduced to two.

In this section, the terms reflectivity and transmissivity are used to refer to the flux of reflected and transmitted Poynting vectors (Poynting vector gives the magnitude and propagation direction of electromagnetic energy) respectively, through the scatterer surface, expressed as percentage of flux of incident Poynting vector through the same surface. Diffraction efficiency refers to the flux of Poynting vector associated with a particular diffraction order, through a surface normal to it's direction of propagation. Diffraction efficiency is also expressed as the percentage of flux of incident Poynting vector through the scatterer surface.

8.2. *Structural Color of Butterfly Wings*

Some species of moths and butterflies have bright, iridescent wing colors, which result from the scales that coat their wings. The scales are flat plates of $\sim 100\mu$m in width, arranged in an overlapping roof tile pattern [24]. A scale's color, to a large extent, is of structural origin. The scales have intricate subwavelength stratification, voids, and complex groove shapes on the externally visible surfaces, and display multilayer interference, scattering, and diffraction effects [25] - [28]. The optical properties of these micro-scale structures depend on the minimum feature size, the periodicity, and the type of arrangement, and also depend strongly on the wavelength, polarization state, the incidence angle, the shape of the illuminating wavefront, and the viewing angle.

In this section, the NS-FDTD simulation of a certain butterfly wing-scale microstructure is presented. The scattered field form of Maxwell's equations (397) and (406) are solved in two-dimensions. Following ref. [29] the complex refractive index ($\hat{n}$) of the material of the scale is taken to be $\hat{n} = 1.56 + i0.06$. The material is a transparent, polymer. From eqs. (418) and (419) one finds that $\varepsilon = 2.43$, and that $\sigma = 0.187\,\omega$. From (120) we have $\alpha = 0.0385\,\omega$. The effective index of refraction is $n_{\mathrm{eff}} = \sqrt{\varepsilon} = 1.559$. Since the exact nature of frequency dependence of ε and σ is unknown, these quantities are assumed to remain unchanged over the range of wavelengths investigated.

Figure 10(A) shows the scanning electron microscope (SEM) image of wing-scales of the Morpho Didius butterflies (ref. [29]). Two layers called the "ground scales," and the "glass scales," give rise to the butterfly's color. The transmission electron micrograph of Fig. 10(B) shows a transverse cross-section containing both layers of scales. The upper surface has quasi-periodic "ridges" crossed by "lamellae" at right angles. The "lamellae" are the (nearly) horizontal dark lines in Fig. 10(B), while the ridges are the more or less vertical dark lines. The cross-sectional structure of any one of the scales can be imagined to be formed by repeating tree-like microstructures in space.

Figure 11 shows a computer-generated model of the wing structure, and shows the geometry for the simulation, including the actual sizes of various structural features. This model abstracts the essential features of the transverse cross-section of a scale from the wing of Morpho Didius. The base of the scale is represented by a simple bar. A "shape factor," f, is defined as the ratio of the width of the top lamella to the bottom one. In (A) the tree-like structures are not tapered so $f = 1$, but in (B) $f = 0.349$. In this calculation simple staircase approximation is adopted for structure representation on the grid. Hence the limits on practical choice of f comes from the value of space discretization (λ/h). Figure 10(A) depicts the approximate directions of propagation for the incident, reflected and transmitted waves for an arbitrary angle of incidence θ. It also indicates the sign conventions adopted for angle measurement. All angles are measured from the normal. An angle is positive if it is taken clockwise and is negative otherwise. The plane of the paper serves as the x-y plane and the quasi-periodic ridges run parallel to the z-axis. In TE mode, electric field (E_z)

is parallel to the ridges and in TM mode, electric field is perpendicular to ridges. Figure 11(B) gives the physical sizes used in these calculations. The ridge-spacing, $0.83\mu m$, is same as that of a ground scale of Morpho Didius butterfly [29], the other feature sizes are chosen following ref. [30].

The "offset" (difference in vertical position) between the left and right lamellae on the same ridge is 0.04 μm. Taking this to be the minimum feature size, the maximum wavelength to be 0.7 μm (red light), and allowing one grid spacing, h to represent this minimum feature size on the grid, one finds that $\lambda/h \geq 11$. Taking $\lambda/h = 11$, the width, w of the smallest lamella is given by $2h \leq w \leq 4h$ (for the tapered structure) in the wavelength range $0.3 \leq \lambda \leq 0.7$ μm. The sizes of features on the grid are thus scaled to the wavelength, and the size of the computational domain is ultimately determined by the wavelength.

Computing Reflection and Transmission Spectrum

The results presented in this section are calculated using the model of 11. Figure 12 shows the reflected intensity distribution due to a plane wave of width 20λ incident from the left. The shape factor here is $f = 1$ (no taper), $\lambda/h = 11$ and the FDTD is iterated for 20 waveperiods. To facilitate better space management, Figs. 12(A) and (B) are rotated by 90 degrees. The computational domain size is $N_x h \times N_y h$, where $N_x = 32\lambda/h$, and $N_y = 20\lambda/h$. The structure is terminated 6λ from each computational boundary in the x-direction leaving 20λ to represent it on the grid in the x-direction, or about 12 ridge-spacings (average spacing between two consecutive tree-like structures). Thus the beam width is the same as the total extent of the structure in x direction. The base of the structure is placed 2λ from the nearest computational boundary (along y axis) leaving a space of 14λ to visualize the reflected electromagnetic fields. The intensity visualization is done by normalizing the scattered intensities from zero to one and by mapping them into 256 shades of grey.

In order to study the dependence of reflected energy on incidence and viewing angles, the reflected intensity distribution is calculated for a series of different incidence angles in TM and TE modes. The simulation parameters (λ/h, size of computational space, total timesteps) are same as above. The results of these calculations are given in Fig. 13. The reflected intensity for incident beam of arbitrary polarization is calculated by taking average of the corresponding TM and TE intensity values. The angular spectrum of reflected intensity at any incidence angle is calculated by interpolating the intensity values along a half-circle of radius 13λ, centered at the mid-point of the base of the structure. From the figure it is clear that at high incidence angles more reflected light goes into grazing angles. Also, increasing the incidence angles decreases the amount of reflected light over all viewing angles.

Figures 14 and 15 plot the reflectivity and transmissivity spectra in TE and TM modes for $f = 1$ and $f = 0.349$ structures respectively. These plots are computed using $\lambda/h = 11$, timesteps $= 10$waveperiods, and a total grid size of $10\lambda \times 22\lambda$ (10λ along y -direction and 22λ along x direction). For the $f = 1$ structure the reflectivity of both polarizations (TE and TM) is highest in the wavelength range 0.42 μm $\leq \lambda \leq$ 0.55 μm, while transmissivity is the lowest. The maximum reflectivity of the TM mode is $\sim 70\%$ at $\lambda = 0.488$ μm. Hence the $f = 1$ structure would appear to be blue-green in reflection, when irradiated normally with an infinite plane wave. On the other hand, for $f = 0.349$ structure, the overall reflectivity is low for both the modes and the reflectivity peaks are shifted more towards the blue end of the spectrum.

However, these calculations alone are not enough to determine the appearance of the butterfly at a far-field point (few thousands of optical wavelengths away). This is because the size of the computational domain in any direction is about few tens of wavelengths at the most. Hence the complex near-fields calculated from NS-FDTD simulations are used to compute the far-field intensities employing a transformation method described in [8]. Figures 16(A) and (B) show the angular dependence of the far-field reflected intensity for various wavelengths in the TE and TM polarizations respectively. These plots are calculated for $f = 1$ structure illuminated by parallel beam of light at normal incidence.

The calculations use the minimum space discretization ($\lambda/h = 11$), with just one grid unit to smallest feature. To check the accuracy of these results, the calculations of Fig. 14 are repeated with $\lambda/h = 20$. In order to reduce the computational burden, the space between the strucure and the computational domain boundaries are reduced to a minimum, scaling all the other sizes in accordance with $\lambda/h = 20$. This ensures that the overall width of the structure in both x and y directions remain the same as in Fig. 14 for all wavelengths. The reflectivities and transmissivities found this way are fairly close to the corresponding values given in Fig. 14 with the maximum difference being 5% at most.

This analysis uses a simplified model of the optical structure of a single wing-scale of the Morpho Didius butterfly. Effects of pigmentation is not taken into account, and the base of the structure which is actually quite complicated is represented by a simple bar. While a real butterfly wing-scale may contain tens of thousands of tree-like structures of Fig. 11, the simulation uses only 6-15 of them. Nevertheless, even with a simple staircase representation, results seem to be quite good even at $\lambda/h = 11$, and they are independent of such details as the size of computational space. Though the actual appearance of the butterfly to a human observer depends on details of the visual perception system (not part of this model), but the simulation seems to at least qualitatively describe structural color of the Morpho Didius butterfly wing.

8.3. *Improving Light Transmission Across Interfaces with Subwavelength Structures*

When light travels from a medium with high refractive index to a medium with low refractive index, a significant amount of incident light gets reflected from the interface back into the high index medium. The total transmitted output through such an interface is determined by the characteristic critical angle of total reflection for the two media. If a light ray is incident at an angle of θ_i to the normal to the interface from an optically dense medium (refractive index $n > 1$) and is transmitted into a lighter medium (refractive index one) at an angle of θ_t to the normal, then Snell's law of refraction is given by

$$\sin \theta_t = n \sin \theta_i. \tag{420}$$

Since $0 \le \sin \theta_t \le 1$, the maximum possible incidence angle (critical angle), θ_c, at which light can be transmitted through the interface is given by

$$\sin \theta_c = \frac{1}{n}. \tag{421}$$

For $\theta_i \ge \theta_c$, 100% of the light is reflected at the interface -this is called "total internal reflection." This phenomenon makes it difficult to extract light from devices such as light emitting diodes into the air.

For high index dielectric materials such as silicon ($n = 3.5$ at $\lambda = 1.55$ μm) the critical angle is as low as 16.4°. Such an interface transmits only about 12% of the total light flux incident upon it. Recently various groups [31]-[34], have tried to enhance the light transmission by using subwavelength grating structures on the interface. Without such measures, the luminous efficiency of most commercially available light emitting diodes is lower than that of fluorescent lamps.

However, the overall transmissivity of such subwavelength structured interfaces depend on different design parameters, such as, profile, groove depth, and spatial period of surface corrugation, wavelength of light and refractive index of the dielectric material. In this section, the overall transmissivity and light extraction effciency of interfaces corrugated with sub-wavelength rectangular and conical profiles are investigated. For this purpose, the NS-FDTD algorithms are used to solve Maxwell's equations with current source (equation (391)) and taking $\sigma = 0$, and $\mu = \mu_0$.

Let an infinite plane wave impinge on an infinite planar interface between two media at an angle θ_i to the normal. The plane of the paper is assumed to be the plane of incidence, which contains the propagation vectors of the incident and transmitted waves. The intersection of the plane of incidence and the interface is a straight line, ℓ on the plane of the paper. There are two possibilites: $\mathbf{E}$ perpendicular to ℓ defines the TE mode, while $\mathbf{E}$ parallel to ℓ defines the TM mode. Thus by solving the two-dimensional forms of Maxwell's equations the amplitude of transmitted wave can be determined in each polarization. Taking the incident amplitude of $\mathbf{E}$ to be E_i, and the transmitted amplitude to be E_t, the normalized amplitude of transmitted wave $A = E_t/E_i$ is given by the Fresnel formulas [23] for the two modes as

$$A_{\mathrm{TM}} = \frac{2n\cos\theta_i}{\cos\theta_i + n\cos\theta_t} \tag{422}$$

$$A_{\mathrm{TE}} = \frac{2n\cos\theta_i}{n\cos\theta_i + \cos\theta_t}. \tag{423}$$

The Fresnel formulas are derived by enforcing the boundary conditions (127) and (128). Notice that at $\theta_i = \theta_c$, the amplitude of transmitted wave does not vanish. There is an electromagnetic field on the other side of the interface, but it is a non-propagating evanescent mode and no energy is transmitted through the interface. Transmissivity, the ratio of the normal components of the incident and the transmitted Poynting vectors, is given by

$$T_{\mathrm{TM}} = \frac{\cos\theta_t}{n\cos\theta_i}\,|A_{\mathrm{TM}}|^2, \tag{424}$$

$$T_{\mathrm{TE}} = \frac{\cos\theta_t}{n\cos\theta_i}\,|A_{\mathrm{TE}}|^2.$$

In Fig. 17 T_{TM} and T_{TE} as obtained from equations (424) and (425) are plotted as functions of θ_i for $n = 3.5$. The figure compares the theoretically determined angular spectra of T_{TM} and T_{TE} with those obtained from NS-FDTD calculations for the same interface.

Figure 18 shows absolute map of incident and reflected intensities when parallel beam of light is incident on a flat interface at a large angle $(-43°)$.

The incident beam is polarized along y-axis and propagating along $-x$-axis. Incident light is generated inside the dense medium (refractive index $n = 2.48$) using two current source arrays with opposing phases and separated by two grid units. Almost all of the incident energy is reflected back into the high-index medium.

Figure 19 shows transmission through an interface structured with sub-wavelength rectangular profile for different incidence angles. Figure 19(A) shows absolute map of incident and reflected intensities when parallel beam of light is incident on the structured interface at a large angle $(-43°)$. Figure 19(B) presents the NS-FDTD computations of transmitted intensities for various incidence angles for the same interface. Refractive index of the optically dense medium is 2.48.

Figure 20 shows transmission through an interface structured with sub-wavelength conical groove profile for different incidence angles. Figure 20(A) shows absolute map of incident and reflected intensities when parallel beam of light is incident on the structured interface at a large angle $(-43°)$. Figure 20(B) presents the NS-FDTD computations of transmitted intensities for various incidence angles for the same interface. Refractive index of the optically dense medium is 3.5.

Figures 19 and 20 indicate that the subwavelength structured interfaces can couple incident light out to propagating transmitted modes in the optically lighter medium even at large incidence angles, higher than the critical angle. Light extraction efficiency for an interface is defined as the ratio of the normal components of transmitted Poynting vectors for that interface and a flat interface between the identical pair of media. For the rectangular grooves of Fig. 19 the average extraction efficiency of TM and TE modes is 2.51. For the conical grooves of Fig. 20 the extraction efficiency is 1.13. Hence the rectangular grooves of Fig. 19 are more effective in coupling light through the interface.

8.4. *Classical Diffraction by Conducting Grating*

Rigorous analytical methods [35]–[38] are applied to solve Maxwell's equations for gratings of infinite extent and simple, periodic groove profile. But for finite, aperiodic, arbitrary grating geometries one has to take recourse to numerical methods. In this section, the NS-FDTD methods are used to determine the classical diffraction characteristics of a highly conducting grating of finite width used in a geometry employing high angle of incidence

160   *J. B. Cole*

(60°). Classical diffraction is observed when the incident and diffracted (reflected, in this case) beams are in a plane normal to the grating grooves. This fact enables us to analyze the gratings in two dimensions.

The diffraction characteristics of the grating are calculated by solving scattered field forms of Maxwell's equations. The NS-FDTD results are compared with finite element method (FEM) results. The NS-FDTD is implemented with NS-Mur-ABC. However, to minimize reflections from the left and right boundaries at an angle of incidence of 60°, u_1 and u_2^2 in equation (231) are chosen as

$$u_1 = \frac{\tan\left(\omega_0/2\right)}{\tan\left(k/2\right)} \tag{425}$$

$$u_2^2 = \frac{2\left(\cos\left(\frac{k}{4}\right) - \cot\left(\frac{k}{2}\right)\sin\left(\frac{k}{4}\right)\right)\sin^2\left(\omega_0/2\right)}{\sin^2\left(k\frac{\sqrt{3}}{4}\right)\cos\left(\frac{k}{4}\right)}. \tag{426}$$

The grating profile is approximated by a sinusoidal curve of the form

$$y(x) = \left[1 + \cos\left(\frac{2\pi}{p}x\right)\right]\frac{d}{2}, \tag{427}$$

where the spatial period, p and the groove depth, d are 1.111 μm, and 0.5 μm respectively. The grating profile is discretized using staircase representation scheme. Wavelength of incident light in vacuum (λ) is 1.55μm. Bulk gold (refractive index $0.63 + i\,9.54$, at $\lambda = 1.55\mu$m) shows high reflectivity (95% in TM polarization and 99% in TE polarization). Hence such a material can be approximated by one with very high conductivity. For TE (electric field parallel to the grating grooves) and TM (electric field perpendicular to the grating grooves) modes, NS-FDTD for absorbing media (Sects. 3 and 4) is employed with $\sigma = 140\,\omega$ [39]. This value of σ satisfies (89).

Figures 21(A) and 22(A) are the scattered intensity plots for the TE and TM mode diffraction patterns respectively. The figures are calculated using a numerical grid of size $15\lambda \times 20\lambda$, $\lambda/h = 16$ and total timesteps equal to 20 waveperiods. Figures 21(C) and 22(C) show the angular spectra of the intensity of diffracted light corresponding to Figs. 21(A) and 22(A) respectively.

The diffraction efficiencies and diffraction angles for the two orders as obtained from FEM simulations are given in Table 1. The diffraction efficiencies and angles of Figs. 21 and 22 are not exactly the same as FEM ones. For more accurate calculation of diffraction pattern, complex near fields are calculated using larger values of λ/h on a small numerical grid. The far-field transformation methods are used to determine the far-field diffraction pattern. Figure 23 contain the far-field diffraction data for TE and TM modes.

Table 1.   Diffraction data obtained from FEM calculations

Order	FEM		
	TM		TE
	Angle	Efficiency	Efficiency
0th	-60^0	2.5%	83.9%
+1st	$+32^0$	92.7%	10.8%

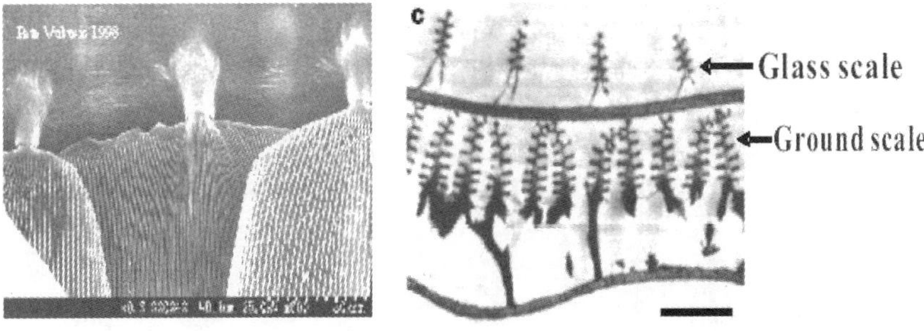

Fig. 10.   (A) Scanning electron micrograph of one ground scale (middle) and two glass scales (left and right) from the wings of the Morpho Didius butterfly. (B) Transmission electron micrograph of a transverse cross-section through the scales. The quasi-periodic array of tree-like microstructures in each of the two layers: glass scales (above) and ground scales (below) are shown. The combination of glass and ground layers give rise to strong diffraction effects causing the wing to appear diffusely colored. The length scale is indicated in the lower right of (B) by a dark line segment of length 1.3 μm. Taken from ref [29].

     J. B. Cole

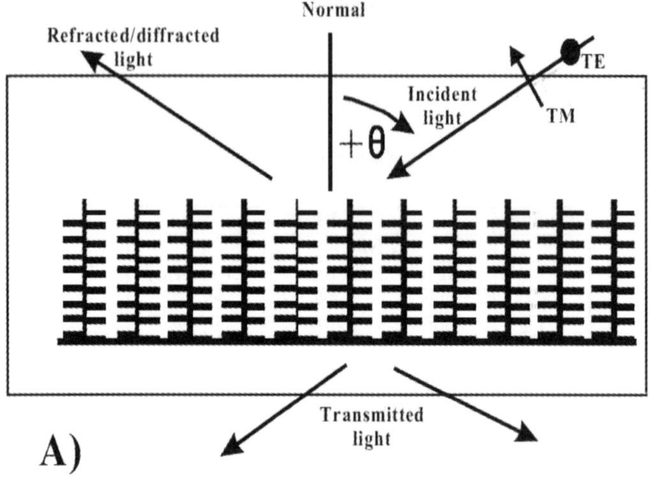

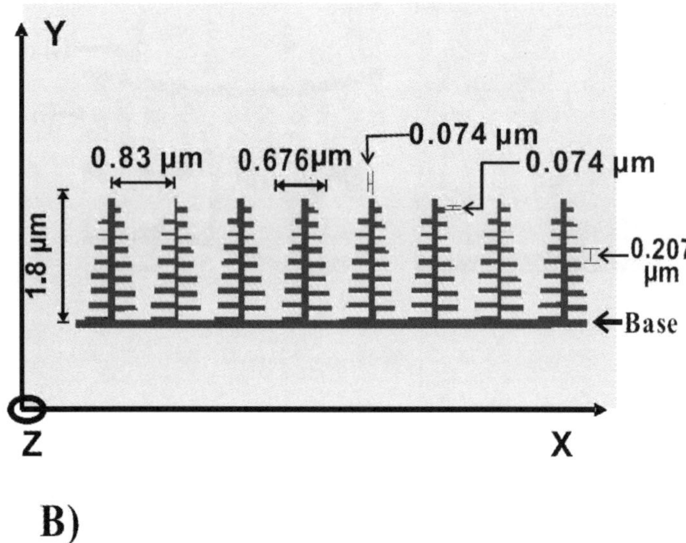

Fig. 11. Two-dimensional computer-generated model of the quasi-periodic structure of a butterfly wing-scale. (A) Non-tapered structure ($f = 1$) showing directions of incident, reflected and transmitted waves. Conventions for angle measurement and the directions of electric fields in each polarization states are indicated. (B) Tapered structure ($f = 0.349$). Actual sizes of various features are indicated. The offset between the left and right lamellae is 0.04μm

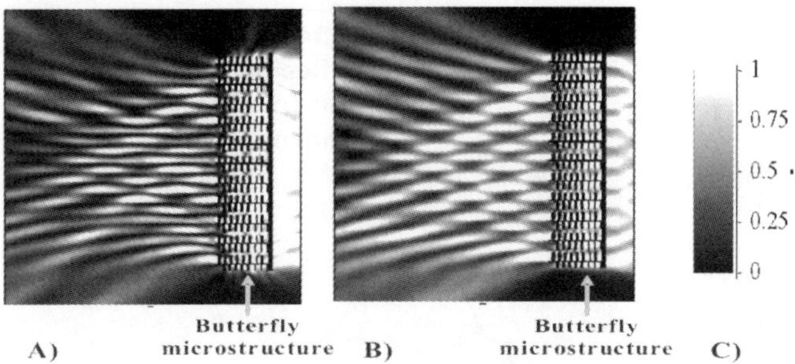

Fig. 12.   Scattered intensity due to a parallel beam of light normally incident (from the left) onto the butterfly structure (right) in the TE mode (A) and the TM mode (B). The incident light is not shown. The grey scale is shown in (C). Shape factor is $f = 1$. Wavelength of incident light (λ) is 0.5 μm.

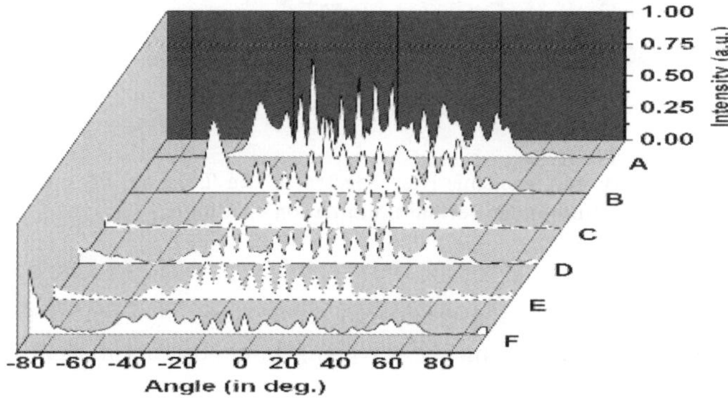

Fig. 13.   Reflected intensity versus viewing angle, plotted for different angles of incidence. The reflected intensity is plotted along the vertical axis and corresponding viewing angles are plotted along the horizontal axis. A: angle of incidence = 10°, B: angle of incidence = 20°, C: angle of incidence = 30°, D: angle of incidence = 40°, E: angle of incidence = 50°, F: angle of incidence = 60°. Calculated using incident light of random polarization, $f = 1$ structure, λ=0.5 μm.

9. FDTD Algorithms for Electromagnetic Fields with Surface Plasmons*

9.1. *Introduction*

With the rise of nanotechnology the optical properties of nanostructures is a research topic of great interest. The properties of nanostructures differ

---

*Contributed by Michael I. Haftel Naval Research Laboratory, USA

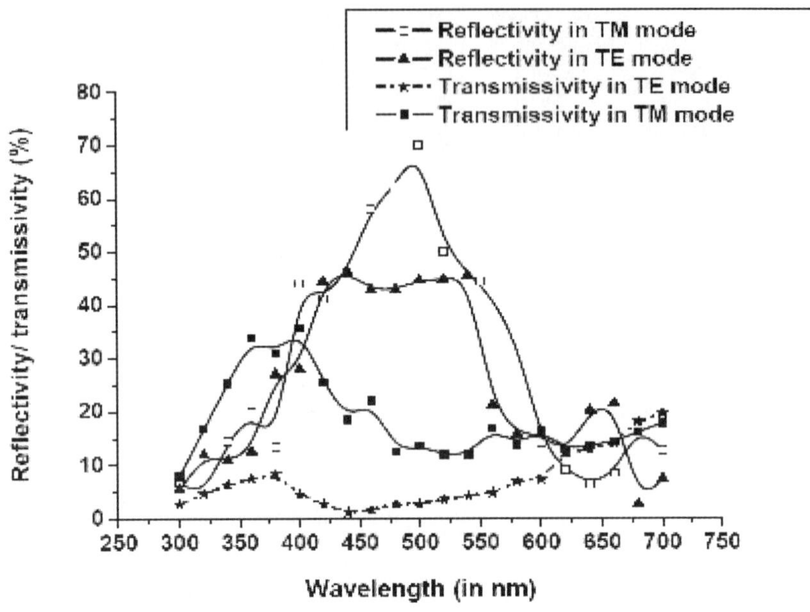

Fig. 14.　Reflectivity and transmissivity spectra in the TM and TE modes for non-tapered ($f = 1$) structure under normal illumination condition.

markedly from the bulk material, and since they have features that are comparable to or smaller than the wavelength, nanostructures can influence light propagation in unusual ways.

In the rapidly developing field of plasmonics the presence of surface plasmons (oscillating electric fields propagating longitudinally along the surface of metals) [16], has been associated with the extraordinary transmission of light through nanometer-size gratings and apertures [17] and the two-dimensional localization [18] and guiding [19] of optical fields. Moreover, there has recently been much interest in using surface plasmons to manipulate light propagation in optical devices. Hence there is great interest in using the FDTD method to compute electromagnetic fields when surface plasmons are present.

Well established solid-state physics theory tells us that the real part of the dielectric constant of a metal becomes zero when the frequency of light is equal to the plasma frequency of the material [20]. The plasma frequency is determined by the free electron density. It is precisely at the plasma frequency that Maxwell's equations allow longitudinally propagating electric

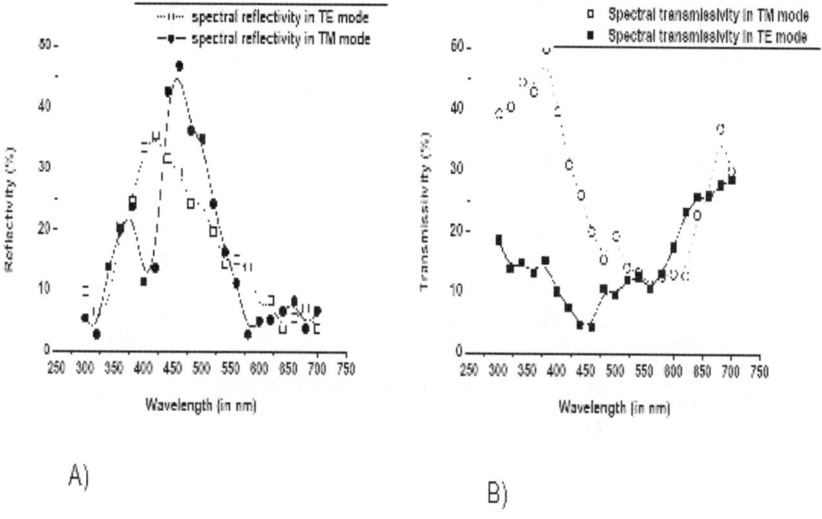

Fig. 15.   Reflectivity and transmissivity spectra in the TM and TE modes for tapered ($f = 0.349$) structure under normal illumination condition.

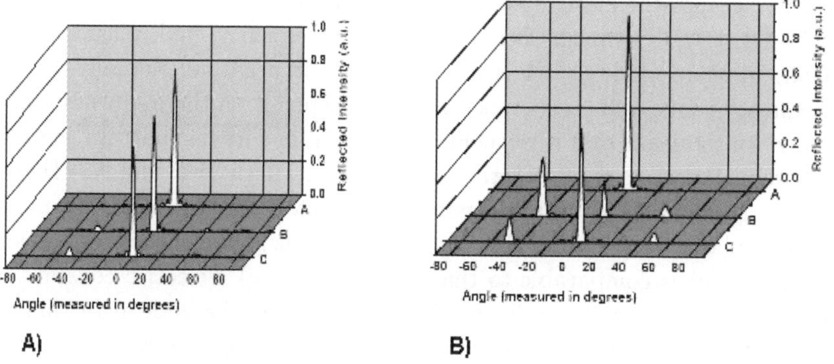

Fig. 16.   (A) Far-field reflection spectra for the TE mode at normal incidence. Normalized reflected intensity is plotted as a function of viewing angle for different wavelengths. A: $\lambda = 0.488$ μm, B: $\lambda = 0.55$ μm, C: $\lambda = 0.62$ μm, (B)Far-field reflection spectra for the TM mode at normal incidence. Normalized reflected intensity is plotted as a function of viewing angle for different wavelengths. A: $\lambda = 0.488$ μm, B: $\lambda = 0.55$ μm, C: $\lambda = 0.62$ μm.

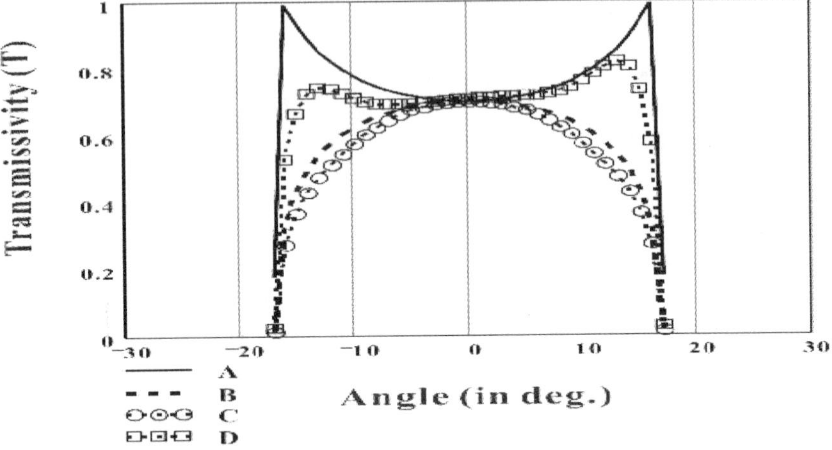

Fig. 17.   Transmissivity through an interface from an optically dense to a lighter medium as a function of incidence angles. A: TM mode transmissivity calculated using Fresnel formulas, B: TE mode transmissivity calculated using Fresnel formulas, C: TE mode transmissivity calculated using NS-FDTD, D: TM mode transmissivity calculated using NS-FDTD, $n = 3.5$ for all the plots and angles of incidence are measured on both sides of the normal to the interface..

fields, because the electric field displacement ($\mathbf{D} = \varepsilon\mathbf{E}$) and its divergence ($\nabla \bullet \mathbf{D}$), trivially vanish. Below the plasma frequency, bulk plasmons appear. In this regime $\mu \cong 1$ is positive, but ε is negative. Actually, in the frequency range of interest, ε is dispersive (depends on the frequency), and lossy, which means that it is a complex number, and its real part is negative. As we have seen in Sect. 6.4, the Yee algorithm (both SFD and NSFD versions), is numerically unstable, when ε is negative. For problems involving surface plasmons (SP), the absolute value of the negative dielectric constant of the metal is comparable to that of the (positive) dielectric constant of the dielectric material interfacing the metal. Thus in the same algorithm we need to handle both positive and negative ε. In this section we will present and assess different methods for avoiding numerical instability in this case.

Kunz and Luebbers [21] describe the recursive convolution method (RC) for the FDTD calculations in lossy, dispersive media. Starting with a convolution integral expression for the displacement they derive a FDTD scheme to update $\mathbf{E}$ and $\mathbf{H}$ by using an "accumulation variable," which is linear in $\mathbf{E}$ and is also updated.

An alternate method has been more recently been proposed by Gray and Kupka [22]. Taking the time dependence of the electromagnetic fields

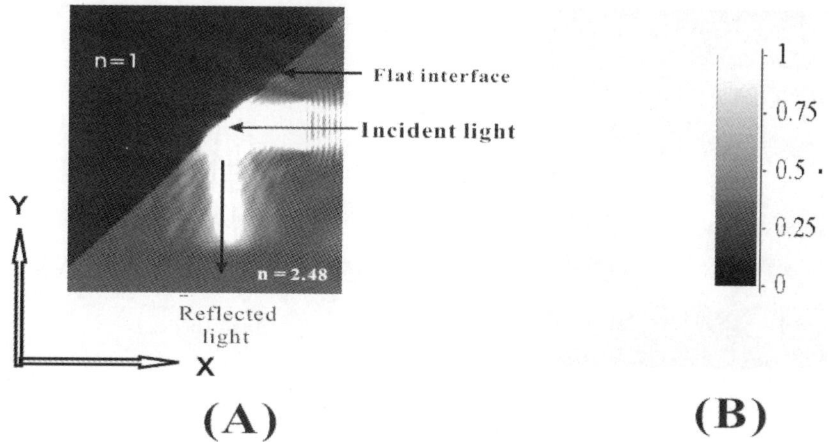

Fig. 18.   (A) Parallel beam of light impinges on a flat inteface at $-43°$. All the radiation is reflected. The electric field is in the TM mode (polarized parallel to the y-axis). Beam is generated by a current source array on the right. (B) The grey scale of intensity visualization used in (A).

to be $e^{-i\omega t}$ the equivalence $-i\omega \leftrightarrow \partial_t$ can be used to derive updates of $\mathbf{E}$ and $\mathbf{H}$ by using an effective current, $\mathbf{J}$, which is also linear in $\mathbf{E}$ and which is updated at each time step. We call this the effective current (EC) method.

While the RC and EC methods have similar features, one of which is that they both require a particular form of the frequency dependence of $\varepsilon(\omega)$, the update algorithms are different and the two methods have slightly different numerical stability properties.

A third method, the hybrid frequency-time domain (FTD) method [3], has also been used in FDTD simulations of surface plasmons. Its advantage, unlike the other methods, is that it does not require any particular form of the frequency dependence of $\varepsilon(\omega)$ or any supplementary variables.

In this section we assess the numerical stability of the RC, EC, and FTD methods in lossy, dispersive media with negative dielectric constants. We briefly outline the derivation of the RC and EC methods, since more detail appears in the literature [7,8]. A longer description of the FTD method is given since it has not yet been published. To be specific we will concentrate on the dielectric constant of gold (Au) for optical wavelengths of around 900 nm. Out of this analysis will come some guidelines for applying one or the other of these methods for surface plasmon problems or more generally for problems with negative, complex, frequency-dependent dielectric constants.

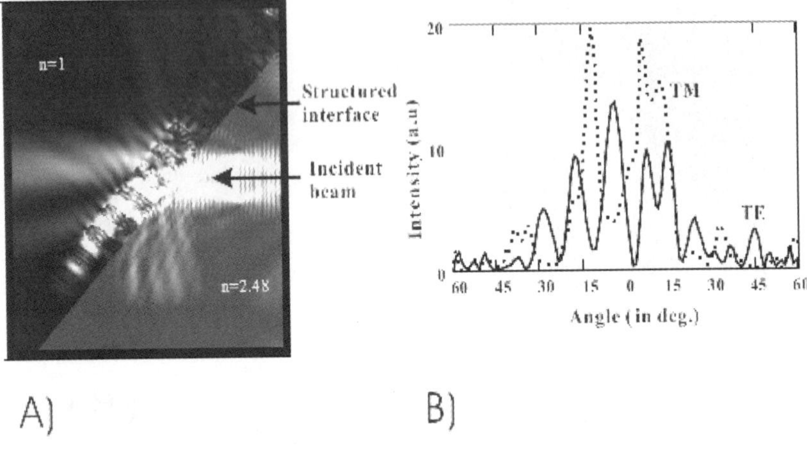

Fig. 19. (A) Transmission through an interface structured with sub-wavelength rect-angular grooves in TM mode for an incidence angle of $-43°$. Groove pitch $= 0.5\ \mu$m, groove depth $= 0.5\ \mu$m. Wavelength of the incident light is $0.45\ \mu$m. The grey-scale intensity visualization is the same one used in Figure 18. (B) Transmitted intensity as a function of incidence angles for TM and TE modes; dotted: TM, solid: TE. Angles of incidence are measured on both sides of the normal to the interface.

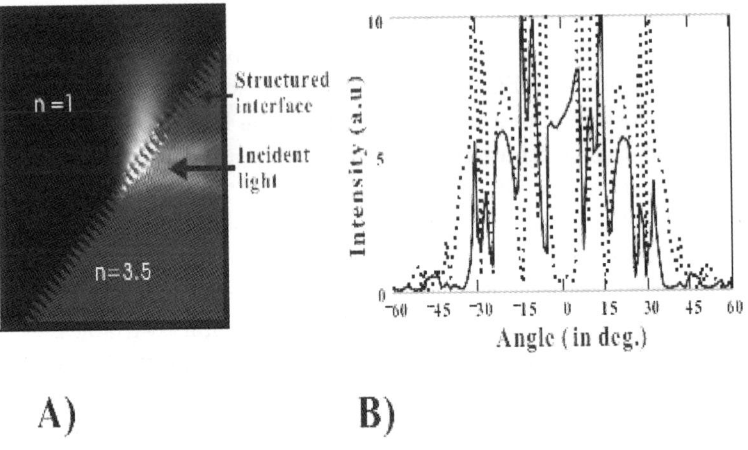

Fig. 20. (A) Transmission through an interface structured with sub-wavelength conical grooves in TM mode for an incidence angle of $-43°$. Groove pitch $= 0.32\ \mu$m, groove depth $= 0.43\ \mu$m. Wavelength of the incident light is $0.632\ \mu$m. The grey-scale intensity visualization is the same one used in Figure 18. (B) Transmitted intensity as a function of incidence angles for TM and TE modes; dotted: TE, solid: TM. Angles of incidence are measured on both sides of the normal to the interface.

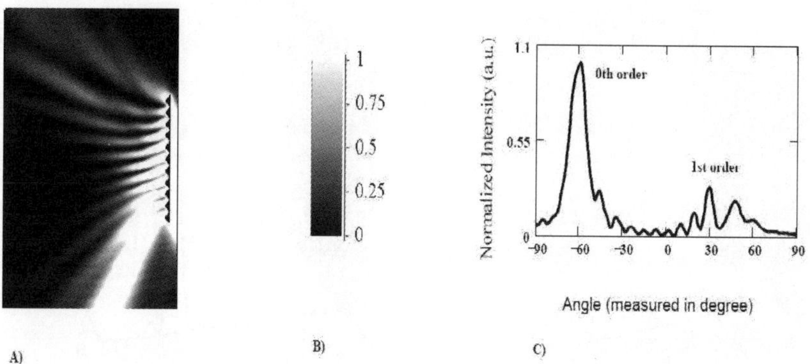

Fig. 21.   (A) Grey-scale visualization of diffracted (reflected) intensity pattern in TE mode. (B) Grey-scale of visualization. (C) Angular spectrum of the normalized diffracted intensity. The intensity of diffracted light in (A) is interpolated over a semicircle of radius $13\,p$ and centered on the midpoint of the back end of the grating. The diffraction efficiencies and angles for the two diffraction orders are 72.8%, $-60°$ and 24.2%, 30° respectively.

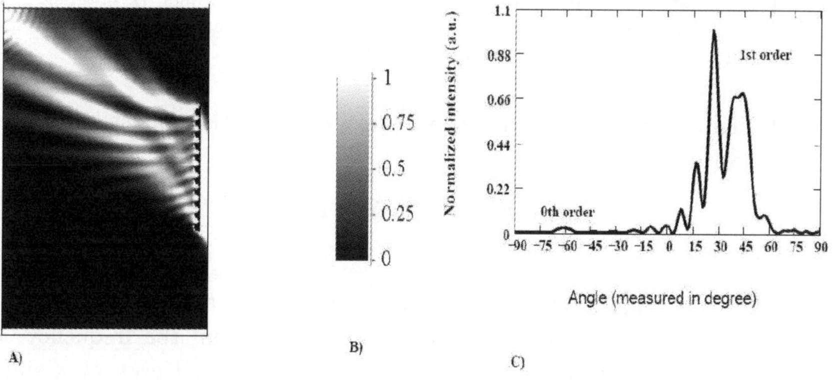

Fig. 22.   (A) Grey-scale visualization of diffracted (reflected) intensity pattern in TM mode. B) Grey-scale of visualization. C) Angular spectrum of the normalized diffracted intensity. The intensity of diffracted light in (A) is interpolated over a semicircle of radius $13\,p$, and centered on the midpoint of the back end of the grating. The diffraction efficiencies and angles for the two diffraction orders are 1.9%, $-62°$ and 91.1%, 28° respectively.

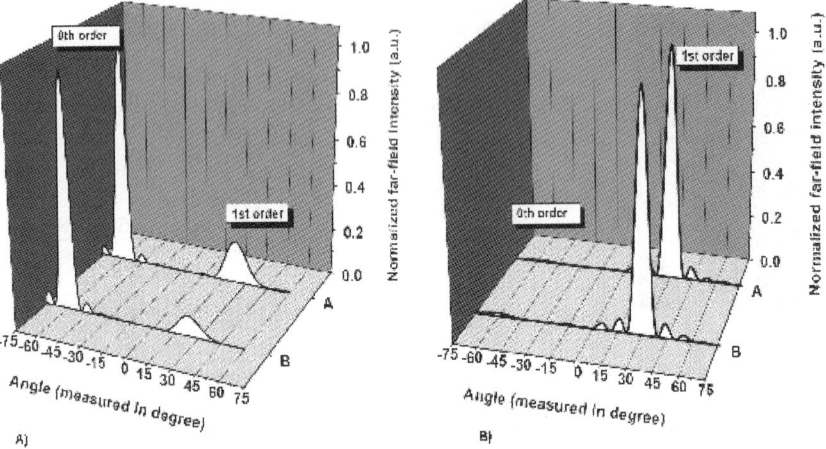

Fig. 23. (A) TE mode far-field diffraction pattern obtained from NS-FDTD simulations. A: obtained with $\lambda/h = 30$ and timesteps $= 10$ waveperiods. The diffraction efficiencies and angles are 70.9%, $-60°$ and 25.0%, $21°$ respectively for $0th$ and $1st$ orders. B: obtained with $\lambda/h = 60$ and timesteps $= 10$ waveperiods. The diffraction efficiencies and angles are 79.9%, $-60°$ and 14.3%, $23°$ respectively for $0th$ and $1st$ orders. (B) TM mode far-field diffraction pattern obtained from NS-FDTD simulations. A: obtained with $\lambda/h = 30$ and timesteps $= 10$ waveperiods, the diffraction efficiencies and angles are 1.6%, $-58°$ and 97.4%, $30°$ respectively for $0th$ and $1st$ orders. B: obtained with $\lambda/h = 40$ and timesteps $= 10$ waveperiods, the diffraction efficiencies and angles are 2.5, $-60°$ and 91.7%, $30°$ respectively for $0th$ and $1st$ orders.

When ε is frequency-dependent, Maxwell's equations must be written in the form

$$\mu\partial_t \mathbf{H}(\mathbf{x}, t) = -\nabla \times \mathbf{E}(\mathbf{x}, t), \tag{428}$$

$$\partial_t \tilde{\mathbf{D}}(\mathbf{x}, t) = \nabla \times \mathbf{H}(\mathbf{x}, t). \tag{429}$$

Whenever a quantity is frequency dependent, it is also time-dependent. This can be understood by fourier-transforming from the frequency- to the time-domain. Therefore $\varepsilon = \varepsilon(\omega) \Rightarrow \varepsilon = \varepsilon(t)$, and even at a single frequency, $\tilde{\mathbf{D}}(t) \neq \varepsilon(t)\mathbf{E}(t)$. The correct relationship between $\tilde{\mathbf{D}}(t)$ and $\mathbf{E}(t)$ is given by the integrodifferential equation (436) below. Thus when ε becomes dispersive, the solution of Maxwell's equations becomes far more difficult. The SFD model of Maxwell's equations is

$$d_t \mathbf{H}(\mathbf{x}, t) = -\left(\frac{1}{\mu}\right)\left(\frac{\Delta t}{h}\right) \mathbf{D}_1 \times \mathbf{E}(\mathbf{x}, t), \tag{430}$$

$$d_t \tilde{\mathbf{D}}(\mathbf{x}, t + \Delta t/2) = \left(\frac{\Delta t}{h}\right) \mathbf{D}_1 \times \mathbf{H}(\mathbf{x}, t + \Delta t/2). \tag{431}$$

To proceed further we need to evaluate $d_t\tilde{\mathbf{D}}$ and extract $d_t\mathbf{E}$. We now discuss different methods to do this.

Editor's Note

The author's original notation and units have been changed to conform with the rest of this chapter. The author used gaussian units in which Maxwell's equations are

$$\mu\partial_t\mathbf{H}(\mathbf{x},t) = -c\nabla \times \mathbf{E}(\mathbf{x},t), \tag{432}$$

$$\partial_t\tilde{\mathbf{D}}(\mathbf{x},t) = c\nabla \times \mathbf{H}(\mathbf{x},t), \tag{433}$$

where c denotes the vacuum velocity of light, and the magnetic permeability of free space is unity. Where the author writes $4\pi\sigma$, we write σ. For example, in (477) the author wrote "$4\pi\sigma = -\varepsilon_2/\tau$." In (479), in the figures, and other places where numerical values are important, we denote the author's numerical value of conductivity by σ^*.

9.2. *The Recursive Convolution Algorithm*

General Development

The recursive convolution method can be used to construct a finite-difference algorithm to solve Maxwell's equations with a frequency-dependent dielectric constant. The electric displacement is given $\tilde{\mathbf{D}}$ is defined in the frequency domain by

$$\tilde{\mathbf{D}}(\mathbf{x},\omega) = \varepsilon(\mathbf{x},\omega)\mathbf{E}(\mathbf{x},\omega), \tag{434}$$

where $\tilde{\mathbf{D}}(\mathbf{x},\omega)$, $\mathbf{E}(\mathbf{x},\omega)$, and $\varepsilon(\mathbf{x},\omega)$ are the Fourier transforms of $\tilde{\mathbf{D}}(\mathbf{x},t)$, $\mathbf{E}(\mathbf{x},t)$, and $\varepsilon(\mathbf{x},t)$, respectively.

To simplify the notation, we now suppress the position dependence, thus $\mathbf{E}(\mathbf{x},\omega) \to \mathbf{E}(\omega)$, etc. The dielectric constant can be expressed as the sum of a frequency-independent part and a frequency-dependent part in the form

$$\varepsilon(\omega) = \varepsilon_1 + \chi(\omega). \tag{435}$$

Maxwell's equations can be cast back into the time domain by using the convolution theorem for the product of Fourier transforms [21] on (434). We obtain

$$\tilde{\mathbf{D}}(t) = \varepsilon_1\mathbf{E}(t) + \int_0^t \mathbf{E}(t-t')\chi(t')dt'. \tag{436}$$

 J. B. Cole

The integral in eq. (436) can now be approximated by discretizing time in the form, $t = 0, \Delta t, 2\Delta t, \cdots, t' = 0, \Delta t, 2\Delta t, \cdots$. For notational simplicity let us write $\mathbf{E}(t) = \mathbf{E}^t$, $\mathbf{E}(t \pm \Delta t) = \mathbf{E}^{t\pm 1}$. For small Δt the $\mathbf{E}$-field is regarded as constant during Δt, and we have

$$\int\limits_{t}^{t+\Delta t} \mathbf{E}(t - t')\chi(t')dt' \cong \mathbf{E}(t) \int\limits_{t}^{t+\Delta t} \chi(t)dt. \tag{437}$$

We can now approximate the convolution integral by

$$\int\limits_{0}^{t} \mathbf{E}(t - t')\chi(t')dt' \cong \sum_{t'=0}^{t-1} \mathbf{E}^{t-t'}\chi^{t'} \tag{438}$$

where,

$$\chi^{t'} = \int\limits_{t}^{t+\Delta t} \chi(t)dt'. \tag{439}$$

Thus in the time domain we now have

$$\tilde{\mathbf{D}}(t) \cong \varepsilon_1 \mathbf{E}(t) + \sum_{t'=0}^{t-1} \mathbf{E}^{t-t'}\chi^{t'}. \tag{440}$$

To construct the FDTD algorithm we need, a FD expression for $\partial_t\tilde{\mathbf{D}}$. Evaluating $\tilde{\mathbf{D}}(t) - \tilde{\mathbf{D}}(t + \Delta t)$ using (440) we have,

$$\tilde{\mathbf{D}}(t + \Delta t) - \tilde{\mathbf{D}}(t) \cong \varepsilon_1 \left(\mathbf{E}^{t+1} - \mathbf{E}^t\right) + \mathbf{E}^{t+1}\chi^0 + \Psi^t, \tag{441}$$

where we have defined

$$\Psi^t = \sum_{t'=0}^{t-1} \mathbf{E}^{t-t'}\Delta\chi^{t'}, \tag{442}$$

and

$$\Delta\chi^t = \chi^{t+1} - \chi^t. \tag{443}$$

The quantity Ψ^t is called the "accumulation field." Now approximating $\partial_t\tilde{\mathbf{D}}$ with a central finite difference, we have,

$$\partial_t\tilde{\mathbf{D}}(t + \Delta t/2) \cong \frac{\varepsilon_1}{\Delta t}\left(\mathbf{E}^{t+1} - \mathbf{E}^t\right) + \frac{1}{\Delta t}\left[\mathbf{E}^{t+1}\chi^0 + \Psi^t\right]. \tag{444}$$

We can now construct an FDTD algorithm to solve Maxwell's equations with frequency-dependent electric permittivity. Solving (430) for $\mathbf{H}(t + \Delta t/2) = \mathbf{H}^{t+1/2}$, and (431) for $\mathbf{E}^{t+1}$ using (444), we have

$$\mathbf{H}^{t+1/2} = \mathbf{H}^{t-1/2} - \frac{1}{\mu}\left(\frac{\Delta t}{h}\right)\mathbf{D} \times \mathbf{E}^t, \tag{445}$$

$$\mathbf{E}^{t+1} = \left(\frac{1}{\varepsilon_1 + \chi^0}\right)\left[\varepsilon_1 \mathbf{E}^t + \left(\frac{\Delta t}{h}\right)\mathbf{D} \times \mathbf{H}^{t+1/2} - \Psi^t\right]. \tag{446}$$

First Order Debye Model

As it stands, this algorithm is costly to implement because the summation, Ψ^t, runs over the $\mathbf{E}$-fields at all time steps. We would like to have some way to find Ψ^{t+1} in terms of recent field values only. For example, if we had

$$\Delta\chi^{t+1} = K\Delta\chi^t, \tag{447}$$

we would have

$$\Psi^{t+1} = \Delta\chi^0 \mathbf{E}^{t+1} + K\Psi^t. \tag{448}$$

For example, in the first order Debye model

$$\chi(\omega) = \frac{\varepsilon_2}{1 - i\omega\tau}. \tag{449}$$

Fourier-transforming to the time domain, we find

$$\chi(t) = \frac{\varepsilon_2}{\tau}e^{-t/\tau}, \qquad \text{for} \quad t > 0. \tag{450}$$

Now evaluating (439) we find

$$\chi^t = \varepsilon_2\left(1 - e^{-\Delta t/\tau}\right)e^{-t/\tau}, \tag{451}$$

$$\Delta\chi^t = -\varepsilon_2\left(1 - e^{-\Delta t/\tau}\right)^2 e^{-t/\tau}, \tag{452}$$

thus in (447), we have

$$K = e^{-\Delta t/\tau}. \tag{453}$$

Extension to the Conducting Maxwell Equations

We can extend these developments to include finite conductivity. Taking the time dependence of the electromagnetic fields to be $e^{-i\omega t}$, in the frequency domain (429) becomes

$$-i\omega\varepsilon\mathbf{E} = \nabla \times \mathbf{H} - \sigma\mathbf{E}, \tag{454}$$

which, using (435) can be transformed into,

$$-i\omega \left(\varepsilon_1 + \chi(\omega) + i\frac{\sigma}{\omega} \right) \mathbf{E} = \nabla \times \mathbf{H}. \tag{455}$$

Incorporating the conductivity into χ, $\chi \to \chi_c$, where

$$\chi_c(\omega) = \chi(\omega) + i\frac{\sigma}{\omega}. \tag{456}$$

Replacing $\chi(\omega)$ by $\chi_c(\omega)$ in (442) and (446) we obtain an FDTD algorithm for dispersive electric permittivity and finite conductivity.

Conductivity can be incorporated into the Debye model, and (456) becomes

$$\chi_c(\omega) = \frac{\varepsilon_2}{1 - i\omega\tau} + i\frac{\sigma}{\omega}. \tag{457}$$

Equation (457) is known as the extended Drude model. With the substition $\chi \to \chi_c$, the FDTD algorithm to solve Maxwell's equations with conductivity and frequency-dependent ε becomes

$$\mathbf{H}^{t+1/2} = \mathbf{H}^{t-1/2} - \frac{1}{\mu} \left(\frac{\Delta t}{h} \right) \mathbf{D} \times \mathbf{E}^t, \tag{458}$$

$$\mathbf{E}^{t+1} = \left(\frac{1}{\varepsilon_1 + \chi_c^0} \right) \left[\varepsilon_1 \mathbf{E}^t + \left(\frac{\Delta t}{h} \right) \mathbf{D} \times \mathbf{H}^{t+1/2} - \Psi^t \right], \tag{459}$$

$$\Psi^{t+1} = \Delta\chi_c^0 \mathbf{E}^{t+1} + K\Psi^t. \tag{460}$$

9.3. *Stability Analysis*

The methods of Sect. 6 can be extended to analyze stability of the above algorithm. For a vector field, $\mathbf{V}$, we can represent $\mathbf{D} \times \mathbf{V}$ in matrix form as,

$$\mathbf{D} \times \mathbf{V} = \begin{pmatrix} 0 & -d_z & d_y \\ d_z & 0 & -d_x \\ -d_y & d_x & 0 \end{pmatrix} \begin{pmatrix} V_x \\ V_y \\ V_z \end{pmatrix}. \tag{461}$$

If $\mathbf{V} = \mathbf{V}_0 e^{i\mathbf{k}\bullet\mathbf{x}}$, where $\mathbf{V}_0$ is a constant vector, then

$$\mathbf{D} \times \mathbf{V} = 2iC\mathbf{V}, \tag{462}$$

where

$$C = \begin{pmatrix} 0 & -s_z & s_y \\ s_z & 0 & -s_x \\ -s_y & s_x & 0 \end{pmatrix} \begin{pmatrix} V_x \\ V_y \\ V_z \end{pmatrix}, \tag{463}$$

where $s_m = \sin(k_m h/2)$ $(m = x, y, z)$. Note that $d_m e^{i\mathbf{k}\bullet\mathbf{x}} = 2is_m e^{i\mathbf{k}}$.

Taking the spatial dependence of the electromagnetic fields to be $e^{i\mathbf{k}\bullet\mathbf{x}}$, the FDTD algorithm (458), and (460) now becomes

$$\mathbf{H}^{t+1/2} = \mathbf{H}^{t-1/2} - 2iC\frac{1}{\mu}\left(\frac{\Delta t}{h}\right)\mathbf{E}^t, \tag{464}$$

$$\mathbf{E}^{t+1} = \left(\frac{1}{\varepsilon_1 + \chi_c^0}\right)\left[\varepsilon_1\mathbf{E}^t + 2iC\left(\frac{\Delta t}{h}\right)\mathbf{H}^{t+1/2} - \Psi^t\right], \tag{465}$$

$$\Psi^{t+1} = \Delta\chi_c^0\mathbf{E}^{t+1} + K\Psi^t. \tag{466}$$

Now we can arrange the fields into a vector defined by

$$\mathbf{F}^t = \begin{pmatrix} \mathbf{H}^{t-1/2} \\ \mathbf{E}^t \\ \Psi^t \end{pmatrix}, \tag{467}$$

and the update equations (464)-(466) can be expressed in matrix form as

$$\mathbf{F}^{t+1} = \mathbf{M}\mathbf{F}^t + \mathbf{N}\mathbf{F}^{t+1}, \tag{468}$$

where

$$\mathbf{M} = \begin{pmatrix} 1 & -\frac{2iC}{\mu}\frac{\Delta t}{h} & 0 \\ 0 & \frac{\varepsilon_1}{\varepsilon_1'} & -\frac{1}{\varepsilon_1'} \\ 0 & 0 & K \end{pmatrix}, \tag{469}$$

$$\mathbf{N} = \begin{pmatrix} 0 & 0 & 0 \\ \frac{2iC}{\varepsilon_1'}\frac{\Delta t}{h} & 0 & 0 \\ 0 & \Delta\chi_c^0 & 0 \end{pmatrix}, \tag{470}$$

and we have made the abbreviation $\varepsilon_1 + \chi_c^0 = \varepsilon_1'$. The update equation (468) can be rewritten as

$$\mathbf{F}^{t+1} = (\mathbf{I} - \mathbf{N})^{-1}\mathbf{M}\mathbf{F}^t, \tag{471}$$

where $\mathbf{I}$ is the unit matrix. Knowing $\mathbf{F}^0$ we can now compute the electromagnetic fields at all future times. Abbreviating $\mathbf{Q} = (\mathbf{I} - \mathbf{N})^{-1}\mathbf{M}$, we have

$$\mathbf{F}^t = \mathbf{Q}^t\mathbf{F}^0. \tag{472}$$

Following the methods of Sect. 6, we find the eigenvalues, λ_i, and normalized eigenvectors, $\mathbf{\Lambda}_i$, of $\mathbf{Q}$, where $i = 1, 2, 3$. We can then express $\mathbf{F}^0$ in the form,

$$\mathbf{F}^0 = c_1\mathbf{\Lambda}_1 + c_2\mathbf{\Lambda}_2 + c_3\mathbf{\Lambda}_3, \tag{473}$$

and (472) becomes

$$\mathbf{F}^t = c_1\lambda_1^t\mathbf{\Lambda}_1 + c_2\lambda_2^t\mathbf{\Lambda}_2 + c_3\lambda_3^t\mathbf{\Lambda}_3. \tag{474}$$

For the fields to remain finite, the eigenvalues of $\mathbf{Q}$ must satisfy

$$|\lambda_i| \le 1, \quad i = 1, 2, 3. \tag{475}$$

Defining $|\Lambda| = \max_{\mathbf{k}}(|\lambda_1|, |\lambda_2|, |\lambda_3|)$, with respect to all possible values of $\mathbf{k}$, (475) becomes

$$|\Lambda| \le 1. \tag{476}$$

Stability of the Recursive Convolution Algorithm

In surface plasmon calculations we often take

$$\sigma = -\frac{\varepsilon_2}{\tau} \tag{477}$$

in the extended Drude model (457), which gives

$$\chi_c(\omega) = \frac{\varepsilon_2}{1 - i\omega\tau} - i\frac{\varepsilon_2}{\tau}. \tag{478}$$

Equation (478) is known as the Drude model.

We now assess the numerical stability of recursive convolution algorithm using the the extended Drude model in one dimension. The one-dimensional form of Maxwell's equations are given by (144) and (145). In this and all subsequent analyses we take $\mu = 1$, and investigate stability about the point $p = (\varepsilon_1, \varepsilon_2, \tau, \sigma, \lambda)$ in parameter space. We define p by

$$\varepsilon_1 = 7.919,$$
$$\varepsilon_2 = -14395.10,$$
$$\sigma^* = 127.27\,\text{fs}^{-1},$$
$$\tau = 9.0\,\text{fs},$$
$$\lambda = 942\,\text{nm}.$$

These values were determined from a fit to the dielectric constant of gold at wavelengths of 600-1000 nm, where experimentally the real part of ε is negative. The parameter values chosen here are close to those used in various studies of surface plasmons in the 600-1000 nm wavelength range. In this wavelength region the wave period is $2 \le T \le 3.33$ fs.

In Fig. 24 we take $h = 5.0$ nm $\Rightarrow 120 \le \lambda/h \le 200$, and plot $|\Lambda|$ (solid line) as a function of Δt. We also plot $\Lambda' = 1 + 100(|\Lambda| - 1)$ (dotted line) which is a magnification of the deviation of $|\Lambda|$ from one. The RC algorithm is stable up to $\Delta t = 0.046$ fs, where the usual SFD algorithm is also stable with $\varepsilon = \varepsilon_1$ (that is real ε). For $\Delta t \geq 0.046$ fs, the value of $|\Lambda|$ rises very rapidly (vertical line on the right of the graph).

One usually expects that the smaller Δt, the more stable the algorithm, but there is a range of small Δt for which it is actually unstable. As the figure shows, $|\Lambda|$ is very slightly greater than unity for $\Delta t < 0.0004$fs. It linearly increases (see plot of Λ') until $\Delta t \cong 0.0004$ fs. After that Λ drops, and the RC algorithm becomes stable at $\Delta t \cong 0.0007$ fs. While technically the algorithm is an unstable for $\Delta t < 0.0004$ fs, such a small time step is never actually used in most FDTD simulations. Furthermore, the instability is so slight that the calculation does not diverge unless it is carried out for millions iterations.

The Yee algorithm given in Section 4 assumes that ε is real, so we cannot directly compare its stability with the RC algorithm, but we can define a point $p' = (\varepsilon, \sigma, \lambda)$ in the parameter space of the Yee algorithm which is equivalent to p. The values of conductivity and and electrical permittivity for the Yee algorithm which correspond to p are given by

$$\varepsilon = -36.37,$$

$$\sigma^* = 0.392 \, \text{fs}^{-1},$$

$$\lambda = 942 \, \text{nm}.$$

The utility of the RC algorithm over the S-Yee algorithm (108) and(109) is shown in Fig. 25. We plot the minimum value of σ for which the RC algorithm becomes conditionally stable (i.e., is stable over some range of Δt) as a function of ε_2, about point p. We make the same plot about p' for the S-Yee algorithm. The parameter space divides into two regions. In the "stable" region there exists a range of Δt for which the algorithm is stable, while in the "unstable" region there is no range of Δt for which the algorithm is stable.

Thus, while the RC algorithm is conditionally stable at p, the S-Yee algorithm is quite unstable at p' and hence cannot be used to simulate surface plasmons in the typical wavelength range in which they occur.

9.4. *Effective Current Algorithm*

Including conductivity in Maxwell's equations, eq. (429) becomes

$$\partial_t \tilde{\mathbf{D}}(\mathbf{x}, t) = \nabla \times \mathbf{H}(\mathbf{x}, t) - \sigma \mathbf{E}(\mathbf{x}, t). \tag{479}$$

Taking the electrical permittivity to be given by (436), (479) becomes

$$\varepsilon_1 \partial_t \mathbf{E}(t) + \partial_t \mathbf{I}(t) = \nabla \times \mathbf{H}(t) - \sigma \mathbf{E}, \tag{480}$$

where

$$\mathbf{I}(t) = \int_0^t \mathbf{E}(t - t')\chi(t')dt', \tag{481}$$

and position dependence is omitted for brevity.

Let the convolution of two arbitrary functions, f and g, be

$$h(t) = \int_0^t f(t - t')g(t')dt'. \tag{482}$$

Using partial integration it is easy to show that

$$\frac{d}{dt}h(t) = f(t)g(0) + \int_0^t f(t - t')\frac{d}{dt'}g(t')dt'. \tag{483}$$

Thus (480) becomes

$$\varepsilon_1 \partial_t \mathbf{E}(t) = \nabla \times \mathbf{H}(t) - [\sigma + \chi(0)]\,\mathbf{E}(t) - \mathbf{J}(t), \tag{484}$$

where $\mathbf{J} = \partial_t \mathbf{I}$, is called the "effective current." Applying (483) to $\partial_t \mathbf{J}$ we find

$$\partial_t \mathbf{J}(t) = \mathbf{E}(t)\partial\chi(0) + \int_0^t \mathbf{E}(t - t')\frac{\partial^2}{\partial t'^2}\chi(t')dt'. \tag{485}$$

Returning to the Debye model, let χ be given by (450) for which $\partial_t \chi = -\chi/\tau$. Thus (485) becomes

$$\partial_t \mathbf{J}(t) = -\frac{1}{\tau}\chi(0)\mathbf{E}(t) - \frac{1}{\tau}\mathbf{J}(t). \tag{486}$$

Using the Debye model, we now have a complete system of equations which can be solved with a finite-difference method,

$$\mu\partial_t \mathbf{H}(t) = -\nabla \times \mathbf{E}(t), \tag{487}$$

$$\varepsilon_1 \partial_t \mathbf{E}(t) = \nabla \times \mathbf{H}(t) - [\sigma + \chi(0)]\,\mathbf{E} - \mathbf{J}(t), \tag{488}$$

$$\partial_t \mathbf{J}(t) = -\frac{1}{\tau}\chi(0)\mathbf{E}(t) - \frac{1}{\tau}\mathbf{J}(t). \tag{489}$$

The EC algorithm becomes

$$\mathbf{H}^{t+1/2} = \mathbf{H}^{t-1/2} - \frac{1}{\mu}\left(\frac{\Delta t}{h}\right)\mathbf{D} \times \mathbf{E}^t, \tag{490}$$

$$\mathbf{E}^{t+1} = \frac{A_-}{A_+}\mathbf{E}^t + \frac{1}{A_+}\frac{1}{\varepsilon_1}\frac{\Delta t}{h}\mathbf{D} \times \mathbf{H}^{t+1/2} - \frac{1}{A_+}\frac{\Delta t}{\varepsilon_1}\mathbf{J}^{t+1/2}, \tag{491}$$

$$\mathbf{J}^{t+3/2} = \left(1 - \frac{\Delta t}{\tau}\right)\mathbf{J}^{t+1/2} - \frac{\Delta t}{\tau}\chi(0)\mathbf{E}^{t+1/2}, \tag{492}$$

where

$$A_\pm = 1 \pm \left(\frac{\sigma + \chi(0)}{2\varepsilon_1}\right). \tag{493}$$

Stability of the Effective Current Algorithm

The stability analysis for the EC algorithm proceeds in analogy with that of the RC algorithm. Taking the position dependence of all field quantities to be $e^{i\mathbf{k}\bullet\mathbf{x}}$ and using (462), the $\mathbf{H}$-field update equation (490) is the same as that of the RC algorithm (465), while (491) becomes

$$\mathbf{E}^{t+1} = \frac{A_-}{A_+}\mathbf{E}^t + 2iC\frac{1}{A_+}\frac{1}{\varepsilon_1}\frac{\Delta t}{h}\mathbf{H}^{t+1/2} - \frac{1}{A_+}\frac{\Delta t}{\varepsilon_1}\mathbf{J}^{t+1/2}. \tag{494}$$

The form of (492) remains the same. The EC algorithm can then be put into the form of (472), and its stability is determined by the eigenvalues of $\mathbf{Q}$.

Figures 26 and 27 depict the stability characteristics of the EC algorithm in one dimension. In one dimension the stability of the EC algorithm is worse than that of the RC algorithm. About point p (479) in parameter space the EC algorithm is unconditionally unstable. For any given value of ε_2 the minimum conductivity needed to ensure conditional stability is higher than for the RC method. As 27 shows, p lies slightly outside the conditional stability line, but in three-dimensional surface plasmon calculations we have found that the EC algorithm is stable at p.

Both the EC and RC method have been used in surface plasmon simulations. Also, for both methods the realistic Drude model parameters are very close to the stability line, so a slight change could completely change the stability of the solutions. We now present a third method, the hybrid frequency-domain time domain (FTD) method, which is always conditionally stable.

9.5. *Hybrid Frequency-Domain Time-Domain Algorithm*

In many FDTD simulations one turns on a monochromatic incident field over many wave periods, and then lets the system approach a steady state solution with the same frequency as the incident field. Taking the time-dependence of the electromagnetic fields to be $e^{-i\omega t}$, the **E**-field is given by (454). Using FD approximations for the derivatives we have

$$(\sigma - i\omega)\,\mathbf{E}^t = \frac{1}{h}\mathbf{D} \times \mathbf{H}^t. \tag{495}$$

In the hybrid frequency-domain time-domain (FTD) method we update the **H**-field in the time domain using (445), but compute the **E**-field in the frequency domain using (495).

$\mathbf{H}(t + \Delta t/2)$ is determined from $\mathbf{H}(t - \Delta t/2)$ and $\mathbf{E}(t)$ in (445), but $\mathbf{E}(t)(t+\Delta t)$ is not given by (495) in terms of $\mathbf{E}(t)$. However we can extrapolate $\mathbf{H}(t)$ to $\mathbf{H}(t + \Delta t)$ using

$$f(t + \Delta t) \cong f(t + \Delta t/2) + \frac{\Delta t}{2}f'(t + \Delta t/2), \tag{496}$$

$$= \frac{3}{2}f(t + \Delta t/2) - \frac{1}{2}f(t - \Delta t/2),$$

where we have used $f'(t + \Delta t/2) \cong d_t f(t)/\Delta t$. The FTD algorithm thus becomes

$$\mathbf{H}^{t+1/2} = \mathbf{H}^{t-1/2} - \frac{1}{\mu}\left(\frac{\Delta t}{h}\right)\mathbf{D} \times \mathbf{E}^t, \tag{497}$$

$$\mathbf{E}^{t+1} = -\left(\frac{1}{\sigma - i\omega}\right)\frac{1}{h}\left[\frac{3}{2}\mathbf{H}^{t+1/2} - \frac{1}{2}\mathbf{H}^{t-1/2}\right]. \tag{498}$$

Stability of the Hybrid Frequency-Domain Time-Domain Algorithm

Again we limit the analysis to one dimension. Using the one dimensional form of Maxwell's equations in Gaussian units we obatin an eigenvalue equation of the form

$$(-i\omega\varepsilon + 4\pi\sigma)\left(\lambda^2 - \lambda\right) = -2c^2\frac{\Delta t^2}{h^2}(3\lambda - 1)\sin^2(kh/2), \tag{499}$$

which becomes

$$\lambda^2 + (3b - 1)\lambda - b = 0, \tag{500}$$

where

$$b = 2c^2\frac{\Delta t^2}{h^2}\left(\frac{\sin^2(kh/2)}{4\pi\sigma - i\omega\varepsilon}\right). \tag{501}$$

We find

$$\lambda = \frac{1 - 3b \pm \sqrt{9b^2 - 2b + 1}}{2}.$$

$$(502)$$

For small $\mathrm{Re}(b)$ the stability condition reduces to $\mathrm{Re}(b) > 0$, but the algorithm becomes unstable when $|b|$ is sufficiently large. Figure 28 indicates values of b in the complex plane for which the algorithm is stable.

The advantage of the FTD method is that it is always conditionally stable, i.e., there is always a Δt small enough to give numerical stability for positive σ. A disadvantage is that one is restricted to a fixed frequency, whereas the RC and EC methods can handle a range of frequencies. Another disadvantage is that while the algorithm is conditionally stable, the maximum permissible time step can be quite small. This is a result of the $\Delta t/h^2$ term in eq. (501). If h is reduced by a factor of f $(h \to h/f)$ to maintain stability Δt must be reduced by a factor of f^2 $(\Delta t \to \Delta t/f^2)$ as seen in Fig. 29. While the RC and EC algorithms also have stability thresholds, once the threshold value of conductivity is exceeded, the permissible value of Δt quickly grows, and soon becomes an order of magnitude greater than for the FTD method. On the other hand, the larger h the better the stability of the FTD algorithm relative to the RC and EC algorithms.

10. Summary

We have assessed the numerical stability of three algorithms to compute electromagnetic propagation in metals with a negative dielectric constant below the plasmon resonance resonance frequency.

The recursive convolution (RC) algorithm and the effective current (EC) algorithm have been used in the surface plasmon FDTD calculations. We find these methods attractive since they allow FDTD calculations over a range of frequencies. These algorithms, however, are based upon an approximate model of of $\varepsilon(\omega)$, and different parameters have to be used over different frequency ranges. This limits the range of frequencies that can be handled in a single calculation.

The commonly-used Drude model for $\varepsilon(\omega)$ seems to be just on the edge of the stable region in parameter space, and with a slight change of parameters these algorithms can become unstable. In one dimension the RC algorithm is stable for Drude model parameters of gold, but the EC algorithm is not. However, the EC algorithm does appear to be stable in three-dimensional calculations.

We have introduced a modification of the Yee algorithm for dispersive materials which we call the hybrid frequency-domain time-domain (FTD) algorithm. It differs from the RC and EC algorithms in that it does not require a particular model of the dielectric constant, and there is no need to introduce auxiliary fields. Even though **E**-field is computed in the frequency dommain, it is updated in the time domain. The FTD algorithm is conditionally stable for any choice of the dielectric constant as long as the conductivity is positive. However the range of Δt for which it is stable is smaller than for the other algorithms, and it must be used at a single frequency.

We have used the FTD algorithm to simulate the fields from an array of apertures [18], and the surface phonons were clearly present in this simulation and agreed quite well with experimental observations.

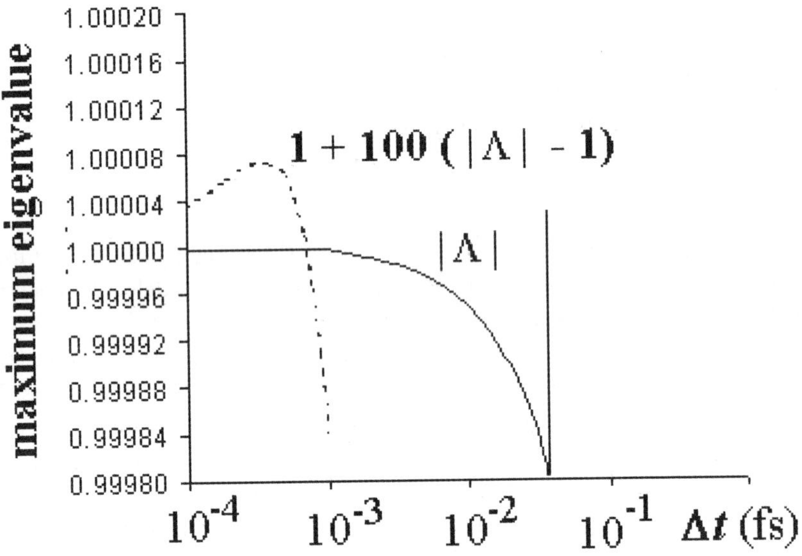

Fig. 24. **Stability of the Recursive Convolution Algorithm as a Function of Time Step.**
$|\Lambda|$ (solid line) and $1 + 100(|\Lambda| - 1)$ (dotted line) are plotted versus Δt at $h= 5.0$ nm at point p in parameter space.

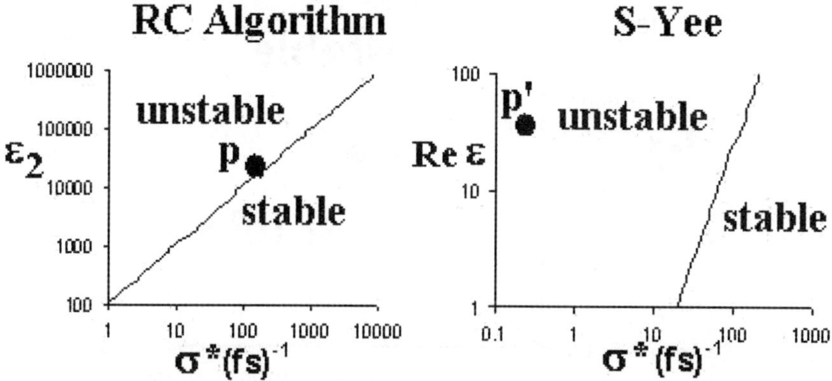

Fig. 25.   **Stability Comparison of the Recursive Convolution and S-Yee Algorithms.**
Left: Stable region as a function of ε_2 and σ^* for the recursive convolution algorithm at $h = 5$ nm. Right: Stable region as a function of ε and σ^* for the S-Yee algorithm at $h = 5$ nm.

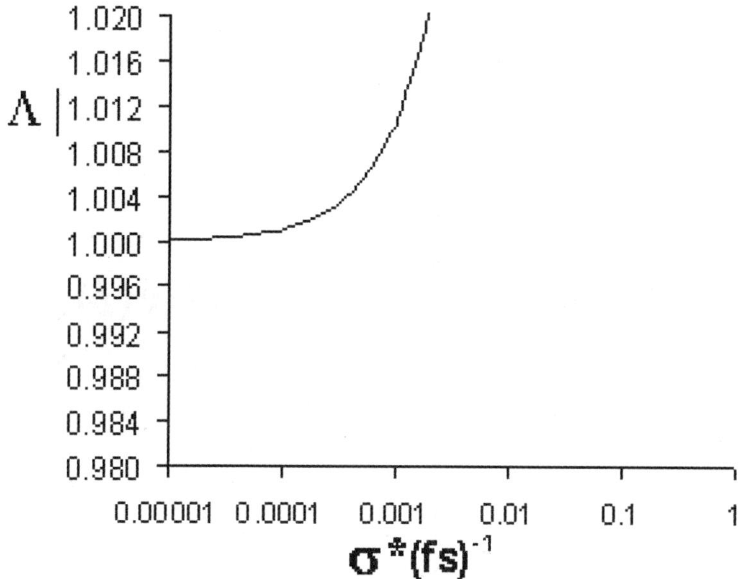

Fig. 26.   **Stability of the EC Algorithm as Function of Conductivity .**
$|\Lambda|$ is plotted as a function of conductivity about point p in parameter space with $h = 5$ nm.

 J. B. Cole

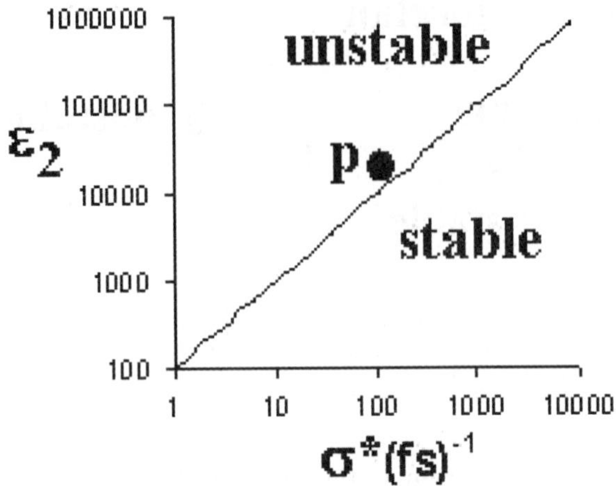

Fig. 27. Stable Region of the EC Algorithm.
Stable region as a function of ε_2 and σ^* about p with with $h = 5$ nm.

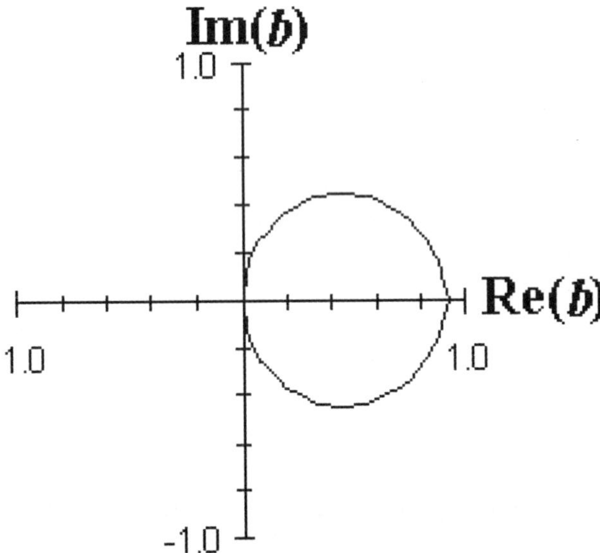

Fig. 28. Stable Region.
Stable region for b in the complex plane.

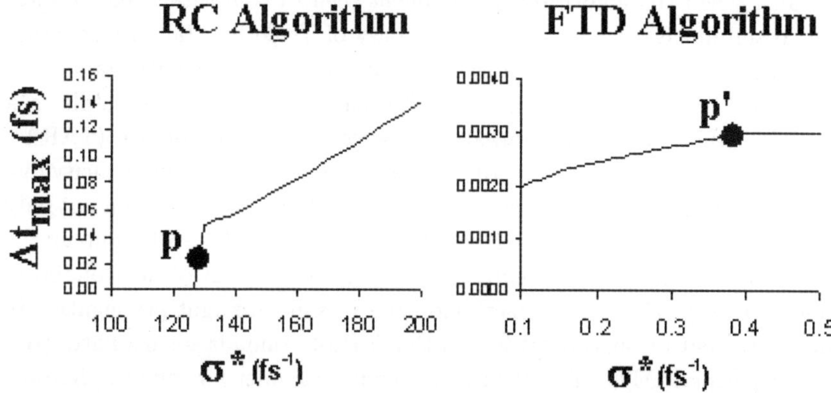

Fig. 29.   **Maximum Stable Time Step for the EC and FTD Algorithms as a Function of Conductivity.**
The maximum time step, Δt_{max}, at $h = 5$ nm. for which the algorithm is stable as a function of σ at p for the EC algorithm, and at p' for the FTD one.

Acknowledgments

"The Devil is in the details." To truly comprehend a concept, it is necessary use it to solve a realistic problem. Wrestling matches with "devils" extends and deepens our understanding, and sometimes fundamental discoveries emerge in the process.

Some of the algorithms presented here were developed about 10 years ago. Although we had every reason to believe that they worked, they were untested. Someone remarked that we had a "solution in search of a problem."

The first useful applications were developed in collaboration with Dr. Neelam Gupta at the Army Research Laboratory (USA), who understood the potential applications to optics. Together we modeled propagation in layered dielectric waveguides and in waveguides with gratings.

Soon afterwards, in collaboration with Dr. Mark Shure at Rohm and Haas Corporation (USA) we developed simulations of Mie scattering of light off colloidal particles. Since there are known analytic solutions for spheres and cylinders (but little else), we could for the first time verify the accuracy of our algorithms in a practical problem [13]. I would like to thank Rohm and Haas Corporation for generous financial support to develop the first working nonstandard FDTD computer programs to compute Mie scattering off particles of arbitrary shape.

Just when the algorithm development seemed complete, I met professor Yoshifumi Katayama at the University of Tsukuba (Japan). Together with Prof. Shigeki Yamada (now at Yokohama City University, Japan) we developed methods to simulate light propation in photonic crystals in a project sponsored by the Tsukuba Advanced Research Alliance (TARA). Prof. Katayama realized that we could also compute the group velocity dispersion of pulses in nanometer scale structures. This collaboration not only stimulated further development of the algorithms, but also taught us the importance of problem modeling (see Sect. 7). As a result of this collaboration [11], [14], [15] we have methodologies to simulate light propagation in sub-wavelength structures. I like to thank Prof. Yamada for his hard work in comparing FDTD calculations with analytical ones for infinite photonic crystals, which verified the correctness of both our algorithms and the numerical models.

The TARA collaboration has turned into a industrial collaboration with Dr. Kensuke Ogawa at the Mitsui Corporation Device Nanotech Research Institute. The goal is to design realistic photonic crystal devices. In three dimensions, where computer memory is limited, clever problem modeling and optimized algorithms are essential, and the results must be effectively visualized and interpreted. It may seem straightforward to generalize from two to three dimensions, but it is not. Dr. Banerjee has devoted much effort to solving the additional problems that have arisen. We have encountered many interesting scientific problems which will be presented in forthcoming papers. We thank Device Nanotech Research Institute for their generous financial support over the past several years.

Dr. Saswatee Banerjee (University of Tsukuba) developed methods to compute the optical properties of conducting diffraction gratings with sub-wavelength features. In collaboration with Prof. Toyohiko Yatagai (University of Tsukuba) and Nippon Sheet Glass Corp. (Japan), Dr. Banerjee has investigated the coupling of light from optically dense dielectric media to less dense ones, and has found ways to reduce reflection at the interface.

We have begun a new collaboration with Dr. Takao Komatsuzaki and Dr. Akitoshi Nishimura of Oscillated Recall Technology Corporation (Japan) with whom we are investigating integrated circuits, and are now addressing scientific problems whose existence we had not even imagined. We thank Oscillated Recall Technology Corp. for their generous financial support.

I thank the University of Tsukuba which has provided a fertile and stimulating environment in which to work. We have received generous research grants ("kakenhi") from the Japanese Ministry of Science and Education, and also internal project grants from the university of Tsukuba. Several students have been involved in these projects, and they have brought new viewpoints to this work. In particular I would like to thank Dr. Rakchnok Rungsawang (now at NTT Basic Research Laboratory, Japan).

I thank my former colleagues and friends at the Naval Research Laboratory (USA). I was first introduced to the FDTD method by Dr. Susan K. Numrich, in whose group I investigated sound propagation in complex ocean environments, with support from the US Office of Naval Research. Dr. Numrich patiently encouraged my fumbling efforts to understand the FDTD method and write working programs. A chance remark by Dr. Rudolph Krutar ("there has be useful information in the diagonal grid points" -rough quotation) inspired my early work in NSFD methods. Dr. Dennis Creamer pointed me to the papers of Professor Ronald Mickens.

Dr. Michael I Haftel and Dr. Mervine Rosen were enthusiastic collaborators in applying NSFD methods to computational electromagnetics problems. I continue to collaborate with Dr. Haftel. I also thank Dr. Ronald Tonucci with whom I have had many stimulating discussions. Dr. Rosenberg of the Visualization Laboratory is an accomplished scientist who is equally competent in solving the problems that arise in large-scale computing, and in the visualization and interpretation of large data sets. I thank him for his enthusiastic help, and especially for giving us his valuable time during a brief visit to Japan.

Last but not least, Professor Mickens is one of the giants, on whose shoulders I stand.

References

1. R. E. Mickens, "Nonstandard Finite Difference Models of Differential Equations," World Scientific, Singapore (1984).
2. J. B. Cole, High Accuracy Yee Algorithm Based on Nonstandard Finite Differences: New Developments and Verifications, IEEE Trans. Antennas and Propagation, vol. 50, no. 9, pp. 1185-1191 (Sept., 2002).
3. J. B. Cole, Application of Nonstandard Finite Differences to Solve the Wave Equation and Maxwellfs Equations, in chap. 3 of "Applications of Nonstandard Finite Difference Schemes," R. E. Mickens, editor, World Scientific, Singapore (2000).
   Note: This paper supercedes the results given in this reference for the conducting Maxwell equations, and the absorbing wave equation.
4. J. B. Cole, High Accuracy FDTD solution of the absorbing wave equation, and conducting Maxwell's equations Based on Nonstandard Finite Difference Model, IEEE Trans. Antennas and Propagation, vol. 53, no. 2, pp. 725-729

(April, 2004).

5. J. B. Cole and S. Banerjee, Applications of Nonstandard Finite Difference Models to Computational Electromagnetics, invited paper in Journal of difference equations and applications, vol. 9, no. 12, pp. 1099-1112 (Dec. 2003).

6. P. W. Barber, S. C. Hill, "Light Scattering by Particles: Computational Methods," World Scientific, Singapore (1990).

7. K. S. Yee, Numerical solution of initial boundary value problems involving Maxwell's equations in isotropic media, IEEE Trans. Antenna Propagation, Vol. AP-14, pp. 302-307 (May, 1966).

8. A. Taflov, "Computational Electrodynamics The Finite-Difference Time-Domain Method," Artech House, Inc., Norwood, MA (1995).

9. G. Mur, Absorbing boundary conditions for the finite-difference approximation of the time domain electromagnetic field equations, IEEE Transactions on Electromagnetic Compatibility, vol. 23, pp. 377-382 (1981).

10. B. Engquist, A. Majda, Absorbing boundary conditions for the numerical simulation of waves, Mathematics of Computation, vol. 31, pp. 629-651 (1977).

11. 17. S. Yamada, Y. Watanabe, Y. Katayama, X. Y. Yan, J. B. Cole, Simulation of Light Propagation in Two-Dimensional Photonic Crystals with a Point Defect by a High Accuracy Finite-Difference Time-Domain Method, Journal of Applied Physics, vol. 92, no. 3, pp. 1181-1184 (1 August 2002).

12. S. K. Godunov, "Difference Schemes," North-Holland, Amsterdam (1987).

13. S. A. Palkar, N. P. Ryde, M. R. Schure, N. Gupta, J. B. Cole, Finite Difference Time Domain Computation of Light Scattering by Multiple Colloidal Particles, Langumir, vol. 14, no. 13, pp. 3484-3492 (June, 1998).

14. J. B. Cole, S. Yamada, Y. Katayama, High Accuracy Finite-Difference Time-Domain Simulation of Light Propagation in Photonic Crystals, invited paper in Laser Research (in Japanese), vol. 30, no. 2, pp. 75-80 (Feb. 2002).

15. S. Yamada, Y. Watanabe, Y. Katayama, J. B. Cole, Simulation of Optical Pulse Propagation in a Two-Dimensional Photonic Crystal Waveguide Using a High Accuracy Finite-Difference Time-Domain Algorithm, Journal of Applied Physics, vol. 93, no. 4, pp. 1859-1864 (15 Feb. 2003).

16. H. Raether, Excitation of Plasmons and Interband Transitions by Electrons, in Chap. 10, Volume 88, "Tracts in Modern Physics," Springer-Verlag, Berlin (1980).

17. T. W. Ebbesen, H. J. Lezec, H. F. Ghaemi, T. Thio, and P. A. Wolff, Extraordinary optical transmission through sub-wavelength hole arrays, Nature 391, pp. 667-669 (1998).

18. D. Egorov, B. S. Dennis, G. Blumberg, and M. I. Haftel, Two-dimensional control of surface plasmons and directional beaming from arrays of subwavelength apertures, Physical Review B, vol. 69, pp. 033404:1-4 (2004).

19. S. A. Maier, M. L. Brongersma, P. G. Kik, S. Meltzer, A. A. G. Requicha, and H. A. Atwater, Plasmonics - A route to nanoscale optical devices, Advanced Materials vol. 13, pp. 1501-1505.

20. G. Burns, p. 51 in "Solid State Physics," Academic Press, New York (1985).

21. K. S. Kunz and R. J. Luebbers, chapter 8 in "The Finite Difference Time

Domain Method for Electromagnetics," CRC Press, New York (1993).

22. S. K. Gray and T. Kupka, Propagation of light in metallic nanowire arrays: Finite-difference time-domain studies of silver cylinders, Physical Review B, vol. 68, pp 045415:1-11 (2003).

23. M. Born, and E. Wolf, "Principles of Optics" Cambridge university press, San Francisco, (1999), 7th (expanded) ed., Chap. XIV p.735-739.

24. C. Lawrence, P. Vukusic, R. Sambles, Grazing incidence iridescence from a butterfly wing, Applied Optics, vol. 41, no.3 pp. 437-441 (2002).

25. H. Ghiradella, D. Aneshansley, T. Eisner, R. E. Silberglied, H. E. Hinton, Science, vol. 178, pp. 1214-1217 (1972).

26. H. Ghiradella, Structure of iridescent Lepidopteran scales: variations on several themes, Annals of the Entomological Society of America, vol. 77, no.6, pp. 637-645 (1984).

27. H. Ghiradella, Structure and development of iridescent lepidopteran scales: the papilionidae as a showcase family. Annals of the Entomological Society of America, vol. 78, no.2, pp. 252-264 (1985).

28. H. Ghiradella, Light and color on the wing: structural colors in butterflies and moths, Applied Optics, vol. 30, pp. 3492-3500 (1991).

29. P. Vukusic, J. R. Sambles, C. R. Lawrence, R. J. Wootton, Quantified interference and diffraction in single Morpho butterfly scales, Proceedings of Royal Society of London B, vol. 266, pp. 1403-1411 (1999).

30. B. Gralak, G. Tayeb, S. Enoch, Morpho butterflies wings color modeled with lamellar grating theory Optics Express Vol.9 no.11, p.1899-1911(2001).

31. S. Fan, P. R. Villeneuve, J. D. Joannopoulos, High extraction efficiency of spontaneous emission from slabs of photonic crystals, Physical Review Letters, vol. 78, no.17, pp. 3294-3297 (1997).

32. A. A. Erchak, et al., Enhanced coupling to vertical radiation using a two-dimensional photonic crystal in a semiconductor light-emitting diode, Applied Physics Letters, vol. 78, no.5, pp. 563-565 (2001)

33. K. Orita, et al., High-extraction efficiency blue light-emitting diode using extended-pitch photonic crystal, Japanese Journal of Applied Physics, vol. 43, no. 8B, pp. 5809-5813 (2004).

34. H. Ichikawa and T. Baba, Efficiency enhancement in a light-emitting diode with a two-dimensional surface grating photonic crystal, Applied Physics Letters, vol. 84, no.2, pp. 457-459 (2004).

35. D. Maystre, "Progress in Optics, ed. Wolf North-Holland, Amsterdam, (1984) Vol. XXI p.1.

36. K. Gaylord, and M. G. Moharam, Proc. IEEE. Vol.73 p.894, 1985.

37. M. G. Moharam and T. K. Gaylord, "Rigorous coupled-wave analysis of metallic surface-relief gratings", J. Opt. Soc. Am. A, vol. 3, No. 11, p.1780, 1986.

38. K. Hirayama, E. N. Glytsis, and T. K. Gaylord, J.Opt.Soc. Am. A Vol. 14, No. 4 p.907 (1997).

39. S. Banerjee, James B. Cole, T. Yatagai, Calculation of diffraction characteristics of subwavelength conducting Gratings using a high accuracy nonstandard finite-difference time-domain method, Optical Review, in press, 2005.

CHAPTER 5

NONSTANDARD FINITE DIFFERENCE SCHEMES FOR SOLVING NONLINEAR MICRO HEAT TRANSPORT EQUATIONS IN DOUBLE-LAYERED METAL THIN FILMS EXPOSED TO ULTRASHORT PULSED LASERS

Weizhong Dai

Mathematics & Statistics
College of Engineering & Science
Louisiana Tech University
Ruston, LA 71272
dai@coes.latech.edu

Ultrashort-pulsed lasers with pulse durations of the order of sub-picosecond to femtosecond domain possess exclusive capabilities in limiting the undesirable spread of the thermal process zone in the heated sample. Parabolic two-step micro heat transport equations have been widely applied to thermal analysis of thin metal films exposed to picosecond thermal pulses. The temperature-dependent thermal property case is a practical and significant problem encountered in microscale heat transfer. Exploration of temperature-dependent thermal properties is absolutely necessary to advance our fundamental understanding of microscale (ultrafast) heat transport. In this chapter, we introduce nonstandard finite difference schemes, by obtaining continuous energy estimates, for solving the parabolic two-step model with temperature-dependent thermal properties in a double-layered micro thin film irradiated by ultrashort-pulsed lasers. Perfect thermal contact and non-perfect thermal contact at interface are considered. The method is illustrated by several examples which are used to investigate the heat transfer in a gold layer on a chromium layer.

1. Introduction

Ultrashort-pulsed lasers with pulse durations of the order of sub-picosecond to femtosecond possess exclusive capabilities in limiting the undesirable spread of the thermal process zone in the heated sample [1]. In addition to demonstrated applications in structural monitoring of thin metal films [2,3], laser micromachining [4] and patterning [5], structural tailoring of mi-

191

crofilms [6], and laser synthesis and processing in thin-film deposition [7], recent applications of ultrashort-pulsed lasers have been in different disciplines such as physics, chemistry, biology, medicine, and optical technology [8,9,10,11,12,13,14]. The non-contact nature of ultrashort-pulsed lasers has made them an ideal candidate for precise thermal processing of functional nanophase material [15,16]. In a recent issue of *Science*, it was reported that the probability and direction of the molecular motion can be understood as a manifestation of strong coupling between the adsorbates' lateral degrees of freedom and the substrate electronic excitation produced by femtosecond laser radiation [17].

Ultrashort-pulsed lasers have been attracting worldwide interest in science and engineering. The trend in their development has been to increase the heating intensity in order to improve the processing power. Success of high-energy ultrashort-pulsed lasers in real applications relies on three factors [1]: (1) well characterized pulse width, intensity and experimental techniques; (2) reliable microscale heat transfer models; and (3) prevention of thermal damage. It should be pointed out here that ultrafast damage induced by sub-picosecond pulses is intrinsically different from that induced by long-pulse or continuous lasers. For the latter, laser damage is caused by the elevated temperatures resulting from the continuous pumping of photon energy into the processed sample. Therefore, the "damage threshold" in heating by long-pulse lasers is often referred to as the laser intensity that drives the heated spot to the melting temperature. Thermal damage induced by ultrashort pulses in the picosecond domain, on the other hand, occurs after the heating pulse is over. Field induced multiphoton ionization produces free electrons that are rapidly accelerated by the laser pulse. These free electrons mobilize and ionize neighboring atoms through high-frequency collisions, which generate more electrons. For microscale heat transfer in metals, the hot electrons transmit thermal energy to lattices through electron-lattice coupling, resulting in a new thermal property, called the electron-lattice coupling factor. This process continues until a critical density of hot electrons is reached. Under a sufficiently high intensity of heating, the ultrafast damage involves shattering of a thin material layer (from the heated surface) without a clear signature of thermal damage by excessive temperature. Rather than the melt damage caused by high temperatures, there exists a new driving force that brings about ultrafast damage, probably in only a few picoseconds after heating is applied [1].

For a high-energy ultrashort-pulsed laser, ultrafast heating involves high-rate heat flow from electrons to lattices in the picosecond domain. Depending on the temperature, electrons have a heat capacity two to three orders of magnitude smaller than that of lattices. When heated by photons (lasers), the laser energy is primarily absorbed by the free electrons that are confined within skin depth during the excitation. Electrons first shoot up to several hundreds or thousands of degrees within a few picoseconds without disturbing the metal lattices. A major portion of the thermal electron energy is then transferred to the lattices. Meanwhile another part of the energy diffuses to the electrons in the deeper region of the target. Because the pulse duration is so short, the laser is turned off before thermal equilibrium between the electrons and lattices is reached. In this time interval, the heat flux is thus essentially limited to the region within the electron thermal diffusion length. This stage is termed non-equilibrium heating due to the large difference of temperatures in electrons and lattices [18]. The energy equations describing the continuous energy flow from hot electrons to lattices during non-equilibrium heating in a metal micro structure can be written as [18,19,20,21,22,23,24,25]:

$$C_e(T_e)\frac{\partial T_e}{\partial t} = \nabla[k_e(T_e, T_l)\nabla T_e] - G(T_e - T_l) + S, \qquad (1.1)$$

$$C_l\frac{\partial T_l}{\partial t} = G(T_e - T_l), \qquad (1.2)$$

where $C_e(T_e) = A_e T_e$, $k_e(T_e, T_l) = k_0(\frac{T_e}{T_l})$, T_e is electron temperature, T_l lattice temperature, k_0 thermal conductivity in thermal equilibrium, C_e and C_l volumetric heat capacity, G electron-lattice coupling factor, S laser heating source, and ∇ the gradient operator. In the classical theory of diffusion, $T_e = T_l$ because thermal equilibrium between the electrons and lattices is reached. Thus, the above two equations can be reduced to the classical heat conduction equation. However, for sub-picosecond pulses and sub-microscale conditions, $T_e > T_l$ during non-equilibrium heating. The significance of the heat transport equations (1.1) and (1.2) as opposed to the classical heat conduction equations has been discussed in [25].

The above coupled Eqs. (1.1) and (1.2), often referred to as parabolic two-step micro heat transport equations, or a modified dual-phase-lag representation of Eqs. (1.1) and (1.2), have been widely applied for thermal analysis of thin metal films exposed to picosecond thermal pulses. For the temperature-dependent thermal property case, a recent report shows that using temperature-dependent conductivity instead of constant conductivity leads to a better agreement between temperature predictions and corresponding measurements for short-pulse laser heating [26]. Consequently, predictions of desorption using temperature-dependent conductivity should be more accurate than predictions using only constant conductivity because desorption is very sensitive to temperature [26]. Since Eq. (1.1) is highly nonlinear, it must be solved numerically. To date, numerical methods that have been employed for solving the temperature-dependent thermal property case use mainly the Crank-Nicholson finite difference scheme. Qiu and Tien [22,24] employed a semi-implicit Crank-Nicholson scheme to solve Eqs. (1.1) and (1.2) in one dimension in a gold thin film and in a double-layered gold and chromium film. Smith *et al.* [27] also employed the Crank-Nicholson scheme to solve the parabolic two-step heat equations for determining the thermal diffusivity of thin films. Tzou and Chiu [28] modified Eqs. (1.1) and (1.2) to a dual-phase-lag model and employed the Crank-Nicholson scheme to study the temperature-dependent thermal lagging in ultrafast laser heating. Antaki [26] used central difference and forward difference approximations for space and time derivatives, respectively, to study the importance of nonequilibrium thermal conductivity during short-pulse laser-induced desorption from metals. Dai and Nassar [29] developed a linearized three-level finite difference scheme for solving a 1D dual-phase-lag equation with temperature-dependent thermal properties. We have seen only two papers using different methods. One paper, Chen and Beraun [18], employed the corrective smoothed particle method [30,31,32] to obtain a numerical solution of ultrashort laser pulse interactions with metal films, the other, Tzou *et al.* [1], used the differential-difference approach (which retains the time derivatives in a partial differential equation and discretizes the spatial derivatives according to the finite difference schemes) and solved the resulting ordinary differential equations by the fifth-order Gear's backward differentiation formula. In this chapter, we introduce nonstandard finite difference schemes for solving nonlinear parabolic two-step heat conduction equations in a double-layered metallic thin film, which is subjected to an ultrashort-pulsed laser irradiation. Multi-layer metal thinfilms are widely used in engineering applications since a single metal layer often cannot satisfy all mechanical, thermal and electronic requirements. For example, high-power infrared-laser systems often use gold-coated metal mirrors because of their extremely high reflectivity - typically over 97%. Even with such high reflectivity, a small but significant portion of the laser energy is still absorbed in the coatings, which can cause excessive heating and thermal damage to the mirrors.

The rest of this chapter is arranged as follows. In section 2, we will set up the governing equations for thermal analysis of a double-layered metal thin film exposed to an ultrashort-pulsed laser. A continuous energy estimate is obtained. In section 3, we develop a nonstandard finite difference scheme based on the continuous energy estimate. The finite difference scheme is showed to satisfy a discrete analogous of the energy estimate. Numerical examples for thermal analysis of a gold layer on a chromium padding layer are illustrated. In section 4, we extend the method to a double-layered thin film case where the interface is not perfectly thermal-contacted. In the conclusion, we summarize this method and suggest the future research work.

2. Governing Equations and an Energy Estimate

Consider a double-layered thin film with thickness L of the order 0.1 μm, which is subjected to a subpicosecond-pulse irradiation. Based on Eqs. (1.1) and (1.2), the governing equations for thermal analysis in the double-layered thin film can be written as follows:

$$C_e^{(m)}\frac{\partial T_e^{(m)}}{\partial t} = \frac{\partial}{\partial x}(k_e^{(m)}\frac{\partial T_e^{(m)}}{\partial x}) - G^{(m)}(T_e^{(m)} - T_l^{(m)}) + S^{(m)}, \qquad (2.1)$$

$$C_l^{(m)}\frac{\partial T_l^{(m)}}{\partial t} = G^{(m)}(T_e^{(m)} - T_l^{(m)}), \qquad (2.2)$$

where $C_e^{(m)} = A_e T_e^{(m)}$, $k_e^{(m)} = k_0 \frac{T_e^{(m)}}{T_l^{(m)}}$, and $0 \leq x \leq \frac{L}{2}$ when $m = 1$, and $\frac{L}{2} \leq x \leq L$ when $m = 2$. The heat source for both layers is chosen to be [25]

$$S^{(m)}(x,t) = 0.94J[\frac{1-R}{t_p\delta}]e^{-\frac{x}{\delta}}I(t), \qquad (2.3)$$

where $I(t)$ is light intensity of the laser beam, δ the penetration depth of laser radiation, J laser fluence, R the radiative reflectivity of the sample to the laser beam, t_p the full-width-at-half-maximum pulse duration. We assume that the interface is perfectly thermal-contacted and hence the interfacial equations are assumed to be

$$k_e^{(1)}\frac{\partial T_e^{(1)}}{\partial x} = k_e^{(2)}\frac{\partial T_e^{(2)}}{\partial x}, \qquad T_e^{(1)} = T_e^{(2)}, \qquad x = \frac{L}{2}. \qquad (2.4)$$

The initial and boundary conditions are assumed to be

$$T_e^{(m)}(x,0) = T_l^{(m)}(x,0) = T_0(= 300 \text{ K}), \qquad (2.5)$$

and

$$\frac{\partial T_e^{(1)}(0,t)}{\partial x} = \frac{\partial T_l^{(1)}(0,t)}{\partial x} = 0, \qquad \frac{\partial T_e^{(2)}(L,t)}{\partial x} = \frac{\partial T_l^{(2)}(L,t)}{\partial x} = 0. \qquad (2.6)$$

Such boundary conditions arise from the fact that there are no heat losses from the film surfaces in the short time response [25].

Assume that $T_e^{(m)}$ and $T_l^{(m)}$ are smooth, and $T_e^{(m)} \geq T_0$ and $T_l^{(m)} \geq T_0$, where $T_0 = 300$ K. We now seek an energy estimate for the above problem. To this end, we first introduce the L^P-norm:

$$\left\| u^{(m)} \right\|_{L^P} = \left(\int_{I^{(m)}} \left| u^{(m)} \right|^p dx \right)^{\frac{1}{p}}, \qquad 1 < p < +\infty, \qquad (2.7)$$

where $I^{(m)}$ represents the intervals $[0, \frac{L}{2}]$ when $m = 1$ and $[\frac{L}{2}, L]$ when $m = 2$, respectively.

Multiplying Eq. (2.1) by $T_e^{(m)}$ and Eq. (2.2) by $T_l^{(m)}$, integrating them over $I^{(m)}$, and then summing the results over m, we obtain

$$\sum_{m=1}^{2} \int_{I^{(m)}} [A_e^{(m)}(T_e^{(m)})^2 \frac{\partial T_e^{(m)}}{\partial t} + C_l^{(m)} T_l^{(m)} \frac{\partial T_l^{(m)}}{\partial t}] dx$$

$$= \sum_{m=1}^{2} \int_{I^{(m)}} \frac{\partial}{\partial x}(k_e^{(m)} \frac{\partial T_e^{(m)}}{\partial x}) T_e^{(m)} dx - \sum_{m=1}^{2} \int_{I^{(m)}} G^{(m)}(T_e^{(m)} - T_l^{(m)})^2 dx$$

$$+ \sum_{m=1}^{2} \int_{I^{(m)}} S^{(m)} T_e^{(m)} dx. \qquad (2.8)$$

Using the integration by parts and Eqs. (2.4) and (2.6), we have

$$\sum_{m=1}^{2} \int_{I^{(m)}} \frac{\partial}{\partial x}(k_e^{(m)} \frac{\partial T_e^{(m)}}{\partial x}) T_e^{(m)} dx = - \sum_{m=1}^{2} \int_{I^{(m)}} k_e^{(m)}(\frac{\partial T_e^{(m)}}{\partial x})^2 dx$$

$$+ k_e^{(1)} \frac{\partial T_e^{(1)}}{\partial x} T_e^{(1)} \Big|_0^{\frac{L}{2}} + k_e^{(2)} \frac{\partial T_e^{(2)}}{\partial x} T_e^{(2)} \Big|_{\frac{L}{2}}^{L}$$

$$= - \sum_{m=1}^{2} \int_{I^{(m)}} k_e^{(m)}(\frac{\partial T_e^{(m)}}{\partial x})^2 dx. \qquad (2.9)$$

Furthermore, we have

$$\int_{I^{(m)}} A_e^{(m)}(T_e^{(m)})^2 \frac{\partial T_e^{(m)}}{\partial t} dx = \frac{d}{dt} \int_{I^{(m)}} \frac{1}{3} A_e^{(m)}(T_e^{(m)})^3 dx$$

$$= \frac{d}{dt}(\frac{1}{3} A_e^{(m)} \left\| T_e^{(m)} \right\|_{L^3}^3), \qquad (2.10)$$

and

$$\sum_{m=1}^{2} \int_{I^{(m)}} C_l^{(m)} T_l^{(m)} \frac{\partial T_l^{(m)}}{\partial t} dx = \frac{d}{dt} \int_{I^{(m)}} \frac{1}{2} C_l^{(m)}(T_l^{(m)})^2 dx$$

$$= \frac{d}{dt}(\frac{1}{2} C_l^{(m)} \left\| T_l^{(m)} \right\|_{L^2}^2). \qquad (2.11)$$

By Young's inequality with ε (*i.e.*, $ab \le \varepsilon a^p + (\varepsilon p)^{-\frac{q}{p}} q^{-1} b^q, \frac{1}{p} + \frac{1}{q} = 1$, [33]), one can obtain, by choosing $\varepsilon = \frac{1}{3} A_e^{(m)}$, $p = 3$, and $q = \frac{3}{2}$,

$$\int_{I^{(m)}} S^{(m)} T_e^{(m)} dx \le \frac{1}{3} A_e^{(m)} \int_{I^{(m)}} (T_e^{(m)})^3 dx + (\frac{1}{3} A_e^{(m)} 3)^{-\frac{1}{2}} \frac{2}{3} \int_{I^{(m)}} (S^{(m)})^{\frac{3}{2}} dx$$

$$= \frac{1}{3} A_e^{(m)} \left\| T_e^{(m)} \right\|_{L^3}^3 + \frac{2}{3\sqrt{A_e^{(m)}}} \left\| S^{(m)} \right\|_{L^{\frac{3}{2}}}^{\frac{3}{2}} . \qquad (2.12)$$

Substituting Eqs. (2.9)–(2.12) into Eq. (2.8) gives

$$\frac{d}{dt} \sum_{m=1}^{2} [\frac{1}{3} A_e^{(m)} \left\| T_e^{(m)} \right\|_{L^3}^3 + \frac{1}{2} C_l^{(m)} \left\| T_l^{(m)} \right\|_{L^2}^2]$$

$$+ \sum_{m=1}^{2} \int_{I^{(m)}} k_e^{(m)} (\frac{\partial T_e^{(m)}}{\partial x})^2 dx + \sum_{m=1}^{2} \int_{I^{(m)}} G^{(m)}(T_e^{(m)} - T_l^{(m)})^2 dx$$

$$\le \sum_{m=1}^{2} \frac{1}{3} A_e^{(m)} \left\| T_e^{(m)} \right\|_{L^3}^3 + \sum_{m=1}^{2} \frac{2}{3\sqrt{A_e^{(m)}}} \left\| S^{(m)} \right\|_{L^{\frac{3}{2}}}^{\frac{3}{2}} . \qquad (2.13)$$

Let $F(t) = \sum_{m=1}^{2} [\frac{1}{3} A_e^{(m)} \left\| T_e^{(m)} \right\|_{L^3}^3 + \frac{1}{2} C_l^{(m)} \left\| T_l^{(m)} \right\|_{L^2}^2]$ and $Q(t) = \sum_{m=1}^{2} \frac{2}{3\sqrt{A_e^{(m)}}} \left\| S^{(m)} \right\|_{L^{\frac{3}{2}}}^{\frac{3}{2}}$. Then, Eq. (2.13) can be simplified as follows:

$$\frac{dF(t)}{dt} \le F(t) + Q(t). \qquad (2.14)$$

By Gronwall's inequality (*i.e.*, $\eta'(t) \leq \phi(t)\eta(t) + \psi(t) \Rightarrow \eta(t) \leq \exp(\int_0^t \phi(s)ds)[\eta(0) + \int_0^t \psi(s)ds]$ [33]), we obtain

$$F(t) \leq e^t[F(0) + \int_0^t Q(s)ds]. \qquad (2.15)$$

Hence, the following theorem has been obtained.

THEOREM 1 [34]. Assume that the solutions of Eqs. (2.1)–(2.6) are smooth, and $T_e^{(m)} \geq T_0$ and $T_l^{(m)} \geq T_0$, where $T_0 = 300$ K. Then the solutions, $T_e^{(m)}$ and $T_l^{(m)}$, satisfy an energy estimate as follows:

$$\sum_{m=1}^{2} [\frac{1}{3} A_e^{(m)} \left\| T_e^{(m)}(t) \right\|_{L^3}^3 + \frac{1}{2} C_l^{(m)} \left\| T_l^{(m)}(t) \right\|_{L^2}^2]$$

$$\leq e^t \{ \sum_{m=1}^{2} [\frac{1}{3} A_e^{(m)} \left\| T_e^{(m)}(0) \right\|_{L^3}^3 + \frac{1}{2} C_l^{(m)} \left\| T_l^{(m)}(0) \right\|_{L^2}^2]$$

$$+ \int_0^t \sum_{m=1}^{2} \frac{2}{3\sqrt{A_e^{(m)}}} \left\| S^{(m)}(s) \right\|_{L^{\frac{3}{2}}}^{\frac{3}{2}} ds \}. \qquad (2.16)$$

3. Nonstandard Finite Difference Scheme

To develop a nonstandard finite difference scheme that satisfies the discrete analogous of the above energy estimate, Eq. (2.16), we denote $(T_e^{(m)})_j^n$ and $(T_l^{(m)})_j^n$ as the numerical approximations of $(T_e^{(m)})(j\Delta x, n\Delta t)$ and $(T_l^{(m)})(j\Delta x, n\Delta t)$, respectively, where Δx and Δt are the x directional spatial and temporal mesh sizes, respectively, and $0 \leq j \leq N + 1$ so that $(N+1)\Delta x = \frac{L}{2}$. We further introduce the l^p-norm and inner product for mesh functions u_j and v_j as follows:

$$(u, v) = \sum_{j=1}^{N} u_j v_j, \qquad \|u\|_{l^P} = (\Delta x \sum_{j=1}^{N} |u_j|^p \, dx)^{\frac{1}{p}}, \qquad 1 < p < +\infty.$$

The first-order forward and backward finite difference operators, ∇_x and $\nabla_{\bar{x}}$, are defined as follows:

$$\nabla_x u_j = \frac{u_{j+1} - u_j}{\Delta x}, \qquad \nabla_{\bar{x}} u_j = \frac{u_j - u_{j-1}}{\Delta x}.$$

It is noted that the term $A_e^{(m)} T_e^{(m)} \frac{\partial T_e^{(m)}}{\partial t}$ in Eq. (2.1) is nonlinear. Based on the definitions given in [35,36], a method is called a *standard* method if it discretizes the nonlinear term using a local representation, and it is a *nonstandard* method if a nonlocal representation is employed. As such, the standard finite difference approximation for this nonlinear term can be written as

$$(A_e^{(m)}T_e^{(m)}\frac{\partial T_e^{(m)}}{\partial t})|_j^{n+\frac{1}{2}} \approx A_e^{(m)}(T_e^{(m)})_j^{n+\frac{1}{2}}((T_e^{(m)})_j^{n+1}-(T_e^{(m)})_j^{n})/\Delta t. \quad (3.1)$$

Hence, a standard finite difference scheme (Crank-Nicholson scheme) for solving Eqs. (2.1)–(2.2) can be expressed as follows:

$$A_e^{(m)}(T_e^{(m)})_j^{n+\frac{1}{2}}\frac{(T_e^{(m)})_j^{n+1} - (T_e^{(m)})_j^{n}}{\Delta t}$$
$$= \nabla_x\left[(k_e^{(m)})_{j-\frac{1}{2}}^{n+\frac{1}{2}}\nabla_{\bar{x}}\frac{(T_e^{(m)})_j^{n+1} + (T_e^{(m)})_j^{n}}{2}\right]$$
$$- G^{(m)}\left[\frac{(T_e^{(m)})_j^{n+1} + (T_e^{(m)})_j^{n}}{2} - \frac{(T_l^{(m)})_j^{n+1} + (T_l^{(m)})_j^{n}}{2}\right]$$
$$+ (S^{(m)})_j^{n+\frac{1}{2}}, \quad (3.2)$$

and

$$C_l^{(m)}\frac{(T_l^{(m)})_j^{n+1} - (T_l^{(m)})_j^{n}}{\Delta t} = G^{(m)}\left[\frac{(T_e^{(m)})_j^{n+1} + (T_e^{(m)})_j^{n}}{2}\right.$$
$$\left. - \frac{(T_l^{(m)})_j^{n+1} + (T_l^{(m)})_j^{n}}{2}\right], \quad (3.3)$$

where $m = 1, 2$. However, it is difficult to obtain the discrete analogous of Eq. (2.16) from Eqs. (3.2)–(3.3), especially, the discrete analogous of the term $\sum_{m=1}^{2}\frac{1}{3}A_e^{(m)}\left\|T_e^{(m)}(t)\right\|_{L^3}^3$ in Eq. (2.16). To avoid this difficulty, we use a nonlocal representation in time to approximate the nonlinear term $A_e^{(m)}T_e^{(m)}\frac{\partial T_e^{(m)}}{\partial t}$ as follows:

$$\left(A_e^{(m)}T_e^{(m)}\frac{\partial T_e^{(m)}}{\partial t}\right)|_j^{n+\frac{1}{2}}$$

$$\approx A_e^{(m)}\frac{\left|(T_e^{(m)})_j^{n+1}\right|^2 + \left|(T_e^{(m)})_j^{n+1}\right|\cdot\left|(T_e^{(m)})_j^{n}\right| + \left|(T_e^{(m)})_j^{n}\right|^2}{3\frac{(1+\theta)(T_e^{(m)})_j^{n+1}+(1-\theta)(T_e^{(m)})_j^{n}}{2}}$$

$$\times \frac{\left|(T_e^{(m)})_j^{n+1}\right| - \left|(T_e^{(m)})_j^{n}\right|}{\Delta t}$$

$$= A_e^{(m)} \frac{\left|(T_e^{(m)})_j^{n+1}\right|^3 - \left|(T_e^{(m)})_j^n\right|^3}{3\Delta t \frac{(1+\theta)(T_e^{(m)})_j^{n+1}+(1-\theta)(T_e^{(m)})_j^n}{2}}, \tag{3.4}$$

where $0 \le \theta \le 1$ is a parameter. Thus, a nonstandard finite difference scheme for solving Eqs. (2.1)–(2.6) can be developed as follows:

$$A_e^{(m)} \frac{\left|(T_e^{(m)})_j^{n+1}\right|^3 - \left|(T_e^{(m)})_j^n\right|^3}{3\Delta t \frac{(1+\theta)(T_e^{(m)})_j^{n+1}+(1-\theta)(T_e^{(m)})_j^n}{2}}$$

$$= \nabla_x [(k_e^{(m)})_{j-\frac{1}{2}}^{n+\frac{1}{2}} \nabla_{\bar{x}} \frac{(1 + \theta)(T_e^{(m)})_j^{n+1} + (1 - \theta)(T_e^{(m)})_j^n}{2}]$$

$$- G^{(m)}[\frac{(1 + \theta)(T_e^{(m)})_j^{n+1} + (1 - \theta)(T_e^{(m)})_j^n}{2}$$

$$- \frac{(1 + \theta)(T_l^{(m)})_j^{n+1} + (1 - \theta)(T_l^{(m)})_j^n}{2}] + (S^{(m)})_j^{n+\frac{1}{2}}, \tag{3.5}$$

$$C_l^{(m)} \frac{(T_l^{(m)})_j^{n+1} - (T_l^{(m)})_j^n}{\Delta t} = G^{(m)}[\frac{(1 + \theta)(T_e^{(m)})_j^{n+1} + (1 - \theta)(T_e^{(m)})_j^n}{2}$$

$$- \frac{(1 + \theta)(T_l^{(m)})_j^{n+1} + (1 - \theta)(T_l^{(m)})_j^n}{2}], \tag{3.6}$$

where $(k_e^{(m)})_{j-\frac{1}{2}}^{n+\frac{1}{2}} = \frac{1}{2}k_0 \left|\frac{(T_e^{(m)})_j^{n+1}+(T_e^{(m)})_j^n}{(T_l^{(m)})_j^{n+1}+(T_l^{(m)})_j^n}\right| + \frac{1}{2}k_0 \left|\frac{(T_e^{(m)})_{j-1}^{n+1}+(T_e^{(m)})_{j-1}^n}{(T_l^{(m)})_{j-1}^{n+1}+(T_l^{(m)})_{j-1}^n}\right|$ and $0 \le \theta \le 1$. The discrete interfacial equations are chosen to be

$$(k_e^{(1)})_{N+\frac{1}{2}}^{n+\frac{1}{2}} \nabla_{\bar{x}} \frac{(1 + \theta)(T_e^{(1)})_{N+1}^{n+1} + (1 - \theta)(T_e^{(1)})_{N+1}^n}{2}$$

$$= (k_e^{(2)})_{\frac{1}{2}}^{n+\frac{1}{2}} \nabla_{\bar{x}} \frac{(1 + \theta)(T_e^{(2)})_1^{n+1} + (1 - \theta)(T_e^{(2)})_1^n}{2}, \tag{3.7a}$$

$$(T_e^{(1)})_{N+1}^n = (T_e^{(2)})_0^n, \tag{3.7b}$$

for any time level n. The initial and boundary conditions are approximated by

$$(T_e^{(m)})_j^0 = (T_l^{(m)})_j^0 = T_0, \tag{3.8}$$

and

$$\nabla_{\bar{x}}(T_e^{(1)})_1^n = \nabla_{\bar{x}}(T_e^{(2)})_{N+1}^n = 0, \quad \nabla_{\bar{x}}(T_l^{(1)})_1^n = \nabla_{\bar{x}}(T_l^{(2)})_{N+1}^n = 0, \tag{3.9}$$

for any time level n.

We now show that the scheme, Eqs. (3.5)–(3.9), satisfies a discrete analogous of the energy estimate, Eq. (2.16). To this end, we multiply Eq. (3.5) by $\Delta x \frac{(1+\theta)(T_e^{(m)})_j^{n+1}+(1-\theta)(T_e^{(m)})_j^n}{2}$ and Eq. (3.6) by $\Delta x \frac{(1+\theta)(T_l^{(m)})_j^{n+1}+(1-\theta)(T_l^{(m)})_j^n}{2}$, sum j over $1 \le j \le N$, and m over 1 and 2, then add the results together. This gives

$$
\frac{1}{\Delta t} \sum_{m=1}^{2} \frac{1}{3} A_e^{(m)} [\left\| (T_e^{(m)})^{n+1} \right\|_{l^3}^3 - \left\| (T_e^{(m)})^n \right\|_{l^3}^3]
$$

$$
+ \frac{1}{\Delta t} \sum_{m=1}^{2} \frac{1}{2} C_l^{(m)} [\left\| (T_l^{(m)})^{n+1} \right\|_{l^2}^2 - \left\| (T_l^{(m)})^n \right\|_{l^2}^2]
$$

$$
+ \frac{1}{\Delta t} \sum_{m=1}^{2} \frac{\theta}{2} C_l^{(m)} \left\| (T_l^{(m)})^{n+1} - (T_l^{(m)})^n \right\|_{l^2}^2
$$

$$
+ \Delta x \sum_{m=1}^{2} G^{(m)} \sum_{j=1}^{N} [\frac{(1+\theta)(T_e^{(m)})_j^{n+1} + (1-\theta)(T_e^{(m)})_j^n}{2}
$$

$$
- \frac{(1+\theta)(T_l^{(m)})_j^{n+1} + (1-\theta)(T_l^{(m)})_j^n}{2}]^2
$$

$$
= -\Delta x \sum_{m=1}^{2} \sum_{j=1}^{N+1} \nabla_x [(k_e^{(m)})_{j-\frac{1}{2}}^{n+\frac{1}{2}} \nabla_{\bar{x}} \frac{(1+\theta)(T_e^{(m)})_j^{n+1} + (1-\theta)(T_e^{(m)})_j^n}{2}]
$$

$$
\cdot \frac{(1+\theta)(T_e^{(m)})_j^{n+1} + (1-\theta)(T_e^{(m)})_j^n}{2}
$$

$$
+ \sum_{m=1}^{2} ((S^{(m)})^{n+\frac{1}{2}}, \frac{(1+\theta)(T_e^{(m)})^{n+1} + (1-\theta)(T_e^{(m)})^n}{2}). \tag{3.10}
$$

The first term on the right-hand-side (RHS) of Eq. (3.10) is equal to, by Eqs. (3.7) and (3.9),

$$
\text{RHS} = \Delta x \sum_{m=1}^{2} \sum_{j=1}^{N} \frac{1}{\Delta x} [(k_e^{(m)})_{j+\frac{1}{2}}^{n+\frac{1}{2}} \nabla_{\bar{x}} \frac{(1+\theta)(T_e^{(m)})_{j+1}^{n+1} + (1-\theta)(T_e^{(m)})_{j+1}^n}{2}
$$

$$
- (k_e^{(m)})_{j-\frac{1}{2}}^{n+\frac{1}{2}} \nabla_{\bar{x}} \frac{(1+\theta)(T_e^{(m)})_j^{n+1} + (1-\theta)(T_e^{(m)})_j^n}{2}]
$$

$$
\cdot \frac{(1+\theta)(T_e^{(m)})_j^{n+1} + (1-\theta)(T_e^{(m)})_j^n}{2}
$$

$$
= \Delta x \sum_{m=1}^{2} \{ \frac{1}{\Delta x} \sum_{j=2}^{N+1} (k_e^{(m)})_{j-\frac{1}{2}}^{n+\frac{1}{2}} \nabla_{\bar{x}} \frac{(1+\theta)(T_e^{(m)})_j^{n+1} + (1-\theta)(T_e^{(m)})_j^n}{2}
$$

$$\cdot \frac{(1+\theta)(T_e^{(m)})_{j-1}^{n+1} + (1-\theta)(T_e^{(m)})_{j-1}^{n}}{2}$$

$$- \frac{1}{\Delta x} \sum_{j=1}^{N} (k_e^{(m)})_{j-\frac{1}{2}}^{n+\frac{1}{2}} \nabla_{\bar{x}} \frac{(1+\theta)(T_e^{(m)})_{j}^{n+1} + (1-\theta)(T_e^{(m)})_{j}^{n}}{2} \cdot$$

$$\frac{(1+\theta)(T_e^{(m)})_{j}^{n+1} + (1-\theta)(T_e^{(m)})_{j}^{n}}{2} \}$$

$$= -\Delta x \sum_{m=1}^{2} \sum_{j=1}^{N+1} (k_e^{(m)})_{j-\frac{1}{2}}^{n+\frac{1}{2}} [\nabla_{\bar{x}} \frac{(1+\theta)(T_e^{(m)})_{j}^{n+1} + (1-\theta)(T_e^{(m)})_{j}^{n}}{2}]^2$$

$$- \Delta x \sum_{m=1}^{2} \frac{1}{\Delta x} (k_e^{(m)})_{\frac{1}{2}}^{n+\frac{1}{2}} \nabla_{\bar{x}} \frac{(1+\theta)(T_e^{(m)})_{1}^{n+1} + (1-\theta)(T_e^{(m)})_{1}^{n}}{2}$$

$$\cdot \frac{(1+\theta)(T_e^{(m)})_{0}^{n+1} + (1-\theta)(T_e^{(m)})_{0}^{n}}{2}$$

$$+ \Delta x \sum_{m=1}^{2} \frac{1}{\Delta x} (k_e^{(m)})_{N+\frac{1}{2}}^{n+\frac{1}{2}} \nabla_{\bar{x}} \frac{(1+\theta)(T_e^{(m)})_{N+1}^{n+1} + (1-\theta)(T_e^{(m)})_{N+1}^{n}}{2}$$

$$\cdot \frac{(1+\theta)(T_e^{(m)})_{N+1}^{n+1} + (1-\theta)(T_e^{(m)})_{N+1}^{n}}{2}$$

$$= -\Delta x \sum_{m=1}^{2} \sum_{j=1}^{N+1} (k_e^{(m)})_{j-\frac{1}{2}}^{n+\frac{1}{2}} [\nabla_{\bar{x}} \frac{(1+\theta)(T_e^{(m)})_{j}^{n+1} + (1-\theta)(T_e^{(m)})_{j}^{n}}{2}]^2 .$$

$$(3.11)$$

By Young's inequality with ε, we have

$$\sum_{m=1}^{2} ((S^{(m)})^{n+\frac{1}{2}}, \frac{(1+\theta)(T_e^{(m)})^{n+1} + (1-\theta)(T_e^{(m)})^{n}}{2})$$

$$\leq \frac{1}{2} \sum_{m=1}^{2} [(1+\theta) \left| ((S^{(m)})^{n+\frac{1}{2}}, (T_e^{(m)})^{n+1}) \right| + (1-\theta) \left| ((S^{(m)})^{n+\frac{1}{2}}, (T_e^{(m)})^{n}) \right|]$$

$$\leq \frac{1}{2} (1+\theta) \sum_{m=1}^{2} [\frac{1}{3} A_e^{(m)} \left\| (T_e^{(m)})^{n+1} \right\|_{l^3}^{3} + \frac{2}{3\sqrt{A_e^{(m)}}} \left\| (S^{(m)})^{n+\frac{1}{2}} \right\|_{l^{\frac{3}{2}}}^{\frac{3}{2}}]$$

$$+ \frac{1}{2} (1-\theta) \sum_{m=1}^{2} [\frac{1}{3} A_e^{(m)} \left\| (T_e^{(m)})^{n} \right\|_{l^3}^{3} + \frac{2}{3\sqrt{A_e^{(m)}}} \left\| (S^{(m)})^{n+\frac{1}{2}} \right\|_{l^{\frac{3}{2}}}^{\frac{3}{2}}]$$

$$\leq \frac{1}{2} \sum_{m=1}^{2} [(1+\theta) \frac{1}{3} A_e^{(m)} \left\| (T_e^{(m)})^{n+1} \right\|_{l^3}^{3} + \frac{1}{3} (1-\theta) A_e^{(m)} \left\| (T_e^{(m)})^{n} \right\|_{l^3}^{3}]$$

$$+ \sum_{m=1}^{2} \frac{2}{3\sqrt{A_e^{(m)}}} \left\| (S^{(m)})^{n+\frac{1}{2}} \right\|_{l^{\frac{3}{2}}}^{\frac{3}{2}} . \tag{3.12}$$

Substituting Eqs. (3.11) and (3.12) into Eq. (3.10), we obtain

$$\frac{1}{\Delta t} \sum_{m=1}^{2} \frac{1}{3} A_e^{(m)} [\left\| (T_e^{(m)})^{n+1} \right\|_{l^3}^{3} - \left\| (T_e^{(m)})^{n} \right\|_{l^3}^{3}]$$

$$+ \frac{1}{\Delta t} \sum_{m=1}^{2} \frac{1}{2} C_l^{(m)} [\left\| (T_l^{(m)})^{n+1} \right\|_{l^2}^{2} - \left\| (T_l^{(m)})^{n} \right\|_{l^2}^{2}]$$

$$+ \frac{1}{\Delta t} \sum_{m=1}^{2} \frac{\theta}{2} C_l^{(m)} \left\| (T_l^{(m)})^{n+1} - (T_l^{(m)})^{n} \right\|_{l^2}^{2}$$

$$+ \Delta x \sum_{m=1}^{2} G^{(m)} \sum_{j=1}^{N} [\frac{(1+\theta)(T_e^{(m)})_j^{n+1} + (1-\theta)(T_e^{(m)})_j^{n}}{2}$$

$$- \frac{(1+\theta)(T_l^{(m)})_j^{n+1} + (1-\theta)(T_l^{(m)})_j^{n}}{2}]^2$$

$$+ \Delta x \sum_{m=1}^{2} \sum_{j=1}^{N+1} (k_e^{(m)})_{j-\frac{1}{2}}^{n+\frac{1}{2}} [\nabla_{\bar{x}} \frac{(1+\theta)(T_e^{(m)})_j^{n+1} + (1-\theta)(T_e^{(m)})_j^{n}}{2}]^2$$

$$= \frac{1}{2} \sum_{m=1}^{2} [(1+\theta) \frac{1}{3} A_e^{(m)} \left\| (T_e^{(m)})^{n+1} \right\|_{l^3}^{3} + \frac{1}{3} (1-\theta) A_e^{(m)} \left\| (T_e^{(m)})^{n} \right\|_{l^3}^{3}]$$

$$+ \sum_{m=1}^{2} \frac{2}{3\sqrt{A_e^{(m)}}} \left\| (S^{(m)})^{n+\frac{1}{2}} \right\|_{l^{\frac{3}{2}}}^{\frac{3}{2}} . \tag{3.13}$$

Taking out the third, fourth and fifth terms on the RHS of Eq. (3.13) and letting $\tilde{F}(n) = \sum_{m=1}^{2} [\frac{1}{3} A_e^{(m)} \left\| (T_e^{(m)})^{n} \right\|_{l^3}^{3} + \frac{1}{2} C_l^{(m)} \left\| (T_l^{(m)})^{n} \right\|_{l^2}^{2}]$ and $\tilde{Q}(n) = \sum_{m=1}^{2} \frac{2}{3\sqrt{A_e^{(m)}}} \left\| (S^{(m)})^{n-\frac{1}{2}} \right\|_{l^{\frac{3}{2}}}^{\frac{3}{2}}$, we simplify Eq. (3.13) as follows:

$$\tilde{F}(n) \leq \frac{1 + \frac{\Delta t}{2}(1-\theta)}{1 - \frac{\Delta t}{2}(1+\theta)} \tilde{F}(n-1) + \frac{\Delta t}{1 - \frac{\Delta t}{2}(1+\theta)} \tilde{Q}(n)$$

$$\leq \frac{1 + \frac{\Delta t}{2}(1-\theta)}{1 - \frac{\Delta t}{2}(1+\theta)} [\frac{1 + \frac{\Delta t}{2}(1-\theta)}{1 - \frac{\Delta t}{2}(1+\theta)} \tilde{F}(n-2) + \frac{\Delta t}{1 - \frac{\Delta t}{2}(1+\theta)} \tilde{Q}(n-1)]$$

$$+ \frac{\Delta t}{1 - \frac{\Delta t}{2}(1+\theta)} \tilde{Q}(n)$$

$$\leq \cdots$$

$$\leq \left(\frac{1+\frac{\Delta t}{2}(1-\theta)}{1-\frac{\Delta t}{2}(1+\theta)}\right)^n \tilde{F}(0) + \frac{\Delta t}{1-\frac{\Delta t}{2}(1+\theta)}\left[1 + \left(\frac{1+\frac{\Delta t}{2}(1-\theta)}{1-\frac{\Delta t}{2}(1+\theta)}\right)\right.$$

$$\left. + \left(\frac{1+\frac{\Delta t}{2}(1-\theta)}{1-\frac{\Delta t}{2}(1+\theta)}\right)^2 + \cdots + \left(\frac{1+\frac{\Delta t}{2}(1-\theta)}{1-\frac{\Delta t}{2}(1+\theta)}\right)^{n-1}\right] \max_{0\leq k\leq n} \tilde{Q}(k)$$

$$\leq \left(\frac{1+\frac{\Delta t}{2}(1-\theta)}{1-\frac{\Delta t}{2}(1+\theta)}\right)^n \tilde{F}(0) + \frac{\Delta t}{1-\frac{\Delta t}{2}(1+\theta)}\left[\frac{1-\left(\frac{1+\frac{\Delta t}{2}(1-\theta)}{1-\frac{\Delta t}{2}(1+\theta)}\right)^n}{1-\left(\frac{1+\frac{\Delta t}{2}(1-\theta)}{1-\frac{\Delta t}{2}(1+\theta)}\right)}\right] \max_{0\leq k\leq n} \tilde{Q}(k)$$

$$\leq \left(\frac{1+\frac{\Delta t}{2}(1-\theta)}{1-\frac{\Delta t}{2}(1+\theta)}\right)^n \tilde{F}(0) - \left[1 - \left(\frac{1+\frac{\Delta t}{2}(1-\theta)}{1-\frac{\Delta t}{2}(1+\theta)}\right)^n\right] \max_{0\leq k\leq n} \tilde{Q}(k)$$

$$\leq \left(\frac{1+\frac{\Delta t}{2}(1-\theta)}{1-\frac{\Delta t}{2}(1+\theta)}\right)^n [\tilde{F}(0) + \max_{0\leq k\leq n} \tilde{Q}(k)]. \tag{3.14}$$

Using the inequalities $(1+\varepsilon)^n \leq e^{n\varepsilon}$ for $\varepsilon > 0$, and $(1-\varepsilon)^{-1} \leq e^{2\varepsilon}$ when $0 < \varepsilon \leq \frac{1}{2}$, we obtain, for a sufficiently small $\Delta t > 0$,

$$\tilde{F}(n) \leq e^{n\frac{\Delta t}{2}(1-\theta)} \cdot e^{n\Delta t(1+\theta)}[\tilde{F}(0) + \max_{0\leq k\leq n} \tilde{Q}(k)]$$

$$\leq e^{2n\Delta t}[\tilde{F}(0) + \max_{0\leq k\leq n} \tilde{Q}(k)]. \tag{3.15}$$

Hence, the following theorem has been obtained.

THEOREM 2. Suppose that $(T_e^{(m)})_j^n$ and $(T_l^{(m)})_j^n$ are the solutions of the scheme, Eqs. (3.5)–(3.9). Then the solutions satisfy

$$\sum_{m=1}^{2}\left[\frac{1}{3}A_e^{(m)}\left\|(T_e^{(m)})^n\right\|_{l^3}^3 + \frac{1}{2}C_l^{(m)}\left\|(T_l^{(m)})^n\right\|_{l^2}^2\right]$$

$$\leq e^{2n\Delta t}\left\{\sum_{m=1}^{2}\left[\frac{1}{3}A_e^{(m)}\left\|(T_e^{(m)})^0\right\|_{l^3}^3 + \frac{1}{2}C_l^{(m)}\left\|(T_l^{(m)})^0\right\|_{l^2}^2\right]\right.$$

$$\left. + \max_{0\leq k\leq n}\sum_{m=1}^{2}\frac{2}{3\sqrt{A_e^{(m)}}}\left\|(S^{(m)})^{k-\frac{1}{2}}\right\|_{l^{\frac{3}{2}}}^{\frac{3}{2}}\right\}. \tag{3.16}$$

It can be seen that Eq. (3.16) is a discrete analogous of Eq. (2.16).

The advantage of the present scheme can be seen by a preliminary test, which involves investigating the temperature rise in a double-layered thin film, namely a gold layer on a chromium padding layer, where the gold layer and the chromium padding layer are 0.05 μm in thickness. The values of thermal properties for both gold and chromium are chosen based on the following table.

Table 3.1. Thermal properties for both gold and chromium [25].

Parameters	Gold($m = 1$)	Chromium($m = 2$)
k_0 (J/mmKps)	3.15×10^{-13}	9.4×10^{-14}
C_l (J/mm^{3}K)	2.5×10^{-3}	3.3×10^{-3}
G (J/mm^{3}Kps)	2.6×10^{-5}	42×10^{-5}
A_e (J/mm^{3}K)	7.0×10^{-8}	1.933×10^{-7}

In this test, we first choose the heat source to be a single Gaussian pulse, $S(x,t) = 0.94J[\frac{1-R}{t_p\delta}]\, e^{-\frac{x}{\delta}-2.77(\frac{t-2t_p}{t_p})^2}$, or a repetitive pulse, $S(x,t) = 0.94J[\frac{1-R}{t_p\delta}]e^{-\frac{x}{\delta}}[e^{-2.77(\frac{t-2t_p}{t_p})^2}+e^{-2.77(\frac{t-4t_p}{t_p})^2}]$, with the fluence $J = 13.4\ \frac{J}{m^2}$, $\Delta t = 0.005$ ps, and $\Delta x = 2.5 \times 10^{-8}$ mm. We compare the present scheme ($\theta = 0$) with the popular method (Crank-Nicholson scheme, Eqs. (3.2)–(3.3)) [22,24,26,27,28]. Figures 3.1 and 3.2 show the change in temperature ($\frac{\Delta T_e}{(\Delta T_e)_{\max}}$) on the surface of the gold layer. It can be seen that there are no differences between these two solutions. We then choose the heat source to be a single step pulse, $S(x,t) = 0.94J[\frac{1-R}{t_p\delta}]e^{-\frac{x}{\delta}}$, when 20 ps $\leq t \leq 22$ ps, and $S(x,t) = 0$ at other times, with the fluence $J = 13.4\ \frac{J}{m^2}$, $\Delta t = 0.5$ ps, and $\Delta x = 2.5 \times 10^{-8}$ mm. Again, we compare the present scheme ($\theta = 0.75$) with the Crank-Nicholson scheme. Figure 3.3 shows the change in temperature ($\frac{\Delta T_e}{(\Delta T_e)_{\max}}$) on the surface of the gold layer. It can be seen that the Crank-Nicholson scheme produces an oscillatory solution. A similar result is obtained for a repetitive step pulse, $S(x,t) = 0.94J[\frac{1-R}{t_p\delta}]e^{-\frac{x}{\delta}}$ when 20 ps $\leq t \leq 22$ ps and 30 ps $\leq t \leq 32$ ps, and $S(x,t) = 0$ at other times, as shown in Figure 3.4. Hence, we conclude that in general the present scheme is better than the Crank-Nicholson scheme.

Figures 3.5 and 3.6 give the electron temperature profiles and lattice temperature profiles along the x-axis for time $t = 0.2, 0.25,$ and 0.5 ps, respectively, when $J = 13.4\ \frac{J}{m^2}$. Results are obtained using a time increment of 0.005 ps and grid points of 100 for each layer.

We further employ the present scheme ($\theta = 0$) to solve the parabolic two-step heat transport equations with other fluencies [34]. Figures 3.7 and 3.8 plot the electron temperature profiles and lattice temperature profiles when $J = 10.0\ \frac{J}{m^2}$. Results in the gold film are similar to those obtained in [18] (see Figures 1 and 2 in [18]). To compare with those obtained in [24], we set the initial condition at $-2t_p$. Figures 3.9 and 3.10 plot the electron temperature profiles and lattice temperature profiles with $J = 500.0\ \frac{J}{m^2}$, respectively. Results have no significant differences from those obtained by the parabolic two-step model (see Figure 6 in [24]).

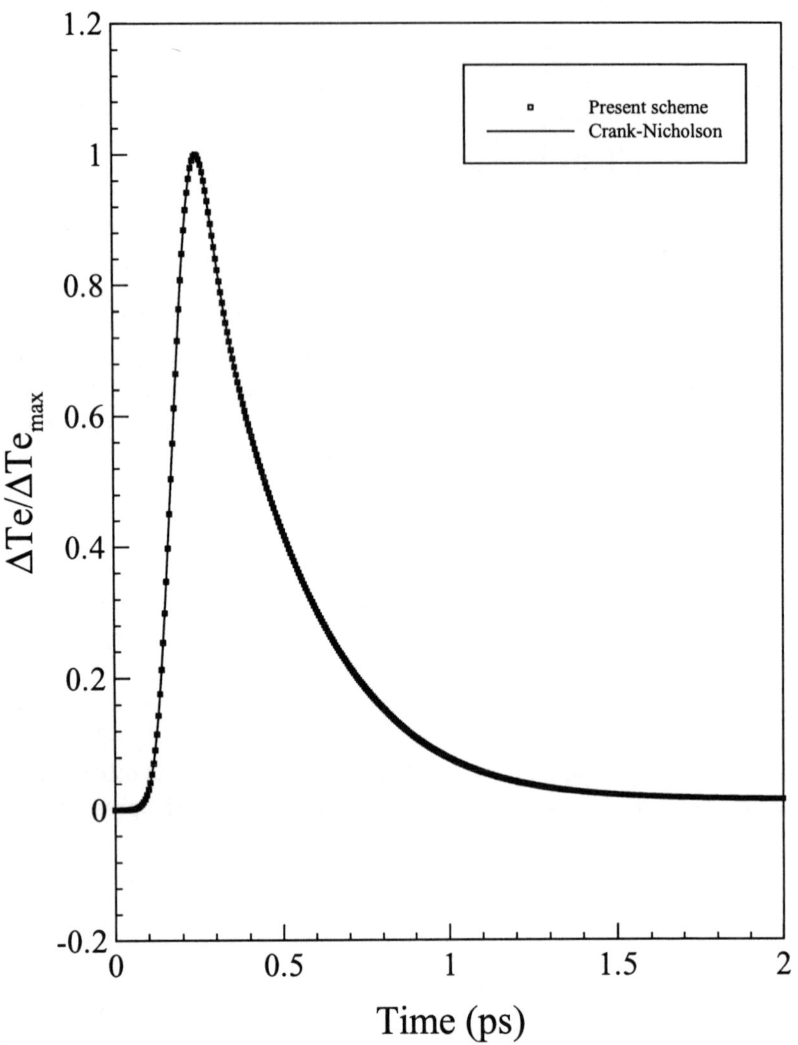

Figure 3.1. Normalized electron temperatures at the front surface irradiated with a Gaussian laser pulse.

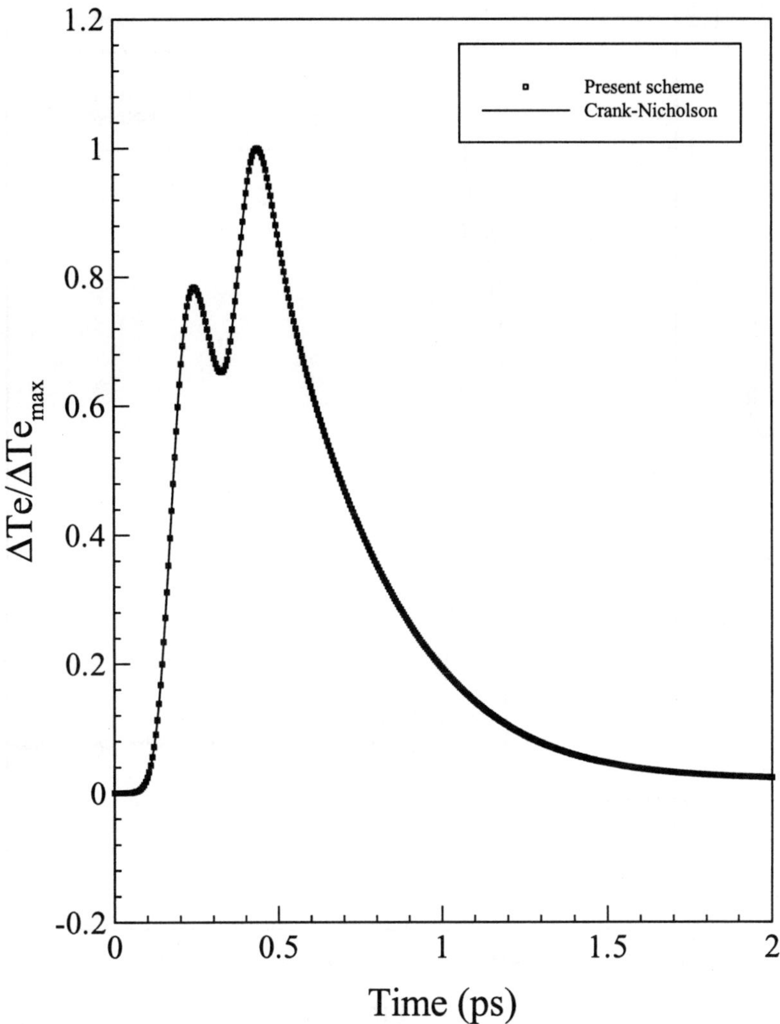

Figure 3.2. Normalized electron temperatures at the front surface irradiated with a repetitive Gaussian laser pulse.

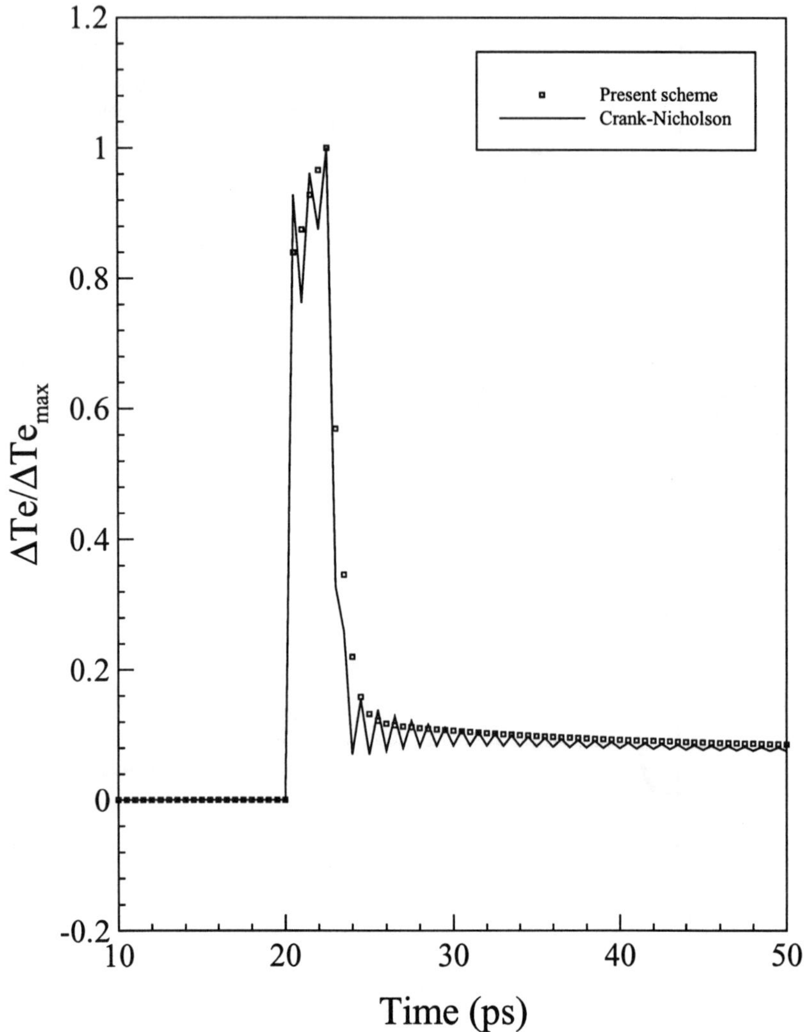

Figure 3.3. Normalized electron temperatures at the front surface irradiated with a single step laser pulse.

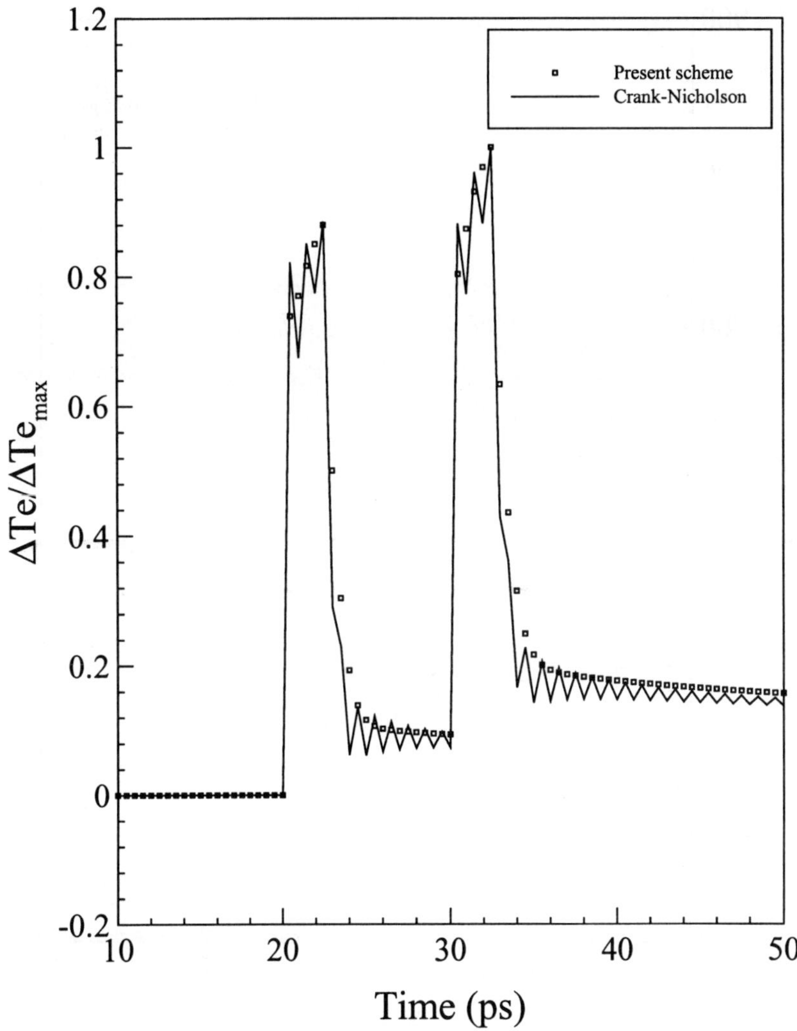

Figure 3.4. Normalized electron temperature at the front surface irradiated with a repetitive step pulse.

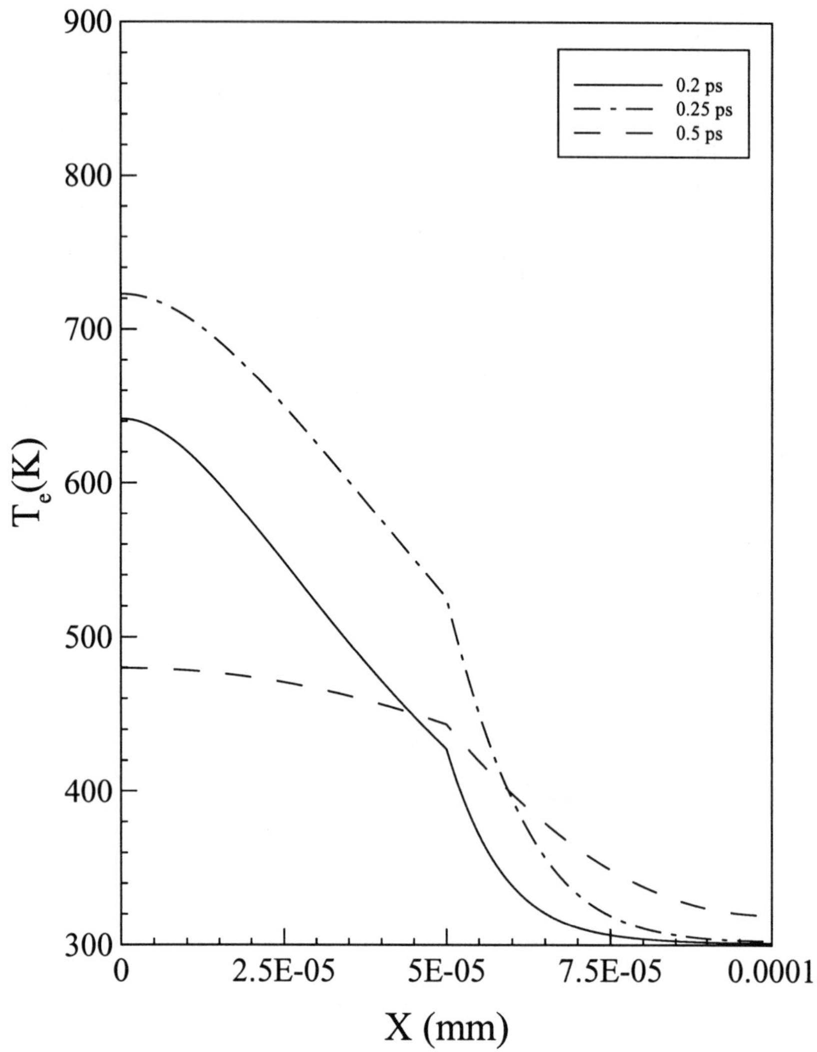

Figure 3.5. Calculated electron temperature profiles for a 100-nm gold and chromium thin film irradiated with a 0.1 ps laser pulse at a fluence of 13.4 J/m^2.

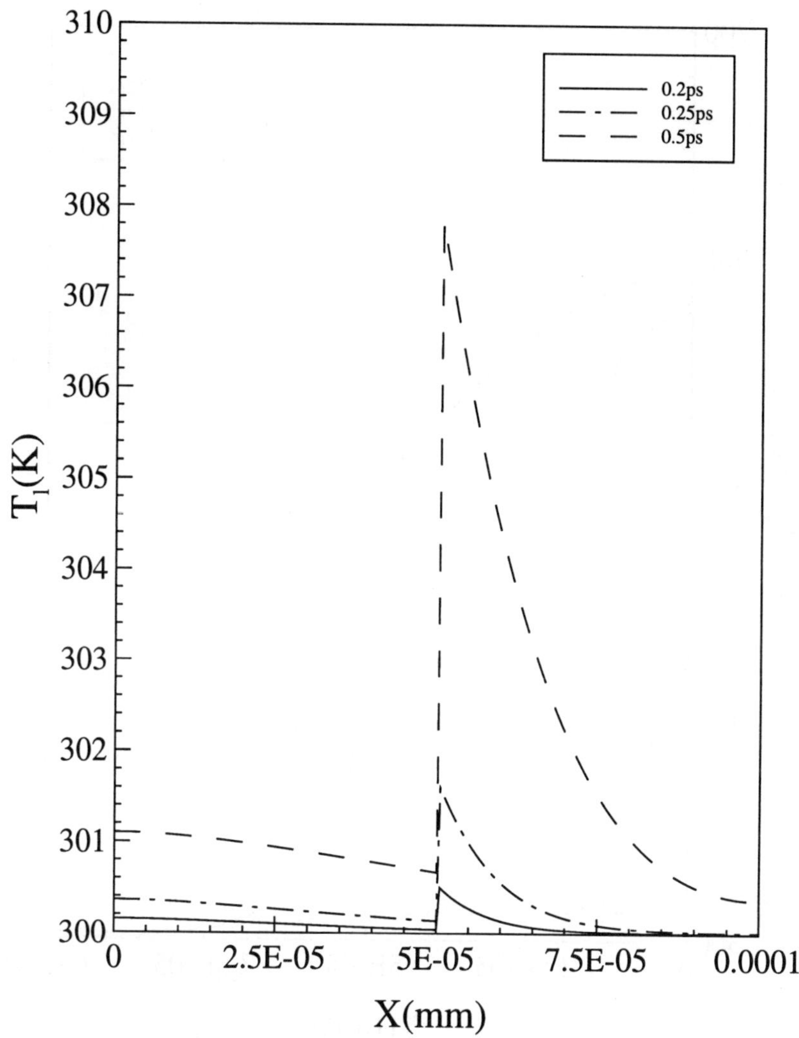

Figure 3.6. Calculated lattice temperature profiles for a 100-nm gold and chromium thin film irradiated with a 0.1 ps laser pulse at a fluence of 13.4 J/m².

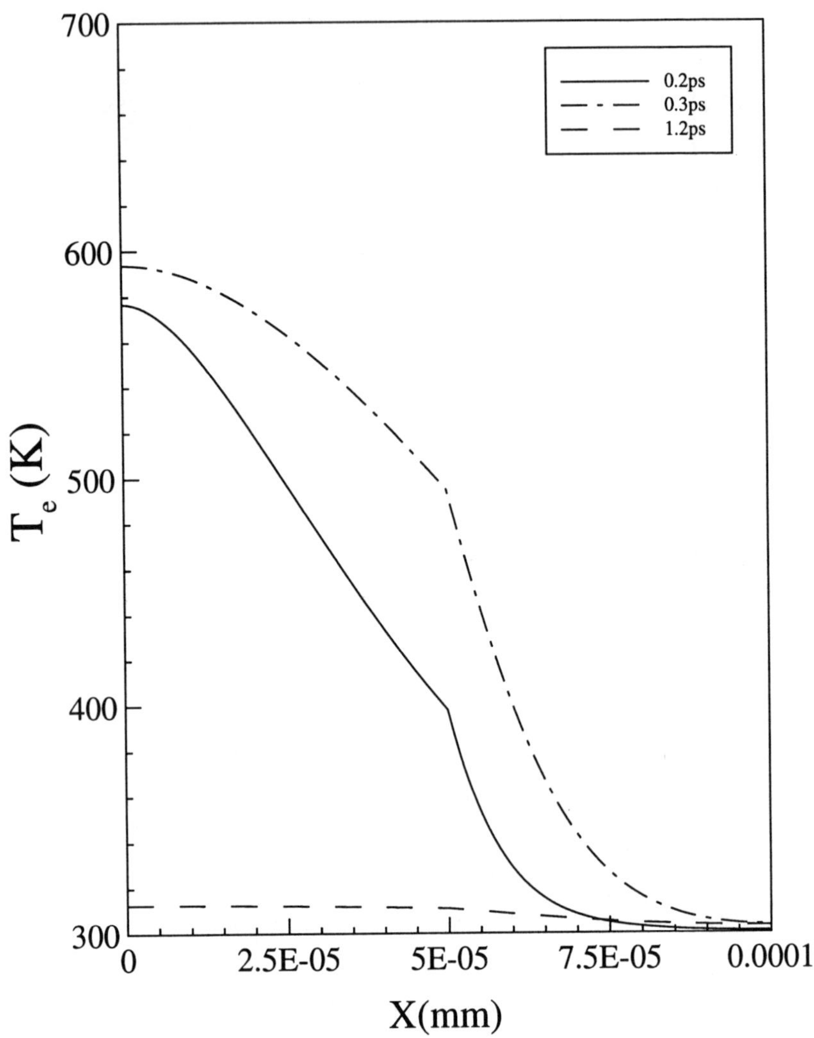

Figure 3.7. Calculated electron temperature profiles for a 100-nm gold and chromium thin film irradiated with a 0.1 ps laser pulse at a fluence of 10.0 J/m^2.

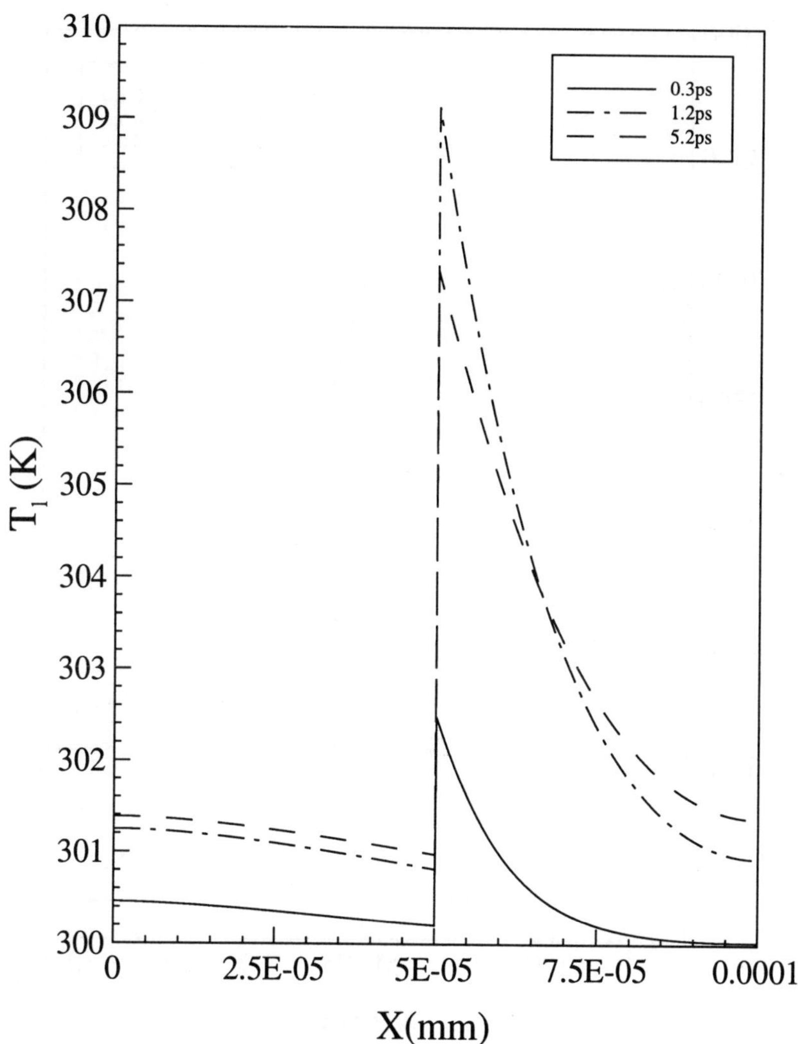

Figure 3.8. Calculated lattice temperature profiles for a 100-nm gold and chromium thin film irradiated with a 0.1 ps laser pulse at a fluence of 10.0 J/m^2.

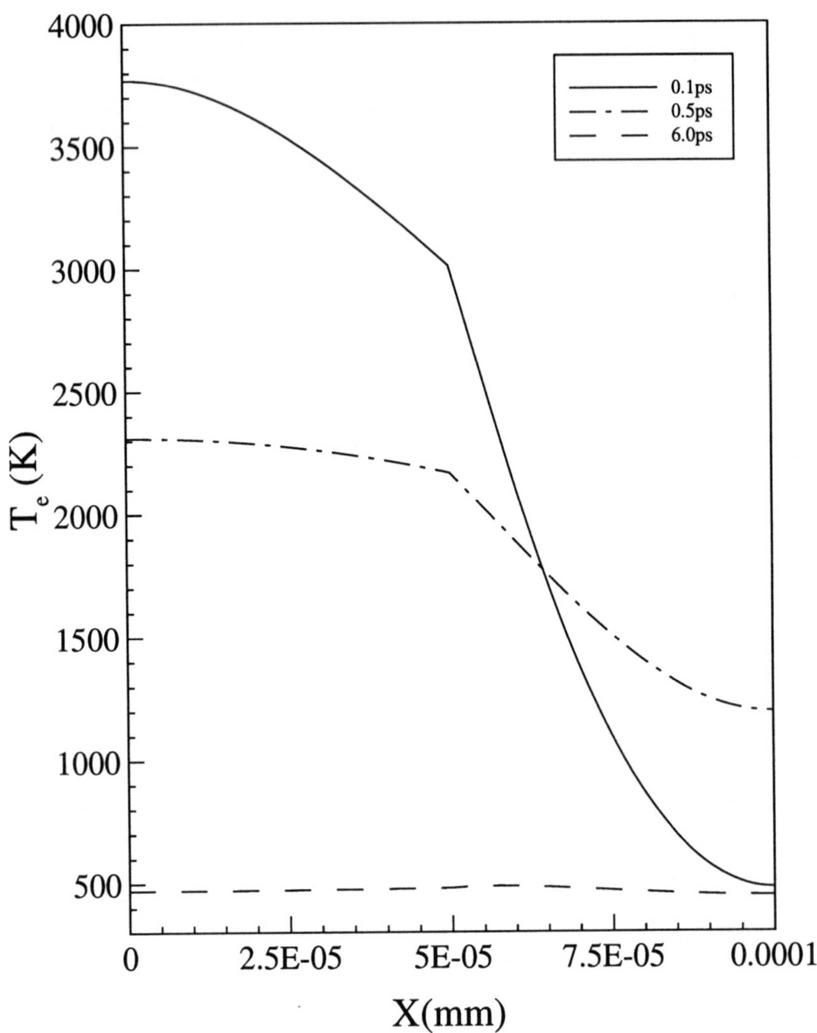

Figure 3.9. Calculated electron temperature profiles for a 100-nm gold and chromium thin film irradiated with a 0.1 ps laser pulse at a fluence of 500.0 J/m^2.

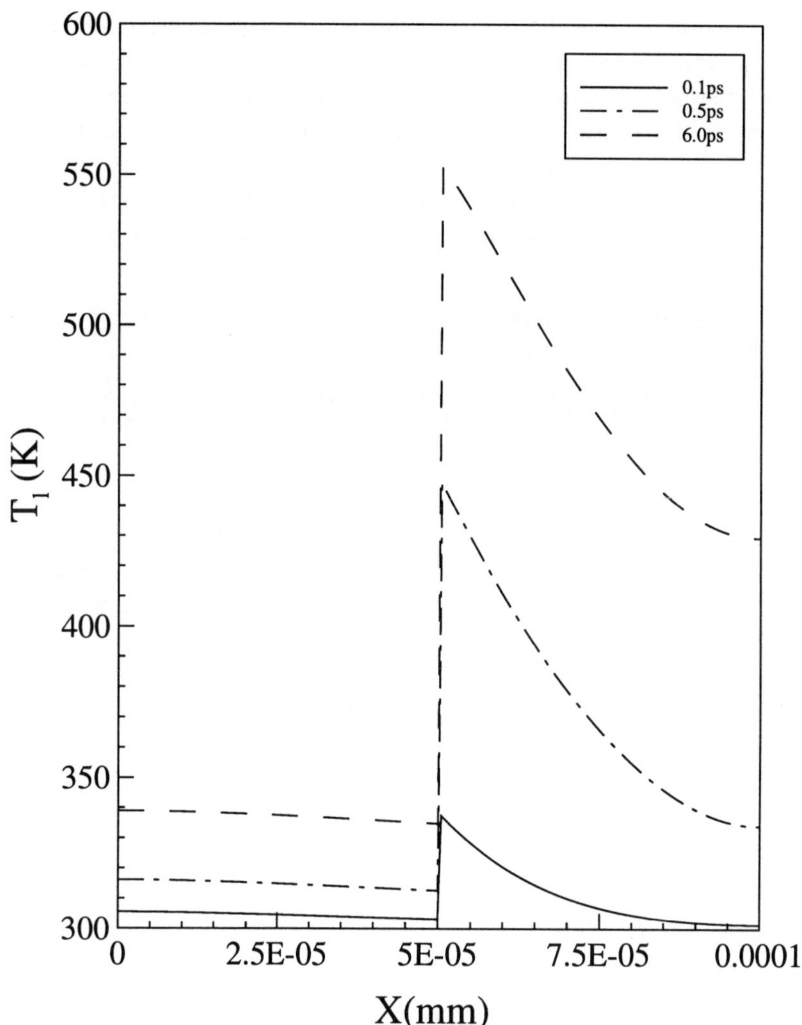

Figure 3.10. Calculated lattice temperature profiles for a 100-nm gold and chromium thin film irradiated with a 0.1 ps laser pulse at a fluence of 500.0 J/m^2.

4. Nonlinear Interfacial Problem

In this section, we consider the parabolic two-step model with temperature-dependent thermal properties in a double-layered micro thin film with non-perfect thermal contact at interface. This case is important for studying the effects of acoustic and diffuse phonon mismatch across the interface [37,38,39], because such an effect will generate additional nonlinear behavior in the interfacial area. For this case, the governing equations are the same as Eqs. (2.1)–(2.6), except that Eq. (2.4) is replaced by the following interfacial equations [37,38,39]

$$-k_e^{(1)}\frac{\partial T_e^{(1)}}{\partial x} = -k_e^{(2)}\frac{\partial T_e^{(2)}}{\partial x}$$

$$= k_c[(T_e^{(1)})^4 - (T_e^{(2)})^4], \qquad T_e^{(1)} \neq T_e^{(2)}, \qquad x = \frac{L}{2}. \qquad (4.1)$$

Here, k_c is a positive constant.

Again, we assume that $T_e^{(m)}$ and $T_l^{(m)}$ are smooth, and $T_e^{(m)} \geq T_0$ and $T_l^{(m)} \geq T_0$, where $T_0 = 300$ K. We would like to obtain an energy estimate for this problem. To this end, we multiply Eq. (2.1) by $T_e^{(m)}$ and Eq. (2.2) by $T_l^{(m)}$, integrating them over $I^{(m)}$, and then summing the results over m. This gives

$$\sum_{m=1}^{2}\int_{I^{(m)}} [A_e^{(m)}(T_e^{(m)})^2\frac{\partial T_e^{(m)}}{\partial t} + C_l^{(m)}T_l^{(m)}\frac{\partial T_l^{(m)}}{\partial t}]dx$$

$$= \sum_{m=1}^{2}\int_{I^{(m)}} \frac{\partial}{\partial x}(k_e^{(m)}\frac{\partial T_e^{(m)}}{\partial x})T_e^{(m)}dx - \sum_{m=1}^{2}\int_{I^{(m)}} G^{(m)}(T_e^{(m)} - T_l^{(m)})^2dx$$

$$+ \sum_{m=1}^{2}\int_{I^{(m)}} S^{(m)}T_e^{(m)}dx. \qquad (4.2)$$

Using the integration by parts and Eqs. (2.6) and (4.1), we obtain

$$\sum_{m=1}^{2}\int_{I^{(m)}} \frac{\partial}{\partial x}(k_e^{(m)}\frac{\partial T_e^{(m)}}{\partial x})T_e^{(m)}dx$$

$$= -\sum_{m=1}^{2}\int_{I^{(m)}} k_e^{(m)}(\frac{\partial T_e^{(m)}}{\partial x})^2dx + k_e^{(1)}\frac{\partial T_e^{(1)}}{\partial x}T_e^{(1)}|_0^{\frac{L}{2}}$$

$$+ k_e^{(2)}\frac{\partial T_e^{(2)}}{\partial x}T_e^{(2)}|_{\frac{L}{2}}^{L}$$

$$= -\sum_{m=1}^{2} \int_{I^{(m)}} k_e^{(m)} \left(\frac{\partial T_e^{(m)}}{\partial x} \right)^2 dx$$

$$- k_c[(T_e^{(1)})^4 - (T_e^{(2)})^4] \cdot [T_e^{(1)} - T_e^{(2)}]\big|_{\frac{L}{2}}$$

$$= -\sum_{m=1}^{2} \int_{I^{(m)}} k_e^{(m)} \left(\frac{\partial T_e^{(m)}}{\partial x} \right)^2 dx$$

$$- k_c[(T_e^{(1)})^2 + (T_e^{(2)})^2] \cdot [T_e^{(1)} + T_e^{(2)}] \cdot [T_e^{(1)} - T_e^{(2)}]^2\big|_{\frac{L}{2}} . \tag{4.3}$$

Using a similar argument as before, Eq. (4.3) can be simplified as follows.

$$\frac{d}{dt} \sum_{m=1}^{2} \left[\frac{1}{3} A_e^{(m)} \left\| T_e^{(m)} \right\|_{L^3}^3 + \frac{1}{2} C_l^{(m)} \left\| T_l^{(m)} \right\|_{L^2}^2 \right]$$

$$+ \sum_{m=1}^{2} \int_{I^{(m)}} k_e^{(m)} \left(\frac{\partial T_e^{(m)}}{\partial x} \right)^2 dx + \sum_{m=1}^{2} \int_{I^{(m)}} G^{(m)} (T_e^{(m)} - T_l^{(m)})^2 dx$$

$$+ k_c[(T_e^{(1)})^2 + (T_e^{(2)})^2] \cdot [T_e^{(1)} + T_e^{(2)}] \cdot [T_e^{(1)} - T_e^{(2)}]^2\big|_{\frac{L}{2}}$$

$$\leq \sum_{m=1}^{2} \frac{1}{3} A_e^{(m)} \left\| T_e^{(m)} \right\|_{L^3}^3 + \sum_{m=1}^{2} \frac{2}{3\sqrt{A_e^{(m)}}} \left\| S^{(m)} \right\|_{L^{\frac{3}{2}}}^{\frac{3}{2}} . \tag{4.4}$$

Taking out the third, fourth and fifth terms on the RHS of Eq. (4.4) since they are positive, and letting $F(t) = \sum_{m=1}^{2} [\frac{1}{3} A_e^{(m)} \left\| T_e^{(m)} \right\|_{L^3}^3 + \frac{1}{2} C_l^{(m)} \left\| T_l^{(m)} \right\|_{L^2}^2]$ and $Q(t) = \sum_{m=1}^{2} \frac{2}{3\sqrt{A_e^{(m)}}} \left\| S^{(m)} \right\|_{L^{\frac{3}{2}}}^{\frac{3}{2}}$, we simplify Eq. (4.4) as follows:

$$\frac{dF(t)}{dt} \leq F(t) + Q(t). \tag{4.5}$$

By Gronwall's inequality, we obtain

$$F(t) \leq e^t [F(0) + \int_0^t Q(s)ds]. \tag{4.6}$$

Hence, the following theorem has been obtained.

THEOREM 3 [34]. Assume that the solutions of the problem are smooth, and $T_e^{(m)} \geq T_0$ and $T_l^{(m)} \geq T_0$, where $T_0 = 300$ K. Then the solutions satisfy an energy estimate as follows:

$$\sum_{m=1}^{2}[\frac{1}{3}A_e^{(m)}\left\|T_e^{(m)}(t)\right\|_{L^3}^{3}+\frac{1}{2}C_l^{(m)}\left\|T_l^{(m)}(t)\right\|_{L^2}^{2}]$$

$$\leq e^t\{\sum_{m=1}^{2}[\frac{1}{3}A_e^{(m)}\left\|T_e^{(m)}(0)\right\|_{L^3}^{3}+\frac{1}{2}C_l^{(m)}\left\|T_l^{(m)}(0)\right\|_{L^2}^{2}]$$

$$+\int_0^t\sum_{m=1}^{2}\frac{2}{3\sqrt{A_e^{(m)}}}\left\|S^{(m)}(s)\right\|_{L^{\frac{3}{2}}}^{\frac{3}{2}}\,ds\}. \tag{4.7}$$

It can be seen that Eq. (4.7) is the same as Eq. (2.16). This is true because the energy does not lose to the surroundings whether or not the film is perfect or non-perfect thermal contact at the interface. It is noted that the term $k_c[(T_e^{(1)})^4-(T_e^{(2)})^4]$ in Eq. (4.1) is nonlinear. A standard finite difference approximation for this term at $(n+\frac{1}{2})\Delta t$ can be expressed as $k_c\{[(T_e^{(1)})^4]^{n+\frac{1}{2}}-[(T_e^{(2)})^4]^{n+\frac{1}{2}}\}$. We have found that it is difficult to obtain an discrete analogous of the above continuous energy estimate, Eq. (4.7). Thus, we employ a nonstandard finite difference approximation based on the idea in [40],

$$k_c[(T_e^{(1)})^4-(T_e^{(2)})^4]^{n+\frac{1}{2}}\approx k_c[(\frac{(1+\theta)(T_e^{(1)})_{N+1}^{n+1}+(1-\theta)(T_e^{(1)})_{N+1}^{n}}{2})^2$$

$$+(\frac{(1+\theta)(T_e^{(2)})_{0}^{n+1}+(1-\theta)(T_e^{(2)})_{0}^{n}}{2})^2]$$

$$\cdot|\frac{(1+\theta)(T_e^{(1)})_{N+1}^{n+1}+(1-\theta)(T_e^{(1)})_{N+1}^{n}}{2}$$

$$+\frac{(1+\theta)(T_e^{(2)})_{0}^{n+1}+(1-\theta)(T_e^{(2)})_{0}^{n}}{2}|$$

$$\cdot[\frac{(1+\theta)(T_e^{(1)})_{N+1}^{n+1}+(1-\theta)(T_e^{(1)})_{N+1}^{n}}{2}$$

$$-\frac{(1+\theta)(T_e^{(2)})_{0}^{n+1}+(1-\theta)(T_e^{(2)})_{0}^{n}}{2}]. \tag{4.8}$$

Hence, a nonstandard finite difference scheme for solving the nonlinear interfacial problem can be written as follows:

$$A_e^{(m)}\frac{\left|(T_e^{(m)})_j^{n+1}\right|^3-\left|(T_e^{(m)})_j^{n}\right|^3}{3\Delta t\frac{(1+\theta)(T_e^{(m)})_j^{n+1}+(1-\theta)(T_e^{(m)})_j^{n}}{2}}$$

$$= \nabla_x[(k_e^{(m)})_{j-\frac{1}{2}}^{n+\frac{1}{2}}\nabla_{\bar{x}}\frac{(1+\theta)(T_e^{(m)})_j^{n+1}+(1-\theta)(T_e^{(m)})_j^n}{2}]$$

$$-G^{(m)}[\frac{(1+\theta)(T_e^{(m)})_j^{n+1}+(1-\theta)(T_e^{(m)})_j^n}{2}$$

$$-\frac{(1+\theta)(T_l^{(m)})_j^{n+1}+(1-\theta)(T_l^{(m)})_j^n}{2}]+(S^{(m)})_j^{n+\frac{1}{2}}, \qquad (4.9)$$

$$C_l^{(m)}\frac{(T_l^{(m)})_j^{n+1}-(T_l^{(m)})_j^n}{\Delta t}=G^{(m)}[\frac{(1+\theta)(T_e^{(m)})_j^{n+1}+(1-\theta)(T_e^{(m)})_j^n}{2}$$

$$-\frac{(1+\theta)(T_l^{(m)})_j^{n+1}+(1-\theta)(T_l^{(m)})_j^n}{2}], \qquad (4.10)$$

where $0 \le \theta \le 1$,

$$(k_e^{(m)})_{j-\frac{1}{2}}^{n+\frac{1}{2}}=\frac{1}{2}k_0|\frac{(T_e^{(m)})_j^{n+1}+(T_e^{(m)})_j^n}{(T_l^{(m)})_j^{n+1}+(T_l^{(m)})_j^n}|+\frac{1}{2}k_0|\frac{(T_e^{(m)})_{j-1}^{n+1}+(T_e^{(m)})_{j-1}^n}{(T_l^{(m)})_{j-1}^{n+1}+(T_l^{(m)})_{j-1}^n}|,$$

and $m = 1, 2$. The interfacial equation is approximated by

$$-(k_e^{(1)})_{N+\frac{1}{2}}^{n+\frac{1}{2}}\nabla_{\bar{x}}\frac{(1+\theta)(T_e^{(1)})_{N+1}^{n+1}+(1-\theta)(T_e^{(1)})_{N+1}^n}{2}$$

$$=-(k_e^{(2)})_{\frac{1}{2}}^{n+\frac{1}{2}}\nabla_{\bar{x}}\frac{(1+\theta)(T_e^{(2)})_1^{n+1}+(1-\theta)(T_e^{(2)})_1^n}{2}$$

$$=k_c[(\frac{(1+\theta)(T_e^{(1)})_{N+1}^{n+1}+(1-\theta)(T_e^{(1)})_{N+1}^n}{2})^2$$

$$+(\frac{(1+\theta)(T_e^{(2)})_0^{n+1}+(1-\theta)(T_e^{(2)})_0^n}{2})^2]$$

$$\cdot|\frac{(1+\theta)(T_e^{(1)})_{N+1}^{n+1}+(1-\theta)(T_e^{(1)})_{N+1}^n}{2}$$

$$+\frac{(1+\theta)(T_e^{(2)})_0^{n+1}+(1-\theta)(T_e^{(2)})_0^n}{2}|$$

$$\cdot[\frac{(1+\theta)(T_e^{(1)})_{N+1}^{n+1}+(1-\theta)(T_e^{(1)})_{N+1}^n}{2}$$

$$-\frac{(1+\theta)(T_e^{(2)})_0^{n+1}+(1-\theta)(T_e^{(2)})_0^n}{2}], \qquad (4.11)$$

for any time level n. The initial and boundary conditions are chosen to be

$$(T_e^{(m)})_j^0=(T_l^{(m)})_j^0=T_0, \qquad (4.12)$$

220   *W. Dai*

and

$$\nabla_{\bar{x}}(T_e^{(1)})_1^n = \nabla_{\bar{x}}(T_e^{(2)})_{N+1}^n = 0, \quad \nabla_{\bar{x}}(T_l^{(1)})_1^n = \nabla_{\bar{x}}(T_l^{(2)})_{N+1}^n = 0, \quad (4.13)$$

for any time level n.

To show that the above scheme satisfies a discrete analogous of Eq. (4.7), we multiply Eq. (4.9) by $\Delta x \dfrac{(1+\theta)(T_e^{(m)})_j^{n+1}+(1-\theta)(T_e^{(m)})_j^n}{2}$ and Eq. (4.10) by $\Delta x \dfrac{(1+\theta)(T_l^{(m)})_j^{n+1}+(1-\theta)(T_l^{(m)})_j^n}{2}$, sum j over $1 \le j \le N$, and m over 1 and 2, then add the results together. This gives

$$\frac{1}{\Delta t}\sum_{m=1}^{2}\frac{1}{3}A_e^{(m)}[\left\|(T_e^{(m)})^{n+1}\right\|_{l^3}^3 - \left\|(T_e^{(m)})^n\right\|_{l^3}^3]$$

$$+ \frac{1}{\Delta t}\sum_{m=1}^{2}\frac{1}{2}C_l^{(m)}[\left\|(T_l^{(m)})^{n+1}\right\|_{l^2}^2 - \left\|(T_l^{(m)})^n\right\|_{l^2}^2]$$

$$+ \Delta x \sum_{m=1}^{2}G^{(m)}\sum_{j=1}^{N}[\frac{(1+\theta)(T_e^{(m)})_j^{n+1}+(1-\theta)(T_e^{(m)})_j^n}{2}$$

$$- \frac{(1+\theta)(T_l^{(m)})_j^{n+1}+(1-\theta)(T_l^{(m)})_j^n}{2}]^2$$

$$= \Delta x \sum_{m=1}^{2}\sum_{j=1}^{N}\nabla_x[(k_e^{(m)})_{j-\frac{1}{2}}^{n+\frac{1}{2}}\nabla_{\bar{x}}\frac{(1+\theta)(T_e^{(m)})_j^{n+1}+(1-\theta)(T_e^{(m)})_j^n}{2}]$$

$$\cdot \frac{(1+\theta)(T_e^{(m)})_j^{n+1}+(1-\theta)(T_e^{(m)})_j^n}{2}$$

$$+ \sum_{m=1}^{2}((S^{(m)})^{n+\frac{1}{2}}, \frac{(1+\theta)(T_e^{(m)})^{n+1}+(1-\theta)(T_e^{(m)})^n}{2}). \quad (4.14)$$

The first term on the RHS of Eq. (4.14) equals to, by Eqs. (4.11) and (4.13),

$$\text{RHS} = \Delta x \sum_{m=1}^{2}\sum_{j=1}^{N}\frac{1}{\Delta x}[(k_e^{(m)})_{j+\frac{1}{2}}^{n+\frac{1}{2}}\nabla_{\bar{x}}\frac{(1+\theta)(T_e^{(m)})_{j+1}^{n+1}+(1-\theta)(T_e^{(m)})_{j+1}^n}{2}$$

$$- (k_e^{(m)})_{j-\frac{1}{2}}^{n+\frac{1}{2}}\nabla_{\bar{x}}\frac{(1+\theta)(T_e^{(m)})_j^{n+1}+(1-\theta)(T_e^{(m)})_j^n}{2}]$$

$$\cdot \frac{(1+\theta)(T_e^{(m)})_j^{n+1}+(1-\theta)(T_e^{(m)})_j^n}{2}$$

$$= \Delta x \sum_{m=1}^{2}\{\frac{1}{\Delta x}\sum_{j=2}^{N+1}(k_e^{(m)})_{j-\frac{1}{2}}^{n+\frac{1}{2}}\nabla_{\bar{x}}\frac{(1+\theta)(T_e^{(m)})_j^{n+1}+(1-\theta)(T_e^{(m)})_j^n}{2}$$

$$\cdot \frac{(1+\theta)(T_e^{(m)})_{j-1}^{n+1} + (1-\theta)(T_e^{(m)})_{j-1}^{n}}{2}$$

$$- \frac{1}{\Delta x} \sum_{j=1}^{N} (k_e^{(m)})_{j-\frac{1}{2}}^{n+\frac{1}{2}} \nabla_{\bar{x}} \frac{(1+\theta)(T_e^{(m)})_{j}^{n+1} + (1-\theta)(T_e^{(m)})_{j}^{n}}{2}$$

$$\cdot \frac{(1+\theta)(T_e^{(m)})_{j}^{n+1} + (1-\theta)(T_e^{(m)})_{j}^{n}}{2} \Big\}$$

$$= -\Delta x \sum_{m=1}^{2} \sum_{j=1}^{N+1} (k_e^{(m)})_{j-\frac{1}{2}}^{n+\frac{1}{2}} [\nabla_{\bar{x}} \frac{(1+\theta)(T_e^{(m)})_{j}^{n+1} + (1-\theta)(T_e^{(m)})_{j}^{n}}{2}]^2$$

$$- \Delta x \sum_{m=1}^{2} \frac{1}{\Delta x} (k_e^{(m)})_{\frac{1}{2}}^{n+\frac{1}{2}} \nabla_{\bar{x}} \frac{(1+\theta)(T_e^{(m)})_{1}^{n+1} + (1-\theta)(T_e^{(m)})_{1}^{n}}{2}$$

$$\cdot \frac{(1+\theta)(T_e^{(m)})_{0}^{n+1} + (1-\theta)(T_e^{(m)})_{0}^{n}}{2}$$

$$+ \Delta x \sum_{m=1}^{2} \frac{1}{\Delta x} (k_e^{(m)})_{N+\frac{1}{2}}^{n+\frac{1}{2}} \nabla_{\bar{x}} \frac{(1+\theta)(T_e^{(m)})_{N+1}^{n+1} + (1-\theta)(T_e^{(m)})_{N+1}^{n}}{2}$$

$$\cdot \frac{(1+\theta)(T_e^{(m)})_{N+1}^{n+1} + (1-\theta)(T_e^{(m)})_{N+1}^{n}}{2}$$

$$= -\Delta x \sum_{m=1}^{2} \sum_{j=1}^{N+1} (k_e^{(m)})_{j-\frac{1}{2}}^{n+\frac{1}{2}} [\nabla_{\bar{x}} \frac{(1+\theta)(T_e^{(m)})_{j}^{n+1} + (1-\theta)(T_e^{(m)})_{j}^{n}}{2}]^2$$

$$+ (k_e^{(1)})_{N+\frac{1}{2}}^{n+\frac{1}{2}} \nabla_{\bar{x}} \frac{(1+\theta)(T_e^{(1)})_{N+1}^{n+1} + (1-\theta)(T_e^{(1)})_{N+1}^{n}}{2}$$

$$\cdot \frac{(1+\theta)(T_e^{(1)})_{N+1}^{n+1} + (1-\theta)(T_e^{(1)})_{N+1}^{n}}{2}$$

$$- (k_e^{(2)})_{\frac{1}{2}}^{n+\frac{1}{2}} \nabla_{\bar{x}} \frac{(1+\theta)(T_e^{(2)})_{1}^{n+1} + (1-\theta)(T_e^{(2)})_{1}^{n}}{2}$$

$$\cdot \frac{(1+\theta)(T_e^{(2)})_{0}^{n+1} + (1-\theta)(T_e^{(2)})_{0}^{n}}{2}$$

$$= -\Delta x \sum_{m=1}^{2} \sum_{j=1}^{N+1} (k_e^{(m)})_{j-\frac{1}{2}}^{n+\frac{1}{2}} [\nabla_{\bar{x}} \frac{(1+\theta)(T_e^{(m)})_{j}^{n+1} + (1-\theta)(T_e^{(m)})_{j}^{n}}{2}]^2$$

$$- k_c \cdot [(\frac{(1+\theta)(T_e^{(1)})_{N+1}^{n+1} + (1-\theta)(T_e^{(1)})_{N+1}^{n}}{2})^2$$

$$+ (\frac{(1+\theta)(T_e^{(2)})_{0}^{n+1} + (1-\theta)(T_e^{(2)})_{0}^{n}}{2})^2]$$

$$\cdot |\frac{(1+\theta)(T_e^{(1)})_{N+1}^{n+1} + (1-\theta)(T_e^{(1)})_{N+1}^n}{2}$$

$$+ \frac{(1+\theta)(T_e^{(2)})_0^{n+1} + (1-\theta)(T_e^{(2)})_0^n}{2} |$$

$$\cdot [\frac{(1+\theta)(T_e^{(1)})_{N+1}^{n+1} + (1-\theta)(T_e^{(1)})_{N+1}^n}{2}$$

$$- \frac{(1+\theta)(T_e^{(2)})_0^{n+1} + (1-\theta)(T_e^{(2)})_0^n}{2}]^2. \tag{4.15}$$

Thus, Eq. (4.14) can be simplified as follows:

$$\frac{1}{\Delta t} \sum_{m=1}^{2} \frac{1}{3} A_e^{(m)} [\left\|(T_e^{(m)})^{n+1}\right\|_{l^3}^3 - \left\|(T_e^{(m)})^n\right\|_{l^3}^3]$$

$$+ \frac{1}{\Delta t} \sum_{m=1}^{2} \frac{1}{2} C_l^{(m)} [\left\|(T_l^{(m)})^{n+1}\right\|_{l^2}^2 - \left\|(T_l^{(m)})^n\right\|_{l^2}^2]$$

$$+ \Delta x \sum_{m=1}^{2} G^{(m)} \sum_{j=1}^{N} [\frac{(1+\theta)(T_e^{(m)})_j^{n+1} + (1-\theta)(T_e^{(m)})_j^n}{2}$$

$$- \frac{(1+\theta)(T_l^{(m)})_j^{n+1} + (1-\theta)(T_l^{(m)})_j^n}{2}]^2$$

$$+ \Delta x \sum_{m=1}^{2} \sum_{j=1}^{N+1} (k_e^{(m)})_{j-\frac{1}{2}}^{n+\frac{1}{2}} [\nabla_{\bar{x}} \frac{(1+\theta)(T_e^{(m)})_j^{n+1} + (1-\theta)(T_e^{(m)})_j^n}{2}]^2$$

$$+ k_c \cdot [(\frac{(1+\theta)(T_e^{(1)})_{N+1}^{n+1} + (1-\theta)(T_e^{(1)})_{N+1}^n}{2})^2$$

$$+ (\frac{(1+\theta)(T_e^{(2)})_0^{n+1} + (1-\theta)(T_e^{(2)})_0^n}{2})^2]$$

$$\cdot |\frac{(1+\theta)(T_e^{(1)})_{N+1}^{n+1} + (1-\theta)(T_e^{(1)})_{N+1}^n}{2}$$

$$+ \frac{(1+\theta)(T_e^{(2)})_0^{n+1} + (1-\theta)(T_e^{(2)})_0^n}{2} |$$

$$\cdot [\frac{(1+\theta)(T_e^{(1)})_{N+1}^{n+1} + (1-\theta)(T_e^{(1)})_{N+1}^n}{2}$$

$$- \frac{(1+\theta)(T_e^{(2)})_0^{n+1} + (1-\theta)(T_e^{(2)})_0^n}{2}]^2$$

$$\leq \frac{1}{2} \sum_{m=1}^{2} [\frac{1}{3}(1+\theta)A_e^{(m)} \left\|(T_e^{(m)})^{n+1}\right\|_{l^3}^3 + \frac{1}{3} A_e^{(m)}(1-\theta) \left\|(T_e^{(m)})^n\right\|_{l^3}^3]$$

$$+ \sum_{m=1}^{2} \frac{2}{3\sqrt{A_e^{(m)}}} \left\| (S^{(m)})^{n+\frac{1}{2}} \right\|_{l^{\frac{3}{2}}}^{\frac{3}{2}} . \tag{4.16}$$

Let $\tilde{F}(n) = \sum_{m=1}^{2} [\frac{1}{3} A_e^{(m)} \left\| (T_e^{(m)})^n \right\|_{l^3}^3 + \frac{1}{2} C_l^{(m)} \left\| (T_l^{(m)})^n \right\|_{l^2}^2]$ and $\tilde{Q}(n) = $

$\sum_{m=1}^{2} \frac{2}{3\sqrt{A_e^{(m)}}} \left\| (S^{(m)})^{n-\frac{1}{2}} \right\|_{l^{\frac{3}{2}}}^{\frac{3}{2}}$. Taking out those positive terms on the left-hand-side of Eq. (4.16), one may obtain

$$(1 - \frac{\Delta t(1+\theta)}{2}) \tilde{F}(n) \leq (1 + \frac{\Delta t(1-\theta)}{2}) \tilde{F}(n-1) + \Delta t \tilde{Q}(n). \tag{4.17}$$

Using a similar argument as that in Eqs. (3.14)–(3.15), we have obtained the following theorem.

THEOREM 4. Suppose that $(T_e^{(m)})_j^n$ and $(T_l^{(m)})_k^n$ are the solutions of the scheme, Eqs. (4.9)–(4.13). For a sufficiently small $\Delta t > 0$, the solutions satisfy

$$\sum_{m=1}^{2} [\frac{1}{3} A_e^{(m)} \left\| (T_e^{(m)})^n \right\|_{l^3}^3 + \frac{1}{2} C_l^{(m)} \left\| (T_l^{(m)})^n \right\|_{l^2}^2]$$

$$\leq e^{2n\Delta t} \{ \sum_{m=1}^{2} [\frac{1}{3} A_e^{(m)} \left\| (T_e^{(m)})^0 \right\|_{l^3}^3 + \frac{1}{2} C_l^{(m)} \left\| (T_l^{(m)})^0 \right\|_{l^2}^2]$$

$$+ \max_{0 \leq k \leq n} \sum_{m=1}^{2} \frac{2}{3\sqrt{A_e^{(m)}}} \left\| (S^{(m)})^{k-\frac{1}{2}} \right\|_{l^{\frac{3}{2}}}^{\frac{3}{2}} \}, \tag{4.18}$$

which is a discrete analogous of Eq. (4.7).

To test the applicability of the above numerical scheme, we investigate the temperature rise in a double-layered thin film, namely a gold layer on a chromium padding layer, where the gold layer and the chromium padding layer are 0.05 μm in thickness. The heat sources are chosen to be

$$(S^{(1)})_j^{n+\frac{1}{2}} = 0.94J \frac{1-R}{t_p \delta} e^{-\frac{j\Delta x}{\delta} - 2.77(\frac{(n+0.5)\Delta t - 2t_p}{t_p})^2} \tag{4.19a}$$

and

$$(S^{(2)})_j^{n+\frac{1}{2}} = 0.94J \frac{1-R}{t_p \delta} e^{-\frac{j\Delta x + 0.00005}{\delta} - 2.77(\frac{(n+0.5)\Delta t - 2t_p}{t_p})^2} , \tag{4.19b}$$

where $t_p = 0.1$ps, $\delta = 15.3$ nm, $R = 0.93$, and J is the laser fluence. In our computation [34], we employ the present scheme with $\theta = 0$.

Figure 4.1 shows the change in temperature $(\frac{\Delta T_e}{(\Delta T_e)_{\max}})$ on the surface of the gold layer when the fluence J is 13.4 $\frac{J}{m^2}$ and when k_c is equal to

$\sigma = 5.669 \times 10^{-8} \frac{\text{W}}{\text{m}^2\text{K}}$ ($= 5.669 \times 10^{-28} \frac{\text{J}}{\text{ps}\cdot(\text{mm})^2\text{K}}$). It can be seen from the figure that the temperature level after the peak is higher than that obtained in [25] (see in [25], Figure 5.18, p. 143). This can be explained by noting that when k_c is very small, only a small portion of heat flow crosses the interface and enters the chromium layer while a large portion of heat still remains in the gold layer. To test for the effect of grid size, we use three different meshes of 100, 200, and 400 grid points. From Figure 4.1, it can be seen that grid size has no significant effect on the solution. Hence, the method is grid independent.

Figures 4.2 and 4.3 give the electron temperature profiles and lattice temperature profiles along the x-axis when $J = 13.4 \frac{\text{J}}{\text{m}^2}$ and times $t = 0.2$, 0.25, and 0.5 ps, respectively. Results are obtained using a time increment of 0.005 ps and 200 grid points for each layer. Figure 4.2 shows that a large temperature difference occurs at the interface, while Figure 4.3 shows that the lattice temperature levels in the gold layer are higher than those in the chromium layer.

To see the difference between the perfect thermal contact and the non-perfect thermal contact at interface, we choose various values of k_c ranging from 10^{-1} to $5.669 \times 10^{-28} \frac{\text{J}}{\text{ps}\cdot(\text{mm})^2\text{K}}$ and plot the change in temperature ($\frac{\Delta T_e}{(\Delta T_e)_{\max}}$) on the surface of the gold layer when $J = 13.4 \frac{\text{J}}{\text{m}^2}$, as shown in Figure 4.4. Here, $(\Delta T_e)_{\max}$ is 426 K, which is obtained based on the perfect thermal contact assumption at the interface. One may see from Figure 4.4 that there is no significant difference in the perfect thermal contact and when k_c is chosen to be 10^{-1}, 10^{-3}, 10^{-5}, and 10^{-10}. However, there is a large difference when k_c is chosen to be 10^{-20} and 5.669×10^{-28}.

Figures 4.5 and 4.6 give the electron temperature profiles and lattice temperature profiles along the x-axis with $J = 13.4 \frac{\text{J}}{\text{m}^2}$ and with various values of k_c ranging from 10^{-1} to $5.669 \times 10^{-28} \frac{\text{J}}{\text{ps}\cdot(\text{mm})^2\text{K}}$ when times $t = 0.2$, 0.25, and 0.5 ps, respectively. It can be seen from Figure 4.5 that there is no significant difference in the perfect thermal contact and when k_c is chosen to be 10^{-1}, 10^{-3}, 10^{-5}, and 10^{-10}. However, a large temperature difference occurs at the interface when k_c is 10^{-20} and 5.669×10^{-28}. Again, Figure 4.6 shows that there is no significant difference in the perfect thermal contact and when k_c is chosen to be 10^{-1}, 10^{-3}, 10^{-5}, and 10^{-10}. However, the lattice temperature levels in the gold layer are higher than those in the chromium layer when k_c is 10^{-20} and 5.669×10^{-28}.

To compare with those obtained in [24], we set the initial condition at $-2t_p$ and chose J to be $500.0 \frac{\text{J}}{\text{m}^2}$. Figures 4.7 and 4.8 plot the electron temperature profiles and lattice temperature profiles with $k_c = 5.669 \times 10^{-28} \frac{J}{ps \cdot (mm)^2 K}$ for times $t = 0.1, 0.5,$ and 6.0 ps, respectively. Again, Figure 4.7 shows that a significant temperature difference occurs at the interface while Figure 4.8 shows that the lattice temperature levels in the gold layer are higher than those in the chromium layer.

Furthermore, we plot in Figures 4.9 and 4.10 the electron temperature profiles and lattice temperature profiles along the x-axis with $J = 500 \frac{J}{m^2}$ and with various values of k_c ranging from 10^{-1} to $5.669 \times 10^{-28} \frac{J}{ps \cdot (mm)^2 K}$ at times $t = 0.1, 0.5,$ and 6.0 ps, respectively. First, it can be seen from both figures that there is no difference between the perfect thermal contact and those obtained in [24] (see Figure 6 in [24]). Second, when k_c is set to be $10^{-1}, 10^{-3}, 10^{-5},$ and 10^{-10}, the electron temperature profiles and lattice temperature profiles overlap together, respectively, and have a large difference in the perfect thermal contact when $t = 0.1$ ps and no difference when $t = 0.5$ and 6.0 ps. Third, a large electron temperature difference occurs at the interface when k_c is 10^{-20} and 5.669×10^{-28}. Fourth, the lattice temperature levels in the gold layer are higher than those in the chromium layer when $t = 6.0$ ps and k_c is 10^{-20} and 5.669×10^{-28}. Results show that there is a significant difference between the perfect thermal contact and the non-perfect thermal contact at the interface when k_c is very small.

5. Conclusion

In this chapter, we develop a type of nonstandard finite difference schemes for solving parabolic two-step model with temperature-dependent thermal properties in a double-layered metal thin film exposed to picosecond thermal pulses. The perfect thermal contact and the non-perfect thermal contact interfacial conditions are considered. The schemes are derived based on the continuous energy estimate obtained from the parabolic two-step model. We have showed that these schemes satisfy a discrete analogous of the energy estimate. Numerical examples for thermal analysis of a gold layer on a chromium padding layer are illustrated.

It should be pointed out that when the laser pulse duration is much shorter than the electron-lattice thermal relaxation time that is the characteristic time for the activation of ballistic behavior in the electron gas, the parabolic two-step model may lose accuracy [23,41]. As Qiu and Tien [24] pointed out, the relaxation time increases dramatically as the temperature decreases from 0.04 ps at room temperature to about 10 ps at 10 K. They [23] developed the hyperbolic two-step heat transport equations based on the macroscopic averages of the electric and heat currents carried by electrons in the momentum space. The hyperbolic two-step model is considered to be a satisfactory extension of the parabolic two-step model. The hyperbolic two-step model for thermal analysis of a double-layered thin metal film can be written as follows [18,24,42]:

$$C_e^{(m)} \frac{\partial T_e^{(m)}}{\partial t} = -\frac{\partial q_e^{(m)}}{\partial x} - G^{(m)}(T_e^{(m)} - T_l^{(m)}) + S^{(m)}, \tag{5.1}$$

$$\tau_e^{(m)} \frac{\partial q_e^{(m)}}{\partial x} + q_e^{(m)} = -k_e^{(m)} \frac{\partial T_e^{(m)}}{\partial x}, \tag{5.2}$$

$$C_l^{(m)} \frac{\partial T_l^{(m)}}{\partial t} = -\frac{\partial q_l^{(m)}}{\partial x} + G^{(m)}(T_e^{(m)} - T_l^{(m)}), \tag{5.3}$$

$$\tau_l^{(m)} \frac{\partial q_l^{(m)}}{\partial x} + q_l^{(m)} = -k_l^{(m)} \frac{\partial T_l^{(m)}}{\partial x}, \tag{5.4}$$

where $C_e^{(m)} = A_e T_e^{(m)}$, $k_e^{(m)} = k_0 \frac{T_e^{(m)}}{T_l^{(m)}}$, and $0 \le x \le \frac{L}{2}$ when $m = 1$, and $\frac{L}{2} \le x \le L$ when $m = 2$. The interfacial condition for perfect thermal contact is given by

$$q_e^{(1)} = q_e^{(2)}, \quad T_e^{(1)} = T_e^{(2)}; \quad q_l^{(1)} = q_l^{(2)}, \quad T_l^{(1)} = T_l^{(2)}; \quad x = \frac{L}{2}. \tag{5.5}$$

While the interfacial condition for non-perfect thermal contact is given by

$$q_e^{(1)} = q_e^{(2)} = k_c^e[(T_e^{(1)})^4 - (T_e^{(2)})^4], \quad T_e^{(1)} \ne T_e^{(2)}; \tag{5.5a'}$$

$$q_l^{(1)} = q_l^{(2)} = k_c^l[(T_l^{(1)})^4 - (T_l^{(2)})^4], \quad T_l^{(1)} \ne T_l^{(2)}. \tag{5.5b'}$$

The initial and boundary conditions are assumed to be

$$T_e^{(m)}(x, 0) = T_l^{(m)}(x, 0) = T_0(= 300 \text{ K}), \tag{5.6}$$

and

$$q_e^{(1)}(0, t) = q_l^{(1)}(0, t) = 0, \quad q_e^{(1)}(L, t) = q_l^{(1)}(L0, t) = 0. \tag{5.7}$$

Here, $q_e^{(m)}$ and $q_l^{(m)}$ are the heat fluxes associated with electrons and the lattice, respectively. $\tau_e^{(m)} = \frac{1}{A_e(T_e^{(m)})^2 + B_l T_l^{(m)}}$ is the electron relaxation time and $\tau_l^{(m)} = \frac{3k_l^{(m)}}{C_l^{(m)}(v_s^{(m)})^2}$ is the lattice relaxation time. It can be seen that if $\tau_e^{(m)}$ and $\tau_l^{(m)}$ are zero, the hyperbolic two-step model will reduce to the parabolic two-step model.

The solution of the above hyperbolic two-step model is difficult to obtain because of the temperature-dependent thermal properties. Only a few works for some specific cases are available in the literature. Chen *et al.* [18,42] employed finite-difference and finite-element methods to solve the hyperbolic two-step model for investigating of thermal response in a single-layered metal thin film caused by pulse laser heating. Al-Nimr and his colleagues [43-47] studied the thermal behavior of thin metal films in the hyperbolic two-step model with constant thermal properties. Future work will focus on the development of nonstandard finite difference schemes for solving the above hyperbolic two-step heat transport equations.

Acknowledgments

This research is supported by a Louisiana Educational Quality Support Fund (LEQSF) grant. Contract No: LEQSF (2002-05)-RD-A-01. The author thanks Professor Ron Mickens for inviting him to contribute to this volume, and for his support.

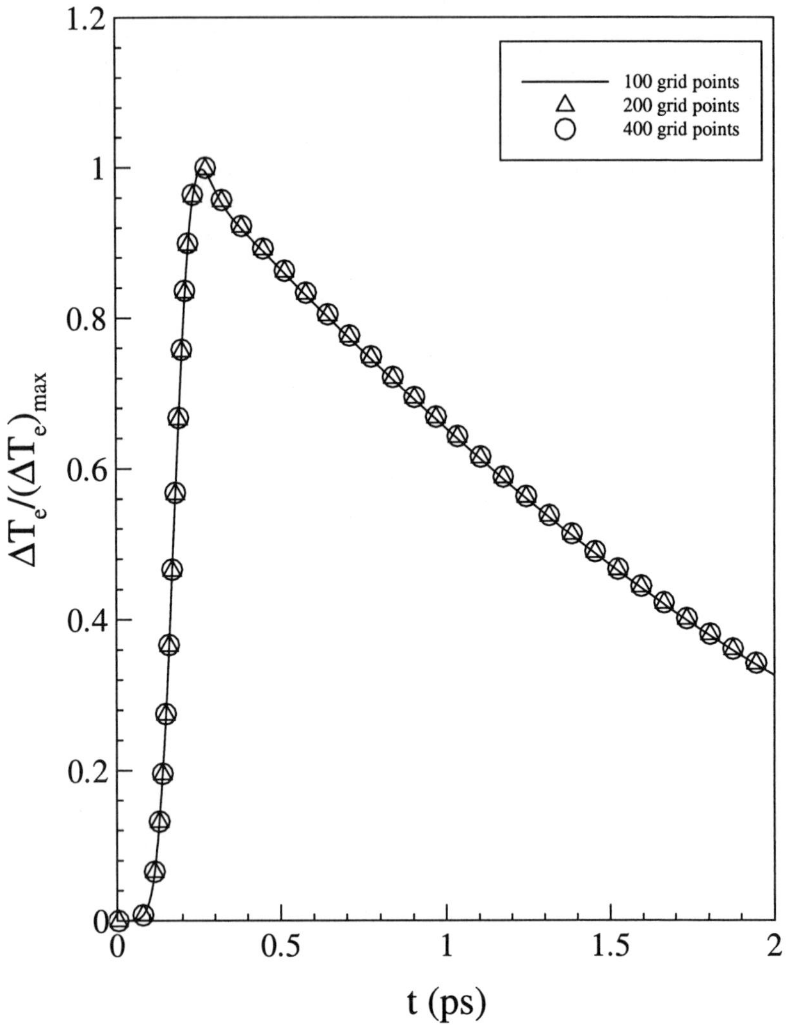

Figure 4.1. Normalized electron temperatures at the front surface of a 100-nm gold and chromium thin film irradiated with a 0.1 ps laser pulse at a fluence of 13.4 J/m^2 and kc = 5.669 x 10^{-8} W/m^{2}K.

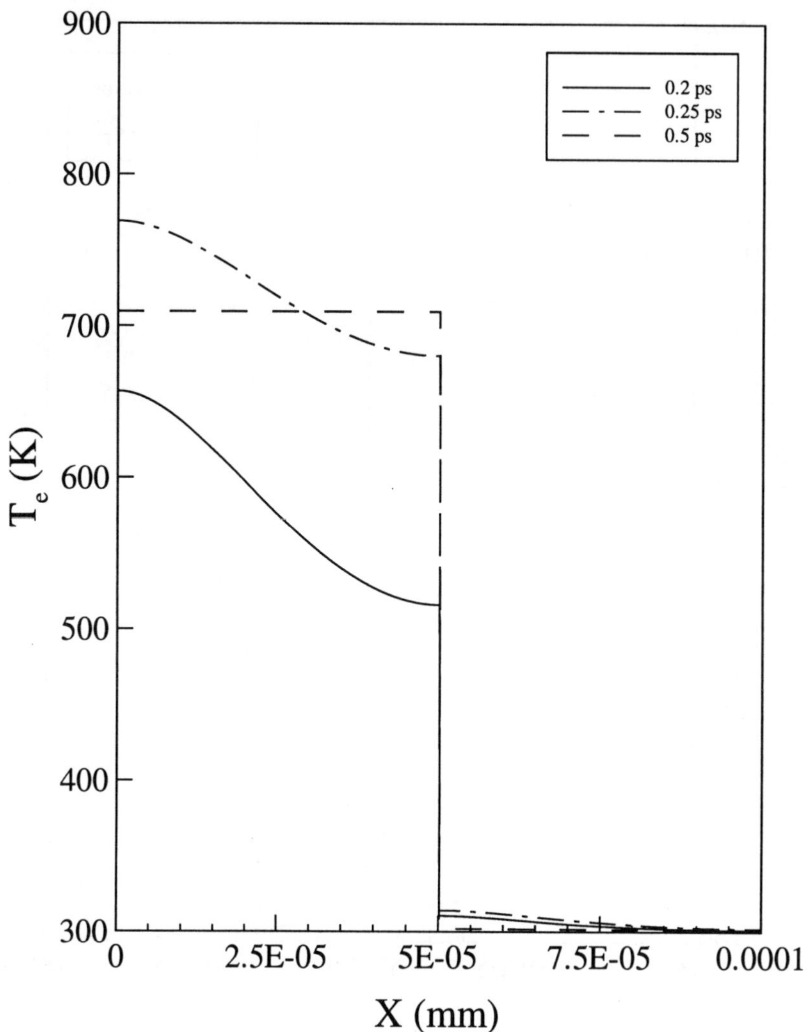

Figure 4.2. Calculated electron temperature profiles for a 100-nm gold and chromium thin film irradiated with a 0.1 ps laser pulse at a fluence of 13.4 J/m^2 and kc $= 5.669$ x 10^{-8} W/m^{2}K.

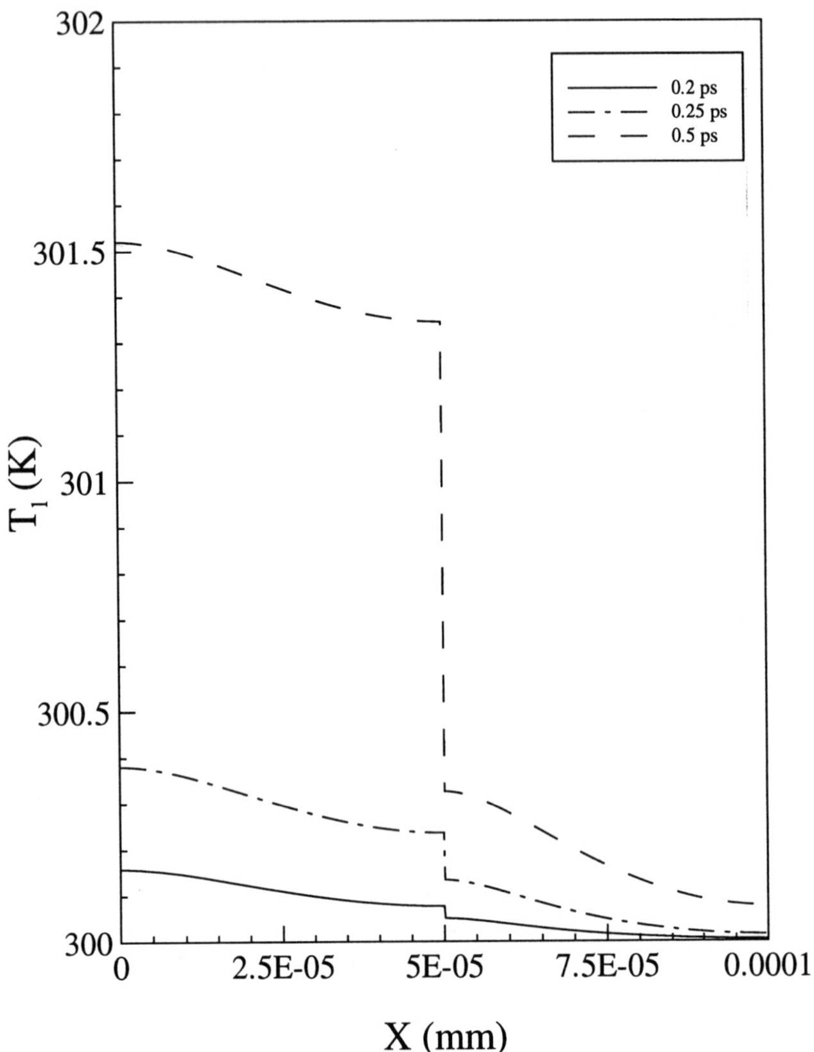

Figure 4.3. Calculated lattice temperature profiles for a 100-nm gold and chromium thin film irradiated with a 0.1 ps laser pulse at a fluence of 13.4 J/m^2 and kc $= 5.669$ x 10^{-8} W/m^{2}K.

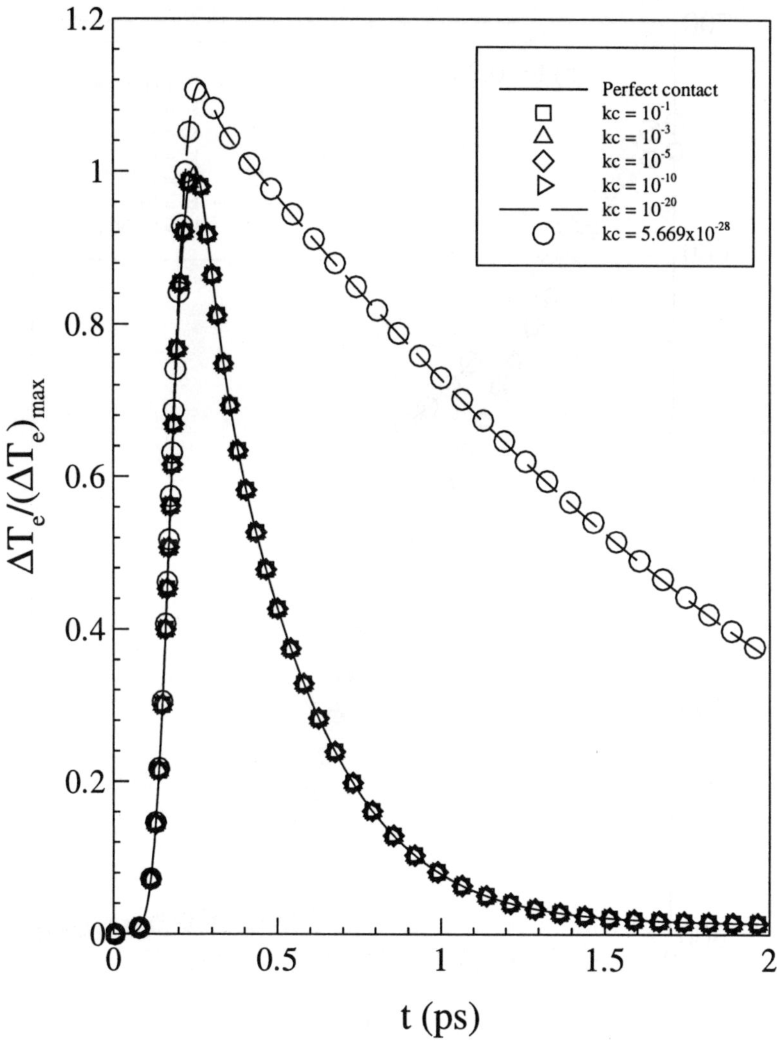

Figure 4.4. Normalized electron temperatures at the front surface of a 100-nm gold and chromium thin film irradiated with a 0.1 ps laser pulse at a fluence of 13.4 J/m^2 and various values of kc.

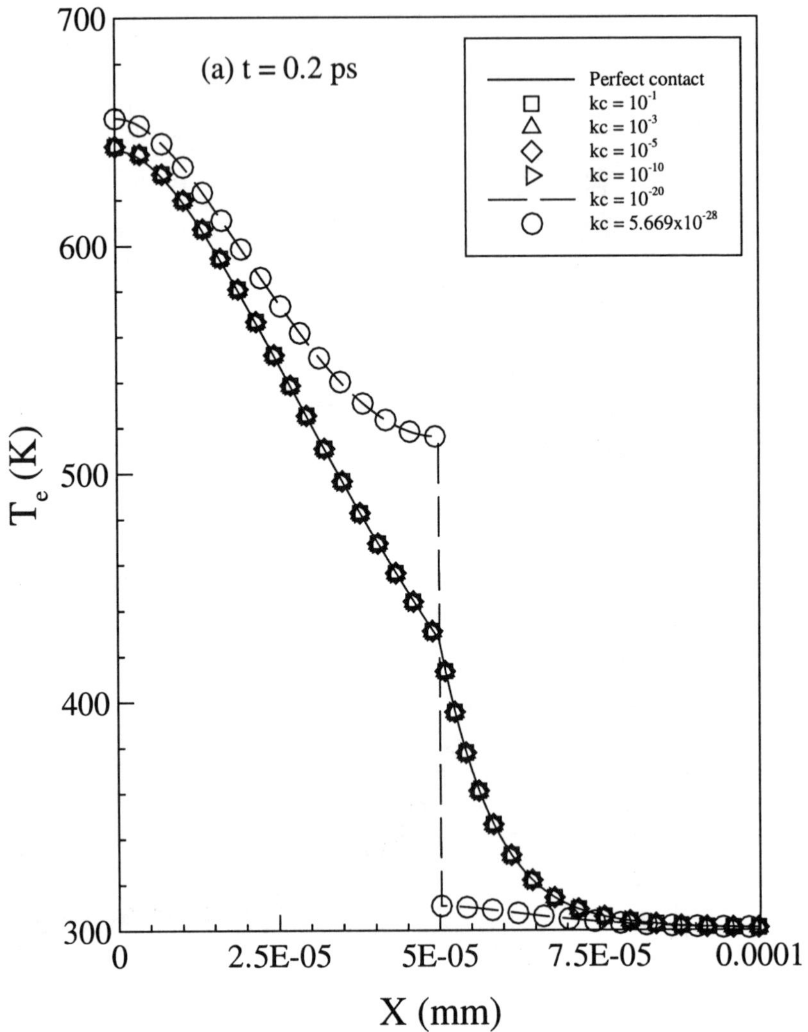

Figure 4.5. Calculated electron temperatures at (a) t = 0.2 ps, (b) t = 0.25 ps, and (c) t = 0.5 ps along the depth of a 100-nm gold and chromium thin film irradiated with a 0.1 ps laser pulse at a fluence of 13.4 J/m² and various values of kc.

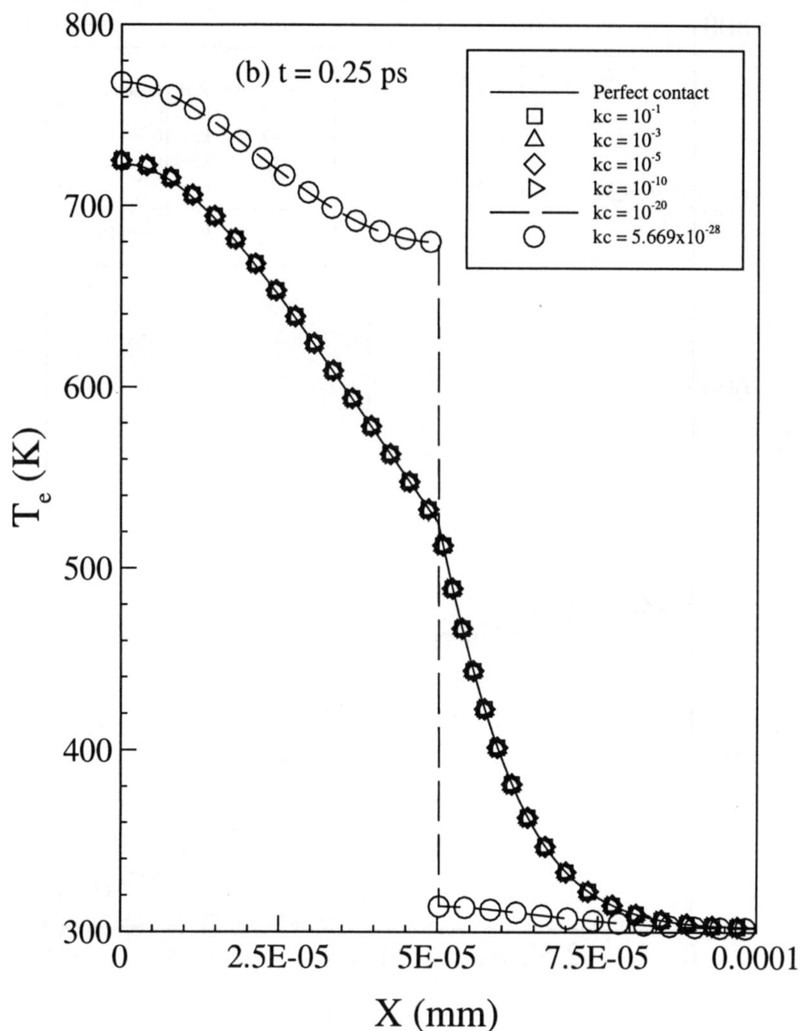

Figure 4.5. Continued.

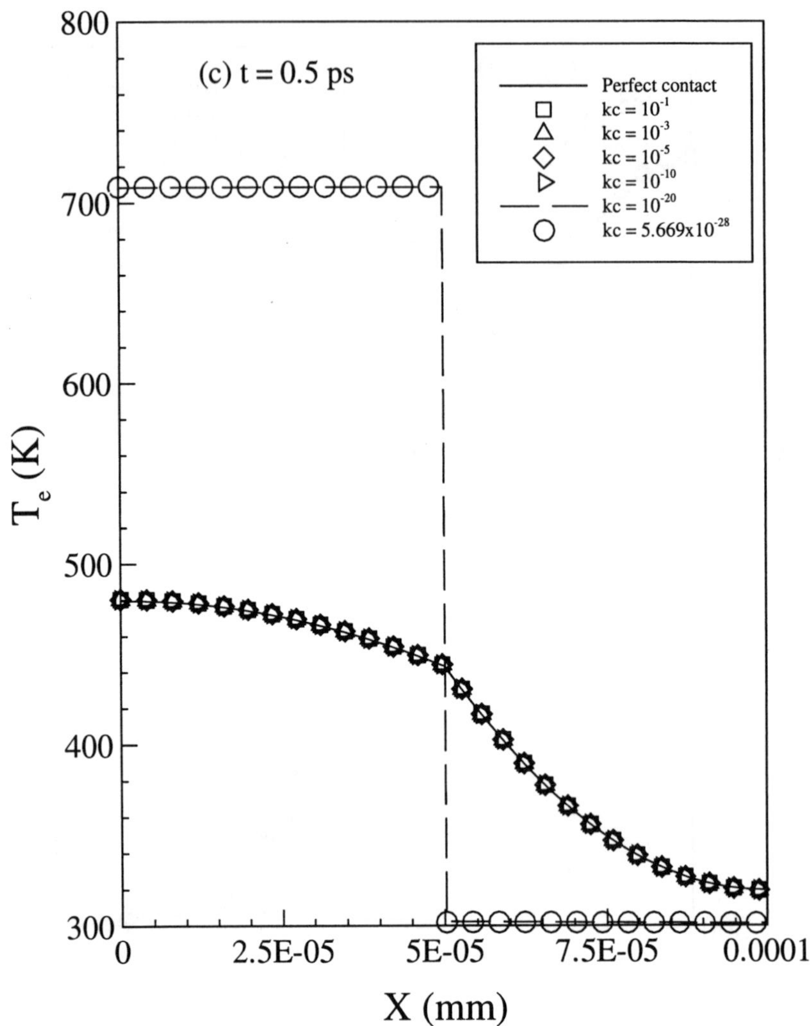

Figure 4.5. Continued.

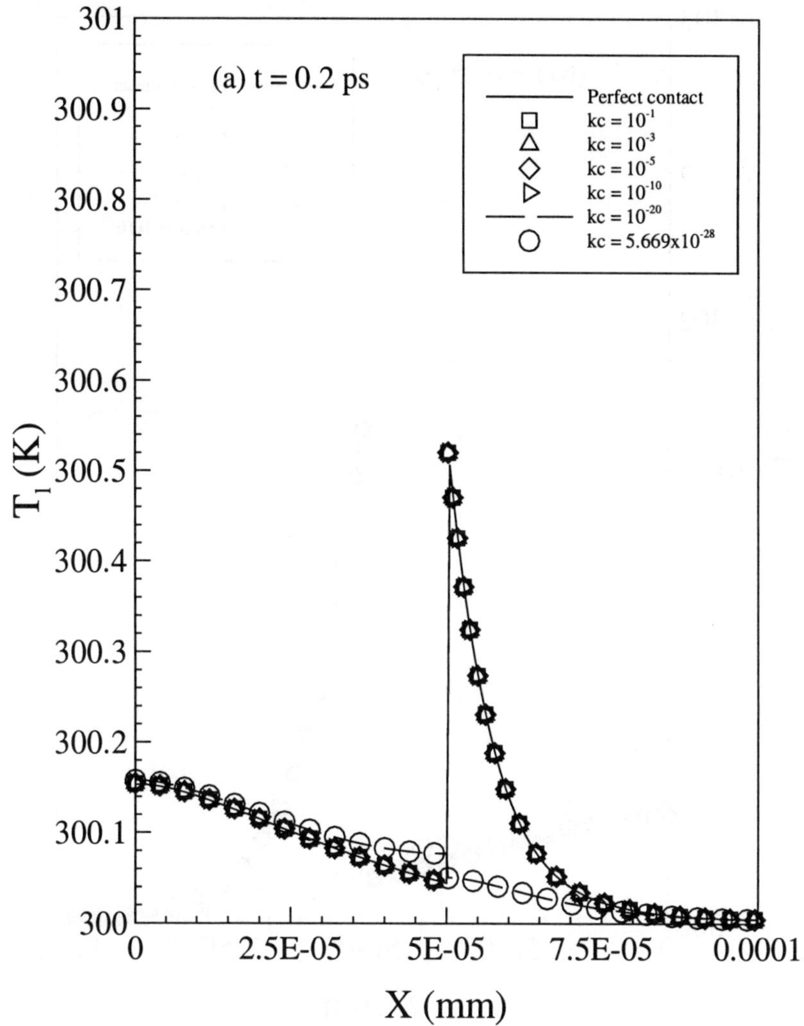

Figure 4.6. Calculated lattice temperatures at (a) $t = 0.2$ ps, (b) $t = 0.25$ ps, and (c) $t = 0.5$ ps along the depth of a 100-nm gold and chromium thin film irradiated with a 0.1 ps laser pulse at a fluence of 13.4 J/m^2 and various values of kc.

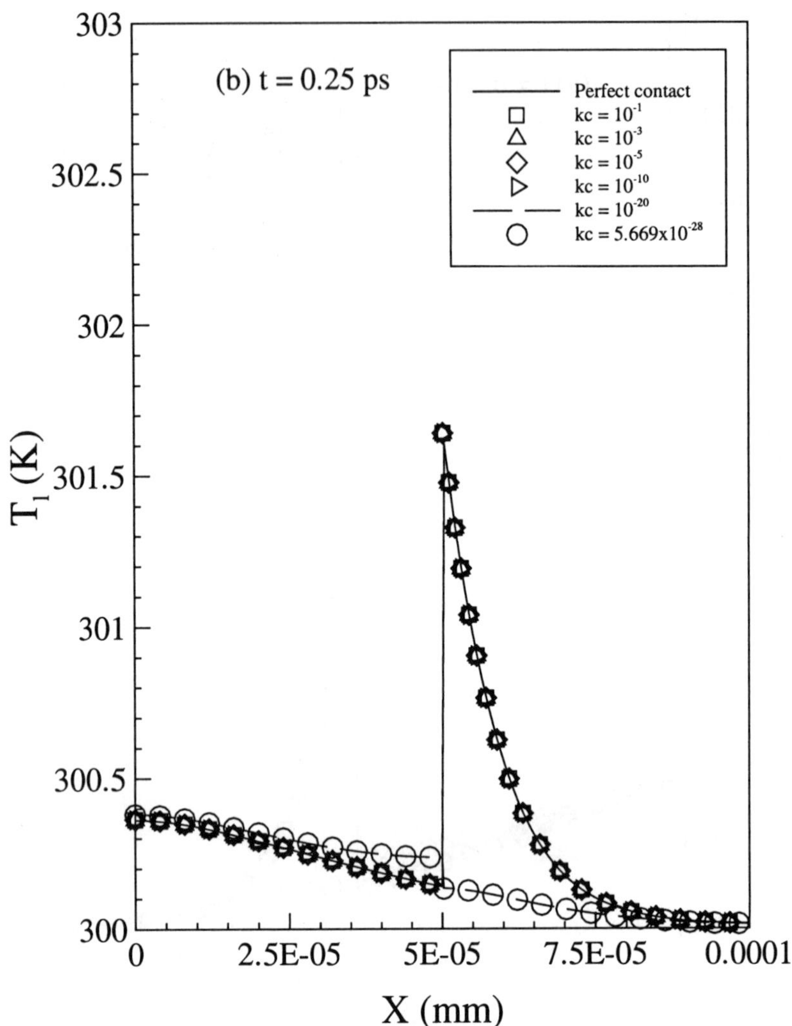

Figure 4.6. Continued.

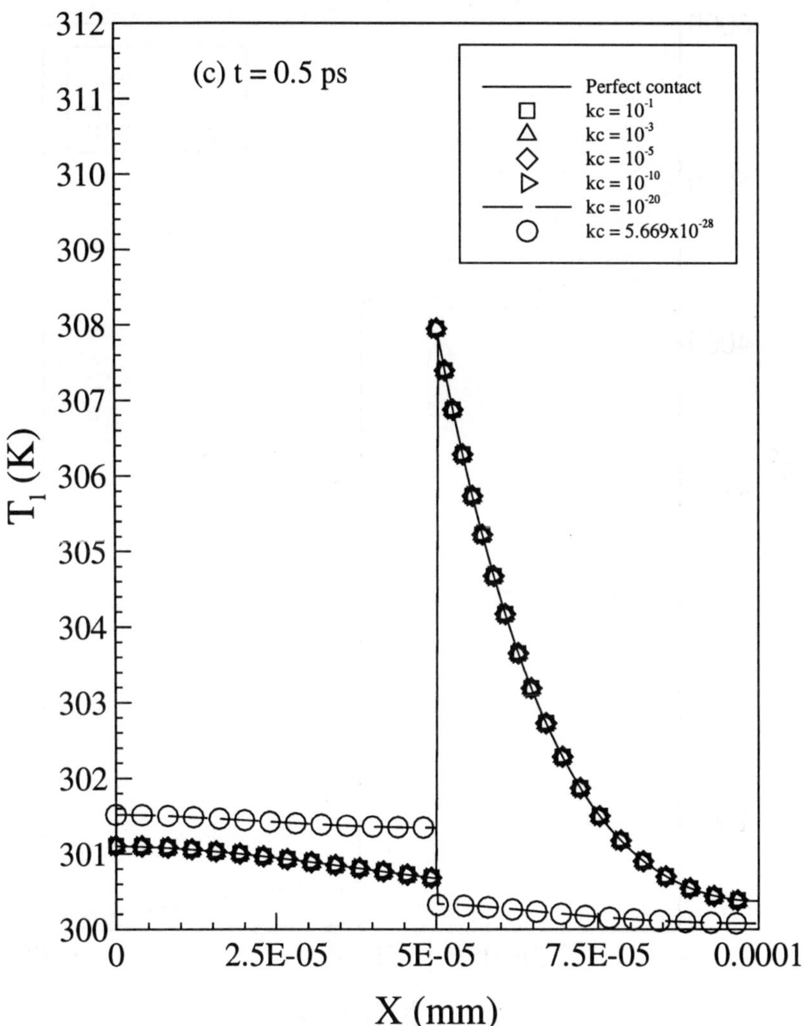

Figure 4.6. Continued.

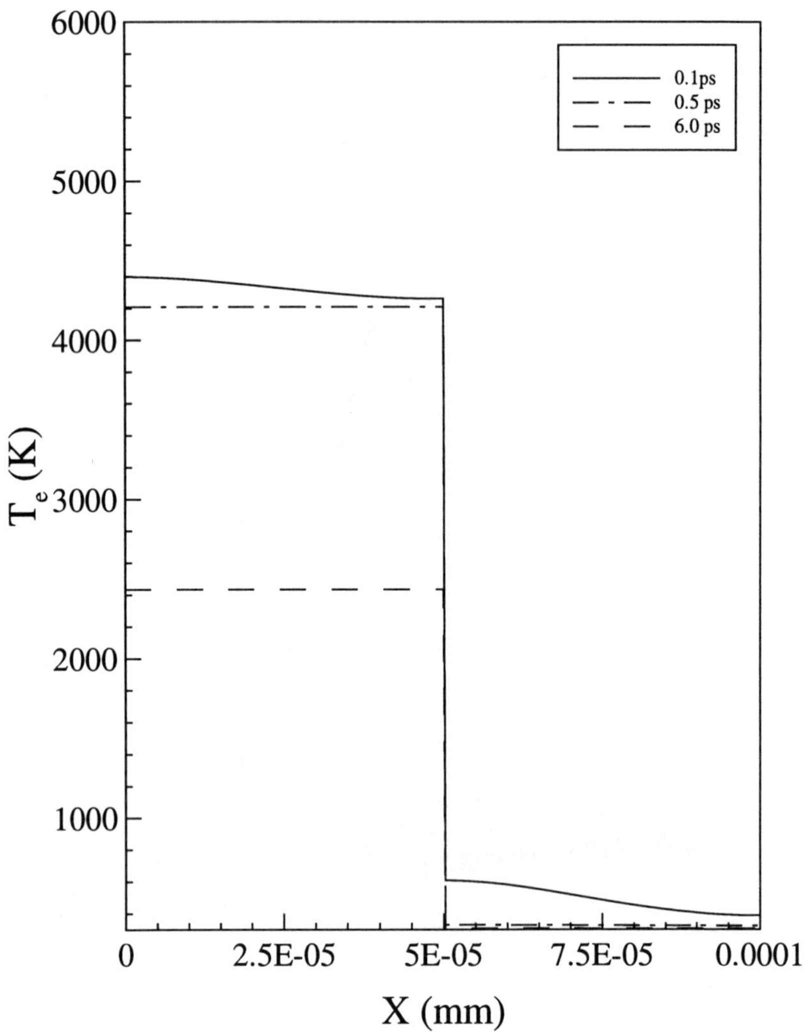

Figure 4.7. Calculated electron temperature profiles for a 100-nm gold and chromium thin film irradiated with a 0.1 ps laser pulse at a fluence of 500.0 J/m^2 and kc = 5.669 x 10^{-8} W/m^{2}K.

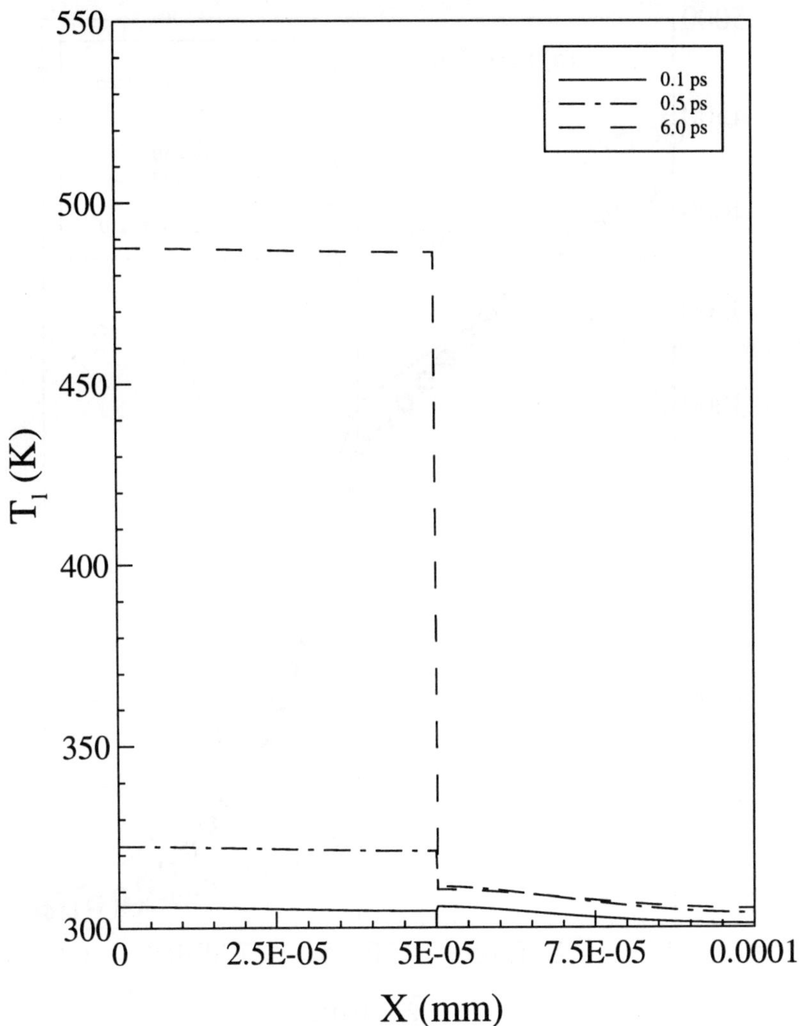

Figure 4.8. Calculated lattice temperature profiles for a 100-nm gold and chromium thin film irradiated with a 0.1 ps laser pulse at a fluence of 500.0 J/m^2 and kc = 5.669 x 10^{-8} W/m^{2}K.

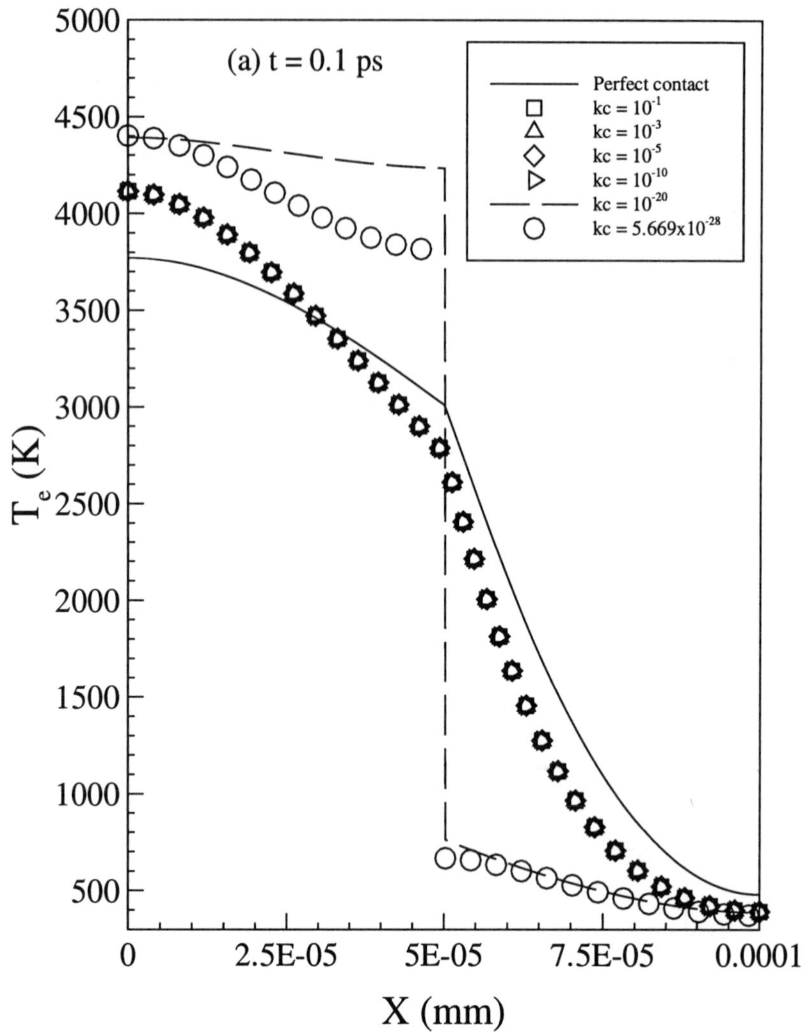

Figure 4.9. Calculated electron temperatures at (a) t = 0.1 ps, (b) t = 0.5 ps, and (c) t = 6.0 ps along the depth of a 100-nm gold and chromium thin film irradiated with a 0.1 ps laser pulse at a fluence of 500 J/m^2 and various values of kc.

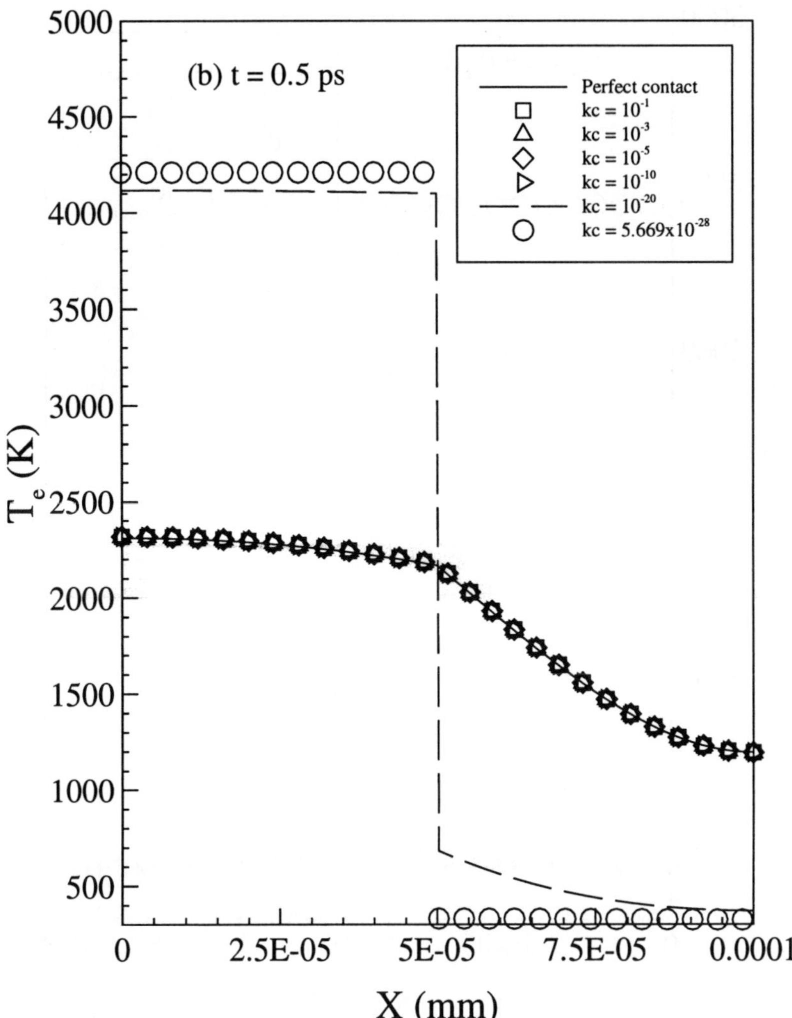

Figure 4.9. Continued.

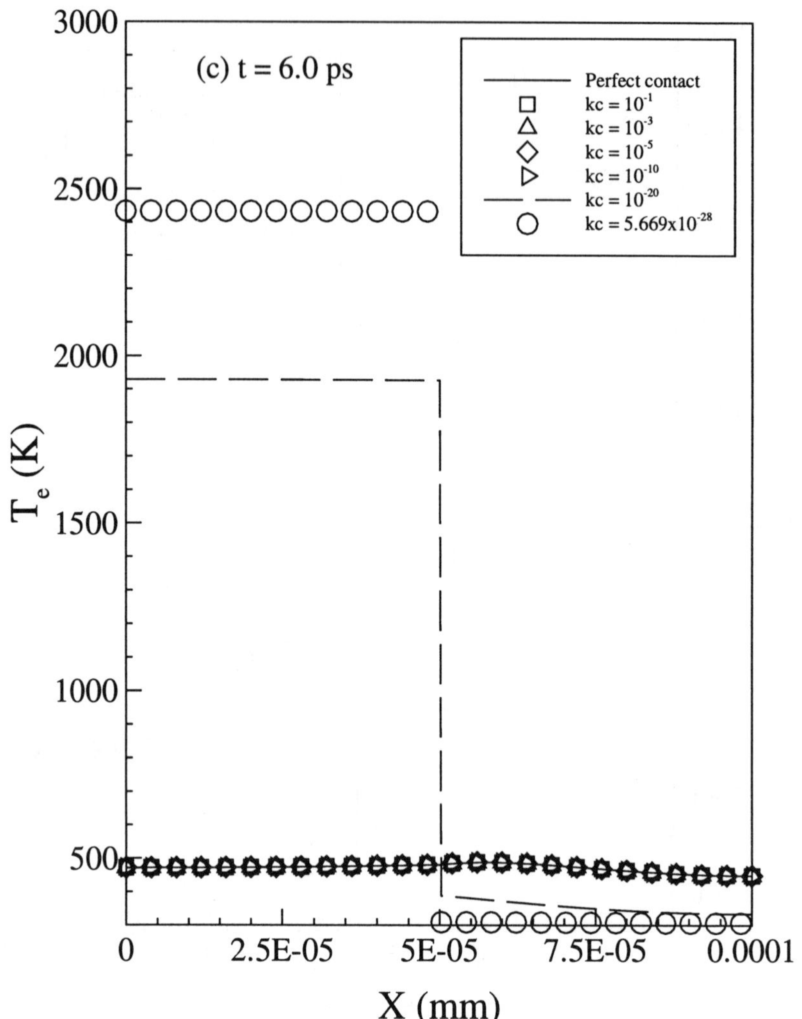

Figure 4.9. Continued.

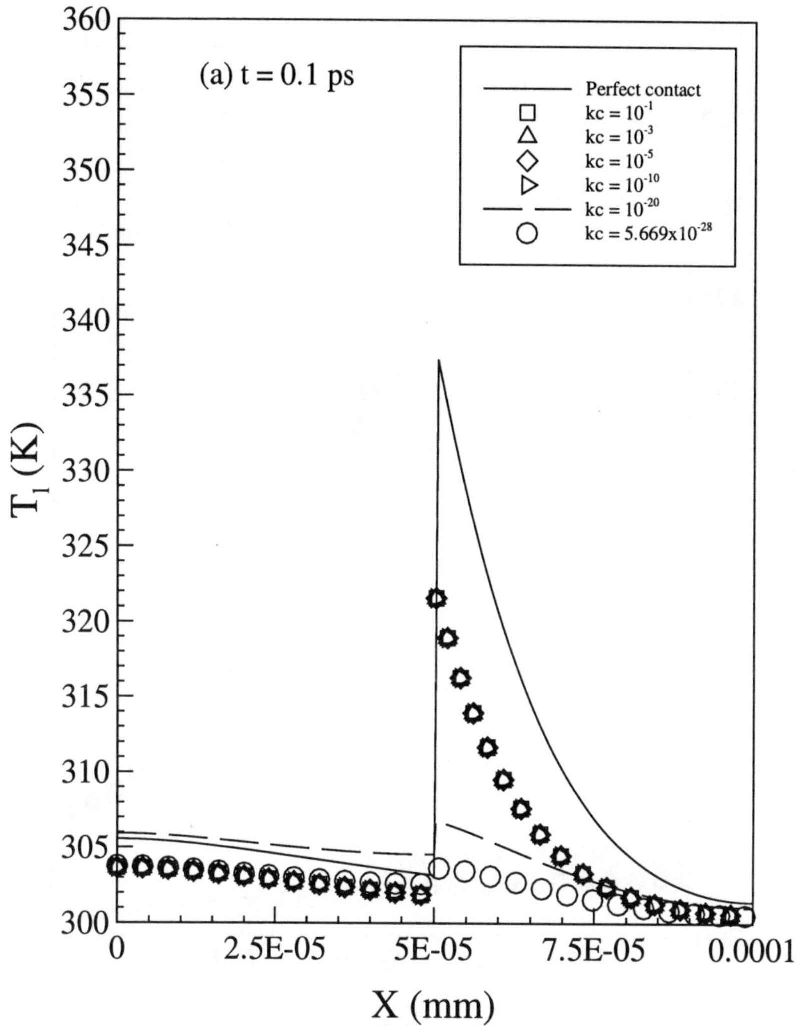

Figure 4.10. Calculated electron temperatures at (a) t = 0.1 ps, (b) t = 0.5 ps, and (c) t = 6.0 ps along the depth of a 100-nm gold and chromium thin film irradiated with a 0.1 ps laser pulse at a fluence of 500 J/m^2 and various values of kc.

W. Dai

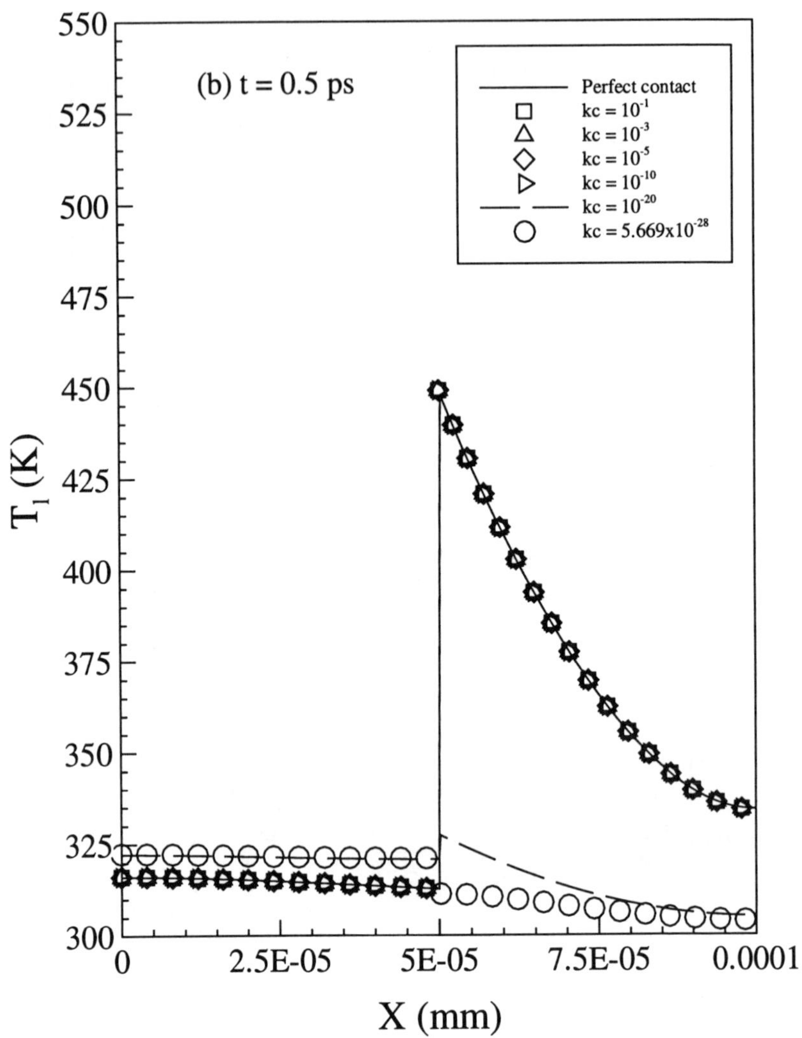

Figure 4.10. Continued.

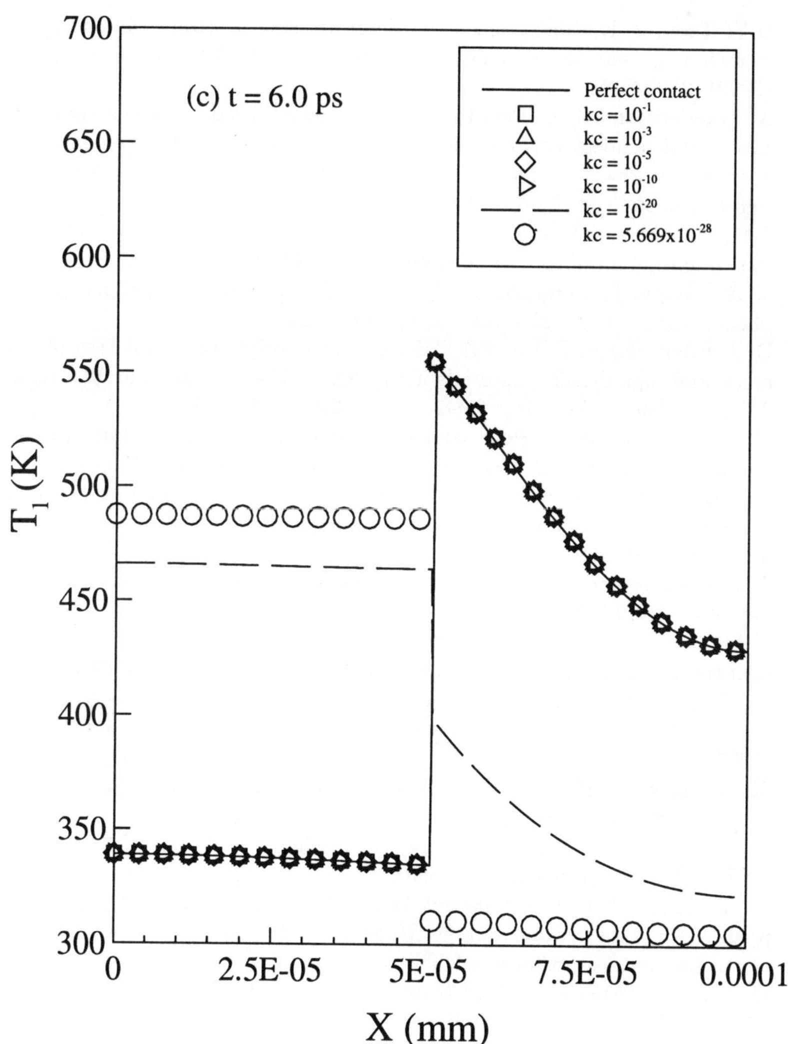

Figure 4.10. Continued.

References

1. D.Y. Tzou, J. K. Chen, and J. E. Beraun, "Hot-electron blast induced by ultrashort-pulsed lasers in layered media," *Int. J. Heat Mass Transfer* **45** (2002), 3369-3382.

2. A. Mandelis and S. B. Peralta, "Thermal-wave based materials characterization and nondestructive evaluation of high-temperature superconductors: a critical review," in *Physics and Materials Science of High Temperature Superconductors II* (Kluwer, Boston, 1992).

3. J. Opsal, "The application of thermal wave technology to thickness and grain size of aluminum films," *SPIE* **1596** (1991), 120-131.

4. J. A. Knapp, P. Borgesen, R. A. Zuhr, "Beam-solid interactions: physical phenomena," *Mater. Res. Soc. Symp. Proc.* **157** (1990)

5. D. J. Elliot and B. P. Piwczyk, "Single and multiple pulse ablation of polymeric and high density materials with excimer laser radiation at 193 nm and 248 nm," *Mater. Res. Soc. Symp. Proc.* **129** (1989), 627-636.

6. C. P. Grigoropoulos, "Heat transfer in laser processing of thin films," in *Annual Review of Heat Transfer V* (Hemisphere, New York, 1994).

7. J. Narayan, V. P. Gosbole, and G. W. White, "Laser method for synthesis and processing of continuous diamond films on nondiamond substrates," *Science* **52** (1991), 416-418.

8. J. M. Hopkins and J. Sibbett, "Ultrashort-pulse lasers: big payoffs in a flash," *Sci. AM.* **283** (2000), 72-79.

9. J. Liu, Preliminary survey on the mechanisms of the wave-like behaviors of heat transfer in living tissues," *Forschung im Ingenieurwesen* **66** (2000), 1-10.

10. C. Momma, S. Nolte, B. N. Chichkov, F. V. Alvensleben, and A. Tunnermann, "Precise laser ablation with ultrashort pulsesm," *App. Surf. Sci.* **109** (1997), 15-19.

11. M. D. Shirk and P. A. Molian, "A review of ultrashort pulsed laser ablation of materials," *J. Laser Applications* **10** (1998), 18-28.

12. D. Y. Tzou, "Ultrafast heat transport: the lagging behavior," 44th *SPIE's Annual Meeting* (Denver, CO, 1999).

13. D. Y. Tzou, "Ultrafast transient behavior in microscale heat/mass transport," *Advanced Photon Source Millennium Lecture Series* (Argonne National Laboratories, Chicago, 2000).

14. D. Y. Tzou, "Microscale heat transfer and fluid flow," 45th *SPIE's Annual Meeting* (San Diego, CA, 2000).

15. DOE (Department of Energy) BES (basic Energy Sciences) Workshop, 1999, Complex Systems - Science for the 21st Century.

16. DOE (Department of Energy) BES (basic Energy Sciences) Workshop, 1999, Nanoscale Science, Engineering, and Technology - Research Direction.

17. L. Bartels, *et al.*, "Real-space observation of molecular motion induced by femtosecond laser pulses," *Science* **305** (2004), 648-651.

18. J. K. Chen and J. E. Beraun, "Numerical study of ultrashort laser pulse interactions with metal films," *Numerical Heat Transfer A* **40** (2001), 1-20.

19. S. I. Anisimov, B. L. Kapeliovich, and T. L. Perel'man, "Electron emission

from metal surfaces exposed to ultrashort laser pulses," *Sov. Phys. JETP* **39** (1974), 375-377.

20. S. D. Brorson, *et al.*, "Femtosecond room-temperature measurement of the electron-phonon coupling constant λ in metallic superconductors," *Phy. Rev. Lett.* **64** (1990), 2172-2175.

21. J. G. Fujimoto, J. M. Liu, and E. P. Ippen, "Femtosecond laser interaction with metallic tungsten and non-equilibrium electron and lattices temperature in thin gold film," *Phys. Rev. Lett.* **53** (1984), 1837-1840.

22. T. Q. Qiu and C. L. Tien, "Short-pulse laser heating on metals," *Int. J. Heat Mass Transfer* **35** (1992), 719-726.

23. T. Q. Qiu and C. L. Tien, "Heat transfer mechanisms during short-pulse laser heating of metals," *ASME Journal of Heat Transfer* **115** (1993), 835-841.

24. T. Q. Qiu and C. L. Tien, "Femtosecond laser heating of multi-layer metals-I. Analysis," *Int. J. Heat Mass Transfer* **37** (1994), 2789-2797.

25. D. Y. Tzou, *Macro To Micro Heat Transfer* (Taylor & Francis, Washington DC, 1996).

26. P. J. Antaki, "Importance of nonequilibrium thermal conductivity during short-pulse laser-induced desorption from metals," *Int. J. Heat Mass Transfer* **45** (2002), 4063-4067.

27. A. N. Smith, J. L. Hosteler, and P. M. Norris, "Nonequilibrium heating in metal films: an analytical and numerical analysis," *Numerical Heat Transfer A* **35** (1999), 859-873.

28. D. Y. Tzou and K. S. Chiu, "Temperature-dependent thermal lagging in ultrafast laser heating," *Int. J. Heat Mass Transfer* **44** (2001), 1725-1734.

29. W. Dai and R. Nassar, "A three level finite difference scheme for solving micro heat transport equations with temperature-dependent thermal properties," *Numerical Heat Transfer B* **45** (2003), 509-523.

30. J. K. Chen, J. E. Beraun, and T. C. Carney, "A corrective smoothed particle method for boundary value problems in heat conduction," *Int. J. Num. Method Eng.* **46** (1999), 231-252.

31. J. K. Chen and J. E. Beraun, "A generalized smoothed particle hydrodynamics method for nonlinear dynamic problems," *Comput. Method Appl. Mech. Eng.* **190** (2000), 225-239.

32. J. K. Chen, J. E. Beraun, and C. J. Jih, "A corrective smoothed particle method for transient elastoplastic dynamics," *Comput. Mech* **27** (2001), 177-187.

33. L. C. Evans, *Partial Differential Equations* (American Mathematical Society, Providence, Rhode Island, 1998).

34. W. Dai, "Nonlinear finite difference schemes for solving the parabolic two-step model with temperature dependent thermal properties in a double-layered thin film heated by ultrashort-pulsed lasers," submitted to *Int. J. Heat Mass Transfer*.

35. R. E. Mickens, *Nonstandard Finite Difference Models of Differential Equations* (World Scientific, Singapore, 1994).

36. R. E. Mickens, *Applications of Nonstandard Finite Difference Schemes* (World Scientific, Singapore, 2000).

37. W. B. Lor and H. S. Chu, "Propagation of thermal waves from a surface or an interface between dissimilar material," *Numerical Heat Transfer* **36** (1999), 681-697.

38. W. B. Lor and H. S. Chu, "Effect of interface thermal resistance on heat transfer in a composite medium using the thermal wave model," *Int. J. Heat Mass Transfer* **43** (2000), 653-663.

39. E. T. Swartz and R. O. Pohl, "Thermal boundary resistance," *Reviews of Modern Physics* **61** (1989), 605-668.

40. W. Dai and S. Su, "A nonstandard finite difference scheme for solving one dimensional nonlinear heat transfer," *Journal of Difference Equations and Applictions* **10** (2004), 1025-1032.

41. D. Y. Tzou, " The generalized lagging response in small-scale and high-rate heating," *Int. J. Heat Mass Transfer* **38** (1995), 3231-3240.

42. J. K. Chen, J. E. Beraun, and C. L. Tham, "Investigation of thermal response caused by pulsed laser heating," *Numerical Heat Transfer A* **44** (2003), 705-722.

43. M. A. Al-Nimr and V. S. Arpaci, "The thermal behavior of thin metal films in the hyperbolic two-step model," *Int. J. Heat Mass Transfer* **43** (2000), 2021-2028.

44. M. A. Al-Nimr, O. M. Haddad, and V. S. Arpaci, "The thermal behavior of metal films - a hyperbolic two-step model," *Heat Mass Transfer* **35** (1999), 459-464.

45. M. Al-Odat, M. A. Al-Nimr, and M. Hamdan, "Thermal stability of superconductors under the effect of a two-dimensional hyperbolic heat conduction model," *Int. J. Numerical Methods for Heat & Fluid Flow* **12** (2002),173-177.

46. M. Naji, M.A. Al-Nimr, and M. Hader, "The validity of using the microscopic hyperbolic heat conduction model under as harmonic fluctuating boundary heating source," *International Journal of Thermophysics* **24** (2003), 545-557.

47. M. A. Al-Nimr and M. K. Alkam, "Overshooting phenomenon in the hyperbolic microscopic heat conduction model," *International Journal of Thermophysics* **24** (2003), 577-583.

CHAPTER 6

RELIABLE FINITE DIFFERENCE SCHEMES WITH APPLICATIONS IN MATHEMATICAL ECOLOGY

Dobromir T. Dimitrov

Department of Mathematics, University of Texas at Arlington
P.O. Box 19408, Arlington, TX 76019-0408, U.S.A.
dobri@uta.edu

Hristo V. Kojouharov

Department of Mathematics, University of Texas at Arlington
P.O. Box 19408, Arlington, TX 76019-0408, U.S.A.
hristo@uta.edu

Benito M. Chen-Charpentier

Department of Mathematics
University of Wyoming
P.O. Box 3036, Laramie, WY 82071, U.S.A.
bchen@uwyo.edu

In this chapter we first develop a new class of nonstandard finite-difference methods for ordinary differential equations with polynomial right-hand sides, based on a combination of exact discretization schemes. Second, we develop a new class of elementary stable nonstandard (ESN) finite-difference schemes, based on the standard Euler and second-order Runge-Kutta methods, for general two-dimensional autonomous dynamical systems. We also develop a class of positive and elementary stable nonstandard (PESN) finite-difference methods for the Rosenzweig-MacArthur predator-prey systems with a logistic intrinsic growth of the prey population and for phytoplankton-nutrient systems with nutrient loss. The proposed new numerical schemes work very well with conservative as well as with non-conservative dynamical systems. Finally, we outline directions for the construction of ESN and PESN methods for general multi-dimensional autonomous dynamical systems.

1. Introduction

Ordinary differential equations and systems of ordinary differential equations appear in many physical, biological and economics applications. The increasing amount of realistic mathematical models helps in understanding the dynamics of analyzed systems. Exact solutions can be obtain in a very few cases and are usually complicated, so good approximations are necessary. Approximate solutions are often the method of choice. Numerical methods based on finite difference approximations, Taylor series expansions and interpolation, such as Euler, Runge-Kutta and Adams methods are widely used (See, for example, [1]). However, their use raises questions such as what the truncation error is, the stability region or even, from a dynamical systems point of view, if there are spurious bifurcation points.

Mickens [2] gave a novel approach for developing new finite difference schemes for ordinary differential equations. His approach consists of solving some differential equations exactly and writing the corresponding finite difference scheme based on this solution. He found that the exact finite difference schemes obtained in this way, which he called nonstandard methods, usually had different denominators in the finite difference approximations for the derivatives and that the right-hand sides were evaluated at both the old and new times. Even though he constructed nonstandard methods for some equations for which there is no exact solution, it remained an open question how to systematically write the nonstandard method for arbitrary equations.

Multi-dimensional dynamical systems in mathematical ecology arise when modeling interspecies interactions and they describe the rates of change of the size of each interacting component [3]. The positivity of the sizes of all populations requires the mathematical models to preserve the invariance of the first quadrant. As in the ODE case, numerical methods that approximate systems of differential equations are expected to be consistent with the original dynamical system, to be zero-stable, convergent and positivity-preserving. General rules for designing nonstandard finite difference methods that preserve the physical properties of the approximated system have been discussed by Mickens [2,4]. Many researchers have worked on developing nonstandard schemes that preserve the stability properties of equilibria, including Anguelov and Lubuma [5], Libuma and Roux [6] and Dimitrov and Kojouharov [7], among others. All of them have designed elementary stable nonstandard (ESN) methods for different classes of dynamical systems. An important number of contributions have also been made to guarantee the positivity of the numerical solution of those nonstandard numerical methods [8,9]. Marcus and Mickens [10] have constructed positive nonstandard methods that suppress numerically induced chaos for a semiconductors system. Piyanwong et.al. [11] and Jansen and Twizell [12] have designed positive and unconditionally stable schemes for the SIR and SEIR models, respectively. Many researchers have also worked on positive and bounded nonstandard finite difference schemes for partial differential equations [13,14].

In this chapter we extend on Mickens' research, in general, and on our previous work [15,7,16], on a research by Lubuma and Roux [6] and on a research Piyanwong et.al. [11], in particular. We first develop a new class of nonstandard finite-difference methods for ordinary differential equations of the form $dx/dt = f(x)$, where f is a polynomial. The new methods are based on a combination of exact discretization schemes, discussed in [15]. We also outline directions for the extension of the methods to ODE's with arbitrary reaction terms. Second, we develop a new class of elementary stable nonstandard (ESN) finite-difference schemes, based on the standard Euler and second-order Runge-Kutta methods, for general two-dimensional autonomous dynamical systems. We also develop a class of positive and elementary stable nonstandard (PESN) finite-difference methods for the Rosenzweig-MacArthur predator-prey systems with a logistic intrinsic growth of the prey population and for phytoplankton-nutrient systems with nutrient loss. The proposed new numerical schemes work very well with conservative as well as with non-conservative dynamical systems. Finally, we outline directions for the construction of ESN and PESN methods for general multi-dimensional autonomous dynamical systems, based on combinations of existing stability preserving methods and Patankar schemes [17].

2. Definitions and Preliminaries

A general n-dimensional autonomous system has the following form:

$$\frac{d\bar{x}}{dt} = f(\bar{x}); \quad \bar{x}(t_0) = \bar{x}_0, \tag{1}$$

where $\bar{x} = [x^1, x^2, \ldots, x^n]^T : [t_0, T) \to \mathbb{R}^n$, the function $f = [f^1, f^2, \ldots, f^n]^T : \mathbb{R}^n \mapsto \mathbb{R}^n$ is differentiable and $\bar{x}_0 \in \mathbb{R}^n$. The equilibrium points of System (1) are defined as the solutions of $f(\bar{x}) = 0$.

Definition 1: Let $\bar{x}^*$ be an equilibrium of System (1), $J(\bar{x}^*)$ be the Jacobian of System (1) at $\bar{x}^*$ and $\sigma(J(\bar{x}^*))$ denotes the spectrum of $J(\bar{x}^*)$.

An equilibrium $\bar{x}^*$ of System (1) is called linearly stable if $Re(\lambda) < 0$ for all $\lambda \in \sigma(J(\bar{x}^*))$ and linearly unstable if $Re(\lambda) > 0$ for at least one $\lambda \in \sigma(J(\bar{x}^*))$.

A numerical scheme with a step size h, that approximates the solution $\bar{x}(t_k)$ of System (1) can be written in the form:

$$D_h(\bar{x}_k) = F_h(f; \bar{x}_k), \tag{2}$$

where $D_h(\bar{x}_k) \approx \left.\dfrac{d\bar{x}}{dt}\right|_{t=t_k}$, $\bar{x}_k \approx \bar{x}(t_k)$, $F_h(f; \bar{x}_k)$ approximates the right-hand side of System (1) and $t_k = t_0 + kh$.

Definition 2: The finite-difference scheme (2) is called exact if its solution satisfies $\bar{x}_k = \bar{x}(t_k)$, where $\bar{x}(t)$ denotes the solution of Equation (1).

Definition 3: Let $\bar{x}^*$ be a fixed point of the scheme (2) and the equation of the perturbed solution $\bar{x}_k = \bar{x}^* + \bar{\epsilon}_k$, where $\bar{\epsilon}_k$ is small, be linearly approximated by

$$D_h \bar{\epsilon}_k = J_h \bar{\epsilon}_k. \tag{3}$$

Here the right-hand side of Equation (3) represents the linear term in $\bar{\epsilon}_k$ of the Taylor expansion of $F_h(f; \bar{x}^* + \bar{\epsilon}_k)$ around $\bar{x}^*$. The fixed point $\bar{x}^*$ is called stable if $\|\bar{\epsilon}_k\| \to 0$ as $k \to \infty$, and unstable otherwise, where $\bar{\epsilon}_k$ is the solution of Equation (3).

Let System (2) be expressed in the following explicit way:

$$\bar{x}_{k+1} = G(\bar{x}_k), \tag{4}$$

where the function $G = [G^1, G^2, \ldots, G^n]^T : \mathbb{R}^n \mapsto \mathbb{R}^n$ is differentiable. If $\bar{x}^*$ is a fixed point of System (4) then the equation for the perturbed solution $\bar{\epsilon}_k$, around $\bar{x}^*$, has the form:

$$\bar{\epsilon}_{k+1} = J(\bar{x}^*)\bar{\epsilon}_k,$$

where $J(\bar{x}^*)$ denotes the Jacobian $\left(\dfrac{\partial G^i(\bar{x}^*)}{\partial x^j}\right)_{1 \leq i,j \leq n}$. The solution $\|\bar{\epsilon}_k\| \to 0$ when $k \to \infty$ if and only if all eigenvalues of $J(\bar{x}^*)$ are less than one in absolute values.

When analyzing the stability of fixed points the following lemma can be useful [18, p.82]:

Lemma 4: *For the quadratic equation $\lambda^2 + \alpha\lambda + \beta = 0$ both roots satisfy $|\lambda_i| < 1, i = 1, 2$ if and only if the following conditions are satisfied:*

- $1 + \alpha + \beta > 0$;
- $1 - \alpha + \beta > 0$; *and*
- $\beta < 1$.

We introduce the next two definitions based on definitions given by Anguelov and Lubuma in [5].

Definition 5: The finite difference method (2) is called elementary stable, if, for any value of the step size h, its only fixed points $\bar{x}^*$ are those of the differential system (1), the linear stability properties of each $\bar{x}^*$ being the same for both the differential system and the discrete method.

Definition 6: The one-step method (2) is called a nonstandard finite-difference method if at least one of the following conditions is satisfied:

- $D_h(\bar{x}_k) = \dfrac{\bar{x}_{k+1} - \bar{x}_k}{\varphi(h)}$, where $\varphi(h) = h + \mathcal{O}(h^2)$ is a nonnegative function;

- $F_h(f; \bar{x}_k) = g(\bar{x}_k, \bar{x}_{k+1}, h)$, where $g(\bar{x}_k, \bar{x}_{k+1}, h)$ is a nonlocal approximation of the right-hand side of System (1).

The following theorem [5] concerns the properties of the general non-standard schemes based on standard finite-difference methods:

Theorem 7: *If the numerical method (2) represents a standard finite-difference scheme that is consistent and zero-stable, then any corresponding nonstandard finite-difference scheme in Definition 6 is necessarily consistent. Furthermore, if the nonstandard scheme is constructed according to the first bullet of Definition 6, then this scheme is zero-stable provided that the operator F_h satisfies, for some $M > 0$ independent of h and for bounded sequences $\{\bar{x}_k\}$ and $\{\tilde{x}_k\}$ in $\mathbb{R}^n$, the Lipschitz condition*

$$\sup_k \|F_h(\bar{x}_k) - F_h(\tilde{x}_k)\| \leq M \sup_k \|\bar{x}_k - \tilde{x}_k\|.$$

3. Numerical Methods for Single-Species Population Models

The following first-order ordinary differential equation is widely used to model the dynamics of single-species populations [3]:

$$\frac{dx}{dt} = f(x); \quad x(t_0) = x_0, \tag{5}$$

where $x(t) : \mathbb{R} \mapsto \mathbb{R}$, the function $f : \mathbb{R} \mapsto \mathbb{R}$ is differentiable and $x_0 \in \mathbb{R}$. The dynamics of Equation (5) depend on the type and the properties of the function $f(x)$.

In this section we derive the necessary and sufficient conditions for the general nonstandard one-step finite-difference method

$$\frac{x_{k+1} - x_k}{\varphi(h)} = g(x_k, x_{k+1}, h), \tag{6}$$

where $\varphi(h) = h + \mathcal{O}(h^2)$ is a nonnegative function, to be a second-order approximation of Equation (5). Based on it we construct reliable finite-difference schemes for Equation (5) with polynomial right-hand sides.

Theorem 8: *If the nonstandard one-step finite-difference method (6) with differentiable right-hand side $g(x, y, h) : \mathbb{R}^3 \mapsto \mathbb{R}$ and twice differentiable denominator function $\varphi(h) = h + \mathcal{O}(h^2)$ is a discretization of Equation (5), then necessary and sufficient conditions for $g(x, y, h)$, that guarantee second-order accuracy of the scheme, are given by:*

- $g(x_k, x_k, 0) = f(x_k)$ *for every k; and*

- $\dfrac{\partial g}{\partial h}(x_k, x_k, 0) = \left(\dfrac{f'(x_k) - \varphi''(0)}{2} - \dfrac{\partial g}{\partial y}(x_k, x_k, 0) \right) f(x_k).$

Proof. To prove that the second-order accuracy of Scheme (6) is equivalent to the set of conditions for the right-hand side $g(x, y, h)$, described in Theorem 8, we compare the Taylor series expansions of $\dfrac{x(h) - x(0)}{\varphi(h)}$ and $g(x(0), x(h), h)$, where $x(t)$ is the solution of Equation (5) with an initial condition $x(0) = x_k$. Since

$$\frac{x(h) - x(0)}{\varphi(h)} = f(x_k) + \frac{f(x_k)(f'(x_k) - \varphi''(0))}{2} h + O(h^2)$$

and

$$g(x(0), x(h), h) = g(x_k, x_k, 0)$$

$$+ \left(\frac{\partial g}{\partial y}(x_k, x_k, 0) f(x_k) + \frac{\partial g}{\partial h}(x_k, x_k, 0) \right) h + O(h^2)$$

then the second-order accuracy of the scheme is guaranteed by the conditions that make the corresponding coefficients of the zero and first degree terms of both expansions equal. $\qquad\square$

If we impose the additional condition of $\dfrac{\partial g}{\partial h}(x_k, x_k, 0) = 0$ on the right-hand side of Scheme (6), which is satisfied in many practical cases, in particular, when the function $g(x, y, h)$ is independent of h, then the conditions in Theorem 8 can be further simplified. We formulate the new conditions in the following Corollary:

Corollary 9: *If the nonstandard one-step finite-difference method (6) with differentiable right-hand side $g(x, y, h) : \mathbb{R}^3 \mapsto \mathbb{R}$, satisfying $\dfrac{\partial g}{\partial h}(x_k, x_k, 0) = 0$, and twice differentiable denominator function $\varphi(h) = h + \mathcal{O}(h^2)$ is a discretization of the equation (5), then necessary and sufficient conditions for $g(x, y, h)$, that guarantee second-order accuracy of the scheme, are given by:*

- $g(x_k, x_k, 0) = f(x_k)$ *for every k; and*

- $\dfrac{\partial g}{\partial y}(x_k, x_k, 0) = \dfrac{f'(x_k) - \varphi''(0)}{2}.$

Based on the results of Theorem 8 and Corollary 9 we formulate the composition rules for nonstandard one-step finite-difference methods of type (6), with denominator functions $\varphi(h) = h$ and $\varphi(h) = \dfrac{e^{ah} - 1}{a}$, that guarantee a second-order accuracy of the resulting schemes.

Lemma 10: *If the nonstandard one-step finite-difference methods*

$$\frac{x_{k+1} - x_k}{h} = g_i(x_k, x_{k+1}, h), \quad i = 1, 2$$

with $g_i(x, y, h)$ satisfying $\dfrac{\partial g_i}{\partial h}(x_k, x_k, 0) = 0$ are second-order approximations of the equations

$$\frac{dx}{dt} = f_i(x), \quad i = 1, 2,$$

respectively, then the combined one-step finite-difference method

$$\frac{x_{k+1} - x_k}{h} = g_1(x_k, x_{k+1}, h) + g_2(x_k, x_{k+1}, h)$$

is a second-order approximation of the equation

$$\frac{dx}{dt} = f_1(x) + f_2(x). \tag{7}$$

Proof. If the additional condition $\dfrac{\partial g}{\partial h}(x_k, x_k, 0) = 0$ on the function $g(x, y, h)$ is imposed, then the second condition in Theorem 8 becomes

$$\frac{\partial g}{\partial y}(x_k, x_k, 0) = \frac{f'(x_k) - \varphi''(0)}{2}.$$

When, in addition, the standard denominator function $\varphi(h) = h$ is used in Scheme (6), the above condition becomes $\dfrac{\partial g}{\partial y}(x_k, x_k, 0) = \dfrac{f'(x_k)}{2}$. Since the schemes

$$\frac{x_{k+1} - x_k}{h} = g_i(x_k, x_{k+1}, h), \quad i = 1, 2$$

are second-order approximations of the equations

$$\frac{dx}{dt} = f_i(x); \quad i = 1, 2,$$

respectively, and both $g_i(x, y, h)$ satisfy $\dfrac{\partial g_i}{\partial h}(x_k, x_k, 0) = 0$ then the conditions of Corollary 9 hold for both $g_i(x, y, h)$. Therefore

$$g_1(x_k, x_k, 0) + g_2(x_k, x_k, 0) = f_1(x_k) + f_2(x_k)$$

and

$$\left. \frac{\partial}{\partial y}\left(g_1(x, y, h) + g_2(x, y, h)\right)\right|_{(x_k, x_k, 0)} = \frac{\left. \dfrac{d}{dx}(f_1(x) + f_2(x))\right|_{x=x_k}}{2},$$

which, according to Corollary 9, implies that the combined numerical scheme is a second-order approximation to Equation (7). $\qquad\square$

Lemma 11: *If the nonstandard one-step finite-difference methods*

$$\frac{x_{k+1} - x_k}{\dfrac{e^{ah} - 1}{a}} = g_i(x_k, x_{k+1}, h) + ax_k, \quad i = 1, 2$$

with $g_i(x, y, h)$ satisfying $\dfrac{\partial g_i}{\partial h}(x_k, x_k, 0) = 0$ are second-order approximations of equations

$$\frac{dx}{dt} = f_i(x) + ax; \quad i = 1, 2,$$

respectively, then the combined one-step finite-difference method

$$\frac{x_{k+1} - x_k}{\dfrac{e^{ah} - 1}{a}} = g_1(x_k, x_{k+1}, h) + g_2(x_k, x_{k+1}, h) + ax_k$$

is a second-order approximation of the equation

$$\frac{dx}{dt} = f_1(x) + f_2(x) + ax. \tag{8}$$

Proof. If $\varphi(h) = \dfrac{e^{ah} - 1}{a}$, then $\varphi''(0) = a$. Since the schemes

$$\frac{x_{k+1} - x_k}{\dfrac{e^{ah} - 1}{a}} = g_i(x_k, x_{k+1}, h) + ax_k, \quad i = 1, 2$$

are second-order approximations of the equations

$$\frac{dx}{dt} = f_i(x) + ax; \quad i = 1, 2,$$

respectively and both $g_i(x, y, h)$, $i = 1, 2$, satisfy $\dfrac{\partial g_i}{\partial h}(x_k, x_k, 0) = 0$ then the conditions of Corollary 9 yield:

- $g_i(x_k, x_k, 0) = f_i(x_k)$; and

- $\dfrac{\partial g_i}{\partial y}(x_k, x_k, 0) = \dfrac{f_i'(x_k)}{2}$

Therefore

$$g_1(x_k, x_k, 0) + g_2(x_k, x_k, 0) + ax_k = f_1(x_k) + f_2(x_k) + ax_k$$

and

$$\frac{\partial}{\partial y}\left(g_1(x, y, h) + g_2(x, y, h) + ax\right)\Bigg|_{(x_k, x_k, 0)} = \frac{f_1'(x_k) + f_2'(x_k)}{2}$$

$$= \frac{\dfrac{d}{dx}\left(f_1(x) + f_2(x) + ax\right)\Big|_{x = x_k} - \varphi''(0)}{2},$$

which, according to Corollary 9, implies that the combined numerical scheme is a second-order approximation to Equation (8). $\qquad\square$

These composition rules help us to construct new second-order accurate nonstandard finite-difference approximations of Equation (5) with an arbitrary polynomial right-hand side $f(x)$. The new methods extend on results from [15], where we considered convection-reaction equations of the form:

$$\frac{\partial c}{\partial t} + v\frac{\partial c}{\partial x} = a_1 c + a_N c^N, \tag{9}$$

for a concentration function $c(x,t)$, and designed the corresponding exact time-discretization schemes:

$$\frac{C^{m+1}(x) - C^m(\tilde{x}^m)}{\dfrac{e^{a_1 \Delta t} - 1}{a_1}} = a_1 C^m(\tilde{x}^m)$$

$$+ a_N \frac{\left(C^{m+1}(x)\right)^{N-1} \left(C^m(\tilde{x}^m)\right)^{N-1}}{\dfrac{\left(C^{m+1}(x)\right)^{N-1} - \left(e^{a_1 \Delta t} C^m(\tilde{x}^m)\right)^{N-1}}{\displaystyle\sum_{k=0}^{N-2} e^{a_1 k \Delta t} \left(C^{m+1}(x) - e^{a_1 \Delta t} C^m(\tilde{x}^m)\right)}}, \qquad (10)$$

where the backtrack point $\tilde{x}^m = x - v\Delta t$ and $C^m(x) \approx c(x, m\Delta t)$.

The schemes (10) are based on the fact that the exact solution of Equation (9) can be obtained by the method of characteristics. A similar approach is applicable to Equation (5) and it is summarized in the following theorem:

Theorem 12: *For the first-order ordinary differential equation (5) the following statements hold:*

(1) If $f(x) = a_n x^n$, $n \geq 2$ is the right-hand side of Equation (1) then the nonstandard finite-difference scheme

$$\frac{x_{k+1} - x_k}{h} = \widehat{g}_n(x_k, x_{k+1}), \qquad (11)$$

where $\widehat{g}_n(x_k, x_{k+1}) = a_n \dfrac{(n-1)x_{k+1}^{n-1} x_k^{n-1}}{\displaystyle\sum_{i=0}^{n-2} x_k^i x_{k+1}^{n-2-i}}$ is exact;

(2) If $f(x) = a_1 x + a_n x^n$, $n \geq 2$ is the right-hand side of Equation (1) then the nonstandard finite-difference scheme

$$\frac{x_{k+1} - x_k}{\dfrac{e^{a_1 h} - 1}{a_1}} = a_1 x_k + \widetilde{g}_n(x_k, x_{k+1}, h), \qquad (12)$$

where $\widetilde{g}_n(x_k, x_{k+1}, h) = a_n \dfrac{x_{k+1}^{n-1} x_k^{n-1}}{\displaystyle\sum_{i=0}^{n-2} e^{i a_1 h} x_k^i x_{k+1}^{n-2-i}} \displaystyle\sum_{i=0}^{n-2} e^{i a_1 h}$ is exact.

Proof. Let us consider the solution $x(t)$ of Equation (5) with the initial condition $x(0) = x_k$.

(1) If $f(x) = a_n x^n$, $n \geq 2$ is the right-hand side of Equation (5) then

$$\int_{x(0)}^{x(h)} \frac{dx}{x^n} = \int_0^h a_n \, dt,$$

which yields that the solution $x(t)$ satisfies

$$\frac{x^{n-1}(h) - x^{n-1}(0)}{(n-1)x^{n-1}(h)x^{n-1}(0)} = a_n h.$$

The above equation can be rewritten in the form:

$$\frac{x(h) - x(0)}{h} = a_n \frac{(n-1)x^{n-1}(h)x^{n-1}(0)}{\displaystyle\sum_{i=0}^{n-2} x^i(0)x^{n-2-i}(h)}$$

which yields that the scheme (11) is exact.

(2) Similarly, if $f(x) = a_1 x + a_n x^n$, $n \geq 2$ is the right-hand side of Equation (1) then the solution $x(t)$ satisfies

$$\frac{x^{n-1}(h)(a_1 + a_n x^{n-1}(0))}{x^{n-1}(0)(a_1 + a_n x^{n-1}(h))} = e^{(n-1)a_1 h}.$$

Therefore

$$a_1\left(x^{n-1}(h) - e^{(n-1)a_1 h}x^{n-1}(0)\right) = a_n x^{n-1}(0)x^{n-1}(h)\left(e^{(n-1)a_1 h} - 1\right)$$

and

$$a_1\left(x(h) - e^{a_1 h}x(0)\right) = \frac{a_n x^{n-1}(0)x^{n-1}(h)}{\displaystyle\sum_{i=0}^{n-2} e^{ia_1 h}x^i(0)x^{n-2-i}(h)}\left(e^{(n-1)a_1 h} - 1\right)$$

which implies that the scheme (12) is exact.

$\square$

Theorem 12 deals with the special cases of polynomial functions $f(x) = a_n x^n$ and $f(x) = a_1 x + a_n x^n$, by designing exact schemes for Equation (5) in those cases. Using the results of Theorem 12 in combination with the composition rules, stated in Lemmas 10 and 11, we develop a new class of second-order nonstandard finite-difference schemes for Equation (5) for general polynomial functions $f(x)$.

Theorem 13: *Let the equation (5) has a polynomial right-hand side $f(x) = a_N x^N + a_{N-1} x^{N-1} + \ldots + a_1 x + a_0$. Then the following statements hold:*

(1) If $a_1 = 0$ then the one-step finite-difference method given by

$$\frac{x_{k+1} - x_k}{h} = \sum_{n=2}^{N} \widehat{g}_n(x_k, x_{k+1}) + a_0,$$

where $\widehat{g}_n(x_k, x_{k+1})$ is defined in Scheme (11), is a second-order approximation of the equation (5);

(2) If $a_1 \neq 0$ then the one-step finite-difference method given by

$$\frac{x_{k+1} - x_k}{\dfrac{e^{a_1 h} - 1}{a_1}} = \sum_{n=2}^{N} \widetilde{g}_n(x_k, x_{k+1}, h) + a_1 x_k + a_0,$$

where $\widetilde{g}_n(x_k, x_{k+1}, h)$ is defined in Scheme (12), is a second-order approximation of the equation (5).

Proof.

(1) Since the nonstandard numerical scheme

$$\frac{x_{k+1} - x_k}{h} = \widehat{g}_n(x_k, x_{k+1})$$

is exact for the equation (1) with right-hand side $f(x) = a_n x^n$ then it clearly is also a second-order approximation scheme. Since $\widehat{g}(x, y)$ does not depend on h, then $\dfrac{\partial \widehat{g}}{\partial h}(x_k, x_k, 0) = 0$ and

$$\frac{x_{k+1} - x_k}{h} = \sum_{n=2}^{N} \widehat{g}_n(x_k, x_{k+1}) + a_0$$

represents a second-order approximation of Equation (5) with a right-hand side $f(x) = a_N x^N + a_{N-1} x^{N-1} + \ldots + a_2 x^2 + a_0$, according to the composition rule in Lemma 10.

(2) Since the nonstandard numerical scheme

$$\frac{x_{k+1} - x_k}{\dfrac{e^{a_1 h} - 1}{a_1}} = \widetilde{g}_n(x_k, x_{k+1}, h) + a_1 x_k + a_0,$$

is exact for the equation (5) with right-hand side $f(x) = a_1 x + a_n x^n$

then it clearly is also a second-order approximation scheme. Since,

$$\frac{\partial}{\partial h}\widetilde{g}_n(x,y,h)\bigg|_{(x_k,x_k,0)} = \frac{\partial}{\partial h}\widetilde{g}_n(x_k,x_k,h)\bigg|_{h=0}$$

$$= \frac{\partial}{\partial h}\left(a_n x_k^n \frac{(1-e^{a_1 h})\left(\sum_{i=0}^{n-2} e^{a_1 ih}\right)}{1-(e^{a_1 h})^{n-1}}\right)\bigg|_{h=0}$$

$$= \frac{\partial}{\partial h}\left(a_n x_k^n\right)\bigg|_{h=0} = 0$$

then the scheme

$$\frac{x_{k+1}-x_k}{\frac{e^{a_1 h}-1}{a_1}} = \sum_{n=2}^{N}\widetilde{g}_n(x_k,x_{k+1},h) + a_1 x_k + a_0$$

represents a second-order approximation of the equation (5) with a right-hand side $f(x) = a_N x^N + a_{N-1}x^{N-1} + \ldots + a_1 x + a_0$, according to the composition rule in Lemma 11. $\qquad\square$

In general, the nonstandard numerical schemes, based on the exact schemes (11) and (12), for Equation (5) with an arbitrary right-hand side function $f(x)$ can be constructed using the following procedure:

- Step 1: Approximate the function $f(x)$ by a polynomial $p(x)$ with at least second-order accuracy;
- Step 2: Approximate the resulting new equation $\dfrac{dx}{dt} = p(x)$ by the nonstandard numerical scheme from Theorem 13.

4. Numerical Methods for Multi-Species Population Models

In structurally complex communities two or more species interact and that affects the dynamics of each species. Mathematical models usually consist of systems of differential equations that represent the rates of change of the size of each interacting species. For example, multi-dimensional epidemi-ological models focus on the transmission dynamics of infectious diseases from individual to individual, from population to population, or from com-munity to community. Multi-species mathematical models are of the form (1), where $\bar{x}$ is the n-dimensional vector of population densities and $f(\bar{x})$ describes the nonlinear interactions between the species. The function $f(\bar{x})$

involves parameters which characterize the various growth and interaction features of the system.

In this section, for simplicity, we concentrate primarily on two-species systems of type (1). However, the results presented here can be easily extended to the multi-dimensional case. We assume that System (1) has a finite number of hyperbolic equilibria, i.e., $Re(\lambda) \neq 0$, for $\lambda \in \Omega$, where $\Omega = \bigcup_{\bar{x}^* \in \Gamma} \sigma(J(\bar{x}^*))$ and Γ represents the set of all equilibria of System (1). In what follows, we also assume that ϕ is a real-valued function on $\mathbb{R}$ that satisfies the property:

$$\phi(h) = h + O(h^2) \text{ and } 0 < \phi(h) < 1 \text{ for all } h > 0, \tag{13}$$

e.g., $\phi(h) = 1 - e^{-h}$.

4.1. *Elementary Stable Nonstandard (ESN) Methods*

The problem of designing elementary stable numerical schemes have been addressed by many researchers. Anguelov and Lubuma [5] have used Mickens' techniques to design nonstandard versions of the explicit and implicit Euler and the second order Runge-Kutta methods, only in the case when all eigenvalues of the Jacobian at each equilibrium of the original differential system are single and real. Libuma and Roux [6] have constructed ESN numerical schemes for systems having all eigenvalues of the equilibria's Jacobians in a specific subregion of the complex plane.

The following theorem extends the above methods and designs ESN schemes for solving general two-dimensional dynamical systems:

Theorem 14: *Let ϕ be a real-valued function on $\mathbb{R}$ that satisfies the property (13). Then the following schemes for solving the two-dimensional system (1) represent ESN methods:*

(a) explicit Euler ESN method given by

$$\frac{\bar{x}_{k+1} - \bar{x}_k}{\varphi(h)} = f(\bar{x}_k); \tag{14}$$

(b) implicit Euler ESN method given by

$$\frac{\bar{x}_{k+1} - \bar{x}_k}{\varphi(h)} = f(\bar{x}_{k+1}); \tag{15}$$

(c) second-order Runge-Kutta ESN method given by

$$\frac{\bar{x}_{k+1} - \bar{x}_k}{\varphi(h)} = \frac{f(\bar{x}_k) + f(\bar{x}_k + \varphi(h)f(\bar{x}_k))}{2}, \tag{16}$$

provided the only fixed points of Scheme (16) are the equilibria of System (1);

where $\varphi(h) = \phi(hq)/q$ for $q > \max\limits_{\Omega}\left(\dfrac{|\lambda|^2}{2|Re(\lambda)|}\right)$.

Proof. Let $\bar{x}^*$ be an equilibrium of System (1), $h_1 = \varphi(h) = \dfrac{\phi(hq)}{q}$ and $J = J(\bar{x}^*)$. Note that since $0 < \varphi(h) < \frac{1}{q}$ then $\varphi(h) < \dfrac{2|Re(\lambda)|}{|\lambda|^2}$ for all $\lambda \in \Omega$.

(a) If $\bar{x}^*$ is an equilibrium of System (1), then the equation (3) for the perturbed solution of the scheme (14) has the form

$$\frac{\bar{\epsilon}_{k+1} - \bar{\epsilon}_k}{h_1} = J\bar{\epsilon}_k. \tag{17}$$

The equation (17) leads to

$$\bar{\epsilon}_{k+1} = (I + h_1 J)\bar{\epsilon}_k, \tag{18}$$

where I represents the 2×2-identity matrix.

If Λ is the Jordan form of J, then $J = S\Lambda S^{-1}$, where S is a nonsingular complex 2×2-matrix. In general, Λ has the form $\begin{pmatrix} \lambda_1 & 0 \\ 0 & \lambda_2 \end{pmatrix}$ or $\begin{pmatrix} \lambda_1 & 1 \\ 0 & \lambda_1 \end{pmatrix}$. After the change of variables $\bar{\epsilon}_k = S\bar{\delta}_k$, the equation (18) becomes $\bar{\delta}_{k+1} = (I + h_1\Lambda)\bar{\delta}_k$. Eigenvalues of $I + h_1\Lambda$ are given by $\mu_i = 1 + h_1\lambda_i, i = 1, 2$. Since $\|\bar{\epsilon}_k\| \to 0$ is equivalent to $\|\bar{\delta}_k\| \to 0$, then $\bar{x}^*$ is a stable fixed point of (14) if $|\mu_i| < 1$ for $i = 1, 2$ and unstable fixed point if at least one $|\mu_i| > 1$.

Let $\bar{x}^*$ be a stable equilibrium of System (1) and $\lambda = a + ib$ be an eigenvalue of J. Then $a < 0$ and $|1 + h_1\lambda|^2 = (1 + h_1 a)^2 + h_1^2 b^2 = 1 + h_1(2a + h_1|\lambda|^2) < 1$. Therefore $\bar{x}^*$ is a stable fixed point of (14).

If $\bar{x}^*$ is unstable then $a > 0$ for some eigenvalue $\lambda = a + ib$ of J. Thus $|1 + h_1\lambda|^2 = (1 + h_1 a)^2 + h_1^2 b^2 > 1$. Therefore $\bar{x}^*$ is an unstable fixed point of (14).

(b) Similarly, the linearized system of the perturbed solution of the method (15) implies that $\bar{\delta}_{k+1} = (I - h_1\Lambda)^{-1}\bar{\delta}_k$. The method (15) is stable at $\bar{x}^*$ if $|\mu| < 1$ for all eigenvalues μ of $(I - h_1\Lambda)^{-1}$, given by $\mu_i = \dfrac{1}{1 - h_1\lambda_i}, i = 1, 2$.

Let $\bar{x}^*$ be a stable equilibrium of System (1) and $\lambda = a + ib$ be an eigenvalue of J. Then $a < 0$ and $|1 - h_1\lambda|^2 = (1 - h_1 a)^2 + h_1^2 b^2 > 1$. Therefore $\left|\dfrac{1}{1 - h_1\lambda}\right| < 1$ and $\bar{x}^*$ is a stable fixed point of (15).

If $\bar{x}^*$ is an unstable equilibrium of System (1), then there exists an eigenvalue $\lambda = a + ib$ of the Jacobian J with $a > 0$. This implies that $|1 - h_1\lambda|^2 = (1 - h_1 a)^2 + h_1^2 b^2 = 1 + h_1(-2a + h_1|\lambda|^2) < 1$. Therefore $\bar{x}^*$ is an unstable fixed point of (15).

(c) For the method (16) the linearized equation about the perturbed solution is as follows:

$$\bar{\epsilon}_{k+1} = \left(I + h_1 J + \frac{h_1^2 J^2}{2} \right) \bar{\epsilon}_k. \tag{19}$$

After the change of variables $\bar{\epsilon}_k = S\bar{\delta}_k$, the equation (19) implies $\bar{\delta}_{k+1} = (I + h_1\Lambda + \frac{h_1^2\Lambda^2}{2})\bar{\delta}_k$. The eigenvalues of $I + h_1\Lambda + \frac{h_1^2\Lambda^2}{2}$ are given by $\mu_i = 1 + h_1\lambda_i + \frac{h_1^2\lambda_i^2}{2}, i = 1, 2$.

Let $\bar{x}^*$ be a stable equilibrium of System (1), $\lambda = a + ib$ be an eigenvalue of J and $\mu = 1 + h_1\lambda + \frac{h_1^2\lambda^2}{2}$. Then $a < 0$ and $|\mu|^2 = (1 + h_1 a + \frac{h_1^2}{2}(a^2 - b^2))^2 + h_1^2 b^2(1 + ah_1)^2$. The condition $|\mu|^2 < 1$ is equivalent to $\alpha(h_1) < 0$, where $\alpha(t) = 2a + 2a^2 t + a(a^2 + b^2)t^2 + \frac{(a^2+b^2)^2}{4}t^3$. The derivative $\alpha'(t) = 2a^2 + 2a(a^2 + b^2)t + \frac{3(a^2+b^2)^2}{4}t^2 = \frac{a^2}{4}\beta(\frac{a^2+b^2}{a}t)$, where $\beta(t) = 8 + 8t + 3t^2$. Since $\beta(t)$ is positive for all t, the derivative $\alpha'(t)$ is positive for all t and $\alpha(t)$ is an increasing function. The inequality $h_1 \leq \frac{-2a}{a^2+b^2}$ implies that $\alpha(h_1) \leq \alpha\left(\frac{-2a}{a^2+b^2}\right) = \frac{2ab^2}{a^2+b^2} < 0$. Therefore $|\mu|^2 < 1$ and $\bar{x}^*$ is a stable fixed point of (16).

If $\bar{x}^*$ is an unstable equilibrium of System (1), then there exists $\lambda = a + ib$, an eigenvalue of J, with $a > 0$. The condition $|\mu|^2 > 1$ is equivalent to $\alpha(h_1) > 0$. Since $\alpha(t)$ is positive for all $t > 0$, then $\bar{x}^*$ is an unstable fixed point of (16). $\qquad\square$

4.2. *Positive and Elementary Stable Nonstandard (PESN) Methods*

Another important characteristic of dynamical systems that model interspecies interactions is that all solutions remain nonnegative in every component. This condition is due to the fact that the size of each interacting population can not be negative. It is natural to extend that positivity property of mathematical models to the numerical methods that approximate those models.

In this section we design PESN methods for the following two classes of predator-prey models:

- Rosenzweig-MacArthur models with a logistic intrinsic growth of the

prey population;
- Phytoplankton-nutrient systems with nutrient loss.

Rosenzweig-MacArthur predator-prey models

The general Rosenzweig-MacArthur predator-prey model [18, p.182] with a logistic intrinsic growth of the prey population has the following form:

$$\frac{dx}{dt} = bx(1 - x) - ag(x)xy; \quad x(0) = x_0 \geq 0,$$

$$\frac{dy}{dt} = g(x)xy - dy; \qquad y(0) = y_0 \geq 0,$$

$$\tag{20}$$

where x and y represent the prey and predator population sizes, respectively, $b > 0$ represents the intrinsic growth rate of the prey, $a > 0$ stands for the capturing rate and $d > 0$ is the predator death rate. In (20) it is reasonable to assume:

$$g(x) \geq 0, \quad g'(x) \leq 0, \quad [xg(x)]' \geq 0, \tag{21}$$

and that $xg(x)$ is bounded as $x \to \infty$. These assumptions express the idea that as prey population increases the consumption rate of prey per predator increases but that the fraction of the total prey population consumed per predator decreases [18].

Depending on the values of the parameters and the functional response $xg(x)$ System (20) has the following equilibria:

(1) $E_0 = (0, 0)$;
(2) $E_1 = (1, 0)$ and
(3) $E^* = (x^*, y^*)$, where x^* is the solution of $xg(x) = d$ and $y^* = \dfrac{bx^*(1 - x^*)}{ad}$. The equilibrium E^* exists if and only if $g(1) > d$.

According to the stability theory for general nonlinear systems [19], the following statements about the stability of the equilibria of System (20) are true:

(1) The equilibrium E_0 is always linearly unstable;
(2) The equilibrium E_1 is linearly stable if $g(1) < d$ and linearly unstable if $g(1) > d$;
(3) The equilibrium E^* is linearly stable if $b + ay^*g'(x^*) > 0$ and linearly unstable if $b + ay^*g'(x^*) < 0$;

PESN methods for solving Rosenzweig-MacArthur predator-prey systems with logistic intrinsic growth of the prey population can be designed, based on the following theorem:

Theorem 15: *Let ϕ be a real-valued function on $\mathbb{R}$ that satisfies the property (13) and the function $g(x)$ satisfies (21). Then the following scheme for solving System (20) represents a PESN method:*

$$\frac{x_{k+1} - x_k}{\varphi(h)} = bx_k - bx_k x_{k+1} - ag(x_k)x_{k+1}y_k;$$

$$\frac{x_{k+1} - x_k}{\varphi(h)} = g(x_k)x_k y_k - dy_{k+1}; \tag{22}$$

where $\varphi(h)$ belongs to the following class of functions:

(1) If $g(1) < d$ then $\varphi(h) = \phi(hq)/q$ for all $q \geq 0$, with $\varphi(h) = h$ for $q = 0$.

(2) If $g(1) > d$ then $\varphi(h) = \phi(hq)/q$ for $q > \dfrac{bd|1 - 2x^*|}{|b + ay^*g'(x^*)|x^*}$, where

$E^* = (x^*, y^*)$ is the interior equilibrium of System (20).

Proof. The explicit expression of the nonstandard scheme (22) has the form:

$$x_{k+1} = \frac{(1 + \varphi(h)b)x_k}{1 + \varphi(h)bx_k + \varphi(h)ag(x_k)y_k}$$

$$y_{k+1} = \frac{(1 + \varphi(h)g(x_k)x_k)y_k}{1 + \varphi(h)d}. \tag{23}$$

Since the constants a, b, d and the function g are all positive then the system (23) is unconditionally positive and its fixed points are exactly the equilibria E_0, E_1 and E^* of System (20). Therefore, Equation (3) for the perturbed solution of the scheme (23) around an equilibrium $\tilde{E} = (\tilde{x}, \tilde{y})$ has the form:

$$\bar{\epsilon}_{k+1} = J(\tilde{E})\bar{\epsilon}_k,$$

where $J(\tilde{E}) = \begin{pmatrix} C(\tilde{E}) & -A(\tilde{E}) \\ B(\tilde{E}) & D(\tilde{E}) \end{pmatrix}$ with

$$A(\tilde{E}) = \frac{(1 + \varphi(h)b)\varphi(h)a\tilde{x}g(\tilde{x})}{(1 + \varphi(h)b\tilde{x} + \varphi(h)a\tilde{y}g(\tilde{x}))^2}, \quad B(\tilde{E}) = \frac{\varphi(h)\tilde{y}(g(\tilde{x}) + \tilde{x}g'(\tilde{x}))}{1 + \varphi(h)d},$$

$$C(\tilde{E}) = \frac{(1 + \varphi(h)b)(1 + \varphi(h)a\tilde{y}(g(\tilde{x}) - g'(\tilde{x})\tilde{x}))}{(1 + \varphi(h)b\tilde{x} + \varphi(h)a\tilde{y}g(\tilde{x}))^2} \quad \text{and}$$

$$D(\tilde{E}) = \frac{1 + \varphi(h)\tilde{x}g(\tilde{x})}{1 + \varphi(h)d}.$$

Since $\begin{pmatrix} C(E_0) & -A(E_0) \\ B(E_0) & D(E_0) \end{pmatrix} = \begin{pmatrix} 1+\varphi(h)b & 0 \\ 0 & \dfrac{1}{1+\varphi(h)d} \end{pmatrix}$ then the Jaco-

bian $J(E_0)$ has eigenvalues $\lambda_1 = 1 + \varphi(h)b$ and $\lambda_2 = \dfrac{1}{1+\varphi(h)d}$. Since $|\lambda_1| > 1$ for $h > 0$ then the unstable equilibrium E_0 is also an unstable fixed point of Scheme (22). The Jacobian $J(E_1) = \begin{pmatrix} C(E_1) & -A(E_1) \\ B(E_1) & D(E_1) \end{pmatrix} =$

$\begin{pmatrix} \dfrac{1}{1+\varphi(h)b} & -\dfrac{\varphi(h)ag(1)}{1+\varphi(h)b} \\ 0 & \dfrac{1+\varphi(h)g(1)}{1+\varphi(h)d} \end{pmatrix}$ has eigenvalues $\lambda_1 = \dfrac{1}{1+\varphi(h)b}$ and $\lambda_2 =$

$\dfrac{1+\varphi(h)g(1)}{1+\varphi(h)d}$. If the equilibrium E_1 is stable then $g(1) < d$ and therefore $|\lambda_1| < 1$ and $|\lambda_2| < 1$ for $h > 0$. Thus E_1 is a stable fixed point of Scheme (22). If the equilibrium E_1 is unstable then $g(1) > d$ and $|\lambda_2| > 1$. Therefore E_1 is an unstable fixed point of Scheme (22). Based on these arguments we conclude that in the case when there is no interior equilibrium E^*, i.e., $g(1) < d$ the scheme (22) is a PESN method for all denominator functions $\varphi(h) = \phi(hq)/q$, where $q \geq 0$.

For the Jacobian $J(E^*) = \begin{pmatrix} C(E^*) & -A(E^*) \\ B(E^*) & D(E^*) \end{pmatrix}$ the following holds:

$$A(E^*) = \frac{\varphi(h)ad}{1+\varphi(h)b}, \quad B(E^*) = \frac{\varphi(h)y^*(g(x^*) + x^*g'(x^*))}{1+\varphi(h)d},$$

$$C(E^*) = 1 - \frac{\varphi(h)x^*(b + ay^*g'(x^*))}{1+\varphi(h)b} \quad \text{and} \quad D(E^*) = 1.$$

Therefore, the eigenvalues λ_1 and λ_2 of $J(E^*)$ are roots of the quadratic equation:

$$\lambda^2 - (C(E^*)+1)\lambda + C(E^*) + A(E^*)B(E^*) = 0. \tag{24}$$

The stability of E^* as a fixed point of Scheme (22) depends of the absolute values of λ_1 and λ_2.

By Lemma 4 we obtain that E^* is a stable fixed point of Scheme (22) if and only if the following conditions are true:

(a) $A(E^*)B(E^*) > 0$;
(b) $2 + 2C(E^*) + A(E^*)B(E^*) > 0$; and

(c) $A(E^*)B(E^*) < 1 - C(E^*)$.

The fixed point E^* is unstable if at least one of the above conditions fails. Since $A(E^*) > 0$ and $B(E^*) = \dfrac{\varphi(h)y^*([xg(x)]'|_{x=x^*})}{1 + \varphi(h)d} > 0$ then the condition (a) is always true. Calculations yield that

$$C(E^*) = \frac{1 + \varphi(h)b(1 - x^*) - \varphi(h)ay^*g'(x^*)x^*}{1 + \varphi(h)b} > 0,$$

because $0 < x^* < 1$ and $g'(x^*) < 0$. Therefore the second condition (b) is always true, as well. The third condition (c) is equivalent to the following inequality:

$$\varphi(h)bd(1 - 2x^*) < x^*(b + ay^*g'(x^*)). \tag{25}$$

Let $\varphi(h) = \dfrac{\phi(hq)}{q}$, where the function ϕ satisfies (13) and the parameter $q > \dfrac{bd|1 - 2x^*|}{|b + ay^*g'(x^*)|x^*}$. Since $0 < \varphi(h) < \dfrac{1}{q}$ then

$$\varphi(h) < \frac{|b + ay^*g'(x^*)|x^*}{bd|1 - 2x^*|}. \tag{26}$$

Assume that $E^* = (x^*, y^*)$ is a stable equilibrium of System (20). Therefore $b + ay^*g'(x^*) > 0$. If $x^* \geq \dfrac{1}{2}$, the inequality (25) is satisfied because the left-hand side is nonpositive, while the right-hand side is positive. If $x^* < \dfrac{1}{2}$, the inequality (25) is satisfied because of Inequality (26). Therefore E^* is a stable fixed point of Scheme (22). If $E^* = (x^*, y^*)$ is an unstable equilibrium of System (20) then $b + ay^*g'(x^*) < 0$. In the case when $x^* \leq \dfrac{1}{2}$, the inequality (25) is not satisfied because the left-hand side is nonnegative, while the right-hand side is negative. If $x^* > \dfrac{1}{2}$, the inequality (25) is not satisfied because of Inequality 26). Therefore E^* is an unstable fixed point of Scheme (22). $\qquad \square$

Phytoplankton-Nutrient Systems with Nutrient Loss

A phytoplankton-nutrient (P-N) model of the surface sea layer, introduced by Taylor et. al. [20], incorporates the nutrient recycling with nutrient loss and has the form:

$$\frac{dx}{dt} = (f(y) - d)x; \qquad x(0) = x_0 \geq 0,$$

$$\frac{dy}{dt} = \gamma(b - f(y))x + c(N_0 - y); \ y(0) = y_0 \geq 0,$$

(27)

where $f(y)$ represents the nutrient-dependent phytoplankton growth rate. The functions $x(t)$ and $y(t)$ represent the concentration of phytoplankton and nutrient, respectively, and N_0 represent the concentrations of the nutrient in the water below the surface layer. The parameters $d = m + \dfrac{v}{h} + \dfrac{k}{h}$, $b = \epsilon m$ and $c = \dfrac{k}{h}$ are positive constants, where m is the phytoplankton mortality rate, v is the sinking speed of phytoplankton cells, h is the thickness of the surface layer and k is the turbulent diffusion coefficient. The parameter γ represents the conversion factor of carbon to chlorophyll ratio and the recycling efficiency ϵ, $\epsilon < 1$, is a parameter which allows for the possibility that not all of the nutrients, from the loss of the phytoplankton cells, remain in the system. The above model assumes [20] that the concentration of the phytoplankton in the water below the considered layer is negligible.

We consider the nutrient-dependent phytoplankton growth rate function $f(y) \in \mathcal{F}$, where $\mathcal{F} = \{f \in C^1(0, \infty), f(0) = 0, f'(y) > 0 \text{ on } (0, \infty)\}$. The class $\mathcal{F}$ contains most of the biologically relevant functions, including the functions of Holling types I, II and III [21].

Depending on the values of the parameters System (27) has the following equilibria:

(1) $E_0 = (0, N_0)$;

(2) $E^* = (x^*, y^*)$ given by $y^* = f^{-1}(d)$ and $x^* = \dfrac{c(N_0 - y^*)}{\gamma(d - b)}$. The equilibrium E^* exists if and only if $f(N_0) > d$.

According the stability analysis of System (27) [22], the following statements about the stability of the equilibria are true:

(1) The equilibrium E_0 is linearly stable if $f(N_0) < d$ and linearly unstable otherwise;
(2) The equilibrium E^* is linearly stable when it exists, i.e., when $f(N_0) > d$.

PESN methods for solving System (27) can be designed, based on the following theorem:

Theorem 16: *Let the function $f(y) \in \mathcal{F}$ and the function ϕ satisfies Condition (13). Denote $g(y) = \dfrac{f(y)}{y}$, $D_1 = 1 + \dfrac{(N_0 - y^*)f'(y^*)}{d - b}$ and $D_2 = \dfrac{dN_0 - by^*}{(d - b)y^*}$, where $E^* = (x^*, y^*)$ is the interior equilibrium of System (27). Then the following numerical scheme for solving System (27) represents a PESN method:*

$$\frac{x_{k+1} - x_k}{\varphi(h)} = f(y_k)x_k - dx_{k+1};$$

$$\frac{y_{k+1} - y_k}{\varphi(h)} = \gamma b x_k - \gamma g(y_k)y_{k+1}x_k + cN_0 - cy_{k+1}, \tag{28}$$

where $\varphi(h)$ belongs to the following class of functions:

(1) If $f(N_0) < d$ or $D_1 \leq 2D_2$ then $\varphi(h) = \phi(hq)/q$ for all $q \geq 0$, with $\varphi(h) = h$ for $q = 0$.

(2) If $f(N_0) > d$ and $D_1 > 2D_2$ then $\varphi(h) = \phi(hq)/q$ for $q > \dfrac{c(D_1 - 2D_2)}{2}$.

Proof. The explicit expression of the nonstandard scheme (28) has the form:

$$x_{k+1} = \frac{(1 + \varphi(h)f(y_k))x_k}{1 + \varphi(h)d};$$

$$y_{k+1} = \frac{y_k + \varphi(h)cN_0 + \varphi(h)\gamma b x_k}{1 + \varphi(h)c + \varphi(h)\gamma g(y_k)x_k}. \tag{29}$$

System (29) has positive solutions for all positive initial conditions and its fixed points are exactly the equilibria of System (27). The analysis of the Jacobian $J(E_0)$ shows that (E_0) is a stable fixed point of System (29) if $f(N_0) < d$ and it is an unstable fixed point if $f(N_0) > d$. The eigenvalues λ_1 and λ_2 of the Jacobian $J(E^*)$ are roots of the quadratic equation:

$$\lambda^2 - (C + 1)\lambda + C + AB = 0,$$

where $A = \dfrac{\varphi(h)f'(y^*)x^*}{1 + \varphi(h)d}$, $B = \dfrac{\varphi(h)\gamma(d - b)}{1 + \varphi(h)cD_2}$ and $C = 1 - \dfrac{\varphi(h)cD_1}{1 + \varphi(h)cD_2}$.

By Lemma 4 we obtain that E^* is a stable fixed point of Scheme (28) if and only if the following is true:

(a) $AB > 0$;
(b) $2 + 2C + AB > 0$; and

(c) $AB < 1 - C$;

and that E^* is an unstable fixed point if at least one of the above conditions fails.

Since $A > 0$ and $B > 0$ then the condition (a) is always true. The condition (c) $AB < 1 - C$ is equivalent to the following inequality:

$$\varphi(h)(f'(y^*)(N_0 - y^* - (D_2 - 1)y^*) - d) < D_1. \tag{30}$$

Since $D_2 - 1 = \dfrac{(N_0 - y^*)d}{(d - b)y^*}$ then $N_0 - y^* - (D_2 - 1)y^* = -\dfrac{(N_0 - y^*)b}{d - b}$.

Therefore, the left-hand side of Inequality (30) is negative and the condition (c) is always true. If $D_1 \leq 2D_2$ then $\dfrac{\varphi(h)cD_1}{1 + \varphi(h)cD_2} < 2$ when $\varphi(h) > 0$.

Therefore the condition (b) is true for $\varphi(h) = \phi(hq)/q$, $q \geq 0$ and the interior equilibrium E^* is a stable fixed point of the Scheme (28).

Let $D_1 > 2D_2$ and $\varphi(h) = \phi(hq)/q$, where $q > \dfrac{c(D_1 - 2D_2)}{2}$. Since $0 < \varphi(h) < \dfrac{1}{q}$ then $\varphi(h) < \dfrac{2}{c(D_1 - 2D_2)}$ for all $h > 0$. Therefore the condition (b) is satisfied and the interior equilibrium E^* is a stable fixed point of the Scheme (28) with $\varphi(h) = \phi(hq)/q$. $\qquad\square$

Let us now consider nutrient-dependent phytoplankton growth rate functions $f(y) \in \mathcal{H}$, defined as follows:

$$f(y) \in \mathcal{H} \iff (f(y) \in \mathcal{F} \text{ and } f''(y) < 0 \text{ for } y > 0).$$

The class $\mathcal{H}$ is a subclass of $\mathcal{F}$ and contains the Holling type II functions. Since $D_1 \leq 2D_2$ is satisfied for System (27) when $f(y) \in \mathcal{H}$ then the following Corollary holds:

Corollary 17: *Let the function $f(y) \in \mathcal{H}$ and the function ϕ satisfies Condition (13). Then the numerical scheme (28) with $\varphi(h) = \phi(hq)/q$ for all $q \geq 0$, where $\varphi(h) = h$ for $q = 0$, represents a PESN method for solving System (27).*

General Multi-Dimensional Systems

The ideas presented in this section can be applied to general n-dimensional systems of the form (1) and can be used to design PESN methods of the form:

$$\frac{\bar{x}_{k+1} - \bar{x}_k}{\varphi(h)} = g(\bar{x}_k, \bar{x}_{k+1}, h), \tag{31}$$

where $\varphi(h)$ satisfies the condition (13). The procedure, based on the construction technique of Theorems 15 and 16, can be summarized as follows:

- Step 1: Select the appropriate non-local approximation $g(\bar{x}_k, \bar{x}_{k+1}, h)$ of the right-hand side function $f(\bar{x})$ that guarantees the unconditional positivity of the discrete solutions of Scheme (31). One possible way to do that is by using the ideas presented by Patankar in [17]. He proposed that if the ith-component of the function $f(\bar{x})$ can be expressed as

$$f^i(\bar{x}) = P^i(\bar{x}) - N^i(\bar{x}), \quad i = 1, 2, \ldots, n,$$

  where $P^i(\bar{x}) \geq 0$ and $N^i(\bar{x}) \geq 0$ for all $\bar{x}$ in the first quadrant, then the ith-component of $g(\bar{x}_k, \bar{x}_{k+1}, h)$ should be selected as follows:

$$g^i(\bar{x}_k, \bar{x}_{k+1}, h) = P^i(\bar{x}_k) - N^i(\bar{x}_k)\frac{x^i_{k+1}}{x^i_k}.$$

  Under the above selection of $g(\bar{x}_k, \bar{x}_{k+1}, h)$, the explicit form of Scheme (31) is given by the system:

$$x^i_{k+1} = \frac{x^i_k + \varphi(h)P^i(\bar{x}_k)}{1 + \varphi(h)\dfrac{N^i(\bar{x}_k)}{x^i_k}}, \quad i = 1, 2, \ldots, n,$$

  which guarantees the unconditional positivity.
- Step 2: Select the denominator function $\varphi(h)$ such that it satisfies the property (13) and makes the Scheme (31) elementary stable. One possible choice for the denominator function is $\varphi(h) = \dfrac{1 - e^{-qh}}{q}$, where the parameter q is big enough to guarantee the elementary stability of the scheme (see Theorem 7).

5. Numerical Examples

5.1. *Single-Species Population Simulations*

To illustrate the efficiency of the combined nonstandard finite difference methods, we examine the following first-order ordinary differential equation:

$$\frac{dx}{dt} = 1 - x^2, \tag{32}$$

subject to the initial condition $x_0 = 0.6$. Mathematical analysis of Equation (32) shows that there exist two equilibria $x = -1$ and $x = 1$, where $x = 1$ is asymptotically stable and $x = -1$ is unstable. We approximate Equation (32) using the forward Euler, the backward Euler and the second-order

Runge-Kutta methods and compare the results with the proposed nonstandard numerical method (Theorem 13):

$$\frac{x_{k+1} - x_k}{h} = 1 - x_k x_{k+1}. \tag{33}$$

In this case, the above combined nonstandard method (33) coincides with the well-known corresponding Rosenbrock-Wanner scheme [23] for solving Equation (32). The exact solution of Equation (32) with an initial condition $x_0 \in (-1, 1)$ is given by $x(t) = \dfrac{e^{2t}\widehat{x}_0 + 1}{e^{2t}\widehat{x}_0 - 1}$, where $\widehat{x}_0 = \dfrac{x_0 + 1}{x_0 - 1}$.

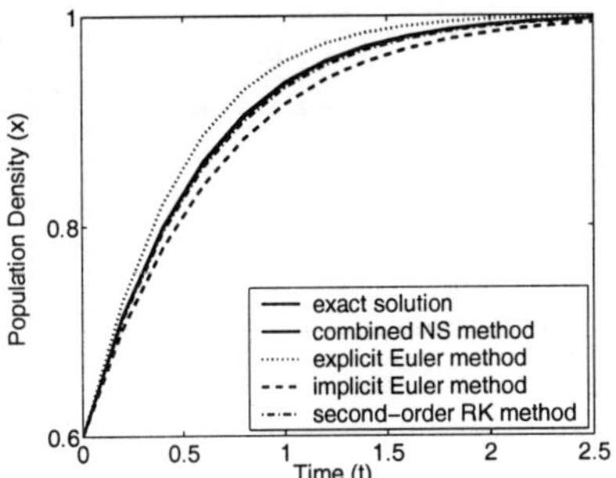

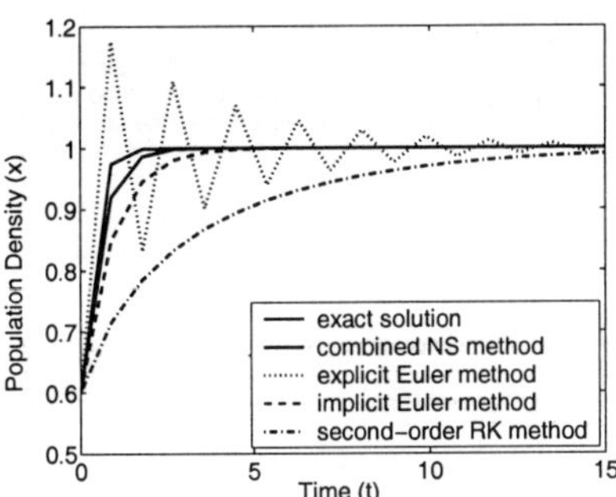

Fig. 1. Numerical approximations of the solution of Equation (32) for $x_0 = 0.6$ with $h = 0.2$ (left) and $h = 0.9$ (right).

Numerical solutions of Equation (32) with the combined nonstandard method (33) versus solutions obtained with standard numerical methods for step-sizes $h = 0.2$ and $h = 0.9$ are presented in Fig. 1. The simulations demonstrate the superior performance, for a wide range of step-sizes h, of the new method (33) over the forward Euler, the backward Euler and the second-order Runge-Kutta methods.

Next we consider the following ordinary differential equation:

$$\frac{dx}{dt} = -\left(x - b - \frac{1}{2}\right)(x - b)\left(x - b + \frac{1}{2}\right). \tag{34}$$

Mathematical analysis of Equation (34) shows that it has an unstable equilibrium at $x = b$ and stable equilibria at $x = b - \frac{1}{2}, b + \frac{1}{2}$, i.e., all initial values x_0 in the interval $(b - \frac{1}{2}, b)$ decay monotonically to the stable equilibrium $x = b - \frac{1}{2}$ and all initial values x_0 in the interval $(b, b + \frac{1}{2})$ grow monotonically to the stable equilibrium $x = b + \frac{1}{2}$.

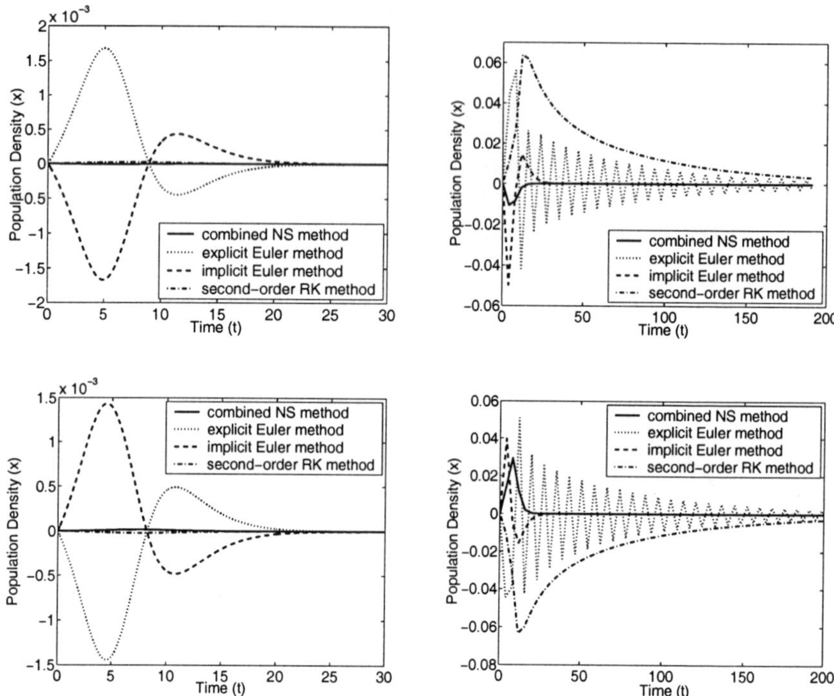

Fig. 2. Numerical errors of approximations of Equation (34) for $x_0 = 0.4$, $b = 1/2$ (top) and $b = 1/(2\sqrt{3})$ (bottom) and $h = 0.1$ (left) and $h = 3.9$ (right).

The exact solution of Equation (34) is given by

$$x(t) = \begin{cases} b - \dfrac{1}{2\sqrt{\widehat{x}_0}}, & x_0 \in (b - \tfrac{1}{2}, b) \\[2ex] b + \dfrac{1}{2\sqrt{\widehat{x}_0}}, & x_0 \in (b, b + \tfrac{1}{2}) \end{cases},$$

where $\widehat{x}_0 = e^{-t/2} \dfrac{\left(x_0 - b + \tfrac{1}{2}\right)\left(b + \tfrac{1}{2} - x_0\right)}{(x_0 - b)^2} + 1$.

First, we approximate Equation (34) for $b = \dfrac{1}{2}$, i.e., a right hand side function $f(x) = -x^3 + \dfrac{3}{2}x^2 - \dfrac{1}{2}x$, using the combined nonstandard method (Theorem 13):

$$\frac{x_{k+1} - x_k}{\varphi(h)} = -\frac{(e^{-h/2} + 1)x_k^2 x_{k+1}^2}{x_{k+1} + e^{-h/2}x_k} + \frac{3}{2}x_k x_{k+1} - \frac{1}{2}x_k. \tag{35}$$

In this case, due to the presence of a first-order term in the right-hand side function f, the denominator function equals $\varphi(h) = \dfrac{e^{-h/2} - 1}{-1/2}$. The right-hand side in (34) for $b = \dfrac{1}{2}$ is a modified logistic-growth reaction term which appears in certain population models involving non trivial extinction levels [24, p. 115].

Second, we approximate Equation (34) for $b = \dfrac{1}{2\sqrt{3}}$, i.e., a right hand side function $f(x) = -x^3 + \dfrac{\sqrt{3}}{2}x^2 - \dfrac{1}{12\sqrt{3}}$, using the combined nonstandard method (Theorem 13):

$$\frac{x_{k+1} - x_k}{h} = -\frac{x_k^2 x_{k+1}^2}{\frac{x_{k+1}+x_k}{2}} + \frac{\sqrt{3}}{2} x_k x_{k+1} - \frac{1}{12\sqrt{3}}. \tag{36}$$

In this case, due to the absence of a first-order term in the right-hand side function f, the denominator function simply equals $\varphi(h) = h$.

The numerical errors in approximations of Equation (34) with $b = 1/2$ and $b = 1/(2\sqrt{3})$, for an initial condition $x_0 = 0.4$ and step-sizes $h = 0.1$ and 3.9, are shown in Fig. 2(top) and Fig. 2(bottom), respectively. The graphs represent the absolute error of the combined nonstandard methods (35) and (36), respectively, versus the absolute errors of the Euler and Runge-Kutta methods. The simulations support the results of Theorem 13, that is, the proposed new methods (35) and (36) approximate the solution of Equation (34) for $b = \frac{1}{2}$ and $b = \frac{1}{2\sqrt{3}}$, respectively, much better than the standard methods for a wide range of the step-sizes h.

5.2. *Multi-Species Population Simulations*

To illustrate the efficiency of the designed ESN finite difference methods, we consider the predator-prey system with Beddington-DeAngelis functional response [25]:

$$\frac{dx}{dt} = x - \frac{axy}{1 + x + y}; \quad x(0) = x_0 \geq 0,$$
$$\tag{37}$$
$$\frac{dy}{dt} = \frac{exy}{1 + x + y} - dy; \quad y(0) = y_0 \geq 0,$$

where x and y represent the prey and predator population sizes and the positive constants $a = 6.0$, $e = 7.5$ and $d = 5.0$ represent the generalized feeding rate, generalized conversion efficiency and generalized mortality rate of the predator, respectively.

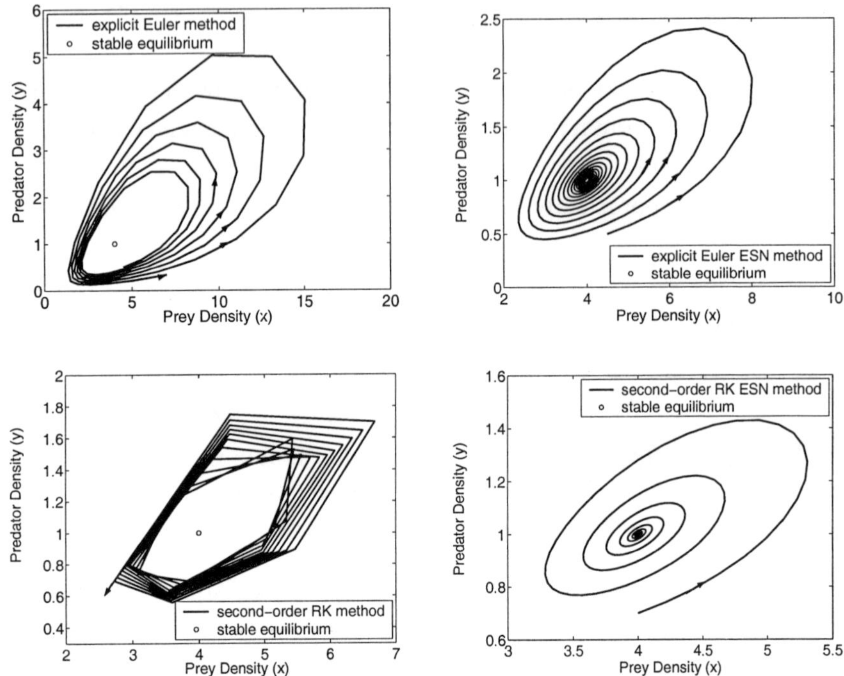

Fig. 3. Numerical approximations of the Beddington-DeAngelis system (37), with initial conditions $x(0) = 4.5$ and $y(0) = 0.5$, for step-sizes $h = 0.45$ (top) and $h = 1.18$ (bottom).

Mathematical analysis of System (37) shows that the system has two equilibria - the equilibrium $E_0 = (0,0)$ which is unstable and the equilibrium $E^* = (4,1)$ which is globally stable in the interior of the first quadrant [25]. The eigenvalues of $J(0,0)$ are given by $\lambda_1 = 1$ and $\lambda_2 = 5$, and the eigenvalues of $J(4,1)$ are given by $\lambda_{3,4} = -\frac{1}{12} \pm i\frac{\sqrt{119}}{12}$. Numerical approximations of the solution of System (37) with initial values $x(0) = 4.5$ and $y(0) = 0.5$ and step-sizes $h = 0.45$ and $h = 1.18$ using the explicit Euler and the second-order Runge-Kutta methods, respectively, (see Fig. 3) support the results of Theorem 14. The ESN methods, with a denominator function $\varphi(h) = \phi(hq)/q = (1 - e^{-hq})/q$, where $q = 5.1$ preserve the global stability of the equilibrium $(4,1)$, while approximations obtained by the standard methods diverge.

In the second example, we consider the following vaccination model with

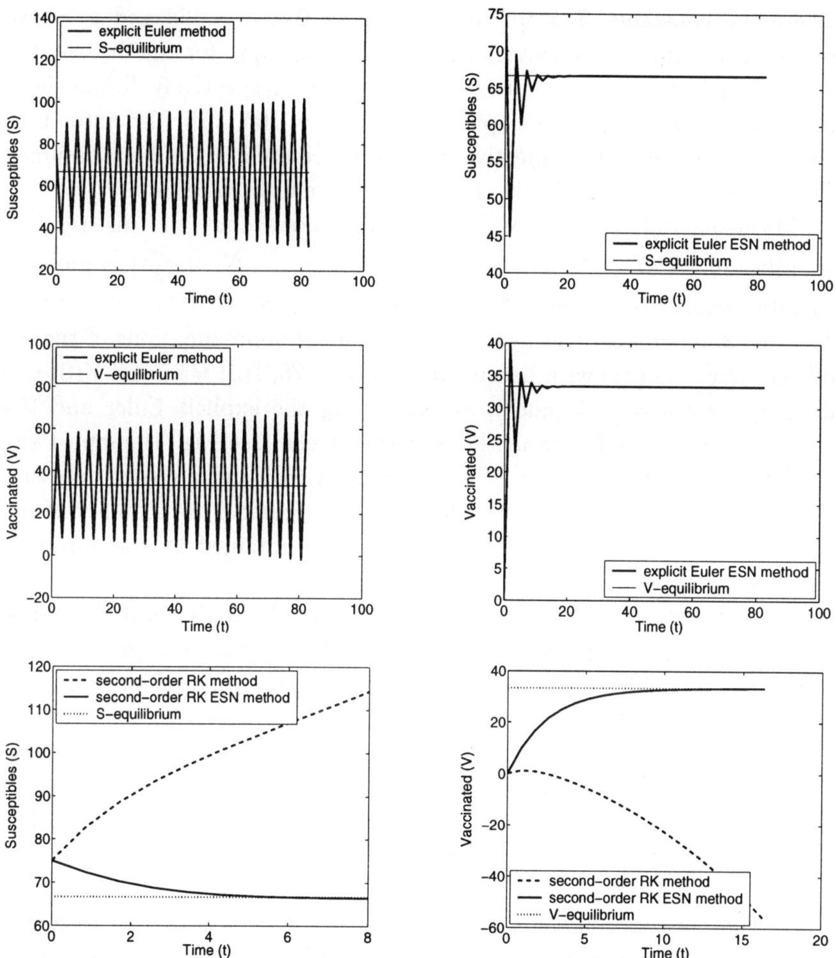

Fig. 4. Numerical approximations of the solution of Equation (38), with initial conditions $S(0) = 75$, $I(0) = 25$ and $V(0) = 0$, for step-sizes $h = 1.68$ (top and middle) and $h = 0.86$ (bottom).

multiple endemic states [26]:

$$\frac{dS}{dt} = \mu N - \beta SI/N - (\mu + w)S + cI + \delta V,$$

$$\frac{dI}{dt} = \beta SI/N - (\mu + c)I, \tag{38}$$

$$\frac{dV}{dt} = wS - (\mu + \delta)V,$$

where the constants $\beta = 0.7$, $c = 0.1$, $\mu = 0.8$, $\delta = 0.8$ and $w = 0.8$ represent the infectious contact rate, the recovery rate for the disease, the natural birth/death rate and the vaccination rate, respectively. In the above model the total (constant) population size $N = 100$ is divided into three classes - susceptibles (S), infectives (I) and vaccinated (V) and it is assumed that the vaccine is completely effective in preventing infection.

Mathematical analysis of System (38) shows that the disease free equilibrium $(S^*, I^*, V^*) = \left(\frac{(\mu+\delta)N}{\mu+\delta+w}, 0, \frac{wN}{\mu+\delta+w} \right) = (\frac{200}{3}, 0, \frac{100}{3})$ is globally asymptotically stable [26]. The eigenvalues of $J(S^*, I^*, V^*)$ are given by $\lambda_1 = -0.8$, $\lambda_2 = -2.4$ and $\lambda_3 = -\frac{13}{30}$. Numerical approximations of the solution of System (38) with initial values $S(0) = 75$, $I(0) = 25$ and $V(0) = 0$ for step-sizes $h = 1.68$ and $h = 0.86$ using the explicit Euler and the second-order Runge-Kutta methods, respectively, (see Fig. 4) support the results of Theorem 14. The ESN methods, with a denominator function $\varphi(h) = \phi(hq)/q = (1 - e^{-hq})/q$, where $q = 1.3$, preserve the stability of the equilibrium (S^*, I^*, V^*), while approximations obtained by the standard methods diverge.

To illustrate the advantages of the designed PESN finite-difference methods, we consider the Rosenzweig-MacArthur predator-prey system (20) with a Holling type II predator functional response of the form $xg(x) = x/(c + x)$ [21]:

$$\frac{dx}{dt} = bx(1 - x) - \frac{axy}{c + x},$$

$$\frac{dy}{dt} = \frac{xy}{c + x} - dy, \tag{39}$$

where the positive constants a, b, c and d represent the feeding rate, the intrinsic growth rate of the prey, the half-saturation level and the mortality rate of the predator, respectively.

We first examine System (39) in the case when the constants are $a = 2.0$, $b = 1.0$, $c = 0.5$ and $d = 6.0$, i.e., $g(1) = \frac{2}{3} < d$. Mathematical analysis of the system shows that there exist two equilibria $E_0 = (0,0)$ and $E_1 = (1,0)$, with the equilibrium $(1,0)$ being globally stable in the interior of the first quadrant. The eigenvalues of $J(0,0)$ are given by $\lambda_1 = 1$ and $\lambda_2 = -6.0$ and the eigenvalues of $J(1,0)$ are given by $\lambda_3 = -1$ and $\lambda_4 = -\frac{16}{3}$. Comparison of numerical approximations of the solution of System (39) with the PESN method (22) using $\varphi(h) = h$, the explicit ESN Euler method (14) using $\varphi(h) = \phi(hq)/q = (1 - e^{-hq})/q$ with $q = 3.1$ and the explicit Euler method supports the results of Theorem 15. The nonstandard (ESN

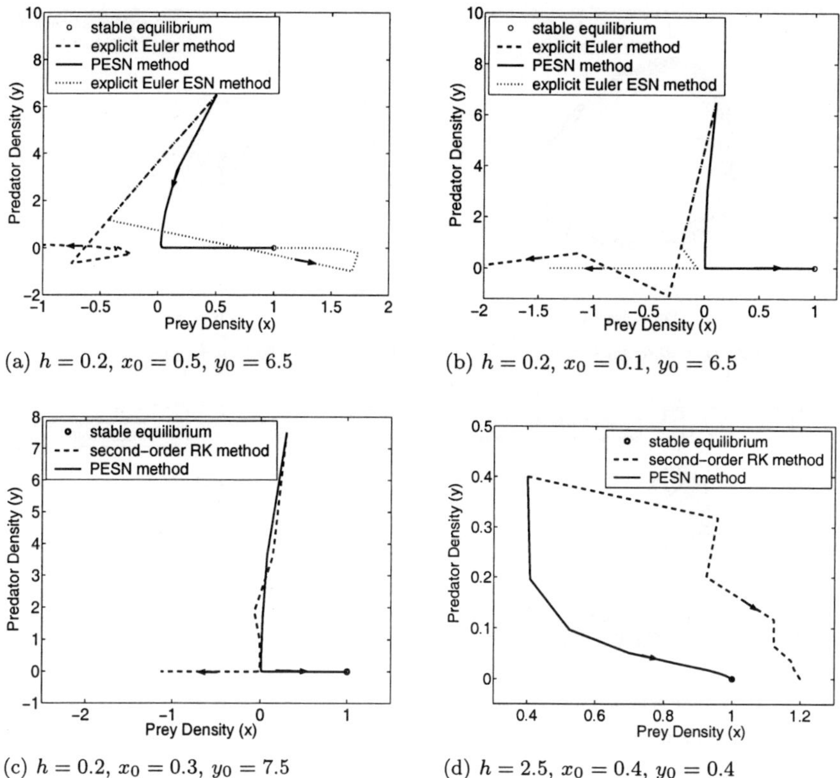

(a) $h = 0.2$, $x_0 = 0.5$, $y_0 = 6.5$

(b) $h = 0.2$, $x_0 = 0.1$, $y_0 = 6.5$

(c) $h = 0.2$, $x_0 = 0.3$, $y_0 = 7.5$

(d) $h = 2.5$, $x_0 = 0.4$, $y_0 = 0.4$

Fig. 5. Numerical approximations of the solutions of System (39) in the case of globally stable equilibrium $(1, 0)$.

and PESN) methods preserve the stability of the equilibrium $(1, 0)$, while the approximation obtained by the standard method diverges (Fig. 5(a)). However, a drawback of the ESN method is that it is not unconditionally positive (Fig. 5(b)). Similar behavior is observed when the standard second order Runge-Kutta method is used to numerically solve System (39) (see Fig. 5(c)). In some cases, for relatively large step-size $h = 2.5$, the Runge-Kutta numerical solution approaches an artificially created non existing equilibrium (Fig. 5(d)).

Next, we examine System (39) in the case when the constants are $a = 2.0$, $b = 1.0$, $c = 1.0$ and $d = 0.2$, i.e., $g(1) = \frac{1}{2} > d$. Mathematical analysis of the system shows that there exist three equilibria $E_0 = (0, 0)$, $E_1 = (1, 0)$ and $E^* = (\frac{1}{4}, \frac{15}{32})$, with the interior equilibrium E^* being globally stable in

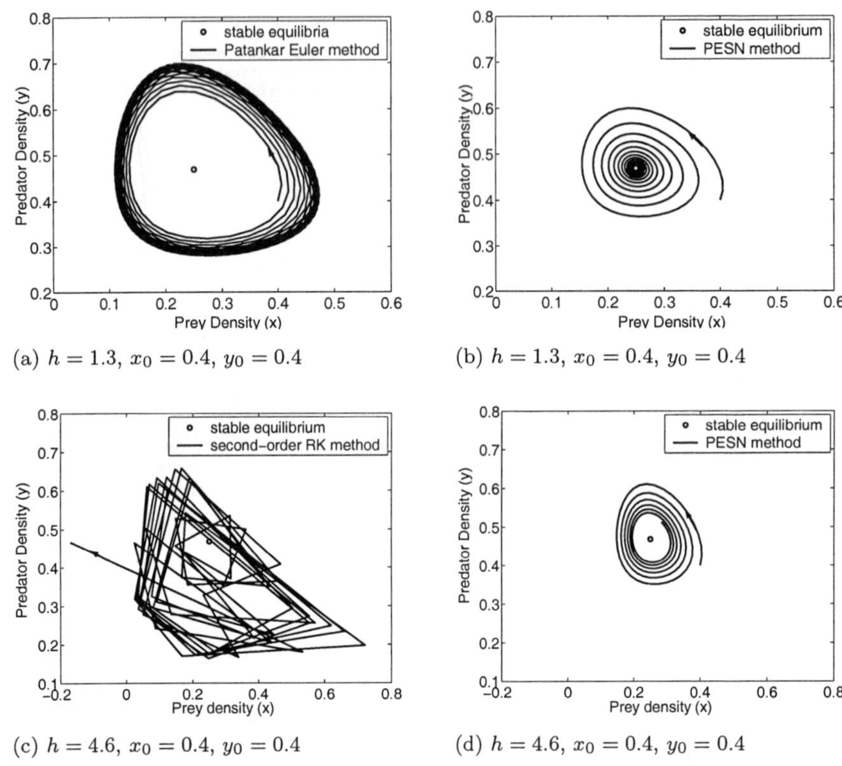

(a) $h = 1.3$, $x_0 = 0.4$, $y_0 = 0.4$

(b) $h = 1.3$, $x_0 = 0.4$, $y_0 = 0.4$

(c) $h = 4.6$, $x_0 = 0.4$, $y_0 = 0.4$

(d) $h = 4.6$, $x_0 = 0.4$, $y_0 = 0.4$

Fig. 6. Numerical approximations of the solutions of System (39) in the case of globally stable interior equilibrium E^*.

the interior of the first quadrant. Comparison of numerical approximations of the solution of System (39) with the PESN method (22) using $\varphi(h) = \phi(hq)/q = (1 - e^{-hq})/q$ with $q = 1.2$, the Patankar Euler scheme [27, p.17] and the second order Runge-Kutta method supports the results of Theorem 15. The PESN method preserves the stability of the equilibrium E^* (Fig. 6(b),(d)), while the approximations obtained by the other two numerical methods diverge (Fig. 6(a),(c)).

Finally, we examine System (27) for two different nutrient-dependent phytoplankton growth rate functions. First, we consider the Holling type II growth rate function $f(y) = \dfrac{\alpha y}{y + \nu}$, where the constants $\alpha = 3.5$, $\nu = d = \gamma = 1.0$, $b = 0.8$, $c = 0.15$ and $N_0 = 3.0$. For this choice of parameters $f(N_0) = 2.625 > d$. The interior equilibrium $E^* = (1.95, 0.4)$ exists and is globally asymptotically stable [22]. The eigenvalues of $J(0, N_0)$ are given by $\lambda_1 = 1.625$ and $\lambda_2 = -0.15$ and the eigenvalues of $J(E^*)$ are given by $\lambda_3 \approx -0.2$ and $\lambda_4 \approx -3.45$. Therefore the scheme (16) is a ESN method for $q > 1.725$. Since $f(y) \in \mathcal{H}$ then the scheme (28) is a PESN method for $\varphi(h) = h$. Figure 7 presents the comparison of the numerical approximations of the solution of System (27) with the PESN method (28)(solid lines) and the second-order Runge-Kutta ESN method (16)(dashed lines) using $\varphi(h) = (1 - e^{-hq})/q$ with $q = 1.8$ and initial conditions $x_0 = 2$ and $y_0 = 4$. The simulations support the results of Theorems 14 and 16. Both ESN and PESN methods preserve the stability of the equilibrium E^* for a relatively small step-size $h = 0.1$ (Fig. 7(top)). The ESN and PESN methods remain stable for $h = 0.6$ and the numerical solutions approach the equilibrium E^* (Fig. 7(bottom)), while the standard Runge-Kutta method diverges. The solution of the ESN method for $h > 0.6$ is no longer positive and as a result of that it diverges and blows up.

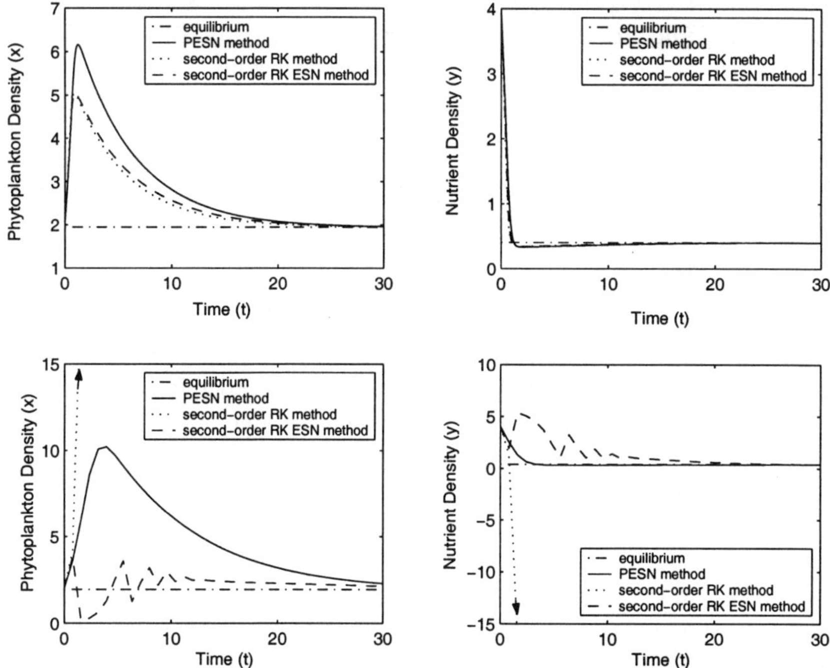

Fig. 7. Numerical approximations of the solution of System (27) with Holling type II growth rate function.

Second, we examine System (27) with $f(y) = \dfrac{\alpha y^3}{y^3 + \nu}$, where the constants $\alpha = 3.5$, $\nu = d = \gamma = 1.0$, $b = 0.8$, $c = 0.15$ and $N_0 = 3.0$. In this case $f(N_0) = 3.375 > d$. The interior equilibrium $E^* = (1.6974, 0.7368)$ is globally stable. The eigenvalues of $J(0, N_0)$ are given by $\lambda_1 = 2.275$ and $\lambda_2 = -0.15$ and the eigenvalues of $J(E^*)$ are $\lambda_3 \approx -0.21$ and $\lambda_4 \approx -4.79$. Therefore the scheme (16) is a ESN method for $q > 2.395$. Since $D_1 - 2D_2 \approx 1.1$ then the scheme (28) is a PESN method for $\varphi(h) = \phi(h\tilde{q})/\tilde{q}$ with $\tilde{q} > 0.083$. Figure 8 presents the comparison of numerical approximations of the solution of System (27) with the PESN method (28)(solid lines) using $\varphi(h) = \phi(h\tilde{q})/\tilde{q} = (1 - e^{-h\tilde{q}})/\tilde{q}$ with $\tilde{q} = 0.1$ and the second-order Runge-Kutta ESN method (16)(dashed lines) using $\varphi(h) = \phi(hq)/q = (1 - e^{-hq})/q$ with $q = 2.5$. The initial conditions are $x_0 = 2$ and $y_0 = 4$. The simulations support the results of Theorems 14 and 16. Both ESN and PESN methods preserve the stability of the equilibrium E^* for a relatively small step-size $h = 0.1$ (Fig. 8(top)). The PESN method remains stable for $h = 0.5905$ and the numerical solution approaches the equilibrium E^* (Fig. 8(bottom)). However, the solutions of the standard and second-order Runge-Kutta ESN methods for $h = 0.5905$ are not positive, which leads to a divergence and solutions blow up.

6. Conclusions

In this chapter we have considered discrete approximations of ecological models by nonstandard numerical methods.

First, we have extended results from [15] to develop second-order accurate nonstandard finite-difference schemes for solving arbitrary first-order differential equation with polynomial right-hand sides. Second, we have designed and analyzed elementary stable nonstandard (ESN) methods, based on the explicit and implicit Euler and the second-order Runge-Kutta for general two-dimensional autonomous dynamical systems. These ESN methods represent generalizations of results obtained earlier by Anguelov and Lubuma [5] and Libuma and Roux [6].

We have also applied the theory of nonstandard numerical methods to develop positive and elementary stable nonstandard (PESN) methods for Rosenzweig-MacArthur predator-prey systems with logistic intrinsic growth of the prey population and phytoplankton-nutrient systems with nutrient loss. These PESN methods preserve essential physical properties of exact solutions of the approximated differential systems. The proposed schemes take into account the major characteristics of the continuous dynamical

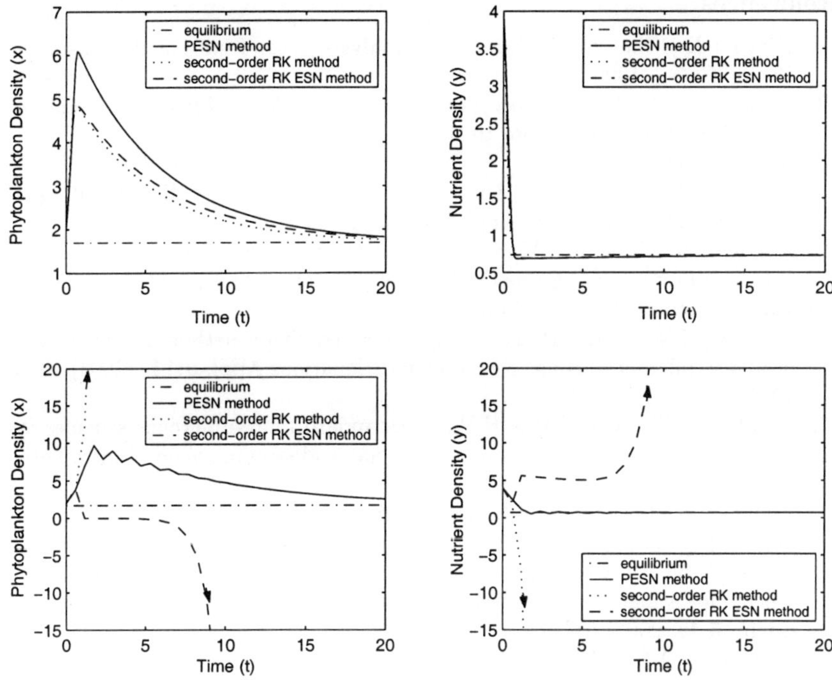

Fig. 8.   Numerical approximations of the solution of System (27) with $f(y) = \dfrac{\alpha y^3}{y^3 + \nu}$.

systems, namely the facts that they are autonomous, nonconservative and unconditionally positive.

Finally, directions for designing new ESN and PESN numerical schemes for general multi-dimensional autonomous dynamical systems were also outlined.

Future research directions include the application of the nonstandard numerical techniques to nonautonomous dynamical systems, the construction of similar nonstandard schemes for general biological systems with non-hyperbolic equilibria and also the development of nonstandard numerical methods that preserve the stability of existing limit cycles.

References

1. Kinkaid D., Cheney W., Numerical Analysis, second edition, Brooks/Cole, Pacific Grove, 1996.
2. Mickens, R. E., *Nonstandard finite difference model of differential equations* (World Scientific, Singapore 1994).
3. Murray, J.D., *Mathematical Biology* (Springer-Verlag, Berlin, 1993).
4. Mickens, R.E., Nonstandard finite difference schemes for differential equations, J. Differ. Equations Appl. **8:9,** (2002) 823-847.
5. Anguelov R., Lubuma J.M.-S., Contributions to the mathematics of the nonstandard finite difference method and applications, *Numer. Methods Partial Diff. Equations,* **17:5** (2001) 518-543.
6. Lubuma, J.M.-S. and Roux A., An improved theta-method for systems of ordinary differential equations, J. Differ. Equations Appl. **9:11,** (2003) 1023-1035.
7. Dimitrov D.T., Kojouharov H.V., Nonstandard finite-difference schemes for general two-dimensional autonomous dynamical systems, Appl. Math. Lett., (2005) in press.
8. Anguelov R., Lubuma J.M.-S., Nonstandard finite difference method by nonlocal approximation, Math. Comput. Simulation **61:3-6,** (2003) 465-475.
9. Gumel A.B., Mickens R.E., Corbett B.D., A non-standard finite-difference scheme for a model of HIV transmission and control, J. Comput. Meth. Sci. Engin. **3:1,** (2003) 91-98.
10. de Markus A.S., Mickens R.E., Suppression of numerically induced chaos with nonstandard finite difference schemes, J. Comput. Appl. Math. **106:2,** (1999) 317-324.
11. Piyawong W., Twizell E.H., Gumel A.B., An unconditionally convergent finite-difference scheme for the SIR model, Appl. Math. Comput. **146,** (2003) 611-625.
12. Jansen H. and Twizell E.H., An unconditionally convergent discretization of the *SEIR* model, Math. Comput. Simulation **58,** (2002) 147-158.
13. Anguelov R., Kama P., Lubuma J.M.-S., On non-standard finite difference models of reaction-diffusion equations, J. Comput. Appl. Math. **175:1,** (2005) 11-29.
14. Mickens, R.E., Relation between the time and space step-sizes in nonstandard finite-difference schemes for the Fisher equation, Numer. Methods Partial Differential Equations **13:1,** (1997) 51-55.
15. Kojouharov H.V., Chen B.M. , Nonstandard Eulerian-Lagrangian Methods for Advection-Diffusion-Reaction Equations, *Applications of Nonstandard Finite Difference Schemes, (R.E. Mickens, Editor)* (World Scientific Publishing, River Edge, NJ, 2000), 55-108.
16. Chen-Charpentier, B.M., Dimitrov, D.T., Kojouharov, H.V., Combined Nonstandard Numerical Methods for ODEs with Polynomial Right-Hand Sides, Applied Numerical Mathematics, (2005) in press.
17. Patankar S.V. *Numerical Heat Transfer and Fluid Flow* (McGraw-Hill, New York 1980).

18. Brauer F., Castillo-Chavez C., *Mathematical Models in Population Biology and Epidemiology* (Springer-Verlag, New York 2001).

19. Coddington E. A., Levinson N., *Theory of Ordinary Differential Equations* (Krieger, Florida 1984).

20. Taylor A.H., Harris J.R.W. and Aiken J., The interaction of physical and biological process in a model of the vertical distribution of phytoplankton under stratification, Mar. Int. Ecohyrd., Jacques C.Nihoul (editor) **42**, (1986) 313-330.

21. Holling C.S., The functional response of predators to prey density and its role in mimicry and population regulation, Mem. Entomol. Soc. Canada **45**, (1965) 1-60.

22. Dimitrov D.T., Kojouharov H.V., Analysis and numerical simulation of phytoplankton-nutrient systems with nutrient loss, Math. Comput. Simulation, (2005) accepted.

23. Hairer E., Wanner G., Solving ordinary differential equations II, Stiff and Differential-Algebraic Problems, Springer, Berlin 1991.

24. Yeargers E.K., Shonkwiler R.W., Herod J.V., An Introduction to the Mathematics of Biology: Computer Algebra Models, Birkhäuser, Boston, 1996.

25. Dimitrov D.T., Kojouharov H.V., Complete mathematical analysis of predator-prey models with linear prey growth and Beddington-DeAngelis functional response, Appl. Math. Comput., **162:2**, (2005) 523-538.

26. Kribs-Zaleta C.M. and Velasco-Hernández J.X., A simple vaccination model with multiple endemic states, Math. Biosci. **164**, (2000) 183-201.

27. Burchard H., Deleersnijder E., Meister A., A high-order conservative Patankar-type discretization for stiff systems of production-destruction equations, Appl. Numer. Math. **47:1**, (2003) 1-30.

CHAPTER 7

APPLICATIONS OF THE NON-STANDARD FINITE DIFFERENCE METHOD IN NON-SMOOTH MECHANICS

Yves Dumont

IREMIA
Université de la Réunion
97715 Saint Denis, France
Yves.Dumont@univ-reunion.fr

We present some applications of the non-standard finite difference method for the numerical approximations of differential inclusions coming from non-smooth mechanics. After a short introduction in the theory and the numerical analysis of differential inclusions, we study several examples of non-smooth dynamical systems, like oscillators subject to dry friction and vibro-impact oscillators. We present some non-standard schemes associated to each of the previous non-smooth systems, and we discuss the results we obtain. Finally, we propose other ways of study in the non-standard finite difference method for the approximation of problems with constraints.

1. Introduction

Non-smooth mechanics involves engineering problems such as vibro-impact systems, gear boxes, suspension bridges, offshore structures, seismic action, heat exchangers, robot walkers .... These problems are non-smooth because the potential force in the gouverning constitutive laws are not differentiable. In non-smooth mechanics, mathematical models are given by variational inequalities, hemivariational inequalities, complementary systems, differential equations with discontinuous right-hand side or differential inclusions [1,2,3,4,5,6].

Indeed, many mechanical systems are made of perfectly rigid bodies, which do not penetrate each other, though they may come into contact or not. As a result, constraints or reaction forces have to be included in the mathematical formulation of such processes. The inclusion of these forces is a source of such mathematical difficulties that unilateral constraints are not sufficiently study in classical mechanics. As a matter of fact, modern tools of (non) convex analysis [7,8] enable to give rigorous mathematical expressions of normal contact laws, Coulomb's friction law, shock laws and others in the form of differential inclusions [9,10].

The dynamics of these systems can be very complex, with both periodic and irregular or even chaotic regime. Moreover, it is possible to observe different regimes for the same problems as parameters vary. It is therefore essential to design numerical methods in order to gain some understanding of these problems. While the focus of most of classical numerical methods is on their convergence, the emphasis of this work is on the design of numerical methods which, apart from being convergent, replicate as much as possible the physical properties of the time-continuous systems. We will use the non-standard finite difference method, developed by R. Mickens in the eighties. The non-standard approach has shown great potential in the design of numerical schemes which preserve qualitative properties of the exact solution. In this regard, we mention the contributions [11,12,13,14] and the references there in. Furthermore, it is worthwhile to refer to the paper [15], which turns out to be the first application of the non-standard approach to non-smooth mechanics.

We will particularly focus on differential inclusions coming from two types of problems. These are:

- frictional problems, including (non)linear stick-slip oscillators,
- perfect unilateral constraint problems, i.e. frictionless constraint problems, such as (non)linear impacting oscillators.

Even such simple systems can develop complex behaviors and we will show that the non-standard method produces reliable results.

The chapter is organized as follows. In section 2, we give a short presentation of existence and/or uniqueness results on differential inclusions. While section 3 is devoted to some classical numerical methods for differential inclusions. In section 4, we design non-standard finite difference schemes for frictional oscillators. A comparative analysis with standard methods is although done in this section. A similar study is considered in section 5, for impacting oscillators. Section 6 deals with concluding remarks and further research orientations on non-standard methods for non-smooth dynamics.

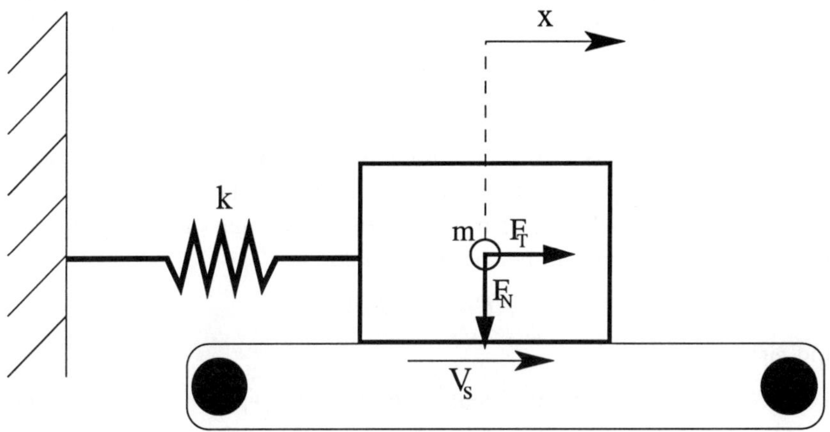

Fig. 1.   The frictional oscillator

2. A Short Introduction to the Theory of Differential Inclusion

For the last forty years, differential inclusions have been studied by engineers and mathematicians because they appear in wide variety of applications such as in control theory, in economic theory, in population dynamics, in the study of differential equation with discontinuous right-hand side, in differential variational inequalities, ... An overview of the theory for differential inclusions can be found in [16,17]. The numerical treatment of differential equations with discontinuous right-hand side or set-valued right-hand side, i.e. differential inclusions, required some specific methods, based on classical finite difference methods (see [18] for a survey, and [19,20]). Some of these methods come directly from standard methods, like Euler, Runge-Kutta methods [18,20], while the others were specifically developed to treat particular differential inclusions, coming from vibro-impact problems [21,10,22,23,24,25].

In order to make the chapter self-contained, we present some important results that will be used in the remaining of the chapter.

Let us first illustrate this section with the following linear stick-slip oscillators. We consider a mass $m > 0$ restrained by a spring with stiffness constant $k > 0$. The mass is riding on a support, that may move with a constant velocity V_s (see Fig. 1). Between the mass and the support, dry friction occurs, with a friction force F_{fr}. The total horizontal force acting on the mass is $F_T + F_{fr}$. The friction force F_{fr} depends on the vertical force F_N exerted by the mass on the support. Thus, the continuous model is also described by

$$m\ddot{x} + kx = F_{fr},$$

with initial conditions $x(0) = x_0$ and $\dot{x}(0) = v_0$. We model the frictional resistance of the support with the classical Coulomb law. Let ν, a positive constant, be the coefficient of friction between the mass and the support. The friction condition can be written as follows

$$|F_{fr}| \leq \nu|F_N|,$$

and either

$$\dot{x} = V_S \quad \text{and} \quad |F_{fr}| \leq \nu|F_N|, \tag{1}$$

or

$$\dot{x} \neq V_S, \quad |F_{fr}| = \nu|F_N| \quad \text{and} \quad F_{fr} = -\nu\mathrm{sgn}(\dot{x} - V_S)|F_N|. \tag{2}$$

Here, we used the following definition of the sgn function

$$\mathrm{sgn}(z) = \begin{cases} +1 & \text{if } z > 0, \\ 0 & \text{if } z = 0, \\ -1 & \text{if } z < 0. \end{cases}$$

In order to rewrite (1) like (2), we have to extend the value for $\mathrm{sgn}(0)$. Indeed, when $\dot{x} = V_S$, (1) indicates that F_T can take any value of magnitude no more than $\nu|F_{fr}|$, which means that we have a set of possible values for $\mathrm{sgn}(0)$, that is "$\mathrm{sgn}(0) = [-1, 1]$." Thus, using set-valued analysis, the Coulomb law can be summarized as follows:

$$F_{fr} \in -\nu\mathrm{SGN}(\dot{x} - V_S)|F_N|,$$

where the set-valued function SGN is given by the graph

$$\mathrm{SGN}(z) = \begin{cases} \{1\} & \text{if } z > 0 \\ [-1, 1] & \text{if } z = 0 \\ \{-1\} & \text{if } z < 0. \end{cases}$$

Finally, the continuous model is described by the following differential inclusion

$$m\ddot{x} \in -kx(t) - \nu mg\mathrm{SGN}(\dot{x} - V_s),$$

which, setting $\omega^2 = k/m$ and $\mu = \nu g$, can be rewritten in the form

$$\begin{cases} \ddot{x} \in -\omega^2 x(t) - \mu\mathrm{SGN}(\dot{x} - V_s), \\ x(0) = x_0, \quad \dot{x}(0) = v_0. \end{cases} \tag{3}$$

We now need some theoretical and numerical tools to study (3).

A general differential inclusion model in mechanics can be formulated as follows:

Find $q : [0, T] \longrightarrow \mathbb{R}^N$, $t \longmapsto q(t)$ such that

$$M\ddot{q}(t) + C\dot{q}(t) + Kq(t) \in f(t) + F(t, q(t), \dot{q}(t)), \quad \text{a.e. } t \in (0, T), \quad (4)$$

where M is the mass matrix, C is the damping matrix, K is the stiffness matrix, and $f : [0, T] \longrightarrow \mathbb{R}^N$ is a vector-valued function related to the given forces acting on the system and $F : [0, T] \times \mathbb{R}^N \longrightarrow \mathcal{P}(\mathbb{R}^N)$ is a set-valued function, i.e. a function from $[0, T] \times \mathbb{R}^N$ onto the set $\mathcal{P}(\mathbb{R}^N)$ of all subsets of $\mathbb{R}^N$, that defines, for each $t \in [0, T]$, a graph in $\mathbb{R}^N$ used to express the unilateral reaction forces. Usual initial conditions such as

$$q(0) = q_0, \qquad \dot{q}(0) = q_1,$$

and impact laws (provided that the system under consideration involves rigid body collisions) are generally introduced to complete the formulation of the model. So far, only some special cases of second order differential inclusions have been studied. See for example the works of Frémond [26], Monteiro Marques [22], Moreau [9,10] and Schatzman [27].

2.1. *A general existence result for differential inclusions*

Let $N \in \mathbb{N}^*$. We consider the general differential inclusion

$$\begin{cases} \dot{q}(t) \in \Phi(q(t), t) & \text{a.e. on } (0, T), \\ q(0) = q_0 \end{cases} \qquad (5)$$

where $q_0 \in \mathbb{R}^N$, q is a function from $[0, T]$ in $\mathbb{R}^N$ and Φ is a map from $\mathbb{R}^N \times [0, T]$ into the set of all subsets of $\mathbb{R}^N$ and differential inclusions of this type have been the subject of many papers and we refer the reader to the books of Aubin-Cellina [16], Filippov [28] and Deimling [17]. The following **general** theorem (see [17]) gives sufficient conditions for existence (and uniqueness)

Theorem 1: *Suppose that Φ satisfies the conditions:*

    (i) Φ *is nonempty, compact and convex-valued on* $[0, T] \times \mathbb{R}^N$,
   (ii) $\Phi(t, \cdot)$ *is upper semicontinuous, for all* $t \in [0, T]$
  (iii) $\Phi(\cdot, x)$ *is measurable, for all* $x \in \mathbb{R}^N$,
  (iv) *there exists constants* k_1 *and* k_2 *such that*

$$\|z\| \leq k_1 \|x\| + k_2, \qquad \forall z \in \Phi(t, x), \ x \in \mathbb{R}^N, \ t \in [0, T].$$

Then Problem (5) has a least one solution, i.e. an absolutely continuous function y that satisfies (5).

If, in addition, the map Φ satisfies a one-sided Lipschitz condition, i.e.

$$(\theta_1 - \theta_2, x_1 - x_2) \le L\|x_1 - x_2\|^2,$$

holds for all x_1 and $x_2 \in \mathbb{R}^N$ and all $\theta_1 \in \Phi(t, x_1)$, $\theta_2 \in \Phi(t, x_2)$ uniformly for all $t \in [0, T]$, then the solution of Problem (5) is unique.

Theorem 1 can be applied to study a great variety of models of unilateral phenomena like dry friction, debonding effects and delamination effects.

Example 2: Consider equation (3), which is equivalent to the following first-order vector differential inclusion

$$\begin{pmatrix} \dot{x} \\ \dot{v} \end{pmatrix} \in \Phi(x, v), \tag{6}$$

with $\Phi(x, v) = \{v\} \times \{-\omega^2 x - \mu \text{SGN}(v - V_s)\}$. It is well known that the set-valued function SGN is bounded, closed, convex-valued and upper semi-continuous on $\mathbb{R}$. Then, we can apply Theorem 1 to prove that problem (6) (and thus (3)) has at least one absolute continuous solution.

2.1.1. *Differential equations with discontinuous right-hand side*

Some dynamical systems may be modeled with a differential equation with single-valued discontinuous right-hand side of the form

$$\dot{q}(t) = f(t, q(t)). \tag{7}$$

Therefore, it is necessary to develop a concept of solution. Let X an arbitrary set, $\text{conv}(X)$ is called the convex hull of X, i.e. the set of all combination $\sum_{i=1}^{P} \lambda_i x_i$ of points $x_i \in X$ where $\lambda_i \ge 0$ for all i and $\sum_{i=1}^{P} \lambda_i = 1$. Filippov [29] was the first to developed a solution concept for differential equations with discontinuous right-hand side, which leads to existence and uniqueness results [28]. He proposed a regularization of the previous discontinuous differential equation, and replace it by a differential inclusion of the form

$$\dot{q}(t) \in \bigcap_{\delta > 0} \bigcap_{P:\mu(P)=0} \overline{\text{conv}\left(f\left(t, \{z \in \mathbb{R}^N : \|z - y(t)\| \le \delta\} - P\right)\right)} \tag{8}$$

for almost $t \in [t_0, T]$, where μ denotes the Lebesgue measure on $\mathbb{R}^N$. Thus, using Theorem 1, we are able to prove the existence (and uniqueness) of a solution of (8) and then of (7). Unfortunately, the direct computation of the right-hand side of (8) can be rather difficult if f is expressed as the product of several functions, for instance.

2.2. *Differential inclusions for friction problems*

In more of Theorem 1, we also present recent results [30] well suited to some problems we are interrested in (see section 4). Indeed, in many cases and in particular when frictional contact may arise, equation (4) can be written in the following formulation

$$M\ddot{q}(t) + C\dot{q}(t) + Kq(t) \in -H_1\partial\Phi\left(H_2\left(\dot{q}(t)\right)\right), \quad \text{a.e.} \ \ t \geq 0,$$

where Φ is a proper convex and lower semicontinuous function from $\mathbb{R}^P \to \mathbb{R}^P$ to $]-\infty, +\infty]$. $H_1 \ \mathbb{R}^{P\times N}$ and $H_2 \in \mathbb{R}^{N\times P}$ are given matrices. The term $H_1\partial\Phi\left(H_2\left(\dot{q}(t)\right)\right)$ has been introduced to model the unilateral contact induced by friction forces and $\partial\Phi$ represents the convex subdifferential of Φ.

Let $(q_0, \dot{q}_0) \in \mathbb{R}^N \times \mathbb{R}^N$, with $H_2\dot{q}_0 \in \mathcal{D}\left(\partial\Phi\right)$, be given, we are looking for a function $t \mapsto q(t)$ $(t \geq 0)$ with $q \in C^1\left([0, +\infty[; \mathbb{R}^N\right)$ which satisfies the following conditions:

$$\begin{cases} \ddot{q} \in L^\infty_{loc}\left([0, +\infty[; \mathbb{R}^N\right), \\ \dot{q} \text{ is right-differentiable on } [0, +\infty[, \\ q(0) = q_0, \\ \dot{q}(0) = q_1, \\ H_2\dot{q}(t) \in \mathcal{D}\left(\partial\Phi\right), \text{ for } t \geq 0, \end{cases} \tag{9}$$

$$M\ddot{q}(t) + C\dot{q}(t) + Kq(t) \in -H_1\partial\Phi\left(H_2\dot{q}(t)\right), \quad \text{a.e.} \ \ t \geq 0. \tag{10}$$

We have the following existence and uniqueness theorem

Theorem 3: [30] *Suppose that the following assumptions are satisfied:*

- *M is non singular*
- *there exists a matrix $R \in \mathbb{R}^N \times \mathbb{R}^N$, symmetric and nonsingular such that*

$$R^{-2}H_2^T = M^{-1}H_1;$$

- *there exists $y_0 = H_2R^{-1}x_0$ ($x_0 \in \mathbb{R}^N$) at which Φ is finite and continuous.*

Let $(q_0, \dot{q}_0) \in \mathbb{R}^N \times \mathbb{R}^N$, with $H_2\dot{q}_0 \in \mathcal{D}\left(\partial\Phi\right)$. Then there exists a unique $q \in C^1\left([0, +\infty[; \mathbb{R}^N\right)$ satisfying (9)-(10).

Remark 4: If M is singular see [31].

Remark 5: The set $\mathcal{S}$ of stationary solutions of (9)-(10) is

$$\mathcal{S} = \left\{ \widetilde{q} \in \mathbb{R}^N \colon K\widetilde{q} \in -H_1 \partial \Phi \left(0 \right) \right\}.$$

Example 6: Consider now equation (3). It enters equation (10) with $\Phi\left(x \right) = \mu \left| x \right|$, that is

$$\ddot{x}\left(t \right) \in -\omega^2 x\left(t \right) - \partial \Phi \left(\dot{x} - V_s \right).$$

Since $V_S = 0$, from Theorem 3 (p. 293), we have the existence of a unique solution $x \in C^1 \left([0, +\infty[, \mathbb{R} \right)$ and the set of stability is $\mathcal{W} = [-\frac{\mu}{\omega^2}, \frac{\mu}{\omega^2}]$. This implies that if the mass is in $\mathcal{W}$ with zero relative velocity, it will remain motionless. In other cases, the mass will oscillate until it stays in $\mathcal{W}$ and the relative velocity becomes zero.

The previous theorem can be generalized to nonsmooth conservative systems. We suppose that the motion is governed by the second-order Lagrange equations:

$$M\ddot{q}\left(t \right) + \Pi'(q\left(t \right)) \in -H_1 \partial \Phi \left(H_1^T \dot{q}\left(t \right) \right), \quad \text{a.e. } t \geq 0, \tag{11}$$

where $\Pi \in C^1 \left(\mathbb{R}^N; \mathbb{R} \right)$ is the potential energy of the system and $H_2 = H_1^T$. Then, we have the following

Theorem 7: [30] *Suppose that the following assumptions are satisfied:*

- *M is symmetric and positive definite,*
- *Π' is lipschitz continous,*
- *there exists $y_0 = H_1^T M^{-1/2} x_0$ $(x_0 \in \mathbb{R}^N)$, at which Φ is finite and continuous.*

Let $t_0 \in \mathbb{R}$, q_0, $\dot{q}_0 \in \mathbb{R}^N$ with $H_1^T \dot{q}_0 \in D\left(\partial \Phi \right)$. Then there exists a unique $q \in C^1 \left([t_0, +\infty[; \mathbb{R}^N \right)$ satisfying conditions (9)-(11).

Remark 8: The set $\mathcal{S}$ of stationary solutions of (9)-(11) is

$$\mathcal{S} = \left\{ \widetilde{q} \in \mathbb{R}^N \colon \Pi'\left(\widetilde{q} \right) \in -H_1 \partial \Phi \left(0 \right) \right\}.$$

2.3. *Differential inclusion in vibro-impacts*

Unfortunately, for problems involving perfect unilateral constraints, Theorem 1 is not useful. Indeed, the assumptions required on Φ in Theorem 1 are too strong to encompass frictionless normal contact laws expressing non-penetration constraints and reactions.

Generally, simple dynamic models or reduced dynamic models involving such unilateral constraints are governed by a system of differential inclusions of the type

$$\mathcal{M}(q)\,\ddot{q} + \partial\Psi_K(q) \ni f(t,q,\dot{q}), \tag{12}$$

where $\mathcal{M}(q)$ is a symmetric matrix, $f : [0,T] \times \mathbb{R}^N \times \mathbb{R}^N \longrightarrow \mathbb{R}^N$ is a single-valued function and $\partial\Psi_K$ denotes the convex subdifferential of the indicator function Ψ_K of some nonempty closed convex set $K \subset \mathbb{R}^N$ defined by the geometric constraints imposed on q [3], i.e

$$\Psi_K(x) = \begin{cases} 0 & \text{if } x \in K \\ +\infty & \text{if } x \notin K. \end{cases}$$

Remark 9: K is usually called the set of admissible positions.

Equation (12) does not contain all necessary information to define a solution: impact laws [2] are also usually considered so as to complete the formulation of the problem in consideration, like

$$\dot{q}(t_+) = -e\dot{q}_N(t_-) + \dot{q}_T(t_-), \qquad \text{if } q(t) \in \partial K, \tag{13}$$

where $\dot{q}_N(t)$ is defined as the normal projection on $\mathbb{R}\nu(q(t))$ of $\dot{q}(t)$ with respect to the local kinetic metric, $\nu(q)$ is the unit exterior normal to ∂K and $\mathbb{R}^+\nu(q(t))$ is the normal cone. The parameter $e \in [0,1]$ is called the restitution coefficient. The impact law (13) states that the normal component of the velocity is reflected and multiplied by e, while the tangential component of the velocity is preserved during the impact. Finally, the solution must satisfy the initial conditions

$$q(0) = q_0, \qquad \dot{q}(0_+) = q_1. \tag{14}$$

Since $\mathcal{M} = \mathrm{Id}_{\mathbb{R}^n}$, the differential inclusion (12) reduces to (5) by setting $y = (q \ \dot{q})^T$, $n = 2N$ and

$$\Phi(t,y) = \begin{pmatrix} y_2 \\ f(t,y_1,y_2) + \partial\Psi_K(y_1) \end{pmatrix}. \tag{15}$$

It is clear that Φ does not satisfy the sublinear growth condition (iv) in Theorem 1. We have indeed

$$\partial\Psi_K(x) = N_K(x),$$

where $N_K(x)$ denotes the normal cone of K at x, that is

$$N_K(x) = \left\{ w \in {}^N : w^T z \leq 0, \quad \forall z \in T_K(x) \right\},$$

and

$$T_K(x) = \overline{\bigcup_{\lambda > 0} \lambda(K - x)},$$

is the tangent cone of K to x [3].

It is also clear that we need specific existence (and uniqueness) theorem for such type of problems. This was first done by Schatzman [27] and then by Buttazo and all [32,33,34] for elastic schocks, i.e. $e = 1$. Monteiro-Marques gave a proof of existence for inelastic shocks, i.e. $e = 0$ [22]. Finally, Paoli, Schatzman and Mabrouk extend these results to the case of partly elastic shocks, i.e. $e \in (0, 1)$. They used two different approaches for the discretization of the dynamics: either a displacement formulation (Paoli and Schatzman) either a velocity formulation (Moreau, Monteiro-Marques and Mabrouk). They also proved the convergence of their schemes, which yield existence results.

We consider the problem (12)-(14). For a trivial inertia operator, i.e. $\mathcal{M} = \mathrm{Id}_{\mathbb{R}^n}$, the following existence result has been proved by Paoli and Schatzman.

Theorem 10: [35,36] *Let K be a closed convex subset of $\mathbb{R}^N$ with non-empty interior and a regular boundary ∂K of class C^2 in the sense that there exists a unique mapping $\nu : \partial K \longrightarrow \mathbb{R}^N$ of class C^1 such that*

$$N_K(x) = \mathbb{R}_+ \nu(x), \qquad \forall x \in \partial K.$$

Let

    (i) *$f : [0, T] \times \mathbb{R}^N \times \mathbb{R}^N \longrightarrow \mathbb{R}^N$ continuous,*
    (ii) *$f(t, \cdot, \cdot)$ Lipschitz continuous for all $t \in [0, T]$.*

Let also $q_0 \in K$, $q_1 \in T_K(q_0)$ and $e \in (0, 1]$ be given. Then there exists $q : [0, T] \longrightarrow \mathbb{R}^N$ Lipschitz continuous such that

    (a) *$\dot{q}$ has bounded variations,*
    (b) *$q(0) = q_0$,*
    (c) *$\dot{q}(0_+) = q_1$,*
    (d) *$q(t) \in K, \quad \forall t \in [0, T]$,*
    (e) *$\langle \ddot{q} - f(t, q, \dot{q}), \varphi - q \rangle \geq 0, \quad \forall \varphi \in C^0([0, T]; K)$.*

Note that the expression $\langle \ddot{q} - f(t, q, \dot{q}), \varphi - q \rangle \geq 0$ (considered in the sense of distributions) constitutes a weak formulation for (12). A generalization of Theorem 10 for a non trivial inertia operator, i.e. $\mathcal{M} \neq \mathrm{Id}_{\mathbb{R}^n}$, as well as for a non-convex set of admissible positions is available in [24,25].

Following [9,10], we can rewrite (12) into the form

$$f(t, q, \dot{q}) - \mathcal{M}(q)\ddot{q} \in \partial\Psi_{T_K(q)}(\dot{q}), \tag{16}$$

where $T_K(z)$ is the tangent cone of K to z.

Since i.e. $\mathcal{M} = \mathrm{Id}_{\mathbb{R}^n}$, Monteiro-Marques [22] proved the existence of a solution of (16) using a time discretization, for inelastic shocks, i.e. $e = 0,$. Mabrouk [23] generalized the result from Monteiro-Marques for $e \in [0, 1]$. Finally, in a more general case of a restitution coefficient depending on t and q, an existence result has been etablished by Ballard [37] when all data are analytic.

3. Numerical Methods for Differential Inclusions

Some difficulties appear if one wants to perform numerical simulations of models as the one formulated in (5). It is necessary to use more advanced numerical methods: one way is to solve differential inclusions by means of appropriate set-valued version of classical difference methods like Euler's method and Runge-Kutta methods ... an other way is to develop new methods for specific differential inclusions. Taubert [38,39] was the first who investigated the convergence of finite difference methods for discontinuous equations and differential inclusions. Then others finite difference schemes were proposed and studied by Niepage [40], Kastner-Maresch [41], Steward [42,43] ... and many others (see [18] and references there in).

For numerical purposes, we replace the differential inclusion $y' \in \Phi(t, y)$ on $[0, T]$ by a sequence of discrete inclusion on the subintervals $t_0 = 0 < t_1 < t_2 < \cdots < t_N = T$ for $N > 0$, with a constant step-size $\Delta t = T/N$. Let a_i, $b_i \in \mathbb{R}$ for $i = 0, \cdots, r$, with $a_r \neq 0$ and $|a_0| + |b_0| > 0$. We are given the starting values $y_j \in \mathbb{R}^n$ for $j = 0, \cdots, r-1$, and the corresponding starting selections $\xi_j \in \Phi(t_i, y_i)$ out of the graph, for $i = 0, \cdots, r-1$. These may be computed by a linear p-step method with $p < r$ or by a one-step method.

Then, for $j = r, \ldots, n$ we compute y_j from

$$\frac{1}{h}\sum_{i=0}^{r} a_i y_{j-r+i} = \sum_{i=0}^{r} b_i \xi_{j-r+i} \qquad \text{with } \xi_i \in \Phi(t_i, y_i).$$

When $b_r \neq 0$ the method is implicit.

The following convergence result for these methods can be found in [18,20,19].

Theorem 11: *Let $D \in \mathbb{R}^n$ and $\Phi : D \times [0, T] \longrightarrow \mathcal{P}(\mathbb{R}^n)$ be a set-valued map. Let the following assumptions be satisfied:*

i) Φ *is a nonempty closed and convex-valued function;*

ii) Φ *is upper semicontinuous in $D \times [0, T]$ and verifies*

$$\|\zeta\| \le c\,(1 + \|x\|)\,,$$

for all $\zeta \in \Phi(x, t)$, $t \in [0, T]$ and $x \in D$ with constant $C \ge 0$;

iii) *All the zeros λ of the polynomial $\sum_{i=0}^{r} a_i \lambda^i$ have absolute value $|\lambda| < 1$ except for the simple zero $\lambda = 1$.*

iv) *The following consistency conditions are satisfied:*

$$\sum_{i=0}^{r} a_i = 0, \qquad \sum_{i=0}^{r} i a_i = \sum_{i=0}^{r} b_i.$$

v) *The coefficients b_i are nonnegative for $i = 0, \cdots, r$.*

vi) *The starting values satisfy*

$$\left\| y_{j+1}^n - y_j^n \right\| \le M \Delta t \qquad (j = 0, \cdots, r - 2)\,.$$

vii) *The approximations of the initial value y_0 satisfy $\lim_{n \to +\infty} y_0^n = y_0$.*

Then, the sequence $(y^n)_{n \in \mathbb{N}}$ of piecewise linear and continuous interpolants of the grid values $(y_0^n, \cdots, y_n^n)$ contains a subsequence which converges uniformly to a solution of the initial value problem (5).

Remark 12: If the Cauchy problem (5) has an unique solution, then the sequence $(y^n)_{n \in \mathbb{N}}$ converges uniformly to this unique solution.

We note that the theorem establishes the convergence of the numerical approximations, however, it provides neither the order of convergence nor any qualitative properties of the limit functions. But, for a special class of right-hand sides, like right-hand sides which satisfy a uniform one-sided lipschitz condition, and under sufficient regularity for the solution, the order of convergence is at least equal to one [41].

As an example and historically because it was the first one to be adapted to solve differential inclusion, the explicit Euler method is here outlined. For $q \in \mathbb{N} \setminus \{0\}$, a grid $0 = t_0 < t_1 < \cdots < t_q = T$ is chosen with stepsize $\Delta t = \frac{T - t_0}{q} = t_j - t_{j-1}$ $(j = 1, \ldots, q)$. Let

$$\eta_0 = y_0,$$

and, for $j = 0, \ldots, q - 1$, the vector η_{j+1} is computed by the formula

$$\eta_{j+1} \in \eta_j + \Delta t \Phi(t_j, \eta_j). \tag{17}$$

Then one sets

$$\eta^q(t) = \eta_j + \frac{1}{\Delta t}(t - t_j)(\eta_{j+1} - \eta_j), \tag{18}$$

for $t_j \leq t \leq t_{j+1}$, $j = 0, \ldots, q - 1$. The piecewise linear function η^q yields an approximation of the solution of Problem (5). The Euler method verifies the assumption of Theorem 11 and thus is convergent.

In section 4, we will propose and use a non-standard version of the previous set-valued Euler scheme.

3.1. *Numerical methods for vibro-impact problems*

This section is devoted to the particular case of the numerical approximations of vibro-impact problems. There exist three main approaches to solve vibro-impact problems:

(1) the event driven approach which consists of solving numerically the unconstrained problem and approximates the impact time. Between each impacts, it is possible to use standard (high order) numerical methods or, if possible, the analytical solution and the impact law to restart the algorithm. The event-driven approach appears to be "conceptually" the most simple and is very convenient when the motion can be decomposed into a finite number of intervals of "free flight." Unfortunately, in the case of "grazing" or impact accumulation, the numerical detection of all the impacts is impossible and a threshold δ must be chosen. In order to have a good approximation of the global dynamics, δ should be chosen small enough and thus we still have to detect a lot of impacts. This approach can become very expensive.

(2) the time-stepping approach: this method takes directly into account the impact law and avoid the detection of impacts. Two main approximations were proposed and studied. The first one, called here after the PS-scheme, was developed by Paoli and Schatzman: it is based on the displacement formulation of the problem. The second one, called here after the MMM-scheme, was given by Moreau, Monteiro Marques and Mabrouk: it is based on a velocity formulation of the problem (see [44] for a survey; [23,22,21,35,36,24,25]). One of the key points in Paoli-Schatzman and Moreau-Monteiro Marques-Mabrouk procedures is that the impact qualitative parameter e is "suitably" incorporated in the scheme, a fact which is consistent with the philosophy of the non-standard finite difference approach.

(3) the compliant approach: this is a common approach for constructing approximations of solutions of non-smooth problems. But, here, the regularization of (12-13) is not obvious [36]. Indeed, for $e \in [0,1]$, we define

$$\alpha = \frac{-\ln e}{\sqrt{\pi^2 + (\ln e)^2}}$$

and

$$G(q,w) = \begin{cases} \frac{(q,w)}{|q|^2}q & \text{if } q \neq 0 \\ 0 & \text{if } q = 0 \end{cases}$$

Then we approximate (12-13) by the penalized problem

$$\ddot{q}_\lambda - \frac{2\alpha}{\sqrt{\lambda}}G(q_\lambda - P_K(q_\lambda), \dot{q}_\lambda) + \frac{1}{\lambda}(q_\lambda - P_K(q_\lambda)) = f(t, q_\lambda, \dot{q}_\lambda), \quad (19)$$

where P_K denotes the projection on K and λ is a small parameter. In (19), we have replaced the rigid boundary of K by a viscoelastic one. The term $\frac{1}{\lambda}(q_\lambda - P_K(q_\lambda))$ is a drawback elastic force, which correspond to the Moreau-Yosida penalization. The term involving G is a viscous damping term which is exerted only when q_λ leaves K. The parameter λ is a positive real number intended to converge to 0. Paoli and Schatzman [36] proved that from any sequence of penalized solutions, it is possible to extract a subsequence which converges strongly in $W^{1,p}([0,T];\mathbb{R}^n)$ for all $p \in [1,+\infty[$ to a solution of (12-13). Equation (19) leads to a very stiff ordinary differential equation. Standard schemes can be used to approximate (19) at the cost of a severe restriction on the step size Δt and on the regularization parameter λ, i.e. $\Delta t = O\left(1/\sqrt{\lambda}\right)$.
In general, it was shown that, numerically, penalty methods could give poor results, like spurious oscillations in the solution, bad periodic regime ... [45,46].

In the remaining of this section, we present the PS-scheme and the MMM-scheme: in [15], we used the PS-scheme and we showed that the non-standard finite difference method can improve the PS-scheme in particular for conservative impacting systems. Later, in section 5, we will present non standard schemes for the MMM-scheme and the PS-scheme.

3.1.1. *The PS-scheme*

The main idea of the PS-scheme is to apply a classical numerical scheme to the ODE $\ddot{q} = f(t, q, \dot{q})$ and to add a discretization of the term $\partial \Psi_K(q)$. Thus, since $\mathcal{M}(q) = Id_{\mathbb{R}^N}$, Paoli and Schatzman proposed and studied [36] the following convergent sequence of discrete solutions (q^k) constructed recursively as follows

$$\begin{cases} q^0 = q_0 \\ q^1 = q_0 + \Delta t p_1 + \dfrac{\Delta t^2}{2} f(0, q_0, p_1) \end{cases} \tag{20}$$

where the impact law (13) permits to set

$$\begin{cases} p_1 = v_0 \ \ \text{if } q_0 \in \overset{\circ}{K}, \\ p_1 = -ev_{0,N} + v_{0,T} \ \ \text{if } q_0 \in \partial K . \end{cases} \tag{21}$$

Thereafter, i.e. for $k \geq 1$, the algorithm is

$$\frac{q^{k+1} - 2q^k + q^{k-1}}{(\Delta t)^2} + \partial \Psi_K \left(\frac{q^{k+1} + eq^{k-1}}{1+e} \right) \ni F^k \tag{22}$$

where F^k is some approximation of $f(t, q, \dot{q})$ at time t_k or t_{k+1}. For example, F^k can be defined by

$$F^k = F\left(\Delta t, \ t_k, \ q^k, \ q^{k-1}, \ \frac{q^{k+1} - q^{k-1}}{2\Delta t} \right) \tag{23}$$

where the function F, satisfying the natural consistency relation

$$F(0, t, x, x, y) = f(t, x, y), \ \ \forall (t, x, y) \in [0, T] \times \mathbb{R} \times \mathbb{R},$$

is continuous in all the arguments and Lipschitz continuous with respect to the last three arguments. In (22), at each time-step the constraint is satisfied by average position $\frac{q^{k+1} + eq^{k-1}}{1+e}$. Paoli and Schatzman proved that the weights involved in this average position yield a correct reflexion of the velocities at impacts. Indeed, at impact, the velocity is reversed in two time-steps and the constraints are also violated during at most two time-steps.

For the implementation of the algorithm, we use the following classical result [7,3]

Lemma 13: *Let K be a closed convex subset of $\mathbb{R}$, λ a positive number and $v \in \mathbb{R}$. Then*

$$q + \lambda \partial \Psi_K(q) \ni v \iff q = P_K(v)$$

where

$$P_K(y) = \begin{cases} y \text{ if } y \in K, \\ 0 \text{ otherwise} \end{cases}$$

is the projection of y on K.

Thus, using Lemma 13, the PS-scheme (22) implies the "workable" formulation

$$q^{k+1} = -eq^{k-1} + P_{(1+e)K}\left(2q^k + (e-1)u^{k-1} + (\Delta t)^2 F^k\right), \quad \text{for } k \geq 1. \tag{24}$$

Remark 14: The velocity of a vibro-impact system is a discontinuous function due to the impact law (13). This discontinuity makes the (global) order of convergence of the PS-scheme to be one for the position and zero for the velocity. However, no theoretical results are available in general about the order convergence of the PS-scheme. The dynamics of vibro-impact systems is often complex. In particular, these systems can be sensitive to initial data and/or have a chaotic behavior. As a result, the order of convergence of the scheme is not that essential since any prescribed accuracy in the computation is lost in finite time.

Remark 15: Recently, Paoli and Schatzman proved [24,25] the convergence of the previous algorithm for a general inertia operator, i.e. $\mathcal{M}(q) \neq Id_{\mathbb{R}^N}$.

3.1.2. *The Moreau-Monteiro Marques-Mabrouk scheme*

Let $\mathcal{M}(q) = Id_{\mathbb{R}^N}$. Contrary to the PS-scheme, we consider a time discretization of (16) written in velocity. The scheme proposed and studied in [9,10,22,23], uses standard discretization for the "ODE" part of (16) and an appropriate discretization for $\partial\psi_{T_K(y)}(v)$ by introducing the restitution coefficient e, i.e.

$$\begin{cases} \frac{y^{n+1}-y^n}{\Delta t} = v^n, \\ -\frac{v^{n+1}-v^n}{\Delta t} + f\left(t_{n+1}, y^{n+1}, v^{n+1}\right) \in \partial\psi_{T_K(y^{n+1})}\left(\frac{v^{n+1}+ev^n}{1+e}\right). \end{cases} \tag{25}$$

Here again, the "introduction" of the restitution coefficient e in the numerical scheme is in the spirit of the nonstandard finite difference method. Note also that (25) is semi-explicit, for numerical stability. Thus, for given

(y_0, v_0), the previous algorithm becomes

$$\begin{cases} y^0 = y_0, \\ -\dfrac{(v^0 - v_0)}{1+e} + \dfrac{\Delta t f(0, y^0, v^0)}{1+e} \in \partial \psi_{T_K(y)} \left(\dfrac{v_0 + e v_0}{1+e} \right) \end{cases} \tag{26}$$

and, for all $n \geq 0$

$$\begin{cases} y^{n+1} = y^n + \Delta t v^n, \\ -\dfrac{(v^{n+1} - v^n)}{1+e} + \dfrac{\Delta t f(t_{n+1}, y^{n+1}, v^{n+1})}{1+e} \in \partial \psi_{T_K(y^{i+1})} \left(\dfrac{v^{n+1} + e v^n}{1+e} \right) \end{cases} \cdot \tag{27}$$

Thus, using convex analysis tools [3], we can rewrite (26)-(27) as follows

$$\begin{cases} y^0 = y_0, \\ v^0 = -e v_0 + (1+e) \operatorname{Proj}_{y_0} \left(T_K(y_0), v_0 + \dfrac{\Delta t}{1+e} f(0, y^0, v^0) \right) \end{cases}$$

and, for all $n \geq 0$

$$\begin{cases} y^{n+1} = y^n + \Delta t v^n, \\ v^{n+1} = -e v^n + (1+e) \operatorname{Proj}_{y^{n+1}} \\ \qquad \left(T_K(y^{n+1}), v^n + \dfrac{\Delta t}{1+e} f(t_{n+1}, y^{n+1}, v^{n+1}) \right), \end{cases} \tag{28}$$

where $\operatorname{Proj}_y (T_K(y), \cdot)$ denotes the projection on $T_K(y)$ with respect to the kinetic metrics. The MMM-scheme is convergent for $e \in [0, 1]$ (see [22] for $e = 0$ and [23] for the general case $e \in [0, 1]$). Note, also that this scheme was principally developed for inelastic impacts, i.e. $e = 0$ [9,21].

Remark 16: In (28), the velocity is also reversed in one time-step at impact (contrary to the PS-scheme). Note also, the scheme is semi-explicit.

Remark 17: A generalization of the result obtained by Mabrouk [23], i.e. since $\mathcal{M}(q) \neq Id_{\mathbb{R}^N}$, is actually under consideration [47].

4. Frictional Oscillators

As a first example of non-smooth dynamical system, we consider linear oscillators subject to dry or Coulomb friction and slip velocity-dependent coefficient of friction. Such systems appear in geophysical systems [48,49] as well as in a wide variety of applications in real-life and in engineering, like in robot grippers and walkers [50], in flexible pin-jointed space structure [51] .... It is known that the presence of dry friction can influence the behavior and performance of these systems [52,53]. Thus the need of mechanical, mathematical and numerical studies in order to improve the performance or the "life" of such systems [54,55,56,57,58]. Dry friction can also be modeled with set-valued models and thus we enter in the theory of differential inclusion. Since the work of Coulomb, numerous frictions laws have been proposed to describe various phenomena that may appear. Here, we present examples of dry friction with constant or velocity-dependent coefficient of friction. Examples of nonlinear oscillators subject to dry friction are given in [59,60]. In general sophisticated methods are used to solve such problems: for instance, in [54], the authors used implicit schemes, such as Adams-Moulton schemes. The implementation of such schemes is rather complicate and can be expansive in CPU-times. The Euler method is generally used because it verifies Theorem 11 and is simple to implement. Unfortunately, it may not give satisfatory results: for instance it can lead to rapid spurious oscillations. In the remaining of the section, we will show that the non-standard method can improve the results obtained with classical method and leads to a "easy-to-use" explicit scheme.

4.1. *The harmonic linear oscillator*

We consider the example presented in the introduction (see Fig. 1), which leads to the following first-order differential inclusion

$$\begin{cases} \dot{x} = v, \\ \dot{v} \in -\omega^2 x - \mu \mathrm{SGN}\left(v - V_s\right). \end{cases} \tag{29}$$

The standard **explicit** Euler scheme associated with (29): for $k \geq 0$

$$\begin{cases} \frac{x^{k+1}-x^k}{\Delta t} = v^k, \\ \frac{v^{k+1}-v^k}{\Delta t} \in -\omega^2 x^k - \mu \mathrm{SGN}(v^k - V_s). \end{cases} \tag{30}$$

In practice, we will use the **semi-explicit** scheme since the explicit ones gives very bad results on the single-valued part, that is

$$\begin{cases} \frac{x^{k+1}-x^k}{\Delta t} = v^k, \\ \frac{v^{k+1}-v^k}{\Delta t} \in -\omega^2 x^{k+1} - \mu \mathrm{SGN}(v^k - V_s). \end{cases} \tag{31}$$

Then, following Mickens [11], we propose the following non-standard scheme associated with (29): for $k \geq 0$

$$\begin{cases} \frac{x^{k+1}-\cos \omega h x^k}{(1/\omega)\sin \omega h} = v^k, \\ \frac{v^{k+1}-\cos \omega h v^k}{(1/\omega)\sin \omega h} \in -\omega^2 x^k - \mu \mathrm{SGN}(v^k - V_s). \end{cases} \tag{32}$$

Remark 18: For the numerical computations, we consider the standard sgn-function because it was used in the modelling.

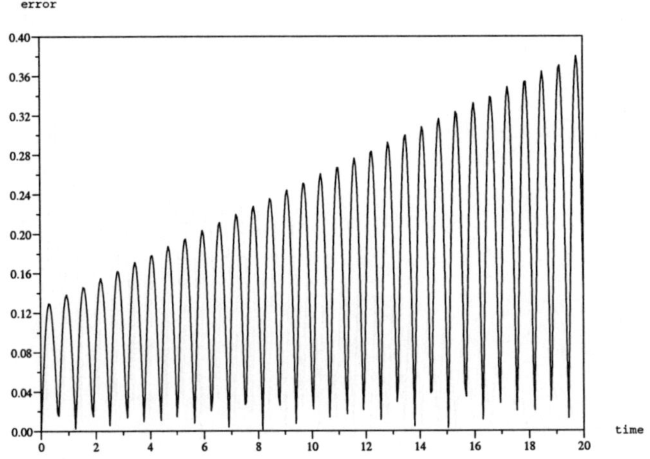

Fig. 2. $\mu = 0.001$ - Numerical error with the standard semi-explicit scheme

Example 19: In order to compare (31) and (32), we consider the following parameters

Δt	ω	x_0	v_0	V_S
5×10^{-3}	5	0	1	0

and different values for the coefficients of friction: $\mu = 0.001$, $\mu = 0.01$ and $\mu = 0.1$. In Figs. 2, 3, 4 and 5 (p. 305 and 306), it appears that the non-standard scheme is powerful as long as the friction coefficient is not "too big". This is not surprising since the oscillatory part of the solution is prevailing. When the coefficient of friction μ is "big," the approximate solutions computed with both methods are similar (Fig. 6, p. 307).

4.2. *Frictional oscillators with velocity-dependent frictions*

Since the works of Rabinowicz [56,57], it is usually admitted that the coefficient of friction varies, rather than being constant, with respect to the velocity, i.e. $\mu = \mu(|\dot{x} - V_S|)$. Let also $\mu : \mathbb{R}_+ \to \mathbb{R}_+$ be a smooth function such that

$$0 < \mu_{min} \leq \mu(r) \leq \mu_{max}, \qquad r \in \mathbb{R}_+.$$

Y. Dumont

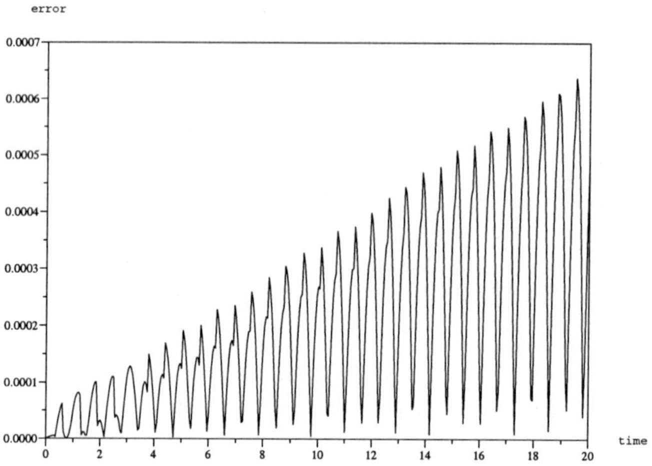

Fig. 3. $\mu = 0.001$ - Numerical error with the non-standard scheme

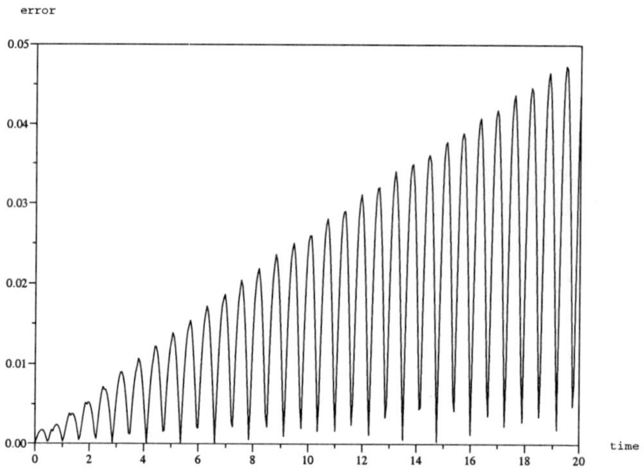

Fig. 4. $\mu = 0.01$ - Numerical error with standard semi-explicit scheme

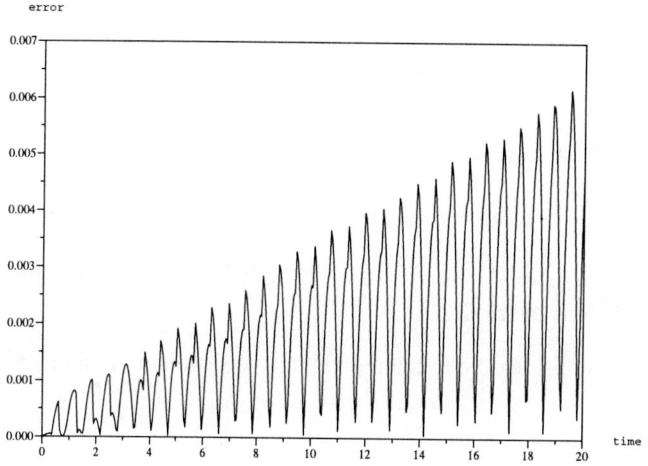

Fig. 5. $\mu = 0.01$ Numerical error with the non-standard scheme

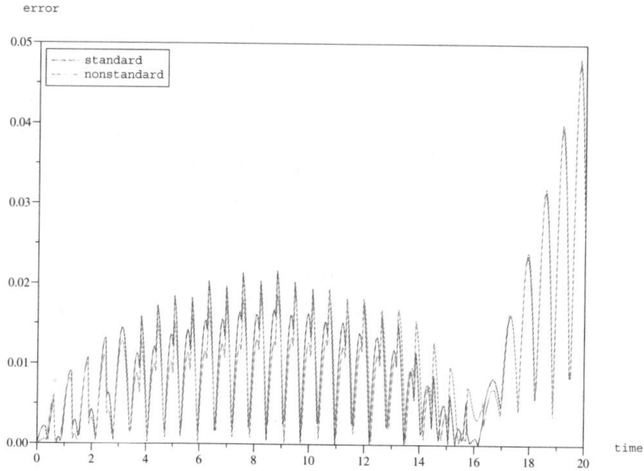

Fig. 6. $\mu = 0.1$ Numerical errors with the non-standard scheme and the standard semi-explicit scheme

Then, the modified problem is to find a function $x : [0, T] \to \mathbb{R}$ such that

$$\begin{cases} \ddot{x} = -\omega^2 x - \mu(|\dot{x} - V_S|) \, \text{SGN}(\dot{x} - V_S) \\ x(0) = x_0, \qquad \dot{x}(0) = v_0. \end{cases} \tag{33}$$

Remark 20: A well-known velocity-dependent friction [61] is

$$\mu(v) = \mu_s + (\mu_{\max} - \mu_s) \exp\left(-\frac{|v|}{V_{cr}}\right), \quad \cdot$$

where V_{cr} is a critical velocity that depends on the material and μ_s is the static coefficient of friction.

Here, we consider the following velocity-dependent coefficient of friction [54]:

$$\mu(v) = \begin{cases} \mu_s + (\mu_{\max} - \mu_s) \frac{v}{V_1}\left(2 - \frac{v}{V_1}\right), \\ \qquad \text{if } 0 \le v \le V_1 \\ \mu_{\max} - (\mu_{\max} - \mu_{\min})\left(\frac{v-V_1}{V_2-V_1}\right)^2\left(3 - 2\frac{v-V_1}{V_2-V_1}\right), \\ \qquad \text{if } V_1 \le v \le V_2 \\ \mu_d + \dfrac{1}{\frac{1}{\mu_d-\mu_{\min}}+b(v-V_2)^2}, \\ \qquad \text{if } V_2 \le v. \end{cases}$$

with $V_1 = 0.05$ m/sec, $V_2 = 1$ m/sec, $\mu_s = 0.8$ (the static friction coefficient), $\mu_d = 0.5$ (the dynamic friction coefficient), $\mu_{\max} = 1$ and $\mu_{\min} = 0.4$ (see Fig. 7, p. 309) Note also that μ is lipschitzian, with lipschitz constant $M_\mu = \frac{3}{2}\frac{\mu_{\max}-\mu_{\min}}{V_2-V_1}$. Thus $\mu(|v-V_s|)\,\text{SGN}(v-V_s)$, in (33) is semi-lipschitzian, which implies that problem (33) has an unique solution, see Theorem 1, p. 291.

The coefficient of friction is based on three monotone parts (see Fig. 7, p. 309) and this particular shape will provide interresting results. Consider the following parameters

Δt	ω	x_0	V_S
5×10^{-3}	2	10^{-4}	4

We will consider different initial velocities, namely $v_0 = 3.411, v_0 = 3.412$ and $v_0 = 3$.

Example 21: We first compare the standard **explicit** scheme with the non-standard one.

Let $v_0 = 3.412$, then we have "similar" phase portrait for both numerical approximations (see Fig. 8, p. 309). But, if we take $v_0 = 3.411$, then

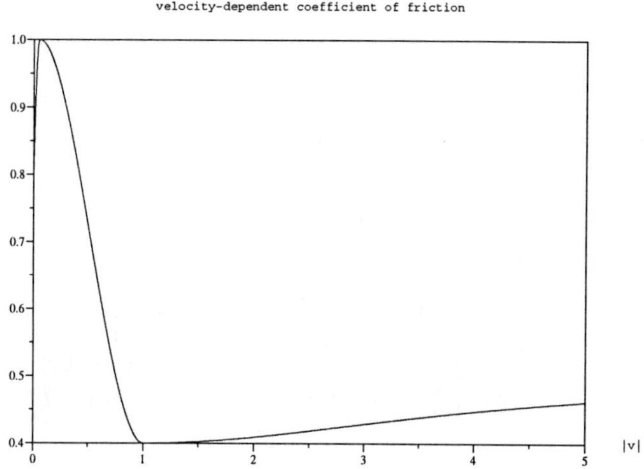

Fig. 7.   Velocity-dependent coefficient of friction

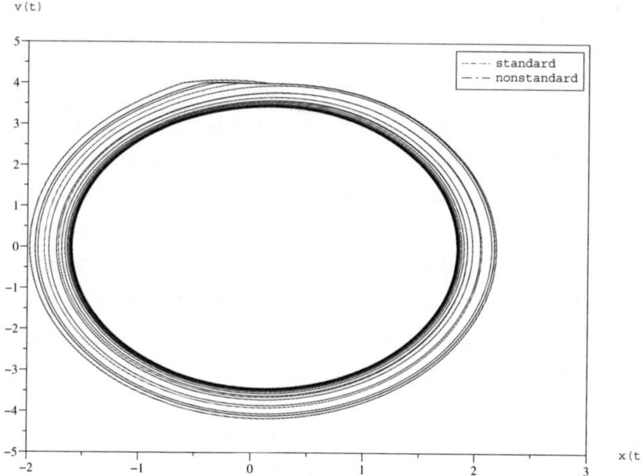

Fig. 8.   Phase portrait by the standard and non-standard scheme with $(x_0, v_0) = (10^{-4}, 3.412)$

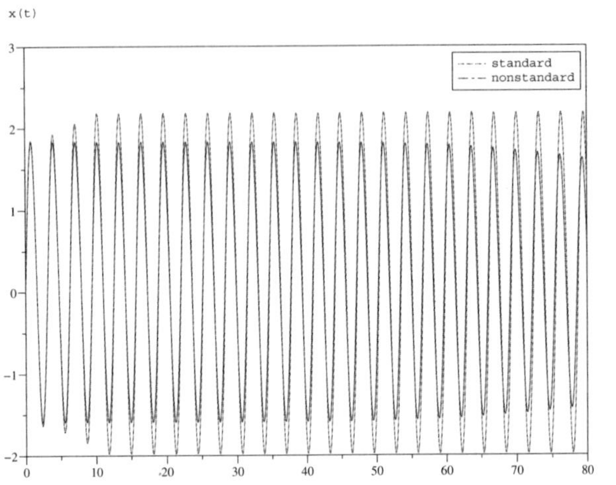

Fig. 9. Displacement by the standard and non-standard schemes with $(x_0, v_0) = (10^{-4}, 3.411)$

the phase portraits are rather different: the non-standard approximation converges to the (stable) stationary point, while the standard approximation converges to a (stable) periodic solution (see Figs. 9 (p. 310) and 10 (p. 311)). The computation requires 18.4 cpu-time for the non-standard and standard schemes. Finally, if $v_0 = 3.3$, the difference remains (even if we choose a lower time-step, for instance $\Delta t = 10^{-3}$ (see Fig. 11, p. 312). Of course, if Δt goes to zero, we recover the same phase portraits as those obtained with the non-standard method, but then the computations become very expansive in CPU-time. The standard explicit scheme does not fit the property of the time-continuous solution, while the non-standard scheme does, even for reasonnable time-step, i.e. $\Delta t = 5 \times 10^{-3}$.

Example 22: Consider now the **semi-explicit** standard scheme: we obtain the same phase portraits for both schemes when $v_0 = 3$ and $v_0 = 3.411$ (see Figs. 12 (p. 313) and 13 (p. 314)). But, when $v_0 = 3.412$, the phase-portrait become rather different (Fig. 14 (p. 315)).

It suffices to choose a lower time-step, i.e. $\Delta t = 10^{-3}$, to obtain similar phase portrait with the standard and non-standard schemes for $v_0 = 3.412$. As expected, the semi-explicit standard scheme improves the approximations with respect to the standard explicit one. But the non-standard

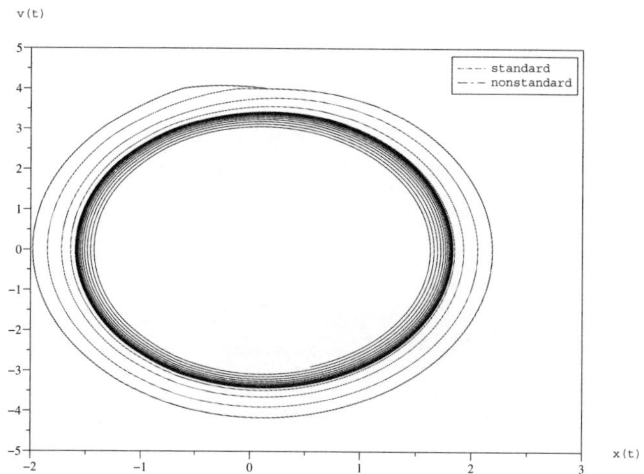

Fig. 10. Phase portrait by the standard and non-standard schemes with $(x_0, v_0) = (10^{-4}, 3.411)$

scheme is preferably to use, because it is explicit and it performs the best results for "usual" time-step in a reasonnable cpu-time.

The non-standard schemes are determined by requesting the same stability properties than the time-continuous system. Here, depending on the initial conditions, the trajectory converges either to a (stable) fixed-point or to a periodic (stable) solution. It is also crucial to use a numerical algorithm that preserves this property: this is very well done by the non-standard approach.

5. Vibro-Impact Oscillators

We now consider non smooth systems that exhibit impacts. They appear in many real life problems, for instance in the models of impact print hammers, gear boxes, offshore structures, ships moored at dockside, rotor-casing systems, control of joint's loose [62,63,64,65,66,67]. Such mechanical systems have been mainly investigated in the single-degree-of freedom case in order to study the behavior of the regime: periodic regime, chaotic regime, bifurcation, transition and global behaviors have been examined [68,2,69,70,71,72]. Of course, some authors studied multi-degree-of-freedom

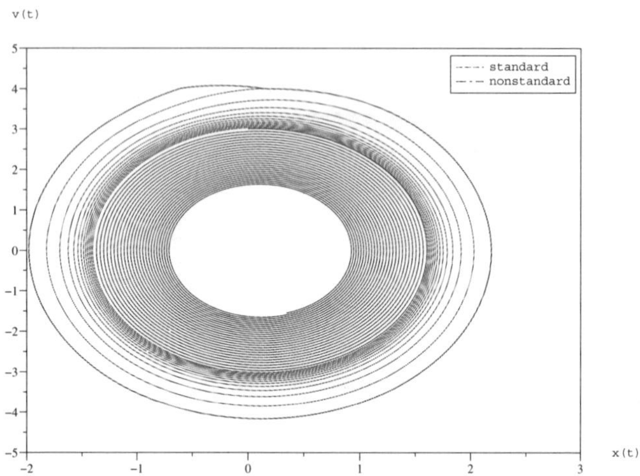

Fig. 11.   Phase portrait by the standard and non-standard schemes with $(x_0, v_0) = (10^{-4}, 3)$

case and models including both impacts and friction. In some cases, authors used classical schemes to solve numerically vibro-impact problems. On the other hand, specific numerical methods have been developed from the mechanical point of view [21,6] and from the mathematical point of view [24,25,65].

In section 3, we have presented the PS- and the MMM-approaches usually used to solve vibro-impact problems. Here, we present their non-standard analogues. A non-standard scheme for the PS-scheme was first proposed in [15] for conservative oscillators.

We consider the following mechanical setup: a mass, subject to a (possibly) external forcing and which can impact against a rigid stop (see Fig. 15). This leads to the following equation for the free flight

$$\ddot{u} + k^2 u = a \sin(\omega t), \tag{34}$$

with $\omega \neq k$ The constraint to be satisfied is $u \geq u_{min}$. Taking into account the constraint, we obtain

$$\begin{cases} \ddot{u} + k^2 u = a \sin(\omega t) + \mu, \\ \qquad u \geq u_{\min}, \\ \mu \geq 0, \qquad \mathrm{supp}\mu \subset \{t : u(t) = u_{\min}\}, \end{cases} \tag{35}$$

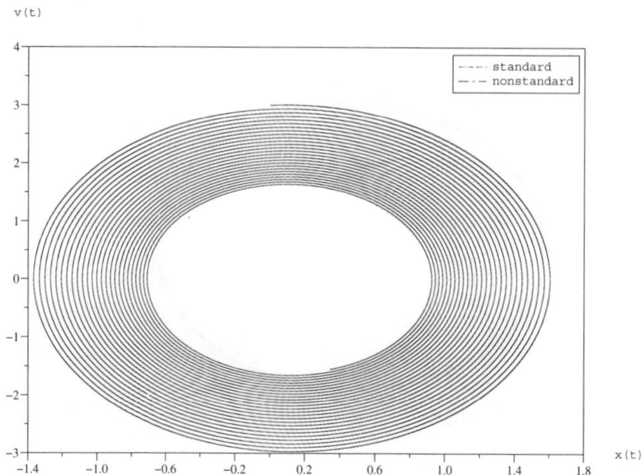

Fig. 12. Phase portrait by the the semi-explicit standard and non-standard schemes with $(x_0, v_0) = (10^{-4}, 3)$

where μ is a measure which describes the reaction force of the system against the obstacle. Using convex analysis tools [7], we rewrite (35) in the form

$$\ddot{u} + k^2 u + \partial \Psi_{[u_{\min}, +\infty[} \ni a\sin(\omega t) \tag{36}$$

or, following the MMM approach

$$-\ddot{u} - k^2 u + a\sin(\omega t) \in \partial \Psi_{T_{[u_{\min}, +\infty[}} \tag{37}$$

We complete the previous differential inclusion, with the following impact's law

$$u(t) = u_{\min} \implies \dot{u}(t+0) = -e\dot{u}(t-0), \tag{38}$$

with $e \in [0, 1]$. Finally, we give initial conditions

$$u(0) = u_0, \qquad \dot{u}(0_+) = u_1,$$

with $u_0 \in [u_{\min}, +\infty[$ and u_1 given.

In [35], Paoli used the PS-scheme, to study the linear damped oscillator case. In particular, she considered different values for $u_{\min}$, showing different behaviors of the solution. Here, we will investigate our linear impacting oscillator with the NSFD method, the PS-scheme and the MMM-scheme, with different values for $u_{\min}$ and e.

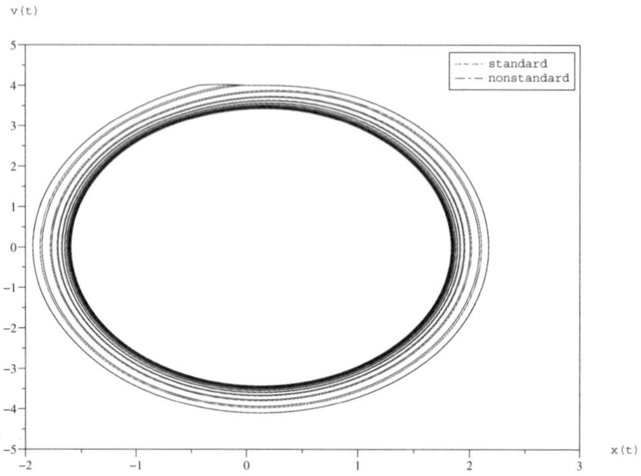

Fig. 13. Phase portrait by the semi-explicit standard and non-standard schemes with $(x_0, v_0) = (10^{-4}, 3.411)$

When no impacts occur and $\omega \neq k$, the exact solution of (34) is: for all $t \geq 0$.

$$u(t) = u_0 \cos kt + \frac{1}{k}\left(u_1 + \frac{a\omega}{\omega^2 - 1}\right) \sin kt + \frac{a}{1 - \omega^2} \sin \omega t. \qquad (39)$$

We compute the exact solution for (34)-(38) as follows: using (39), we consider a dichotomy procedure to approximate $t_{impact} \in \,]t_n, t_{n+1}[$ Because, we have the exact solution between each impact, we approximate t_{impact} up to 10^{-14} in order to have "reasonnable" CPU-time. We obtain u_{impact} and we calculate $v_{impact} = u'(t_{impact})$. Then, we compute: for $i = n+1, \ldots$

$$u(t_i) = u_{\min} \cos k\,(t_i - t_{impact})$$
$$+ \frac{1}{k}\left(-ev_{impact} + \frac{a\omega}{\omega^2 - 1}\right) \sin k\,(t_i - t_{impact})$$
$$+ \frac{a}{1 - \omega^2} \sin \omega\,(t_i - t_{impact}),$$

until the next impact.

In the remaining of this section, we will consider either $a = 0$ (no force occurs: the system is then conservative), either $a \neq 0$. We first give non-

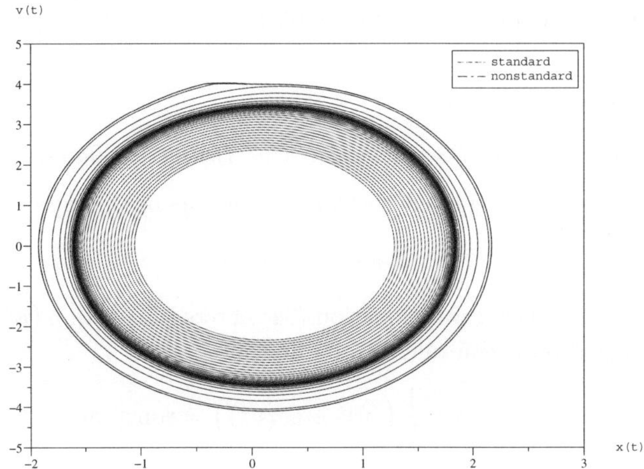

Fig. 14.   Phase portrait by the semi-explicit standard and non-standard schemes with $(x_0, v_0) = (10^{-4}, 3.412)$

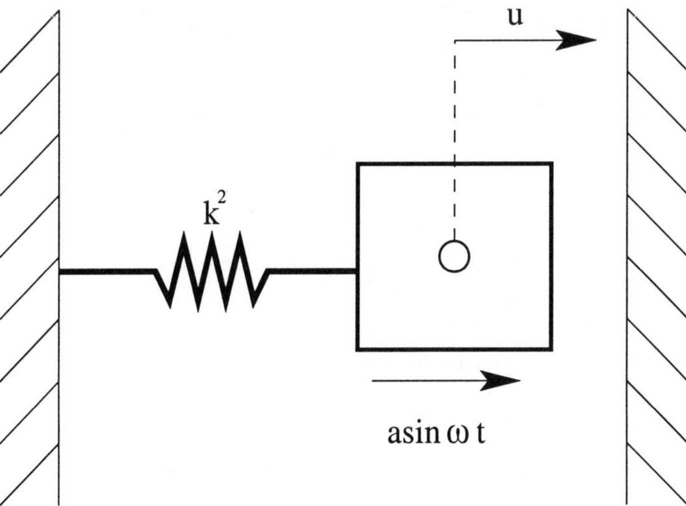

Fig. 15.   The impacting oscillator

standard schemes for the PS- and the MMM-approaches in both cases, then we illustrate through numerical examples.

5.1. *No forcing: $a = 0$*

5.1.1. *A non-standard PS-scheme for conservative oscillators*

In [15], we studied energy conservative oscillators of the type

$$\ddot{u} + k^2 u g(u^2) = 0, \tag{40}$$

with $g(0) = 1$. Since $a = 0$, equation (34) enters (40) with $g(u^2) = 1$. The previous equation is equivalent to its first integral

$$H(u, \dot{u}) = \frac{1}{2}\left((\dot{u})^2 + G(u^2)\right) = \text{constant},$$

with

$$G(z) = \int_0^z g(s)\,ds,$$

where H represents the sum of kinetic energy and potential energy. We also consider a discrete energy of the general form

$$H_{\Delta t}(u^n) := \frac{1}{2}\left\{\left(\frac{u^{n+1} - u^n}{\phi(\Delta t)}\right)^2 + k^2 \tilde{G}_{\Delta t}(u^n)\right\}, \tag{41}$$

and we show that this discrete energy is conserved, i.e.

$$H_{\Delta t}(u^n) = H_{\Delta t}(u^{n+1}) \tag{42}$$

if

$$\tilde{G}_{\Delta t}(u^n) := G(u^n u^{n+1}), \ \forall k \geq 0. \tag{43}$$

Thus we deduced the following important theorem that gives us a non-standard scheme associated with (40):

Theorem 23: [15] *The non-standard finite difference scheme*

$$\frac{u^{n+1} - 2u^n + u^{n-1}}{(4/\omega^2)\sin^2(\omega\Delta t/2)} + k^2 u^n \frac{G(u^n u^{n+1}) - G(u^{n-1} u^n)}{u^n u^{n+1} - u^{n-1} u^n} = 0 \tag{44}$$

for (40) *is equivalent to the discrete principle of conservation of energy* (42) *where the discrete potential energy is given by* (43) *and* $\phi(\Delta t) = (2/k)\sin(k\Delta t/2)$.

Thus, motivated by Theorem 23 and the PS-algorithm, we set

$$G^{n-1,n,n+1} = \frac{G(u^n u^{n+1}) - G(u^{n-1} u^n)}{u^n u^{n+1} - u^{n-1} u^n} \tag{45}$$

and propose the following non-standard impact PS-scheme, call here after NSPS-scheme

$$\frac{u^{n+1} - 2u^n + u^{n-1}}{(4/\omega^2)\sin^2(\omega\Delta t/2)} + \partial\Psi_K\left(\frac{u^{n+1} + eu^{n-1}}{1+e}\right) + k^2 u^n G^{n-1,n,n+1} \ni 0, \tag{46}$$

which, using lemma 13 (p. 301), leads to "workable" formulation

$$u^{n+1} = -eu^{n-1} + P_{(1+e)K}\left(2u^n + (e-1)u^{n-1} - 4u^n \sin^2(k\Delta t/2)G^{n-1,n,n+1}\right). \tag{47}$$

The first starting value u^0 of the NSPS-scheme is as in Eq. (20). The second initial guess u^1 is obtained depending on the initial value u^0. Owing to the condition

$$(2/k)\sin(k\Delta t/2) = \Delta t + O[(\Delta t)^2],$$

the results obtained by the NSPS-scheme and the PS-scheme display similar convergence property for small values of Δt. We have however to emphasize that the non-standard scheme has no restriction on the step size, being unconditionally stable, as a consequence of Theorem 23. For the same reason, the new scheme preserves the principle of conservation of energy between two consecutive impact times. Here is a further important qualitative stability property, valid in the case when impact times are isolated.

Remark 24: Theorem 23 is very general and can even be applied on nonlinear conservative oscillators (see subsection 5.3).

Applying Theorem 23 to (34) (with $a = 0$) leads to the following NSPS-scheme: for $n \geq 1$

$$u^{n+1} = -eu^{n-1} + P_{(1+e)K}\left(2u^n + (e-1)u^{n-1} - 4u^n \sin^2(k\Delta t/2)\right). \tag{48}$$

5.1.2. *A non-standard MMM-scheme*

Let us first explain the MMM-scheme associated with (37)-(38). In fact, it suffices to use (28) with $f(t, u, v) = -k^2 u$: for all $n \geq 0$

$$\begin{aligned}
&u^{n+1} = u^n + \Delta t v^n, \\
&Y = v^n - \Delta t k^2 u^{n+1}/(1+e) \\
&\begin{cases}
\quad \text{if } u^{n+1} > u_{\min}, \text{ then } v^{n+1} = -ev^n + (1+e)Y, \\
\text{if } u^{n+1} \leq u_{\min} \text{ then } v^{n+1} = -ev^n + (1+e)\max(Y, 0)
\end{cases}
\end{aligned} \tag{49}$$

In order to understand how to obtain the non-standard MMM-scheme, called here after the NSMMM-scheme, we have to do some computations. Indeed, using [11] and (28), we have: for all $n \geq 0$

$$\begin{cases} u^{n+1} = \psi\left(\Delta t\right) u^n + \phi\left(\Delta t\right) v^n, \\ -\left(\frac{v^{n+1}-\psi(\Delta t)v^n}{1+e}\right) - \frac{\phi(\Delta t)k^2 u^n}{1+e} \in \partial\psi_{T_K(u^{n+1})}\left(\frac{v^{n+1}+ev^n}{1+e}\right), \end{cases}$$

with $\psi\left(\Delta t\right) = \cos\left(k\Delta t\right)$ and $\phi\left(\Delta t\right) = \sin(k\Delta t)/k$. Then, we rewrite the previous inclusion into the form

$$\begin{cases} u^{n+1} = \psi\left(\Delta t\right) u^n + \phi\left(\Delta t\right) v^n, \\ \frac{e+\psi(\Delta t)}{1+e} v^n - \frac{\phi(\Delta t)k^2 u^n}{1+e} - \left(\frac{v^{n+1}+ev^n}{1+e}\right) \in \partial\psi_{T_K(u^{n+1})}\left(\frac{v^{n+1}+ev^n}{1+e}\right), \end{cases}$$

which is equivalent to

$$\begin{cases} u^{n+1} = \psi\left(\Delta t\right) u^n + \phi\left(\Delta t\right) v^n, \\ v^{n+1} = -ev^n + (1+e)\operatorname{Proj}_{u^{n+1}} \\ \qquad \left(T_K, \left((e+\psi\left(\Delta t\right))v^n - \phi\left(\Delta t\right)k^2 u^n\right)/(1+e)\right), \end{cases}$$

Finally, the NSMMM-scheme is

$$\begin{aligned} & u^{n+1} = \psi\left(\Delta t\right) u^n + \phi\left(\Delta t\right) v^n, \\ & Y = \left((e+\psi\left(\Delta t\right))v^n - \phi\left(\Delta t\right)k^2 u^n\right)/(1+e) \\ & \begin{cases} \quad \text{if } u^{n+1} > u_{\min}, \text{ then } v^{n+1} = -ev^n + (1+e)Y, \\ \text{if } u^{n+1} \leq u_{\min} \text{ then } v^{n+1} = -ev^n + (1+e)\max(Y,0). \end{cases} \end{aligned} \tag{50}$$

When no impacts occur, a simple computation show that the NSMMM-scheme (50) is conservative, i.e. for all $n \geq 0$:

$$\begin{aligned} H\left(u^{n+1}, v^{n+1}\right) &= \frac{1}{2}\left(\left(v^{n+1}\right)^2 + k^2 \left(u^{n+1}\right)^2\right) \\ &= \frac{1}{2}\left(\psi\left(\Delta t\right)^2 + k^2\phi\left(\Delta t\right)^2\right)\left(\left(v^n\right)^2 + k^2 \left(u^n\right)^2\right) \\ &= H\left(u^n, v^n\right). \end{aligned}$$

In contrary, the standard MMM-scheme (49) studied by Mabrouk is not conservative (for a comparison, see Fig. 16 with $k = 100$, $(u_0, v_0) = (1, 0)$ and $\Delta t = 10^{-2}$). Thus, like the NSPS-scheme (46), the NSMMM-scheme preserves the principle of conservation of energy between two consecutive impact times. The NSMMM-scheme and the MMM-scheme display the same convergence property as Δt goes to zero. In addition to be convergent, the NSMMM has no restriction on the step size because it is based on the exact scheme when no impacts occur.

We now consider some test-cases in order to compare the standard schemes and the non-standard schemes.

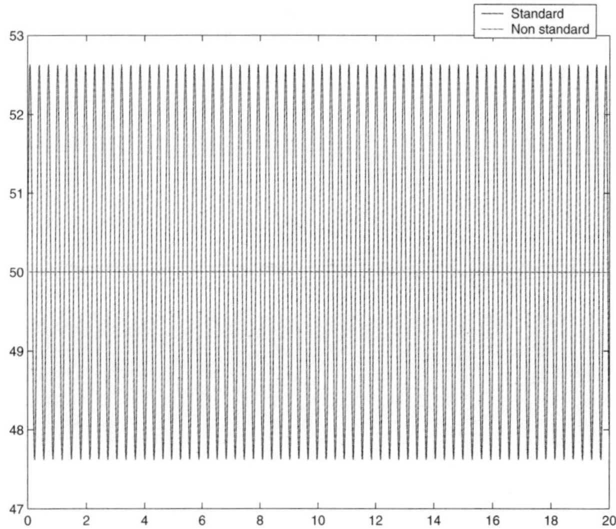

Fig. 16. Total energy: standard and non-standard MMM-schemes

Example 25: We first consider a case, which show another qualitative property hold by the NSPS- and the NSMMM-schemes. Consider the following parameters

Δt	k	x_0	v_0	$x_{\min}$	e
5×10^{-4}	10	1	0	-1.0000001	0.5

Figs. 17 (p. 320) and 18 (p. 320) show the error between the approximations and the exact solution. Moreover, the exact solution verifies the boundedness property $-1 \leq u\,(t_n) \leq 1$ for all $n \geq 0$. The MMM-scheme and the PS-scheme, based on standard finite difference, fail to replicate this property and generate spurious impacts. While the NSPS- and NSMMM-schemes, being exact schemes, are stable with respect to this property.

Example 26: We now consider different values for e, i.e $e = 0.5$ and $e = 1$, and the following parameters

Δt	k	x_0	v_0	$x_{\min}$
5×10^{-4}	1	1	0	-0.5

In Figs. 19 (p. 321) and 20 (p. 321), we show the error between the exact solution and the approximations computed with the PS and MMM-schemes

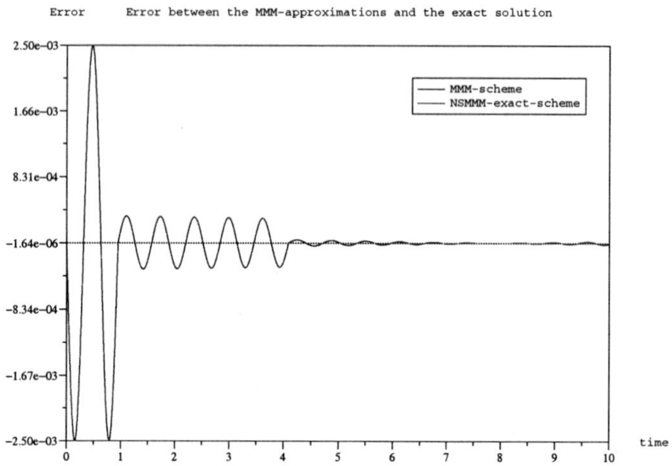

Fig. 17.   Error between the MMM-approximations and the exact solution

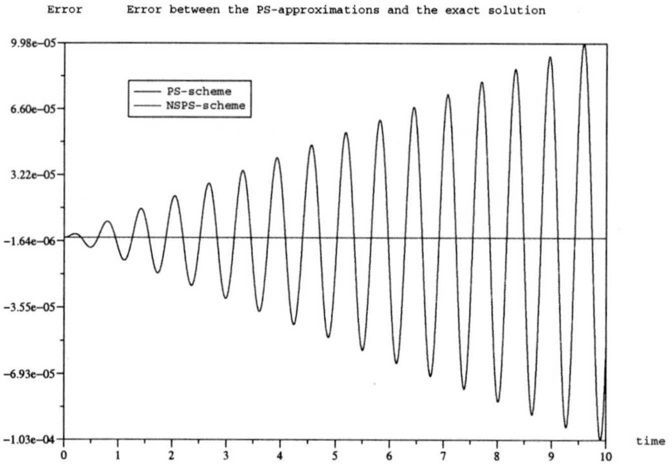

Fig. 18.   Error between the PS-approximations and the exact solution

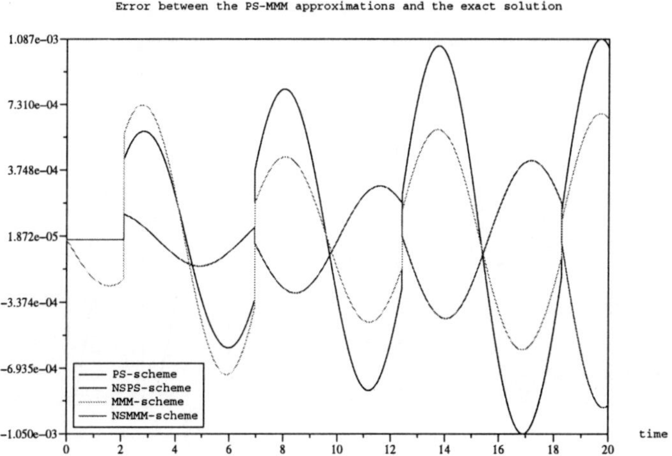

Fig. 19. Error between the PS-MMM approximations and the exact solution for $e = 0.5$

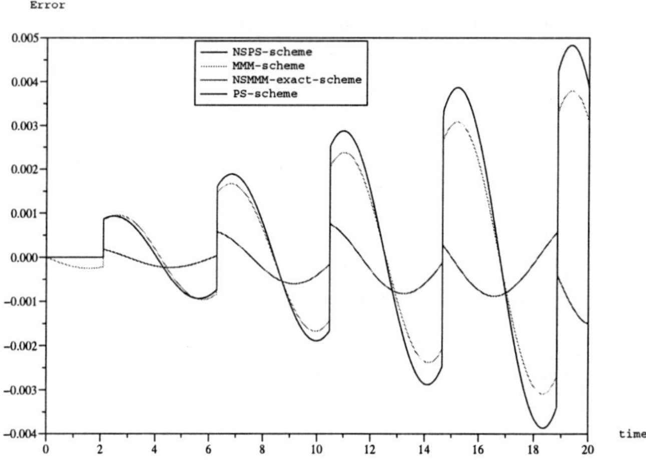

Fig. 20. Error between the PS-MMM approximations and the exact solution for $e = 1$

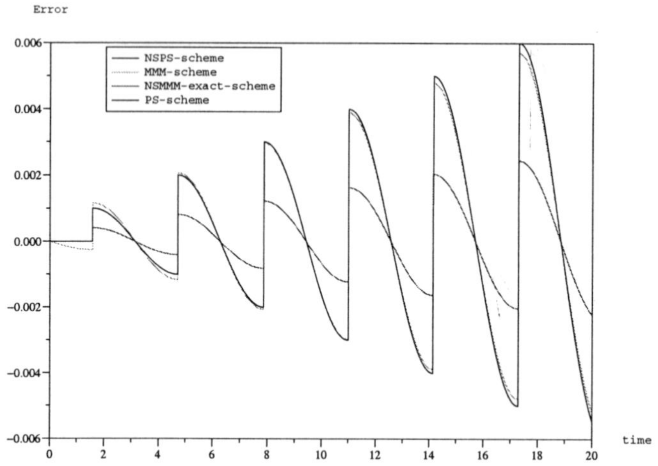

Fig. 21.   Error between the PS- and MMM-approximations and the exact solution

over a long time: the NSMMM-scheme performs the lowest error. Note also
that the NSPS-scheme and the PS-scheme give the same errors. In the next
table, we give the CPU-time for each method

CPU-time	Exact solution	MMM-scheme	PS-scheme	NSMMM-scheme	NSPS-scheme
	423.62	172.63	177.43	173.96	176.67

Note that the CPU times for the PS- and MMM-schemes are approxima-
tively three times less than the CPU-time for the exact solution.

Example 27: We now change the position of the obstacle

Δt	k	x_0	v_0	$x_{\min}$	e
5×10^{-4}	1	1	0	0	0.5

As in the previous example, the NSMMM-exact scheme gives the best
results, see Fig. 21 (p. 322). The PS-schemes (standard and non-standard)
give the same error.

5.2. *The forcing case: $a \neq 0$*

Following (48), we will now consider the following NSPS-scheme associated with (36): for $n \geq 1$

$$y^{n+1} = -ey^{n-1} + P_{(1+e)K}$$
$$\left(2y^n + (e-1)\, y^{n-1} + 4\frac{\sin^2(k\Delta t/2)}{k^2} \left(-k^2 y^n - a\sin\omega t_n\right) \right).$$

In a same manner, following (50), we may propose the following NSMMM-scheme associated with (37)

$$u^{n+1} = \psi\,(\Delta t)\, u^n + \phi\,(\Delta t)\, v^n,$$
$$Y = \left((e + \psi\,(\Delta t))\, v^n + \phi\,(\Delta t) \left(-k^2 u^n + a\sin\omega t_n\right)\right) / (1+e)$$
$$\begin{cases} \quad \text{if } u^{n+1} > u_{\min}, \text{ then } v^{n+1} = -ev^n + (1+e)Y, \\ \text{if } u^{n+1} \leq u_{\min} \text{ then } v^{n+1} = -ev^n + (1+e)\max(Y,0). \end{cases}$$

In more of the previous non-standard schemes, we now propose some non-standard schemes associated with the exact formulation (39). We will show that the use of exact-schemes can seriously improve the results obtained with the non-standard approach.

5.2.1. *The NSMMM-exact-scheme*

Using the previous NSMMM-scheme, we are able to propose the following NSMMM-exact scheme, associated with (37)-(38), with the help of [11]. Indeed, using (39), after some straighforward computations, we have

$$u\,(t + \Delta t) = \cos(k\Delta t)\, u\,(t) + \frac{\sin k\Delta t}{k} u'\,(t) +$$
$$+ \frac{a}{1-\omega^2} \left(\sin\omega\,(t+\Delta t) - \cos(k\Delta t)\sin\omega t - \frac{\sin k\Delta t}{k}\omega\cos\omega t \right)$$
$$u'\,(t+\Delta t) = \cos(k\Delta t)\, u'\,(t) - \frac{\sin k\Delta t}{k}\left(k^2 u\,(t)\right) +$$
$$+ \frac{a}{1-\omega^2}\left(k\sin k\Delta t\sin\omega t - \omega\cos(k\Delta t)\cos\omega t \right.$$
$$\left. + \omega\cos\omega\,(t+\Delta t) \right).$$

Thus, if we rewrite (34) in a first order system

$$\begin{cases} u' = v \\ v' = -k^2 u + a\sin\omega t, \end{cases} \tag{51}$$

the exact-scheme associated with (51) is: for $n \geq 0$

$$\begin{cases} u^{n+1} = \psi\left(\Delta t\right) u^n + \phi\left(\Delta t\right) v^n + F^n, \\ v^{n+1} = \psi\left(\Delta t\right) v^n - \phi\left(\Delta t\right) k^2 u^n + G^n \end{cases} \tag{52}$$

where

$$F^n = \frac{a}{1 - \omega^2} \left(\sin \omega t_{n+1} - \psi\left(\Delta t\right) \sin \omega t_n - \omega \phi\left(\Delta t\right) \cos \omega t_n\right), \tag{53}$$

$$G^n = \frac{a}{1 - \omega^2} \left(k^2 \phi\left(\Delta t\right) \sin \omega t_n - \omega \psi\left(\Delta t\right) \cos \omega t_n + \omega \cos \omega t_{n+1}\right),$$

and $\phi\left(\Delta t\right) = \sin(k\Delta t)/k$ and $\psi\left(\Delta t\right) = \cos\left(k\Delta t\right)$. Thus, having in mind (50), we deduce the following NSMMM-exact scheme associated with (37)-(38): for $n \geq 0$

$$\begin{cases} u^{n+1} = \psi\left(\Delta t\right) u^n + \phi\left(\Delta t\right) v^n + F^n, \\ Y = \left(\left(e + \psi\left(\Delta t\right)\right) v^n - \phi\left(\Delta t\right) k^2 u^n + G^n\right)/(1 + e) \\ \quad\quad \text{if } u^{n+1} > u_{\min}, \text{ then } v^{n+1} = -ev^n + (1 + e)Y, \\ \text{if } u^{n+1} \leq u_{\min} \text{ then } v^{n+1} = -ev^n + (1 + e)\max(Y, 0). \end{cases} \tag{54}$$

5.2.2. *The NSPS-exact-scheme*

In the same way, we are able to propose an exact PS-scheme. Indeed, using (52), it is possible to propose an exact scheme associated with (34). After some straightforward computations, we obtain: for all $n \geq 1$

$$u^{n+1} - 2\psi\left(\Delta t\right) u^n + u^{n-1} = F^n - \psi\left(\Delta t\right) F^{n-1} + \phi\left(\Delta t\right) G^{n-1},$$

or equivalently

$$u^{n+1} - 2u^n + u^{n-1} = -2k^2\phi^2\left(\Delta t/2\right) F^n - \psi\left(\Delta t\right) F^{n-1} + \phi\left(\Delta t\right) G^{n-1}$$

with F^n and G^n given by (53). Thus we can deduce the exact-NSPS-scheme: for $n \geq 1$

$$\begin{aligned} u^{n+1} = -eu^{n-1} + P_{(1+e)K} \big(2u^n + (e - 1) u^{n-1} - 2k^2\phi^2\left(\Delta t/2\right) F^n \\ - \psi\left(\Delta t\right) F^{n-1} + \phi\left(\Delta t\right) G^{n-1}\big). \end{aligned} \tag{55}$$

We now present some illustrative examples.

Example 28: Let us give the following paramaters

Δt	k	a	ω	u_0	u_1	$u_{\min}$	e
5×10^{-4}	10	1	9	1	0	-0.5	0.2

Figs. 22 (p. 325) and 24 (p. 326) show the displacement of the system computed with all methods. As expected, the system oscillates between

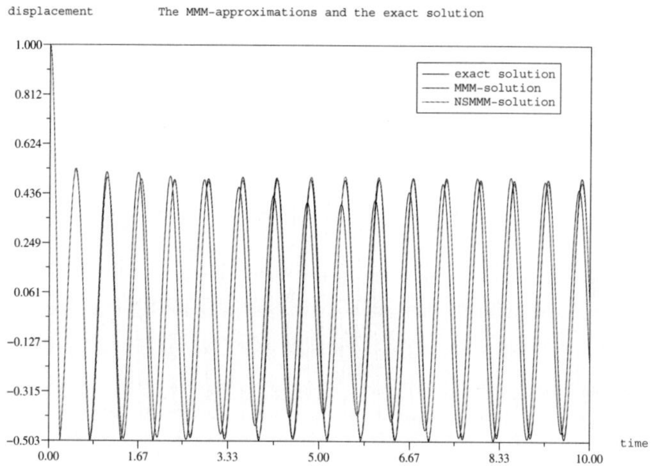

Fig. 22.   Comparison between the MMM-approximations and the exact solution

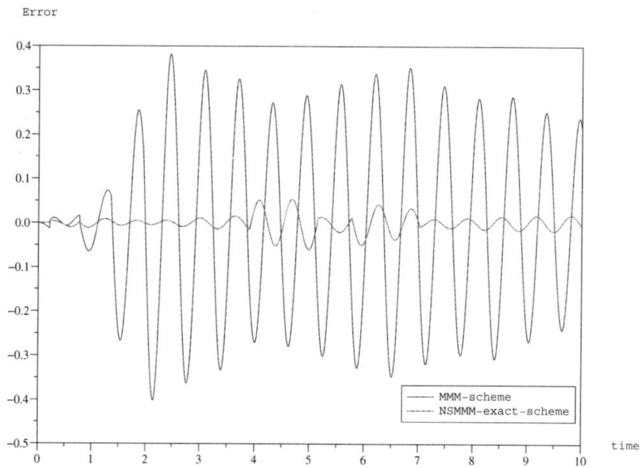

Fig. 23.   Error between the MMM-approximations and the exact solution

     Y. Dumont

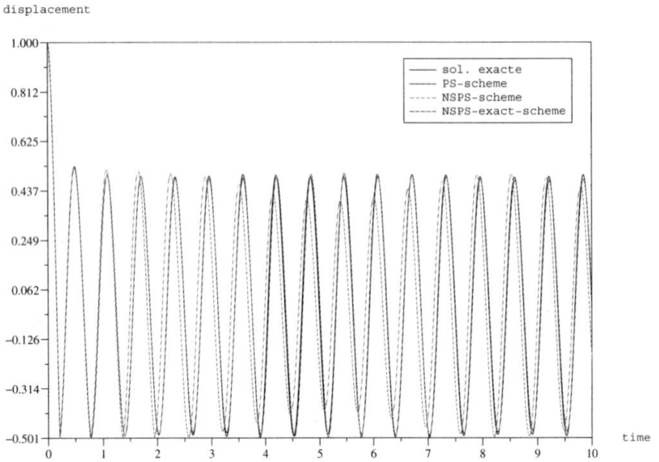

Fig. 24.   Comparison between the PS-approximations and the exact solution

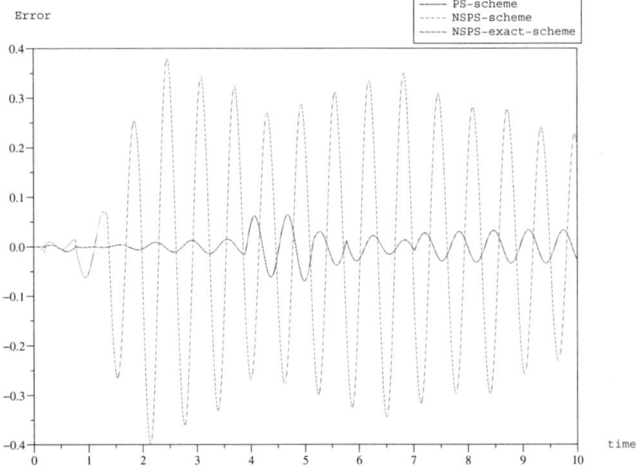

Fig. 25.   Error between the PS-approximations and the exact solution

the fixed point $x = 0$. In Figs. 23 (p. 325) and 25 (p. 326), we show that the NSMMM-exact scheme and the NSPS-exact scheme give rather good approximations of the exact solution, i.e.

$$\max_n |u^n_{NSMMM} - u_{exact}(t_n)| = 5.95 \times 10^{-2},$$

$$\max_n |u^n_{NSPS-exact} - u_{exact}(t_n)| = 5.95 \times 10^{-2}$$

while the standard PS-scheme, the MMM-scheme and even the NSPS-scheme give the worst approximations (see Figs. 22-23-24-25, p. 325-p. 326). The CPU-time for each method is given in the following table

CPU-time	Exact solution	PS-scheme	MMM-scheme	NSPS-scheme	NSMMM--exact-scheme	NSPS--exact-scheme
	82.03	28.59	28.33	28.37	30.55	29.96

Example 29: We now consider a "special" case: ω takes a value (very) near k, i.e.

Δt	k	a	ω	u_0	u_1	u_{min}	e
5×10^{-4}	2	1	2.01	1	0	-0.5	0.5

Here, we show that the NSPS- and NSMMM-exact-schemes give very good result, while the standard PS- and MMM-schemes do not (see Figs. 26-28, p. 328-p. 329). The error between the NSPS- and NSMMM-exact-approximations and the exact solution (see Figs. 27-29, p. 328-p. 329) are

$$\max |u^n_{NSMMM-exact} - u_{exact}(t_n)| = 7.9 \times 10^{-3},$$

$$\max |u^n_{NSPS-exact} - u_{exact}(t_n)| = 8.2 \times 10^{-3}.$$

The CPU-time for each method is given in the following table

CPU-time	Exact solution	PS-scheme	MMM-scheme	NSPS-scheme	NSMMM--exact-scheme	NSPS--exact-scheme
	4357.67	28.59	29.7	28.53	30.54	30.18

Note here the huge CPU-time needed to compute the exact solution.

Example 30: Finally, consider an other value for u_{min}, that is

Δt	k	a	ω	u_0	u_1	u_{min}
5×10^{-4}	1	1	50	0.8	0.1	0.8

Fig. 30 (p. 330) show that the non-standard schemes perform in a similar

Y. Dumont

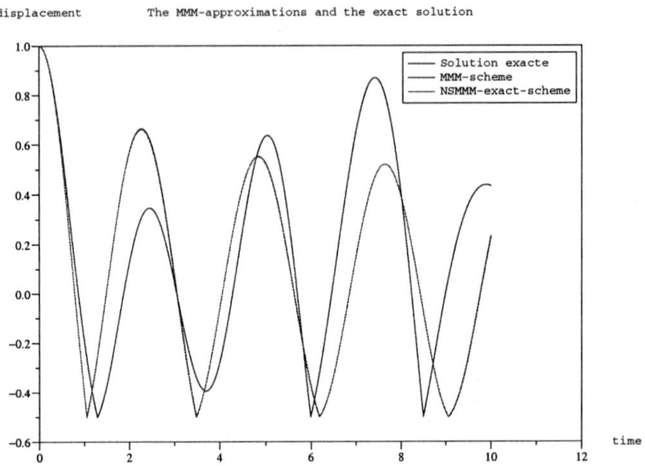

Fig. 26. Comparison between the MMM-approximations and the exact solution ($\omega = 2.01$)

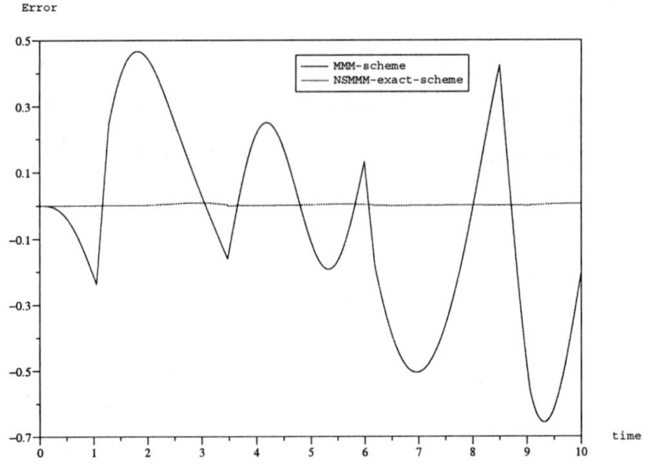

Fig. 27. Error between the MMM-approximations and the exact solution ($\omega = 2.01$)

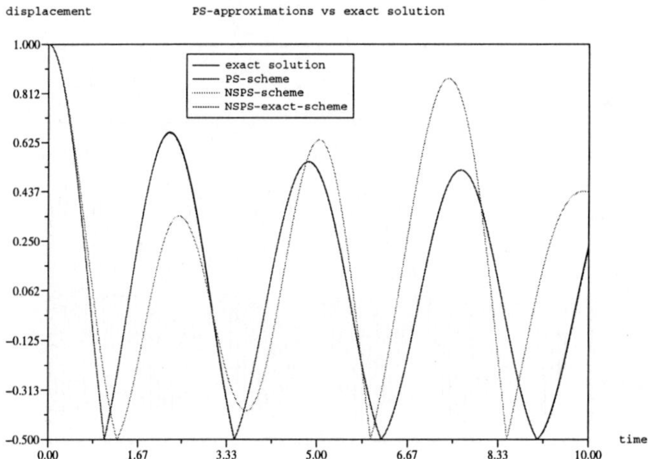

Fig. 28.   PS-approximations and the exact solution

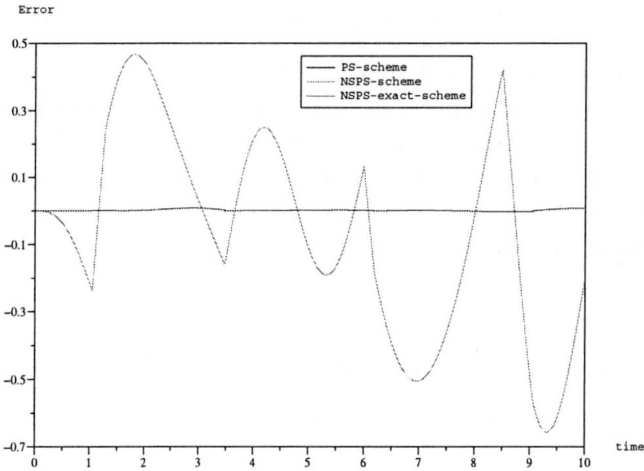

Fig. 29.   Error between the PS-approximations and the exact solution

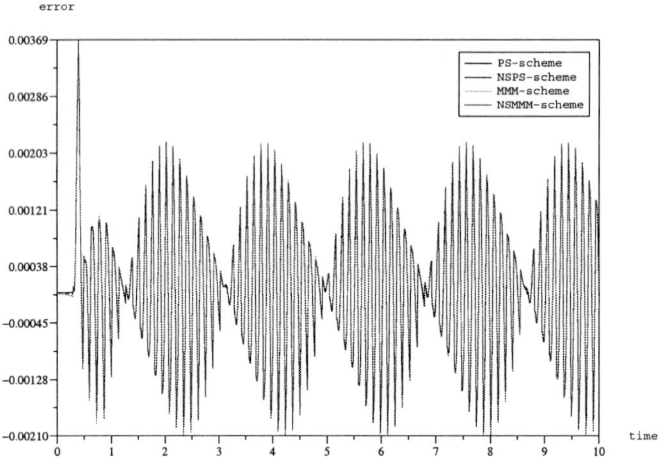

Fig. 30.   Error between all numerical approximations and the exact solution

manner than the standard schemes. This should indicate that the proposed non-standard schemes are not appropriate for these parameters.

All together, the non-standard approach performs better numerical result than the standard approach. In particular, if we consider similar (not equal) values for ω and k, the NSMMM-exact-scheme performs the best results. Further investigations are needed, but, in these cases, the non-standard approach gives very satisfactory results. Finally, the use of exact schemes seems to be very useful, even when impacts occur (see example 29, p. 327).

5.3. *The duffing oscillator with impacts*

Following [15], we consider the duffing oscillator

$$\begin{cases} \ddot{x}(t) + \partial\Psi_K(x) \ni -\omega^2 x(1 + \lambda x^2), \ \lambda > 0, \\ x(0) = x_0, \\ \dot{x}(0) = v_0, \end{cases} \tag{56}$$

where $K = [z_0, +\infty[$. When no impacts occur, we enter equation (40) with $g(z) = 1 + \lambda z$. Between each impact, the duffing oscillator is conservative and thus, the NSPS-scheme associated with (56) is given by Theorem 23 with $G(z) = z(1 + \lambda z/2)$, that is

$$\frac{x^{k+1} - 2x^k + x^{k-1}}{(4/\omega^2)\sin^2(\omega\Delta t/2)} + \partial\Psi_K\left(\frac{x^{k+1} + ex^{k-1}}{1 + e}\right)$$
$$+ \omega^2 x^k\left(1 + \lambda x^k\left(\frac{x^{k+1} + x^{k-1}}{2}\right)\right) \ni 0,$$

For given (x_0, v_0), we are able to deduce x^1 in the absence of collision

$$x^1 = \frac{x_0[\cos(\omega\Delta t) + 4(\lambda/\omega)x_0 v_0 \sin^3(\omega\Delta t/2)] + 2(v_0/\omega)\sin(\omega\Delta t/2)}{1 + 2\lambda(x_0)^2\sin^2(\omega\Delta t/2)} \tag{57}$$

On the contrary, the condition on x_0 in (21_2) means that there is collision, i.e. $x_0 \in \partial K$. In this case, the impact law (13) legitimates the replacement of Eq. (57) with

$$x^1 = \frac{x_0[\cos(\omega\Delta t) - 4e(\lambda/\omega)x_0 v_0 \sin^3(\omega\Delta t/2)] - 2e(v_0/\omega)\sin(\omega\Delta t/2)}{1 + 2\lambda(x_0)^2\sin^2(\omega\Delta t/2)}.$$

Example 31: We apply the non-standard scheme for the parameters in the table below:

ω	λ	x_0	v_0	z_0	Δt
5	15	2	0	-1	0.0001

Fig. 31, p. 332, shows the displacements for different values of e. This figure shows that the system oscillates between the fixed-point $x = 0$, even if the restitution coefficient is not equal to one: the system becomes periodic as displayed by the phase portraits in Fig. 32, p. 332. Moreover, when $0 \le e \le 1$, the total energy of the system converges to a constant. In Figs. 33-34 (p. 333), the total energy computed by the standard method shows some "bad" oscillations while the total energy obtained by the non-standard method is constant. The computation requires 9.93 cpu-time and 9.84 cpu-time for the non-standard and standard algorithms, respectively.

Example 32: We revisit the previous case on changing only the position of the obstacle as indicated in the following table:

ω	λ	x_0	v_0	z_0	Δt
5	15	2	0	-1.7	0.0001

The phase portraits in Figs. 35-36 (p. 334 and p. 335) show a possible chaotic behavior of the Duffing oscillator. There is a clear distinction between the phase portraits computed by the standard and the non-standard

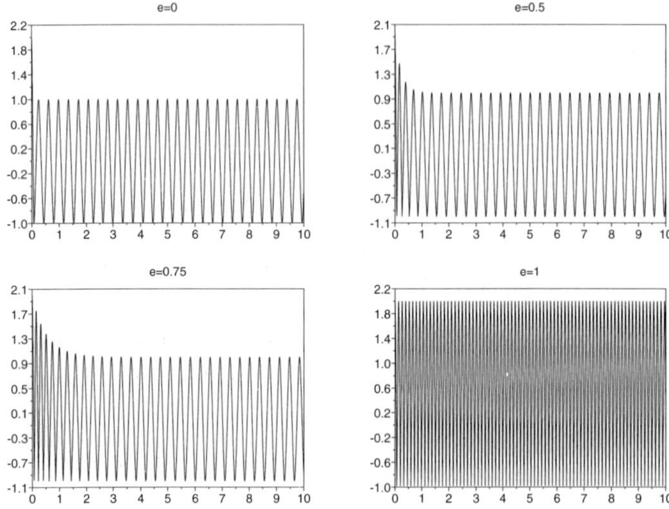

Fig. 31.   Displacements (Duffing) for different values of *e*

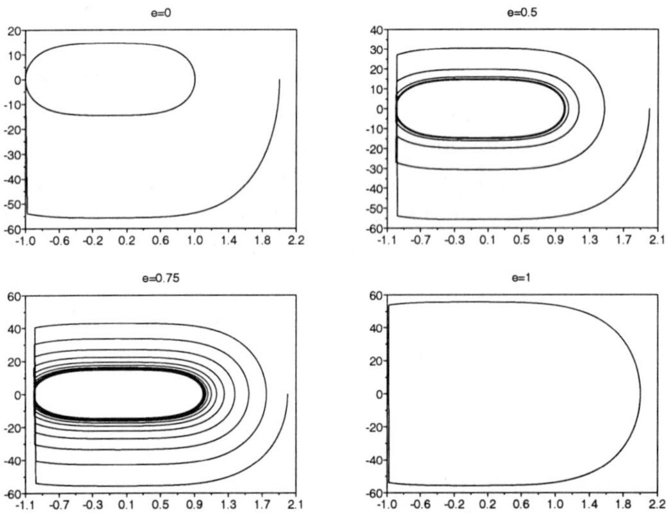

Fig. 32.   Phase portrait (Duffing) by non-standard scheme for different values of *e*

schemes. The chaotic behavior is also confirmed by Figs. 37-38 (p. 335 and p. 336), which show that the total energy computed by both methods (stan-

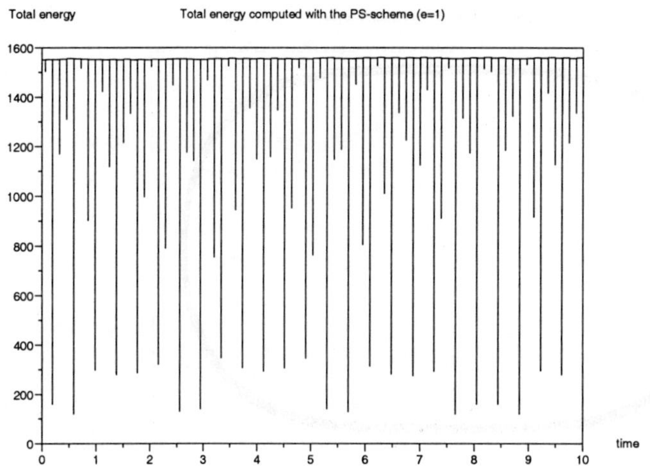

Fig. 33.   Total energy (Duffing) for $e = 1$ by standard scheme

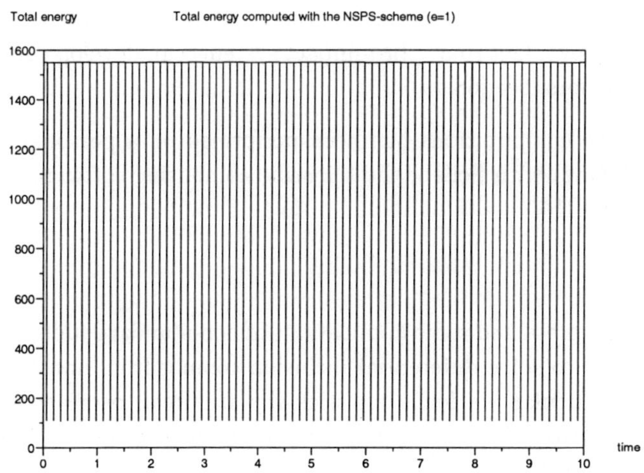

Fig. 34.   Total energy (Duffing) for $e = 1$ by non-standard scheme

dard and non-standard) is not constant: the standard scheme is dissipative whereas the non-standard scheme is not. The non-standard scheme being

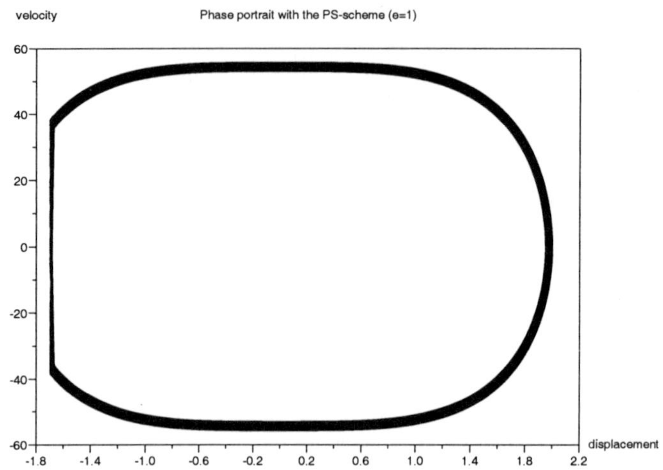

Fig. 35. Chaotic behavior (Duffing) via the phase portrait by standard scheme

more reliable, we believe that the chaotic behavior is due to the mechanical system itself and not to the numerical methods. This statement is consistent with a well known fact that vibro-impact systems can develop complex and chaotic behaviors (see [68,69,71]).

Remark 33: The computations has been performed on a personal computer (PIV, 1.6 Ghz) with Scilab, the scientific computing software developed by ENPC-INRIA.

6. Conclusions

In this chapter, we presented some applications of non-standard finite difference schemes used to solve differential inclusion coming from non-smooth mechanics. In particular we focus on stick-slip frictional oscillators and vibro-impact oscillators. We also introduced non-standard finite difference schemes associated with some numerical schemes used to solve frictional or impacting problems. The application of the non-standard approach in non-smooth mechanics is at a very early stage: as a first step, we follow Micken's rules to construct non-standard approximations. In most of the examples, the non-standard approach improves the numerics. However, in one example, the proposed non-standard schemes perform in a similar manner than the standard schemes. We fill that this is due to the fact that the qualita-

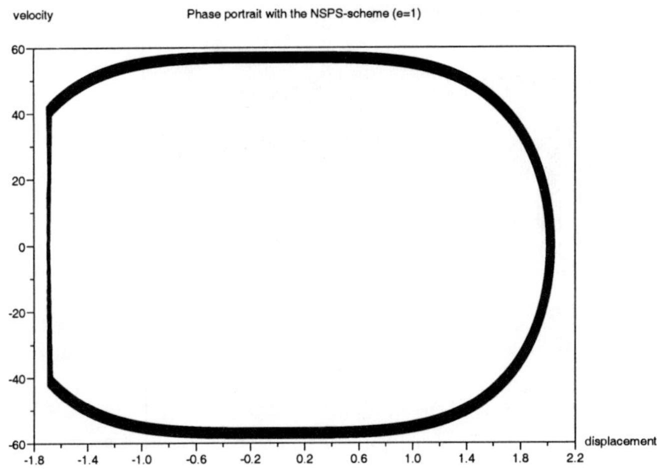

Fig. 36.   Chaotic behavior (Duffing) via the phase portrait by non-standard scheme

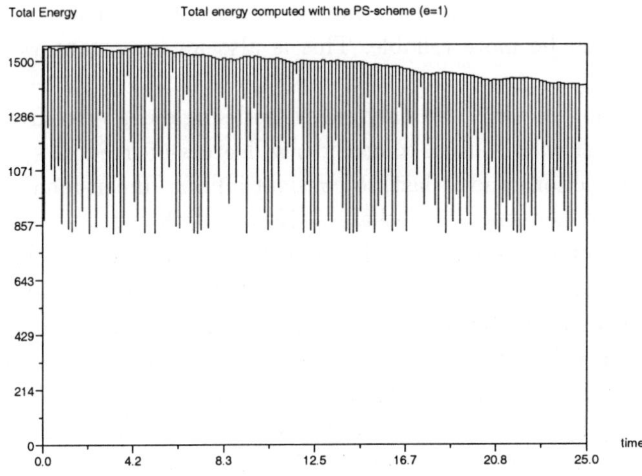

Fig. 37.   Chaotic behavior (Duffing) via the total energy by standard scheme

tive properties of the time-continuous problem (periodic behavior, chaotic behavior, ...) are not suitably incorporated in the proposed non-standard

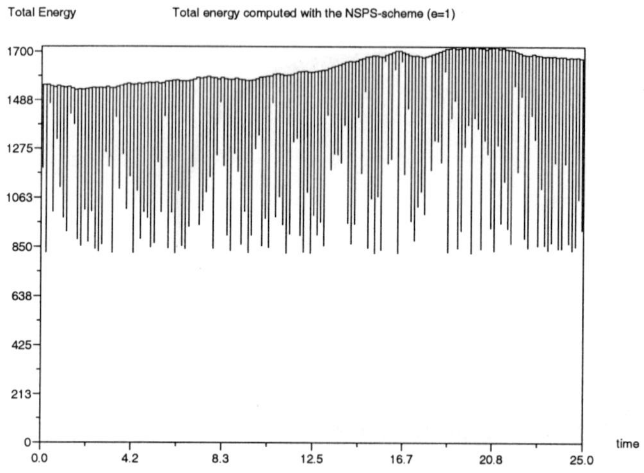

Fig. 38.   Chaotic behavior (Duffing) via the total energy by non-standard scheme

schemes. Therefore, we are investigating the study of new non-standard schemes, which will be more reliable. This is also an important way for further studies.

In the preceding sections, we only considered single-degree-of-freedom systems. In the reality, it is usual to consider n-degrees-of-freedom systems (with impact or friction) and at last, to consider infinite dimensional problems, coming from non-smooth mechanics. Non-smooth continuous problems are very difficult to solve both theoretically and numerically. For instance, the vibrations of a beam that may oscillate between stops. This setting is of great interest in order to understand the dynamic vibrations and the appearance of unwanted noise in mechanical settings. Even if the modeling looks like simple, this problem is very difficult to study theoretically and numerically [73,45,74,35]. Recently, the author and L. Paoli proved the convergence of a full discretization [75], based on finite element for the space discretization and the standard PS-scheme for the time-discretization. Finally, in [76], we present and compare different full discretizations based on the event-driven approach, the normal compliance approach and the time-stepping approach.

Thus, in order to extend the range of applications of the non-standard finite difference method, it is necessary to develop the non-standard approach or exact schemes for n-dimensional system without constraint and then to extend them constrained systems.

Differential inclusions appear in other areas than in mechanics: in economics or in population dynamics, for instance. In population dynamics, recent models take into account certain constraints which may be given by the environment. These constraints include, for instance, individual strategies, food preferences, space constraints [77,78]. These contraints define a so-called "viability" set [79]. It is admited that the only biological plausible solutions are those that belong to the "viability" set at each instant of time. Using the viability theory [79], we have the existence of viable solutions. All together, we deal with differential inclusions and, for numerical simulations, researchers used classical numerical methods. Thus, it would be of some interest to show if it is possible to propose non-standard schemes associated with constrained population dynamics.

Acknowledgements

I would like to thank Jean Lubuma, who introduced me to the nonstandard finite difference method and for many helpful comments that improved this work.

I would like to express my gratitude to Prof. R. Mickens who gave me the opportunity to present some of my works in non-smooth mechanics.

Finally, I would like to dedicate this chapter to my youngest daughter, Solène.

References

1. P.D. Panagiotopoulos, *Inequality problems in mechanics and applications. Convex and nonconvex energy functions.* (Birkhäuser, Massachusetts, 1985)

2. B. Brogliato, *Nonsmooth mechanics: models, dynamics and control, 2nd ed.,* (Springer-Verlag, Communications and Control Engineering Series, London, 2000).

3. D. Goeleven, D. Motreanu, Y. Dumont and M. Rochdi, *Variational and hemivariational inequalities: theory, methods and applications Volume I: Unilateral analysis and unilateral mechanics,* (Kluwer Academic Publishers, Nonconvex Optimization and Its Applications, vol. 69, London, 2003).

4. D. Goeleven and D. Motreanu, *Variational and hemivariational inequalities: theory, methods and applications Volume II: Unilateral problems,* (Kluwer Academic Publishers, Nonconvex Optimization and Its Applications, vol. 70, London, 2003).

5. D. E. Stewart, "Rigid-body dynamics with friction and impact," SIAM Review, **42** (2000), no 1, 3–39.

6. F. Pfeiffer and C. Glocker, *Multibody dynamics with unilateral contacts*, (Wiley: New York, 1996)

7. R.T. Rockafellar, *Convex analysis.* (Princeton University Press, Princeton,1970)

8. F.H. Clarke, *Optimization and nonsmooth analysis* (Wiley-Interscience, New York 1983)

9. J.J. Moreau, "Liaisons unilatérales sans frottement et chocs inélastiques," C.R. Acad. Sc. Paris., **296** (1983), Série II, 1473–1476.

10. J.J. Moreau, *Unilateral contact and dry friction in finite freedom dynamics* in MOREAU, J.J.; PANAGIOTOPOULOS, P.D. (eds.): Nonsmooth Mechanics and Applications, CISM, vol. 302, Springer-Verlag, Wien, New York 1988.

11. R.E. Mickens, *Nonstandard finite difference models of differential equations,* (World Scientific, Singapore 1994)

12. R.E. Mickens, Nonstandard finite difference schemes: a status report. In: Y.C. Teng, E.C. Shang, Y.H. Pao, M.H. Schultz and A.D. Pierce (Editors), *Theoretical and Computational Accoustics 97,* World Scientific, Singapore (1999), pp. 419–428.

13. R.E. Mickens, *Applications of nonstandard finite difference schemes* (World Scientific, Singapore, 2000).

14. R.E. Mickens, "Nonstandard finite difference schemes for differential equations," *J. Diff. Equations Appl.* **8** (2002), 823–847.

15. Y. Dumont and J. Lubuma, Non-standard finite difference methods for vibro-impact problems, Proceedings A of the Royal Society of London, **461** (2005), 1927–1950.

16. J.P. Aubin and A. Celina, *Differential inclusion,* (Grundelheren der mathematischen wissenschaftn 264, Springer-Verlag Berlin Heidelberg New York Tokyo, 1984)

17. K. Deimling, *Multivalued Differential Equations,* (De Gruyter Series in Nonlinear Analysis and Applications, 1992)

18. Dontchev, A. and Lempio, F.: "Difference methods for differential inclusions: A survey," in SIAM Reviews. **4** (1992) 2, 263–294.

19. F. Lempio, Difference methods for differential inclusions, in Lecture notes in Economics and Math. systems, **378** (1992), 236–273.

20. F. Lempio and V. Veliov, "Discrete approximations of Differential Inclusions," Bayreuther Mathematische Shriften, **54** (1998), 149–232.

21. J.J. Moreau, "Dynamique des systèmes à liaisons unilatérales avec frottement sec éventuel," Technical note no. 1-85, LMGC, Université des Sciences et Techniques du Languedoc (1986)

22. M.P.D. Monteiro Marques, *Differential inclusions in non-smooth mechanical problems: shocks and dry friction,* (Progress in nonlinear differential equations, vol. 9, Birkhaüser, 1993)

23. M. Mabrouk, "An unified variational model for the dynamics of perfect unilateral constraints," Eur. J. Mech. A, **17** (1998), 819–842.

24. L. Paoli and M. Schatzman, "A numerical scheme for impact problems I: the one-dimensional case," SIAM J. Numer. Anal. **40** (2002), 702–733.

25. L. Paoli and M. Schatzman, "A numerical scheme for impact problems II:

the multidimensional case," SIAM J. Numer. Anal. **40** (2002), 734–768.

26. M. Frémond, "Rigid bodies collisions", Phys. Lett., Ser. A **204** (1995), 33–41.

27. M. Schatzman, "A class of nonlinear differential equations of second order in time," Nonlinear analysis Anal., Theory, Methods and Applications, **2** (1978), 355–373.

28. A.F. Filippov, *Differential equations with discontinuous righthand side*, (Mathematics and its applications. Kluwer Academic Publishers, Dordrecht, Boston, London 1988).

29. A.F. Filippov, *Differential equations with discontinuous righthand side*, Amer. Math. Soc. Translations, **42**, ser. 2 (1964), 199–231.

30. S. Adly and D. Goeleven, "A stability theory for second-order nonsmooth dynamical systems with applications to friction problems," in J. Math. Pures Appl. **83** (2004),17–51.

31. Y. Dumont, D. Goeleven and M. Rochdi, "Reduction of second order Unilateral singular systems. Applications in mechanics," ZAMM **81** 4 (2001), pp 219–245.

32. G. Buttazzo and D.Percivale, "The bounce problem on n-dimensional Riemannian manifolds," Atti. Accad. Naz. Lincei. Cl. Sci. Fis. Mat. Natur., **70** (1981), 246–250.

33. G. Buttazzo and D. Percivale, "On the approximation of the elastic bounce problem on Riemannian manifolds," J. Diff. Equ. **47** (1983), 227–245.

34. D. Percivale, "Bounce problem with weak hypotheses on regularity," Ann. Mat. Pura. Appl. **4** (1986), 259–274.

35. L. Paoli, *Analyse numérique de vibrations avéc contraintes unilatérales*, Ph.D. Thesis, Université Claude Bernard, Lyon, France, 1993.

36. L. Paoli and M. Schatzman, "Mouvement à un nombre fini de degrés de liberté avec contraintes unilatérales: Cas avec perte d'énergie," *RAIRO, Modélisation Math. Anal. Numér.* **27** (1993), 673–717.

37. P. Ballard, The dynamics of discrete mechanical systems with perfect unilateral constraints, Arch. Rat. Mach. Anal. **154** (2000), 199–274.

38. K. Taubert, Differenzenverfahren für schwingungen mit trockener und zäher reibung und für regelungssytems, Numerische Mathematik, **26** (1976), 379–395.

39. K. Taubert, Converging multisteps methods for initial values problems involving multivalued maps, Computing, **27** (1981), 123–136.

40. H.D. Niepage, Inverse stability and convergence of difference approximations for boundary values problems for differential inclusions, Numr. Funct. Anal. and Optimz., **9** (1987), 1221–1249.

41. A. Kastner-Maresch, Implicit Runge Kutta methods for differential inclusions, Numr. Funct. Anal. and Optimz., **11** (1990-1991), 937–958.

42. D. E. Stewart, A high accuracy method for solving ODEs with discontinuous right-hand side, Numer. Math., **58** (1990), 299–328.

43. D. E. Stewart, Numerical methods for friction problems with multiple contacts, J. Austral. Math. Soc. series B, 37 (1996), no3, 288–308.

44. L. Paoli, "Time discretization of vibro-impact," Philos. Trans. R. Soc. Lond., Ser. A, Math. Phys. Eng. Sci., **359** no. 1789 (2001), 2405–2428.

45. Y. Dumont, *Vibrations of a beam between stops: Numerical simulations and comparison of several numerical schemes*, in Math. Comput. Simul. **60-1-2** (2002), 45–83.

46. L. Paoli and M. Schatzman, "Ill-posedness in vibro-impact and its numerical consequences," in ECCOMAS (Barcelona, 11-14 Sept., 2000).

47. R. Dzonou, A numerical scheme for inelastic contact with non trivial inertia operator, submitted to Proceedings ENOC (2005).

48. W.R. Brace and J.D. Byerlee, Stick-slip as a mechanism for earth-quakes, Science **153** (1996), 990–992.

49. F. Horowitz and A. Ruina, A slip patterns generated in a spatially homogeneous elastic fault model, J. of Geophysical research **94** (1989), 279–298.

50. V.R. Kumar and K.J. Waldron, Force distribution in closed kinematic chains, IEEE J. of Robotics and Automation **4** (1988), 657–664.

51. A.V. Srinivasan, Dynamic friction, in Large Space sturctures: Dynamics and Control (1987). (S.N. Atluri and K.A. Amos editors).

52. B. Feeny, *Chaos and Friction* (Phd Thesis Cornell University, 1990)).

53. B. Feeny and F.C. Moon, Chaos in forced dry-friction oscillator: experiments and numerical modelling, in J. of Sound and Vibration **170** (3) (1994), 303–323.

54. Y. Renard, *Modélisation des instabilités liées au frottement sec des solides élastiques, Aspects théoriques et numériques* Ph. D. Thesis. Grenoble I University 1998.

55. G. Gao and D. Kuhlmann-Wilsdorf, On stick-slip and velocity dependence of friction at low speeds, ASMA journal of Tribology, **112** (1990), 355–360.

56. E. Rabinowicz, Study of the stick-slip process, in Friction and Wear (1949), (Davies, Elsevier, New York), 149–164.

57. E. Rabinowicz, The intrinsic variables affecting the stick-slip process, Proceedings of the Royal Phys. Soc. **71** (1958), 668–675.

58. D. Tabor, Friction: the present state of our understanding, J. of Lubrification Technology, **103** (1981), 169–179.

59. M. Cadivel, D. Goeleven and M. Shillor, Study of a unilateral oscillator with friction, in Math. Comp. Model. **32** (2000), 381–392.

60. Y. Dumont, D. Goeleven, M. Rochdi and M. Shillor, Frictional contact of a nonlinear spring, Math. Model. comp. 31 (2-3) (2000), pp 83–97.

61. Li Chun Bo and D. Pavelescu, "The friction-speed relation and its influence on the critical velocity of stick-slip motion," Wear **82**(1992), 227–289.

62. T. Mc Geer, *Passive dynamic walking*, Int. J. of Robotic Research, vol. **8** (1990), 62–82.

63. A. Kaharaman and R. Singh, *Nonlinear dynamics of a spur gear pair*, Journal of Sound and Vibration, vol. **142** (1990), 49–75.

64. G. Li and M. Paidoussis, Impact phenomena of rotor-casing dynamical systems, Nonlinear dynamics, **5** (1994), 53–70.

65. L. Paoli, M. Schatzman and M. Panet, Vibrations with an obstacle and finite number of degrees of freedom, (EuroMech280, Proceedings of the international symposium on identification of nonlinear mechanical systems from dynamics tests. Balkema: Rotterdam 1992).

66. F. Pfeiffer and W. Prestl, Hammering in diesel engine driveline systems, Nonlinear dynamics, **5** (1994), 477–492.

67. P. C. Tung and S. W. Shaw, *The Dynamics of an Impact Print Hammer*, J. of Vibration, Acoustics, Stress and Reliability in Design, Vol. **110**, pp 193–200, (1988).

68. C. Budd, F. Dux and A. Cliff, The effect of frequency and clearance variations on single-degree-of-freedom impact oscillators, J. of Sound and Vibration, **184** (1995), 475–502.

69. S. Foale and S. Bishop, Bifurcations in impact oscillations, Nonlinear dynamics, **6** (1994), 285–299.

70. M. Frederiksson and A. Nordmark, Bifurcations caused by grazing incidence in many degrees of freedom impact oscillators, Proc. of Royal Soc. of London A, **453** (1997), 1261–1276.

71. F. Peterka and J. Vacik, Transitions to chaotic motion in mechanical systems with impacts, J. of Sound and Vibrations, **154** (1992), 95–115.

72. G. Whiston, Singularities in vibro-impact dynamics, J. of Sound and Vibrations, **152** (1992), 427–160.

73. K. Kuttler, M. Shillor, Vibrations of a beam between two stops, in Dynamics of continuous, discrete and impulsive systems, Series B, Applications and Algorithms, **8** (1998), 93–110.

74. Y. Dumont Some remarks on a vibro-impact scheme, in Numerical Algorithms, **33** (2003), 227–240.

75. Y. Dumont, L. Paoli, Vibrations of a beam between stops. Convergence of a fully discretized approximation, submitted to Numerische Mathematik (2005) (Preprint available:
http://arxiv.org/find/math/1/au:+Dumont_Y/0/1/0/all/0/1).

76. Y. Dumont, L. Paoli, Simulations of beam vibrations between stops: comparison of several numerical approaches, ENOC Conference 7-12 August (2005), Rotterdam (Netherland).

77. A. Kastner-Maresch and V. Krivan, Modelling food preferences and viability constraints, J. of Biological systems, Vol. **3** (2) (1995), 313–322.

78. V. Krivan, Individual behavior and population dynamics, in Lectures Notes on Biomath. and BioInfo. (DATECS Publ. Sofia) (1995), 17–31.

79. J.P. Aubin, *Viability theory*, (Birkhaüser, Boston, 1992).

CHAPTER 8

FINITE DIFFERENCE SCHEMES ON UNBOUNDED DOMAINS

Matthias Ehrhardt

Institut für Mathematik
Technische Universität Berlin
Straße des 17.Juni 136
D–10623 Berlin, Germany
ehrhardt@math.tu-berlin.de

We discuss the nonstandard problem of using the finite difference method to solve numerically a partial differential equation posed on an unbounded domain. We propose different strategies to construct so–called *discrete artificial boundary conditions* (ABCs) and present an efficient implementation by the sum–of–exponential ansatz. The derivation of the ABCs is based on the knowledge of the exact solution, the construction of asymptotic solutions or the usage of a continued fraction expansion to a second–order difference equation. Our approach is explained by means of three different types of partial differential equations arising in option pricing, in quantum mechanics and in (underwater) acoustics. Finally, we conclude with an illustrating numerical example from underwater acoustics showing the superiority of our new approach.

1. Introduction

It is a nonstandard task to solve numerically a partial differential equation posed on an unbounded domain. Usually finite differences are used to discretize the equation and *artificial boundary conditions* (ABCs) are introduced in order to confine the computational domain. If the solution on the computational domain coincides with the exact solution on the unbounded domain (restricted to the finite domain), one refers to these ABCs as *transparent boundary conditions* (TBCs).

However, ad-hoc discretizations of an *analytic TBC* may induce numerical reflections at this artificial boundary and also may destroy the stability properties of the underlying finite difference method. To overcome both problems so–called *discrete ABCs* (or *discrete TBCs*) are derived directly from the fully discretized problem on the unbounded domain. These discrete ABCs/TBCs are already adapted to the inner scheme and therefore the numerical stability is often better–behaved than for a discretized differential TBC. An additional motivation for this discrete approach arises from the fact that the numerical scheme often needs more boundary conditions than the analytical problem can provide (especially hyperbolic equations, systems of equations and high–order schemes).

In the literature the discrete approach did not gain much attention yet. The first *discrete derivation of artificial boundary conditions* was presented in [1, Section 5]. This discrete approach was also used by Schmidt and Deuflhard [2] for the Schrödinger equation, in [3], [4], [5] for linear hyperbolic systems and in [6] for the wave equation in one dimension, also with error estimates for the reflected part. In [4] a discrete (nonlocal) solution operator for general difference schemes (strictly hyperbolic systems, with constant coefficients in 1D) is constructed. Lill generalized in [7] the approach of Engquist and Majda [1] to boundary conditions for a convection–diffusion equation and drops the standard assumption that the initial data is compactly supported inside the computational domain.

In this work we will propose different strategies to construct these discrete ABCs by using the Z–transformation and exact or asymptotic solutions to the second–order linear difference equation:

$$\Delta^2 y_j - p(j)\, y_j = 0, \quad j \in \mathbb{Z}. \tag{1}$$

Here, $\Delta^2 y_j = y_{j+1} - 2\, y_j + y_{j-1}$ denotes the standard second–order difference operator.

We consider (1) with three different discrete potential terms:

A) constant coefficients: $p(j) = d, \quad d \in \mathbb{C}$,
B) Coulomb–type term: $p(j) = d + c/j, \quad c, d \in \mathbb{C}$,
C) affin–linear term: $p(j) = d + c\,j, \quad c, d \in \mathbb{C}$.

Equation A) can easily be solved explicitly. For the other two model equations it is not clear a-priori whether one can find *explicit solutions*. However, it is a standard task [8, Chapter 7] to determine *asymptotic solutions* if the difference equation is of *Poincaré type*, i.e. the coefficient $p(j)$ in equation (1) must approach a constant value as $j \to \infty$. This is the case for the equation B) with the Coulomb–type term, but the difference equation of case C) (a general discrete Airy equation) does not satisfy this condition.

In §2 we will present the fields of applications of these three cases, namely the Black–Scholes equation for American options, a time–dependent Schrödinger equation with a Coulomb–like potential and with a linearly varying potential. Finite difference schemes are introduced in §3 to solve numerically these partial differential equations. For the derivation of the ABCs we apply the Z–transformation technique and need to solve in the sequel difference equations of the form (1). Afterwards, in §4, §5 and §6, we outline general procedures to construct ABCs and present different techniques to obtain exact and asymptotic solutions of these three model equations.

Since the discrete TBC includes a convolution with respect to time with a weakly decaying kernel, its numerical evaluation becomes very costly for large–time simulations. As a remedy we construct in §7 an *approximate discrete ABC* with a kernel having the form of a finite sum–of–exponentials, which can be evaluated by a very efficient recursion formula.

The Schrödinger equation with a linear varying potential term arises in (underwater) acoustics and we will present at the end of this chapter a concrete numerical example in §8 which will show the superiority of the new (approximated) discrete TBC.

2. Fields of Applications

2.1. *The Black–Scholes Equation for American Options*

The famous Black–Scholes equation is an effective model for option pricing. It was named after the pioneers Black, Scholes and Merton who suggested it 1973 [9], [10]. A derivation of the Black–Scholes equation can be found in [11] and for a more complete discussion in the context of discrete TBCs we refer the interested reader to [12].

An option is the right to buy ('call option') or to sell ('put option') an asset (typically a stock or a parcel of shares of a company) for a price E by the date T. European options can only be exercised at the expiration date T. For *American options* exercise is permitted at any time until the expiry date. While for European options the Black–Scholes equation results (after a standard transformation) in a boundary value problem, for American options it results in a free boundary problem (FBP) for the heat equation. In general, closed–form solutions do not exist (especially for American options) and the solution has to be computed numerically. The standard approach for solving the Black–Scholes equation for American options consists in transforming the original equation to a heat equation posed on a semi-unbounded domain with a free boundary [11].

The Black–Scholes Equation. Here we consider an *American call.* V denotes the value of an option and depends on the current value S of the underlying asset, and time t: $V = V(S, t)$.

$$\frac{\partial V}{\partial t} + \frac{\sigma^2}{2} S^2 \frac{\partial^2 V}{\partial S^2} + (r - D_0)S \frac{\partial V}{\partial S} - rV = 0, \quad 0 < S < S_f(t), \tag{2a}$$

$0 \le t < T$, where σ is the volatility of the asset price, r is the risk-free interest rate and T is the expiry date. We assume that dividends are paid with a continuous yields of constant level D_0. $S_f(t)$ denotes the free boundary ('early exercise boundary') separating the *holding region* $(S < S_f(t))$ and the *exercise region* $(S > S_f(t))$.

The final condition (*'payoff condition'*) at the expiry $t = T$ is

$$V(S, T) = (S - E)^+, \quad 0 \le S < S_f(T), \tag{2b}$$

with the notation $f^+ = \max(f, 0)$, $E > 0$ denotes the *exercise price* or 'strike', and $S_f(T) = \max(E, rE/D_0)$.

The asset–price boundary conditions at $S = 0$, and $S = S_f(t)$ are

$$V(0, t) = 0, \quad 0 \le t \le T, \tag{2c}$$

$$V(S_f(t), t) = (S_f(t) - E)^+, \quad \frac{\partial V}{\partial S}(S_f(t), t) = 1, \quad 0 \le t \le T, \tag{2d}$$

i.e. at $S = 0$ the option is worthless. Note that we need two conditions at the free boundary $S = S_f(t)$. One condition is necessary for the solution of (2a) and the other is needed for determining the position of the *free boundary* $S = S_f(t)$ itself. At $S = S_f(t)$ one requires that $V(S, t)$ touches the payoff function tangentially.

The transformation to the heat equation. In the sequel we shall show how to transform (2a) into a diffusion equation (cf. [11, § 5.4]). First it is convenient to transform (2a)–(2d) to a forward in time equation by the change of variable $t = T - 2\tau/\sigma^2$. The new time variable τ stands for the *remaining life time* of the option (up to the scaling by $\sigma^2/2$). We denote the new variables by:

$$\widetilde{V}(S, \tau) = V(S, t) = V\left(S, T - \frac{2\tau}{\sigma^2}\right), \quad \widetilde{S}_f(\tau) = S_f\left(T - \frac{2\tau}{\sigma^2}\right),$$

$$\tilde{r} = \frac{2}{\sigma^2} r, \quad \widetilde{D}_0 = \frac{2}{\sigma^2} D_0, \quad \widetilde{T} = \frac{\sigma^2}{2} T.$$

The forward equation then reads:

$$\frac{\partial \widetilde{V}}{\partial \tau} = S^2 \frac{\partial^2 \widetilde{V}}{\partial S^2} + (\tilde{r} - \widetilde{D}_0)S \frac{\partial \widetilde{V}}{\partial S} - \tilde{r}\widetilde{V}, \quad 0 < S < \widetilde{S}_f(\tau), \tag{3a}$$

$0 \leq \tau < \widetilde{T}$, with the initial condition

$$\widetilde{V}(S,0) = (S - E)^+, \qquad 0 \leq S < \widetilde{S}_f(0) = S_0, \tag{3b}$$

and the boundary conditions

$$\lim_{S \to 0} \widetilde{V}(S,\tau) = 0, \ 0 \leq \tau \leq \widetilde{T}, \tag{3c}$$

$$\widetilde{V}(\widetilde{S}_f(\tau),\tau) = (\widetilde{S}_f(\tau) - E)^+, \ \frac{\partial \widetilde{V}}{\partial S}(\widetilde{S}_f(\tau),\tau) = 1, \ 0 \leq \tau \leq \widetilde{T}. \tag{3d}$$

The right hand side of (3a) is a well–known *Euler's differential equation* and therefore it is standard practice to transform (3a) to the heat equation. To do so, we let

$$\alpha = -\frac{1}{2}(\tilde{r} - \widetilde{D}_0 - 1), \quad \beta = -\frac{1}{4}(\tilde{r} - \widetilde{D}_0 + 1)^2 - \tilde{r},$$

and use the change of variables

$$S = Ee^x, \quad \widetilde{V}(S,\tau) = Ee^{\alpha x + \beta \tau} v(x,\tau).$$

Then problem (3a)–(3d) is equivalent to the free boundary problem for the heat equation:

$$\frac{\partial v}{\partial \tau} = \frac{\partial^2 v}{\partial x^2}, \quad -\infty < x < x_f(\tau), \quad 0 \leq \tau < \widetilde{T}, \tag{4a}$$

with the initial condition

$$v(x,0) = \left(e^{\frac{1}{2}(\tilde{r} - \widetilde{D}_0 + 1)x} - e^{\frac{1}{2}(\tilde{r} - \widetilde{D}_0 - 1)x} \right)^+, \quad x < x_f(0), \tag{4b}$$

and the boundary conditions

$$\lim_{x \to -\infty} v(x,\tau) = 0, \quad 0 \leq \tau \leq \widetilde{T}, \tag{4c}$$

$$v(x_f(\tau),\tau) = g(x_f(\tau),\tau), \quad 0 \leq \tau \leq \widetilde{T}, \tag{4d}$$

$$e^{(\alpha-1)x - \beta\tau}\left(\alpha v(x_f(\tau),\tau) + \frac{\partial v(x_f(\tau),\tau)}{\partial x} \right) = 1, \ 0 \leq \tau \leq \widetilde{T}, \tag{4e}$$

where

$$g(x,\tau) = e^{-\alpha x - \beta \tau}(e^x - 1)^+.$$

It is known that the free boundary given by $x_f(\tau) = \ln(\widetilde{S}_f(\tau)/E)$ has the property $x_f(\tau) > 0$ for $0 \leq \tau \leq \widetilde{T}$.

2.2. *The Schrödinger–Poisson System*

The second example arises in quantum mechanics and details concerning the computation of solutions on unbounded domains can be found in [13]. In many applications one wants to calculate the evolution of an ensemble of particles over long time. These computations include the solution of the single particle Schrödinger equation obtained from a mean field approximation using Coulomb potentials [14]. The transient Schrödinger–Poisson problem describes the time evolution of the wave function ψ under the force of the self–consistent potential V caused by the charged electrons. It is an appropriate model for semiconductor heterostructures (cf. [14] and the references therein).

The Schrödinger–Poisson system. The transient *Schrödinger–Poisson system* (SPS) associated with a single particle system in vacuum reads for the complex–valued wave function $\psi(x,t)$ and the electrostatic potential $V(x,t)$:

$$i\hbar\partial_t\psi = -\frac{\hbar^2}{2m}\Delta_x\psi + V\psi, \qquad x \in \mathbb{R}^3, \quad t > 0, \tag{5a}$$

$$\Delta_x V = -\gamma n, \qquad x \in \mathbb{R}^3, \quad t > 0, \tag{5b}$$

where $n = |\psi(x,t)|^2$ denotes the particle density for a pure quantum state and $\gamma > 0$ (repulsive case) or $\gamma < 0$ (attractive case) depending on the considered type of Coulomb force. Here $\hbar$ denotes the Planck constant and m is the particle mass. Throughout this application we will be interested in the attractive case. Equations (5) are supplied with some initial data $\psi(x,0) = \psi^I(x)$ and the decay conditions

$$\lim_{|x|\to\infty} \psi(x,t) = 0, \qquad \lim_{|x|\to\infty} V(x,t) = 0.$$

The spherically symmetric Schrödinger–Poisson system. Since we want to keep the numerical effort to a minimum we only consider the case of a spherically symmetric *initial condition*: $\psi(x,0) = \psi^I(r)$. It can be shown that $\psi(x,t)$ is invariant under rotations and therefore a radial function at any time. For convenience we introduce the *reduced wave function* $u(r,t)$ by

$$\psi(x,t) = \frac{1}{\sqrt{4\pi}}\frac{u(r,t)}{r}, \tag{6}$$

and define the *effective charge* $\phi(r,t) = rV(x,t)$. The *SPS* reduces then to

$$ i\hbar \partial_t u = -\frac{\hbar^2}{2m} \partial_r^2 u + \frac{\phi}{r} u, \qquad r > 0, \quad t > 0, \tag{7a} $$

$$ \partial_r^2 \phi = -\frac{\gamma}{4\pi} \frac{|u|^2}{r}, \qquad r > 0, \quad t > 0, \tag{7b} $$

together with the *homogeneous Dirichlet conditions at the origin*

$$ u(0, t) = 0, \qquad \phi(0, t) = 0, $$

and the *decay conditions*

$$ \lim_{r \to \infty} u(r, t) = 0, \qquad \lim_{r \to \infty} \phi(r, t) = \frac{\gamma}{4\pi}. $$

2.3. *The Standard "Parabolic Equation"*

The third example, a Schrödinger equation with a linear varying potential can be used for *standard "parabolic equation"* (SPE) [15] simulations in (underwater) acoustics and for radiowave propagation in the troposphere. Details about this example can be found in [16]. Here we focus on the application to underwater acoustics.

The standard parabolic equation in underwater acoustics. A standard task in oceanography is to calculate the acoustic pressure $p(z, r)$ emerging from a time–harmonic point source located in the water at $(z_s, 0)$. Here, $r > 0$ denotes the radial range variable and $0 < z < z_b$ the depth variable (assuming a cylindrical geometry). The water surface is at $z = 0$, and the (horizontal) sea bottom at $z = z_b$. We denote the local sound speed by $c(z, r)$, the density by $\rho(z, r)$, and the attenuation by $\alpha(z, r) \geq 0$. The *complex refractive index* is given by $N(z, r) = c_0/c(z, r) + i\alpha(z, r)/k_0$ with a reference sound speed c_0 and the reference wave number $k_0 = 2\pi f/c_0$, where f denotes the frequency of the emitted sound.

The *SPE* in cylindrical coordinates (z, r) reads:

$$ 2ik_0\psi_r(z, r) + \rho\, \partial_z(\rho^{-1}\partial_z)\psi(z, r) + k_0^2(N^2(z, r) - 1)\,\psi(z, r) = 0, \tag{8} $$

where ψ denotes the (complex valued) *outgoing acoustic field*

$$ \psi(z, r) = \sqrt{k_0 r}\, p(z, r)\, e^{-ik_0 r}, \tag{9} $$

in the far field approximation ($k_0 r \gg 1$). This *Schrödinger equation* (8) is an evolution equation in r and a reasonable description of waves with a propagation direction within about $15°$ of the horizontal.

Here, the physical problem is posed on the unbounded z–interval $(0, \infty)$ and one wishes to restrict the computational domain in the z–direction by introducing an artificial boundary at the water–bottom interface ($z = z_b$), where the wave propagation in water has to be coupled to the wave propagation in the the bottom. At the water surface one usually employs a Dirichlet ("pressure release") BC: $\psi(0, r) = 0$.

Since the density is typically discontinuous at the water–bottom interface ($z = z_b$), one requires continuity of the pressure and the normal particle velocity:

$$\psi(z_b-, r) = \psi(z_b+, r), \tag{10a}$$

$$\frac{\psi_z(z_b-, r)}{\rho_w} = \frac{\psi_z(z_b+, r)}{\rho_b}, \tag{10b}$$

where $\rho_w = \rho(z_b-, r)$ is the water density just above the bottom and ρ_b denotes the constant density of the bottom. This situation is sketched in Fig. 1.

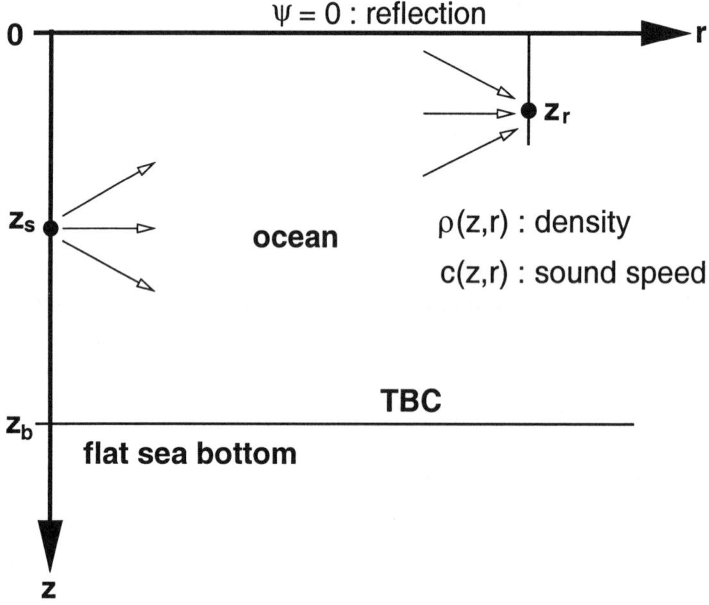

Fig. 1.   Underwater sound propagation in cylindrical coordinates.

In this application we are especially interested in the case of a *linear squared refractive index* in the bottom region. For most underwater acoustics (and also radiowave propagation) problems the squared refractive index in the exterior domain increases with z. However, the usual TBC (see e.g.

[17]) was derived for a homogeneous medium (i.e. all physical parameters are constant for $z > z_b$). This TBC is not matched to the behaviour of the refractive index and spurious reflections will occur. Instead we will derive a TBC that matches the squared refractive index gradient at $z = z_b$. We denote the physical parameters in the bottom with the subscript b and assume that the squared refractive index N_b below $z = z_b$ can be written as

$$N_b^2(z,r) = 1 + \beta + \mu(z - z_b), \quad z > z_b, \tag{11}$$

with real parameters β and $\mu \neq 0$, i.e. no attenuation in the bottom: $\alpha_b = 0$. All other physical parameters are assumed to be constant in the bottom. Here, the slope $\mu > 0$ corresponds to a downward-refracting bottom (energy loss) and $\mu < 0$ represents the upward-refracting case, i.e. energy is returned from the bottom.

3. The Finite Difference Equations

In this section we derive the *discrete ABCs/TBCs* of the fully discretized problems based on a finite difference discretization. This strategy helps to minimize any numerical reflections at the boundary since the discrete ABC/TBC is matched to the finite difference scheme in the interior domain. Moreover, the stability of the resulting scheme is often better behaved (compared to the discretized analytic TBC).

While a uniform spatial grid is necessary in the exterior domain, the interior grid may be nonuniform in space. To derive the discrete ABCs/TBCs we make the *basic assumption* that the initial data is supported inside the computational domain. We note that a strategy to overcome this restriction could be found in [18].

The basic tool for the derivation of the discrete ABCs/TBCs is the *Z–transformation* of a series $\{f_j^{(n)}\}_{n\in\mathbb{N}_0}$ (with j fixed):

$$\mathcal{Z}\{f_j^{(n)}\} = \hat{f}_j(z) := \sum_{n=0}^{\infty} f_j^{(n)} z^{-n}, \quad z \in \mathbb{C}, \quad |z| > 1.$$

The Z–transformation is the discrete analogue of the Laplace–transformation and a collection of the most important properties is given in the Appendix.

3.1. *The Black–Scholes Equation for American Options*

With the uniform grid points $x_j = a + j\Delta x$, $\tau_n = n\Delta\tau$ and the approximation $v_j^{(n)} \approx v(x_j, \tau_n)$ the *Crank-Nicolson scheme* for solving the heat equation (4a) is

$$v_j^{(n+1)} - v_j^{(n)} = \rho \left(v_{j+1}^{(n+\frac{1}{2})} - 2v_j^{(n+\frac{1}{2})} + v_{j-1}^{(n+\frac{1}{2})} \right), \tag{12}$$

with the time averaging $v_j^{(n+\frac{1}{2})} = (v_j^{(n+1)} + v_j^{(n)})/2$ and the parabolic mesh ratio $\rho = \Delta\tau/(\Delta x)^2$.

We obtain the discrete TBC by solving the *discrete exterior problem*, i.e. (12) for $j \leq 1$. To do so, we apply the Z–transformation to solve (12) for $j \leq 1$ explicitly. We assume for the initial data, $v_j^{(0)} = 0$, $j \leq 2$ and obtain the *transformed exterior scheme*

$$\Delta^2 \hat{v}_j(z) - \frac{2}{\rho} \frac{z-1}{z+1} \hat{v}_j(z) = 0, \quad j \leq 1. \tag{13}$$

Obviously, (13) is a difference equation of the form (1) Case A).

3.2. *The Schrödinger–Poisson System*

For simplicity we use the uniform grid points

$$u_j^{(n)} \sim u(r_j, t_n), \quad \phi_j^{(n)} \sim \phi(r_j, t_n), \quad r_j = j\Delta r, \ t_n = n\Delta t,$$

with $0 \leq j \leq J$, $n \geq 0$. The *discretized SPS* (7) reads

$$i\hbar D_t^+ u_j^{(n)} = -\frac{\hbar^2}{2m} D_r^2 u_j^{(n+\frac{1}{2})} + \frac{\phi_j^{(n+\frac{1}{2})}}{r_j} u_j^{(n+\frac{1}{2})}, \quad j \geq 1, \tag{14a}$$

$$D_r^2 \phi_j^{(n+1)} = -\frac{\gamma}{4\pi} \frac{\left| u_j^{(n+1)} \right|^2}{r_j}, \quad j \geq 1, \tag{14b}$$

together with the discrete boundary conditions

$$u_0^{(n)} = 0, \qquad \lim_{j\to\infty} u_j^{(n)} = 0, \qquad \phi_0^{(n)} = 0, \qquad \phi_J^{(n)} = \frac{\gamma}{4\pi}. \tag{14c}$$

In (14) we have used the standard abbreviations for the forward, and second–order difference quotient:

$$D_t^+ u_j^{(n)} = \frac{u_j^{(n+1)} - u_j^{(n)}}{\Delta t}, \qquad D_r^2 u_j^{(n)} = \frac{u_{j+1}^{(n)} - 2u_j^{(n)} + u_{j-1}^{(n)}}{(\Delta r)^2},$$

and the time averaging $u_j^{(n+\frac{1}{2})} = (u_j^{(n+1)} + u_j^{(n)})/2$.

On the unbounded domain $j \geq 0$ the nonlinear method (14) conserves the *discrete mass* and *discrete total energy* (cf. [13]). In order to obtain a mass and energy conserving *linear method* we now proceed to present a predictor–corrector scheme approximating the nonlinear Crank–Nicolson scheme (14). It only requires the solution of linear equations at each step and is of the same order as the nonlinear scheme (14). One step of this scheme will be of the form

$$(u_j^{(n)}, \phi_j^{(n)}) \rightarrow u_j^{(n,1)} \rightarrow \phi_j^{(n,1)} \rightarrow u_j^{(n,2)} \rightarrow \phi_j^{(n,2)} \rightarrow (u_j^{(n+1)}, \phi_j^{(n+1)}),$$

where $u_j^{(n,1)}$, $\phi_j^{(n,1)}$, $u_j^{(n,2)}$, $\phi_j^{(n,2)}$ denote intermediate values. For brevity we define the *difference operators* $D_{t,k}^+ u_j^{(n)} = (u_j^{(n,k)} - u_j^{(n)})/\Delta t$, and the *time averaging* $S_{t,k} u_j^{(n)} = (u_j^{(n,k)} + u_j^{(n)})/2$, $k = 1, 2$.

Given $u_j^{(n)}$, the *predictor step* to compute $u_j^{(n,1)}$, $\phi_j^{(n,1)}$ is then defined as

$$i\hbar D_{t,1}^+ u_j^{(n)} = -\frac{\hbar^2}{2m} D_r^2 S_{t,1} u_j^{(n)} + \frac{\phi_j^{(n)}}{r_j} S_{t,1} u_j^{(n)}, \quad j \geq 1, \tag{15a}$$

$$D_r^2 \phi_j^{(n,1)} = -\frac{\gamma}{4\pi} \frac{\left| u_j^{(n,1)} \right|^2}{r_j}, \quad j \geq 1. \tag{15b}$$

The standard *corrector step* for determining $u_j^{(n,2)}$, $\phi_j^{(n,2)}$ is

$$i\hbar D_{t,2}^+ u_j^{(n)} = -\frac{\hbar^2}{2m} D_r^2 S_{t,2} u_j^{(n)} + \frac{S_{t,1} \phi_j^{(n)}}{r_j} S_{t,2} u_j^{(n)}, \quad j \geq 1, \tag{16a}$$

$$D_r^2 \phi_j^{(n,2)} = -\frac{\gamma}{4\pi} \frac{\left| u_j^{(n,2)} \right|^2}{r_j}, \quad j \geq 1. \tag{16b}$$

It is easily verified that the scheme (15)–(16) is second order consistent in time.

The Modulation Strategy. This predictor–corrector approximation to the Crank–Nicolson scheme preserves mass, but exhibits a spurious gain / loss of the total energy which is of order Δt^3 at each time step. Ringhofer and Soler [19] remedied this situation by *modulating the phase* of the second stage $u_j^{(n,2)}$ of the scheme by setting

$$u_j^{(n+1)} = u_j^{(n,2)} \exp(i\Delta t^3 \omega g_j), \qquad \phi_j^{(n+1)} = \phi_j^{(n,2)}, \quad j \geq 1, \tag{17}$$

where ω is a real parameter and $g_j = g(r_j)$ denotes an appropriate chosen real valued function bounded uniformly for $j \in \mathbb{N}$. Obviously, this correction does not change the discrete ℓ^2-norm of $u_j^{(k,2)}$, and therefore the mass conservation property is retained by this phase correction. Also, adding an order $O(\Delta t^3)$ correction at each step does not destroy the overall second order accuracy of the method. For the detailed choice of the modulation parameters ω and g_j we refer to [13].

Since the problem (7a) is posed on an unbounded domain we have to introduce an artificial boundary at $j = J$ for the numerical solution. Here we use the approach of a discrete TBC first assuming a constant potential term: $V_j^{(n)} = \phi_j^{(n)}/r_j = \text{const}$ for $j \geq J$ (exterior domain). Afterwards we extend these calculations to the case of a Coulomb–type potential, i.e. $\phi_j^{(n)}/r_j \sim \text{const}/r_j$, $j \to \infty$. It will turn out that the discrete TBC for zero potential is the lowest order approximation to the discrete TBC for the Coulomb–type potential. To derive the discrete TBC we assume $u_j^{(0)} = 0$, $j \geq J - 1$, and rewrite the scheme (14a) in the form:

$$-i\rho(u_j^{(n+1)} - u_j^{(n)}) = \Delta^2(u_j^{(n+1)} + u_j^{(n)}) + w\frac{\phi_j^{(n+\frac{1}{2})}}{r_j}(u_j^{(n+1)} + u_j^{(n)}), \quad (18)$$

with the mesh ratio ρ and the abbreviation w given by

$$\rho = \frac{4m}{\hbar}\frac{\Delta r^2}{\Delta t}, \qquad w = -\frac{2m}{\hbar^2}\Delta r^2.$$

i) Constant potential term outside the computational domain. We start with assuming that $V_j^{(n)} = \phi_j^{(n)}/r_j = \text{const}$ for $j \geq J$ (exterior domain). The Z–transformed finite difference scheme (18) for $j \geq J$ reads

$$\Delta^2 \hat{u}_j(z) + i\rho\Big[\frac{z-1}{z+1} + i\kappa\Big]\hat{u}_j(z) = 0, \quad \kappa = \frac{\Delta t}{2}\frac{V_R}{\hbar}, \quad (19)$$

i.e. (19) represents a difference equation of the form (1) case A).

ii) Coulomb–type potential term outside the computational domain. We now assume that $\phi_j^{(n)} = \phi_j^{(n,1)} = \phi_j^{(n,2)} = \phi_\infty$, $j \geq J$ and write the discrete Z–transformed exterior problem (18) as

$$\Delta^2 \hat{u}_j(z) + \Big[i\rho\frac{z-1}{z+1} + w\frac{\phi_\infty}{j\Delta r}\Big]\hat{u}_j(z) = 0, \quad j \geq J. \quad (20a)$$

Clearly, (20a) has the form (1) case B) with

$$d = -i\rho\frac{z-1}{z+1}, \qquad c = -w\frac{\phi_\infty}{\Delta r}. \quad (20b)$$

3.3. *The Standard "Parabolic Equation"*

In order to solve the Schrödinger equation (8) numerically we use a *Crank-Nicolson* finite difference scheme which is of second order (in Δz and Δr)

and unconditionally stable. We choose the uniform grid $z_j = jh$, $r_n = nk$ with $h = \Delta z$, $k = \Delta r$ and the approximations $\psi_j^{(n)} \approx \psi(z_j, r_n)$, $\rho_j \approx \rho(z_j)$. The discretized SPE (8) then reads:

$$-iR(\psi_j^{(n+1)} - \psi_j^{(n)}) =$$

$$\rho_j \Delta_z^0 (\rho_j^{-1} \Delta_z^0)(\psi_j^{(n+1)} + \psi_j^{(n)}) + w((N^2)_j^{(n)} - 1)(\psi_j^{(n+1)} + \psi_j^{(n)}), \quad (21)$$

with $\Delta_z^0 \psi_j^{(n)} = \psi_{j+1/2}^{(n)} - \psi_{j-1/2}^{(n)}$, the ratio $R = 4k_0 h^2/k$ and $w = k_0^2 h^2$.

To derive the discrete TBC at $z_b = Jh$ we assume vanishing initial data $\psi_j^{(0)} = 0$, $j \geq J - 1$ and use the linear potential term $(N^2)_j^{(n)} - 1 = \beta + \mu h(j - J)$, and solve the *discrete exterior problem*

$$-iR(\psi_j^{(n+1)} - \psi_j^{(n)}) =$$

$$\Delta^2 \psi_j^{(n+1)} + \Delta^2 \psi_j^{(n)} + w[\beta + \mu h(j - J)](\psi_j^{(n+1)} + \psi_j^{(n)}), \quad (22)$$

$j \geq J$. Hence, the Z–transformed finite difference scheme (22), is a *general discrete Airy equation* of the form (cf. case C)):

$$\Delta^2 y_j - (d + cj)\, y_j = 0, \quad c, d \in \mathbb{C}. \tag{23a}$$

with

$$d = -2i\zeta(z) + \mu k_0^2 h^3 J, \quad c = -\mu k_0^2 h^3, \tag{23b}$$

$$\zeta(z) = \frac{R}{2}\frac{z-1}{z+1} - i\frac{\beta}{2}k_0^2 h^2. \tag{23c}$$

Transparent Boundary Conditions. Here, we review from [20] the derivation of the analytic TBC at $z = z_b$. We assume that the initial data $\psi^I = \psi(z, 0)$ is supported in the *computational domain* $0 < z < z_b$ and use the Laplace transform (8) for $z > z_b$:

$$\hat{\psi}_{zz}(z, s) + [\mu k_0^2(z - \tilde{z}_b) + 2ik_0 s]\hat{\psi}(z, s) = 0, \quad z > z_b, \tag{24}$$

with $\tilde{z}_b = z_b - \beta/\mu$. To solve (24) in the *exterior domain* $z > z_b$ we set $\sigma^3 = -\mu k_0^2$ and $\tau = 2ik_0/\sigma^2$. Then (24) can be written as

$$\hat{\psi}_{zz}(z, s) + \sigma^2[\sigma(z - \tilde{z}_b) + \tau s]\hat{\psi}(z, s) = 0, \quad z > z_b. \tag{25}$$

Introducing the change of variables $\zeta_s(z) = \sigma(z - \tilde{z}_b) + \tau s$, $U(\zeta_s(z)) = \hat{\psi}(z, s)$, we can write (25) in the form of an *Airy equation*:

$$U''(\zeta_s(z)) + \zeta_s(z)\, U(\zeta_s(z)) = 0, \quad z > z_b. \tag{26}$$

The decaying solution of (26) for $z \to \infty$, for fixed s, Re $s > 0$ is

$$\hat{\psi}(z, s) = C_1(s)\, \mathrm{Ai}(\zeta_s(z)), \quad z > z_b, \tag{27}$$

if we define the physically relevant branch of σ to be

$$\sigma = \begin{cases} (\mu k_0^2)^{1/3} e^{-i\pi/3}, & \mu > 0, \\ (-\mu k_0^2)^{1/3}, & \mu < 0. \end{cases} \tag{28}$$

Elimination of $C_1(s)$ gives

$$\hat{\psi}(z, s) = \hat{\psi}(z_b+, s) \, \frac{\mathrm{Ai}(\zeta_s(z))}{\mathrm{Ai}(\zeta_s(z_b))}, \qquad z > z_b. \tag{29}$$

Finally, differentiation w.r.t. z yields with the matching conditions (10) the *transformed analytic TBC* at $z = z_b$:

$$\hat{\psi}_z(z_b-, s) = \frac{\rho_w}{\rho_b} s \, \hat{\psi}(z_b-, s) \, W(s), \qquad W(s) = \sigma \, \frac{\mathrm{Ai}'(\zeta_s(z_b))}{s \, \mathrm{Ai}(\zeta_s(z_b))}, \tag{30}$$

i.e. the *analytic TBC* at $z = z_b$ reads:

$$\psi_z(z_b, r) = \frac{\rho_w}{\rho_b} \int_0^r \psi_r(z_b, r') \, g_\mu(r - r') \, dr'. \tag{31}$$

The kernel g_μ is obtained by an inverse Laplace transformation of $W(s)$ (cf. [21]):

$$g_\mu(r) = \sigma \left\{ \frac{\mathrm{Ai}'(\zeta_0(z_b))}{\mathrm{Ai}(\zeta_0(z_b))} + \sum_{j=1}^{\infty} \frac{e^{-(a_j - \zeta_0(z_b))r/\tau}}{a_j - \zeta_0(z_b)} \right\}, \tag{32}$$

where $\zeta_0(z_b) = \sigma\beta/\mu$ and the (a_j) are the zeros of the Airy function Ai which are all located on the negative real axis. This TBC is *nonlocal* in the range variable r and can be discretized, e.g. in conjunction with a finite difference scheme for (8). The constant term in g_μ acts like a Dirac function and the infinite series represents the continuous part. As Levy noted in [20] the kernel g_μ decays extremely fast for $\mu > 0$ and for negative μ it decays slowly at short ranges and then oscillates.

Discretization of the continuous TBC. To incorporate the analytic TBC (31) in a finite difference scheme we make the approximation that $\psi_r(z_b, r')$ is constant on each subinterval $r_n < r' < r_{n+1}$ and integrate the kernel g_μ exactly. In the following we review the discretization from [20] and start with the discretization in range:

$$\psi_z(z_b, r_n) = \frac{\rho_w}{\rho_b} \sum_{m=0}^{n-1} \frac{\psi_b^{(n-m)} - \psi_b^{(n-m-1)}}{k} \, G_m, \tag{33}$$

where we set $\psi_b^{(n)} = \psi(z_b, r_n)$ and G_m is given by

$$
\begin{aligned}
G_m &= \int_{r_m}^{r_{m+1}} g_\mu(\eta)\, d\eta \\
&= k\sigma \frac{\mathrm{Ai}'(\zeta_0(z_b))}{\mathrm{Ai}(\zeta_0(z_b))} + \frac{2ik_0}{\sigma} \sum_{j=1}^{\infty} \frac{e^{-(a_j - \zeta_0(z_b))r/\tau}}{(a_j - \zeta_0(z_b))^2} \Big|_{r=r_m}^{r=r_{m+1}}.
\end{aligned}
\tag{34}
$$

This leads after rearranging to

$$
k\frac{\rho_b}{\rho_w}\psi_z(z_b, r_n) = -\psi_b^{(0)} G_n + \psi_b^{(n)} G_0 + \sum_{m=1}^{n-1} \psi_b^{(n-m)}(G_m - G_{m-1}).
\tag{35}
$$

In [20] Levy used an *offset grid* in depth, i.e. $\tilde{z}_j = (j+\frac{1}{2})h$, $\tilde{\psi}_j^{(n)} \approx \psi(\tilde{z}_j, r_n)$, $j = -1, \ldots, J$, where the water–bottom interface lies between the grid points $j = J - 1$ and J:

$$
\psi_b^{(m)} = \psi(z_b, r_m) \approx \frac{\tilde{\psi}_J^{(m)} + \tilde{\psi}_{J-1}^{(m)}}{2}, \quad \psi_z(z_b, r_n) \approx \frac{\tilde{\psi}_J^{(n)} - \tilde{\psi}_{J-1}^{(n)}}{h}.
\tag{36}
$$

This finally yields (recall that $\tilde{\psi}_J^{(0)} = \tilde{\psi}_{J-1}^{(0)} = 0$) the following *discretized TBC* for the SPE:

$$
(1 - b_0)\tilde{\psi}_J^{(n)} - (1 + b_0)\tilde{\psi}_{J-1}^{(n)} = \sum_{m=1}^{n-1} b_m(\tilde{\psi}_J^{(n-m)} + \tilde{\psi}_{J-1}^{(n-m)}),
\tag{37}
$$

with

$$
b_0 = \frac{1}{2}\frac{h}{k}\frac{\rho_w}{\rho_b}G_0, \quad b_m = \frac{1}{2}\frac{h}{k}\frac{\rho_w}{\rho_b}(G_m - G_{m-1}).
\tag{38}
$$

Note that the constant term in (34) enters only b_0. Since $a_j \sim -\left(\frac{3\pi}{8}(4j - 1)\right)^{2/3}$ for $j \to \infty$ the series (34) defining G_m has good convergence properties for positive range r but for $r = 0$ the convergence is very slow. To overcome this problem we use the identity

$$
\sum_{j=1}^{\infty} \frac{1}{(a_j - \zeta_0(z_b))^2} = \left(\frac{\mathrm{Ai}'(\zeta_0(z_b))}{\mathrm{Ai}(\zeta_0(z_b))}\right)^2 - \zeta_0(z_b),
\tag{39}
$$

which can be derived analogously to the one in [20].

In a numerical implementation one has to limit the summation in (32) and therefore the TBC is no more fully transparent. Moreover, the stability of the resulting scheme is not clear since the discretized TBC (37) is not matched to the finite difference scheme (21) in the interior domain.

4. Discrete TBCs via Exact Solutions

We will now show how to find exact solutions for the presented difference equations in order to formulate the TBCs.

It is well–known how to solve second–order linear difference equations with *constant coefficients* (this is the case for the transformed Crank–Nicolson scheme (13) for solving the Black–Scholes equation for American options). In contrast, second–order linear difference equations with *variable coefficients* cannot be solved in closed form in most cases. However, if one wants to solve a difference equation with *polynomial coefficients*, one approach is to find the solution by the *"method of generating functions"*; i.e., a generating function for a solution of the difference equation can be shown to satisfy a differential equation, which may be solvable in terms of known functions.

4.1. *The Black–Scholes Equation for American Options*

The two linearly independent solutions of the resulting *second–order difference equation* (13) take the form $\hat{v}_j(z) = v_{1,2}^j(z)$, $j \leq 1$, where $v_{1,2}(z)$ are the solutions of the quadratic equation

$$v^2 - 2\left[1 + \frac{1}{\rho}\frac{z-1}{z+1}\right]v + 1 = 0. \tag{40}$$

Since we are seeking decreasing modes as $j \rightarrow -\infty$ we have to require $|v_1| > 1$ and obtain the Z–transformed *discrete TBC* as

$$\hat{v}_1(z) = v_1(z)\,\hat{v}_0(z). \tag{41}$$

It only remains to inverse Z–transform $v_1(z)$ in order to obtain the discrete TBC from (41). This can be performed explicitly (cf. [22]) and the discrete TBC becomes:

$$v_1^{(n)} = \ell^{(n)} * v_0^{(n)} = \sum_{k=1}^{n} \ell^{(n-k)} v_0^{(k)}, \quad n \geq 1, \tag{42}$$

with convolution coefficients $\ell^{(n)}$ given in [22]. Since the asymptotical behaviour $\ell^{(n)} \sim 4(-1)^n/\rho$ of the convolution coefficients may lead to subtractive cancellation in (42) we prefer to use the following *summed coefficients* in the implementation

$$s^{(n)} := \ell^{(n)} + \ell^{(n-1)}, \quad n \geq 1, \quad s^{(0)} := \ell^{(0)}. \tag{43}$$

The *discrete TBC* then reads

$$v_1^{(n)} - s^{(0)} v_0^{(n)} = \sum_{k=1}^{n-1} s^{(n-k)} v_0^{(k)} - v_1^{(n-1)}, \quad n \geq 1 \tag{44}$$

with the convolution coefficients

$$s^{(0)} = 1 + \frac{1 + \sqrt{1 + 2\rho}}{\rho}, \quad s^{(1)} = 1 - \frac{1}{\rho} - \frac{1}{\rho\sqrt{1 + 2\rho}},$$

$$s^{(n)} = -\frac{\sqrt{1 + 2\rho}}{\rho} \frac{\widetilde{P}_n(\mu) - \lambda^{-2}\widetilde{P}_{n-2}(\mu)}{2n - 1}, \quad n \geq 2, \tag{45}$$

where $\widetilde{P}_n(\mu) := \lambda^{-n} P_n(\mu)$ denotes the *"damped" Legendre polynomials* ($\widetilde{P}_0 \equiv \lambda^{-1}$, $\widetilde{P}_{-1} \equiv 0$). The parameters λ, μ are given by

$$\lambda = \frac{\sqrt{1 + 2\rho}}{\sqrt[4]{1 - 2\rho}}, \quad \mu = \frac{1}{\sqrt{1 + 2\rho}\sqrt[4]{1 - 2\rho}}. \tag{46}$$

Alternatively, the convolution coefficients can be computed by the recursion formula

$$s^{(n+1)} = \frac{2n - 1}{n + 1} \mu\lambda^{-1} s^{(n)} - \frac{n - 2}{n + 1} \lambda^{-2} s^{(n-1)}, \quad n \geq 2, \tag{47}$$

after calculating $s^{(n)}$, $n = 0, 1, 2$ by formula (45).

For a derivation of the discrete TBC for a class of difference schemes for a general convection diffusion equation we refer to [22, Chapter 2].

4.2. *The Schrödinger–Poisson System*

Unfortunately, the exact solution to the discrete Schrödinger equation with a Coulomb–type potential (case ii)) is not known explicitly. However, in the case of a constant potential (case i)) we can easily write down an explicit solution:

The two linearly independent solutions of the second–order difference equation (19) are $\hat{u}_j(z) = \nu_{1,2}^j(z)$, $j \geq J$, where $\nu_{1,2}(z)$ solve

$$\nu^2 - 2\left[1 - \frac{i\rho}{2}\left(\frac{z - 1}{z + 1} + i\kappa\right)\right]\nu + 1 = 0. \tag{48}$$

For the decreasing mode (as $j \to \infty$) we have to require $|\nu_1(z)| < 1$ and obtain the *Z–transformed discrete TBC* as

$$\hat{u}_{J-1}(z) = \nu_1^{-1}(z)\,\hat{u}_J(z). \tag{49}$$

It only remains to inverse transform (49) and in a tedious calculation this can be achieved explicitly [23]. However, since the magnitude of $\ell^{(n)} :=$

$\mathcal{Z}^{-1}\left\{\nu_1^{-1}(z)\right\}$ does not decay as $n \to \infty$ (Im $\ell^{(n)}$ behaves like const $\cdot(-1)^n$ for large n), it is more convenient to use a modified formulation of the discrete TBC (cf. [18]). Therefore we introduce the *summed coefficients*

$$s^{(n)} = \mathcal{Z}^{-1}\left\{\hat{s}(z)\right\}, \quad \text{with } \hat{s}(z) := \frac{z+1}{z}\hat{\ell}(z), \tag{50}$$

which satisfy

$$s^{(0)} = \ell^{(0)}, \quad s^{(n)} = \ell^{(n)} + \ell^{(n-1)}, \quad n \geq 1.$$

The *discrete TBC* for the discretization (18) now reads (cf. [23]):

$$u_{J-1}^{(n)} - s^{(0)}u_J^{(n)} = \sum_{k=1}^{n-1} s^{(n-k)}u_J^{(k)} - u_{J-1}^{n-1}, \quad n \geq 1, \tag{51}$$

with

$$s^{(n)} = \left[1 - i\frac{\rho}{2} + \frac{\sigma}{2}\right]\delta_n^0 + \left[1 + i\frac{\rho}{2} + \frac{\sigma}{2}\right]\delta_n^1 + \alpha\, e^{-in\varphi}\,\frac{P_n(\mu) - P_{n-2}(\mu)}{2n-1},$$

$$\varphi = \arctan\frac{2\rho(\sigma+2)}{\rho^2 - 4\sigma - \sigma^2}, \quad \mu = \frac{\rho^2 + 4\sigma + \sigma^2}{\sqrt{(\rho^2+\sigma^2)(\rho^2 + [\sigma+4]^2)}}, \tag{52}$$

$$\sigma = -wV_R, \quad \alpha = \frac{i}{2}\sqrt[4]{(\rho^2+\sigma^2)(\rho^2 + [\sigma+4]^2)}\, e^{i\varphi/2}.$$

P_n denotes the Legendre polynomials ($P_{-1} \equiv P_{-2} \equiv 0$) and δ_n^j the Kronecker symbol. The P_n only have to be evaluated at one value $\mu \in \mathbb{R}$, and hence the numerically stable recursion formula for the Legendre polynomials can be used. Using asymptotic properties of the Legendre polynomials one finds the decay rate $s^{(n)} = O(n^{-3/2})$.

4.3. The Standard "Parabolic Equation"

We show that in the case of the discrete Airy equation (23a) the exact solution can be found explicitly by the method of generating functions. We define the *generating function* to be

$$g(\xi) = \sum_{j=-\infty}^{\infty} y_j\xi^j. \tag{53}$$

We multiply (23a) with ξ^{j-1} and sum it up for $j \in \mathbb{Z}$:

$$\sum_{j=-\infty}^{\infty} y_j\xi^{j-2} - (2+d)\sum_{j=-\infty}^{\infty} y_j\xi^{j-1} + \sum_{j=-\infty}^{\infty} y_j\xi^j - c\sum_{j=-\infty}^{\infty} jy_j\xi^{j-1} = 0.$$

This results in the following ordinary differential equation for g:

$$g'(\xi) - \frac{1 - (2+d)\xi + \xi^2}{c\xi^2}\, g(\xi) = 0,$$

for which the solution is

$$g(\xi) = \xi^{-\frac{2+d}{c}}\, e^{(\xi - \frac{1}{\xi})/c} = \xi^{-\frac{2+d}{c}} \sum_{\nu=-\infty}^{\infty} J_\nu(\tfrac{2}{c})\xi^\nu.$$

Hence, the exact decaying solution of (23a) is the *Bessel function* $J_\nu(\frac{2}{c})$ (regarded as function of its order ν), i.e. the discrete Airy equation is nothing else but the recurrence relation for $J_\nu(\frac{2}{c})$. It is well–known [24] that the recurrence equation for the Bessel functions

$$J_{\nu+1}(z) - 2\,\frac{\nu}{z}\, J_\nu(z) + J_{\nu-1}(z) = 0, \tag{54}$$

still holds for complex orders ν and complex arguments z.

Thus the decaying solution to (23a) can be represented as (cf. [24, Chapter 3.1]):

$$y_j = J_{j+\frac{2+d}{c}}\left(\tfrac{2}{c}\right) = \frac{1}{c^{j+\frac{2+d}{c}}} \sum_{n=0}^{\infty} \frac{(-1)^n}{c^{2n}\, n!\,\Gamma(j + \frac{2+d}{c} + n + 1)}, \quad c, d \in \mathbb{C}. \tag{55}$$

We also observe that (53) is not a generating function in the strict sense but a *Laurent series*, which is uniformly convergent, i.e. differentiating each term is permissible (cf. [24]). Note that this generating function approach is not suitable for determining the growing solution of (23a) for $j \to \infty$. This solution is the so–called *"Neumann–Function"* (or Bessel function of the second kind) which is also known to satisfy the recursion equation of the Bessel functions.

Remark. A difference equation more general than (23a) was examined by Barnes [25] in 1904. He also considered (23a) and found (through a different construction) the solution (55).

Comparing (23a) with the recurrence relation of the Bessel function $J_\nu(\sigma)$ yields the condition

$$\frac{\nu}{\sigma} = 1 - i\zeta(z) - \mu\frac{k_0^2}{2} h^3(j - J) = \frac{j + \text{offset}}{\sigma}, \tag{56}$$

and we conclude that the exact solution of (23a) is

$$\hat{\psi}_j(z) = J_{\nu_j(z)}(\sigma), \tag{57}$$

with

$$\nu = \nu_j(z) = \sigma(1 - i\zeta(z)) + j - J, \quad \sigma = -\left(\mu\frac{k_0^2}{2}h^3\right)^{-1} \in \mathbb{R}. \tag{58}$$

From (57) we obtain the *transformed discrete TBC* at $z_b = Jh$:

$$\hat{\psi}_{J-1}(z) = \hat{g}_{\mu,J}(z)\hat{\psi}_J(z) \tag{59a}$$

with

$$\hat{g}_{\mu,J}(z) = \frac{J_{\nu_{J-1}(z)}(\sigma)}{J_{\nu_J(z)}(\sigma)} = \frac{J_{\sigma(1-i\zeta(z))-1}(\sigma)}{J_{\sigma(1-i\zeta(z))}(\sigma)}. \tag{59b}$$

Finally, an inverse Z–transformation yields the *discrete TBC*

$$\psi_{J-1}^{(n)} - \ell_J^{(0)}\psi_J^{(n)} = \sum_{m=1}^{n-1} \psi_J^{(n-m)}\,\ell_J^{(m)}, \tag{60}$$

with $\ell_J^{(n)} = \mathcal{Z}^{-1}\{\hat{g}_{\mu,J}(z)\}$ given by

$$\ell_J^{(n)} = \frac{\tau^n}{2\pi}\int_0^{2\pi} \hat{g}_{\mu,J}(\tau e^{i\varphi})e^{in\varphi}\,d\varphi, \quad n \in \mathbb{Z}_0, \quad \tau > 0. \tag{61}$$

Since this inverse Z–transformation cannot be done explicitly, we use a numerical inversion technique based on FFT (cf. [18]); for details of this routine we refer the reader to [22]. Note that the Bessel functions in (59) with complex order and (possibly large) real argument can be evaluated numerically by special software packages (see e.g. [26]).

Analogously, to §4.2 we introduce the *summed coefficients*

$$s_J^{(n)} = \mathcal{Z}^{-1}\{\hat{s}_J(z)\}, \quad \text{with } \hat{s}_J(z) := \frac{z+1}{z}\hat{\ell}_J(z), \tag{62}$$

i.e. in physical space, the discrete TBC is:

$$\psi_{J-1}^{(n)} - s_J^{(0)}\psi_J^{(n)} = \sum_{m=1}^{n-1} s_J^{(n-m)}\psi_J^{(m)} - \psi_{J-1}^{(n-1)}, \quad n \geq 1. \tag{63}$$

Remark. For brevity of the presentation, we omit the discussion of an adequate discrete treatment of the typical density shock at $z = z_b$ and refer the reader to [17] for a detailed discussion of various strategies.

5. Discrete ABCs through Asymptotic Solutions

If the exact solution is not known or too complicated (i.e. too expensive for an efficient numerical calculation) then one can use asymptotic solutions of the second–order difference equation (1) (cf. [8, Chapter 7] and the references therein).

If the coefficient $p(j)$ in (1) has the finite limit $p = \lim_{j\to\infty} p(j)$ one calls (1) a *Poincaré difference equation* and

$$\Phi(t) = t^2 - (2+p)t + 1 \tag{64}$$

the *characteristic polynomial* of (1). The idea of Poincaré [27]is now that the solutions of a Poincaré difference equation behave asymptotically for large j similar to the solutions of the corresponding constant coefficent difference equation

$$\Delta^2 y_j - p\, y_j = 0, \quad j \in \mathbb{Z}.$$

We formulate the classical *Theorem of Poincaré* (for the special case of this second–order difference equation):

Theorem 1: (Poincaré Theorem [8]) *Suppose that the zeros t_1, t_2 of the characteristic polynomial (64) have distinct moduli. Then for any nontrivial solution y_j of (1)*

$$\lim_{j\to\infty} \frac{y_{j+1}}{y_j} = t_k$$

for $k = 1$ or $k = 2$.

This theorem was improved 1921 by Perron [28]:

Theorem 2: (Perron Theorem, [8]) *Assume that $p(j) \neq 0$ for all $j \in \mathbb{N}_0$. Then under the assumptions of Theorem 1, Equation (1) has a fundamental set of solutions $\{y_j^{(1)}, y_j^{(2)}\}$ with the property*

$$\lim_{j\to\infty} \frac{y_{j+1}^{(k)}}{y_j^{(k)}} = t_k, \quad k = 1, 2$$

Remark. If equation (1) has characteristic roots with equal moduli then Poincaré's Theorem may fail (cf. Example of Perron [8, Example 7.12]).

In the case of the Schrödinger–Poisson problem we have a difference equation of Poincaré–type. However, the applicability of the above two theorems is limited; they only yield the first term approximation of the asymptotic solution. Therefore we apply standard perturbation techniques of asymptotic analysis to the equation

$$\Delta^2 y_j = 0, \quad j \in \mathbb{Z}.$$

Remark. We remark that the same approach is given in [8, Chapter 7.3], but the Theorem [8, Theorem 7.17.] does not apply to our case of the Coulomb–type potential.

If the classic theorems of Poincaré and Perron cannot be applied to (23a) it is not straight forward to obtain information about the asymptotic behaviour of the solutions to this equation. This is the case for the discrete Airy equation. One ansatz is the one of Mickens [29] and another approach is due to Wong and Li [30]. They obtain asymptotic solutions to the second–order difference equation

$$y_{j+2} + j^p\, a(j)\, y_{j+1} + j^q\, b(j)\, y_j = 0, \tag{65}$$

where p, q are integers and $a(j)$, $b(j)$ have expansions of the form

$$a(j) = \sum_{s=0}^{\infty} \frac{a_s}{j^s}, \qquad b(j) = \sum_{s=0}^{\infty} \frac{a_s}{j^s}, \tag{66}$$

with nonzero leading coefficients: $a_0 \neq 0$, $b_0 \neq 0$.

5.1. *The Schrödinger–Poisson System*

We consider the difference equation (20a) which is the discrete Z–transformed exterior problem. Motivated by (49), we want to obtain the *transformed discrete TBC* in the form:

$$\hat{u}_{J-1}(z) = \hat{\ell}(z)\, \hat{u}_J(z). \tag{67}$$

In the sequel we will construct some expressions for $\hat{\ell}(z)$ by determining asymptotic solutions to (20a) through different approaches.

Approach of Mickens. Following the approach [31], the asymptotic solution of (20a) written as

$$\hat{u}_{j+1}(z) - 2\left[A_0 + \frac{A_1}{j}\right]\hat{u}_j(z) + \hat{u}_{j-1}(z) = 0, \quad j \geq J, \tag{68}$$

takes the form

$$\hat{u}_j(z) \sim j^\theta e^{B_0 j}\left[1 + \sum_{k=1}^{\infty} \frac{B_k}{j^k}\right], \tag{69}$$

where the parameters θ and B_k are expressible in terms of $A_0 = A_0(z)$, A_1. The parameters θ, B_0, B_1 can be obtained by

$$\cosh(B_0) = A_0, \text{ i.e. } B_0 = \ln\left(A_0 \pm \sqrt{A_0^2 - 1}\right), \tag{70a}$$

$$\theta = \frac{A_1}{\sinh(B_0)}, \tag{70b}$$

$$B_1 = \frac{\theta(\theta - 1)}{2}\coth(B_0). \tag{70c}$$

In our case we obtain

$$e^{B_0} = \nu_1(z), \tag{71a}$$

$$\theta = \frac{2m\Delta r}{\hbar^2} \frac{\phi_\infty}{\nu_1(z) - \nu_1^{-1}(z)}, \tag{71b}$$

$$B_1 = \frac{\theta(\theta - 1)}{2} \frac{\nu_1(z) + \nu_1^{-1}(z)}{\nu_1(z) - \nu_1^{-1}(z)}, \tag{71c}$$

where $\nu_1(z)$ is the solution to (48) for the Schrödinger equation with zero potential (i.e. $\kappa = 0$) with $|\nu_1(z)| < 1$.

Approach of Wong & Li. Alternatively, one can use the *approach of Wong and Li* [32] to obtain a formula for the asymptotic behaviour of the solutions to this second–order difference equation of Poincaré type. To do so, we rewrite (20a) in the form

$$\hat{u}_{j+2} + a(j)\,\hat{u}_{j+1} + \hat{u}_j = 0, \quad j \geq J, \tag{72}$$

with $a(j) = -2[A_0 + A_1/(j+1)]$. Now $a(j)$ has a power expansion

$$a(j) = \sum_{k=0}^{\infty} \frac{a_k}{j^k},$$

with coefficients:

$$a_0 = -2A_0, \quad a_k = 2A_1(-1)^k, \quad k \geq 1.$$

Then the decaying asymptotic solution (cf. [32]) is of the form

$$\hat{u}_j \sim \nu_1(z)^j j^\alpha \sum_{k=0}^{\infty} \frac{c_k}{j^k}, \quad j \to \infty, \tag{73}$$

where α can be calculated as

$$\alpha = \frac{a_1 \nu_1(z)}{a_0 \nu_1(z) + 2} = \frac{A_1 \nu_1(z)}{A_0 \nu_1(z) - 1} = \frac{2A_1}{\nu_1(z) - \nu_1^{-1}(z)}. \tag{74}$$

Without loss of generality, we assume that $c_0 = 1$ and determine the values of the coefficients $c_1, c_2, \ldots$ by formula (2.3) in [32] or more illustrative by substituting the solution (73) in (72):

$$\nu_1^2 \left(1 + \frac{2}{j}\right)^\alpha \sum_{k=0}^{\infty} \frac{c_k}{(j+2)^k} + a(j)\nu_1 \left(1 + \frac{1}{j}\right)^\alpha \sum_{k=0}^{\infty} \frac{c_k}{(j+1)^k} + \sum_{k=0}^{\infty} \frac{c_k}{j^k} = 0.$$

We now obtain after a Taylor expansion in $1/j$ and setting all the linearly independent terms equal to zero, by a lengthy calculation

$$c_1 = \frac{\alpha^2 + A_0 A_1 \alpha - \alpha - A_1 \nu_1 + A_1^2}{2(A_0 \nu_1 - 1)}, \tag{75a}$$

$$c_2 = \frac{c_1^2}{2} + \frac{A_0 \nu_1 - \alpha}{2(A_0 \nu_1 - 1)} c_1 + \frac{1 - A_1 \nu_1 - A_0 A_1 - A_0^2}{3(A_0 \nu_1 - 1)} c_1 \tag{75b}$$

$$+ \frac{(A_0^3 c_1 + \nu_1 - A_0)A_1}{6(A_0 \nu_1 - 1)} + \frac{(A_0^2 - A_1 \nu_1 + A_1 A_0 - 3)A_1^2}{12(A_0 \nu_1 - 1)},$$

etc.

This result can be checked easily with a symbolic package like MAPLE. After some basic manipulations one observes that these two approaches lead to the same asymptotic solution of the equation (20a).

5.2. *The Standard "Parabolic Equation"*

It is a nontrivial task to determine asymptotic solutions of the discrete Airy equation (23a) since equation (23a) is not of Poincaré type.

Approach of Wong & Li. We increase the index of (23a) by one to put it in the form of (65) and make the following identifications:

$$a_0 = -c, \quad a_1 = -(2 + c + d), \quad a_s = 0, \ s \geq 2,$$
$$b_0 = 1, \quad b_s = 0, \ s \geq 1, \quad p = 1, \quad q = 0.$$

Then the two formal series solutions (cf. [30]) are given by

$$y_j^{(1)} = \frac{c^{-j}}{(j-2)!} j^{-2-(2+d)/c} \sum_{s=0}^{\infty} \frac{c_s^{(1)}}{j^s}, \tag{76a}$$

$$y_j^{(2)} = (j-2)! \, c^j \, j^{1+(2+d)/c} \sum_{s=0}^{\infty} \frac{c_s^{(2)}}{j^s}. \tag{76b}$$

To determine the values of the coefficients $c_1^{(1)}, c_2^{(1)}, c_3^{(1)}, \ldots$ we substitute the decaying solution $y_j^{(1)}$ in (65):

$$\frac{1}{cj}\left(\frac{j+1}{j+2}\right)^{\theta} \sum_{s=0}^{\infty} \frac{c_s^{(1)}}{(j+2)^s} + c(j-1)\left(\frac{j+1}{j}\right)^{\theta} \sum_{s=0}^{\infty} \frac{c_s^{(1)}}{j^s}$$

$$= \big(c(\theta - 1) + cj\big) \sum_{s=0}^{\infty} \frac{c_s^{(1)}}{(j+1)^s}, \tag{77}$$

with $\theta = 2 + (2+d)/c$. We now obtain after a Taylor expansion in $1/j$ and setting all the linearly independent terms equal to zero, by a lengthy but elementary calculation, the results:

$$c_1^{(1)} = -c_0^{(1)}\left[c^{-2} - \theta + \frac{\theta}{2}(\theta - 1)\right], \tag{78a}$$

$$c_2^{(1)} = \frac{c_0^{(1)}}{2}\left[\theta c^{-2} + \frac{\theta}{2}(\theta - 1) - \frac{\theta}{6}(\theta - 1)(\theta - 2)\right]$$
$$- \frac{c_1^{(1)}}{2}\left[c^{-2} - 2 + \frac{\theta}{2}(\theta - 1)\right], \tag{78b}$$

$$c_3^{(1)} = -\frac{c_0^{(1)}}{3}\left[\left(2\theta + \frac{\theta}{2}(\theta - 1)\right)c^{-2} - \frac{\theta}{6}(\theta - 1)(\theta - 2)\right.$$
$$\left. - \frac{\theta}{24}(\theta - 1)(\theta - 2)(\theta - 3)\right]$$
$$+ \frac{c_1^{(1)}}{3}\left[(2 + \theta)c^{-2} - 2 + \theta + \frac{\theta}{2}(\theta - 1) - \frac{\theta}{6}(\theta - 1)(\theta - 2)\right]$$
$$- \frac{c_2^{(1)}}{3}\left[c^{-2} - 5 + \theta + \frac{\theta}{2}(\theta - 1)\right], \tag{78c}$$

etc.

Similarly we obtain for the increasing solution $y_j^{(2)}$ the first three coefficients

$$c_1^{(2)} = c_0^{(2)}\left[c^{-2} - \eta + \frac{\eta}{2}(\eta - 1)\right], \tag{79a}$$

$$c_2^{(2)} = -\frac{c_0^{(2)}}{2}\left[(\eta - 1)c^{-2} - \eta + \eta(\eta - 1) - \frac{\eta}{6}(\eta - 1)(\eta - 2)\right] \tag{79b}$$
$$+ \frac{c_1^{(2)}}{2}\left[c^{-2} + 3 - 2\eta - \frac{\eta}{2}(\eta - 1)\right],$$

$$c_3^{(2)} = \frac{c_0^{(2)}}{3}\left[\left(1 + \frac{\eta}{2}(\eta - 1)\right)c^{-2} - \eta + \frac{3}{2}\eta(\eta - 1)\right.$$
$$\left. - \frac{\eta}{2}(\eta - 1)(\eta - 2) + \frac{\eta}{24}(\eta - 1)(\eta - 2)(\eta - 3)\right]$$
$$- \frac{c_1^{(2)}}{3}\left[(\eta - 1)c^{-2} - 7 - 6\eta + 2\eta(\eta - 1) - \frac{\eta}{6}(\eta - 1)(\eta - 2)\right]$$
$$+ \frac{c_2^{(2)}}{3}\left[c^{-2} + 9 - 3\eta + \frac{\eta}{2}(\eta - 1)\right], \tag{79c}$$

with $\eta = 1 + (2+d)/c$. Here $c_0^{(1)}$, $c_0^{(2)}$ denote arbitrary constants.

We recall that the Z–transformed scheme in the exterior $j \geq J - 1$, given by (23b), is a discrete Airy equation of the form (23a) with

$$c = 2\sigma^{-1}, \quad d = -2i\zeta(z) - cJ, \text{ i.e. } \theta = 2 + \frac{2+d}{c} = 2 + \nu_0.$$

Using the asymptotic solution $y_j^{(1)}$ from (76a) we thus obtain the *transformed discrete ABC*

$$\hat{\psi}_{J-1}(z) = \hat{k}_{\mu,J}(z)\hat{\psi}_J(z), \tag{80a}$$

with

$$\hat{k}_{\mu,J}(z) = \frac{y_{J-1}^{(1)}}{y_J^{(1)}} = c(J-2)\Big(\frac{J}{J-1}\Big)^{\theta} \frac{\sum\limits_{s=1}^{\infty} \frac{c_s^{(1)}}{(J-1)^s}}{\sum\limits_{s=1}^{\infty} \frac{c_s^{(1)}}{J^s}}. \tag{80b}$$

Asymptotic Expansion of Explicit Solution. Another approach is to use an asymptotic expansion for the exact solution. We will explain this using the discrete Airy equation (23). We use the following asymptotic representation of Bessel functions for large values of the order ν (cf. [24]):

$$J_\nu(z) \approx \frac{e^{\nu+\nu \log(z/2)-(\nu+1/2)\log\nu}}{\sqrt{2\pi}}, \quad |\nu| \to \infty, \ |\arg\nu| \leq \pi - \delta. \tag{81}$$

Using the formula (81) leads to the *transformed discrete ABC*

$$\hat{\psi}_{J-1}(z) = \hat{h}_{\mu,J}(z)\hat{\psi}_J(z), \tag{82a}$$

with

$$\hat{h}_{\mu,J}(z) = \frac{2}{e}\sigma^{-1}\sqrt{\nu_J(\nu_J - 1)}\Big(\frac{\nu_J}{\nu_J - 1}\Big)^{\nu_J}, \tag{82b}$$

and ν_J, σ given by (58).

6. The Continued Fraction Approach

Finally, for a third approach to construct a discrete ABC, we use a formulation as a *continued fraction*. This approach is suitable for general second–order difference equations, since the exact solvability of the difference equation is not necessary. One can deduce such an expression for the quotient of two spatially neighboured Z–transformed solutions as a continued fraction directly from the difference scheme (i.e. without knowing the solution). This approach is often better than evaluating the quotient of two asymptotic solutions (obtained by any of the previous approaches) at two neighboured grid points. For the numerical implementation one can use the *modified Lentz's method* [33] which is an efficient general method for evaluating continued fractions. The calculations in [13] and [16] showed that the numerical evaluation of the continued fraction formula is stable for all considered parameter values.

6.1. *The Schrödinger–Poisson System*

If we rewrite the transformed discrete exterior problem (20a) as

$$\frac{\hat{u}_{J-1}(z)}{\hat{u}_J(z)} = 2\left[A_0 + \frac{A_1}{j}\right] - \frac{1}{\dfrac{\hat{u}_J(z)}{\hat{u}_{J+1}(z)}},$$

it is obvious that we have the following continued fraction

$$\frac{\hat{u}_{J-1}(z)}{\hat{u}_J(z)} = 2\left[A_0 + \frac{A_1}{J}\right] - \frac{1}{2[A_0 + \frac{A_1}{J+1}] -} \cdots - \frac{1}{2[A_0 + \frac{A_1}{J+M}] -} \frac{\hat{u}_{J+M+1}(z)}{\hat{u}_{J+M}(z)}.$$

For decreasing solutions the last quotient may be neglected if $M \to \infty$, i.e. we obtain the expansion

$$\hat{\ell}(z) = 2\left[A_0 + \frac{A_1}{J}\right] - \frac{1}{2[A_0 + \frac{A_1}{J+1}] -} \frac{1}{2[A_0 + \frac{A_1}{J+2}] -} \cdots. \tag{83}$$

This continued fractions formula (83) offers another way to evaluate the quotient $\hat{\ell}(z)$ needed in the transformed discrete TBC (49).

Finally we want to end with a short note about the implementation of the discrete TBC using the asymptotic expansions or the continued fraction approach. As in the case for the constant potential in the exterior domain (cf. (50)) it is favourable to use

$$\hat{s}(z) := \frac{z+1}{z}\,\hat{\ell}(z). \tag{84}$$

An inverse Z–transformation yields finally the *discrete TBC*

$$u_{J-1}^{(n)} - s^{(0)} u_J^{(n)} = \sum_{k=1}^{n-1} s^{(n-k)} u_J^{(k)} - u_{J-1}^{(n-1)}, \quad n \geq 1 \tag{85}$$

with

$$s^{(n)} = \mathcal{Z}^{-1}\{\hat{s}(z)\} = \frac{\tau^n}{2\pi} \int_0^{2\pi} \hat{s}(\tau e^{i\varphi}) e^{in\varphi}\, d\varphi, \quad n \in \mathbb{Z}_0, \quad \tau > 0. \tag{86}$$

This inverse Z–transformation (86) must be performed numerically (cf. §4.3).

Remark. We remark that this discrete TBC (85) can also used for both the predictor (15a) and the corrector step (16a) for the Schrödinger equation. In the exterior domain they are

$$i\hbar D_{t,k}^{+} u_j^{(n)} = -\frac{\hbar^2}{2m} D_r^2 S_{t,k} u_j^{(n)} + \frac{\phi_\infty}{r_j} S_{t,k} u_j^{(n)}, \quad j \geq J, \qquad (87)$$

$k = 1, 2$, i.e. after a (slightly modified) Z–transformation they are of the form (20a) and a discrete TBC analogue to (85) can be applied.

6.2. *The Standard "Parabolic Equation"*

Following [24, Section 5.6] we can easily deduce an expression for the quotient of Bessel functions as a continued fraction from the recurrence formula (54). If we rewrite (54) as

$$\frac{J_{\nu-1}(z)}{J_\nu(z)} = 2\nu z^{-1} - \frac{1}{\frac{J_\nu(z)}{J_{\nu+1}(z)}},$$

it is obvious that

$$\frac{J_{\nu-1}(z)}{J_\nu(z)} = 2\nu z^{-1} - \frac{1}{2(\nu+1)z^{-1} - \dots - } \frac{1}{2(\nu+M)z^{-1} - } \frac{J_{\nu+M-1}(z)}{J_{\nu+M}(z)}.$$

This holds for general values of ν and it can be shown, with the help of the theory of Lommel polynomials [24, Section 9.65], that for $M \to \infty$, the last quotient may be neglected, so that

$$\frac{J_{\nu-1}(z)}{J_\nu(z)} = 2\nu z^{-1} - \frac{1}{2(\nu+1)z^{-1} - } \frac{1}{2(\nu+2)z^{-1} - \dots}. \qquad (88)$$

This continued fractions formula offers another way to evaluate the quotient of two Bessel functions needed in the transformed discrete TBC.

In the sequel we will focus exclusively on this example of the Standard "parabolic equation". However, the approximation technique described in the following §7 applies generally to boundary conditions of convolution type.

7. The Approximation by the Sum of Exponentials

An ad-hoc implementation of the discrete convolution (63) with convolution coefficients $s_j^{(n)}$ from (62) (or obtained by any of the above approaches) has still one disadvantage. The boundary condition is non–local and therefore computationally expensive. In fact, the evaluation of (63) is as expensive as for an discretization of the TBC (31). As a remedy, we proposed in [34] the sum–of–exponentials ansatz (for a comparison of the computational efforts see Fig. 7). In the sequel we will briefly review this approach which can also be used for more general "parabolic equations" [35].

In order to derive a fast numerical method to calculate the discrete convolutions in (63), we approximate the coefficients $s_J^{(n)}$ by the following (*sum of exponentials*):

$$s_J^{(n)} \approx \tilde{s}_J^{(n)} := \begin{cases} s_J^{(n)}, & n = 0, 1 \\ \sum_{l=1}^{L} b_l \, q_l^{-n}, & n = 2, 3, \dots, \end{cases} \tag{89}$$

where $L \in \mathbb{N}$ is a fixed number. Evidently, the approximation properties of $\tilde{s}_J^{(n)}$ depend on L, and the corresponding set $\{b_l, q_l\}$. Below we propose a deterministic method of finding $\{b_l, q_l\}$ for fixed L.

Remark. The "split" definition of $\{\tilde{s}_J^{(n)}\}$ in (89) is motivated by the fact that the implementation of the discrete TBCs (63) involves a convolution sum with k ranging only from 1 to $m = n - 1$. Since the first coefficient $s_J^{(0)}$ does not appear in this convolution, it makes no sense to include it in our sum–of–exponential approximation, which aims at simplifying the evaluation of the convolution. The "special form" of $s_J^{(0)}$ and $s_J^{(1)}$ (in the case of a constant potential, cf. (52)) suggests to even exclude $s_J^{(1)}$ from this approximation.

Let us fix L and consider the formal power series:

$$g(x) := s_J^{(2)} + s_J^{(3)} x + s_J^{(4)} x^2 + \dots, \qquad |x| \leq 1. \tag{90}$$

If there exists the $[L - 1 | L]$ *Padé approximation*

$$\tilde{g}(x) := \frac{P_{L-1}(x)}{Q_L(x)} \tag{91}$$

of (90), then its Taylor series

$$\tilde{g}(x) = \tilde{s}_J^{(2)} + \tilde{s}_J^{(3)} x + \tilde{s}_J^{(4)} x^2 + \dots \tag{92}$$

satisfies the conditions

$$\tilde{s}_J^{(n)} = s_J^{(n)}, \qquad n = 2, 3, \dots, 2L + 1, \tag{93}$$

due to the definition of the Padé approximation rule.

Theorem 3: ([34]) *Let $Q_L(x)$ have L simple roots q_l with $|q_l| > 1$, $l = 1, \dots, L$. Then*

$$\tilde{s}_J^{(n)} = \sum_{l=1}^{L} b_l \, q_l^{-n}, \qquad n = 2, 3, \dots, \tag{94}$$

where

$$b_l := -\frac{P_{L-1}(q_l)}{Q'_L(q_l)} \, q_l \neq 0, \qquad l = 1, \ldots, L. \tag{95}$$

It follows from (93) and (94) that the set $\{b_l, q_l\}$ defined in Theorem 3 can be used in (89) at least for $n = 2, 3, .., 2L + 1$. The main question now is: Is it possible to use these $\{b_l, q_l\}$ also for $n > 2L + 1$? In other words, what quality of approximation

$$\tilde{s}_J^{(n)} \approx s_J^{(n)}, \qquad n > 2L + 1 \tag{96}$$

can we expect?

The above analysis permits us to give the following description of the approximation to the convolution coefficients $s_J^{(n)}$ by the representation (89) if we use a $[L-1|L]$ Padé approximant to (90): the first $2L$ coefficients are reproduced exactly, see (93); however, the asymptotic behaviour of $s_J^{(n)}$ and $\tilde{s}_J^{(n)}$ (as $n \to \infty$) differs strongly (algebraic versus exponential decay). A typical graph of $|s_J^{(n)} - \tilde{s}_J^{(n)}|$ versus n for $L = 27$ is shown in Fig. 3 in §8.

Fast Evaluation of the Discrete Convolution. Let us consider the approximation (89) of the discrete convolution kernel appearing in the DTBC (63). With these "exponential" coefficients the convolution

$$C^{(n)} := \sum_{m=1}^{n-1} \tilde{s}_J^{(n-m)} \psi_J^{(m)}, \quad \tilde{s}_J^{(n)} = \sum_{l=1}^{L} b_l \, q_l^{-n}, \tag{97}$$

$|q_l| > 1$, of a discrete function $\psi_J^{(m)}$, $m = 1, 2, \ldots$, with the kernel coefficients $\tilde{s}_J^{(n)}$, can be calculated by recurrence formulas, and this will reduce the numerical effort significantly (cf. Fig. 7 in §8).

A straightforward calculation (cf. [34]) yields:

Theorem 4: ([34]) *The value $C^{(n)}$ from (97) for $n \geq 2$ is represented by*

$$C^{(n)} = \sum_{l=1}^{L} C_l^{(n)}, \tag{98}$$

where

$$C_l^{(1)} \equiv 0,$$

$$C_l^{(n)} = q_l^{-1} C_l^{(n-1)} + b_l \, q_l^{-1} \psi_J^{(n-1)}, \tag{99}$$

$n = 2, 3, \ldots \; l = 1, \ldots, L.$

Finally we summarize the approach by the following *algorithm*:

1. calculate $s_J^{(n)}$, $n = 0, \ldots, N - 1$, via an explicit formula or a numerical inverse Z–transformation;
2. calculate $\tilde{s}_J^{(n)}$ via Padé–algorithm;
3. the corresponding coefficients b_l, q_l are used for the efficient calculation of the discrete convolutions.

Remark. We note that the Padé approximation must be performed with high precision ($2L - 1$ digits mantissa length) to avoid a 'nearly breakdown' by ill conditioned steps in the Lanczos algorithm. If such problems still occur or if one root of the denominator is smaller than 1 in absolute value, the orders of the numerator and denominator polynomials are successively reduced.

8. Numerical Example

In the example of this section we will consider the SPE for comparing the numerical result from using our new (approximated) discrete TBC to the solution using the discretized TBC of Levy [20]. We used the environmental test data from [21] and the Gaussian beam from [17] as starting field ψ^I. Below we present the so–called *transmission loss* $-10\log_{10}|p|^2$, where the acoustic pressure p is calculated from (9). We computed a *reference solution* on a three times larger computational domain confined with the discrete TBC from [17].

Example. As an illustrating example we chose the typical *downward refracting case* (i.e. energy loss to the bottom): $\mu = 2 \cdot 10^{-4}\,\mathrm{m}^{-1}$. The source at $z_s = 91.44\,\mathrm{m}$ is emitting sound with a frequency $f = 300\,\mathrm{Hz}$ and the receiver is located at the depth $z_r = 27.5\,\mathrm{m}$. The TBC is applied at $z_b = 152.5\,\mathrm{m}$ and the discretization parameters are given by $\Delta r = 10\,\mathrm{m}$, $\Delta z = 0.5\,\mathrm{m}$. It contains no attenuation: $\alpha = 0$. We consider a range-independent situation for $0 < r < 50\,\mathrm{km}$, i.e. 5000 range steps. The sound speed varies linearly from $c(0\,\mathrm{m}) = 1536.5\,\mathrm{m\,s}^{-1}$ to $c(152.5\,\mathrm{m}) = 1539.24\,\mathrm{m\,s}^{-1}$. The reference sound speed c_0 is chosen to be equal to $c(z_b)$ such that $\beta = 0$ in (11).

For this choice of parameters the mesh ratio becomes $R \approx 0.12246$ and the parameter $\sigma \approx -53345.32$; that is, the value of ν_J defined in (58) is much too large for the routines like COULCC [26] for evaluating Bessel functions. On the other hand, using the asymptotic formula (82) is not advisable since for large ν_J we have $\hat{h}_{\mu,J}(z) \sim 2(1 - i\zeta(z))$ which is only the first term in the continued fraction expansion (88). Therefore, we decided to evaluate the ratio of the two Bessel functions in (59) by the continued fraction formula (88) together with the sum–of–exponentials ansatz (89). We note that all approaches fulfilled for moderate choices of ν_J the growth condition (101) needed for stability.

We computed the first 1000 terms in the expansion (88) and used a radius $\tau = 1.04$ with 2^{10} sampling points for the numerical inverse Z–transformation (61). The choice of an appropriate radius τ is a delicate problem: it may not be too close to the convergence radius of (88) due to the approximation error and τ too large raises problems with rounding errors during the rescaling process. For a discussion of that topic we refer the reader to [34, Section 2] and [36]. In order to calculate the convolution coefficients b_n (discretized TBC of Levy) we used the MATLAB routine from [37] to compute the first 100 zeros of the Airy function. Alternatively, using precomputed values from the call `evalf(AiryAiZeros(1..100));` in MAPLE with high precision yielded indistinguishable results.

First we examine the convolution coefficients of the two presented approaches. Fig. 2 shows a comparison of the coefficients b_n from the discretized TBC (37) with the coefficients $s_J^{(n)}$ from the approximated discrete TBC. The coefficients b_n decay even faster than the coefficients $s_J^{(n)}$. In Fig. 3 we plot both the exact convolution coefficients $s_J^{(n)}$ and the error $|s_J^{(n)} - \tilde{s}_J^{(n)}|$ versus n for $L = 27$ (observe the different scales).

Now we investigate the *stability* of our numerical scheme for the SPE (21) along with a surface condition and the discrete TBC (60):

$$
\begin{cases}
-iR(\psi_j^{(n+1)} - \psi_j^{(n)}) = \rho_j \Delta_z^0(\rho_j^{-1}\Delta_z^0)(\psi_j^{(n+1)} + \psi_j^{(n)}) \\
\qquad\qquad + w\big[(N^2)_j^{(n)} - 1\big](\psi_j^{(n+1)} + \psi_j^{(n)}), \\
\qquad\qquad\qquad\qquad j = 1, \ldots, J-1, \\
\psi_j^{(0)} = \psi^I(z_j), \ j = 0, 1, 2, \ldots, J-1, J; \\
\text{with } \psi_{J-1}^{(0)} = \psi_J^{(0)} = 0, \\
\psi_0^{(n)} = 0, \\
\hat{\psi}_{J-1}(z) = \hat{g}_{\mu,J}(z)\hat{\psi}_J(z),
\end{cases}
\tag{100}
$$

where $\hat{g}_{\mu,J}(z)$ is given by (59). The following theorem bounds the exponential growth of solutions to the numerical scheme for a fixed discretization:

Theorem 5: ([16]) *Let the boundary kernel $\hat{g}_{\mu,J}$ satisfy*

$$
\operatorname{Im} \hat{g}_{\mu,J}(\gamma e^{i\varphi}) \le 0, \quad \forall 0 \le \varphi \le 2\pi,
\tag{101}
$$

for some (sufficiently large) $\gamma \ge 1$ (i.e. the system is dissipative). Assume

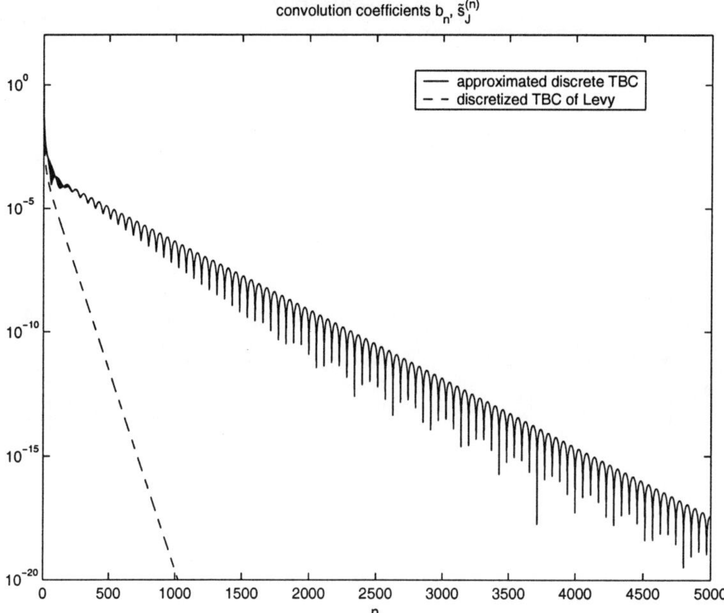

Fig. 2. Comparison of the convolution coefficients b_n of the discretized TBC (37) and $\tilde{s}_J^{(n)}$ from the approximated discrete TBC (with $L = 27$).

also that $\hat{g}_{\mu,J}(z)$ is analytic for $|z| \geq \gamma$. Then, the solution of (100) satisfies the a-priori estimate

$$\|\psi^{(n)}\|_2 \leq \|\psi^0\|_2 \, \gamma^n, \quad n \in \mathbb{N}, \tag{102}$$

where

$$\|\psi^{(n)}\|_2^2 := h \sum_{j=1}^{J-1} |\psi_j^{(n)}|^2 \rho_j^{-1}, \tag{103}$$

denotes the discrete weighted ℓ^2–norm.

Remark. Above we have assumed that the Z–transformed boundary kernel $\hat{g}_{\mu,J}(z)$ is analytic for $|z| \geq \beta$. Hence its imaginary parts is a harmonic functions there. Since the average of $\hat{g}_{\mu,J}(z)$ on the circles $z = \beta e^{i\varphi}$ equals $g_{\mu,J}^{(0)} = \hat{g}_{\mu,J}(z = \infty)$, condition (101) implies Im $\hat{g}_{\mu,J}(z = \infty) \leq 0$. Then we have the following simple consequence of the maximum principle for the Laplace equation:

If condition (101) holds for some γ_0, it also holds for all $\gamma > \gamma_0$.

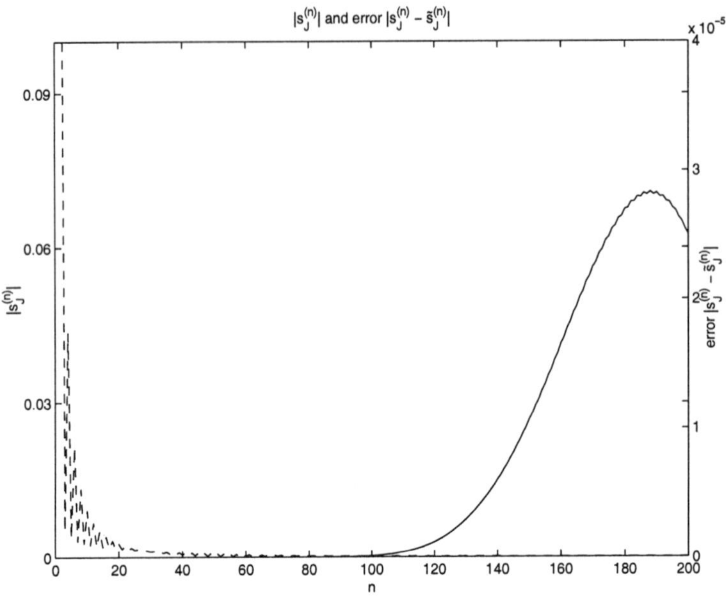

Fig. 3.   Convolution coefficients $s_J^{(n)}$ (left axis, dashed line) and error $|s_J^{(n)} - \tilde{s}_J^{(n)}|$ of the convolution coefficients (right axis); ($L = 27$).

We want to check the *growth condition* (101) for this example. For $\gamma = 1$ we have $\max\{\mathrm{Im}\,(\hat{g}_{\mu,J}(\gamma\,e^{i\varphi})\} = 0.153$ and, with $\gamma = 1.01$, we obtain $\max\{\mathrm{Im}\,(\hat{g}_{\mu,J}(\gamma\,e^{i\varphi})\} = -0.002$ (see Fig. 4). Hence, the Z–transformed kernel $\hat{g}_{\mu,J}(\gamma\,e^{i\varphi})$ of the approximated discrete TBC satisfies the condition (101) for $\gamma \geq 1.01$ (for this discretization).

In Fig. 5 and Fig. 6 we compare the transmission loss results for the discretized TBC and the approximated discrete TBC in the range from 0 to 50 km. The transmission loss curve of the solution using the approximated discrete TBC is indistinguishable from the one of the reference solution while the solution with the discretized TBC still deviates significantly from it (and is more oscillatory) for the chosen discretization. The result in Fig. 6 does not change if we compute more zeros of the Airy function.

Evaluating the convolution appearing in the discretized TBC (37) is quite expensive for long-range calculations. Therefore we extended the range interval up to 250 km and shall now illustrate the difference in the computational effort for both approaches in Fig. 7: The computational effort for the discretized TBC is quadratic in range, since the evaluation of

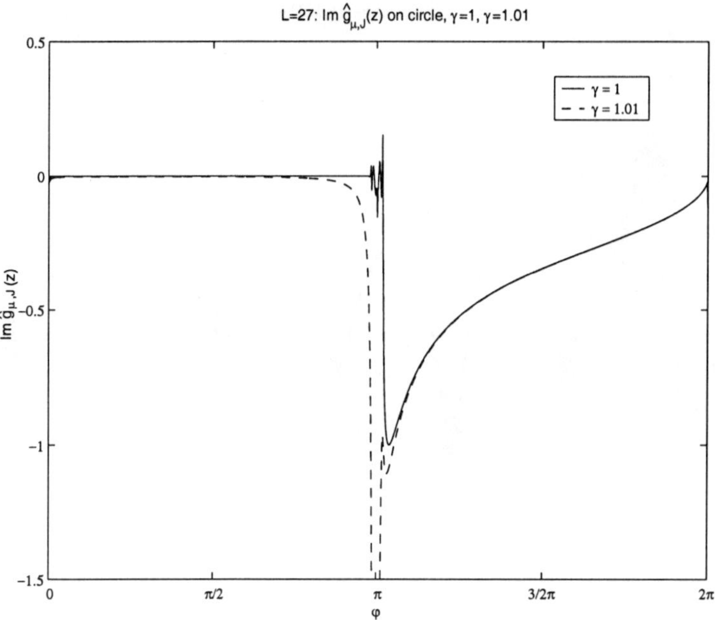

Fig. 4. Growth condition $\hat{g}_{\mu,J}(z)$, $z = \gamma e^{i\varphi}$ $(L = 27)$.

the boundary convolutions dominates for large ranges. On the other hand, the effort for the approximated discrete TBC only increases linearly. The line (- - -) does not change considerably for different values of L since the evaluation of the sum–of–exponential convolutions has a negligible effort compared to solving the PDE in the interior domain.

Conclusion

We have proposed a variety of general strategies to derive discrete ABCs/TBCs for the Black-Scholes equation for American options and the Schrödinger equation with a linear or Coulomb–type potential term in the exterior domain. The derivation was based on the knowledge of the exact solution, the construction of asymptotic solutions or the usage of a continued fraction expansion. Our approach has two advantages over the standard approach of discretizing the continuous TBC: higher accuracy and efficiency; while discretized TBCs have usually quadratic effort, the sum–of–exponential approximation to discrete ABCs/TBCs has only linear effort. Moreover, we have provided in the case of the standard "parabolic

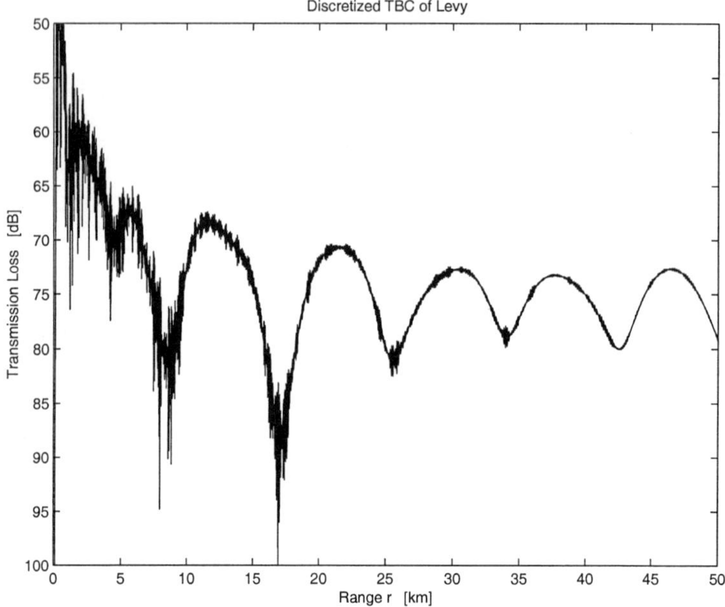

Fig. 5.   Transmission loss at $z_r = 27.5\,\mathrm{m}$.

equation" a simple criteria to check the stability of our method and gave an illustrating numerical example from underwater acoustics showing the superiority of our new approach.

9. Future Directions

It can easily be seen that the solutions to this discretization (23) do not have the same asymptotic properties as the solutions of the continuous Airy equation (26) which motivated the construction of a *nonstandard discretization scheme* (cf. [29], [38]). The logical consequence would be to study discrete TBC to the nonstandard discretization. This will be a topic of future work.

Acknowledgments

The author acknowledges support by the DFG Research Center "Mathematics for key technologies" MATHEON in Berlin.

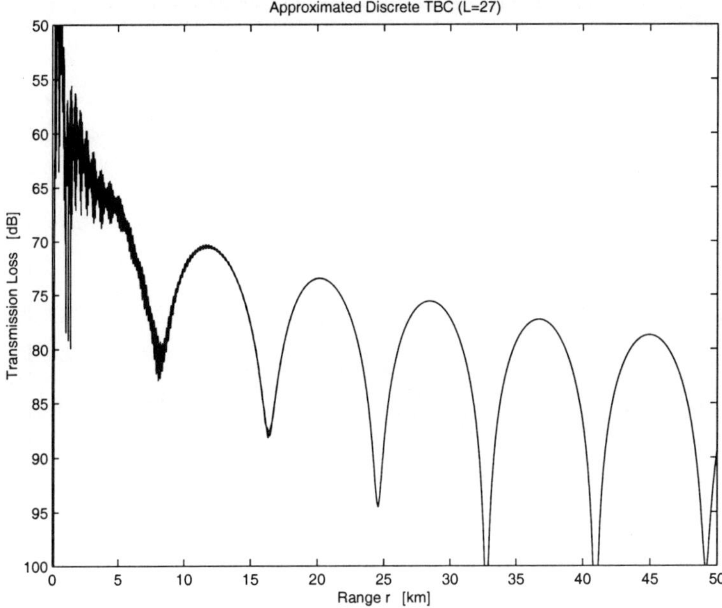

Fig. 6.   Transmission loss at $z_r = 27.5\,\mathrm{m}$.

Appendix A.   The Z–Transformation

The main tool of this work is the Z–transformation which is the discrete
analogue of the Laplace–transformation. The Z–transformation can be applied to the solution of linear difference equations in order to reduce the
solutions of such equations into those of algebraic equations in the complex
z–plane. In this work we used it to solve the finite difference schemes in
the exterior domain in order to construct the discrete ABCs/TBCs. The
Z–transformation is described in more detail in [39]. It is defined in the
following way:

Definition A1: (*Z–transformation* [39]) The formal connection between a sequence and a complex function given by the correspondence

$$\mathcal{Z}\{f_n\} = \hat{f}(z) := \sum_{n=0}^{\infty} f_n\, z^{-n}, \quad z \in \mathbb{Z}, \quad |z| > R_{\hat{f}}, \tag{A1}$$

is called Z–transformation. The function $\hat{f}(z)$ is called Z–transformation
of the sequence $\{f_n\}$, $n = 0, 1, \ldots$ and $R_{\hat{f}} \geq 0$ denotes the radius of convergence.

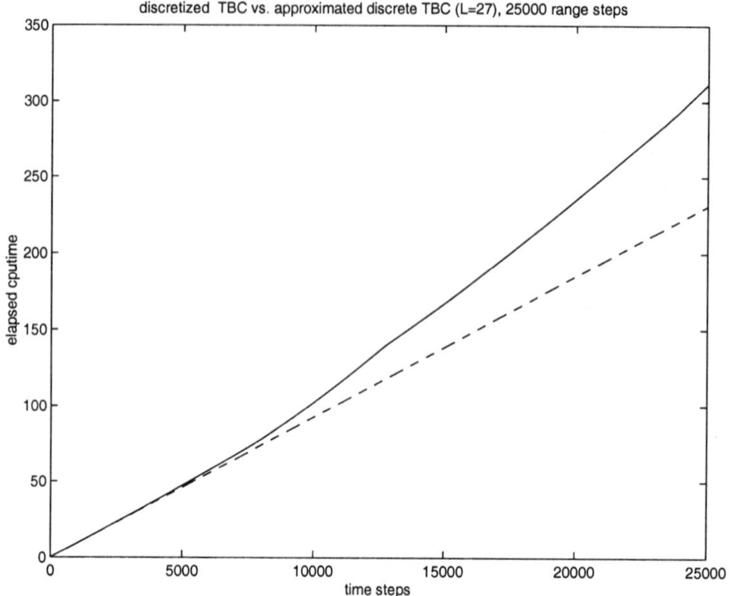

Fig. 7.   Comparison of CPU times: the discretized TBC of Levy (37) has quadratic effort (—), while the sum–of–exponential approximation to the discrete TBC has only linear effort (- - -).

The discrete analogue of the Differentiation Theorem for the Laplace transformation is the *shifting theorem*:

Theorem A6: (Shifting Theorem [39]) *If the sequence $\{f_n\}$ is exponentially bounded, i.e. there exist $C > 0$ and c_0 such that*

$$|f_n| \le C\,e^{c_0 n}, \qquad n = 0, 1, \ldots,$$

then the Z–transformation $\hat{f}(z)$ is given by the Laurent series (A1) and for the shifted sequence $\{g_n\}$ with $g_n = f_{n+1}$ holds

$$\mathcal{Z}\{f_{n+1}\} = z\hat{f}(z) - zf_0. \tag{A2}$$

The initial values enter into the transformation of the shifted sequence. As a useful consequence of the shifting theorem we have:

$$\mathcal{Z}\{f_{n+1} \pm f_n\} = (z \pm 1)\hat{f}(z) - zf_0. \tag{A3}$$

The convolution $f_n * g_n$ of two sequences $\{f_n\}$, $\{g_n\}$, $n = 0, 1, \ldots$ is defined by $\sum_{k=0}^{n} f_k\,g_{n-k}$. For the Z–transformation of a convolution of two sequences we formulate the following theorem:

Theorem A7: (Convolution Theorem [39]) *If $\hat{f}(z) = \mathcal{Z}\{f_n\}$ exists for $|z| > R_{\hat{f}} \geq 0$ and $\hat{g}(z) = \mathcal{Z}\{g_n\}$ for $|z| > R_{\hat{g}} \geq 0$, then there also exists $\mathcal{Z}\{f_n * g_n\}$ for $|z| > \max(R_{\hat{f}}, R_{\hat{g}})$ with*

$$\mathcal{Z}\{f_n * g_n\} = \hat{f}(z)\,\hat{g}(z). \tag{A4}$$

Note that (A4) is nothing else but an expresssion for the Cauchy product of two power series.

Now we present two basic rules for calculating the *inverse Z–transformation* which are essential for formulating the discrete TBCs.

Theorem A8: (Inverse Z–transformation [39]) *If $\{f_n\}$ is an exponentially bounded sequence and $\hat{f}(z)$ the corresponding Z–transformation then the inverse Z–transformation is given by*

$$f_n = \mathcal{Z}^{-1}\left\{\hat{f}(z)\right\} = \frac{1}{2\pi i} \oint_C \hat{f}(z)\, z^{n-1} dz, \qquad n = 0, 1, \ldots, \tag{A5}$$

where C is a circle around the origin with sufficiently large radius.

Other inversion formulas can be obtained by using the fact that $\hat{f}(z^{-1})$ is a Taylor series or if $\hat{f}(z)$ is a rational function of z, analytic at ∞. The most important formula is the *inverse Z–transformation of a product*:

$$\mathcal{Z}^{-1}\left\{\hat{f}(z)\,\hat{g}(z)\right\} = f_n * g_n = \sum_{k=0}^{n} f_k\, g_{n-k}. \tag{A6}$$

Theorem A9: (Initial Value Theorem [39]) *If $\hat{f}(z) = \mathcal{Z}\{f_n\}$ exists then*

$$f_0 = \lim_{z \to \infty} \hat{f}(z). \tag{A7}$$

z can tend to ∞ on the real axis or on an arbitrary path, since $\hat{f}(z)$ is analytic at $z = \infty$.

This theorem, when repeatedly applied to $\hat{f}(z)$, $\hat{f}(z) - f_0$, $\hat{f}(z) - f_0 - f_1 z^{-1}$, etc., yields a method for the inversion of the Z–transformation:

$$f_n = \lim_{z \to \infty} z^n \left[\hat{f}(z) - \sum_{k=0}^{n-1} f_k z^{-k}\right], \qquad n = 0, 1, 2, \ldots. \tag{A8}$$

References

1. B. Engquist and A. Majda, "Radiation Boundary Conditions for Acoustic and Elastic Wave Calculations," *Comm. Pure Appl. Math.* **32** (1979), 313–357.

2. F. Schmidt and P. Deuflhard, "Discrete transparent boundary conditions for the numerical solution of Fresnel's equation," *Comput. Math. Appl.* **29** (1995), 53–76.

3. C.W. Rowley and T. Colonius, "Discretely Nonreflecting Boundary Conditions for Linear Hyperbolic Systems," *J. Comp. Phys.* **157** (2000), 500–538.

4. L. Wagatha, "On Boundary Conditions for the Numerical Simulation of Wave Propagation," *Appl. Num. Math.* **1** (1985), 309–314.

5. J.C. Wilson, "Derivation of boundary conditions for the artificial boundaries associated with the solution of certain time dependent problems by Lax–Wendroff type difference schemes," *Proc. Edinb. Math. Soc.* **II. Ser. 25** (1982), 1–18.

6. L. Halpern, "Absorbing Boundary Conditions for the Discretization Schemes for the One–Dimensional Wave Equation," *Math. Comp.* **38** (1982), 415–429.

7. G. Lill, *Diskrete Randbedingungen an künstlichen Rändern* (Ph.D. Thesis, Technische Hochschule Darmstadt, 1992).

8. S.N. Elaydi, *An introduction to difference equations* (Springer, New York, 1996).

9. F. Black and M. Scholes, "The pricing of options and corporate liabilities," *J. Polit. Econ.* **81** (1973), 637–659.

10. R.C. Merton, "Theory of rational option pricing," *Bell J. Econ. Manag. Sci.*, **4** (1973), 141–183.

11. P. Wilmott, S. Howison, and J. Dewynne, *The Mathematics of Financial Derivatives, A Student Introduction* (Cambridge University Press, 2002).

12. M. Ehrhardt and R.E. Mickens, "Discrete Artificial Boundary Conditions for the Black–Scholes Equation of American Options," (in preparation).

13. M. Ehrhardt and A. Zisowsky, "Fast Calculation of Energy and Mass preserving solutions of Schrödinger–Poisson systems on unbounded domains," *Preprint No. 162 of the DFG Research Center* MATHEON *Berlin, 2004*, (submitted to J. Comp. Appl. Math.).

14. P. Markowich, C. Ringhofer and C. Schmeiser, *Semiconductor equations* (Springer, New York, 1990).

15. F.D. Tappert, "The parabolic approximation method," in: *Wave Propagation and Underwater Acoustics*, Lecture Notes in Physics 70, eds. J.B. Keller and J.S. Papadakis, (Springer, New York, 1977), 224–287.

16. M. Ehrhardt and R.E. Mickens, "Solutions to the Discrete Airy Equation: Application to Parabolic Equation Calculations," *J. Comp. Appl. Math.* **172** (2004), 183–206.

17. A. Arnold and M. Ehrhardt, "Discrete transparent boundary conditions for wide angle parabolic equations in underwater acoustics," *J. Comp. Phys.* **145** (1998), 611–638.

18. M. Ehrhardt and A. Arnold, "Discrete Transparent Boundary Conditions for

the Schrödinger Equation," *Riv. Mat. Univ. Parma* **6** (2001), 57–108.

19. C. Ringhofer and J. Soler, "Discrete Schrödinger–Poisson Systems preserving Energy and Mass," *Appl. Math. Lett.* **13** (2000), 27–32.

20. M.F. Levy, *Parabolic equation models for electromagnetic wave propagation* (IEE Electromagnetic Waves Series 45, 2000).

21. T.W. Dawson, D.J. Thomson and G.H. Brooke, "Non–local boundary conditions for acoustic PE predictions involving inhomogeneous layers," *Proceedings of the Third European Conference on Underwater Acoustics*, FORTH–IACM, Heraklion, Greece, 1996, 183–188.

22. M. Ehrhardt, *Discrete Artificial Boundary Conditions* (Ph.D. Thesis, Technische Universität Berlin, 2001).

23. A. Arnold, "Numerically Absorbing Boundary Conditions for Quantum Evolution Equations," *VLSI Design* **6** (1998), 313–319.

24. G.N. Watson, *A treatise on the theory of Bessel functions* (Cambridge University Press, Cambridge, 1966).

25. E.W. Barnes, "On the homogeneous linear difference equation of the second order with linear coefficients," *Messenger* **34** (1904), 52–71.

26. I.J. Thompson and A.R. Barnett, "Coulomb and Bessel Functions of Complex Arguments and Order," *J. Comp. Phys.* **46** (1986), 490–509.

27. H. Poincaré, "Sur Les Equation Linéaires aux Differentielles Ordinaires et aux Différence Finies," *Amer. J. Math.* **7** (1885), 203–258.

28. O. Perron, "Über Summengleichungen und Poincarésche Differenzengleichungen," *Math. Annalen* **84** (1921), 1–15.

29. R.E. Mickens, "Asymptotic properties of solutions to two discrete Airy equations," *J. Difference Equ. Appl.* **3** (1998), 231–239.

30. R. Wong and H. Li, "Asymptotic expansions for second–order linear difference equations. II," *Stud. Appl. Math.* **87** (1992), 289–324.

31. R.E. Mickens, "Asymptotic properties of solutions to discrete Coulomb equations," *Comput. Math. Appl.* **36** (1998), 285–289.

32. R. Wong and H. Li, "Asymptotic expansions for second–order linear difference equations," *J. Comput. Appl. Math.* **41** (1992), 65–94.

33. W.J. Lentz, "Generating Bessel Functions in Mie Scattering Calculations Using Continued Fractions," *Appl. Opt.* **15** (1976), 668–671.

34. A. Arnold, M. Ehrhardt and I. Sofronov, "Discrete transparent boundary conditions for the Schrödinger equation: Fast calculation, approximation, and stability," *Comm. Math. Sci.* **1** (2003), 501–556.

35. M. Ehrhardt and A. Arnold, "Discrete Transparent Boundary Conditions for Wide Angle Parabolic Equations: Fast Calculation and Approximation," *Proceedings of the Seventh European Conference on Underwater Acoustics*, July 3–8, 2004, TU Delft, The Netherlands, 9–14.

36. A. Zisowsky, *Discrete Transparent Boundary Conditions for Systems of Evolution Equations* (Ph.D. Thesis, Technische Universität Berlin, 2003).

37. S. Zhang and J. Jin, *Computation of special functions: with over 100 computer programs in FORTRAN* (John Wiley & Sons, New York, 1996). (MATLAB routines: http://ceta.mit.edu/comp_spec_func/)

38. R.E. Mickens, "Asymptotic solutions to a discrete Airy equation," *J. Diffe-*

rence Equ. Appl. **7** (2001), 851–858.

39. G. Doetsch, *Anleitung zum praktischen Gebrauch der Laplace-Transformation und der Z-Transformation* (R. Oldenburg Verlag München, Wien, 1967).

CHAPTER 9

ASYMPTOTICALLY CONSISTENT NON-STANDARD FINITE-DIFFERENCE METHODS FOR SOLVING MATHEMATICAL MODELS ARISING IN POPULATION BIOLOGY

A. B. Gumel

Department of Mathematics
University of Manitoba
Winnipeg, Manitoba, R3T 2N2, Canada
gumelab@cc.umanitoba.ca

K. C. Patidar

Department of Mathematics and Applied Mathematics
University of Pretoria
Pretoria 0002, South Africa

R. J. Spiteri

Department of Computer Science
University of Saskatchewan
Saskatoon, Saskatchewan, S7N 5C9, Canada

Ever since the pioneering work of Kermack and McKendrick in the 1930s, numerous compartmental mathematical models have been used to help gain insights into the transmission and control mechanisms of many human diseases. These models are often of the form of systems of non-linear differential equations, whose closed-form solutions are not easily obtainable (if at all), necessitating the use of numerical methods for their approximate solutions. Easy-to-use standard explicit finite-difference methods, such as the forward Euler and explicit Runge-Kutta methods, have often been used to solve these models. Unfortunately, these methods may suffer spurious behaviours, which are not the features of the continuous model being approximated, when certain values of the associated discretization and model parameters are used in the simulations. The aim of this chapter is to investigate a class of finite-difference methods, designed *via* the non-standard framework of Mickens, for solving systems of differential equations arising in population biology. It will be shown that this class of methods can often give numerical results that are asymptotically consistent with those of the corresponding continuous model. This fact is illustrated using a number of case studies arising from population biology (human epidemiology and ecology).

1. Introduction

Following the successes of Kermack and McKendrick in modelling a malaria epidemic in the 1930s [11], many mathematical models have been developed and used to study the transmission and control dynamics of numerous emerging and re-emerging human diseases (see, for instance, [2,3,5,10] and the references therein). These models, often non-linear and deterministic in nature, are generally formulated by subdividing the total population into a number of mutually exclusive compartments. The non-linear and multi-dimensional nature of these models often necessitates the use of numerical integrators for their solutions (because closed-form solutions, expressible in terms of a combination of elementary functions, may not exist or be obtainable).

For many decades, easy-to-use standard numerical methods, such as the explicit forward Euler and higher-order Runge-Kutta methods, have been frequently used to solve non-linear initial-value problems (IVPs) arising from the mathematical modelling of many real-life phenomena (such as those arising in disease transmission and control). A number of studies have shown that the use of such schemes to solve real-life models may lead to scheme-dependent instabilities and/or convergence to spurious solutions (see, for instance, [5,6,22]). In other words, the use of some standard numerical methods may lead to numerical solutions with artifacts that do not correspond to features of the solutions of the continuous model.

Although the afore-mentioned drawbacks can, in general, be circumvented by using small step-sizes [5,13,14,23], the computing costs associated with using such step-sizes in monitoring the long-term dynamics of population models can be substantial. Thus, there is a need to construct numerical schemes that allow the use of the largest possible step-sizes (that are consistent with stability) in the numerical simulations. One other well-known method of getting around the stability drawbacks associated with the use of standard explicit methods is to opt for implicit formulations for the solution of non-linear IVPs. Unfortunately, although they are more robust than the explicit ones (in terms of stability), it will be shown in this chapter that some of these implicit methods are not free of spurious behaviour.

The main aim of this chapter is to introduce the non-standard family of finite-difference methods [5,6,14,15,22] and to show that, by generally preserving essential qualitative properties of the continuous (population) models being investigated, they are more robust in capturing the correct asymptotic dynamics of the models in comparison to the standard explicit methods as well as implicit methods that do not preserve these properties. Models from mathematical epidemiology and ecology will be used to illustrate this.

2. SIS Model

A basic model for the spread of a disease in a population of size $N = N(t)$ subdivided into compartments of susceptible ($S(t)$) and infected ($I(t)$) sub-populations is given by (see [3,10,11]):

$$\frac{dS}{dt} = \Pi - \beta SI - \mu S + \gamma I,$$
$$\frac{dI}{dt} = \beta SI - \mu I - dI - \gamma I, \tag{1}$$

where Π is the recruitment rate (by birth or immigration) of individuals into the population (assumed susceptible), β is the effective contact rate, μ is the natural death rate, γ is the recovery rate, and d is the disease-induced death rate. This model assumes that an average infective makes contact sufficient to transmit infection with βN others *per* unit time.

As noted earlier, because the SIS model (1) monitors human populations, it is assumed that all its state variables and parameters are non-negative for all $t \geq 0$. Adding the two equations in (1) gives $dN/dt = \Pi - \mu N - dI$. Consequently, in the absence of infection, $N \to \Pi/\mu$ as $t \to \infty$ and Π/μ is an upper bound of $N(t)$ provided that $N(0) \leq \Pi/\mu$. Thus, the following feasible region:

$$\mathcal{D} = \left\{ (S,I) \in \mathbb{R}^2_+ : S + I \leq \Pi/\mu \right\},$$

is positively invariant. It is therefore sufficient to consider the solutions of (1) in $\mathcal{D}$. In this region, the usual existence and uniqueness results hold for the system.

2.1. *Qualitative features*

Before designing any numerical method for solving (1), it is imperative that the essential qualitative features of the model are determined. These (analytical results) can then be used as yardsticks for measuring the suitability and/or competitiveness of numerical methods for solving (1). Numerical methods that fail to faithfully mimic the qualitative features of the continuous model can then be termed as "inappropriate".

2.1.1. *Disease-free equilibrium*

The equilibria of the model are obtained by setting the right-hand sides of the equations in (1) to zero. It follows that in the absence of the disease ($I = 0$), the model has a disease-free equilibrium given by $\mathcal{E}_0 = (S^*, I^*) = (\Pi/\mu, 0)$. It can be shown that the eigenvalues of the Jacobian of (1) evaluated at $\mathcal{E}_0$ have negative real parts provided $\mathcal{R}_0 = \beta\Pi/[\mu(\mu + d + \gamma)] < 1$. This result is summarized below.

Lemma 1: *The disease-free equilibrium $\mathcal{E}_0$ is locally asymptotically stable if $\mathcal{R}_0 < 1$ and unstable if $\mathcal{R}_0 > 1$.*

This threshold quantity $\mathcal{R}_0$ is the basic reproduction number of infection [2,10]. It measures the average number of new cases generated by a single infected individual in a completely susceptible population. Lemma 1 shows that if $\mathcal{R}_0 < 1$, then the disease can be eliminated from the population if the initial sizes of the sub-populations of the model are in the basin of attraction of $\mathcal{E}_0$. The following result guarantees disease elimination regardless of the initial sizes of the sub-populations of the model:

Theorem 1: *The disease-free equilibrium is globally asymptotically stable if $\mathcal{R}_0 < 1$.*

Proof: The disease-free equilibrium, $\mathcal{E}_0$, is the only biologically feasible equilibrium in $\mathcal{D}$ whenever $\mathcal{R}_0 < 1$ (see Section 2.1.2 below). Define the following Lyapunov function $V = I$ with Lyapunov derivative (using dot for time derivative)

$$
\begin{aligned}
\dot{V} &= \beta SI - \mu I - dI - \gamma I \\
&\leq \left(\frac{\beta\Pi}{\mu} - \mu - d - \gamma\right) I \quad \text{(because } S \leq \Pi/\mu \text{ in } \mathcal{D}) \\
&= (\mu + d + \gamma)(\mathcal{R}_0 - 1)I \leq 0 \quad \text{for} \quad \mathcal{R}_0 \leq 1.
\end{aligned}
$$

Furthermore, $\dot{V} = 0$ if and only if $I = 0$. Since the largest compact invariant set on the line $I = 0$ is $\mathcal{E}_0$, it follows by the Lasalle invariance principle (see, for instance, [9, Theorem 3.1]), every solution to the equations in (1) with initial conditions in $\mathbb{R}_+^2$ approaches $\mathcal{E}_0$ as $t \to \infty$. Thus, $\mathcal{E}_0$ is globally asymptotically stable whenever $\mathcal{R}_0 < 1$. $\qquad\square$

2.1.2. *Endemic equilibrium*

The SIS model (1) has a unique endemic equilibrium ($I \neq 0$) given by

$$\mathcal{E}_1 = (S^*, I^*) = \left(\frac{\mu + d + \gamma}{\beta}, \frac{\mu(\mu + d + \gamma)(\mathcal{R}_0 - 1)}{\beta(\mu + d)} \right). \qquad (2)$$

It follows from (2) that the unique endemic equilibrium exists only if $\mathcal{R}_0 > 1$ (the model has no endemic equilibrium when $\mathcal{R}_0 < 1$, since, in this case, $I^* < 0$ which is biologically meaningless). It can be shown that $\mathcal{E}_1$ is locally asymptotically stable if $\mathcal{R}_0 > 1$. Furthermore, using the Dulac criterion [9,17] with multiplier $1/I$, it follows that

$$\frac{\partial}{\partial S} \left(\frac{\Pi - \beta SI - \mu S + \gamma I}{I} \right) + \frac{\partial}{\partial I} \left(\frac{\beta SI - \mu I - dI - \gamma I}{I} \right) = -(\beta + \mu/I) < 0.$$

Thus, there are no periodic solutions in $\mathcal{D}$.

Define $\mathcal{D}_0 = \{(S, I) \in \mathcal{D} : I = 0\}$ (the stable manifold of the disease-free equilibrium). The following result can be proved:

Theorem 2: *The unique endemic equilibrium $\mathcal{E}_1$ is globally asymptotically stable in $\mathcal{D} \setminus \mathcal{D}_0$ whenever $\mathcal{R}_0 > 1$.*

The proof is based on the fact that the disease-free equilibrium is globally asymptotically stable whenever $\mathcal{R}_0 < 1$ and unstable if $\mathcal{R}_0 > 1$ (Theorem 1 and Lemma 1), and that there are no periodic solutions in $\mathcal{D}$ (using the Dulac criterion). Thus, the local stability of $\mathcal{E}_1$ (when $\mathcal{R}_0 > 1$) implies its global stability.

In summary, the above analyses show that the SIS model (1) has the following qualitative features:

(i) a globally stable disease-free equilibrium ($\mathcal{E}_0$) whenever $\mathcal{R}_0 < 1$;
(ii) a unique globally stable endemic equilibrium ($\mathcal{E}_1$) whenever $\mathcal{R}_0 > 1$.

Consequently, the disease will be eliminated if $\mathcal{R}_0 < 1$, and will persist if $\mathcal{R}_0 > 1$. In the following section, a number of numerical methods will be constructed for solving (1), aimed at determining which one(s) would have precisely the same asymptotic dynamical features as the ones summarized above (for the continuous SIS model (1)).

2.2. Finite-difference methods

2.2.1. Forward Euler

A forward Euler scheme can be constructed by replacing the derivatives in (1) with their respective forward-difference approximations, and the non-derivative terms approximated at the base time level. This gives:

$$\frac{S_{n+1} - S_n}{\ell} = \Pi - \beta S_n I_n - \mu S_n + \gamma I_n,$$
$$\frac{I_{n+1} - I_n}{\ell} = \beta S_n I_n - \mu I_n - d I_n - \gamma I_n, \tag{3}$$

where $\ell > 0$ is an increment in time (the time-step or step-size) and the interval $t \geq 0$ at the points $t_n = n\ell$ $(n = 0, 1, 2, ...)$ is discretized in the standard way. The theoretical solution of (1) at t_n is denoted by $S(t_n)$ and $I(t_n)$, and the corresponding numerical solution denoted by S_n and I_n. The equations in (3) can be rearranged to give:

$$S_{n+1} \equiv f_1(S_n, I_n) = S_n + \ell \left(\Pi - \beta S_n I_n - \mu S_n + \gamma I_n \right),$$
$$I_{n+1} \equiv f_2(S_n, I_n) = I_n + \ell \left(\beta S_n I_n - \mu I_n - d I_n - \gamma I_n \right). \tag{4}$$

2.2.2. Analysis of fixed points

The expressions for S_{n+1} and I_{n+1} are of the forms $S_{n+1} = f_1(S_n, I_n)$ and $I_{n+1} = f_2(S_n, I_n)$ respectively. It is easy to verify that the fixed points of the Euler method (4) are

$$(S_n^*, I_n^*) = (\Pi/\mu, 0) \quad \text{and} \quad (S_n^*, I_n^*) = \left(\frac{\mu + d + \gamma}{\beta}, \frac{\mu(\mu + d + \gamma)(\mathcal{R}_0 - 1)}{\beta(\mu + d)} \right),$$

which correspond to the two equilibria ($\mathcal{E}_0$ and $\mathcal{E}_1$) of the continuous SIS model (1). A crucial point of interest is to determine whether these fixed points have the same asymptotic stability properties as the corresponding equilibrium solutions of the continuous model (1). Consider, first of all, the functions

$$f_1(S, I) = S + \ell \left(\Pi - \beta S I - \mu S + \gamma I \right)$$
$$f_2(S, I) = I + \ell \left(\beta S I - \mu I - d I - \gamma I \right). \tag{5}$$

The Jacobian of the system (5) is given by

$$J(S, I) = \begin{bmatrix} 1 - \ell \left(\beta I + \mu \right) & \ell \left(-\beta S + \gamma \right) \\ \ell \beta I & 1 + \ell \left(\beta S - (\mu + d + \gamma) \right) \end{bmatrix}. \tag{6}$$

Evaluating the Jacobian at the fixed point $(S^*, I^*) = (\Pi/\mu, 0)$ gives

$$J(\Pi/\mu, 0) = \begin{bmatrix} 1 - \mu\ell & \ell\,(\gamma - \beta\Pi/\mu) \\ 0 & 1 + \ell\,(\Pi\beta/\mu - (\mu + d + \gamma)) \end{bmatrix}, \tag{7}$$

with eigenvalues

$$\lambda_1 = 1 - \mu\ell \quad \text{and} \quad \lambda_2 = 1 + \ell(\mu + d + \gamma)(\mathcal{R}_0 - 1).$$

It can be shown that the two eigenvalues are less than unity in magnitude provided $\mathcal{R}_0 < 1$ and $0 < \ell < \dfrac{2}{(\mu + d + \gamma)(1 - \mathcal{R}_0)} = \ell_c$. Thus, we have the following result:

Lemma 2: *The Euler method* (4) *will converge to the disease-free equilibrium, $\mathcal{E}_0$, whenever $\mathcal{R}_0 < 1$ and $0 < \ell < \ell_c$.*

Similarly, for $\mathcal{R}_0 > 1$, it can be shown that the Euler method will converge to the fixed point $\mathcal{E}_1$ whenever a certain step-size restriction is satisfied (which can be obtained by evaluating the Jacobian (6) at the endemic equilibrium $\mathcal{E}_1$ and ensuring that the resulting eigenvalues are all less than unity in magnitude.) This (step-size dependency) underlines the conditional stability property of the explicit forward Euler method (as expected). Since population biology models, such as the SIS model (1), are typically monitored over a long time period, it is desirable that an approximating numerical method allows the largest possible step-size (that is consistent with stability of the method).

To illustrate the convergence properties of the Euler method (4), numerous simulations are carried out using the following set of parameter values: $\Pi = 2000$, $\mu = 1/70$, $\gamma = 25$ and $d = 25$ (see [2,3,5,6,10,21] and the references therein for further description and estimation of these parameters) and initial values $S(0) = \Pi/\mu + 5$, $I(0) = 2$ with various values of $\beta \in (0, 1)$. The results obtained, based on using a step-size of $\ell = 0.002$, are tabulated in Table 1, from which it is evident that for the step-size used, the Euler method gives numerical results that are consistent with the theoretical analysis of Section 2 (convergence to the globally stable disease-free equilibrium, $\mathcal{E}_o$, whenever $\mathcal{R}_0 < 1$; and convergence to the endemic equilibrium, $\mathcal{E}_1$, whenever $\mathcal{R}_0 > 1$).

It should be noted that for the simulations corresponding to $\mathcal{R}_0 < 1$ in Table 1, it can be shown (using Lemma 2) that the step-size used ($\ell = 0.002$) is less than the corresponding threshold step-size values (ℓ_c). The effect of varying the step-size is monitored by simulating the case with $\beta = 0.000004$ using various step-sizes. In this case, $\ell_c = 0.0404$ (and, from Lemma 2, convergence to $\mathcal{E}_0$ is expected if $\mathcal{R}_0 < 1$ and $\ell < \ell_c$). The results obtained, tabulated in Table 2, show that the Euler method converges to the correct equilibrium solution ($\mathcal{E}_0$) for values of step-sizes in the range $0 < \ell < 0.03$. Oscillatory convergence, involving negative transient values for the state variables $S(t)$ and $I(t)$, are recorded for $0.03 < \ell < 0.04045$. The method undergoes a period-doubling oscillation (of period two) for $0.04045 \leq \ell < 0.06$ and finally diverges (method fails) for $\ell > 0.06$. In summary, as is typically the case, the convergence of the forward Euler method to a correct steady-state solution of (1) is dependent on the magnitude of the step-size used in the simulations. It induces a scheme-dependent numerical instability (such as period-doubling oscillations leading to divergence) when the step-size restriction is violated. Furthermore, even for step-sizes that are within the allowable threshold but large enough (e.g. $0.03 < \ell < 0.04 < \ell_c$), the Euler method gives profiles that oscillate between negative and positive values during the transient stage. This phenomenon (negative solutions) is alien to the SIS model being solved (which requires the solution profiles to always be non-negative for all initial conditions in $\mathcal{D}$). It is therefore important to design numerical methods that not only allow the largest possible step-size but will also guarantee the preservation of the positivity property of the SIS model (1).

2.2.3. *Second-order implicitly-derived explicit method*

In this section, a second-order method for solving S and I will be constructed.

Two implicit methods for solving S are given below

$$
\begin{aligned}
M_S^{(1)} &: \frac{S_{n+1} - S_n}{\ell} = \Pi - \beta S_{n+1} I_n - \mu S_{n+1} + \gamma I_n, \\
M_S^{(2)} &: \frac{S_{n+1} - S_n}{\ell} = \Pi - \beta S_n I_{n+1} - \mu S_n + \gamma I_{n+1}.
\end{aligned}
\tag{8}
$$

The local truncation errors of $M_S^{(1)}$ and $M_S^{(2)}$, denoted by $L_S^{(1)}$ and $L_S^{(2)}$, are

$$
L_S^{(1)} = \left(\frac{1}{2}\ddot{S} + \beta \dot{S} I + \mu \dot{S} \right) \ell^2 + O(\ell^3)
$$

and

$$
L_S^{(2)} = \left(\frac{1}{2}\ddot{S} + \beta S \dot{I} - \gamma \dot{I} \right) \ell^2 + O(\ell^3),
$$

respectively, as $\ell \to 0$, where all derivatives are evaluated at time t_n. It follows that the linear combination

$$L_c = L_S^{(1)} + L_S^{(2)} = \left(\dddot{S} + \beta \dot{S} I + \beta S \dot{I} + \mu \dot{S} - \gamma \dot{I} \right) \ell^2 + O(\ell^3), \quad \text{as } \ell \to 0. \quad (9)$$

Substituting for $\ddot{S}$ in (9) shows that $L_c = O(\ell^3)$. Thus, a second-order method for solving S in (1) can be constructed by taking the linear combination $M_S^{(1)} + M_S^{(2)}$. This implicit (second-order) method is given by:

$$S_{n+1} = \frac{[2 - \ell(\mu + \beta I_{n+1})]\, S_n + \ell\, [2\Pi + \gamma(I_n + I_{n+1})]}{2 + \ell(\mu + \beta I_n)}. \quad (10)$$

It should be noted that this implicit method involves I_{n+1} which has not yet been determined.

Using a similar approach, it can be shown that the following is a second-order method for solving the equation for I in (1):

$$I_{n+1} = \frac{[2 + \ell(\beta S_{n+1} - \mu - d - \gamma)]\, I_n}{2 + \ell(\gamma + d + \mu - \beta S_n)}. \quad (11)$$

Although the second-order method $\{(10), (11)\}$ is implicit by construction, the solution for S and I can be obtained explicitly *via* solving (10) and (11) simultaneously. This gives:

$$S_{n+1} = \{((2 - \mu\ell)S_n + \ell(2\Pi + \gamma I_n))\, [2 + \ell(\mu + d + \gamma - \beta S_n)]$$
$$+ [\ell(\gamma - \beta S_n)(2 - \ell(\mu + d + \gamma))I_n]\} / F(S_n, I_n), \quad (12)$$

and

$$I_{n+1} = \{((2 + \ell(\mu + \beta I_n))[2 - \ell(\mu + d + \gamma)]I_n)$$
$$+ \ell\beta I_n\, [(2 - \mu\ell)S_n + \ell(2\Pi + \gamma I_n)]\} / F(S_n, I_n), \quad (13)$$

where

$$F(S_n, I_n) = (2 + \ell(\mu + \beta I_n))\, [2 + \ell(\mu + d + \gamma - \beta S_n)] - \ell^2(\gamma - \beta S_n)\beta I_n.$$

It is easy to show that the fixed points of the second-order method $\{(12), (13)\}$ are the same as the corresponding critical points of the SIS model (1). Although this method is generally more robust than the Euler method, in terms of accuracy and asymptotic stability properties by allowing relatively larger step-sizes in the simulations (see Table 3), it also suffers some scheme-dependent instabilities for certain choices of parameters and step-size. For instance, simulating the method with $\mu = 1/70, \Pi = 2000, \gamma = d = 25, \beta = 0.000004$ (so that $\mathcal{R}_0 = 0.0112$) results in negative transient profiles whenever a step-size of $\ell \geq 0.05$ is used in the simulation, albeit the method eventually converges to the correct equilibrium solution. Figures 1 and 2 depict the profiles for $I(t)$ using step-sizes $\ell = 2$ and $\ell = 10$ respectively. Overall, it can be concluded that, despite its superior asymptotic stability property and accuracy (in comparison to the Euler method), the fact that the second-order method $\{(12), (13)\}$ introduces negative solution profiles when some reasonably small step-sizes are used in the simulations, renders it generally ineffective for use to monitor the dynamics of the SIS model (1).

The SIS model was also simulated using Matlab's ODE45 solver with variable step-size. For $\beta = 0.000004$, this method fails when a minimum step-size of $\ell_{\min} > 0.0125$ is used in the simulations (i.e., the solver is told that the minimum acceptable step-size is $\ell = 0.0125$). Furthermore, for $\beta = 0.005$, ODE45 fails once $\ell_{\min} > 0.00113$.

In summary, the failure of these methods (Euler, second order and ODE45) necessitate the design of a new numerical integrator that can preserve the essential properties of the model (1) whilst allowing the use of the largest possible step-size in the simulations.

2.3. *Asymptotically consistent finite-difference method*

It is important to note, first of all, that both the Euler method ((4)) and the second-order method $\{(12), (13)\}$ have negative terms on their right - hand sides, enabling the possibility of generating negative solution profiles. Furthermore, the ODE45 method for solving (1) also admits negative terms on its right-hand sides. Thus, these methods violate the positivity property of the continuous model (1). For models that require positivity, such as the SIS model (1), it was noted by Mickens [15] that any finite-difference method that allows negative transient solutions (characterized by the presence of negative terms on the right-hand sides of the methods) will exhibit spurious behaviour. The failure of the three methods above, in faithfully capturing the proper dynamical behaviour of the model (1) (owing to the absence of positivity preservation), further supports this claim. The new non-standard method to be developed below will be especially designed so that it preserves this important positivity property of the SIS model (1).

Consider the following first-order implicit methods for solving (1):

$$\frac{S_{n+1} - S_n}{\ell} = \Pi - \beta S_{n+1} I_n - \mu S_{n+1} + \gamma I_n,$$

$$\frac{I_{n+1} - I_n}{\ell} = \beta S_{n+1} I_n - (\mu + d + \gamma) I_{n+1}, \tag{14}$$

so that

$$S_{n+1} = \frac{S_n + \ell(\Pi + \gamma I_n)}{1 + \ell(\beta I_n + \mu)} \quad \text{and} \quad I_{n+1} = \frac{I_n + \ell\beta S_{n+1} I_n}{1 + \ell(\mu + d + \gamma)}. \tag{15}$$

It is worth noting that although the method (14) is implicit by construction, the equations in (15) enable the method to be implemented explicitly in a sequential manner (where S_{n+1} is computed first, and then I_{n+1}). Since all the terms on the right-hand sides of the method (15) are positive, it follows that this method will preserve the positivity property of the SIS model (1). In other words, starting the numerical simulation of the method with non-negative initial conditions will generate a sequence of iterates that are non-negative (i.e., $S_0 > 0, I_0 > 0$ implies $S_n > 0, I_n > 0$ for all $n = 1, 2, \ldots$). Furthermore, because the denominators of the right-hand sides of the equations in (15) are never zero, and that $S + I \leq \Pi/\mu$ in $\mathcal{D}$, it follows that the numerical solutions generated by the positivity-preserving method (15) are bounded for all t.

Like the two previous standard methods (Euler and second-order), the fixed points of the method (15) are the same as the critical points of the continuous model (1). Furthermore, it can be shown that the eigenvalues of the Jacobian of the method (15) at the fixed point $(S^*, I^*) = (\Pi/\mu, 0)$ are

$$\lambda_1 = \frac{1}{1 + \mu\ell} \quad \text{and} \quad \lambda_2 = \frac{1 + \ell\beta\Pi/\mu}{1 + \ell(\mu + d + \gamma)},$$

from which it follows that $|\lambda_1| < 1$ (since $\mu\ell > 0$) and $|\lambda_2| < 1$ whenever $\mathcal{R}_0 < 1$. Thus, we have established the following result.

Lemma 3: *The asymptotically consistent method* (15) *will converge to the disease-free equilibrium,* $\mathcal{E}_0$, *whenever* $\mathcal{R}_0 < 1$.

Similarly, it can be shown that

Lemma 4: *The asymptotically consistent method* (15) *will converge to the endemic equilibrium,* $\mathcal{E}_1$, *whenever* $\mathcal{R}_0 > 1$.

Lemmas 3 and 4 show that, unlike the Euler and second-order methods, the new positivity-preserving method (15) will converge to the correct equilibrium solution in accordance with the theoretical results in Section 2 (Theorems 1 and 2) regardless of the magnitude of the step-size (ℓ) used in the simulations. In other words, the method (15) is unconditionally stable and is free of any scheme-dependent numerical instability. Extensive numerical simulations show that the method converges to $\mathcal{E}_0$ whenever $\mathcal{R}_0 < 1$, and to $\mathcal{E}_1$ whenever $\mathcal{R}_0 > 1$ irrespective of the size of ℓ and that the method always generates non-negative solution profiles for all non-negative initial conditions (see Table 4, Table 5). This clearly shows that the method (15) is better suited for use to capture the asymptotic dynamics of the SIS model than the Euler method and the second-order method.

2.4. *Preservation of total population*

Recalling that for $d = 0$, the model (1) gives the rate of change of the total population $dN/dt = \Pi - \mu N$ so that $N \to \Pi/\mu$ as $t \to \infty$. Thus, for the case $d = 0$, the curve $N = \Pi/\mu$ is invariant. Consequently, simulating a numerical method (for this case) with initial conditions $S(0)$ and $I(0)$ such that $S(0) + I(0) = \Pi/\mu$ should generate profiles with $S(t) + I(t) = \Pi/\mu$ for all $t > 0$. In other words, for $d = 0$, any initial conditions that start at $S(0) + I(0) = \Pi/\mu$ stay there for all $t > 0$. The three methods ((4), {12, 13} and (15)) are simulated to determine whether they capture this essential property.

Whilst the Euler method and the second-order method gave results that are in line with the above theory (see Figure 3, where the total population is preserved for all t), the positivity-preserving method (15) does not generally preserve this property (Figure 4). Although the method (15) seems to do very well for the cases where $\mathcal{R}_0 < 1$, it fails to preserve the total population during the early transient stage for cases where $\mathcal{R}_0 > 1$. This constitutes a situation where the positivity-preserving scheme fails to satisfy a crucial qualitative feature of the model. Clearly further research is required into the construction of non-standard finite-difference schemes that generally preserve *linear invariants*; i.e., linear combinations of solution components that remain invariant even though the components themselves may be time-varying. It should be noted that standard methods such as Runge–Kutta or linear multistep methods automatically preserve linear invariants [4,20].

3. Transmission Dynamics of Two HIV Subtypes

Since its emergence in the 1980s, the human immuno-deficiency virus (HIV) remains a major global public health menace, accounting for over 42 million infections and 20 million HIV-related deaths. Although significant progress has been made in designing effective therapeutic drugs, a definitive cure and/or vaccine for the disease remain elusive. One of the main reasons for such failure is the existence of numerous subtypes of HIV, which differ significantly in geographical distribution, cell tropism, and transmission efficiencies [18]. Typically, an invading subtype is brought into a community in which another subtype is already endemic.

This section presents a modified vaccination model for the transmission dynamics of two HIV subtypes in a given community. The total population $(N(t))$ is subdivided into the sub-populations of wholly susceptible individuals $(X(t))$, vaccinated susceptible individuals $(V(t))$, individuals infected with an endemic HIV-Subtype-1 $(Y_1(t))$ and individuals infected with an invading HIV-Subtype-2 $(Y_2(t))$. The model contains the following equations [5]:

$$
\begin{aligned}
\frac{dX}{dt} &= \Pi(1-\rho) - \mu X - \frac{1}{N}\beta_1 cXY_1 - \frac{1}{N}\beta_2 cXY_2, \\
\frac{dV}{dt} &= \Pi\rho - \mu V - \frac{1}{N}(1-\xi_1)\beta_1 cVY_1 - \frac{1}{N}(1-\xi_2)\beta_2 cVY_2, \\
\frac{dY_1}{dt} &= \frac{1}{N}\beta_1 cXY_1 + \frac{1}{N}(1-\xi_1)\beta_1 cVY_1 - (\mu + \gamma_1 + \tau)Y_1, \\
\frac{dY_2}{dt} &= \frac{1}{N}\beta_2 cXY_2 + \frac{1}{N}(1-\xi_2)\beta_2 cVY_2 - (\mu + \gamma_2 + \tau)Y_2,
\end{aligned}
\tag{16}
$$

where $N(t) = X(t)+V(t)+Y_1(t)+Y_2(t)$ and the parameters are as described in Table 6. This four-dimensional model is a modified version of the model proposed in [18]. As in Section 2, the dynamical features of the model (16) will be determined, first of all, before attempting to construct numerical methods for solving the model.

3.1. *Existence and stability of equilibria*

It can be shown that the equilibria of (16) are:

(a) The disease-free equilibrium given by
$$
(X^*, V^*, Y_1^*, Y_2^*) = \left(\frac{(1-\rho)\Pi}{\mu}, \; \frac{\rho\Pi}{\mu}, \; 0, \; 0\right);
$$
(b) Subtype-1-only equilibrium (where the endemic subtype persists while the invading subtype is eliminated),
(c) Subtype-2-only equilibrium (the invading subtype dominates while the endemic subtype is eliminated), and
(d) Co-existence equilibrium (both subtypes co-exist).

The last three equilibria cannot be expressed cleanly in closed form. It can be shown that the eigenvalues of the Jacobian associated with the disease-free equilibrium are given by

$$\lambda_1 = \lambda_2 = -\mu, \lambda_3 = \beta_1 c(1 - \rho\xi_1) - (\mu + \gamma_1 + \tau), \lambda_4$$
$$= \beta_2 c(1 - \rho\xi_2) - (\mu + \gamma_2 + \tau). \tag{17}$$

Since all model parameters are positive, it follows that $\lambda_1 < 0$ and $\lambda_2 < 0$. Furthermore, $\lambda_3 < 0$ provided $R_\rho^{(1)} = \dfrac{\beta_1 c\,(1 - \rho\xi_1)}{\mu + \gamma_1 + \tau} < 1$ and $\lambda_4 < 0$ whenever $R_\rho^{(2)} = \dfrac{\beta_2 c\,(1 - \rho\xi_2)}{\mu + \gamma_2 + \tau} < 1$. Thus, we have the following result.

Lemma 5: *The disease-free equilibrium of the system* (16) *is locally asymptotically stable if* $\mathcal{R}_1 = \max\{R_\rho^{(1)}, R_\rho^{(2)}\} < 1$.

Lemma 5 shows that a small influx of HIV-infected individuals into the community would not result in a major epidemic provided $\mathcal{R}_1 < 1$ (i.e., this occurs when both reproduction numbers $R_\rho^{(1)}$ and $R_\rho^{(2)}$ are less than unity simultaneously). On the other hand, HIV will persist (or be established in the community) if and only if at least one of the two eigenvalues λ_3 or λ_4 has a positive real part (so that $\mathcal{R}_1 > 1$). Further analysis shows that when both reproduction numbers exceed unity, the subtype with the higher reproduction number eventually overcomes the other (and becomes the only existing HIV subtype at steady-state). The two subtypes co-exist if the two reproduction numbers are equal and greater than unity [18].

3.2. *Finite-difference methods*

Two implicit methods for solving the system (16) will be designed.

3.2.1. *Method 1*

Consider the following first-order implicit methods for solving the model (16)

$$\frac{X^{n+1} - X^n}{\ell} = \Pi(1 - \rho) - \mu X^{n+1} - \frac{\beta_1 c Y_1^n X^{n+1} + \beta_2 c Y_2^n X^{n+1}}{X^n + V^n + Y_1^n + Y_2^n},$$

$$\frac{V^{n+1} - V^n}{\ell} = \Pi\rho - \mu V^{n+1}$$

$$- \frac{(1 - \xi_1)\beta_1 c V^{n+1} Y_1^n + (1 - \xi_2)\beta_2 c V^{n+1} Y_2^n}{X^{n+1} + V^n + Y_1^n + Y_2^n},$$

$$\frac{Y_1^{n+1} - Y_1^n}{\ell} = \frac{\beta_1 c X^{n+1} Y_1^{n+1} + (1 - \xi_1)\beta_1 c V^{n+1} Y_1^{n+1}}{X^{n+1} + V^{n+1} + Y_1^n + Y_2^n}$$
$$- (\mu + \gamma_1 + \tau)Y_1^{n+1},$$

$$\frac{Y_2^{n+1} - Y_2^n}{\ell} = \frac{\beta_2 c X^{n+1} Y_2^{n+1} + (1 - \xi_2)\beta_2 c V^{n+1} Y_2^{n+1}}{X^{n+1} + V^{n+1} + Y_1^{n+1} + Y_2^n}$$
$$- (\mu + \gamma_2 + \tau)Y_2^{n+1}, \tag{18}$$

so that

$$X^{n+1} = \frac{X^n + \Pi\ell(1 - \rho)}{1 + \ell\left[\mu + \dfrac{c\,(\beta_1 Y_1^n + \beta_2 Y_2^n)}{X^n + V^n + Y_1^n + Y_2^n}\right]},$$

$$V^{n+1} = \frac{V^n + \rho\Pi\ell}{1 + \ell\left\{\mu + \dfrac{c\,[(1 - \xi_1)\beta_1 Y_1^n + (1 - \xi_2)\beta_2 Y_2^n]}{X^{n+1} + V^n + Y_1^n + Y_2^n}\right\}},$$

$$Y_1^{n+1} = \frac{Y_1^n}{1 + \ell\left\{\mu + \gamma_1 + \tau - \dfrac{\beta_1 c\left[X^{n+1} + (1 - \xi_1)V^{n+1}\right]}{X^{n+1} + V^{n+1} + Y_1^n + Y_2^n}\right\}},$$

$$Y_2^{n+1} = \frac{Y_2^n}{1 + \ell\left\{\mu + \gamma_2 + \tau - \dfrac{\beta_2 c\left[X^{n+1} + (1 - \xi_2)V^{n+1}\right]}{X^{n+1} + V^{n+1} + Y_1^{n+1} + Y_2^n}\right\}}. \tag{19}$$

Like in Section 2, although (18) is implicit by construction, the equations in (19) allow the solution of the model (16) to be computed explicitly *via* the following Gauss-Seidel-like sequential process: Compute X^{n+1}, then V^{n+1}, then Y_1^{n+1} and, finally, Y_2^{n+1}.

Furthermore, it is worth noting that the method for Y_1^{n+1} in (19) has negative terms in its denominator resulting from the approximation of the terms (in the model (16))

$$\frac{\beta_1 c X Y_1}{N} \longrightarrow \frac{\beta_1 c X^{n+1} Y_1^{n+1}}{X^{n+1} + V^{n+1} + Y_1^n + Y_2^n},$$

and

$$\frac{(1 - \xi_1)\beta_1 c V Y_1}{N} \longrightarrow \frac{(1 - \xi_1)\beta_1 c V^{n+1} Y_1^{n+1}}{X^{n+1} + V^{n+1} + Y_1^n + Y_2^n}$$

in (18). Similarly, the method for Y_2^{n+1} in (19) also has negative terms. As noted earlier, the presence of such negative terms is a pre-cursor for spurious behaviour and scheme-dependent instabilities.

3.2.2. *Method 2: Non-standard positivity-preserving*

Here, the aim is to construct a finite-difference scheme that is free of the afore-mentioned instabilities. A crucial fact to note is that the positivity property of the state variables $(X, V, Y_1,$ and $Y_2)$ of the model (16) must be preserved. Since the methods for X and V, given in (19), do not admit negative solutions (for $0 < \xi_1, \xi_2 < 1$), we need only to ensure that the methods for Y_1 and Y_2 do not have negative terms on their right-hand sides. To construct schemes for Y_1 and Y_2 satisfying the positivity property of the model, the variables Y_1 and Y_2 in the first two terms of the equations for dY_1/dt and dY_2/dt in (16) are, respectively, approximated using their non-local representations given below:

$$Y_1 \to 2Y_1^n - Y_1^{n+1} \text{ and } Y_2 \to 2Y_2^n - Y_2^{n+1}. \tag{20}$$

Doing so gives the following methods for Y_1 and Y_2:

$$\frac{Y_1^{n+1} - Y_1^n}{\ell} = \frac{\beta_1 c X^{n+1}(2Y_1^n - Y_1^{n+1}) + (1 - \xi_1)\beta_1 c V^{n+1}(2Y_1^n - Y_1^{n+1})}{X^{n+1} + V^{n+1} + Y_1^n + Y_2^n} \\ - (\mu + \gamma_1 + \tau)Y_1^{n+1},$$

$$\tag{21}$$

and

$$\frac{Y_2^{n+1} - Y_2^n}{\ell} = \frac{\beta_2 c X^{n+1}(2Y_2^n - Y_2^{n+1}) + (1 - \xi_2)\beta_2 c V^{n+1}(2Y_2^n - Y_2^{n+1})}{X^{n+1} + V^{n+1} + Y_1^{n+1} + Y_2^n} \\ - (\mu + \gamma_2 + \tau)Y_2^{n+1}.$$

$$\tag{22}$$

In the terminology of Mickens [14,15], these methods are "non-standard" because non-local discretizations have been used. Upon rearrangement of the discretizations for Y_1^{n+1} and Y_2^{n+1}, the non-standard method (*Method 2*) for solving the HIV model (16) consists of the equations:

$$X^{n+1} = \frac{X^n + \Pi\ell(1 - \rho)}{1 + \ell\left[\mu + \dfrac{c\left(\beta_1 Y_1^n + \beta_2 Y_2^n\right)}{X^n + V^n + Y_1^n + Y_2^n}\right]},$$

$$V^{n+1} = \frac{V^n + \rho\Pi\ell}{1 + \ell\left\{\mu + \dfrac{c\left[(1 - \xi_1)\beta_1 Y_1^n + (1 - \xi_2)\beta_2 Y_2^n\right]}{X^{n+1} + V^n + Y_1^n + Y_2^n}\right\}},$$

$$Y_1^{n+1} = \frac{\left\{1 + \dfrac{2\ell\beta_1 c\left[X^{n+1} + (1-\xi_1)V^{n+1}\right]}{X^{n+1} + V^{n+1} + Y_1^n + Y_2^n}\right\} Y_1^n}{1 + \ell\left\{\dfrac{\beta_1 c\left[X^{n+1} + (1-\xi_1)V^{n+1}\right]}{X^{n+1} + V^{n+1} + Y_1^n + Y_2^n} + \mu + \gamma_1 + \tau\right\}},$$

$$Y_2^{n+1} = \frac{\left\{1 + \dfrac{2\ell\beta_2 c\left[X^{n+1} + (1-\xi_2)V^{n+1}\right]}{X^{n+1} + V^{n+1} + Y_1^{n+1} + Y_2^n}\right\} Y_2^n}{1 + \ell\left\{\dfrac{\beta_2 c\left[X^{n+1} + (1-\xi_2)V^{n+1}\right]}{X^{n+1} + V^{n+1} + Y_1^{n+1} + Y_2^n} + \mu + \gamma_2 + \tau\right\}}. \tag{23}$$

It follows from (23) that *Method 2* admits no negative solutions (note that $0 < \xi_1, \xi_2 < 1$), hence it is expected to preserve the positivity property of the model (16). It is worth noting that *Method 2*, like *Method 1*, is first-order accurate.

Methods 1 and 2 are compared *via* extensive simulations using the following parameter and initial values: $\mu = \frac{1}{32}$, $\Pi = 2000$, $\beta_1 = 0.3$, $\beta_2 = 0.35$, $c = 4$, $\gamma_1 = 0.1$, $\gamma_2 = 0.1$, $\xi_1 = 0.3$, $\xi_2 = 0.4$, $\rho = 0.5$, $\tau = 0.4$, $X(0) = 8000$, $V(0) = 800$, $Y_1(0) = 200$, and $Y_2(0) = 300$. For convenience, only the steady-state values of the infected components (Y_1^* and Y_2^*) are compared. Simulations with the standard Runge-Kutta method of order 4 (RK4) are also carried out for further comparisons.

With the above parameter values, it can be shown that $R_\rho^{(1)} = 1.920$ and $R_\rho^{(2)} = 2.108$ (so that $\mathcal{R}_1 = 2.108$). Thus, Subtype-2 would be expected to invade Subtype-1. The exact steady-state values for Y_1 and Y_2 are $\mathcal{E}_Y = (Y_1^*, Y_2^*) = (0, 3552)$. The effect of variations in step-size (ℓ) is monitored by using various values of ℓ in the simulations. The results, tabulated in Table 7, show that both *Method 1* and the RK4 are affected by the size of time-step used in the computation. *Method 1*, for instance, exhibits contrived oscillations (involving negative transient values) for $2 \le \ell \le 3$ and converged to spurious zeros for $\ell > 3$. Figure 5 depicts the profiles of Y_2 generated using *Method 1* with $\ell = 2.5$. This figure shows that although the method eventually converged to the correct steady-state solution, it did give negative transient values of Y_2 during the first 200 units of time. The presence of such negative values is obviously not consistent with the original HIV model (16) (which requires all the dependent variables to be non-negative at all times). Furthermore, the profiles generated by *Method 1* for $\ell = 6$ (Figure 6) show convergence to the wrong steady-state solution (with $Y_2^* = 0$ instead of $Y_2^* = 3552$). Similarly, the RK4 converged to spurious (wrong) results for $\ell \geq 5$, before subsequently diverging when $\ell \geq 7$. Simulations with ODE45 method were also carried out, generating profiles that are consistent with the theoretical findings above for reasonably small values of the minimum allowable step-size (Figure 7). This method fails, however, when the minimum allowable step-size exceeds $\ell_{\min} = 1.32$.

While Table 7 and Figures 5-7 clearly illustrate the spurious behaviour and scheme-dependent numerical instabilities associated with the use of standard schemes like *Method 1* and the RK4, *Method 2*, on the other hand, seem to always give numerical results that converge to the correct steady-state solutions $\mathcal{E}_Y = (0, 3552)$ for any arbitrary large value of $\ell > 0$ and/or parameter value used in the simulations. Such robustness of *Method 2* in capturing the correct asymptotic behaviour of the model is clearly attributable to its positivity-preserving property. Further simulations are carried out with *Method 2* to assess the effect of the size of the reproduction numbers, $R_\rho^{(i)}$, by using various combinations of $R_\rho^{(1)}$ and $R_\rho^{(2)}$ in the numerical simulations. The results generated are tabulated in Table 8 from which it follows that the subtype with the higher reproduction number dominates the other (in line with the theory). When both reproductive numbers are less than unity (simultaneously), so that $\mathcal{R}_1 < 1$, the method converges to the disease-free equilibrium as expected.

4. A Predator-Prey Model

Predator-prey models typically exhibit oscillatory dynamics involving limit cycles. The aim here is to investigate the competitiveness of a positivity-preserving numerical scheme in capturing the essential dynamics of such a model. Consider the following predator-prey model [16]

$$\frac{dx}{dt} = rx\left(1 - \frac{x}{K}\right) - \frac{xy}{a + x^2},$$

$$\frac{dy}{dt} = y\left(\frac{\mu x}{a + x^2} - D\right), \tag{24}$$

where x and y represent the population densities of prey and predator, respectively; the parameter $K > 0$ represents the maximum carrying capacity of the prey, $D > 0$ is the mortality rate of the predator, $\mu > 0$ is the rate of conversion of prey into predator, $r > 0$ is growth rate of prey in the absence of predator.

4.1. *Existence and stability of equilibria*

The dynamics of the predator-prey model (24) is considered in the feasible (positively invariant) region

$$\mathcal{D}_1 = \{(x, y) : x \geq 0, \, y \geq 0\}.$$

The model (24) has two boundary equilibria namely $(x_1, y_1) = (0, 0)$ and $(x_2, y_2) = (K, 0)$. To find the interior equilibria of the model (where $x > 0$ and $y > 0$), Ruan and Xiao [19] considered the following three cases:

Case 1: $\mu^2 - 4aD^2 < 0$. In this case, the model has no interior equilibria;

Case 2: $\mu^2 - 4aD^2 = 0$ and $\frac{\mu}{2D} < K$. In this case, the model has a unique interior equilibrium given by:

$$(x_3, y_3) = \left(\frac{\mu}{2D}, \, r\left(1 - \frac{x_3}{K}\right)\left(a + x_3^2\right) \right);$$

Case 3: $\mu^2 - 4aD^2 > 0$. Here, the model has at most two interior equilibria denoted by

$$(x_4, y_4) = \left(\frac{\mu - \sqrt{\mu^2 - 4aD^2}}{2D}, \, r\left(1 - \frac{x_4}{K}\right)\left(a + x_4^2\right) \right),$$

$$(x_5, y_5) = \left(\frac{\mu + \sqrt{\mu^2 - 4aD^2}}{2D}, \, r\left(1 - \frac{x_5}{K}\right)\left(a + x_5^2\right) \right).$$

It can be shown that $(x_1, y_1) = (0, 0)$ is a saddle point (hence, unstable) and (x_2, y_2) is globally asymptotically stable if $\delta = \frac{\mu K}{a + K^2} - D < 0$. Since the main focus of this subsection is on the design and simulation of a suitable numerical method for solving the system (24), the dynamics of the associated interior equilibria $((x_i, y_i)$ for $i = 3, 4, 5)$ will not be reported (the reader may refer to [19] for further theoretical details).

4.2. *Positivity-preserving finite-difference method*

The method is constructed based on the discretization below [24]:

$$\frac{X_{n+1} - X_n}{\ell} = rX_n\left(1 - \frac{X_{n+1}}{K}\right) - \frac{X_{n+1}Y_n}{a + X_n^2},$$

$$\frac{Y_{n+1} - Y_n}{\ell} = \frac{\mu X_{n+1}Y_n}{a + X_{n+1}^2} - DY_{n+1} \tag{25}$$

so that,

$$X_{n+1} = \frac{(1+r\ell)X_n}{1+\ell\left(\dfrac{X_n r}{K} + \dfrac{Y_n}{a+X_n^2}\right)}, \qquad Y_{n+1} = \frac{\left(1+\dfrac{\mu\ell X_{n+1}}{a+X_{n+1}^2}\right)Y_n}{1+D\ell}. \tag{26}$$

Like the methods (15) and (23) in earlier sections, it is clear that the first-order method (26) has no negative terms on its right-hand side and, hence, it will be expected to preserve the positivity property of the predator-prey model (24). Furthermore, even-though the construction of the method is implicit (Eq. (25)), the equations in (26) allow the solution of the predator-prey model to be obtained explicitly.

It can be verified that the fixed points of the method (26) are the same as the equilibrium points of the predator-prey model (24). It can further be shown that the fixed point $(0,0)$ is unstable; similarly, the fixed point $(K,0)$ is stable provided

$$\left| \frac{1+\dfrac{\mu\ell K}{a+K^2}}{1+D\ell} \right| < 1,$$

which holds whenever

$$\delta = \frac{\mu K}{a+K^2} - D < 0.$$

Thus, the discrete model (26) has the same standard properties (for the equilibria $(0,0)$ and $(K,0)$) as the continuous predator-prey model (24).

4.3. *Numerical simulations*

Numerous simulations are carried out to test the stability and convergence properties of the positivity-preserving method (26), and comparisons are made with RK4. The main focus of these simulations is to investigate the existence and stability of limit cycles associated with the use of the method (26) to solve the model equations in (24). Consequently, parameter values for simulating the model have to be chosen such that none of the two equilibria $((0,0)$ and $(K,0))$ is stable. This requires making $\delta = \frac{\mu K}{a+K^2} - D > 0$. Furthermore, for the case $\mu^2 - 4aD^2 > 0$, it is shown that [19]:

(i) the model has no interior equilibrium if $K < x_4$;

(ii) the model has exactly one interior equilibrium (x_4, y_4) if $x_4 < K < x_5$. In this case, there may exist one, two, or more limit cycles surrounding (x_4, y_4) ;

(iii) the model has two interior equilibria (x_4, y_4) and (x_5, y_5) if $K > x_5$, where (x_4, y_4) is a hyperbolic saddle, and (x_5, y_5) is a focus or node.

The positivity-preserving method (26) is simulated using the following set of parameter values: $r = 0.2$, $\mu = 4.5$, $K = 10$, $a = 30$, $D = 0.3$, $\ell = 1$. For this choice of parameter values, $\delta = 0.0461 > 0$, $\mu^2 - 4aD^2 = 9.45 > 0$ and the equilibrium $(x_4, y_4) = (2.37652, 5.43521)$ is a stable focus. Simulating the method with initial conditions very close to this equilibrium gives profiles that converge to it. For initial conditions relatively far away from this equilibrium, such as $(X(0), Y(0)) = (12, 2)$, the method converged to a limit cycle (Figure 8). Furthermore, using an initial condition inside the limit cycle such as $(X(0), Y(0)) = (3, 5)$, the method gave solution profiles that converged to a stable limit cycle (Figure 9). In other words, the method shows that the limit cycle is stable since trajectories from inside and outside it always approach it as $t \to \infty$. Furthermore, since the equilibrium (x_4, y_4) is a stable focus, the method (26) suggest the existence of yet another limit cycle (which is unstable) in a small neighbourhood of (x_4, y_4) (see also [19]).

An estimate of the period of oscillation is obtained by evaluating the Jacobian of the continuous predator-prey model (24) at an arbitrary point on the limit cycle, such as $(x, y) = (3.279503, 11.3001)$, resulting in eigenvalues $-4 \times 10^{-9} \pm 0.20869i$. It follows that the period of oscillation (T) is approximated by $T = \frac{2\pi}{0.20869} \approx 30.12$ (see also [24]). Although, in general, numerical integrators must always be simulated with step-sizes lower than the period of oscillation (or smallest associated time-scale), the positivity-preserving method (26) seems unaffected by the size of ℓ. For instance, using initial conditions $(X(0), Y(0)) = (12, 2)$ and step-sizes that exceed the period of oscillation, such as $\ell = 10, 100$ or 10^6, the method (26) gave profiles that converged to the stable limit cycle (in agreement with the theoretical results).

For comparison purposes, simulations with RK4 were carried out to determine the effect of step-size on its convergence using the above parameter values with various values of ℓ and initial conditions $(X(0), Y(0)) = (12, 2)$. The results obtained are summarized below:

(i) For relatively small step-sizes, such as $\ell = 0.5$ or 3, the RK4 method gives the same phase portrait as those generated using the method (26);

(ii) For step-sizes such as $\ell = 5$ or 7, the RK4 converges to the stable focus (x_4, y_4), instead of converging to the stable limit cycle. Thus, by converging to (x_4, y_4) in this case, the RK4 has introduced scheme-dependent instability (convergence to spurious solution);

(iii) The RK4 method diverged (overflow) for $\ell \geq 10$.

Thus, the RK4 fails to capture the true dynamics of the model even when a step-size below the period of oscillation (e.g. $5 \leq \ell < T \approx 30.12$) is used.

It is to be noted that the ODE45 fails for this model when $\ell_{\min} > 1.91885$.

4.4. *A modified predator-prey model*

Consider the following modified mathematical model [12]

$$
\begin{aligned}
\frac{dx}{dt} &= x\left(1 - \frac{x}{\gamma}\right) - \frac{xy}{\dfrac{x^2}{\alpha} + x + 1}, \\
\frac{dy}{dt} &= \frac{\beta\delta xy}{\dfrac{x^2}{\alpha} + x + 1} - \delta y,
\end{aligned}
\tag{27}
$$

where γ, α, β, δ are parameters and x, y are population densities of prey and predator, respectively. A suitable positivity-preserving method for solving (27) is given by

$$
\begin{aligned}
\frac{X_{n+1} - X_n}{\ell} &= X_n\left(1 - \frac{X_{n+1}}{\gamma}\right) - \frac{X_{n+1}Y_n}{\dfrac{X_n^2}{\alpha} + X_n + 1}, \\
\frac{Y_{n+1} - Y_n}{\ell} &= \frac{\beta\delta X_{n+1}Y_n}{\dfrac{X_{n+1}^2}{\alpha} + X_{n+1} + 1} - \delta Y_{n+1},
\end{aligned}
\tag{28}
$$

so that

$$
X_{n+1} = \frac{(1+\ell)X_n}{1 + \ell\left(\dfrac{X_n}{\gamma} + \dfrac{Y_n}{\dfrac{X_n^2}{\alpha} + X_n + 1}\right)},
$$

$$
Y_{n+1} = \frac{Y_n + \dfrac{\ell\beta\delta X_{n+1}Y_n}{\dfrac{X_{n+1}^2}{\alpha} + X_{n+1} + 1}}{1 + \delta\ell}.
\tag{29}
$$

Table 9 depicts the results obtained using the method (29) with $\alpha = 5.2, \beta = 2.0, \delta = 2.5, \ell = 1$ and various values of γ. These results are consistent with those reported in [12]. In particular, the formation of a stable limit cycle (*via* a supercritical Hopf bifurcation) when $\gamma = 3.5$ and the global bifurcation, where the limit cycle runs into a saddle point (forms a homoclinic orbit) and disappears, when $\gamma = 4.6$. Overall, the method (29) seems to capture the essential dynamics of the modified predator-prey model (27) for all initial values and step-sizes.

5. Concluding Remarks and Challenges

This chapter considers the problem of designing appropriate finite-difference methods for solving systems of non-linear differential equations arising in mathematical epidemiology and ecology. It is shown that by failing to preserve the positivity property of such models, standard numerical integrators (such as the explicit RK methods) generally exhibit scheme-dependent instabilities (including period-doubling oscillations, giving negative profiles during transient stages, divergence) and often converge to wrong solutions. Furthermore, this chapter shows that implicit schemes that fail to preserve the positivity property of the population biology model under consideration can, in general, be expected to suffer from spurious behaviour and scheme-dependent instabilities.

Although the above drawbacks can often be obviated by using small step-sizes, the computing costs associated with using small step-sizes in monitoring the long-term dynamics of the corresponding continuous models can be substantial. Thus, for population biology models (such as those considered in this chapter), the need to design numerical schemes that allow the use of the largest possible step-size is paramount. This chapter shows that standard numerical integrators may not always be suited for solving real-life models arising in population biology (such as those arising in human epidemiology and ecology). These phenomena (of scheme-dependent instabilities and convergence to spurious zeros demonstrated by some standard numerical methods) further emphasize the need to, first of all, carry out rigorous qualitative analysis of the given continuous model before any numerical methods are designed. This is especially necessary for validating the asymptotic behaviour of the numerical methods.

Based on the models considered, this chapter further shows that finite-difference schemes that preserve the positivity property of the population model under consideration generally give numerical results that are free of scheme-dependent numerical instabilities and always converge to the correct steady-state solution. In other words, such methods tend to always capture the essential (asymptotic) dynamics of the continuous model being solved regardless of the size of the step-size and parameters used in the simulations. This suggests that positivity-preserving finite-difference formulations are more suited for epidemic and ecology models if capturing asymptotic dynamics is the key objective. These schemes, however, are not without their key challenges. These include:

(i) **Transient dynamics**: The positivity-preserving methods (e.g. (15),(23), (26) and (29)) are only first-order accurate. Thus, unless very small step-sizes are used in the simulations, they may not be able to faithfully capture the transient dynamics (albeit they capture the asymptotic dynamics well). Because fitting continuous and discrete models to data is often a key desirable aspect of mathematical epidemiology and ecology, it is essential that a class of higher-order positivity-preserving schemes be designed. The approach for constructing second-order schemes in Section 2, based on taking linear combinations of appropriate first-order methods, may be the way to proceed.

(ii) **Preservation of linear invariants**: The positivity-preserving scheme often fails to preserve the linear invariants of the given population biology model. This means, starting a simulation with initial conditions that add up to the total population (e.g., $S(0) + I(0) = \Pi/\mu = N(0)$ for the SIS model (1) when $d = 0$) implies that the sum $S(t) + I(t)$ generated by the positivity-preserving method (e.g., (15)) is not always constant. This violates one of the properties of the model, since $N = \Pi/\mu$ is invariant.

(iii) **Oscillatory systems**: Not much work has been done in the nonstandard finite-difference methods community towards designing appropriate numerical schemes for oscillatory systems arising in the natural and engineering sciences. Recent studies show that positivity-preserving schemes tend to be less competitive in capturing the asymptotic dynamics of engineering systems with limit cycles, except if very small step-sizes are used. For relatively modest step-sizes, these methods often converge to an unstable equilibrium solution when it should be converging to a stable limit cycle. An example is the diffusion-free Brusselator system given by the kinetic equations [1,7,8]:

$$\frac{du}{dt} = B + u^2 v - (A+1)u,$$

$$\frac{dv}{dt} = Au - u^2 v, \tag{30}$$

where $u = u(t)$, $v = v(t)$ are chemical concentrations and A and B are positive real constants. The system (30) has a unique equilibrium solution given by $(u^*, v^*) = (B, \frac{A}{B})$, which is stable whenever $1 - A + B^2 > 0$. A stable limit cycle emerges (*via* a Hopf bifurcation) for $1 - A + B^2 < 0$. A typical positivity-preserving scheme for the Brusselator system is given by:

$$\frac{U_{n+1} - U_n}{\ell} = B + U_n^2 V_n - (A+1)U_{n+1},$$

$$\frac{V_{n+1} - V_n}{\ell} = AU_{n+1} - U_{n+1}^2 V_{n+1}. \tag{31}$$

Choosing $A = 3$ and $B = 1$ (so that $1 - A + B^2 = -1 < 0$) and a step-size $\ell = 1$, the method (31) converges to the unstable fixed point $(u^*, v^*) = (1, 3)$ instead of converging to the stable limit cycle. The method converges to the limit cycle for smaller step-sizes such as $\ell = 0.1$. This shows that, for oscillatory problems such as the Brusselator system, a relatively small step-size (less than the smallest timescale of the system) must be used in the simulations. Otherwise, the positivity-preserving scheme will also fail to capture the essential dynamics of the model.

Overall, these positivity-preserving non-standard methods offer significant hope towards finding robust numerical integrators for population biology models. Based on the studies carried out so far, these methods seem to be robust in wide classes of problems arising in the natural and engineering sciences. More examples need to be found in other disciplines to further test the robustness (and applicability) of the non-standard positivity-preserving finite-difference methods.

Acknowledgments: A.B.G. and R.J.S. acknowledge the support of the Natural Sciences and Engineering Research Council of Canada (NSERC) and Mathematics of Information Technology and Complex Systems (MITACS) of Canada. K.C.P. acknowledges the support of the University of Pretoria, South Africa. The authors are grateful to Professor Edward Twizell (Brunel University, England) for his editorial comments.

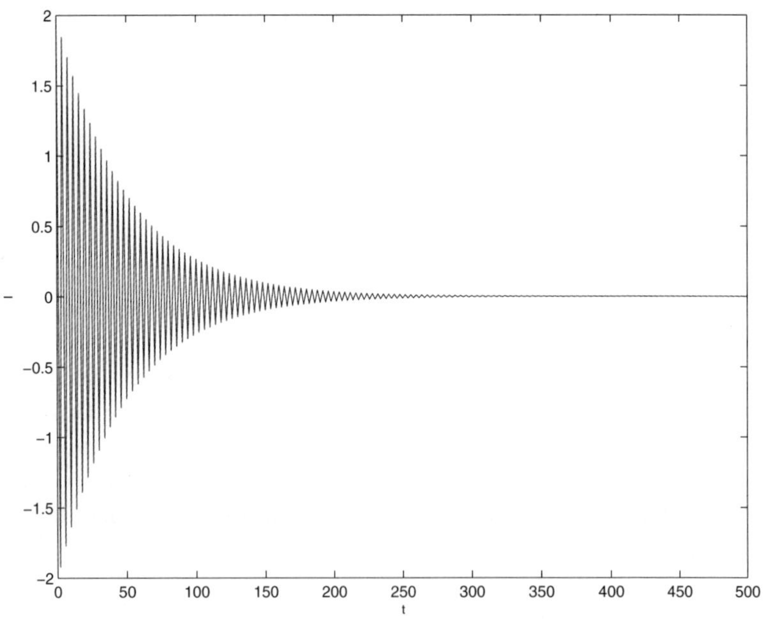

Fig. 1. Profiles of $I(t)$ generated using the second-order method $\{(12), (13)\}$ with $\mu = 1/70$, $\Pi = 2000$, $\gamma = 25$, $d = 25$, $\beta = 0.000004$, $S(0) = \Pi/\mu + 5$, $I(0) = 2$ and $\ell = 2$.

Table 1. Effect of β on the convergence of the Euler method (4) for solving the SIS model (1) using $\ell = 0.002$, $\mu = 1/70$, $\Pi = 2000$, $\gamma = 25$, $d = 25$, $S(0) = \Pi/\mu + 5$, $I(0) = 2$

β	$\mathcal{R}_0$	Equilibrium solutions	Comment
0.000005	0.013996	$\mathcal{E}_0 = (140000, 0)$, $\mathcal{E}_1 = (10002857, -5633)$	Convergence to $\mathcal{E}_0$
0.00005	0.13996	$\mathcal{E}_0 = (140000, 0)$, $\mathcal{E}_1 = (1000286, -491)$	Convergence to $\mathcal{E}_0$
0.00035	0.97972	$\mathcal{E}_0 = (140000, 0)$, $\mathcal{E}_1 = (142892, -2)$	Convergence to $\mathcal{E}_0$
0.00036	1.00771	$\mathcal{E}_0 = (140000, 0)$, $\mathcal{E}_1 = (138929, 1)$	Convergence to $\mathcal{E}_1$
0.0005	1.3996	$\mathcal{E}_0 = (140000, 0)$, $\mathcal{E}_1 = (100029, 23)$	Convergence to $\mathcal{E}_1$
0.005	13.996	$\mathcal{E}_0 = (140000, 0)$, $\mathcal{E}_1 = (10003, 74)$	Convergence to $\mathcal{E}_1$

Table 2. Effect of ℓ on the convergence of the Euler method (4) for solving the SIS model (1) using $\beta = 0.000004$, $\mu = 1/70$, $\Pi = 2000$, $\gamma = 25$, $d = 25$, $S(0) = \Pi/\mu + 5$, $I(0) = 2$

ℓ	$\mathcal{R}_0$	Equilibrium solutions	Comment
0.002	0.0112	$\mathcal{E}_0 = (140000, 0)$, $\mathcal{E}_1 = (12503571, -7061)$	Convergence to $\mathcal{E}_0$
0.02	0.0112	$\mathcal{E}_0 = (140000, 0)$, $\mathcal{E}_1 = (12503571, -7061)$	Convergence to $\mathcal{E}_0$
0.03	0.0112	$\mathcal{E}_0 = (140000, 0)$, $\mathcal{E}_1 = (12503571, -7061)$	Oscillatory convergence to $\mathcal{E}_0$ involving negative values
0.04	0.0112	$\mathcal{E}_0 = (140000, 0)$, $\mathcal{E}_1 = (12503571, -7061)$	Oscillatory convergence to $\mathcal{E}_0$ involving negative values
0.041	0.0112	$\mathcal{E}_0 = (140000, 0)$, $\mathcal{E}_1 = (12503571, -7061)$	Period-doubling oscillation involving negative values
0.061	0.0112	$\mathcal{E}_0 = (140000, 0)$, $\mathcal{E}_1 = (12503571, -7061)$	Method diverges (fails)

Table 3.　Effect of ℓ on the convergence of the second-order method $\{(12),(13)\}$ for solving the SIS model (1) using $\beta = 0.000004$, $\mu = 1/70$, $\Pi = 2000$, $\gamma = 25$, $d = 25$, $S(0) = \Pi/\mu+5$, $I(0) = 2$

ℓ	$\mathcal{R}_0$	Equilibrium solutions	Comment
0.002	0.0112	$\mathcal{E}_0 = (140000, 0)$, $\mathcal{E}_1 = (12503571, -7061)$	Convergence to $\mathcal{E}_0$
0.02	0.0112	$\mathcal{E}_0 = (140000, 0)$, $\mathcal{E}_1 = (12503571, -7061)$	Convergence to $\mathcal{E}_0$
0.05	0.0112	$\mathcal{E}_0 = (140000, 0)$, $\mathcal{E}_1 = (12503571, -7061)$	Oscillatory convergence to $\mathcal{E}_0$ involving negative values
1	0.0112	$\mathcal{E}_0 = (140000, 0)$, $\mathcal{E}_1 = (12503571, -7061)$	Oscillatory convergence to $\mathcal{E}_0$ involving negative values
10	0.0112	$\mathcal{E}_0 = (140000, 0)$, $\mathcal{E}_1 = (12503571, -7061)$	Oscillatory convergence to $\mathcal{E}_0$ involving negative values

Table 4.　Effect of ℓ on the convergence of the positivity-preserving method (15) for solving the SIS model (1) using $\beta = 0.000004$, $\mu = 1/70$, $\Pi = 2000$, $\gamma = 25$, $d = 25$, $S(0) = \Pi/\mu + 5$, $I(0) = 2$

ℓ	$\mathcal{R}_0$	Equilibrium solutions	Comment
0.002	0.0112	$\mathcal{E}_0 = (140000, 0)$, $\mathcal{E}_1 = (12503571, -7061)$	Convergence to $\mathcal{E}_0$
0.02	0.0112	$\mathcal{E}_0 = (140000, 0)$, $\mathcal{E}_1 = (12503571, -7061)$	Convergence to $\mathcal{E}_0$
0.2	0.0112	$\mathcal{E}_0 = (140000, 0)$, $\mathcal{E}_1 = (12503571, -7061)$	Convergence to $\mathcal{E}_0$
1	0.0112	$\mathcal{E}_0 = (140000, 0)$, $\mathcal{E}_1 = (12503571, -7061)$	Convergence to $\mathcal{E}_0$
10	0.0112	$\mathcal{E}_0 = (140000, 0)$, $\mathcal{E}_1 = (12503571, -7061)$	Convergence to $\mathcal{E}_0$
10^6	0.0112	$\mathcal{E}_0 = (140000, 0)$, $\mathcal{E}_1 = (12503571, -7061)$	Convergence to $\mathcal{E}_0$

Table 5. Effect of ℓ on the convergence of the positivity-preserving method (15) for solving the SIS model (1) using $\beta = 0.005$, $\mu = 1/70$, $\Pi = 2000$, $\gamma = 25$, $d = 25$, $S(0) = \Pi/\mu + 5$, $I(0) = 2$

ℓ	$\mathcal{R}_0$	Equilibrium solutions	Comment
0.002	13.996	$\mathcal{E}_0 = (140000, 0)$, $\mathcal{E}_1 = (10003, 74)$	Convergence to $\mathcal{E}_1$
0.02	13.996	$\mathcal{E}_0 = (140000, 0)$, $\mathcal{E}_1 = (10003, 74)$	Convergence to $\mathcal{E}_1$
0.2	13.996	$\mathcal{E}_0 = (140000, 0)$, $\mathcal{E}_1 = (10003, 74)$	Convergence to $\mathcal{E}_1$
1	13.996	$\mathcal{E}_0 = (140000, 0)$, $\mathcal{E}_1 = (10003, 74)$	Convergence to $\mathcal{E}_1$
10	13.996	$\mathcal{E}_0 = (140000, 0)$, $\mathcal{E}_1 = (10003, 74)$	Convergence to $\mathcal{E}_1$
10^6	13.996	$\mathcal{E}_0 = (140000, 0)$, $\mathcal{E}_1 = (10003, 74)$	Convergence to $\mathcal{E}_1$

Table 6. Description of parameters for the HIV model (16)

Parameter	Interpretation
Π	annual recruitment rate of individuals into the sexually active community
ρ	fraction of susceptible individuals vaccinated
μ	rate of cessation of sexual activity
c	number of sexual partners
β_i	probability of *per* partnership transmission
ξ_i	vaccine-induced immunity against subtype$-i$
τ	*per capita* rate of treatment coverage

Table 7.   Effect of time-step ℓ on the convergence of *Methods 1,2* and RK4 for solving the HIV model (16)

ℓ	Method 1	RK4	Method 2
1	Convergence to $\mathcal{E}_Y$	Convergence to $\mathcal{E}_Y$	Convergence to $\mathcal{E}_Y$
2	Convergence to $\mathcal{E}_Y$	Convergence to $\mathcal{E}_Y$	Convergence to $\mathcal{E}_Y$
5	Convergence to $(Y_1^*, Y_2^*) = (0,0)$	Convergence to $(Y_1^*, Y_2^*) = (0, 2768)$	Convergence to $\mathcal{E}_Y$
6	Convergence to $(Y_1^*, Y_2^*) = (0,0)$	Convergence to $(Y_1^*, Y_2^*) = (0, 2362)$	Convergence to $\mathcal{E}_Y$
7	Convergence to $(Y_1^*, Y_2^*) = (0,0)$	Divergence	Convergence to $\mathcal{E}_Y$
10	Convergence to $(Y_1^*, Y_2^*) = (0,0)$	Divergence	Convergence to $\mathcal{E}_Y$
1000	Convergence to $(Y_1^*, Y_2^*) = (0,0)$	Divergence	Convergence to $\mathcal{E}_Y$

Table 8.   Effect of reproduction numbers, $R_p^{(i)}$, using *Method 2* to solve the HIV model (16)

$R_\rho^{(1)}$	$R_\rho^{(2)}$	X^*	V^*	Y_1^*	Y_2^*	Comment
0.8698	0.9406	25600	38400	0	0	Both subtypes eliminated
1.7397	0.9406	792	1674	1700	0	Subtype-1 persists
0.8698	1.7467	761	1866	0	1695	Subtype-2 persists
2.9706	3.2121	826	2021	0	5769	Subtype-2 persists
3.4657	2.7532	788	1667	5806	0	Subtype-1 persists
3.4657	3.4657	788	1667	2507	3299	Both subtypes co-exist

Table 9. Effect of γ on the convergence of the positivity-preserving method (29) for solving the modified predator-prey model (27) using $\alpha = 5.2, \beta = 2.0, \delta = 2.5, \ell = 1$

γ	Comment
1.1	Convergence to $E_1 = (\gamma, 0) = (1.1, 0)$
2.5	Convergence to $(1.351, 1.416)$
3.5	Convergence to limit cycle
4.1	Convergence to limit cycle or $E_1 = (4.1, 0)$ depending on initial conditions
4.6	Convergence to $E_1 = (4.6, 0)$ (global bifurcation of limit cycle)

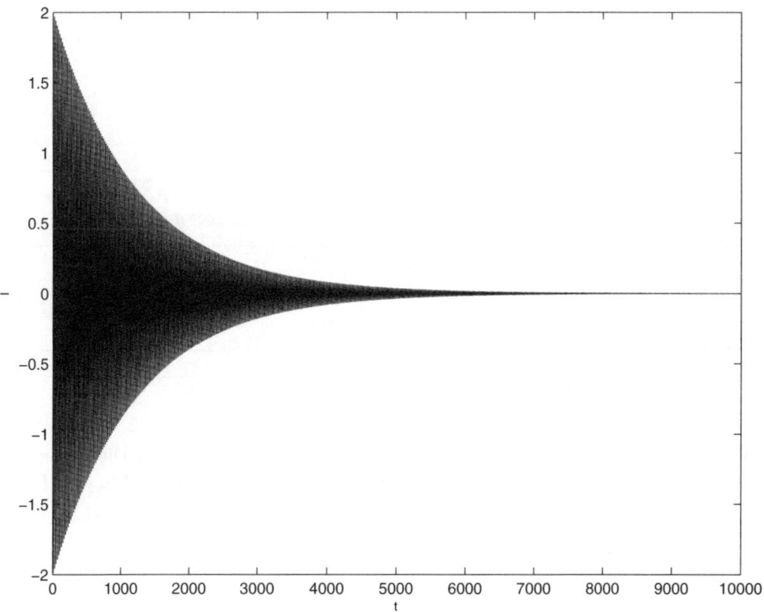

Fig. 2. Profiles of $I(t)$ generated using the second-order method $\{(12), (13)\}$ with $\mu = 1/70, \Pi = 2000, \gamma = 25, d = 25, \beta = 0.000004, S(0) = \Pi/\mu + 5, I(0) = 2$ and $\ell = 10$.

    A. B. Gumel, K. C. Patidar and R. J. Spiteri

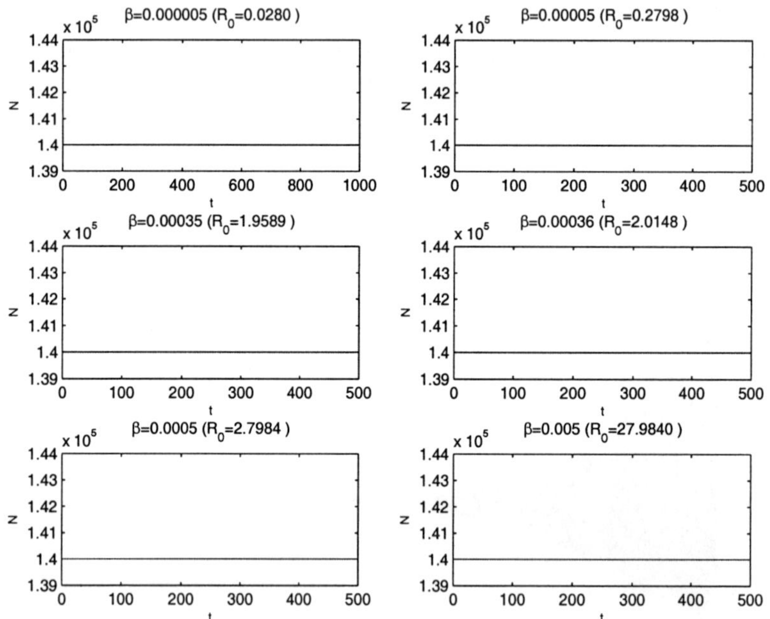

Fig. 3. Profiles of $N(t)$ generated using the forward Euler (4) and the second-order method $\{(12),(13)\}$ with $\mu = 1/70, \Pi = 2000, \gamma = 25, d = 0, S(0) = \Pi/\mu - 2, I(0) = 2$ and $\ell = 0.002$.

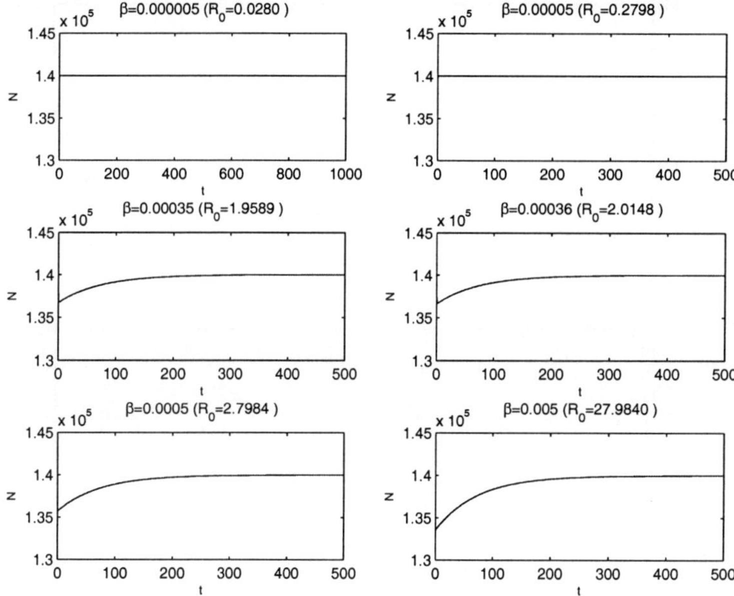

Fig. 4. Profiles of $N(t)$ generated using the non-standard method $((15))$ with $\mu = 1/70$, $\Pi = 2000$, $\gamma = 25$, $d = 0$, $S(0) = \Pi/\mu - 2$, $I(0) = 2$ and $\ell = 0.002$.

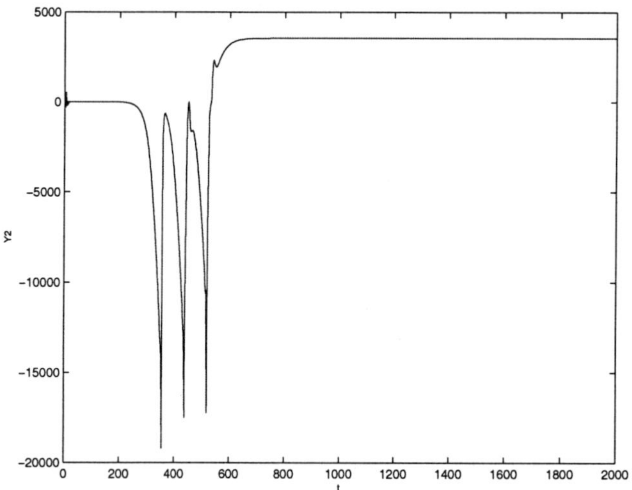

Fig. 5. Profiles of Y_2 generated using *Method 1* $((19))$ with $\mu = 1/32$, $\Pi = 2000$, $\beta_1 = 0.3$, $\beta_2 = 0.35$, $c = 4$, $\gamma_1 = 0.1$, $\gamma_2 = 0.1$, $\xi_1 = 0.3$, $\xi_2 = 0.4$, $\rho = 0.5$, $\tau = 0.4$, $X(0) = 8000$, $V(0) = 800$, $Y_1(0) = 200$, $Y_2(0) = 300$ and $\ell = 2.5$.

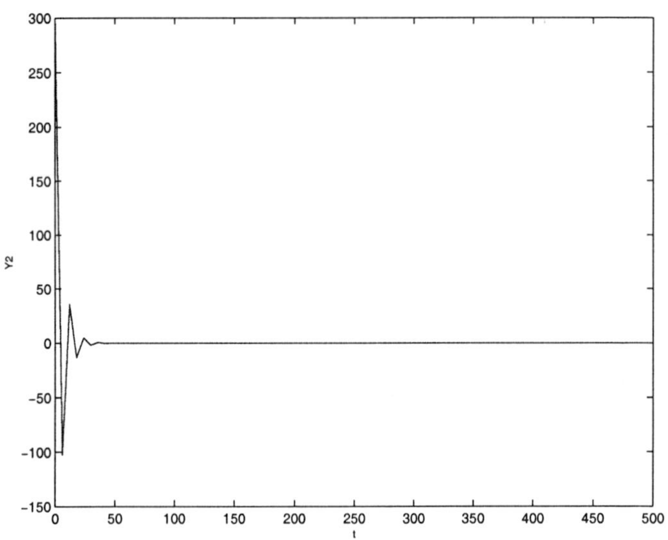

Fig. 6. Profiles of Y_2 generated using *Method 1* ((19)) with $\mu = 1/32$, $\Pi = 2000$, $\beta_1 = 0.3$, $\beta_2 = 0.35$, $c = 4$, $\gamma_1 = 0.1$, $\gamma_2 = 0.1$, $\xi_1 = 0.3$, $\xi_2 = 0.4$, $\rho = 0.5$, $\tau = 0.4$, $X(0) = 8000$, $V(0) = 800$, $Y_1(0) = 200$, $Y_2(0) = 300$ and $\ell = 6$.

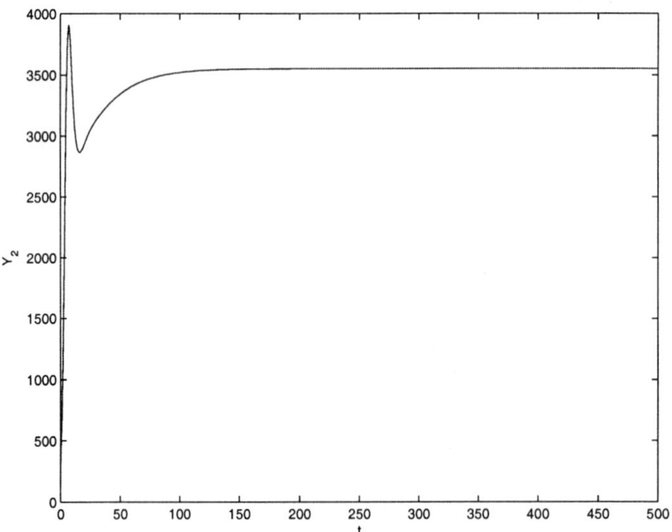

Fig. 7. Profile of Y_2 generated using ODE45 for the HIV model ((16)) with $\mu = 1/32$, $\Pi = 2000$, $\beta_1 = 0.3$, $\beta_2 = 0.35$, $c = 4$, $\gamma_1 = 0.1$, $\gamma_2 = 0.1$, $\xi_1 = 0.3$, $\xi_2 = 0.4$, $\rho = 0.5$, $\tau = 0.4$, $X(0) = 8000$, $V(0) = 800$, $Y_1(0) = 200$, $Y_2(0) = 300$ and $\ell = 1.32$

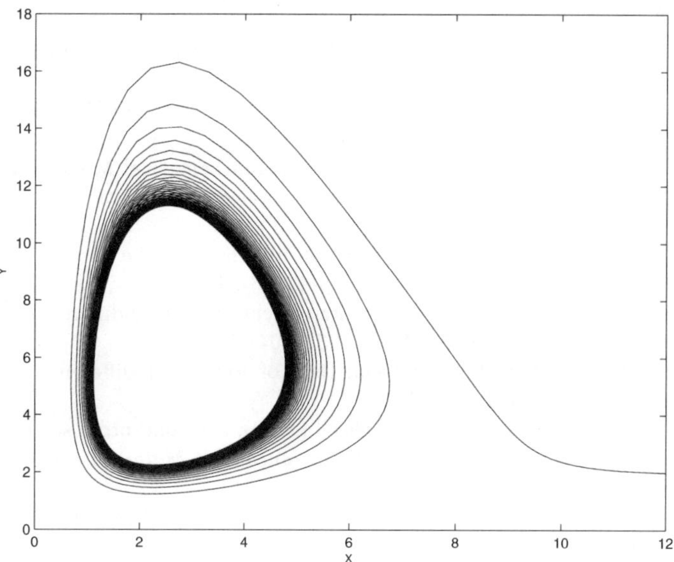

Fig. 8. Profiles of Y *versus* X generated using the positivity-preserving method ((26)) with $r = 0.2$, $\mu = 4.5$, $K = 10$, $a = 30$, $D = 0.3$, $\ell = 1$ and $X(0) = 12$, $Y(0) = 2$.

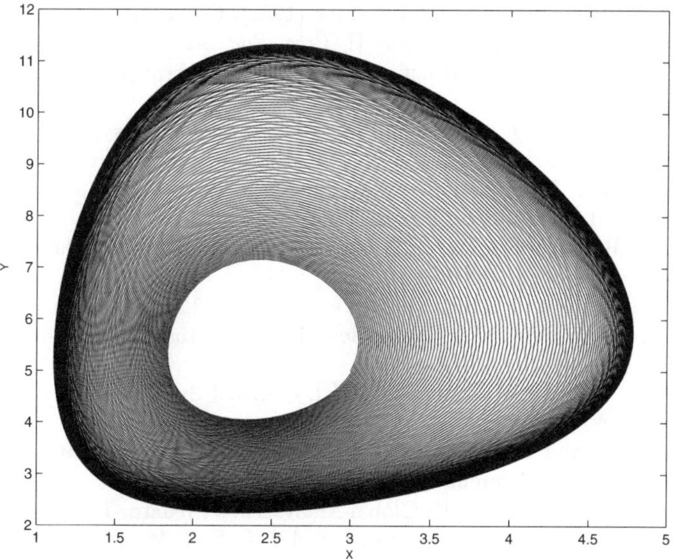

Fig. 9. Profiles of Y *versus* X generated using the positivity-preserving method ((26)) with $r = 0.2$, $\mu = 4.5$, $K = 10$, $a = 30$, $D = 0.3$, $\ell = 1$ and $X(0) = 3$, $Y(0) = 5$.

References

1. Adomian, G. (1995). The diffusion Brusselator equation. *Computers Math. Applic.* **29**(5):1–3.

2. Anderson, R.M. and R.M. May (1991). Infectious Diseases of Humans. Oxford University Press, London/New York.

3. Brauer, F. and C. Castillo-Chavez (2000). Mathematical Models in Population Biology and Epidemiology. Texts in Applied Mathematics Series, Volume 40, Springer-Verlag, New York.

4. Gear, W.C (1986). Maintaining solution invariants in the numerical solution of ODEs, *SIAM J. Sci. Statist. Comput.* **7**(3): 734–743.

5. Gumel, A.B. (2002). Removal of contrived chaos in finite-difference methods. *Intern. J. Compuetr Math.* **79**(9): 1033–1041.

6. Gumel A.B. (editor)(2003). Special Issue of Journal of Difference Equations and Applications, **vol 9** (11,12).

7. Gumel, A.B., Q. Cao, and E.H. Twizell (1999). A second-order scheme for the Brusselator reaction-diffusion system. *Journal of Mathematical Chemistry.* **26**: 297–316.

8. Gumel, A.B., W.F. Langford, E.H. Twizell and J. Wu (2000). Numerical solutions for a coupled non-linear oscillator. *Journal of Mathematical Chemistry.* **28**(4): 325–340.

9. Hale, J.K. (1969). Ordinary Differential Equations, John Wiley & Sons, New York

10. Hethcote, H. W. (2000). The mathematics of infectious diseases. *SIAM Review* **42**(4): 599–653.

11. Kermack, W.O. and A.G. McKendrick (1932). Contributions to the mathematical theory of epidemics, part II. *Proc. Roy. Soc. Lond.* **138**: 55–83.

12. Kot, M. Elements of Mathematical Ecology. Cambridge University Press, 2001.

13. J.D. Lambert. Numerical methods for ordinary differential systems: the initial value problem. John Wiley and Sons, Chichester, England, 1991.

14. Mickens, R.E. Non-standard Finite-difference Models of Differential Equations. World Scientific, Singapore, 1994.

15. Mickens, R.E. Applications of non-standard finite-difference schemes. World Scientific, Singapore, 2000.

16. Mischaikow, K. and G.S.K. Wolkowicz (1990). A Predator-Prey System Involving Group Defense: A Connection Matrix Approach, *Nonlinear Analysis* **14**, 955–969.

17. Perko, L. Differential Equations and Dynamical Systems (1991). Springer-Verlag, New York.

18. Porco, T.C. and S.M. Blower (1998). Designing HIV vaccination policies: subtypes and cross-immunity. *Interfaces* **28**(3): 167–190.

19. Ruan, S. and D. Xiao (2001). Global Analysis in a Predator-Prey System with Nonmonotonic Functional Response, *SIAM J. Appl. Math.* **61**, 1445–1472.

20. Shampine, L.F. (1986). Conservation laws and the numerical solution of ODEs, *Comput. Math. Appl.* **12** Part B (5/6): 1287–1296.

21. van den Driessche, P. and J. Watmough (2002). Reproduction numbers and sub-threshold endemic equilibria for compartmental models of disease transmission. *Math. Bios.* **180**: 29–48.

22. Serfaty de Markus, A. and R.E. Mickens (1999). Suppression of numerically induced chaos with non-standard finite difference schemes. *J. Comp. Appl. Math.* **106**: 317–324.

23. Twizell, E.H., W. Wang and P.G. Price (1990). Chaos-free numerical solutions of reaction-diffusion equations. *Proc. R. Soc. Lond.* **430**: 541-576.

24. Zhen, C., A.B. Gumel, and S.M. Moghadas (2005). A semi-explicit numerical scheme for a predator-prey model with non-monotonic functional response. *Journal of Computational Methods in Science and Engineering.* To appear.

CHAPTER 10

NONSTANDARD FINITE DIFFERENCE METHODS AND
BIOLOGICAL MODELS

Sophia R.-J. Jang

Department of Mathematics
University of Louisiana at Lafayette
Lafayette, LA 70504-1010
jang@louisiana.edu

Continuous-time biological models including epidemic models and a competition population model are presented. Nonstandard finite difference methods are then applied to approximate these continuous-time systems. The resulting difference equations are studied and comparisons between continuous-time and discrete-time models are made. In particular, conditions for the existence and stability of steady states are investigated. Numerical simulations of the discrete approximations are also presented.

1. Introduction

The concept of mathematical modeling in biology was in large part inherited from the very successful modeling process in physics. But the natures of the modeling process in these two areas are quite different. In physics, one largely achieves mathematical models that describe the real world very precisely, based for example on Newtonian paradigm [1]. Experiments can also be performed to test for theory relatively easy. However, no such precision is possible for many biological phenomena. The biological world is too complex and unpredictable. There are usually no laws associated with biological systems with a few exceptions such as for example the Hardy-Weienberg law in population genetics [2]. In addition, it is very hard if not impossible to test for hypotheses by experiments. Nevertheless, it is very important to build relevant mathematical models so that some conclusions about the biology can be drawn.

Many ecological and epidemic models have been constructed and ana-

423

lyzed by researchers in a diversified disciplines to help us understand and interpret biological problems in ecology and epidemics. Among these is the Malthus equation, which can be viewed as the simplest way to model population growth,

$$\frac{dx}{dt} = \lambda x, \tag{1}$$

where $\lambda > 0$, independent of population size and time, is the growth rate of the population, and $x(t)$ is the population density or size at time t. Under the simple assumption that population growth rate is a constant, the population will always grow to unboundedly large over time as long as the initial population size is positive. Therefore the equation does not capture the long time realistic population growth phenomenon.

One way to modify the above biological assumption is to incorporate density dependence into the growth parameter λ. The well-known continuous-time logistic equation

$$\frac{dx}{dt} = rx(1 - \frac{x}{K}) \tag{2}$$

has been used very frequently in modeling a single population growth prior to its interaction with other populations. In (2), parameter $r > 0$ is the intrinsic growth rate of the population and $K > 0$ is the carrying capacity of the environment. In other words, the growth parameter λ in equation (1) is replaced by $r(1 - \frac{x(t)}{K})$, which depends on both population density at time t and the carrying capacity K of the environment. This equation possesses a very simple asymptotic dynamics: all solutions with positive initial conditions will eventually approach the carrying capacity K. Therefore, population size will eventually be stabilized to K in the long run even if population dynamics initially either overshoot or undershoot the carrying capacity.

If we use a simple forward Euler scheme [3,4] to approximate solutions of (2) by letting h denote the step size of the approximation and replace $\frac{dx}{dt}$ by

$$\frac{x(t+h) - x(t)}{h}$$

and denote $x(nh)$ by x_n, we then have

$$x_{n+1} = \bar{r}x_n(1 - \frac{rh}{K\bar{r}}x_n), \tag{3}$$

where $\bar{r} = 1 + rh$. Although the continuous-time logistic equation (2) has only equilibrium dynamics, this discrete counterpart, the well known discrete logistic equation (3), exhibits period doubling bifurcation cascade to chaos [5,6,7,8]. Even if the step size h is chosen to be very small, the discrepancy between the ordinary differential equation and its difference approximation is still inevitable as we increase the intrinsic growth rate r. Therefore, the forward Euler approximation is not appropriate for the simple model presented.

Another example is to consider the well-studied Lotka-Volterra ordinary differential equations of two interacting populations. Ushiki [9] presented a forward Euler approximation with step size h. It was demonstrated in their paper that the discrete model possesses period-doubling bifurcation route to chaos. Consequently, the discrete approximation by using forward Euler method is not in agreement with its continuous counterpart. Several researchers used piecewise constant arguments to obtain a discrete analogue of the Lotka-Volterra equation. For example, Jiang and Rogers [10] studied the competitive case and Krawcewicz and Rogers [11] discussed the cooperative case. Both studies showed dynamical inconsistency between the continuous-time and discrete-time models.

On the other hand, Liu and Elaydi [12], Al-Kahby, Dennan and Elaydi [13] used Mickens nonstandard discretization method [3] to derive a discrete version of the two dimensional continuous-time Lotka-Volterra model. They proved dynamical consistency between the continuous and discrete-time models in these two pioneering papers. Since then several researchers such as Roeger and Allen [14], Roeger [15,16,17] have applied the method to continuous-time three dimensional May-Leonard competitive systems and showed similar dynamics between the discrete and continuous-time models.

More recently, Jang and Elaydi [18] also used Mickens nonstandard finite difference method to approximate a continuous-time epidemic model proposed by Brauer and van den Driessche [19] and proved similarity between two types of models. In particular, the existence criteria of the steady states between the continuous-time and discrete-time models are the same, and both models also have the same equilibria. However, unlike the continuous-time SIS model for which global asymptotical stability of the steady state can be easily established by using well known theory of Dulac criterion and Poincaré-Bendixson Theorem [20], the global asymptotic stability of the steady state for the discrete-time model derived in [18] is not trivial.

The purpose of this chapter is to present some interesting continuous-time biological models including epidemic models and a competition model, and to introduce nonstandard discretization methods to approximate these systems. We will then study the resulting systems of difference equations and, in particular, we will compare these two types of models with respect to existence and stability of the steady states, and their asymptotic dynamics whenever it is possible. Section 2 focuses on several epidemic models and Section 3 presents a competition model. Numerical simulations for the systems of approximations using nonstandard finite difference methods are provided. The final section gives a brief summary.

2. Epidemic Models

The study of epidemics has a long history with a vast variety of models and explanations for the spread and cause of epidemic outbreaks. The investigation of epidemic models may help us understand how to eradicate and control the infectious diseases. Most of the earlier epidemic models are continuous-time models, and in particular, they are expressed in terms of ordinary differential equations. This is probably in large part because the theory of ordinary differential equations are well developed and therefore they can be easily applied to study continuous-time systems derived from mathematical modelling. However, data collected from transmitted diseases are usually discrete. Therefore, it is necessary and important to study discrete time epidemic models.

In a pioneering paper by Allen [21], discrete-time epidemic models are proposed and comparisons between their continuous-time counterparts are made. In this section, we shall first review some of the basic epidemic models presented in [21]. We then describe some more simple models for the population dynamics of disease agents and use Mickens nonstandard finite difference method to approximate these models. Comparisons between continuous-time models and their discrete counterparts will be made.

2.1. *SI and SIS models*

In this subsection, some of the models presented in [21] will be reviewed and the Mickens nonstandard finite difference scheme [3] will be developed. We will then demonstrate that the discrete-time models derived from Mickens method have the same asymptotic dynamics as their corresponding continuous-time models.

As in most of the literature in epidemics, we let S denote the susceptible, who can catch the disease; I be the infectives, who have the disease and can transmit it; and R, the removed class, namely those who have either had the disease, or are recovered, immune or isolated until recovered. By assuming total population size of a population is a constant and equals to N, a continuous-time SI model can be written in terms of a single equation in terms of I as

$$\begin{cases} \dfrac{dI}{dt} = \dfrac{\alpha}{N}(N - I)I \\ 0 \le I(0) \le N, \end{cases} \tag{4}$$

where $\alpha > 0$ is the contact rate [21]. Equation (4) can be easily solved explicitly yielding

$$I(t) = \frac{I(0)N}{I(0) + e^{-\alpha t}(N - I(0))}$$

and thus

$$\lim_{t \to \infty} I(t) = N \text{ if } I(0) > 0.$$

Using a forward Euler approximation with step size $\Delta t > 0$ and replace $\dfrac{dI}{dt}$ by

$$\frac{I(t + \Delta t) - I(t)}{\Delta t}$$

and denote $I(n\Delta t)$ by I_n, we have [21]

$$\begin{cases} I_{n+1} = I_n(1 + \alpha\Delta t - \dfrac{\alpha\Delta t}{N}I_n). \\ 0 \le I_0 \le N. \end{cases} \tag{5}$$

Notice that solutions of the above nonlinear difference equation may not remain nonnegative. In [21], it is assumed that

$$\alpha\Delta t \le 1. \tag{6}$$

Consequently, solutions of the above equation (5) remain nonnegative. Furthermore, since $I_n \le N$ for all n we have

$$\alpha\Delta t - \frac{\alpha\Delta t}{N}I_n \ge 0 \text{ for } n \ge 0$$

and thus

$$I_{n+1} \ge I_n \text{ for } n \ge 0.$$

As a result, $\lim_{n \to \infty} I_n = \bar{I} > 0$ exists and we immediately conclude that $\bar{I} = N$ if $I_0 > 0$. Therefore, both the continuous-time and the discrete-time models have the same asymptotic dynamics. However, it is necessary to impose an extra condition (6) on the parameter α or step size Δt.

Consider a continuous-time SIS model given in [21] by assuming constant population size N again. Moreover, individuals that are recovered after infection do not develop permanent immunity and so they are immediately susceptible again. Let γ denote the rate of becoming susceptible. Since the total population is a constant, the SIS model can be written in terms of I equation only with

$$\begin{cases} \dfrac{dI}{dt} = (\dfrac{\alpha}{N}(N - I) - \gamma)I \\ 0 \leq I(0) \leq N. \end{cases} \tag{7}$$

Define the threshold

$$R = \frac{\alpha}{\gamma}.$$

Then since (7) is only a one-dimensional autonomous equation, we immediately have

$$\lim_{t \to \infty} I(t) = 0 \text{ if } R < 1$$

and

$$\lim_{t \to \infty} I(t) = \frac{(\alpha - \gamma)N}{\alpha} \text{ if } R > 1 \text{ and } I(0) > 0.$$

Using a forward Euler approximation with step size Δt and replace $\dfrac{dI}{dt}$ by

$$\frac{I(t + \Delta t) - I(t)}{\Delta t},$$

and denote $I(n\Delta t)$ by I_n, we have [21]

$$\begin{cases} I_{n+1} = I_n(1 - \gamma\Delta t + \alpha\Delta t - \dfrac{\alpha\Delta t}{N}I_n). \\ 0 \leq I_0 \leq N. \end{cases} \tag{8}$$

Under the assumption

$$\gamma\Delta \leq 1 \text{ and } \alpha\Delta t < (1 + \sqrt{\gamma\Delta t})^2,$$

it was proved in [21] that solutions of (8) remain nonnegative. Clearly, if $R < 1$, then $\{I_n\}_{n=0}^{\infty}$ is monotonically decreasing and so it converges to a fixed point of the equation, which is easily seen to be 0. If $R > 1$, then numerical simulations performed in [21] demonstrate the existence of periodic solutions. Therefore, in addition to the positivity problem associated with the approximation by using Euler's method, the discrete-time model is not dynamically consistent with its corresponding continuous-time model. Since the nonlinear scalar difference equation (8) is quadratic, it is believed that the equation will undergo period doubling bifurcations to chaos.

We now apply Mickens nonstandard finite difference method [3] to the continuous-time model (7). Let h denote the step size of our approximation and replace $\dfrac{dI}{dt}$ by

$$\frac{I(t+h) - I(t)}{\phi(h)},$$

where $\phi(h)$ satisfies

$$\phi(h) = h + O(h^2), 0 < \phi(h) < 1, \tag{9}$$

and approximate I by $I(t)$, I^2 by $I(t)I(t+h)$. Letting $I_n = I(nh)$. We then have the following scalar difference equation

$$\begin{cases} I_{n+1} = \dfrac{(1 + \alpha\phi)I_n}{1 + \dfrac{\alpha\phi}{N}I_n} \\ 0 \le I_0 \le N. \end{cases} \tag{10}$$

The equation only has two fixed points 0 and N. Since $I_n \le N$ for $n \ge 0$, we have

$$I_{n+1} \ge I_n \text{ for } n \ge 0.$$

Therefore solutions of (10) with $I_0 > 0$ converge to the positive steady state N. This asymptotic dynamics is exactly same as its continuous-time model (7) and is also the same as equation (8), where a first order Euler approximation is used. However, there is no restriction on the parameter values under Mickens nonstandard method.

We next proceed to approximate (7) by using Mickens nonstandard finite difference scheme. Similar to model (4), we let h denote the step size and assume $\phi(h)$ satisfy (9). With the same spirit as in the derivation of (10), we have

$$\begin{cases} I_{n+1} = \dfrac{(1 + \alpha\phi)I_n}{1 + \dfrac{\alpha\phi}{N}I_n + \gamma\phi} \\ 0 \le I_0 \le N. \end{cases} \tag{11}$$

Notice if $R < 1$, i.e, if $\dfrac{\alpha}{\gamma} < 1$, then equation (11) only has a trivial fixed point 0. On the other hand if $R > 1$, then in addition to 0, there is a unique positive steady state

$$\bar{I} = \frac{N}{\alpha}(\alpha - \gamma).$$

If $R < 1$, then $I_{n+1} \le I_n$ for all n implies $\lim_{n \to \infty} I_n \ge 0$ exists and is a fixed point of the map associated with the equation. Since 0 is the only fixed point, we conclude that solutions of (11) converge to 0.

Suppose now $R > 1$. Let g denote the map induced by (11), i.e.,

$$g(x) = \frac{(1 + \alpha\phi)x}{1 + \dfrac{\alpha\phi}{N}x + \gamma\phi}.$$

We have $g'(x) > 0$ for all $x \ge 0$ and $g(0) = 0$. Since

$$\frac{1 + \alpha\phi}{1 + \dfrac{\alpha\phi}{N}\bar{I} + \gamma\phi} = 1,$$

it is clear if $0 < I_0 < \bar{I}$, then

$$I_1 = \frac{1 + \alpha\phi}{1 + \dfrac{\alpha\phi}{N}I_0 + \gamma\phi}I_0 > I_0$$

and

$$I_1 = g(I_0) < g(\bar{I}) = \bar{I}.$$

As a result, $\{I_n\}$ is a monotone increasing sequence of real numbers and bounded above by $\bar{I}$. Therefore if must converge to the fixed point $\bar{I}$. A symmetric argument can be applied to the case when $I_0 > \bar{I}$. We can thus conclude that solutions of equation (11) with $I_0 > 0$ converge to $\bar{I}$ if $R > 1$. We summarize our discussion of equation (11) below.

Theorem 2.1 *Dynamics of scalar equation* (11) *depends on the threshold* $R = \dfrac{\alpha}{\gamma}$.

    (a) If $R < 1$, then solutions of (11) *all converge to 0.*
    (b) If $R > 1$, then solutions of (11) *with $I_0 > 0$ converge to $\bar{I} = \dfrac{N}{\alpha}(\alpha - \gamma)$.*

We now conclude from above discussion that Mickens nonstandard finite difference scheme results in discrete-time models that share the same global asymptotic dynamics as their corresponding continuous-time models. Moreover, solutions to the equations also remain nonnegative without any additional conditions imposed on the parameters of the equations.

2.2. *A simple SIR model*

In this section, we will examine a simple classical SIR model given in [22]. Suppose the total population is again a constant. Consider a disease which confers immunity after recovery. The population can then be divided into three distinct classes: susceptible, infectives, and removed. For simplicity, it is assumed that the gain in the infective class is at a rate proportional to the number of infectives and susceptible, that is, according to rSI, where $r > 0$ is a constant. The susceptibles are lost at the same rate. The rate of removal of infectives to the removed class is proportional to the number of infectives, that is, aI, where $a > 0$ is a constant. The incubation period is short enough to be negligible. Therefore, a susceptible who contracts the disease is infective right away. Under these assumptions, the disease dynamics can be described by the following differential equations involving only S and I.

$$\begin{cases} \dfrac{dS}{dt} = -rSI \\[2mm] \dfrac{dI}{dt} = rSI - aI \\[2mm] S(0) = S_0 > 0, \ I(0) = I_0 > 0, \end{cases} \tag{12}$$

where $r > 0$ is the infection rate and $a > 0$ is the removal rate of infectives.

Note that the system has no interior steady state and there are uncountably many steady states on the nonnegative S-axis. If $S_0 < a/r$, then $\lim_{t\to\infty} S(t) = 0$, and $\lim_{t\to\infty} S(t) = S^* > 0$ if $S_0 > a/r$, where S^* depends on the initial condition. Therefore we can define the threshold

$$\rho = \frac{a}{r}, \tag{13}$$

where ρ can be interpreted as the relative removal rate as $\dfrac{1}{a}$ is the average infectious period [22].

We now apply Mickens nonstandard discretization method [3] to the above system and show that the resulting system of difference equations has the same dynamical behavior as its continuous counterpart. Let h denote the step size of our approximation. Apply Mickens nonstandard discretization method by using the following scheme:

$$\phi_i(h) = h + O(h^2), 0 < \phi_i(h) < 1 \text{ for } i = 1, 2,$$

and replace $\dot{S}$ by

$$\frac{S(t+h) - S(t)}{\phi_1(h)},$$

$\dot{I}$ by

$$\frac{I(t+h) - I(t)}{\phi_2(h)},$$

SI by $S(t+h)I(t)$ and by $S(t)I(t)$ in $\dot{S}$ and $\dot{I}$ respectively, and I by $I(t+h)$ in $\dot{I}$. There are several different ways to discretize SI and I. We choose the preceding procedure so that solutions of the resulting difference equations remain nonnegative as can be seen below. Indeed, under the above discretization method if we denote $S(nh)$ by S_n and $I(nh)$ by I_n, we then have

$$\begin{cases} \dfrac{S_{n+1} - S_n}{\phi_1(h)} = -rS_{n+1}I_n \\[2mm] \dfrac{I_{n+1} - I_n}{\phi_2(h)} = rS_nI_n - aI_{n+1} \\[2mm] S_0 > 0, \ I_0 > 0. \end{cases} \tag{14}$$

For simplicity, we write $\phi_i(h)$ by ϕ_i for $i = 1, 2$. Then the discretization of system (14) yields the following system of difference equations

$$\begin{cases} S_{n+1} = \dfrac{S_n}{1 + r\phi_1 I_n} \\[3mm] I_{n+1} = \dfrac{(1 + r\phi_2 S_n)I_n}{1 + a\phi_2} \\[3mm] S_0 > 0, \ I_0 > 0. \end{cases} \tag{15}$$

Since parameters are positive, we immediately see from the above system that solutions remain nonnegative for all $n \geq 0$.

System (15) has the same steady states as its continuous counterpart (12). Indeed, steady state $E_0 = (0, 0)$ always exists for all parameter values and there are also steady states of the form $(S, 0)$ for any $S > 0$. Consequently, steady states of (15) are not isolated and there is no interior steady state. We conclude that the existence conditions of steady states of the system of difference equations derived from Mickens discretization method and its original system of ordinary differential equations are the same.

We next investigate asymptotic behavior of the solutions. Clearly for $n \geq 0$ we have

$$S_{n+1} \leq S_n$$

and

$$I_{n+1} < I_n \text{ if and only if } S_n < \rho,$$

where ρ is defined as in (13). Therefore if $S_0 < \rho$, we have $\lim\limits_{n \to \infty} I_n = I^* \geq 0$ exists as solutions are bounded below by zero. If $I^* > 0$, then by the second equation of (15) we have

$$\lim_{n \to \infty} \frac{1 + r\phi_2 S_n}{1 + a\phi_2} = 1,$$

i.e., $\lim\limits_{n \to \infty} S_n = \rho$. But this is impossible as $S_{n+1} \leq S_n < \rho$ for $n \geq 0$. Therefore we can conclude that $\lim\limits_{n \to \infty} (S_n, I_n) = (S^*, 0)$, where $(S^*, 0)$ is a steady state that depends on S_0. In this case, there is no epidemic as $I_n < I_0$ for $n \geq 1$.

Suppose now $S_0 > \rho$. Then $I_1 > I_0$ and there is an epidemic. If $S_n > \rho$ for all $n \geq 1$, then $\lim\limits_{n \to \infty} S_n \geq \rho$ exists, and $I_{n+1} > I_n$ for all $n \geq 0$ implies $\lim\limits_{n \to \infty} I_n > 0$ exists. Using the first equation of (15), we arrive at

$$1 = \lim_{n \to \infty} \frac{1}{1 + r\phi_1 I_n},$$

i.e., $\lim\limits_{n \to \infty} I_n = 0$ and obtain a contradiction. Therefore there exists $k > 0$ such that $S_k \leq \rho$. We let $k_0 > 0$ be the smallest such k. Consequently,

$$S_n \leq \rho \text{ and } I_{n+1} \leq I_n \text{ for } n \geq k_0,$$

and we can conclude that $\lim\limits_{n \to \infty} I_n = 0$. That is, the epidemic will eventually die out. However, the eradication of the disease is due to a lack of infectives and not from a lack of susceptibles. Moreover, since

$$I_{k_0} > I_{k_0-1} \text{ and } I_{k_0+1} \leq I_{k_0}$$

we see that the number of infectives reaches a maximum of I_{k_0} during the course of epidemics. For the continuous-time model (12), $I_{\max}$ occurs when $\dot{I} = 0$, that is, when $S = \rho$. Integrating

$$\frac{dI}{dS} = -1 + \frac{\rho}{S},$$

yields

$$I + S - \rho \ln S = I_0 + S_0 - \rho \ln S_0$$

and thus

$$I_{\max} = I_0 + S_0 - \rho + \rho \ln(\frac{\rho}{S_0}). \tag{16}$$

However, no such a relation can be obtained for the discrete model (15). Moreover, the total number of infectives can be seen to satisfy

$$I_{\text{total}} = I_0 + S_0 - S_\infty,$$

where $S_0 - S_\infty$ is the number of susceptible individuals that become infected during the course of epidemic. Since $S_\infty = \lim_{n \to \infty} S_n$ depends on initial condition (S_0, I_0), we conclude that the severity of the epidemic depends on initial data and parameters. This conclusion is the same as the continuous-time model (12). The above discussion for system (15) can be summarized below.

Theorem 2.2 *Dynamics of system* (15) *are summarized below.*

    (a) If $S_0 < \rho$, then $I_{n+1} < I_n$ for $n = 0, 1, 2, \cdots$, and $\lim_{t \to \infty} S_t = S^* > 0$ and $\lim_{t \to \infty} I_t = 0$, where $(S^*, 0)$ is a boundary steady state.

    (b) If $S_0 > \rho$, then there exists $k_0 > 0$ such that $I_{n+1} > I_n$ for $n = 0, 1, \cdots, k_0 - 1$ and $I_{n+1} \le I_n$ if $n \ge k_0$. Moreover, $\lim_{n \to \infty} S_n = S^* > 0$ and $\lim_{n \to \infty} I_n = 0$, where $(S^*, 0)$ is a boundary steady state.

We next use a numerical example to simulate solutions of system (15), i.e., to approximate solutions of (12) by using Mickens method. Specifically, we choose $\phi_i(h) = h$ for $i = 1, 2$, $r = 0.5$ and $a = 0.7$. Therefore, with these parameter values we have $\rho = a/r = 1.4$ and there is an epidemic according to our earlier discussion. Figure 1 provides two particular solutions, one with initial condition $(S_0, I_0) = (100, 10)$ and $h = 0.01$. Using equation (16) we see that $I_{\max} = 102.6238$, which is very close to the maximum number of infectives estimated from the Mickens discretization method even when $h = 0.01$, which is not very small. The other solution with initial condition $(300, 20)$ is also plotted. Using (16), we have $I_{\max} = 311.08$ and the approximation is also very close. See Figure 1 for these two particular solutions of (15).

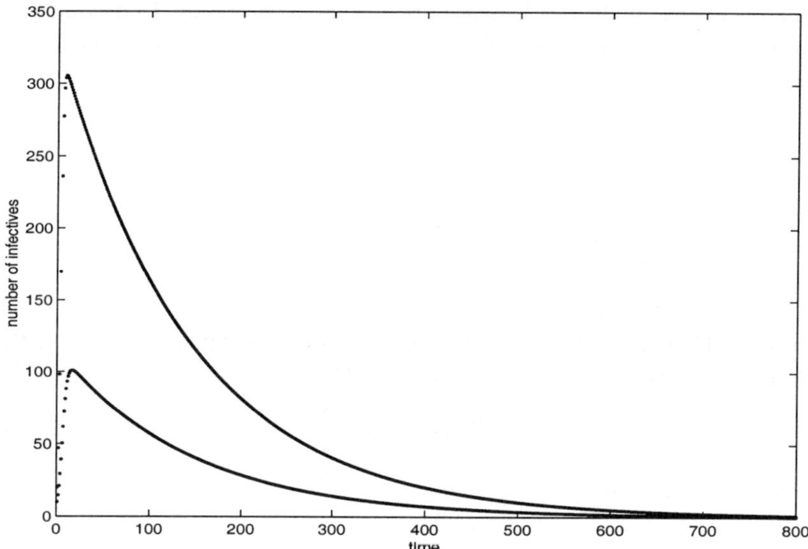

Fig. 1. The I-component of two solutions of the discrete-time SIR model (15) are plotted with $\phi_i(h) = h = 0.01$ for $i = 1, 2$. One initial condition $(S_0, I_0) = (100, 10)$. It can be seen that $I_{\max}$ approximated from system (15) is very close to the exact solution 102.6238 given by equation (16). The other solution has initial condition $(300, 20)$. In this case, $I_{\max} = 311.08$ by using equation (16). The approximated value from Mickens finite difference method is very close to it.

2.3. *A crisscross disease model*

The increasing incidence of sexually transmitted diseases (STD), such as gonorrhea, chlamydia, syphilis, and AIDS, is a major health problem in both developed and developing countries. For example, it is believed that more than 2 million people contract gonorrhea annually. To help us understand such a disease dynamics, we next consider a crisscross infectious disease such as malaria and bilharzia. Bilharzia is an infectious disease between humans and a particular type of snail, while malaria is transmitted between mosquitoes and humans. Since the incubation period for venereal diseases is usually very short, we will use an SI model to model the disease dynamics.

Because the diseases are transmitted between two populations, for simplicity, we let I and I^* denote the male and female infective populations, respectively. Therefore, the disease dynamics that we wish to capture may include AIDS epidemics. However, these two populations may represent mosquitoes and humans populations, respectively. In particular, we assume that total populations are constants and let N and N^* be the total population sizes for the male and female populations, respectively. Consequently, it is enough to consider the following two dimensional system of ordinary differential equations [22].

$$\begin{cases} \dfrac{dI}{dt} = rI^*(N - I) - aI \\[2mm] \dfrac{dI^*}{dt} = r^*I(N^* - I^*) - a^*I^* \\[2mm] I(0) = I_0 > 0, \ I^*(0) = I_0^* > 0, \end{cases} \qquad (17)$$

where $r, r^* > 0$ are the infection rates and $a, a^* > 0$ are the removal rates of infectives. They have the same biological meanings as given in model (12).

Similar to the previous SIR model, we let

$$\rho = \frac{a}{r} \text{ and } \rho^* = \frac{a^*}{r^*}. \qquad (18)$$

The local dynamics of system (17) can be summarized below. There exists a trivial steady state $E_0 = (0, 0)$ for all parameter values. A simple calculation shows that E_0 is locally asymptotically stable if

$$NN^* - \rho\rho^* < 0. \qquad (19)$$

In this case there exists no interior steady state. However, if the above inequality is reversed,

$$NN^* - \rho\rho^* > 0, \qquad (20)$$

then the system has a unique positive steady state $E_1 = (\bar{I}, \bar{I}^*)$ which is moreover locally asymptotically stable, where

$$\bar{I} = \frac{NN^* - \rho\rho^*}{\rho + N^*} \text{ and } \bar{I}^* = \frac{NN^* - \rho\rho^*}{\rho^* + N}.$$

The global asymptotic behavior of solutions of (17) is not studied here.

We proceed to use Mickens nonstandard discretization method to approximate the continuous-time model (17). Similar to the previous model we adopt the following scheme:

$$\phi_i(h) = h + O(h^2), 0 < \phi_i(h) < 1 \text{ for } i = 1, 2,$$

and replace $\dot{I}$ by

$$\frac{I(t + h) - I(t)}{\phi_1(h)},$$

$\dot{I}^*$ by

$$\frac{I^*(t + h) - I^*(t)}{\phi_2(h)},$$

and approximate II^* by $I^*(t)I(t+h)$ in $\dot{I}$ and by $I(t)I^*(t+h)$ in $\dot{I}^*$. By writing ϕ_i for $\phi_i(h)$ for $i = 1, 2$, and I_n for $I(nh)$ and I_n^* for $I^*(nh)$, we arrive at the following discrete-time system

$$\begin{cases} I_{n+1} = \dfrac{I_n + \phi_1 r N I_n^*}{1 + \phi_1 r I_n^* + \phi_1 a} \\[3mm] I_{n+1}^* = \dfrac{I_n^* + \phi_2 r^* N^* I_n}{1 + \phi_2 r^* I_n + \phi_2 a^*} \\[3mm] I_0 > 0, \ I_0^* > 0. \end{cases} \tag{21}$$

Similar to its continuous-time model, system (21) always has a trivial steady state $(0,0)$. A direct computation yields the Jacobian matrix of the system at $(0,0)$ given below

$$J_0 = J(0,0) = \begin{pmatrix} \dfrac{1}{1 + \phi_1 a} & \dfrac{\phi_1 r N}{1 + \phi_1 a} \\[3mm] \dfrac{\phi_2 r^* N^*}{1 + \phi_2 a^*} & \dfrac{1}{1 + \phi_2 a^*} \end{pmatrix}.$$

Apply the Jury condition [21] it can be easily verified that E_0 is locally asymptotically stable if (19) is true. Indeed,

$$tr J_0 = \frac{1}{1 + \phi_1 a} + \frac{1}{1 + \phi_2 a^*}$$

and

$$\det J_0 = \frac{1 - \phi_1 \phi_2 N N^* r r^*}{(1 + \phi_1 a)(1 + \phi_2 a^*)},$$

where

$$|tr J_0| < 1 + \det J_0 \text{ if and only if } aa^* - NN^* rr^* > 0,$$

i.e., if and only if (19) is true. Furthermore, it is easy to see that $|\det J_0| < 1$ if (19) holds. Therefore, we conclude that the trivial steady state $E_0 = (0,0)$ is locally asymptotically stable if (19) holds. We shall show in this case that system (21) has no interior steady state.

Toward this end, an interior steady state (I_s, I_s^*) of system (21) must satisfy

$$\begin{cases} I = \dfrac{I + \phi_1 N I^* r}{1 + \phi_1 r I^* + \phi_1 a} \\[3mm] I^* = \dfrac{I^* + \phi_2 N^* I r^*}{1 + \phi_2 r^* I + \phi_2 a^*}. \end{cases}$$

As a result, we have I and I^* components of the interior steady state given below

$$I_s = \frac{NN^* - \rho\rho^*}{\rho + N^*} \text{ and } I_s^* = \frac{NN^* - \rho\rho^*}{\rho^* + N}.$$

Consequently, we see that an interior steady state of (21) exists if and only if (20) holds. In this case, the systems also have the same steady state. Moreover, if E_0 is locally asymptotically stable, then system (21) has no interior steady state by the above discussion.

For the ordinary differential equations model (17), it is known that the interior steady state is always locally asymptotically stable provided it exists. We will demonstrate this finding for the resulting discrete model only by numerical simulations. Specifically, we choose parameters $a = 10^5$, $r = 0.0098$, $a^* = 10^6$, $r^* = 0.0115$, $N = 10^7$ and $N^* = 10^8$. Consequently, (20) holds and there is a unique interior steady state $(10.225 \times 10^5, 11.6225 \times 10^5)$. Similar to the previous discrete-time SIR model, we use $\phi_i(h) = h = 0.01$ for $i = 1, 2$. Figure 2 plots solutions of the I^*-component with initial conditions $(10^4, 2000)$ and $(10^6, 10000)$ respectively. We see that solutions converge to the unique interior steady state of (21). Indeed, the I^*-component of the solutions asymptotically approach 11.6225×10^5. Therefore one may conclude that the continuous-time model and its discrete-time counterpart by using Mickens finite difference method have the same local dynamics.

2.4. *An SIS model with immigration of infectives*

In the previous several epidemic models it was assumed that the total population size is a constant. In particular, there is no immigration and migration between populations. In this subsection we shall consider a case when there is a constant flow of immigrants coming to the population, where a small proportion of the immigrants carrying the disease.

As before, we let $S(t)$ and $I(t)$ denote the number of susceptible and infective of a population at time t. It is assumed that there is a constant flow of A new members into the population, of which a fraction p $(0 \le p \le 1)$ is infective. Let $d > 0$ be the per capita natural death rate of the population. The disease related death rate is denoted by $\alpha \ge 0$. In this model, a simple mass action βSI is used to model disease transmission, where β is a positive constant, and a fraction $\gamma \ge 0$ of these infectives recovers. We refer the reader to Brauer and van den Driessche [19] for more biological discussion

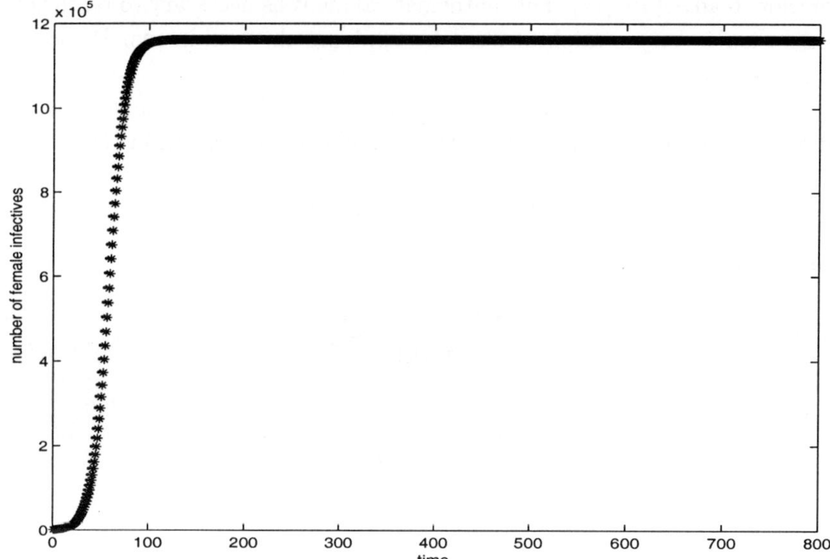

Fig. 2.   Two solutions of the discrete-time model (21) are plotted with $\phi_i(h) = h = 0.01$ for $i = 1, 2$ and initial conditions $(I_0, I_0^*) = (10^4, 2000)$ and $(10^6, 10000)$. It can be seen that the I^*-component of both solutions go to approximately 11.6×10^5.

about the continuous-time model presented below. Epidemic models with different transmission assumptions were also discussed in their work.

The continuous-time *SIS* model studied by Brauer and van den Driessche [19] can be written in terms of the following two-dimensional ordinary differential equations.

$$\begin{cases} \dot{S} = (1 - p)A - \beta SI - dS + \gamma I \\ \dot{I} = pA + \beta SI - (d + \gamma + \alpha)I \\ S(0), I(0) \geq 0. \end{cases} \tag{22}$$

Notice that birth and death rates are not modeled into the system for previous models. System (22) however takes these forces into consideration. Consequently, the disease under consideration in this section may take a longer time to evolve than the diseases considered earlier.

The dynamics of model (22) can be understood very easily. In fact since solutions are bounded, all solutions of (22) converge to the steady state when the system has only a single equilibrium. When there are two equilibria, one will be on the boundary and the other will be in the interior. Solutions with positive initial condition always asymptotically approach the interior steady state [19]. The mathematical analysis necessary to reach this conclusion is very straightforward by applying the well-known Poincare-Bendixson Theorem [20].

We now describe Mickens nonstandard discretization procedure. Much of our presentation in this subsection is similar to that given in [18]. Let

$$\phi_i(h) = h + O(h^2), 0 < \phi_i(h) < 1 \text{ for } i = 1, 2.$$

We replace $\dot{S}$ by

$$\frac{S(t+h) - S(t)}{\phi_1(h)},$$

$\dot{I}$ by

$$\frac{I(t+h) - I(t)}{\phi_2(h)},$$

and approximate SI by $S(t+h)I(t)$ in $\dot{S}$ and by $S(t)I(t)$ in $\dot{I}$. Notice similar to the previous two models that there are several different ways to discretize SI. We choose $S(t+h)I(t)$ in $\dot{S}$ and $S(t)I(t)$ in $\dot{I}$ so that solutions of the resulting difference equations will remain nonnegative as can be seen below. Substituting these into equation (22), and letting $S_n = S(nh)$ and $I_n = I(nh)$, yields

$$\begin{cases} \dfrac{S_{n+1} - S_n}{\phi_1(h)} = (1 - p)A - \beta S_{n+1}I_n - dS_{n+1} + \gamma I_n \\[2mm] \dfrac{I_{n+1} - I_n}{\phi_2(h)} = pA + \beta S_n I_n - (d + \gamma + \alpha)I_{n+1}. \end{cases} \tag{24}$$

For simplicity, we write ϕ_i instead of $\phi_i(h)$ for $i = 1, 2$. Then the system of difference equations is given below.

$$\begin{cases} S_{n+1} = \dfrac{S_n + (1 - p)\phi_1 A + \gamma\phi_1 I_n}{1 + \beta\phi_1 I_n + d\phi_1} \\[3mm] I_{n+1} = \dfrac{I_n + \phi_2 pA + \beta\phi_2 S_n I_n}{1 + (d + \gamma + \alpha)\phi_2} \\[3mm] S_0, I_0 \geq 0. \end{cases} \tag{25}$$

We first study the special case when $\beta = 0$, i.e., when there is no disease transmission between individuals in the population. In this case the only new infectives are coming from immigration, and system (24) now takes the following form

$$\begin{cases} S_{n+1} = \dfrac{S_n + (1-p)\phi_1 A + \gamma\phi_1 I_n}{1 + d\phi_1} \\[3mm] I_{n+1} = \dfrac{I_n + \phi_2 p A}{1 + (d + \gamma + \alpha)\phi_2} \\[3mm] S_0, I_0 \geq 0. \end{cases} \qquad (25)$$

System (25) is a special case of system (24) when $\beta = 0$. We see that system (25) always has a unique steady state $E_0 = (S_0^*, I_0^*)$, where

$$S_0^* = \frac{(1-p)A}{d} + \frac{\gamma}{d} I_0^*$$

and

$$I_0^* = \frac{pA}{d + \gamma + \alpha}.$$

Since the equation for I_n can be decoupled from S_n, the global dynamics of (25) can be easily understood. See [18] for the proof.

Theorem 2.3 *If $\beta = 0$, then every solution of (24) converges to E_0.*

Theorem 2.3 illustrates that the discrete-time model derived from Mickens nonstandard finite difference method [3] has the same asymptotic dynamics as the original continuous-time model when $\beta = 0$. Suppose now $\beta > 0$ and $p = 0$, that is, all the immigrants are susceptible but there is disease transmission within the population. System (24) then takes the following form.

$$\begin{cases} S_{n+1} = \dfrac{S_n + \phi_1 A + \gamma\phi_1 I_n}{1 + \beta\phi_1 I_n + d\phi_1} \\[3mm] I_{n+1} = \dfrac{I_n + \beta\phi_2 S_n I_n}{1 + (d + \gamma + \alpha)\phi_2} \\[3mm] S_0, I_0 \geq 0. \end{cases} \qquad (26)$$

We rewrite system (26) as

$$(S_{n+1}, I_{n+1}) = H(S_n, I_n),$$

where

$$S_{n+1} = F(S_n, I_n), I_{n+1} = G(S_n, I_n) \text{ and } H = (F, G).$$

A steady state (S, I) of (26) must satisfy

$$(d + \gamma + \alpha)I = \beta I \frac{A + \gamma I}{\beta I + d}.$$

Clearly, one solution is $I = 0$ and the other solution is

$$I^* = \frac{\beta A - d(d + \gamma + \alpha)}{(d + \alpha)\beta}. \tag{28}$$

Define a threshold σ,

$$\sigma = \beta A - d(d + \gamma + \alpha).$$

Then $(\frac{A}{d}, 0)$ is the only feasible steady state of (26) if $\sigma < 0$, and in addition a nontrivial steady state (S^*, I^*) exists if $\sigma > 0$, where I^* is given by (27) and

$$S^* = \frac{A + \gamma I^*}{\beta I^* + d} = \frac{d + \gamma + \alpha}{\beta}.$$

Consider the case when $\sigma < 0$. Then (26) has only boundary steady state $(\frac{A}{d}, 0)$. It can be easily shown that the trivial steady state is globally asymptotically stable. Our proof presented here follows that of [18].

Theorem 2.4 *If $\beta > 0$, $p = 0$ and $\sigma < 0$, then every solution of (24) converges to $(\frac{A}{d}, 0)$.*

Proof. This is trivial if $I_0 = 0$. As $I_n = 0$ for $n \geq 0$ and thus $\lim_{n \to \infty} S_n = \frac{A}{d}$. We may assume $I_0 > 0$. Then $I_n > 0$ for $n \geq 0$. If there exists $k = 0, 1, \cdots$ such that

$$S_k \leq \frac{d + \gamma + \alpha}{\beta}$$

then

$$S_{k+1} \leq \frac{\dfrac{d + \gamma + \alpha}{\beta} + \phi_1 A + \gamma \phi_1 I_k}{1 + \beta \phi_1 I_k + d\phi_1} < \frac{d + \gamma + \alpha}{\beta} \quad \text{as } \sigma < 0.$$

Therefore $S_n < \dfrac{d + \gamma + \alpha}{\beta}$ for $n > k$. As a result,

$$I_{n+1} = \frac{(1 + \beta \phi_2 S_n)I_n}{1 + (d + \gamma + \alpha)\phi_2} < I_n$$

for $n > k$, which implies

$$\lim_{n \to \infty} = \bar{I} \geq 0 \text{ exits .}$$

Notice that if $\bar{I} = 0$, then for any $\epsilon > 0$ there exists $n' > 0$ such that $-\epsilon < I_n < \epsilon$ for $n \geq n'$. Thus

$$S_{n+1} \geq \frac{S_n + A\phi_1}{1 + \beta\phi_1\epsilon + d\phi_1} \quad \text{for} \quad n \geq n',$$

and we have

$$\liminf_{n \to \infty} S_n \geq \frac{A}{d}.$$

Similarly it can be proven that

$$\limsup_{n \to \infty} S_n \leq \frac{A}{d}.$$

Hence $\lim_{n \to \infty} S_n = \dfrac{A}{d}$ and the assertion is shown. Suppose now $\bar{I} > 0$. Then

$$1 = \lim_{n \to \infty} \frac{I_{n+1}}{I_n} = \lim_{n \to \infty} \frac{1 + \beta\phi_2 S_n}{1 + (d + \gamma + \alpha)\phi_2}$$

implies

$$\lim_{n \to \infty} S_n = \frac{d + \gamma + \alpha}{\beta}.$$

Consequently, solutions of (26) converge to

$$(\frac{d + \gamma + \alpha}{\beta}, \bar{I}),$$

a fixed point of H. Since $(\dfrac{A}{d}, 0)$ is the only fixed point of H, we obtain a contradiction and the result follows.

Suppose on the other hand that

$$S_n > \frac{d + \gamma + \alpha}{\beta} \text{ for } n = 0, 1, \cdots.$$

Then $I_{n+1} > I_n$ for $n = 0, 1, \cdots$ and thus $\lim_{n \to \infty} I_n > 0$ exists (maybe ∞). Notice that

$$S_{n+1} \leq \frac{S_n}{1 + d\phi_1} + a \text{ for } n \geq 0,$$

and hence

$$\limsup_{n \to \infty} S_n \leq \frac{a(1 + d\phi_1)}{d\phi_1}.$$

Consequently, if

$$\lim_{n\to\infty} I_n = \infty,$$

then from the first equation of (26), we have

$$\lim_{n\to\infty} S_{n+1} = \frac{\gamma}{\beta} < \frac{d+\gamma+\alpha}{\beta} \le \liminf_{n\to\infty} S_n$$

and obtain a contradiction. Therefore $\lim_{n\to\infty} I_n = \hat{I}$ is a positive real number. As a consequence, $\lim_{n\to\infty} S_n = \dfrac{d+\gamma+\alpha}{\beta}$. But since $\sigma < 0$,

$$\frac{d+\gamma+\alpha}{\beta} = \lim_{n\to\infty} S_{n+1} = \frac{\dfrac{d+\gamma+\alpha}{\beta} + \phi_1 A + \gamma\phi_1\hat{I}}{1 + \beta\phi_1\hat{I} + d\phi_1}$$

$$< \frac{(d+\gamma+\alpha)(1 + d\phi_1 + \beta\phi_1\hat{I})}{\beta(1 + \beta\phi_1\hat{I} + d\phi_1)} = \frac{d+\gamma+\alpha}{\beta}.$$

We thus arrive at a contradiction. Therefore, there must exist $k_0 \ge 0$ such that $S_{k_0} \le \dfrac{d+\gamma+\alpha}{\beta}$ and hence $(\dfrac{A}{d}, 0)$ is globally asymptotically stable by our earlier analysis. $\qquad\qquad\qquad\qquad\qquad\qquad\qquad\qquad\qquad\square$

It follows from Theorem 2.4 that the system of difference equations and its corresponding system of ordinary differential equations have the same asymptotic behavior when $\beta > 0$, $p = 0$ and $\sigma < 0$. Therefore, Mickens nonstandard finite difference scheme preserves the same asymptotic dynamics as its continuous counterpart for these special cases.

Suppose now $\beta > 0$, $p = 0$ and $\sigma > 0$. Then system (24) has two steady states $(A/d, 0)$ and (S^*, I^*). Their local stability can be obtained rather straightforwardly and are summarized below. The proof of the following theorem follows similarly as in [18].

Theorem 2.5 *If $\beta > 0$, $p = 0$ and $\sigma > 0$, then $(\dfrac{A}{d}, 0)$ is unstable and (S^*, I^*) is locally asymptotically stable. Moreover,*

$$\frac{\gamma}{\beta} \le \liminf_{n\to\infty} S_n \le \limsup_{n\to\infty} S_n \le \frac{A}{d}$$

for any solution (S_n, I_n) of (24) with $S_0, I_0 > 0$.

Proof. We first verify that (S^*, I^*) is locally asymptotically stable. The linearization of system (24) about the steady state yields the following Jacobian matrix

$$J = \begin{pmatrix} \dfrac{1}{1 + \beta\phi_1 I^* + d\phi_1} & \dfrac{\phi_1(\gamma + d\gamma\phi_1 - \beta S^* - A\beta\phi_1)}{(1 + \beta\phi_1 I^* + d\phi_1)^2} \\ \dfrac{\beta\phi_2 I^*}{1 + (d + \gamma + \alpha)\phi_2} & 1 \end{pmatrix}.$$

Notice

$$\gamma + d\gamma\phi_1 - \beta S^* - A\beta\phi_1 < 0 \tag{28}$$

implies

$$\det J = \frac{1}{1 + \beta\phi_1 I^* + d\phi_1} - \frac{(\beta\phi_1\phi_2 I^*)(\gamma + d\gamma\phi_1 - S^*\beta - A\beta\phi_1)}{[1 + (d + \gamma + \alpha)\phi_2](1 + \beta\phi_1 I^* + d\phi_1)^2} > 0,$$

and also

$$tr J = 1 + \frac{1}{1 + \beta\phi_1 I^* + d\phi_1} > 0.$$

Applying the Jury conditions given in [6,8], we have eigenvalues λ of J satisfying $|\lambda| < 1$ if and only if

$$\det J < 1 \text{ and } tr J < 1 + \det J.$$

Clearly, $tr J < 1 + \det J$ is true by (28), and $\det J < 1$ holds if and only if

$$\frac{-\beta\phi_2 I^*(\gamma + d\gamma\phi_1 - \beta S^* - A\beta\phi_1)}{[1 + (d + \gamma + \alpha)\phi_2][1 + \beta\phi_1 I^* + d\phi_1]} < \beta I^* + d.$$

Since $A\beta = d^2 + d\gamma + d\alpha + (\alpha + d)\beta I^*$, a straightforward calculation shows that the latter inequality is indeed true. Hence the steady state (S^*, I^*) is locally asymptotically stable.

For the boundary steady state $(\dfrac{A}{d}, 0)$, the linearization of system (24) about steady state yields the following Jacobian matrix

$$J_0 = \begin{pmatrix} \dfrac{1}{1 + d\phi_1} & \dfrac{\phi_1(\gamma + \phi_1 d\gamma - \frac{A}{d}\beta - A\beta\phi_1)}{(1 + d\phi_1)^2} \\ 0 & \dfrac{1 + \phi_2\beta\frac{A}{d}}{1 + (d + \gamma + \alpha)\phi_2} \end{pmatrix}.$$

Since J_0 is upper triangular, the eigenvalues of J_0 are $\lambda_1 = \dfrac{1}{1 + d\phi_1}$ and $\lambda_2 = \dfrac{1 + \beta\frac{A}{d}\phi_2}{1 + (d + \gamma + \alpha)\phi_2}$. Since $\sigma > 0$, we have $\lambda_2 > 1$ and $(\dfrac{A}{d}, 0)$ is thus unstable.

It remains to prove the two inequalities stated in the theorem. Since

$$\frac{\gamma}{\beta} < \frac{d+\gamma+\alpha}{\beta} < \frac{A}{d},$$

therefore if $S_n < \frac{\gamma}{\beta}$ for $n = 0, 1, \cdots$, then $I_{n+1} < I_n$ for $n \geq 0$ and thus $\lim_{n\to\infty} I_n = \hat{I} \geq 0$ exists. If $\hat{I} = 0$, then we have $\liminf_{n\to\infty} S_n \geq \frac{A}{d}$ and obtain a contradiction. If $\hat{I} > 0$, then by using a similar argument as in the proof of Theorem 2.4, we have $\lim_{n\to\infty} S_n = \frac{d+\gamma+\alpha}{\beta}$. But this again is impossible as $S_n < \frac{\gamma}{\beta}$ for $n \geq 0$. We therefore conclude that $S_{n*} \geq \frac{\gamma}{\beta}$ for some $n^* \geq 0$. It is then straightforward to show that $S_{n*+1} > \frac{\gamma}{\beta}$. As a consequence $S_n \geq \frac{\gamma}{\beta}$ for all n large and thus $\liminf_{n\to\infty} S_n \geq \frac{\gamma}{\beta}$ is shown.

On the other hand if $S_n > \frac{A}{d}$ for $n = 0, 1, \cdots$, then $I_{n+1} > I_n$ for $n \geq 0$ and thus $\lim_{n\to\infty} I_n = \bar{I} > 0$ exists. By using similar arguments as in the proof of Theorem 2.4, we have $\lim_{n\to\infty} S_{n+1} = \frac{\gamma}{\beta}$ if $\bar{I} = \infty$ and $\lim_{n\to\infty} S_n = \frac{d+\gamma+\alpha}{\beta}$ if $\bar{I}$ is a real number. In any case we obtain a contradiction. Hence $S_{k'} \leq \frac{A}{d}$ for some $k' \geq 0$ and

$$S_{k'+1} \leq \frac{A + dA\phi_1 + \gamma d\phi_1 I_{k'}}{d(1 + \beta\phi_1 I_{k'} + d\phi_1)} < \frac{A + dA\phi_1 + \beta A\phi_1 I_{k'}}{d(1 + \beta\phi_1 I_{k'} + d\phi_1)} = \frac{A}{d}.$$

Therefore $S_n \leq \frac{A}{d}$ for all n large and $\limsup_{n\to\infty} S_n \leq \frac{A}{d}$ is proved. $\qquad\square$

Finally, suppose $\beta > 0$ and $p > 0$. Then steady state (S, I) of (24) must satisfy

$$\beta(d+\alpha)I^2 - \sigma I - dpA = 0,$$

where $\sigma = \beta A - d(d+\gamma+\alpha)$. Clearly, one root is negative and the other root is

$$\bar{I} = \frac{\sigma + \sqrt{\sigma^2 + 4\beta dpA(d+\alpha)}}{2\beta(d+\alpha)} > 0.$$

Consequently, system (24) has a unique steady state $(\bar{S}, \bar{I})$, where

$$\bar{S} = \frac{A + \gamma\bar{I}}{\beta\bar{I} + d}.$$

Similar to the continuous-time model, it can be shown that $\bar{I} > I_0^*$, where I_0^* is the I-component of the steady state of (24) when $\beta = 0$. The linearization of (24) at $(\bar{S}, \bar{I})$ yields the following Jacobian matrix

$$
J = \begin{pmatrix}
\dfrac{1}{1 + \beta\phi_1\bar{I} + d\phi_1} & \dfrac{\phi_1(\gamma + d\gamma\phi_1 - \beta\bar{S} - (1-p)A\beta\phi_1)}{(1 + \beta\phi_1\bar{I} + d\phi_1)^2} \\[4mm]
\dfrac{\beta\phi_2\bar{I}}{1 + (d + \gamma + \alpha)\phi_2} & \dfrac{1 + \beta\phi_2\bar{S}}{1 + (d + \gamma + \alpha)\phi_2}
\end{pmatrix}.
$$

Unlike the continuous model for which global asymptotic stability of the positive steady state can be easily proved by using the Dulac criterion and the Poincaré-Bendixson Theorem [20], the local stability of the steady state for the discrete model requires a lot of tedious computations. By using the Jury condition, the authors in [18] verified that $(\bar{S}, \bar{I})$ is locally asymptotically stable for (24) if

$$
\phi_1 < \frac{d + \alpha}{d\gamma - A\beta}
$$

when $d\gamma > A\beta$ is true. If $d\gamma \leq A\beta$, then the inequality imposed on ϕ_1 as given above is unnecessary. These results are summarized in the following proposition [18].

Proposition 2.6 *If $d\gamma \leq A\beta$, then $(\bar{S}, \bar{I})$ is locally asymptotically stable for model (24). If $d\gamma > A\beta$, then $(\bar{S}, \bar{I})$ is locally asymptotically stable if* $\phi_1 < \dfrac{d + \alpha}{d\gamma - A\beta}.$

We now use a numerical example to simulate model (24). Choose $d = 0.01$, $\gamma = 0.2$, $A = 200$, $\beta = 0.001$, $\alpha = 0.05$, $p = 0.0001$ and $\phi_i(h) = h = 0.01$ for $i = 1, 2$. Then the unique steady state of the resulting system (24) is $(3290, 260)$. We plot three solutions with initial conditions $(S_0, I_0) = (10000, 2000)$, $(100, 20)$ and $(500, 100)$ respectively. Figure 3 plots the I-component of these solutions. We see that they all asymptotically approach approximately $\bar{I} = 3290$. Therefore, it is strongly suspected that the unique steady state is globally asymptotically stable at least for the parameter values chosen here.

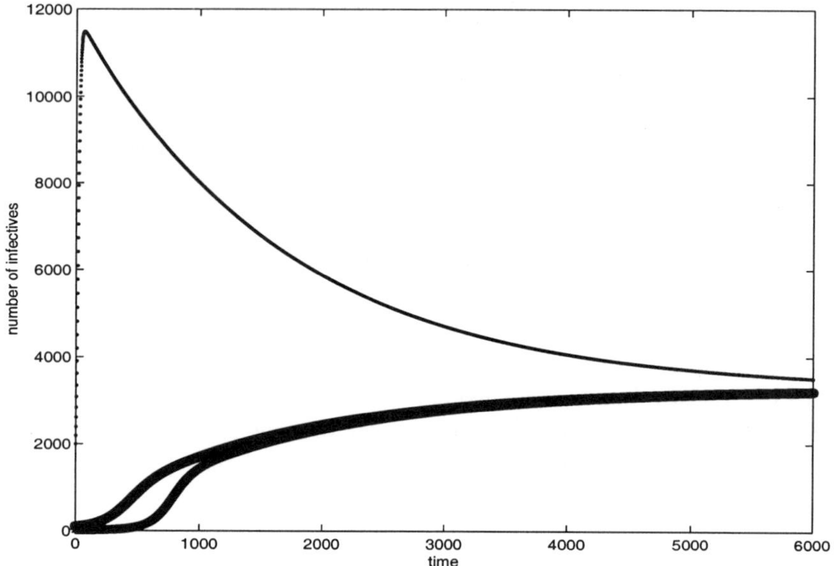

Fig. 3.   The I-component of the solutions of system (24) are plotted with three different initial conditions. If is clear from the figure that they all converge to approximately 3290.

3. A Multiple Populations Competition Model

Classical competition theory states that the number of populations competing for the same limiting resources that can survive indefinitely cannot exceed the number of distinct resources. Such a statement is often referred to as the *competitive exclusion principle* [23]. In this section, we shall present a continuous-time competition model and adopt nonstandard discretization method to obtain a system of difference equations and compare dynamics between the continuous-time and discrete-time systems.

Our presentation in this section follows similarly as that in [24]. We are interested in the following distributed rate model developed by Ackleh et al. [25] which is given by the following integro-differential equation

$$\begin{cases} \dfrac{\partial x(t,q)}{\partial t} = x(t,q)\left[q_1 - q_2 X(t)\right], \\ x(0,q) = x_0(q). \end{cases} \tag{29}$$

Here the parameter $q = (q_1, q_2)$ lies in the set $Q = [\alpha_1, \beta_1] \times [\alpha_2, \beta_2]$, where $\alpha_1, \alpha_2, \beta_1, \beta_2 \in \mathbb{R}_+$. The function $x(t,q)$ is the density of the subpopulation having parameter q. Also, $X(t) = \int_Q x(t,q)dq$ is the total population at time $t \geq 0$. The initial data $x_0(q)$ is a nonnegative continuous function satisfying

$x_0(q^*) > 0$ with $q^* = (\beta_1, \alpha_2)$. Using the theory of weak convergence of probability measures, the authors in [25] elegantly showed that solutions to model (29) satisfy

$$x(t, q) \to c\delta_{q^*}(q) \text{ as } t \to \infty,$$

where δ_{q^*} denotes the Dirac delta function concentrated at q^*, and the constant c is the ratio $\dfrac{\beta_1}{\alpha_2}$. Therefore, all subpopulations with parameters $q \neq q^*$ go to extinction as $t \to \infty$ while the subpopulation with parameter q^* persists. Notice in this model that the resource is not modeled explicitly. Nevertheless, the model validates the well established classical competitive exclusion principle successfully.

We now proceed to discretize the system by choosing a partition V of the parameter space Q. Let $V = V_1 \times V_2$, where

$$V_1 : \alpha_1 = p_1^0 < p_1^1 < \cdots < p_1^{n_1} = \beta_1,$$
$$V_2 : \alpha_2 = p_2^0 < p_2^1 < \cdots < p_2^{n_2} = \beta_2.$$

Let $n = n_1 \times n_2$, and

$$\{Q^i\}, i = 1, ..., n,$$

to be the family of subrectangles of Q resulting from the above defined partition. Let

$$q^i = \left(q_1^i, q_2^i\right)$$

be the midpoint of the rectangle $Q^i, i = 1, ..., n$.

With these at hand, we approximate the continuous-time model (29) by the following system of ordinary differential equations

$$\begin{cases} \dfrac{dz^i(t)}{dt} = z^i(t) \left(q_1^i - q_2^i \sum_{j=1}^{n} \mu(Q^j) z^j(t) \right), \\ z^i(0) = \dfrac{1}{\mu(Q^i)} \displaystyle\int_{Q^i} x_0(q) dq, \end{cases} \tag{30}$$

where $\mu(Q^i)$ denotes the Lebesgue measure of the rectangle Q^i. Making the substitution $y^i = \mu(Q^i)z^i$, (30) becomes

$$\begin{cases} \dfrac{dy^i(t)}{dt} = y^i(t) \left(q_1^i - q_2^i \sum_{j=1}^{n} y^j(t) \right), \\ y^i(0) = \displaystyle\int_{Q^i} x_0(q) dq. \end{cases} \tag{31}$$

We now approximate $\dfrac{dy^i(t)}{dt}$ by

$$\frac{y^i(t+h) - y^i(t)}{h}$$

and use a mixed implicit explicit approximation (i.e., a nonstandard finite difference method) for the quadratic term in (31) to obtain the following fully discretized version of (29)

$$\begin{cases} \dfrac{y^i(t+h) - y^i(t)}{h} = q_1^i y^i(t) - q_2^i y^i(t+h) \displaystyle\sum_{j=1}^{n} y^j(t), \\ y^i(0) = \displaystyle\int_{Q^i} x_0(q)dq. \end{cases} \tag{32}$$

Rearranging terms we have

$$\begin{cases} y^i(t+h) = \dfrac{(1+hq_1^i)y^i(t)}{1+hq_2^i \sum_{j=1}^{n} y^j(t)}, \\ y^i(0) = \displaystyle\int_{Q^i} x_0(q)dq. \end{cases} \tag{33}$$

Letting $h = 1$, $a_i = 1 + q_1^i > 1$ and $b_i = q_2^i$, results in the following system.

$$\begin{cases} x_i(t+1) = \dfrac{a_i x_i(t)}{1+b_i \sum_{j=1}^{n} x_j(t)} \\ x_i(0) \geq 0 \\ i = 1, 2, \cdots, n, \end{cases} \tag{34}$$

where parameters $a_i > 0$ and $b_i > 0$ are assumed to satisfy the following conditions

$$a_i > 1 \text{ for } i = 1, 2, \cdots, n \tag{35}$$

and

$$\frac{a_1 - 1}{b_1} > \frac{a_i - 1}{b_i} \text{ for } i = 2, \cdots, n. \tag{36}$$

Notice that solutions of the resulting system remain nonnegative. Moreover, under the assumption of (35), we see that each individual population can persist in the absence of other populations, and population $i = 1$ has the largest fitness by condition (36).

In the following, we shall study system (34). It is easy to see for the following first order scalar difference equation

$$y(t+1) = \frac{ay(t)}{1+by(t)}, \tag{37}$$

all solutions with $y(0) > 0$ converge to the steady state $\dfrac{a-1}{b}$ if $a > 1$ and $b > 0$ [8]. Using the asymptotic dynamics of equation (37) and condition (35), it can be easily shown that

$$\limsup_{t\to\infty} x_i(t) \le \frac{a_i - 1}{b_i}$$

for $i = 1, 2, \cdots, n$ for any solution of (34). On the other hand, letting

$$a_0 = \min\{a_1, a_2, \cdots, a_n\}, \, b_0 = \max\{b_1, b_2, \cdots, b_n\}$$

and

$$P(t) = \sum_{i=1}^{n} x_i(t),$$

we have for $t \ge 0$

$$P(t+1) \ge \frac{a_0 P(t)}{1 + b_0 P(t)}. \tag{38}$$

Consequently, solutions $(x_1(t), x_2(t), \cdots, x_n(t))$ of (34) with $P(0) > 0$ satisfy

$$\liminf_{t\to\infty} P(t) \ge \frac{a_0 - 1}{b_0}.$$

The above discussion regarding the discrete-time model is summarized in the following lemma.

Lemma 3.1 *Solutions of system* (34) *exist and remain nonnegative for* $t > 0$. *Moreover, let*

$$m = \frac{a_0 - 1}{b_0} > 0$$

and

$$M = \frac{\hat{a} - 1}{\hat{b}} > 0$$

where $\hat{a} = \max\{a_1, a_2, \cdots, a_n\}$ *and* $\hat{b} = \min\{b_1, b_2, \cdots, b_n\}$, *then*

$$m \le \liminf_{t\to\infty} P(t) \le \limsup_{t\to\infty} P(t) \le M$$

if $P(0) > 0$, *where* $P(t) = \sum_{i=1}^{n} x_i(t)$ *and* $(x_1(t), x_2(t), \cdots, x_n(t))$ *is the solution of* (34).

We next proceed to discuss equilibria of system (34). Clearly, (34) always has a trivial steady state $E_0 = (0, 0, \cdots, 0)$ which exists for all parameter values. The nontrivial equilibrium equations satisfy

$$\sum_{j=1}^{n} x_j = \frac{a_i - 1}{b_i} \quad \text{for } i = 1, 2, \cdots, n.$$

As a result, there are no interior equilibria, and in addition to E_0, there are n boundary equilibria: $E_1 = (\frac{a_1 - 1}{b_1}, 0, \cdots, 0)$, $E_2 = (0, \frac{a_2 - 1}{b_2}, 0, \cdots, 0)$, $\cdots$, and $E_n = (0, \cdots, 0, \frac{a_n - 1}{b_n})$. The local stability of each of these steady states can be determined easily. Using conditions (35) and (36) imposed on the parameters we have, E_0 is a repeller, E_1 is locally asymptotically stable, and E_i is a saddle point for $i = 2, \cdots, n$.

For the special case when $n = 2$, the global asymptotic dynamics of the system can be understood as theory of planar competitive maps are well developed,. In the following we prove that when $n = 2$ solutions of (34) with $x_1(0) > 0$ converge to $E_1 = (\frac{a_1 - 1}{b_1}, 0)$. Notice that system (34) is competitive and it is known that bounded solutions of planar competitive systems either converge to a steady state or have a 2-cycle as its ω-limit set [26, Theorem 4.2]. An explicit calculation was carried out in [27] to eliminate the existence of an interior 2-cycle and one can thus conclude that the interior steady state is globally asymptotically stable. Our analysis performed here is different from that given in [27].

System (34) with $n = 2$ can be written explicitly as

$$\begin{cases} x_1(t+1) = \dfrac{a_1 x_1(t)}{1 + b_1(x_1(t) + x_2(t))} \\[2ex] x_2(t+1) = \dfrac{a_2 x_2(t)}{1 + b_2(x_1(t) + x_2(t))} \\[2ex] x_1(0), x_2(0) \geq 0, \end{cases} \qquad (39)$$

where $a_1, a_2 > 1$, $b_1, b_2 > 0$ and $\dfrac{a_1 - 1}{b_1} > \dfrac{a_2 - 1}{b_2}$. Specifically, we will use a result of [26] to assert competitive exclusion for two populations. The asymptotic dynamics of (39) is given below.

Theorem 3.2 *Solutions $(x_1(t), x_2(t))$ of system (39) with $x_1(0) > 0$ converge to the steady state $E_1 = (\frac{a_1 - 1}{b_1}, 0)$.*

Proof. Let $f : \mathbb{R}^2 \to \mathbb{R}^2$ denote the map induced by system (39), i.e., $f(x_1, x_2) = (f_1(x_1, x_2), f_2(x_1, x_2))$, where

$$f_1(x_1, x_2) = \frac{a_1 x_1}{1 + b_1(x_1 + x_2)} \quad \text{and} \quad f_2(x_1, x_2) = \frac{a_2 x_2}{1 + b_2(x_1 + x_2)}.$$

The Jacobian matrix associated with the system is given below.

$$J = \begin{pmatrix} \dfrac{a_1 + a_1 b_1 x_2}{[1 + b_1(x_1 + x_2)]^2} & \dfrac{-a_1 b_1 x_1}{[1 + b_1(x_1 + x_2)]^2} \\[2em] \dfrac{-a_2 b_2 x_2}{[1 + b_2(x_1 + x_2)]^2} & \dfrac{a_2 + a_2 b_2 x_1}{[1 + b_2(x_1 + x_2)]^2} \end{pmatrix}.$$

Define a partial ordering $\leq_K$ on $\mathbb{R}^2_+$ by

$$(x_1, x_2) \leq_K (z_1, z_2) \text{ if } x_1 \leq z_1 \text{ and } x_2 \geq z_2.$$

Then the map J is K-positive as its diagonal entries are nonnegative and its off diagonal entries are non-positive on $\mathbb{R}^2_+$, and the determinant of J is also positive on $\mathbb{R}^2_+$. Moreover, $\mathbb{R}^2_+$ contains order intervals and is $\leq_K$-convex. It remains to verify that the map f is one-to-one by [26, Corollary 4.4].

Let $(y_1, y_2) \in \mathbb{R}^2_+$ be given arbitrarily. If $y_1 = y_2 = 0$, then it has a unique preimage $(0, 0)$. If $y_1 = 0$, then $(0, y_2)$ has at most one preimage as $\hat{g}(x) = \dfrac{a_2 x}{1 + b_2 x}$ is strictly increasing. The same is true if $y_2 = 0$. Suppose now $y_1, y_2 > 0$. If there are $(\bar{x}_1, \bar{x}_2) \neq (\hat{x}_1, \hat{x}_2)$ such that

$$(y_1, y_2) = f(\bar{x}_1, \bar{x}_2) = f(\hat{x}_1, \hat{x}_2),$$

we may assume $\bar{x}_2 < \hat{x}_2$ as the other cases can be proved similarly. Let x_2^* with $\bar{x}_2 \leq x_2^* \leq \hat{x}_2$ be given. Notice

$$y_1 = \frac{a_1 x_1}{a + b_1(x_1 + x_2^*)}$$

has a unique solution $x_1^* > 0$, where

$$\frac{a_1 \bar{x}_1}{1 + b_1(\bar{x}_1 + x_2^*)} \leq \frac{a_1 \bar{x}_1}{1 + b_1(\bar{x}_1 + \bar{x}_2)} = y_1$$

and

$$\frac{a_1 \hat{x}_1}{1 + b_1(\hat{x}_1 + x_2^*)} \geq \frac{a_1 \hat{x}_1}{1 + b_1(\hat{x}_1 + \hat{x}_2)} = y_2.$$

Hence $\bar{x}_1 \leq x_1^* \leq \hat{x}_1$. Denote this x_1^* by $X_1^*(x_2^*)$, $\bar{x}_2 \leq x_2^* \leq \hat{x}_2$. Since

$$\frac{a_1 X_1^*(x_2^*)}{1 + b_1(X_1^*(x_2^*) + x_2^*)} = y_1,$$

we have $\dfrac{dX_1^*(x_2^*)}{dx_2^*} = \dfrac{b_1 X_1^*(x_2^*)}{1 + b_1 x_2^*} > 0$. Define

$$Y_2(x_2^*) = \frac{a_2 x_2^*}{1 + b_2(X_1^*(x_2^*) + x_2^*)}$$

for $\bar{x}_2 \leq x_2^* \leq \hat{x}_2$. Using $\dfrac{dX_1^*}{dx_2*}$ given above, a direct computation yields

$$\frac{dY_2}{dx_2^*} = \frac{a_2 + a_2 b_1 x_2^* + a_2 b_2 X_1^*(x_2^*)}{1 + b_2 x_2^*} > 0.$$

Since $\bar{x}_2 \leq x_2^* \leq \hat{x}_2$ and $\bar{x}_2 < \hat{x}_2$, we must have

$$y_2 = Y_2(\bar{x}_2) < Y_2(\hat{x}_2) = y_2$$

and obtain a contradiction. Therefore, f is one-to-one and solutions of (39) with $x_1(0) > 0$ converge to the steady state E_1 as solutions are bounded and E_1 is locally asymptotically stable [26, Corollary 4.4]. $\square$

It is strongly suspected for $n > 2$ that the equilibrium E_1 is globally asymptotically stable provided that conditions (35) and (36) are satisfied. A numerical example was given in [27], which confirmed our observation. We shall use another numerical example to demonstrate this conjecture. Specifically, parameter values are $n = 3$, $a_1 = 1.5, a_2 = 2, a_3 = 2.75$, $b_1 = 0.25$, $b_2 = 0.6$ and $b_3 = 1.8$. Therefore population one has the largest fitness than other two populations. Numerical simulation presented in Figure 4 with initial population size $x_1(0) = 3, x_2(0) = 10$ and $x_3(0) = 25$ shows that only population one survives while other populations become extinct. Simulations with the same parameter values and different initial conditions also result in the same asymptotic behavior. Therefore, we conclude that the nonstandard finite difference method yields discrete approximations with the same local dynamics. For $n = 2$, it was shown that the two systems even have the same global asymptotic dynamics.

4. Discussion

It has been known that discrete approximations of continuous-time systems introduce an extra parameter, the step size of the approximation, and thus the resulting difference equations usually have different properties as their original continuous-time equations. From the models presented in this

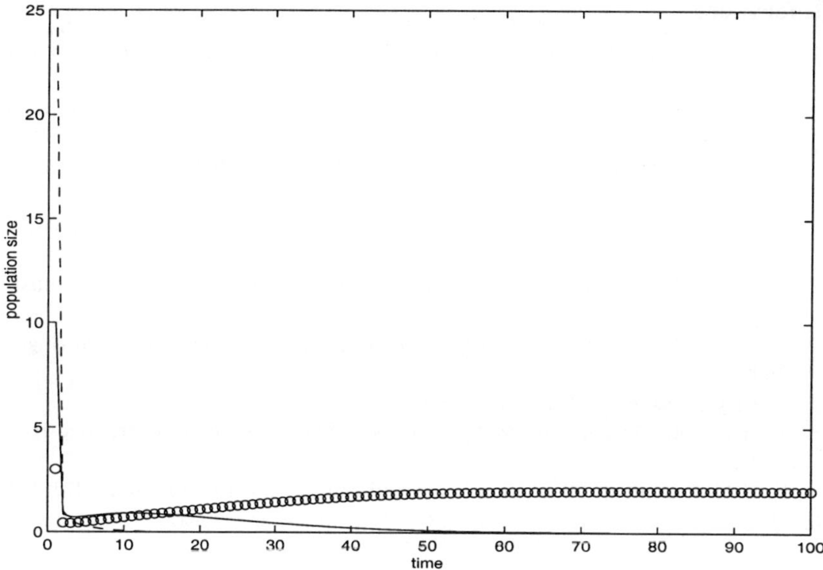

Fig. 4. The figure plots population dynamics of three competing populations. We can see from the graph that population 1 with the largest fitness drives the other populations to extinction.

chapter and the equations studied by other researchers, we see that Mickens nonstandard finite difference method results in difference equations that share the same dynamics as their original continuous-time equations. However, it is still an open question as to why the nonstandard discretization schemes work. This is a fundamental problem that needs to be resolved by numerical analysts.

Moreover, there are some types of equations such as partial differential equations, periodic ordinary differential equations, and delay differential equations for which the methods of nonstandard finite difference schemes have not been applied to. However, many biological models are described by the above mentioned equations or systems of equations. Therefore, it would be interesting to see whether the methods would work on these types of equations successfully.

References

1. S. B. Palmer, M. S. Rogalski, *Advanced Univeristy Physics*, Gordon and Breach, 1996.
2. W. J. Ewens, *Mathematical Population Genetics I: Theoretical Introduction*, Springer-Verlag, 2004.
3. R. Mickens, *Nonstandard Finite Difference Methods of Differential Equations*, World Scientific, Singapore, 1994.
4. R. Mickens, *Applications of Nonstandard Finite Difference Schemes*, World Scientific, Singapore, 2000.
5. R. Devaney, *A First Course in Chaotic Dynamical Systems,* Perseus Publishing Co., 1992.
6. S. Elaydi, *An Introduction to Difference Equations*, Second Edition, Springer, 1999.
7. S. Elaydi, *Discrete Chaos*, CRC, 2000.
8. L. Allen, *An Introduction to Deterministic Models in Biology*, Prentice-Hall, 2005.
9. S. Ushiki, Central difference scheme and chaos, *Physica D* **4**(1982), 407-424.
10. H. Jiang and T. Rogers, The discrete dynamics of symmetric competiiton in the plane, *J. Math. Biol.* **25**(1987), 573–596.
11. W. Krawcewicz and T. Rogers, Perfect harmony: The discrete dynamics of cooperation, *J. Math. Biol.* **28**(1990), 383–410.
12. P. Liu and S. Elaydi, Discrete competitive and cooperative models of Lotka-Volterra type, *J. Comput. Anal. Appli.* **3**(2001), 53–73.
13. H. Al-Kahby, F. Dannan and S. Elaydi, Non-standard discretization methods for some biological models, *Appl. of Nonstandard Finite Diff. Schemes*, edited by R. Mickens, 155–178, 2000.
14. L-I. Roeger and L. Allen, Discrete May-Leonard competitive models I, *J. Diff. Equ. Appl.*, **10**(2004), 77-98.
15. L-I. Roeger, Discrete May-Leonard competitive models II, *Disc. Cont. Dyn. System, SerB*, to appear.
16. L-I. Roeger, Disctete May-Leonard competitive models III, *J. Diff. Equ. Appl.*, **10**(2004), 773–790.
17. L-I. Roeger, Three-dimensional disctete competitive Lotka-Volterra system that preserves Hopf bifurcations, submitted.
18. S. R.-J. Jang and S. Elaydi, Difference equations derived from discretization of a continuous epidemic model with immigration of infectives, *Canadian Applied Math. Quarterly*, to appear.
19. F. Brauer and P. van den Driessche, Models for transmission of disease with immigration of infectives, *Math. Biosci.* **171**(2001), 143–54.
20. E. Coddington and N. Levinson, *Theory of Ordinary Differential Equations*, New York: McGraw Hill, 1955.
21. L. Allen, Some discrete-time SI, SIR, and SIS epidemic models, *Math. Biosci.*, **124**(1994), 83–105.
22. J. D. Murray, *Mathematical Biology*, Spring-Verlag, 1998.
23. M. Begon, J. Harper and C. Jownsend, *Ecology: Individuals, Populations and*

Communities, Blackwell Science Ltd, NY, 1996.

24. A. S. Ackleh, Y. M. Dib and S. R.-J. Jang, A discrete-time Beverton-Holt competition model, *Proceedings of the 9th International Conference on Difference Equations and Applications*, to appear.

25. A. S. Ackleh, D. Marshall, H. Heatherly and B. G. Fitzpatrick, Survival of the fittest in a generalized logistic model, *Mathematical Models and Methods in Applied Sciences*, **9**(1999), 1379–1391.

26. H. L. Smith, Planar competitive and cooperative difference equations, *J. Diff. Eqns. Appl.* **3**(1998), 335–357.

27. S. R.-J. Jang and A.S. Ackleh, Discrete-time, discrete stage-structured predator-prey models, *J. Diff. Equ. Appl.*, to appear.

CHAPTER 11

ROBUST DISCRETIZATIONS VERSUS INCREASE OF THE TIME STEP FOR CHAOTIC SYSTEMS

Christophe Letellier

CORIA UMR 6614 - Université de Rouen
Av. de l'Université, BP 12
F-76801 Saint-Etienne du Rouvray cedex, France

Eduardo M. A. M. Mendes

Universidade Federal de Minas Gerais
Av. Antonio Carlos 6627
Belo Horizonte 31270-901, Brazil

When continuous systems are discretized, their solutions depend on the time step chosen *a priori*. Such solutions are not necessarily spurious in the sense that they can still correspond to a solution of the differential equations but with a displacement in the parameter space. Consequently, it is of great interest to obtain discrete equations which are robust even when the discretization time step is large. Using non standard schemes, different discretizations of three chaotic systems are discussed versus the values of the discretization time step. It is shown that the sets of difference equations proposed are more robust versus increases of the time step than conventional discretizations built with standard schemes such as the forward Euler or the backward Euler. The non-standard schemes used here are Mickens' scheme and Monaco and Normand-Cyrot's scheme. The solution to the discretization only becomes spurious when the time step is greater than the time related to the so-called Nyquist time.

1. Introduction

Until recently, most of the physical processes have been modeled by differential equations where the processes are assumed to be evolving continuously. When these differential equations are non linear, there is very often no analytical solution and only numerical integration techniques can provide accurate numerical solutions to the original differential equations. With the advent of digital computers, this is easily done using a fourth-order Runge-Kutta integration scheme. Nevertheless, sometimes it is necessary to replace the set of differential equations with a continuous dependence on time by a set of difference equations with a discrete time variable. When standard schemes as the forward Euler, backward Euler or central difference schemes is used, the discrete system has solutions which are equivalent to those of the continuous counterpart only for sufficiently small discretization time step. With standard schemes, the upper value of the time step for which the solutions are equivalent to the continuous counterpart is significantly smaller than the sampling time used for recording the time evolution of a physical quantity. The question of a possible equivalence between differential equations for quite large time step has thus emerged as well as a new scientific theory advocating the discreteness of time [1,2].

This equivalence is particularly important to address when a comparison between different global modeling techniques is attempted. Typically, a global modeling technique is used for obtaining a set of equations that captures the dynamics described by a recorded time series. Such a technique is based on the following procedure. First, a phase space is reconstructed from the measured time series using either delay or derivatives coordinates [3]. If derivative coordinates are used then we obtain a system of ordinary differential equations whose dimensions are equal to the dimensions of the original phase space [4]. On the other hand, if delay coordinates are used then a set of difference equations is obtained. Most of time, the dimensions of the obtained models are significantly greater than those of the original systems [5,6]. Such a feature was never considered as a major problem since Takens proposed an existence theorem which ensures the existence of a reconstructed phase space that is diffeomorphically equivalent to the original phase space as long as the dimension of the former is sufficiently large [7]. Indeed, Takens' criterion corresponds to a dimension significantly greater than the dimension of the original phase space. Tempkin and Yorke [8] showed that for almost any choice of measurements (in the sense of prevalence), there exists a scalar difference equation that describes the evolution of the sequence of measurements whose dimension is larger than twice the box-counting dimension of the dynamical system's attractor.

Nevertheless, according to recent works [9,10,11] where both types of techniques were applied successfully to the same data sets, the dimensions and the number of terms of the obtained global models are different and very much dependent on the techniques used. No direct comparison between the difference and the differential models was therefore possible. Moreover, the obtained global discrete model strongly depends on the discretization time-step [12]. Consequently, it becomes an important task to compare the global models provided by these modeling techniques. It then appears obvious that, before being able to address correctly this problem, it is necessary to have a clear idea of the possible equivalence between differential equations with time step comparable to the sampling time.

For all the above-mentioned reasons, we decided to reinvestigate the possible equivalence between difference and differential equations. This is obviously connected to the important problem of finding a discretization of a set of ordinary differential equations [13,14,15]. In this context, it is well known that the discretization of continuous equations may have solutions which depend on the discretization time-step h and numerical instabilities may be encountered. Numerical instabilities are solutions to the discrete finite difference equations that do not correspond to any solution to the original differential equations. Such a feature occurs mainly when the time step h is too large [16]. This may also be encountered when the order of the difference equations is larger than that of the differential equations [13].

A simple example of numerical instabilities is observed in the discretization of the logistic map using the Euler scheme. Indeed, the elementary nonlinear ordinary differential equation

$$\dot{x}(t) = x(t)\left[1 - x(t)\right] \tag{1}$$

has two fixed points $x_1^* = 0$ and $x_2^* = 1$. The first fixed point is unstable while the second is stable. Thus, every solution of $x(t)$ with $x(0) > 0$ is asymptotically stable. Using the Euler-type discretization scheme

$$\dot{x}_n \mapsto \frac{x_{n+1} - x_n}{h}, \tag{2}$$

where $h > 0$ denotes the time step and x_n the value of $x(t)$ for $t = nh$. The elementary equation (1) thus becomes

$$x_{n+1} = hx_n\left(1 + \frac{1}{h} - x_n\right), \tag{3}$$

which is a slightly modified logistic map. This difference equation has two fixed points which are still $x_1^* = 0$ and $x_2^* = 1$. Nevertheless, the fixed point x_2^* is stable only for $0 < h < 2$. When the time step h is increased, a period-doubling cascade is observed as well as various chaotic attractors and periodic solutions usually encountered in the logistic map. Consequently, it appears that the discretization of equation (1) is only valid, from the asymptotic behavior point of view, for a finite interval of the time step. In fact, equation (3) may be rewritten as

$$x_{n+1} = \delta_x x_n + \delta_{x^2} x_n^2, \tag{4}$$

where $\delta_x = h+1$ and $\delta_{x^2} = -h$. If we take $\delta_x = \lambda$ and $\delta_{x^2} = -\lambda$, we obtain the usual logistic map, where λ is the bifurcation parameter. It appears that the discretization of equation (1) leads to a discrete equation which has the same structure as the logistic equation since h and λ are closely related. The discretization time step h may be viewed as the bifurcation parameter of the equation.

In this chapter, we will investigate the discretization of three different sets of ordinary differential equations possessing chaotic solutions. This will lead us to discuss the solution to the possible equivalence between a set of ordinary differential equations and the solution to its discretization. We will limit ourselves to the cases of three-dimensional systems to take advantage of the topological analysis for accurately characterizing the nature of the solutions. Although in a rigorous way, difference and differential equations are said to have the same general solution if and only if $u_n = u(t_n)\,|_{t=n\,h}$ for $h > 0$, a weaker equivalence is required since the time step h is necessarily a bifurcation parameter of the system, that is, the solution may change in its nature under bifurcation when h is varied. Thus, it obviously appears that the rigorous equivalence could only be obtained for specific value of h. But since the considered solutions are chaotic, the condition $u_n = u(t_n)$ cannot be verified beyond the Lyapunov time, that is, the upper limit before which two solutions from two slightly different initial conditions diverge significantly. An obvious question then comes : *which kind of equivalence are we really looking for since an exact solution is not attainable when a chaotic attractor is considered?*

Moreover, as soon as h is varied, the solution to the discretization differs from the solution to the corresponding ordinary differential equations but this does not necessarily mean that the obtained solution is no longer a solution to the differential equation. Another important point is therefore to determine in which way a solution to a discretization model could remain a solution to the corresponding differential equation although both are different. The domain where a solution to the discretization is also a solution of the corresponding differential equations will be discussed.

The rest of this chapter is organized as follows. Section 2 briefly describes the topology analysis of chaotic solutions to the Rössler which is used as an example. The motivation for the analysis performed in this chapter is laid out in Section 3. In Section 4, the nonstandard Mickens discretization scheme is used and solutions to the discretisations obtained are investigated versus the time step. In Section 5, another nonstandard scheme, the so-called Normand and Monaco-Cyrot scheme is investigated and the quality of the solutions obtained are compared to those obtained with the Mickens scheme. Our conclusions are given in Section 6.

2. Topological Analysis of Chaotic Systems

2.1. *Poincaré section and bifurcation diagram*

As a benchmark system for introducing the main concepts used in topological analysis, let us introduce the Rössler system [17] reading as:

$$
\begin{cases}
\dot{x} = -y - z \\
\dot{y} = x + ay \\
\dot{z} = b + xz - cz
\end{cases}
\tag{5}
$$

where (a, b, c) are the bifurcation parameters. The Rössler system has two fixed points given by

$$
\begin{cases}
x_\pm = \dfrac{c \pm \sqrt{c^2 - 4ab}}{2} \\[2mm]
y_\pm = -\dfrac{c \pm \sqrt{c^2 - 4ab}}{2a} \\[2mm]
z_\pm = \dfrac{c \pm \sqrt{c^2 - 4ab}}{2a} \, .
\end{cases}
\tag{6}
$$

For $a = 0.432$, $b = 2$ and $c = 4$, the Rössler system has a chaotic attractor for solution (Fig. 1a). According to Rössler [18], we designate this attractor as the *spiral* attractor. This attractor is characterized by a first-return map to the Poincaré section

$$
P \equiv \left\{ (y_n, z_n) \in \mathbb{R}^2 \mid x_n = x_-, \dot{x}_n > 0 \right\},
\tag{7}
$$

which is constituted by an increasing monotonic branch and a decreasing branch separated by the critical point located at the maximum (Fig. 1b). A symbol is associated with each branch. Chaotic trajectories and the periodic orbits constituting their skeleton are thus encoded over the symbol set $\{0, 1\}$. The symbol "0" is associated with the increasing branch and symbol "1" with the decreasing branch. Periodic orbits may thus be encoded by symbolic strings. For instance, a period-2 orbit having one intersection with the Poincaré section located on the branch "0" and one located in the branch

"1" is designated by the sequence (10). A period-3 orbit would have three symbols, and so on. The population of periodic orbits embedded within the attractor solution to system (5) for $a = 0.432$ is reported in Table 1. Thus, for $a = 0.432$, most of periodic orbits encoded with two symbols are embedded within the attractor. In fact, the increasing branch is very close to the bisecting line and, consequently, the symbolic dynamics is almost complete. A symbolic dynamics defined on the symbol set $\{0, 1\}$ is *complete* when all periodic orbits, which can be encoded with these two symbols, are solutions to the Rössler system.

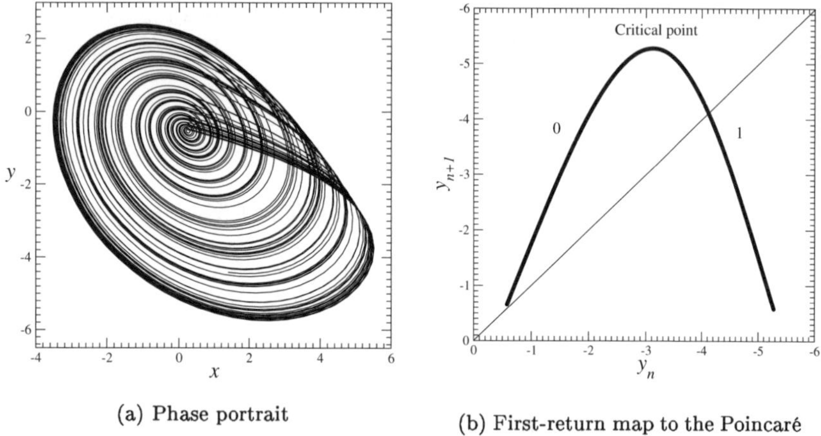

(a) Phase portrait

(b) First-return map to the Poincaré section P

Fig. 1.   Spiral attractor solution to the Rössler system (5) with the parameters $(a, b, c) = (0.432, 2, 4)$.

When the bifurcation parameter a is increased, new periodic orbits are created and the chaotic attractor increases in size (Fig. 2b). The corresponding first-return map is constituted by more than two branches and, for $a = 0.556$, up to eleven monotonous branches may been identified [19]. The corresponding attractor is designated as the *screw* attractor [18]. For a greater than 0.556, there is metastable chaos, that is, the trajectory visits the neighborhood of the unstable periodic orbits solution to the Rössler attractor before being ejected to infinity [19]. The dynamics of the Rössler system can therefore be investigated for $a < 0.556$, b and c remaining constant.

A bifurcation diagram synthesizes the evolution of the dynamics under

Orbit	Orbit	Orbit
1	100	100010
10	100101	100011
1011	10010	10001
101110	10011	10000
101111	100111	100001
10111	100110	100000
10110	1001	
101	1000	

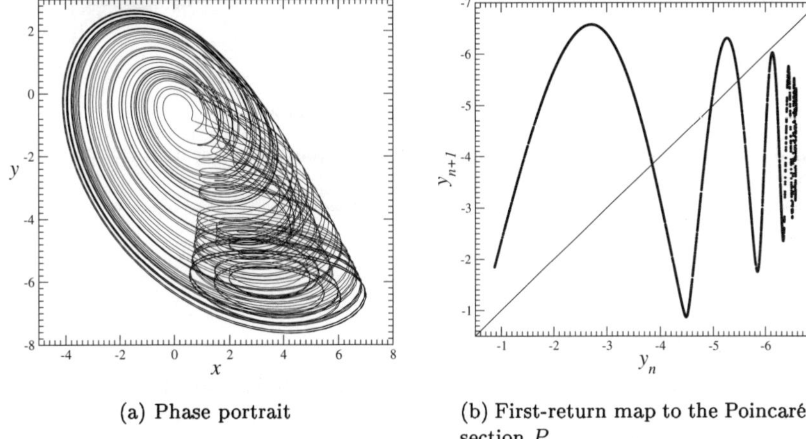

(a) Phase portrait

(b) First-return map to the Poincaré section P

Fig. 2.   Screw attractor solution to the Rössler system (5) with the bifurcation parameters $(a, b, c) = (0.556, 2, 4)$.

the change of the bifurcation parameter a (Fig. 3). It is built as saving an hundred of intersections of the trajectory with the Poincaré section for each value of the parameter. In this diagram, one may identify a period-doubling cascade between $a = 0.2$ and $a = 0.385$, then a chaotic solution is observed. The chaotic solution arises when the orbit with period 2^n with $n \rightarrow \infty$ has been created. Still increasing parameter a, various periodic windows are observed. In each of them, a new period-doubling cascade is observed. Orbits have period equal to $2^n p$ where p is the period of the orbit at the left of the window. For instance, a period-3 window is observed around $a = 0.410$. We will show that quite similar dependences of the dynamics on the bifurcation parameter is recovered when the time step h of the discretization of the Rössler system is increased.

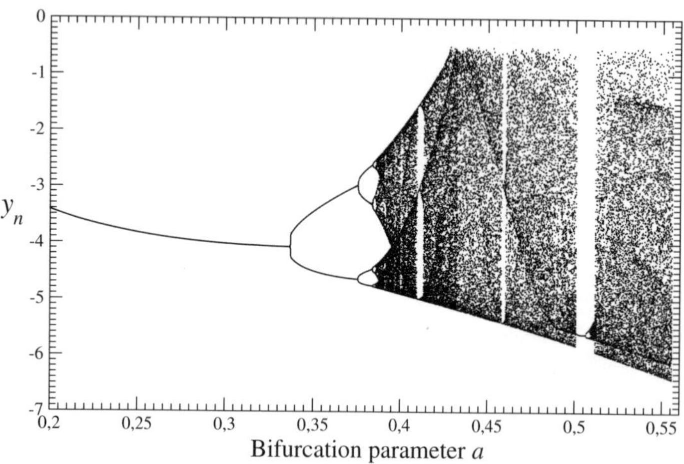

Fig. 3. Bifurcation diagram versus parameter a of the Rössler system (5). Parameter values : $a \in [0.2 \, ; 0.556]$ and $(b, c) = (2.0, 4.0)$.

2.2. *Template and linking numbers*

When the system is described in a three dimensional phase space, it possible to describe the flow by a two-dimensional branched manifold where the two dimensions describe the direction of the flow and the direction of stretching [20]. The number of branches of the template is equal to the number of branches in the first-return map. The increasing branches are preserving order and decreasing branches are reversing order [21]. The spiral attractor (Fig. 1) solution to the Rössler system is thus divided in one preserving order strip and one reversing order strip. A preserving order strip represents an even number of half-turns while a reversing order strip represents an odd number of half-turns. Consequently, the corresponding template will be composed of two strips as shown in Fig. 4.

An adequate template must predict topological invariants like linking numbers between pairs of periodic orbits. A periodic orbit is here considered as a knot. Periodic orbits embedded within the attractor can be approximated by segments of the chaotic time series that mimic the behavior of nearby unstable periodic orbits. A 'close return' method [19] is applied to

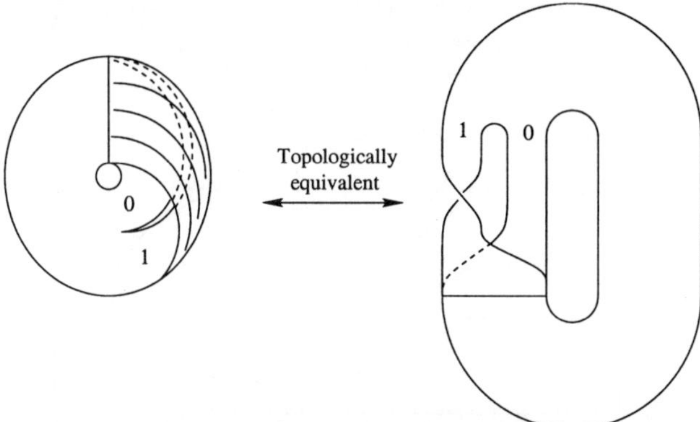

Fig. 4.　Two topologically equivalent representations of the template with two strips which encodes the topological properties of the spiral attractor of the Rössler system.

the Poincaré section to extract them. The linking numbers are ambient isotopy invariant defined as follows.

Definition 1: Let α and β be two knots defining a link L in $\mathbb{R}^3$. Let σ denotes the set of crossings of α with β. Then the linking number reads :

$$lk(\alpha, \beta) = \frac{1}{2} \sum_{p \in \sigma} \epsilon(p) \tag{8}$$

where ϵ is the sign of each crossing p with the usual convention, that is

The linking number $lk(\alpha, \beta)$ between two periodic orbits α and β is the half of the algebraic sum of all crossings between α and β (ignoring self-crossings). In practice, the linking numbers are counted in plane regular projections of orbit pairs by using the third coordinate to define the sign of crossings. For instance, orbits (10) and (101) are depicted in Fig. 5. This example is very simple and the linking number is found to be equal to -2 since five negative crossings and one positive crossing are identified.

All the linking numbers between orbits embedded within the attractor are well predicted by the template (Fig. 6a) where periodic orbits (10) and (101) are drawn. Four negative crossings are identified. The linking number $\tilde{lk}(10, 101)$ predicted by the template is thus -2 as for the plane projection

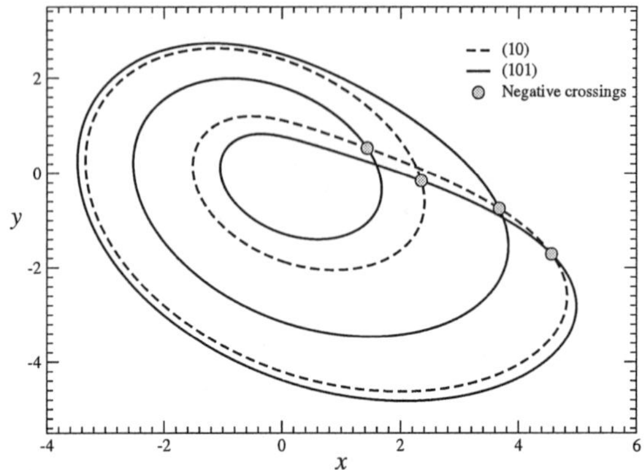

Fig. 5.   A link made of two periodic orbits encoded by (10) and (101), respectively. Four negative crossings are identified. The linking number $lk(10, 101)$ is therefore equal to -2.

of orbits shown in Fig. 5. Such a template can be described by the linking matrix

$$M_{ij} = \begin{bmatrix} 0 & -1 \\ -1 & -1 \end{bmatrix} \qquad (9)$$

according to the standard insertion convention [22]. The diagonal elements M_{ii}'s are equal to the number of π-twists of the ith strip and off-diagonal elements M_{ij}, $(i \neq j)$ are given by the algebraic number of intersections between the ith and jth strips. Further details for such a topological characterization procedure are extensively discussed in [23,19,20].

As previously explained, the screw attractor can have many branches. For instance, for $a = 0.52385$, the first-return map is made of four branches and the template describing the topology of the corresponding attractor has four strips as shown in Fig. 6. This template remains valid until $a \in$ $]0.496\ ; 0.525[$, that is, when the first-return map has four branches. In a general way, all the Rössler attractors are well described by templates defined by the general linking matrix reading as [19] :

$$M_{ij,n} = \begin{bmatrix} 0 & -1 & -1 & \dots & -1 & -1 & -1 \\ -1 & -1 & -2 & \dots & -2 & -2 & -2 \\ -1 & -2 & -2 & \dots & \dots & \dots & \dots \\ \vdots & \vdots & \vdots & \vdots & \vdots & \vdots & \vdots \\ -1 & -2 & \dots & \dots & -(n-2) & -(n-1) & -(n-1) \\ -1 & -2 & \dots & \dots & -(n-1) & -(n-1) & -n \\ -1 & -2 & \dots & \dots & -(n-1) & -n & -n \end{bmatrix} \qquad (10)$$

where n is the number of branches. Thus, two chaotic attractors will be said "*topologically equivalent*" when both are described by the same linking matrix — or template — and both have the same population of periodic orbits.

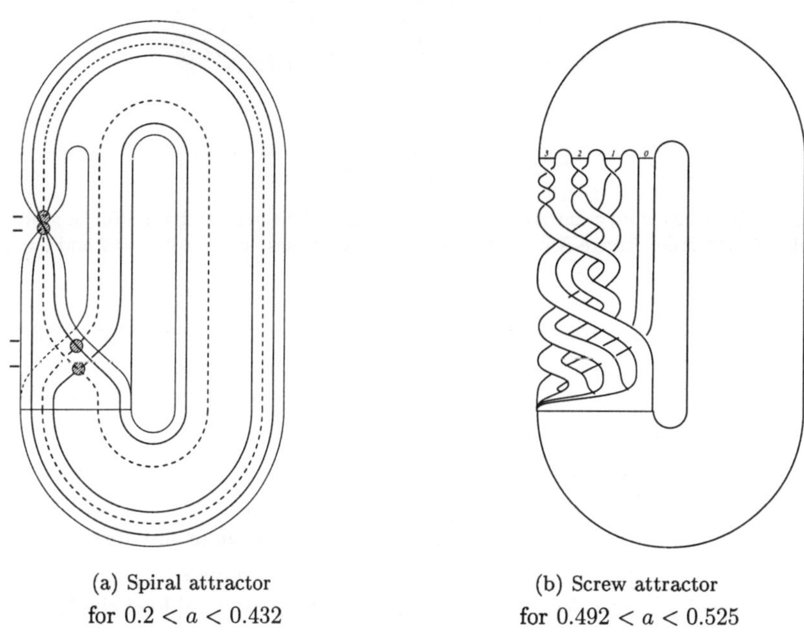

(a) Spiral attractor
for $0.2 < a < 0.432$

(b) Screw attractor
for $0.492 < a < 0.525$

Fig. 6.   Two templates for solutions to the Rössler system. Periodic orbits (10) and (101) are drawn in the template for the spiral attractor. From this template, the linking number $\tilde{l}k(10, 101)$ is equal to -2 as counted in the plane projection shown in Fig. 5.

3. Motivations for Nonstandard Schemes

Since existence of chaotic solutions is linked to the lack of any analytical solution, the use of numerical integration techniques that provide accurate numerical solutions to the original differential equations is necessary. Most of the time, this is easily done using a fourth-order Runge-Kutta integration scheme. For instance, when the Rössler system is investigated, the time step for numerical integration can be varied over a quite wide interval without presenting any particular bifurcation as revealed by the bifurcation diagram depicted in Fig. 7.

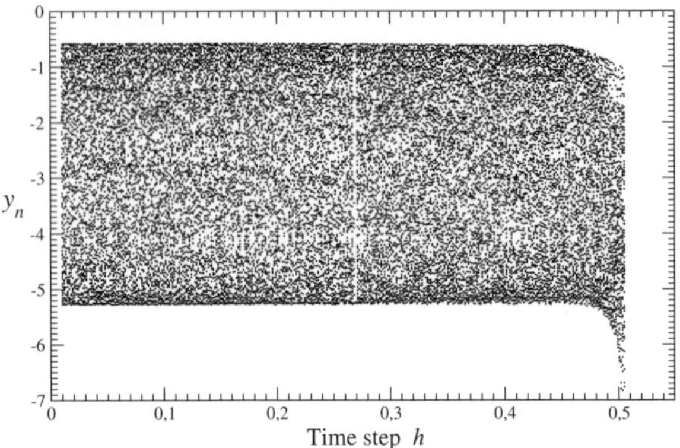

Fig. 7.   Bifurcation diagram versus the time step used for integrating the Rössler system with a fourth-order Runge-Kutta scheme. Parameter values : $a = 0.432$, $b = 2$ and $c = 4$.

In fact, the trajectory is ejected to infinity when h is greater than 0.506 s. First, it is interesting to compare this threshold value to the time associated with the Nyquist criterion. This criterion states that all the information required for a proper description of the dynamics can only be recovered when the sampling time is less than $T_N = \frac{1}{2f_{\max}}$ where $f_{\max}$ is the largest frequency estimated using a Fourier spectrum. In the present case, this would mean that the largest time step cannot exceed T_N without any lost of information on the dynamics, that is, without a serious change in the nature of the observed solution. The Fourier spectrum of the y-variable of the Rössler system suggests $f_{\max}$ equal to 1.0 Hz (Fig. 8). This frequency corresponds to a time step equal to 0.5 s. Since the main frequency (corresponding to the highest peak in the power spectrum plot) f_0 is equal to 0.1587 Hz, the main time period is around 6.3 s. Consequently, around $12 = \left(\frac{2f_{\max}}{f_0}\right)$ points per oscillation are required to have a proper description of the dynamics. The "Nyquist time" is quite close to the largest time step which can be safely used for integrating the Rössler system without numerical instabilities.

Although valid numerical solutions can be found, there is no possibility to obtain a set of difference equations using the Runge-Kutta technique. The simplest finite difference scheme to achieve such a task is the Euler scheme (2). In this case, the Rössler system (5) becomes :

$$\begin{cases} x_{k+1} = x_k + h\left[-y_k - z_k\right] \\[2mm] y_{k+1} = y_k + h\left[x_k + ay_k\right] \\[2mm] z_{k+1} = z_k + h\left[b + z_k\left(x_k - c\right)\right]. \end{cases} \tag{11}$$

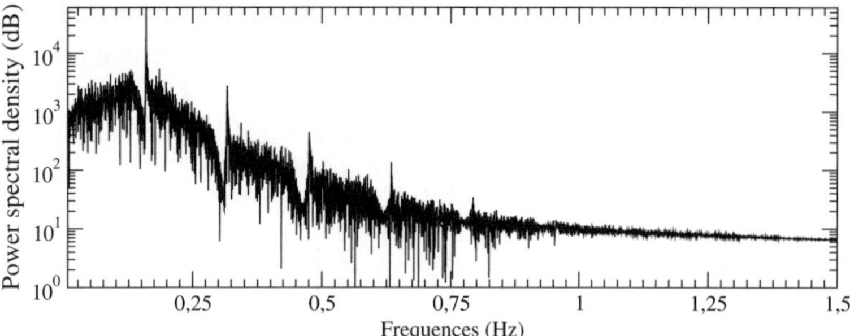

Fig. 8.   Fourier spectrum of the Rössler system with $(a, b, c) = (0.432, 2, 4)$. The main frequency f_0 is equal to 0.1587 Hz which corresponds to a pseudo-period equal to 6.3 s. The highest significative frequency $f_{\max} \approx 1.0$ Hz.

When the time step h is smaller than 0.011 s, there is no significant departure from the original dynamics shown in Fig. 1a. Nevertheless, a bifurcation diagram versus h (Fig. 9a) shows that when h is greater than 0.011 s, different bifurcations and their associated periodic-windows occur.

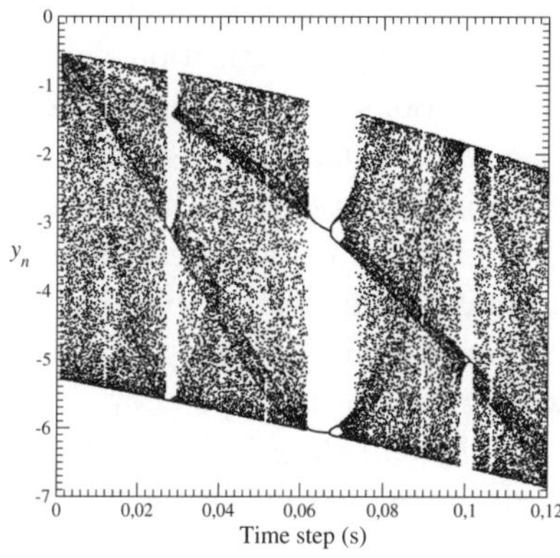

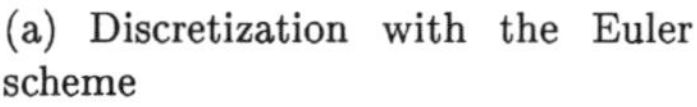

(a) Discretization with the Euler scheme

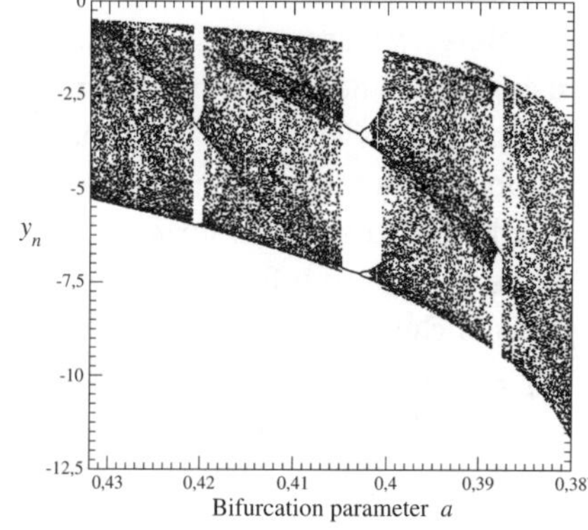

(b) Continuous Rössler system

Fig. 9.   Two similar bifurcation diagrams. (a) Bifurcation diagram versus the time step h for the Rössler system discretized with the Euler scheme. Parameter values : $a = 0.432$, $b = 2$ and $c = 4$. (b) Bifurcation diagram for the continuous Rössler system versus parameter a with $b = \alpha a + \beta$ where $\alpha = \frac{2.0 - 0.15}{0.432 - 0.380}$ and $\beta = 2.0 - 0.432\alpha$. Parameter c is equal to 4.0.

For $h < 0.12$ s, the solution to the simple discretization (11) of the Rössler system is still equivalent to solutions to the continuous Rössler system (5) but with parameters $(a, b, c) = (0.380, 0.15, 4.0)$. This is shown in Fig. 10. Both attractors have a topology described by the linking matrix

$$M_{ij} = \begin{bmatrix} 0 & 0 & 0 \\ 0 & -1 & -1 \\ 0 & -1 & -2 \end{bmatrix} \tag{12}$$

and similar populations of periodic orbits. Both solutions are therefore topologically equivalent. One could want two solutions with the same geometrical domain, that is, with all the variable fluctuating over the same intervals. This could be obtained with a more accurate parameter search or by rescaling the dynamical variables. When equation (11) is rewritten as

$$\begin{cases} x_{k+1} = a_1 x_k + a_2 y_k + a_3 z_k \\ y_{k+1} = b_1 x_k + b_2 y_k \\ z_{k+1} = c_0 + c_1 x_k + c_2 y_k + c_3 z_k z_k \end{cases} \tag{13}$$

an automatic search for the nine parameter values could be performed using a multi-variate global modeling technique adapted from [4]. By performing this search, the same scaling for the dynamical variables is obtained.

Here, we are more concerned to show that the discretization derived using the Euler scheme exhibits the same solutions — from the topological point of view — as the solutions of the continuous Rössler system. An important feature has to be detailed : the discretization does not provide any bounded solution for $h > 0.12$ s due to numerical instabilities; the solutions diverge to infinity exactly as the original Rössler system solutions do. This is checked by comparing the bifurcation diagram between the attractor solution to the Rössler system (5) with $a = a_1 = 0.432$, $b = b_1 = 2.0$, $c = c_1 = 4.0$ and with $a = a_2 = 0.380$, $b = b_2 = 0.15$, $c = c_2 = 4.0$. The diagram of the original Rössler system is thus computed versus a, and b is varied as indicated in the caption of Fig. 9. This allows to follow a path in the parameter space which is similar as those followed for the discretization as shown in Fig. 9a.

The two diagrams show similar periodic windows and both diagrams are interrupted by a boundary crisis (right part of the diagram) between the chaotic attractor and the boundary of the attraction basin. This means that the Euler discretization has a strong dependence on the time step which quickly sends the solution to infinity. Since 0.12 s is less than the Nyquist

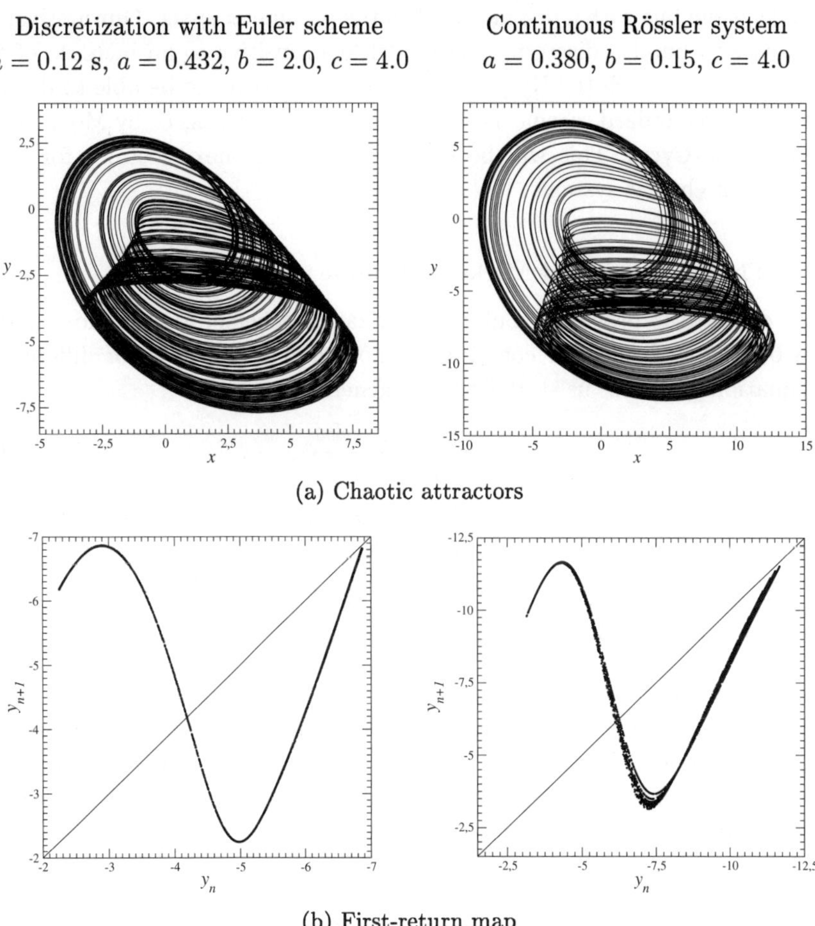

(a) Chaotic attractors

(b) First-return map

Fig. 10. Solutions to the discretization of the Rössler system with the Euler scheme (left column) and to the continuous Rössler system (right column). Note that the scales are different, but both solutions are topologically equivalent.

time, the obtained solution to the Euler discretization is still a solution to the original Rössler system (5) with different set of parameters and with the time step acting as a bifurcation parameter. Therefore numerical instabilities cannot be blamed when the Euler discretization for $h \in]0.0\,;0.12]$ is used, but a displacement in the parameter space is responsible for the boundary crisis. Thus the solution obtained with the Euler discretization is still a solution (at least) topologically equivalent to a solution to the original system but with a displacement in the parameter space. Our motivation is now to build a discretization model less sensitive to an increase of the time step to reach the Nyquist time. One can expect to be able to do that using nonstandard scheme as those proposed by Mickens or by Monaco and Normand-Cyrot. This will be investigated in the next sections for three different chaotic systems.

4. The Nonstandard Mickens Scheme

In order to improve the Euler discretization scheme, Mickens formulated a basic set of rules for constructing nonstandard schemes for differential equations [24]. Let us start from the general form

$$\dot{u} \mapsto \frac{u_{k+1} - \Psi u_k}{\varphi},\tag{14}$$

where Ψ and φ depends on the time step h and other parameters occurring in the differential equations. Functions Ψ and φ should satisfy the conditions

$$\begin{cases} \Psi = 1 + O(h) \\ \varphi = h + O(h^2) \end{cases}\tag{15}$$

and may vary from one equation to another. Unfortunately, there is no clear prior set of guidelines for determining them. In most applications, Ψ is usually selected to be $\Psi = 1$, and φ is determined by the requirement of having the correct stability properties for special solutions to the differential equations. A general choice for φ may be

$$\varphi_i(h, g_i) = \frac{1 - e^{-g_i h}}{g_i},\tag{16}$$

where g_i's are determined as follows. Let $\boldsymbol{u}_j$ denote the coordinates of the jth fixed point of the continuous system to discretize. The function g_i can be chosen

$$g_i = \max \left(\left| \frac{\partial f_i}{\partial u_i} \right|_{\boldsymbol{u}=\boldsymbol{u}_j} \right).\tag{17}$$

The search over all fixed points allows to identify the fastest time scale and to "normalize" the time according to it. Functions φ_i can be interpreted as a "normalized" or a rescaled time step such that its value is never larger than the smallest time scale of the system. Since many of the mechanisms that lead to the occurrence of numerical instabilities have their origin in using a time step that is greater than some relevant physical time scale, this method for selecting φ_i's reduces these types of instabilities. In other words, the use of functions φ_i, rather than just h, allows the value of h to be much larger than the one normally selected because it is the effective time step φ_i that determines the of the system stability and not the actual time step h. Another issue of great importance is that, in general, nonlinear terms are modeled by discrete expressions that are non local on the computational grid [24]. For instance, a u^2 term should be replaced by $u_{k+1}u_k$ in the finite difference scheme, whereas conventional methods would use the local form u_k^2. An important rule to build discretization is that the discrete equations should be equal to the order of the corresponding derivatives of the differential equations, otherwise spurious solutions (numerical instabilities) can occur [13]. The fundamental reason for the existence of numerical instabilities is that the discrete models of differential equations have a larger parameter space than the corresponding differential equations, as previously discussed. This is easily argued by the fact that the time step h can be regarded as an additional parameter. Even if $\{y(\lambda)\}$ and $\{y_k(\lambda, h)\}$ are "close" to each other for a particular value of h, say $h = h_1$, if h is changed to a new value, say $h = h_2$, the possibility exists that $y_k(\lambda, h_2)$ differs greatly from $y_k(\lambda, h_1)$ both qualitatively and quantitatively [13].

Our aim is to show how this nonstandard discretization scheme may influence the quality of the model and its sensitivity to the choice of the discretization time step h. This will be investigated for three different chaotic systems in the next section.

4.1. *A finite difference model for the Rössler system*

A discrete model of the Rössler system (5) is obtained according to the Mickens' guideline, that is, using the transformation [25]

- first equation: $(x_k, y_k, z_k) \mapsto (x_k, y_k, z_k)$
- second equation: $(x_k, y_k, z_k) \mapsto (x_{k+1}, y_k, z_k)$
- third equation: $(x_k, y_k, z_k) \mapsto (x_{k+1}, y_{k+1}, z_k)$.

The nonlinearity xz in the third equation of the Rössler system is replaced by the non local term $x_{k+1}z_k$. This choice has also the advantage of preserving a polynomial form for the discretization model. Function Ψ is equal to 1. Thus, we obtain

$$\begin{cases} x_{k+1} = x_k - \varphi_1(y_k + z_k) \\[2mm] y_{k+1} = (1 + a\varphi_2)y_k + \varphi_2 x_{k+1} \\[2mm] z_{k+1} = b\varphi_3 + [1 + \varphi_3(x_{k+1} - c)]\,z_k\,. \end{cases} \tag{18}$$

The function φ_i is chosen according to the on-diagonal elements of the Jacobian matrix of the original continous system (5)

$$J_{ij} = \begin{bmatrix} 0 & -1 & -1 \\ 1 & a & 0 \\ z & 0 & x-c \end{bmatrix}. \tag{19}$$

Note that since $J_{11} = 0$, there is no natural time scale associated with the autoregressive part of the first equation. We therefore choose the function φ_1 equal to h, thus expressing this lack of time scale. According to equations (16) and (17), the second function is

$$\varphi_2 = \frac{1 - e^{-ah}}{a} \tag{20}$$

and the third one is

$$\varphi_3 = \frac{1 - e^{-x_c h}}{x_c} \tag{21}$$

where $x_c = \frac{-c + \sqrt{c^2 - 4ab}}{2}$ is estimated using the fixed point coordinates and equation (17).

This discrete model can be iterated with a discretization time step h over the range $(0, 0.6194)$. The bifurcation diagram versus h (Fig. 11) is very similar to the diagram computed with the continuous Rössler system versus the parameter a (Fig. 3). Note, however, that increasing h is similar to decreasing the parameter a. This can be explained by replacing x_{k+1} in the second equation of system (18) by the first equation of system (18)

$$y_{k+1} = [1 + \varphi_2(a - \varphi_1)]\,y_k + \varphi_2 x_k - \varphi_1\varphi_2 z_k \tag{22}$$

where in the term $[1 + \varphi_2(a - \varphi_1)]\,y_k$, $\varphi_1 = h$ tends to balance the effect of parameter a. The bifurcation diagram (Fig. 11) shows that all solutions to the discretization of the Rössler system are topologically equivalent to a solution to the continuous counterpart with appropriate values of the bifurcation parameters. Note that, when $h \in\,]0, 0.05]$, very few bifurcations are observed and, consequently, the dependence of the solution of the discrete model (18) on the time step h is weaker than those obtained with the Euler

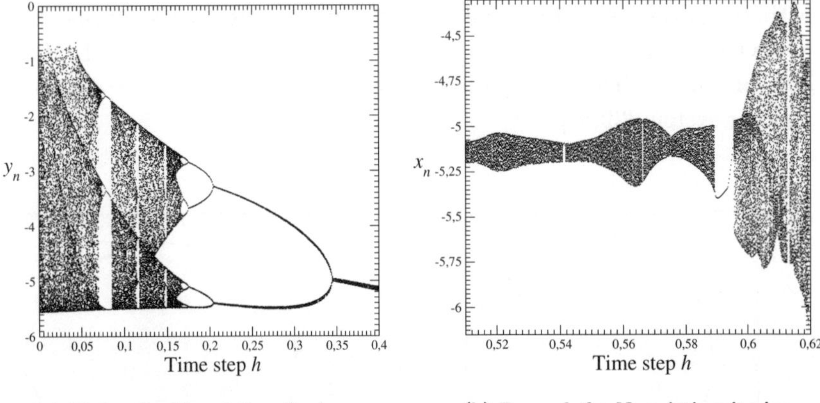

(a) Under the Nyquist's criterion   (b) Beyond the Nyquist's criterion

Fig. 11.   Bifurcation diagram versus the time step h for the optimized discrete model (18) with the functions φ_i given in relations (20) and (21). Numerical instabilities occur when the discretization time step is greater than the Nyquist time (b).

scheme. This model is therefore much more robust against increasing the time step h.

It is interesting to note that completely different dynamic behaviors, which are no longer topologically equivalent to any solution to the original Rössler system, appear for discretization time step h greater than the Nyquist time.

For instance, for $h = 0.5$ s, a period-1 limit cycle (Fig. 12a), which is still obviously topologically equivalent to the period-1 limit cycle of the original Rössler system, is observed. Increasing again the time step h, the limit cycle becomes progressively constrained by the heterocline connections between the periodic points of an unstable period-11 orbit (Figs. 12b and 12c). Such a type of bifurcation has been evidenced in a non autonomous 5D system [26] and in a glow discharge experiments [27] for a Poincaré section of a torus. This means that by taking too large a step ($h > T_N$), the discretization becomes a kind of return map to a Poincaré section and, therefore, there is no longer a connection to the original dynamics. With $h = 0.589$ s, the "limit cycle" is shown with the unstable periodic orbit (Fig. 12c). These behaviors are never observed in the continuous Rössler system. Consequently, they result from numerical instabilities. When the time step h is further increased, the "limit cycle" presents apparent cusps which are actually points of extreme curvature as observed in the scenario depicted in

[26,27]. Unfortunately, the discrete model is not sufficiently stable to be used with larger discretization time steps but we expect that spurious chaotic behavior could be observed in such a case. This was observed by Lorenz [28] and in another investigation using a different scheme for discretizing another system [29,30].

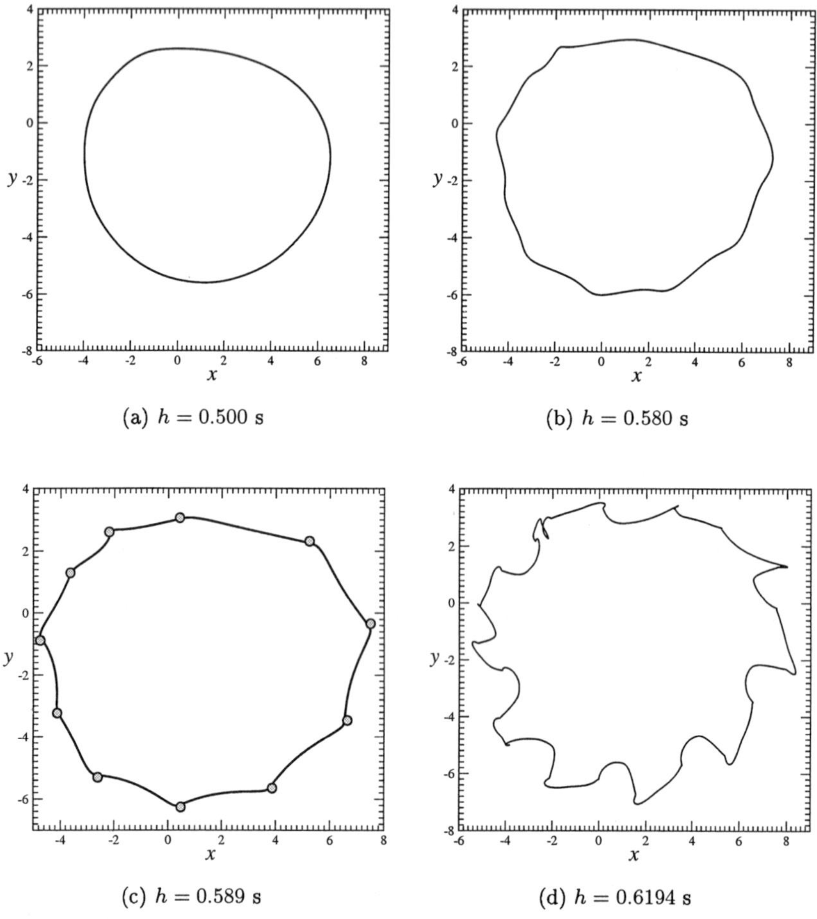

(a) $h = 0.500$ s   (b) $h = 0.580$ s

(c) $h = 0.589$ s   (d) $h = 0.6194$ s

Fig. 12. Solutions to the optimized discrete model of the Rössler system for different values of h. A bifurcation similar to a torus breaking is observed.

Note that numerical instabilities appear when the discretization time step h is greater than the Nyquist time. We conjecture that when a discretization of a continuous system is built and iterated with a discretization time step h greater than the Nyquist time, the dynamics can no longer be recovered without any damage. Consequently, one can expect solutions that do not correspond to any solution to the original differential system, even with a displacement in the parameter space. This is directly related to a comment made by Mickens [24] who stated that numerical instabilities can occur when the finite difference equations do not satisfy a condition that is of importance for the corresponding differential equations. The main time scale of the dynamics is definitely an important characteristic and choosing the time step without considering it can lead to spurious results. In summary, we have obtained difference equations which have the best equivalence with the Rössler dynamics, that is, which have solution topologically equivalent to some solution to the Rössler system with a possible displacement in the parameter space.

4.2. *The Genesio and Tesi system*

We choose now to apply the discretization scheme to the system introduced by Genesio & Tesi in [31] and represented by the following differential equation

$$\dddot{x} + a\ddot{x} + b\dot{x} + x(1 + x) = 0 \tag{23}$$

which may be rewritten as a set of first-order ordinary differential equations under the form :

$$\begin{cases} \dot{x} = y \\ \dot{y} = z \\ \dot{z} = -az - by - x(1 + x) \end{cases} \tag{24}$$

where a and b are the bifurcation parameters. This system has two fixed points: one, F_0, is located at the origin of the phase space and the other one, F_1, is located at $(-1, 0, 0)$. For $b = 1.1$, the point F_0 becomes unstable when a Hopf bifurcation for $a = 0.9091$ occurs. When $a < 0.9091$, the fixed point F_0 is a saddle-focus with two complex conjugated eigenvalues with positive real parts and F_1 is also saddle-focus but with complex eigenvalues with negative real parts. Such a configuration with two fixed points is very similar to the configuration of the Rössler system [17]. The asymptotic behavior of the Genesio and Tesi system settles down onto a limit cycle or a chaotic attractor as long as the trajectory does not cross the boundary of the attraction basin associated with the stable manifold of the fixed point F_1 [19]. Indeed, as soon as the trajectory crosses this manifold, it is ejected to infinity by the unstable manifold corresponding to the fixed point F_1. A typical chaotic attractor solution to system (24) is shown in Fig. 13a.

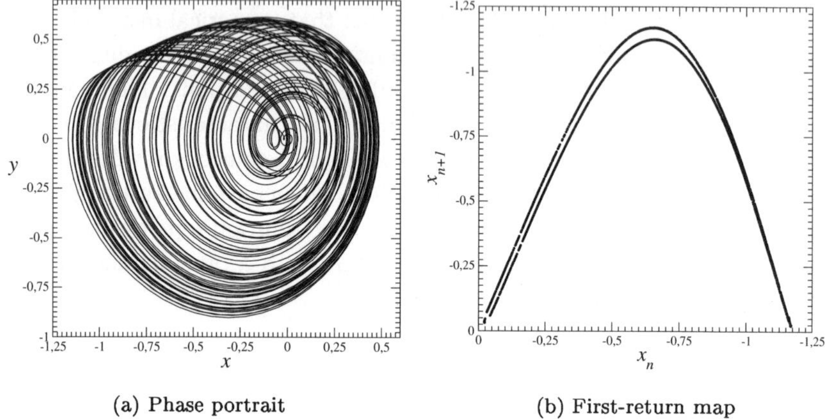

(a) Phase portrait　　　　　　　　　　(b) First-return map

Fig. 13.　Chaotic behavior solution to system (24) with $a = 0.446$ and $b = 1.1$.

A first-return map to a Poincaré section for system (24) was computed (Fig. 13b). This is a parabola with a layered structure resulting from the small damping rate of the dynamics. The damping is not sufficient for fully squeezing the folding in one cycle around the fixed point F_0. Such a feature is very characteristic of low dissipative system. Nevertheless, the dynamics still has the main characteristics of the class defined by the discrete logistic map. Thus, a period-doubling cascade is observed in the bifurcation diagram as well as many periodic windows (Fig. 14). Basically, after the period-doubling cascade, system (24) may generate a *spiral* attractor associated with a map mainly constituted by one increasing and one decreasing branch.

Using the same rules as those used for the Rössler system, we obtain the discretization model as follows :

$$\begin{cases} x_{k+1} = x_k + \varphi_1 y_k \\ y_{k+1} = y_k + \varphi_2 z_k \\ z_{k+1} = z_k + \varphi_3 \left[-a z_k - b y_{k+1} - x_{k+1} \left(1 + x_k \right) \right] \end{cases} \qquad (25)$$

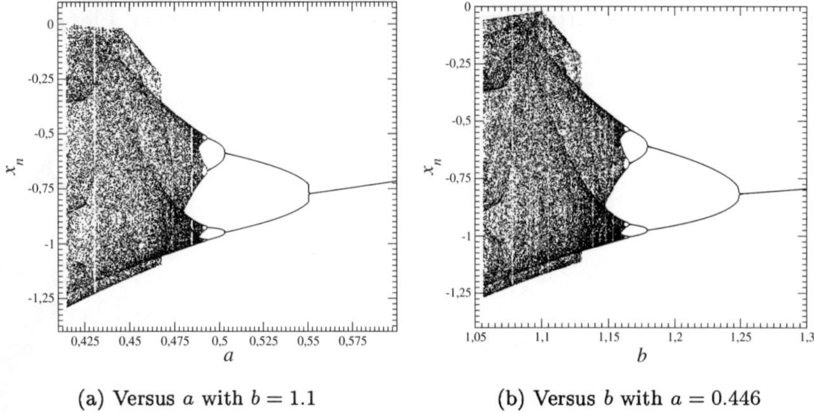

(a) Versus a with $b = 1.1$ (b) Versus b with $a = 0.446$

Fig. 14.   Bifurcation diagrams of system (24). The period-doubling as a route to chaotic attractor ensures us that the spiral attractor corresponds to the class of chaotic behaviors associated with the discrete logistic map.

where

$$\left|\begin{array}{l} \varphi_1 = h \\[2mm] \varphi_2 = h \\[2mm] \varphi_3 = \dfrac{1-e^{-ah}}{a}\,. \end{array}\right. \tag{26}$$

When iterated with a small time step, the asymptotic behavior is topologically equivalent to the attractor shown in Fig. 13. When the time step is increased, various bifurcations arise up to a boundary crisis which ejects the trajectory to infinity for $h = 0.358$ s. This is significantly less than the upper value of the time step which corresponds to the Nyquist time ($T_N \approx 0.75$ s). The impossibility to obtain a discretization of the Genesio and Tesi system given by equation (24) which exhibits solutions topologically equivalent to some solutions to the original system does not result from the weakness of this discretization but from the fact that, in the parameter space, $a = 0.446$ and $b = 1.1$ are parameters for which the investigated behavior is quite close to the boundary crisis which ejects the trajectory to infinity.

This can be checked by computing the value of the time step at each point where the boundary crisis occurs versus the value of the bifurcation parameter a (Fig. 15). The curve reveals that when the a-value is such that the attractor is very close to the boundary of the attraction basin ($a = 0.446$), the largest time step is very close to zero. Contrary to this, the limit cycle just after the Hopf bifurcation ($a = 0.9091$), is very far from the boundary of the attraction basin and h can be varied over a long range before the boundary crisis.

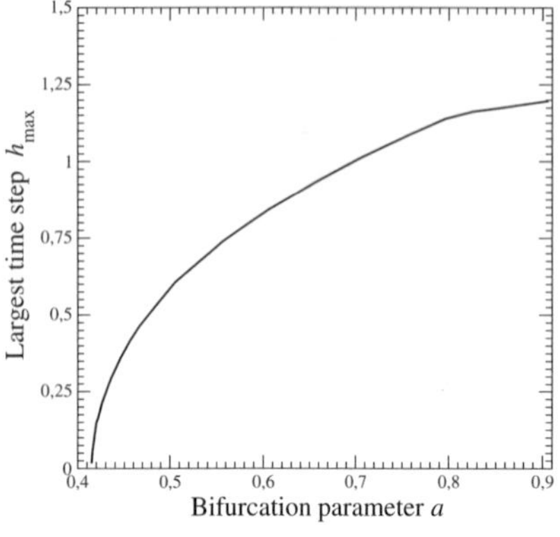

Fig. 15. The largest time step, which can be used before the boundary crisis occurs, depends on the bifurcation parameter. Increasing the time step thus induces a development of the dynamics toward an ejection of the trajectory to infinity. Parameter: $b = 1.1$.

Fig. 16. Bifurcation diagram versus the time step for the discretization of the Genesio and Tesi system using the Mickens nonstandard scheme. Parameter: $a = 0.796$ and $b = 1.1$.

For instance, for $a = 0.796$, the attractor solution to the continuous Genesio and Tesi system (24) is a small period-1 limit cycle, just after a Hopf bifurcation. When discretization (25) is iterated, the solution is equivalent to a solution to the continuous system for all values of h that do not exceed the Nyquist time which is around 1 s (Fig. 15). The bifurcation diagram presents many bifurcations (Fig. 16). The fact that this bifurcation diagram does not exactly correspond to those diagrams associated with the continuous system when a is varied just means that the time step does not act exactly as the a-parameter in the continuous system or, at least, not with the same b-value. Nevertheless, the general tendency to develop the dynamics from the Hopf bifurcation toward a boundary crisis is verified.

When the time step becomes greater than the Nyquist time (1.0 s), the solution to the discretization no longer corresponds to solutions to the original system in equation (24). After an inverse cascade of period-doubling bifurcations ended by a period-2 limit cycle ($h = 1.01$ s), the chaotic attractors have some structures which are never observed in the original dynamics (Fig. 17). These chaotic solutions are numerical instabilities as expected when the time step is greater than the Nyquist time.

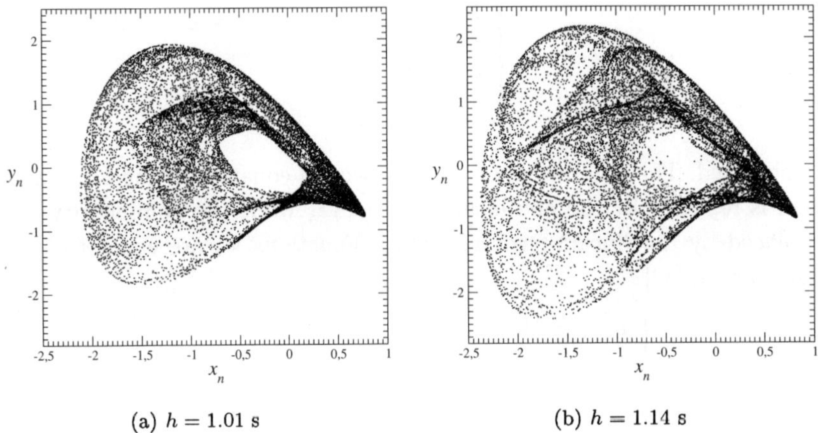

(a) $h = 1.01$ s   (b) $h = 1.14$ s

Fig. 17.   Two attractors which are not topologically equivalent to any attractor solution to the continuous system (24). This is mainly due to the structures occurring in the middle of the attractor. Parameter: $a = 0.796$ and $b = 1.1$.

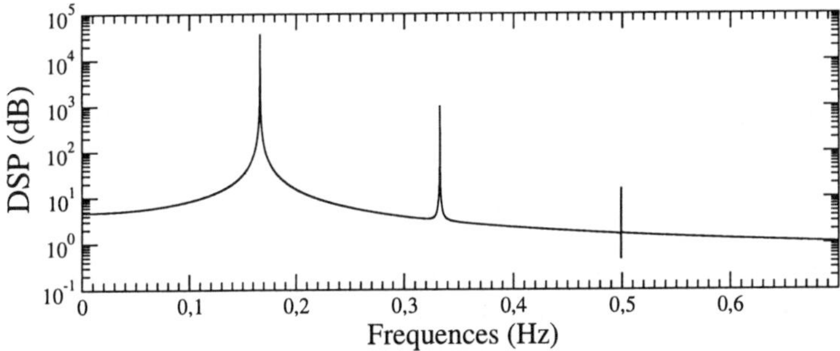

Fig. 18.   Fourier spectrum computed from the x-variable of the Genesio and Tesi system. The parameters used for simulation are $a = 0.796$ and $b = 1.1$. The attractor settles in a limit cycle. The largest frequency is equal to 0.5 Hz, thus leading to a Nyquist time around 1.0 s.

4.3. *Discretization of the Lorenz system*

Using the same rules as for the two previous systems, a discretization of
the Lorenz system

$$
\begin{cases}
\dot{x} = \sigma(y - x) \\[1mm]
\dot{y} = Rx - y - xz \\[1mm]
\dot{z} = xy - bz
\end{cases}
\tag{27}
$$

is obtained. The nonlinearity xz in the second equation of the Lorenz system is replaced by the non local term $x_{k+1}z_k$ while the nonlinearity xy is
replaced by the "local" term $x_{k+1}y_{k+1}$. Functions Ψ_i are equal to 1. We
thus obtain [32] :

$$
\begin{cases}
x_{k+1} = (1 - \varphi_1\sigma)x_k + \varphi_1\sigma y_k \\[1mm]
y_{k+1} = \varphi_2 Rx_{k+1} + (1 - \varphi_2)y_k - \varphi_2 x_{k+1}z_k \\[1mm]
z_{k+1} = (1 - \varphi_3 b)z_k + \varphi_3 x_{k+1}y_{k+1}
\end{cases}
\tag{28}
$$

with

$$
\begin{cases}
\varphi_1 = \dfrac{1 - e^{-\sigma h}}{\sigma} \\[3mm]
\varphi_2 = 1 - e^{-h} \\[3mm]
\varphi_3 = \dfrac{1 - e^{-bh}}{b} \, .
\end{cases}
\tag{29}
$$

This means that $g_1 = \sigma$, $g_2 = 1$ and $g_3 = b$, that is, the three on-diagonal
elements of the Jacobian matrix of the Lorenz system. Note that this discretization is equivariant like the Lorenz system, that is, it obeys the following relation

$$
\Gamma \cdot F(x_k) = F(\Gamma \cdot x_k)
\tag{30}
$$

where x_k is the state vector at time kh, F is the discretized vector field and
Γ is a 3×3 matrix defining the symmetry property. In the Lorenz case, this
matrix reads as

$$
\Gamma =
\begin{bmatrix}
-1 & 0 & 0 \\
0 & -1 & 0 \\
0 & 0 & 1
\end{bmatrix}
\tag{31}
$$

and defines a rotation by π around the z-axis. This means that the coordinates (x_k, y_k, z_k) are mapped to $(-x_k, -y_k, z_k)$ under the action of Γ.

Discretized model (28) of the Lorenz system is iterated with $(R, \sigma, b) = (28, 10, 8/3)$. When the discretization time step is very small, say $h = 0.002$ s, the phase portrait is the usual Lorenz attractor. The discretization time step h is then increased up to the greatest value for which the behavior is still oscillating. For time steps greater than 0.071 s, the trajectory is ejected to infinity. A bifurcation diagram (Fig. 19) versus h is computed using the Poincaré section

$$\mathcal{P} \equiv \left\{ (x_n, y_n) \in \mathbb{R}^2 \mid z_n = R - 1, \dot{z}_n > 0 \right\}. \tag{32}$$

Note that for h values of less than 0.04 s, the Poincaré section does not change too much, confirming that discretization (28) is rather robust against changes in the time step h. The chaotic attractor (Fig. 20) obtained for this value of the time step is still very close to the usual Lorenz attractor.

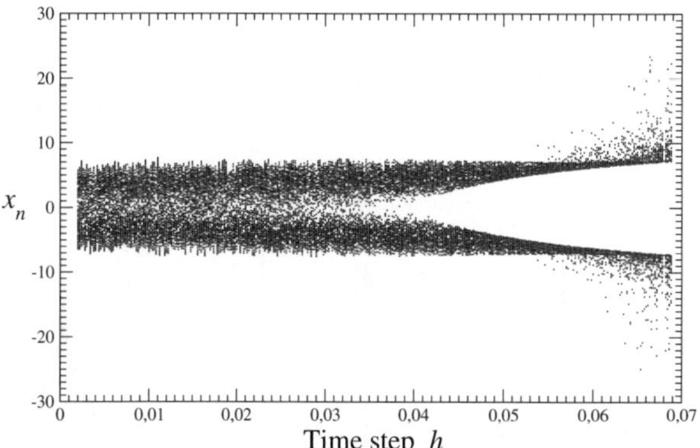

Fig. 19.   Bifurcation diagram of the discretization of the Lorenz system using the non-standard Mickens' scheme with $(R, \sigma, b) = (28, 10, 8/3)$.

Although the first-return map has a layered increasing branch (Fig. 21a), it can be shown that this attractor is topologically equivalent to the usual Lorenz attractor [33]. In fact, a similar map (Fig. 21b) can be obtained from the original Lorenz system with $R \approx 40.0$. An increase of h thus corresponds to an increase of R. Decreasing the R-value to 17.0 (rather than 28.0) therefore balances the influence of the large time step. Such an R-change implies that the population of periodic orbits embedded within the attractor of the discrete model (28) is the same as the population extracted from the original Lorenz attractor. This is confirmed by the first-return map which is very close to the Lorenz map with the characteristic cusp

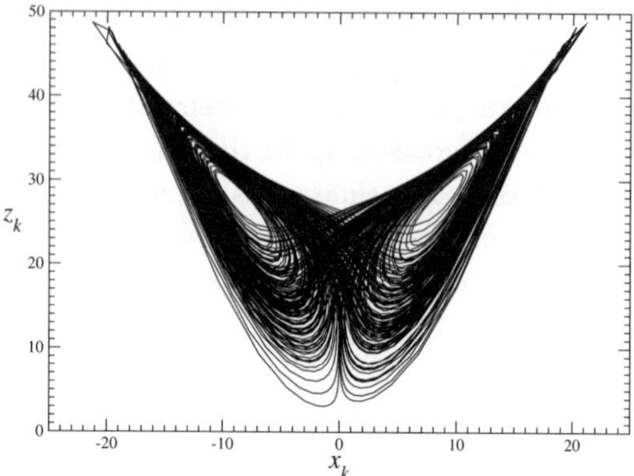

Fig. 20. Chaotic attractor solutions to the discretization of the Lorenz system using the non-standard Mickens' scheme with $(R, \sigma, b) = (28, 10, 8/3)$. The discretization time step is $h = 0.040$ s.

(Fig. 21b). Nevertheless, the upper parts of the wings remain sharp (Fig. 20), a dynamical signature not observed in the original Lorenz attractor.

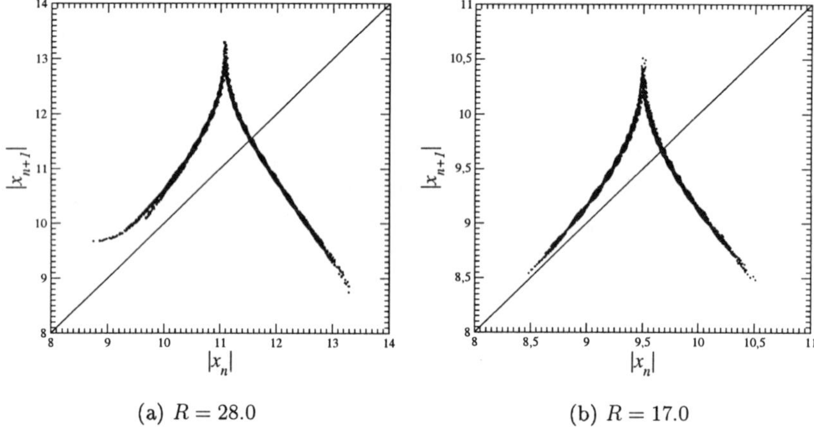

(a) $R = 28.0$ (b) $R = 17.0$

Fig. 21. First-return map of discrete model (28) for two different values of the R-parameter. Other parameters are $(\sigma, b) = (10, 8/3)$ and the discretization time step is $h = 0.040$ s.

A Fourier spectrum (Fig. 22) computed from the z-variable of the Lorenz system reveals that the largest frequency $f_{\max}$ is around 12.5 Hz and therefore the Nyquist time is $T_N = \frac{1}{2 f_{\max}} \approx 0.040$ s. This is exactly the value beyond which the chaotic attractor of discretization (28) becomes very different from any solution to the original Lorenz system. For instance, reinjections from one wing to the other (Fig. 23) do not follow the paths that can be observed in solutions to the original Lorenz system, even for parameter values different from the usual values. In fact, only well-defined domains of the phase portrait are deformed under the effect of the increase in the time step. Such a feature was not observed for discretizations of the Rössler system or discretizations of the Genesio and Tesi system. Around the Nyquist time these two systems exhibit a periodic window, and then a limit cycle with points of extreme curvature. The root of such a departure could result from the very large band observed in the Fourier spectrum (Fig. 22) and the fact that the velocity of the trajectory may significantly vary from one region of the phase space to another. For instance, around the saddle fixed point located at the origin of the phase space, the derivatives are rather small, and even with a time step $h \approx 0.071$ s, the attractor is still "smoothly" visited (Fig. 23). However when the trajectory visits the stable manifold of one of the symmetry related saddle-focus fixed points, the trajectory is jagged. This is a signature of significantly larger velocities. In this region, the time step is already too large to describe the attractor smoothly. Consequently, the topology of the attractor is affected in a well-defined region of the phase portrait, whereas elsewhere the dynamics is preserved. Numerical instabilities therefore occur from a local lack of definition of the dynamics. Such a feature was not identified in the two systems previously investigated for which the dynamics is more homogeneous.

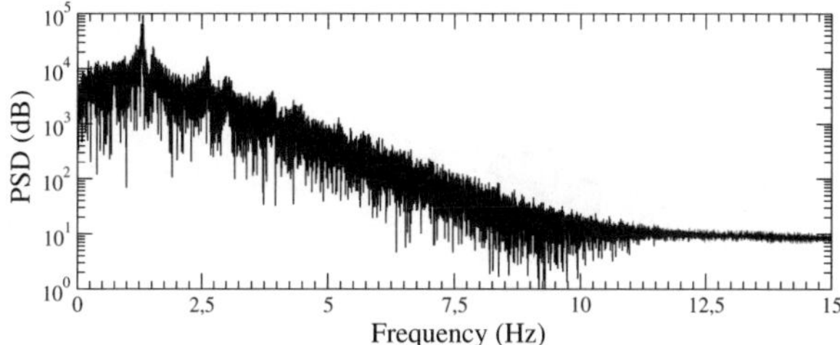

Fig. 22.   Fourier spectrum computed from the z-variable of the Lorenz system with $(R, \sigma, b) = (28, 10, 8/3)$.

Despite all of that, we can still consider that discretization (28) is an optimized set of discrete difference equations for the Lorenz system, since the time step h may be increased up to T_N, which is an upper bound, before

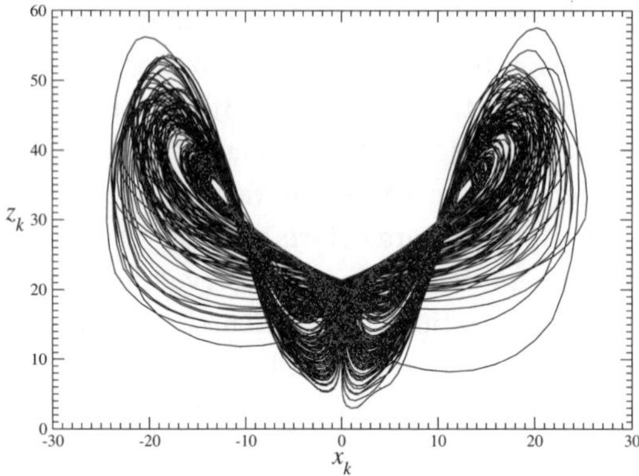

Fig. 23. Chaotic attractor solutions to the discretization of the Lorenz system using the non-standard Mickens' scheme with $(R, \sigma, b) = (28, 10, 8/3)$. The discretization time step is $h = 0.071$ s. The data are interpolated to give a better representation of the topological structure of the dynamics.

numerical instabilities show up. With $(R, \sigma, b) = (28, 10, 8/3)$, the time step h may be increased up to 0.071 s. For this value significantly larger than T_N, the chaotic attractor (Fig. 23) which is a solution to discretization (28) does not correspond to any solution to the Lorenz system, even with a displacement in the parameter space.

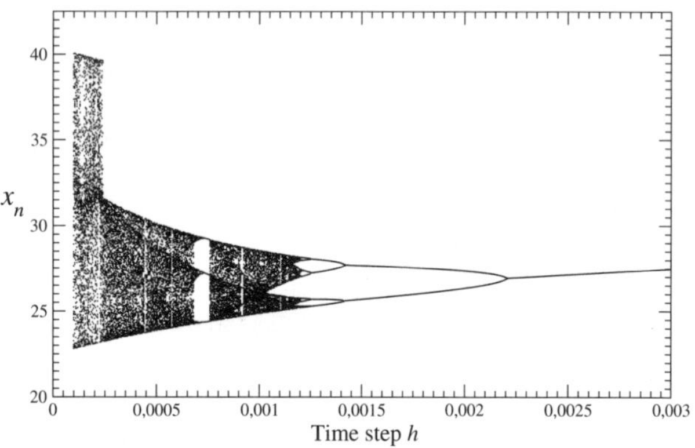

Fig. 24. Bifurcation diagram of the discretization of the Lorenz system using the non-standard Mickens scheme with $(R, \sigma, b) = (200, 10, 8/3)$ for small values of the time step h.

Now we choose a point in the parameter space where the dynamics is rather different ($R = 200$), keeping unchanged the other parameters, b and

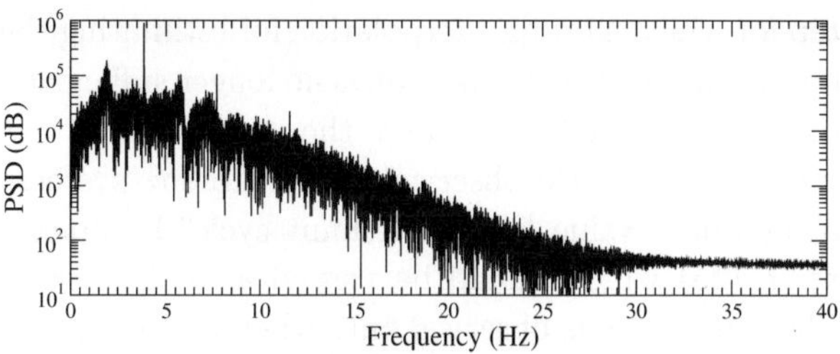

Fig. 25. Fourier spectrum computed from the z-variable of the Lorenz system with $(R, \sigma, b) = (200, 10, 8/3)$.

σ. The chaotic attractor is located just before (when the parameter R is decreased) a merging attractor crisis [34] occurring for $R = 203.04$. After such a crisis, the chaotic attractor is no longer globally invariant under the action of the Γ-matrix. There are, therefore, two symmetry-related attractors co-existing in the phase space. The dynamics is thus very different. This particularity in the bifurcation diagram will allow us to exhibit small dynamical changes versus the value of the discretization time step h. The first bifurcation diagram is computed for h varying from 0.0001 s to 0.003 s (Fig. 24). The inverse merging attractor crisis is clearly identified ($h \approx 0.00025$ s) as well as the inverse period-doubling cascade ($0.0012 < h < 0.0025$). All these bifurcations occur for very small values of the discretization time step. For $h \approx 0.000235$ s, the period-1 limit cycle occurs just after the inverse cascade. In the original Lorenz system, this limit cycle is observed after the final period-doubling bifurcation occurring at $R = 229.48$. All these solutions to discretization (28) are therefore close to solutions to the corresponding differential equations.

Fig. 25 shows the Fourier spectrum computed from the z-variable of the Lorenz system with $R = 200$. The main frequency is located at $f_0 = 3.86\,\mathrm{Hz}$. The pseudo-period T_0 is thus equal to 0.26 s, which is roughly a third of the pseudo-period for $R = 28$ ($T_0 = 0.754$ s). The largest frequency is $f_{\max} \approx 35$ Hz. The Nyquist time is therefore around $T_N = 0.014$ s, which is also roughly a third of the Nyquist time for $R = 28$.

With $h = 0.01$ s for which the asymptotic behavior is a limit cycle, the bifurcation diagram shown in Fig. 26 is computed up to the h-value where the trajectory is ejected to infinity, that is, up to $h_{\max} \approx 0.023$ s ($> T_N = 0.014$ s). Discretization (28) is therefore an optimized set of discrete difference equations for the Lorenz system. Note that the interval visited in the Poincaré section is rather small. This is mostly the effect of a large time step for which a linear interpolation for estimating the location of the period-1 point in the Poincaré section is no longer sufficient to obtain an accurate estimation. Around $h = 0.12$ s, the period-1 limit cycle becomes different from any limit cycle observed in the Lorenz system. For values of h close to the upper value $h_{\max}$, the "limit cycle" becomes the solution to discretization (28) which should be viewed as a "Poincaré map". This means that the limit cycle is observed only when a large number of points are plotted (and not connected by segments). For $h = 0.0221$ s, a period-18 orbit is observed (Fig. 27). This means that in fact, the period-1 "limit cycle" must be considered as the Poincaré section of a quasi-periodic regime since the time step is no longer small enough when compared to the pseudo-period T_0 (compare 0.0221 s to $T_0 = 0.26$s). The values of h between T_N and T_0 thus correspond to a critical interval over which discretization (28) switches from a description of the flow of the original Lorenz system to a Poincaré map which cannot correspond to the original dynamics (it does not seems logical to imagine that the same equations could describe a flow and its associated Poincaré map). There is, therefore, always an interval, typically $T_N < h < T_0$, over which the dynamics is necessarily spurious and cannot correspond to any solution to the original continuous system. The assumption that numerical instabilities occur roughly for discretization time steps larger than the Nyquist time is still observed for the Lorenz system.

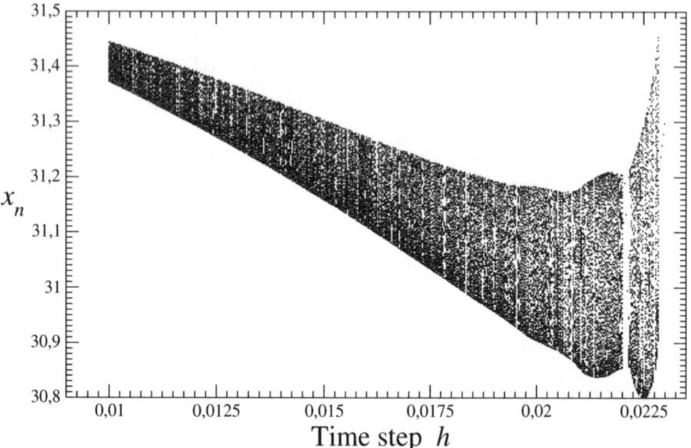

Fig. 26. Bifurcation diagram of the discretization of the Lorenz system using the non-standard Mickens' scheme with $(R, \sigma, b) = (200, 10, 8/3)$.

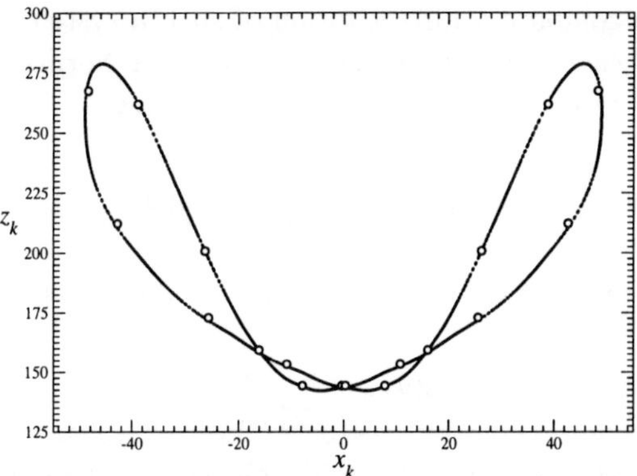

Fig. 27.   Spurious solutions to the discretization of the Lorenz system using the non-standard Mickens' scheme with $(R, \sigma, b) = (200, 10, 8/3)$. If discretization (28) is considered as a Poincaré map, the solution is a limit cycle (drawn with $\circ$) for $h = 0.0221$ s and a quasi-periodic regime for $h = 0.0222$ s.

5. The Normand and Monaco-Cyrot Scheme

The nonstandard Mickens scheme works with quite simple rules and provides rather simple difference equations as discretizations of a nonlinear set of ordinary differential equations. In previous section, we showed that varying the time step provokes a displacement in the parameter space. In order to obtain discretizations less sensitive to the increase of the time step, one could expect to have more complicated difference equations. This statement will be investigated in the light of the so-called Monaco and Normand-Cyrot scheme in this section.

5.1. *The scheme*

Consider a dynamical system

$$\dot{\boldsymbol{x}} = \boldsymbol{f}(\boldsymbol{x}) \tag{33}$$

where $\boldsymbol{x} \in \mathbb{R}^m$ is the vector made of the dynamical variables and, $\boldsymbol{f}$ are analytic functions of appropriate dimension. The aim is to obtain a discretization of equation (33) for a relatively large value of the discretization time step h. Such a discretization reads as :

$$\boldsymbol{x}_{k+1} = \boldsymbol{g}(\boldsymbol{x}_k, h) \tag{34}$$

where $\boldsymbol{x}_k \in \mathbb{R}^m$ are the dynamical variables at time $t = t_0 + kh$. h is the discretization time step.

In order to compute a discretization of such a system, we will use the discretization scheme introduced by Monaco and Normand-Cyrot [35,36] based on a Lie expansion of equation (33) as follows :

$$\boldsymbol{x}_{k+1} = \boldsymbol{x}_k + \sum_{n=1}^{\eta} \frac{h^n}{n!} \mathcal{L}_{\boldsymbol{f}}^n(\boldsymbol{x}_k) \tag{35}$$

where η is the order of the expansion. The Lie derivative is given by

$$\mathcal{L}_{\boldsymbol{f}}(\boldsymbol{x}_k) = \sum_{j=1}^{m} f_j \frac{\partial \boldsymbol{x}}{\partial x_j} \tag{36}$$

where f_j designates the jth component of the vector field. The higher order derivatives are given recursively

$$\mathcal{L}_{\boldsymbol{f}}^n(\boldsymbol{x}_k) = \mathcal{L}_{\boldsymbol{f}}\left(\mathcal{L}_{\boldsymbol{f}}^{n-1}(\boldsymbol{x}_k)\right). \tag{37}$$

We will show that such a Lie expansion (35) can be truncated at order η to avoid an excessive number of terms. The dependence of the discretization robustness on the time step increase will be investigated up to $\eta = 4$. The series truncated at the first order ($\eta = 1$) corresponds to the Euler scheme. It has been shown that the discretization model of the first order obtained using Monaco and Normand-Cyrot's scheme preserves the number and location of fixed points of its original differential equations [29]. For any higher-order discretization ($\eta > 1$), the location of fixed points of the differential equations is preserved but the location and number of fixed points of spurious fixed points introduced by the higher-order terms depend on the discretization time step h.

The subsequent part of this section is devoted to four different discretized models obtained by applying Monaco and Normand-Cyrot's scheme to the Genesio and Tesi system and to the Lorenz system. They are analysed as the time step is increased.

5.2. *The Genesio and Tesi system*

Applying Monaco & Normand-Cyrot's scheme to system (24) ($\eta = 1$) leads to :

$$\begin{cases} x_{k+1} = x_k + h y_k \\ y_{k+1} = y_k + h z_k \\ z_{k+1} = z_k - h\left(a z_k + b y_k + x_k(1 + x_k)\right). \end{cases} \tag{38}$$

For sufficiently small time step h, the solution of this discretization is topologically equivalent to the attractors solution of continuous system (24). This means that the attractor is characterized by the same template as the spiral attractor of the continuous counterpart. Only the population of periodic orbits is slightly changed. Nevertheless, as usually observed for any type of discretization scheme, the nature of the solution depends on the value of the discretization time step h. In order to have a global overview of these solutions, a bifurcation diagram is computed versus h (Fig. 28a). The discretization time step may be varied over the range $]0\,;0.0358]$. For larger values, the trajectory is ejected to infinity. Over this range, discretized model (38) exhibits either a chaotic attractor or a limit cycle. It has been verified that the chaotic attractors are topologically equivalent to the attractor shown in Fig. 13. The largest value of the discretization time step corresponds to $\frac{T_0}{165}$, which is a particular small value. Note that the period-4 window is here clearly observed with $h \approx 0.035$ s in the bifurcation diagram (Fig. 28a). The boundary crisis therefore occurs in conditions very similar to those of system (24). Increasing h is similar to decreasing the parameter a. Consequently, all solutions of the discretization of system (24) are topologically equivalent to solutions of its continuous counterpart. The discretization scheme, even for the first order, is already very efficient as long as only the nature of the solution and sufficiently small value of the discretization time step are considered.

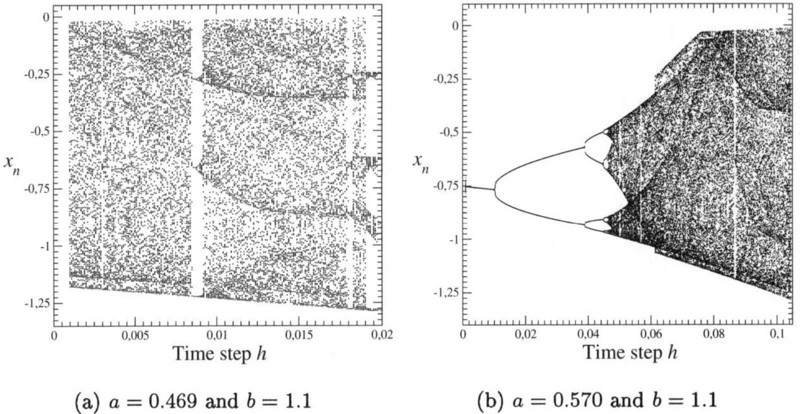

(a) $a = 0.469$ and $b = 1.1$ (b) $a = 0.570$ and $b = 1.1$

Fig. 28. Bifurcation diagram versus the discretization time h.

When the highest value of the discretization time step is considered ($h = 0.0358$ s), the funnel attractor is observed with $a = 0.469$. The spiral attractor can be recovered when the bifurcation parameter a is changed to 0.504. This is just a consequence of the interplay between the time step h and the bifurcation parameter a. Indeed, the effect of the discretization time step h can be balanced by an action on the value of the parameter a. Such a feature was also observed in a discretization of the Rössler system using non standard Mickens' schemes [10]. Applying such a shift in the bifurcation parameters allows the recovery of the period-doubling cascade which is not observed in the bifurcation diagram computed with $a = 0.469$ and $b = 1.1$ (Fig. 28a). Choosing $a = 0.570$ with b unchanged, the bifurcation diagram versus the discretization time step presents the period-doubling cascade (Fig. 28b) as observed in the original system (Fig. 14). Note that, in this case, the time step can be increased up to 0.1046, that is, over a range roughly three time larger than for the original values of the bifurcation parameters.

The topology of the attractor solution of the first-order discretization is investigated for different values of h. When $a = 0.469$ and $b = 1.1$, the spiral attractor is obtained for h smaller than 0.012 s. For this particular value, the first-return map is made of two monotonic branches (Fig. 29a). With larger values of h, a third branch occurs and the funnel attractor is observed. With $h = 0.012$ s, the population of periodic orbits embedded within the attractor is the same as the population extracted from system (24) with $a = 0.446$ and $b = 1.1$ (compare the first-return maps shown in Fig. 29a and Fig. 13b). The linking number $lk(10, 101)$ is equal to -2 as observed for the spiral attractor solution of system (24) (compare Figs. 29b and Fig. 5). This topological equivalence between the first-order discretized model and system (24) may be obtained for any other value of h if the a value for system (24) is properly tuned, since a change of the time step in the discrete model can be balanced by a change of the value of parameter a in system (24).

The second order discretization of system (24) is now considered. It reads as :

$$\begin{cases} x_{k+1} = x_k + h y_k + \frac{h^2}{2} z_k \\[2mm] y_{k+1} = y_k + h z_k - \frac{h^2}{2} \left(a z_k + b y_k + x_k(1 + x_k) \right) \\[2mm] z_{k+1} = z_k - h \left(a z_k + b y_k + x_k(1 + x_k) \right) \\[2mm] \qquad - \frac{h^2}{2} \left[(1 + 2x_k) y_k + b z_k - a \left(a z_k + b y_k + x_k(1 + x_k) \right) \right]. \end{cases} \tag{39}$$

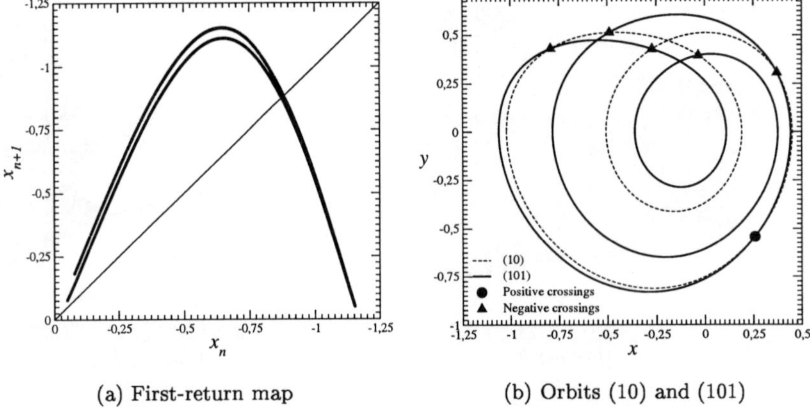

<table>
<tr><td>(a) First-return map</td><td>(b) Orbits (10) and (101)</td></tr>
</table>

Fig. 29.   First-return map to a Poincaré section of the attractor solution to the first-order discrete model with $a = 0.469$, $b = 1.1$, and $h = 0.012$ s. The link made of periodic orbits (10) and (101) shows that the linking number $lk(10, 101)$ is left unchanged, that is equal to -2.

The first obvious advantage presented by this second order discretization is that the discretization time step may be varied over the larger range $]0 \,; 0.4733]$. The largest time step corresponds now to $\frac{T_0}{12}$. More than a factor ten is gained with this second order. Such a feature shows the efficiency of Monaco & Normand-Cyrot's scheme. Indeed, most of discretization schemes found in the literature cannot provide discretized models which remain valid over a very large range of time step h. It should be also noted that when the discretization time step is less than 0.1, the bifurcation diagram does not show too many changes in the chaotic behavior. This means that over this interval, the dynamics is less sensitive to the increase of the time step h. When discretized model (39) is iterated with the largest time step $h = 0.473$ s, twelve points are obtained per cycle. This is still quite far from the Nyquist time (1.0 s) although 12 points per cycle is already below the limit to perform a topological analysis without an interpolation procedure.

When the value of the bifurcation parameter a is changed to 0.560, the bifurcation diagram (Fig. 30b) exhibits a period-doubling cascade that is also observed in the original system given by equation (24). The largest discretization time step is equal to 0.723 s, that is $\frac{T_0}{8}$. Such a value for the time step does not allow to have accurate structure of the dynamics as seen in the bifurcation diagram (Fig. 30b). Note that the period-4 window observed for $h \approx 0.62$ s can no longer be detailed nor can the period-

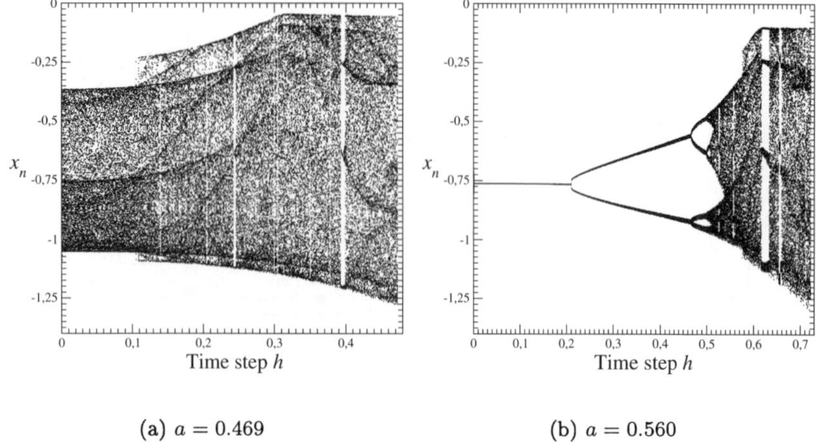

(a) $a = 0.469$ (b) $a = 0.560$

Fig. 30. Bifurcation diagrams of the second-order discretization versus the time step h.

doubling cascade occuring in the beginning of this window be identified. The range over which the dynamics remains topologically equivalent can still be improved as explained in section 5.3.

In the same way that was done for the first-order model, the topology of the attractor solution to the second-order model is investigated for many different values of h. With $a = 0.469$ and $b = 1.1$, the spiral attractor is recovered for $h \leq 0.31$ s. With $h = 0.31$ s, the two branch first-return map (Fig. 31a) is very similar to the map shown in Fig. 29a for the first order discretization with $h = 0.012$ s. The population of periodic orbits is also the same as the population reported in Table 1. Linking numbers are unchanged as exemplified by the link made of orbits (10) and (101) shown in Fig. 31b. The attractor is therefore topologically equivalent to the original attractor. Spurious crossings which were not identified in the previous cases are now observed. They result from the quite large time step h used here. But note that the linking number $lk(10, 101)$ is still equal to -2 and, consequently, the topological properties are not affected. These spurious crossings could be deleted by interpolating the data to avoid the low sampling effect.

We will end this investigation with the third order discretization of

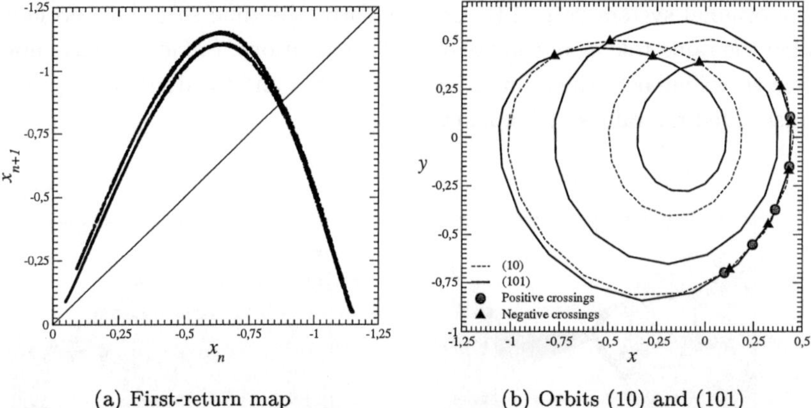

(a) First-return map   (b) Orbits (10) and (101)

Fig. 31.   First-return map to a Poincaré section of the attractor solution to the second-order discrete model with $a = 0.469$, $b = 1.1$, and $h = 0.31$ s. The link made of periodic orbits (10) and (101) shows that the linking number $lk(10, 101)$ is left unchanged, that is equal to $\frac{1}{2}[+5 - 9] = -2$, although spurious crossings due to the low sampling effect are present.

system (24) given by

$$
\begin{cases}
x_{k+1} = x_k + hy_k + \dfrac{h^2}{2}z_k - \dfrac{h^3}{6}\left(az_k + by_k + x_k + x_k^2\right) \\[2ex]
y_{k+1} = y_k + hz_k - \dfrac{h^2}{2}\left(az_k + by_k + x_k + x_k^2\right) \\[2ex]
\qquad + \dfrac{h^3}{6}\left[-(1 + 2x_k)y_k - bz_k + a\left(az_k + by_k + x_k + x_k^2\right)\right] \\[2ex]
z_{k+1} = z_k - h\left(az_k + by_k + x_k + x_k^2\right) \\[2ex]
\qquad + \dfrac{h^2}{2}\left[-(1 + 2x_k)y_k - bz_k + a\left(az_k + by_k + x_k + x_k^2\right)\right] \\[2ex]
\qquad + \dfrac{h^3}{6}\left[(-2y_k + a(1 + 2x_k))y_k - (ab - 1 - 2x_k)z_k \right. \\[1ex]
\qquad \left. + (a^2 - b)\left(az_k + by_k + x_k + x_k^2\right)\right].
\end{cases}
\tag{40}
$$

The number of terms becomes quite large but we will show that this discretization provides solutions which are close to solutions of the continuous counterpart for a quite significant interval of the time step. First, the bifurcation diagram (Fig. 32a) does not look like all diagrams computed for the previous lower order discretizations obtained using the Normand-Cyrot scheme. In particular, an inverse period-doubling cascade is observed as in the original system (Fig. 14). Consequently, the time step does no longer play the same role as in the previous discretizations. In fact, it plays now a role very similar to the bifurcation parameter a and the dynamics is almost unchanged for values of h up to 0.25 s.

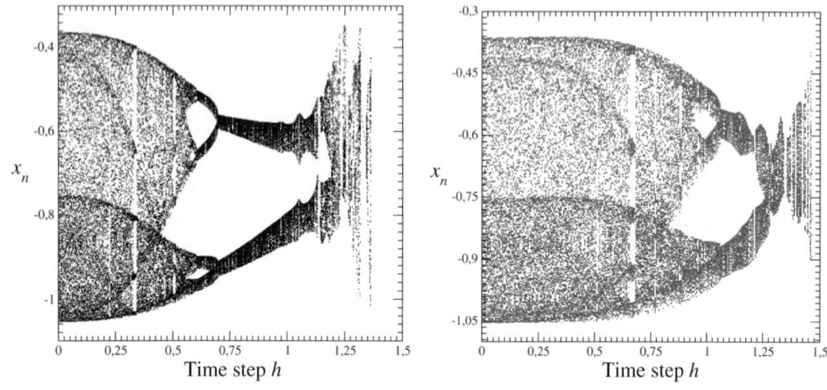

(a) Third-order discretization   (b) Fourth-order discretization

Fig. 32.   Bifurcation diagrams of the third-order (a) and the fourth-order (b) discretizations versus the time step h. Parameters values: $(a, b) = (0.469, 1.1)$.

As shown in Fig. 32a, increasing the value of h does no longer correspond to a development of the dynamics (there is an inverse period-doubling cascade). In order to check the topology of the spiral attractor of the third-order discrete model (40), a is decreased to $a = 0.446$ to obtain a spiral attractor with $h = 0.31$ s. A first-return map (Fig. 33a) is thus made of a two branches as in the case of the map of system (24) for the same value of a (Fig. 13a). Spurious crossings are detected between orbits (10) and (101) as shown in Fig. 33b due to the large time step. Nevertheless, the linking number remains unchanged as for the second-order discrete model (see Fig. 31b). The population of periodic orbits is the same as for system (24) with the same parameter values. This means that, even with a quite large value of h, the topology is not affected.

For larger values of the time step ($h \approx 1.0$ s), an inverse period-doubling cascade ending with a period-2 limit cycle occurs. For such a time step, only 6 points are available per cycle which is far from enough to have a good

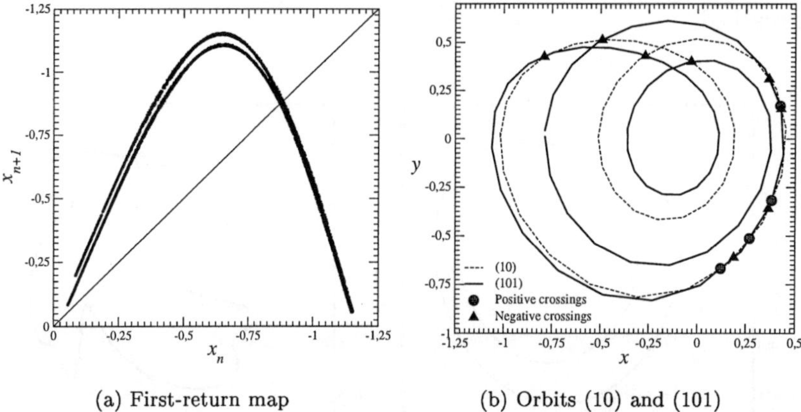

(a) First-return map   (b) Orbits (10) and (101)

Fig. 33. First-return map to a Poincaré section of the attractor solution of the second-order discrete model with $a = 0.445$, $b = 1.1$, and $h = 0.31$ s. The link made of periodic orbits (10) and (101) shows that the linking number $lk(10, 101)$ is equal to -2, although spurious crossings due to the large time step are present.

representation of the dynamics. Indeed, the continuity of the solution can no longer be obtained step by step but only when a quite large number of iterations is taken into account. In addition to that, the time step is larger than the Nyquist time. Despite that, the period-2 limit cycle can be reproduced when the points are not joined by straight segments (Fig. 34a). The blurred aspect of the bifurcation diagram therefore results from the inaccuracy of the Poincaré section and not from the inadequacy of the solution of the third-order discretization. For larger values of the discretization time step, the dynamics is no longer well reproduced as expected. For instance, with $h = 1.25$ s$> T_N = 1.0$ s, the limit cycle observed (Fig. 34b) does not have the same configuration as the solution to system (24) would have. Despite its rough appearance, the limit cycle is still topologically equivalent to a circle, that is, to the limit cycle observed with $h = 1.0$ s. It is worth emphasing that the value of the time is sufficiently large to imply that the discretization becomes quite close to a first-return map to a Poincaré section. Since it is not possible to switch continuously from the attractor to a first-return map to a Poincaré section, there is necessarily a range of the discretization time step over which the solution cannot correspond to a solution to the continuous system. Therefore some additional oscillations appear on the limit cycle. It should be stressed that this happens only for time steps larger than the Nyquist time. This means that with this discretization scheme, the upper limit for the discretization time step is reached.

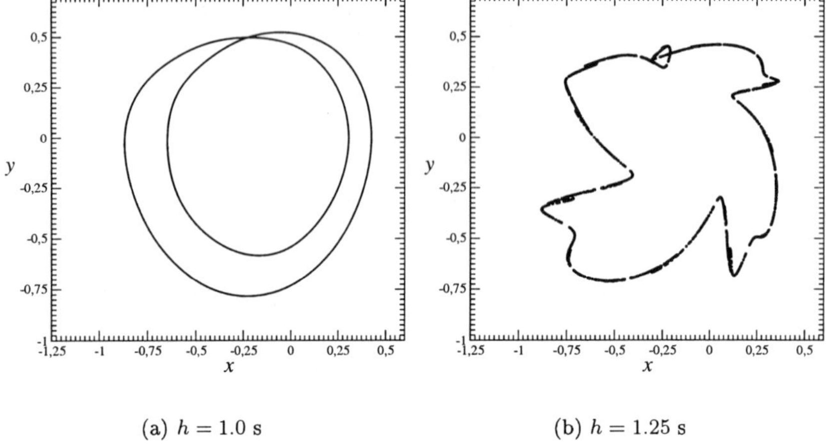

(a) $h = 1.0$ s

(b) $h = 1.25$ s

Fig. 34.   Limit cycle generated by the third order discretization of system (24) for quite large time step. A large number of points is required to obtain a global representation of the dynamics. The continuity can no longer be observed step by step but only at a global point of view. Data points are not connected by segments.

Using higher order helps a little bit to improve the range over which the time step h is varied without any major bifurcation (Fig. 32b). In the case of the fourth-order discretization, not reported here, the time step may be increased up to $h = 0.5$ s without any major modification of the topology of the original attractor. With larger values, periodic orbits are destroyed as seen for the case of the period-5 orbits. These orbits were destroyed just after the window ($h \approx 0.66$ s) where an inverse period-doubling cascade starts leading to the end of the bifurcation diagram. Note that just before the boundary crisis, the solution to the fourth-order discretization with $h = 1.458$ s is a chaotic attractor (Fig. 35) and no longer a limit cycle. The route to chaos followed from the period-1 limit cycle to the chaotic attractor is similar to the route described by Lorenz [28], that is, the period-1 limit cycle start to present a distorted structure as observed in the third-order discretization (Fig. 34). The boundary crisis occurs sufficiently late to allow the attractor to become chaotic. Obviously, the attractor does not have any topological equivalent in the continuous counterpart. This means that getting a discretization with time step around 1.5 s is no longer an attainable goal since the amount of information required per cycle to describe reasonably well the structure of the attractor is not sufficient. This was expected as the Nyquist criterion was violated and therefore such a discretization should not be attempted.

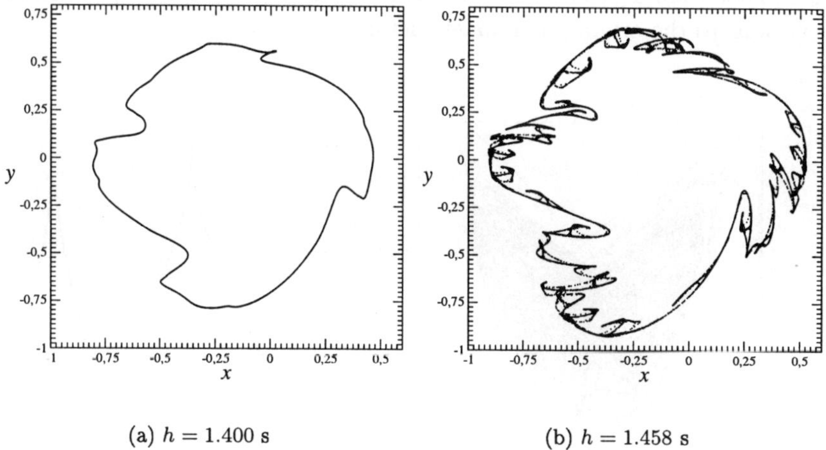

(a) $h = 1.400$ s   (b) $h = 1.458$ s

Fig. 35.   Spurious limit cycle (a) and chaotic attractor (b) observed in the fourth-order discretization of system (24) with two different values of h. This representations are closer to a first-return map to a Poincaré section than to a continuous flow.

5.3. *The Lorenz system*

Applying Monaco and Normand-Cyrot's scheme to the Lorenz system leads to :

$$\begin{cases} x_{k+1} = x_k + h\left(-\sigma x_k + \sigma y_k\right) \\[2ex] y_{k+1} = y_k + h\left(R x_k - y_k - x_k z_k\right) \\[2ex] z_{k+1} = z_k + h\left(x_k y_k - b z_k\right). \end{cases} \tag{41}$$

For a sufficiently small time step h, the solution to this discretization is topologically equivalent to the Lorenz attractor. For $R = 28$, the upper value of the discretization time step is 0.0265 s. Beyond this value, the trajectory is ejected to infinity when the first-order discretization (41) is iterated. The chaotic attractor (Fig. 36a), which is a solution to discretization (41), is close to a solution to the original Lorenz attractor, but with a displacement in the parameter space. One possible set of parameters (R, σ, b)

for which a similar behavior is obtained is (80,3,0.25). The first-return map of the chaotic attractor (Fig. 37) has four monotonic branches as observed in the map (Fig. 36b) of the first-order discrete model (41). Since for $R = 28$ the Nyquist time is 0.04 s, Euler discretization (41) is therefore not very robust versus increases of the time step. But since the largest value of h is smaller than T_N, all the solutions are thus topologically equivalent to solutions to the original Lorenz attractor.

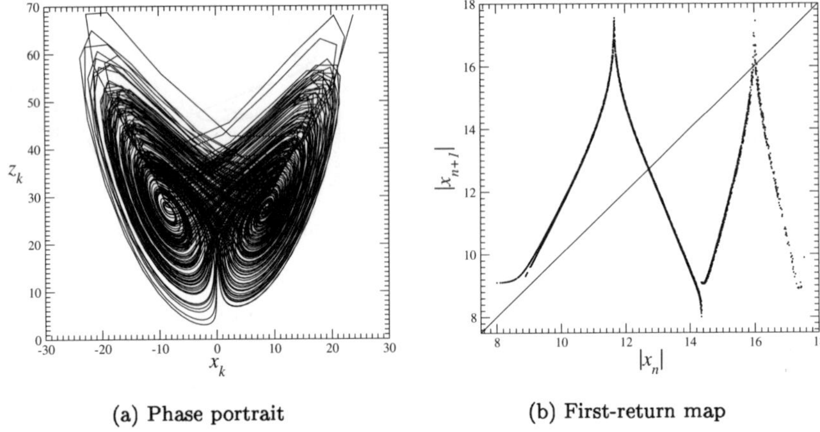

(a) Phase portrait   (b) First-return map

Fig. 36. Chaotic behavior of the first order discretization of the Lorenz system using Monaco and Normand-Cyrot's scheme with $(R, \sigma, b) = (28, 10, 8/3)$. The discretization time step is $h = 0.0265$ s and therefore smaller than T_N.

The second order discretization of the Lorenz system is now considered. It reads as :

$$
\begin{cases}
x_{k+1} = x_k + h\left(-\sigma x_k + \sigma y_k\right) \\[2mm]
\qquad + \dfrac{h^2}{2}\left[-\sigma^2(y - x) + \sigma(x(R - z) - y)\right] \\[2mm]
y_{k+1} = y_k + h\left(Rx_k - y_k - x_k z_k\right) \\[2mm]
\qquad + \dfrac{h^2}{2}\left[(R - z)(\sigma y - x(\sigma + 1)) + y - x(xy - bz)\right] \\[2mm]
z_{k+1} = z_k + h\left(x_k y_k - bz_k\right) \\[2mm]
\qquad + \dfrac{h^2}{2}\left[-(\sigma + 1 + b)xy + x^2(R - z) + \sigma y^2 + b^2 z\right].
\end{cases}
\tag{42}
$$

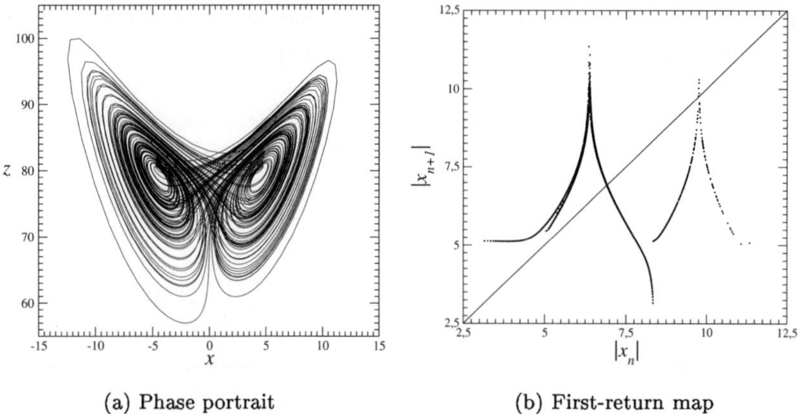

(a) Phase portrait   (b) First-return map

Fig. 37.   Chaotic behavior of the Lorenz system with $(R, \sigma, b) = (80, 3.0, 0.25)$. The attractor is topologically equivalent to the solutions of the first-order discretization shown in Fig. 36.

The first obvious advantage presented by this second order discretization is that the upper value ($h = 0.064$ s) of the discretization time step is larger than the upper value ($h = 0.0265$ s) obtained for the first order discretization. For $h = 0.064$ s, the chaotic attractor (Fig. 38a) presents some departures from the original Lorenz attractor. This topological inequivalence is confirmed by computing a first-return map to a Poincaré section (Fig. 38b). It still presents a reminiscence of the usual Lorenz map, but its structure is blurred by trajectories which do not follow the path usually observed on the Lorenz attractor. These trajectories are not close to any solution to the original Lorenz attractor (such a map has never been observed in the original Lorenz system for any control parameter values). As observed for discretization (28), the numerical instabilities occur through the stable manifold of the two symmetry-related saddle-focus fixed points which are the regions of the phase space where the dynamics is fastest. The region associated with the stable manifold of the saddle fixed point is still well described. Note that the thickness is due to the interpolation used to estimate the intersection between the trajectory and the Poincaré plane. Indeed, the Poincaré section is here defined as

$$\mathcal{P} \equiv \left\{ (x_n, z_n) \in \mathbb{R}^2 \mid x_n = \sqrt{b(R-1)}, \dot{x}_n > 0 \right\}$$
$$\cup \left\{ (x_n, z_n) \in \mathbb{R}^2 \mid x_n = -\sqrt{b(R-1)}, \dot{x}_n < 0 \right\} \tag{43}$$

and the variable z_n is used to compute an invariant first return map. The the first-return map is computed near the saddle fixed point where the dynamics is still defined in a rather smooth way.

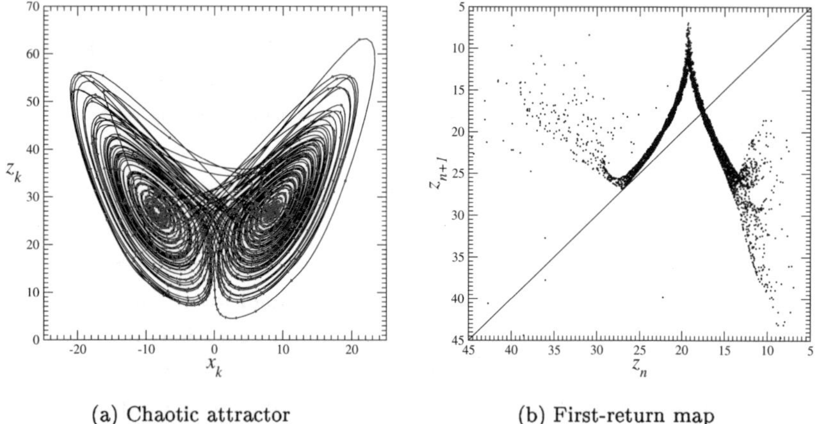

(a) Chaotic attractor   (b) First-return map

Fig. 38. Chaotic attractor of the second order discretization of the Lorenz system using Monaco and Normand-Cyrot's scheme with $(R, \sigma, b) = (28, 10, 8/3)$. The discretization time step is $h = 0.064$ s. Data are interpolated to give a better description of the dynamics.

We end this investigation with the third order discretization of the Lorenz system which has quite a large number of term. We demonstrate that this discretization provides solutions that are close to the solution to the corresponding differential equations for quite a significant interval of the time step. It reads as :

$$
\left\{
\begin{aligned}
x_{k+1} &= x_k + h\left(-\sigma x_k + \sigma y_k\right) + \frac{h^2}{2}\left[-\sigma^2(y-x) + \sigma(x(R-z)-y)\right]\\
&\quad + \frac{h^3}{6}\left[\sigma^2((y-x)(\sigma+R-z) - 2(R-z)x)\right.\\
&\quad \left. + \sigma((\sigma+1-x^2)y + bxz)\right]\\
y_{k+1} &= y_k + h\left(Rx_k - y_k - x_k z_k\right)\\
&\quad + \frac{h^2}{2}\left[(R-z)(\sigma y - x(\sigma+1)) + y - x(xy-bz)\right]\\
&\quad + \frac{h^3}{6}\left[(-(R-z)(\sigma+1) - 3xy + 2bz)\sigma(y-x)\right.\\
&\quad \left. + (\sigma(R-z)+1-x^2)(x(R-z)-y) + (b+1)(xy-bz)x\right]\\
z_{k+1} &= z_k + h\left(x_k y_k - b z_k\right)\\
&\quad + \frac{h^2}{2}\left[-(\sigma+1+b)xy + x^2(R-z) + \sigma y^2 + b^2 z\right]\\
&\quad + \frac{h^3}{6}\left[(3x(R-z) - (\sigma+b+2)y)\sigma(y-x)\right.\\
&\quad \left. + (\sigma y - (b+1)x)(x(R-z)-y) + (b^2-x^2)(xy-bz)\right].
\end{aligned}
\right.
\tag{44}
$$

The discretization time step may be increased up to 0.083 s which is around twice the Nyquist time. Beyond $h = 0.065$ s, the asymptotic behavior settles down onto one of the symmetry-related fixed points which are stable focus-nodes. This means that the large value of the discretization time step induces a displacement to a point of parameter space where the subcritical Hopf bifurcation has not yet destabilized the symmetry-related fixed points. For $h = 0.065$ s, which is a value larger than the Nyquist time, the chaotic attractor (Fig. 39a) does not look very different from the usual Lorenz attractor. This does not seem to be in agreement with our assumption concerning the occurrence of numerical instabilities. Nevertheless, a first-return map to a Poincaré section reveals that there is a spurious structure superimposed on the structure of the usual Lorenz map. Note that the value of the discretization time step is here almost the same as the time step used to compute the first-return map for the second order discretization (Fig. 38b). Obviously the points in this latter map (Fig. 38b) located far from the usual structure of the Lorenz map are no longer observed in the first-return map of the third order discretization (Fig. 39b). But there is a modulation superimposed on the thickness of the map which was never observed on the original Lorenz dynamics. Here, the numerical instabilities

  C. Letellier and E. Mendes

have a reduced effect; they only affect the way in which the trajectory visits the attractor. The reason for this is the thickness of the first-return map which provides more freedom to the trajectory to visit the structure of the attractor. Note that the thickness is not due to the interpolation for estimating the intersection between the trajectory and the Poincaré plane as in the case of the second order discretization.

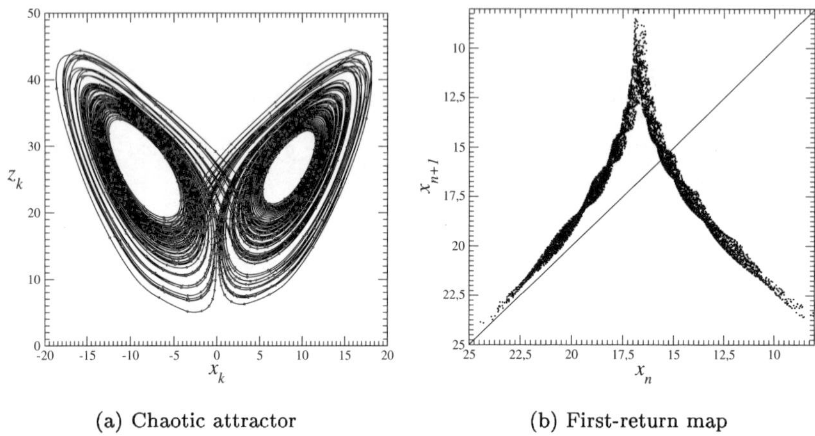

(a) Chaotic attractor   (b) First-return map

Fig. 39.   Chaotic attractor of the third order discretization of the Lorenz system using Monaco and Normand-Cyrot's scheme with $(R, \sigma, b) = (28, 10, 8/3)$. The discretization time step is $h = 0.065$ s. Data are interpolated.

Using higher order discretizations slightly helps to improve the range over which the time step h is varied. In the case of the fourth-order discretization, not reported here, the time step may be increased up to $h = 0.102$ s. The chaotic attractor observed for $h = 0.065$ s does not present any modification compared to the third-order discretization. The third-order and the fourth-order discretizations induce the same first-return map for this value of h. When the discretization time step is increased beyond this value, the fourth-order discretization still has a chaotic attractor as a solution, and not one of the two symmetry-related fixed points. Such a chaotic attractor is shown in Fig. 40a. The shape of the attractor is slightly different from the original Lorenz attractor and a first-return map to a Poincaré section reveals that there are numerous trajectories following paths not observed in the original dynamics (Fig. 40b). As observed with the second-order and the third-order discretizations, when the time step is greater than the Nyquist time, the dynamics has spurious properties which are revealed by the first-return map. Using the fourth-order discretization helps to increase the range of values of the time step over which the asymptotic behavior is a chaotic attractor seemingly induced by a flow.

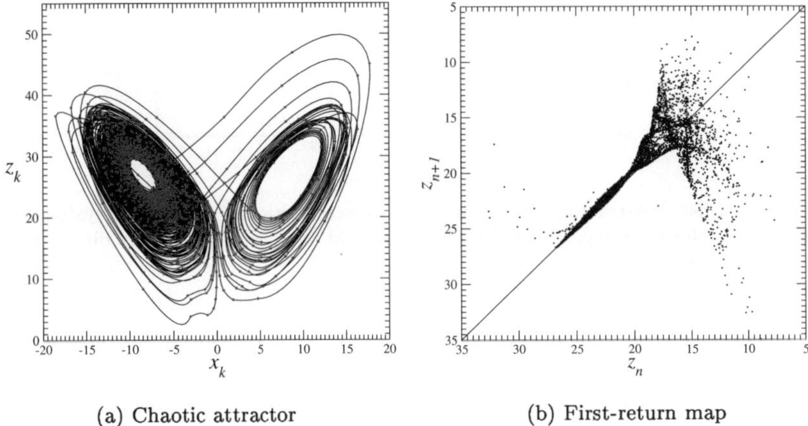

(a) Chaotic attractor   (b) First-return map

Fig. 40.   Chaotic attractor of the fourth order discretization of the Lorenz system using Monaco and Normand-Cyrot's scheme with $(R, \sigma, b) = (28, 10, 8/3)$. The discretization time step is $h = 0.0962$ s. Data are interpolated.

When the time step is further increased, a period-7 torus is observed for $h = 0.102$ s (Fig. 41). This is a behavior which does not correspond to any solution to the original Lorenz system. Using higher order discretization allows us to increase the range over which the time step may be varied before an ejection of the trajectory to infinity. Nevertheless, numerical instabilities always occur around the Nyquist time, but switching from the second-order to the third-order discretization reduces the effect of the numerical instabilities for h around 0.065 s. The fourth-order discretization does not improve the quality of the dynamics for this value. Since numerical instabilities are already superimposed on the original dynamics, we believe that there is no real interest in using the fourth-order discretization.

6. Conclusion

Different sets of discrete difference equations have been proposed for three chaotic systems, namely the Rössler system, the Genesio & Tesi, and the Lorenz system. The discretizations built with the non standard Mickens'

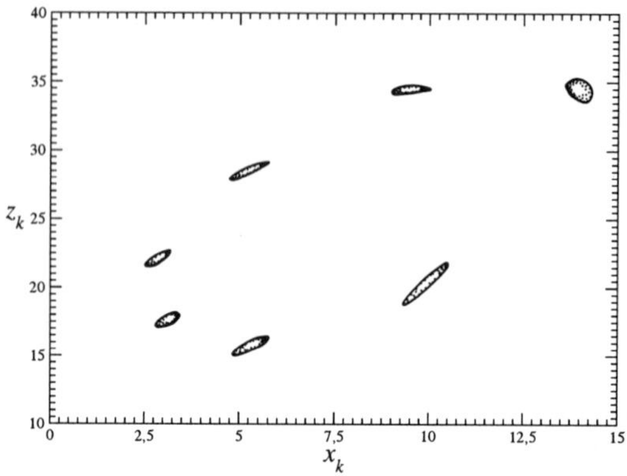

Fig. 41. Chaotic attractor of the fourth order discretization of the Lorenz system using Monaco & Normand-Cyrot's scheme with $(R, \sigma, b) = (28, 10, 8/3)$. The discretization time step is $h = 0.102$ s.

scheme have algebraic structures which are not very complicated. They provide solutions close to solutions of original systems for discretization time step up to the Nyquist time. Nevertheless, the asymptotic behavior remains quite sensitive to the time step which can be considered as a bifuraction parameter.

We showed that Monaco & Normand-Cyrot's scheme based on a truncated Lie expansion of the differential equation that describes the continuous system under investigation can be used very efficiently to obtain quite accurate discretization of continuous nonlinear system. The first order discretizations using Monaco & Normand-Cyrot's scheme, correspond to the Euler's scheme, and are not very robust if large values of the time step are considered. The largest value of the time step is significantly smaller than the Nyquist time. With second-order discretizations, this time can be reached but numerical instabilities significantly affect the topology of the chaotic attractors. The third-order discretization improve the robustness by reducing the deformation of the topology of the attractor. The fourth-order discretizations just help to increase the largest values before which the trajectory is ejected to infinity but do not improve the quality of the chaotic structure for values smaller than the Nyquist time.

It was shown that even with low-order scheme, the solutions of the discretization are topologically equivalent to some solutions of the continuous counterpart. To recover the original dynamics, the bifurcation parameters must be changed. In other words, applying a discretization scheme necessarily induces a displacement in the parameter space when the discretization time step is greater than a threshold which depends on the scheme used. With higher-order Monaco & Normand-Cyrot's schemes difference equations are obtained that can reproduce the solutions of the original continuous system without any displacement in the parameter space and for a quite significant range of the time step. It has been shown that it becomes almost impossible to obtain accurate discretization for time step greater than the Nyquist time. This seems reasonable since for larger values the dynamics can no longer be accurately described. Similar results have been observed with discretizations using Mickens' non standard scheme and we believe that the features here reported are quite general.

From our topological analysis, it may be concluded from this investigation that the discretization tends to converge towards a quite stable topology when the third-order or higher-order discretization are used. Using higher order scheme would slightly improve the range over which the time step may be varied without any major change in the dynamics. However the significant increase of the number of terms involved may not always justify such a choice. For the majority of cases, it seems that the second-order would be a practical choice due to the rather limited amount of terms and the significant range of value for the time step for which the dynamics is reproduced without a significant displacement in the parameter space.

Acknowledgements

We wish to thank Ronald Mickens and Luis A. Aguirre for encouraging this work. A part of this work has been done during stays by C. Letellier at UFMG (Belo Horizonte, Brazil) supported by CNRS and CNPq. E. Mendes acknowledges the support of CNPq under the grant 301313/96-2. Dilys Moscato is thanked for helping us to improve the English of this chapter.

References

1. K. Inagaki, On the discreteness at the edge of chaos, *IPSJ Trans. Mathematical Modeling and its applications*, **40**, SIG 2(TOM 1), 76-81, 1999.
2. S. Elaydi, Is the world evolving discretely ?, *Advances in Applied Mathematics*, (2003).
3. N. H. Packard, J. P. Crutchfield, J. D. Farmer & R. S. Shaw, Geometry from a time series, *Physical Review Letters*, **45** (9), 712-716, 1980.
4. G. Gouesbet & C. Letellier, Global vector field reconstruction by using a multivariate polynomial L_2-approximation on nets, *Physical Review E*, **49** (6), 4955-4972, 1994.

5. I. J. Leontaritis & S. A. Billings, Input-output parametric models for nonlinear systems part II: Stochastic nonlinear systems, *International Journal of Control*, **41** (2), 329-344 ,1985.

6. R. Brown, N. F. Rul'kov & E. R. Tracy, Modeling and synchronizing chaotic systems from time-series data, *Physical Review E*, **49** (5), 3784-3800, 1994.

7. F. Takens, Detecting strange attractors in turbulence, *Lecture Notes in Mathematics*, **898**, 366-381 ,1981.

8. J. A. Tempkin & J. Yorke, Measurements of a physical process satisfy a difference equation. *Journal of Difference Equations and Applications*, **8** (1), 13-24 ,2002.

9. C. Letellier, L. A. Aguirre, J. Maquet & A. Aziz-Alaoui, Should all the species of a food chain be counted to investigate the global dynamics?, *Chaos, Solitons & Fractals*, **13**, 1099-1113, 2002.

10. C. Letellier, L. A. Aguirre, J. Maquet & B. Lefebvre, Analogy between a 10D model for nonlinear wave-wave interaction in a plasma and the 3D Lorenz dynamics, *Physica D*, **179**, 33-52, 2003.

11. J. Maquet, C. Letellier & L. A. Aguirre, Scalar modeling and analysis of a 3D biochemical reaction model, *Journal of Theoretical Biology*, **228** (3), 421-430, 2004.

12. S. A. Billings & L. A. Aguirre, Effects of the sampling time on the dynamics and identification of nonlinear models, *International Journal of Bifurcation & Chaos*, **5** (6), 1541-1556 ,1995.

13. R. E. Mickens, *Nonstandard finite difference models of differential equations*, World Scientific, 1994.

14. P. Liu & S. N. Elaydi, Discrete competitive and cooperative models of Lotka-Volterra type, *J. Computational Anal. Appl.*, **3**, 53-73, 2001.

15. H. Al-Kahby, F. Dannan & S. N. Elaydi, Non-Standard discretization methods for some biological models, *Nonstandard finite difference models of differential equations*, Ed. R. E. Mickens, World Scientific, 155-188, 2000.

16. R. E. Mickens, Genesis of elementary numerical instabilities in finite-difference models of ordinary differential equations, *Proceedings of Dynamic Systems and Applications*, **1**, 251-258, 1994.

17. O. E. Rössler, An equation for continuous Chaos, *Physics Letters A*, **57** (5), 397-398, 1976.

18. O. E. Rössler, Chaos in abstract kinetics : two prototypes, *Bulletin of Mathematical Biology*, **39** (2), 275-289, 1977.

19. C. Letellier, P. Dutertre & B. Maheu, Unstable periodic orbits and templates of the Rössler system: toward a systematic topological characterization, *Chaos*, **5** (1), 271-282, 1995.

20. R. Gilmore, Topological analysis of chaotic dynamical systems, *Reviews of Modern Physics*, **70** (4), 1455-1529, 1998.

21. P. Bergé, Y. Pomeau, Ch. Vidal, *L'ordre dans le chaos*, Hermann, Paris, 1984.

22. P. Melvin & N. B. Tufillaro, Templates and Framed Braids, *Physical Review A*, **44** (6), 3419-3422, 1991.

23. G. B. Mindlin, H. G. Solari, M. A. Natiello, R. Gilmore & X. J. Hou, Topological analysis of chaotic time series data from the Belousov-Zhabotinski

reaction, *Journal of Nonlinear Sciences*, **1**, 147-173, 1991.
24. R. E. Mickens, Nonstandard finite difference schemes for differential equations, *Journal of Difference Equations and Applications*, **8** (9), 823-947, 2002.
25. C. Letellier, S. Elaydi, L. A. Aguirre & Aziz-Alaoui, Difference equations *versus* differential equations, a possible equivalence ? *Physica D*, **195** (1-2), 29-49, 2004.
26. M. S. Baptista & I. L. Caldas, Dynamics of the two frequency torus breakdown in the driven double scroll circuit, *Physical Review E*, **58** (4), 4413-4420, 1998.
27. C. LETELLIER, A. DINKLAGE, H. EL-NAGGAR, C. WILKE & G. BONHOMME, Experimental evidence for a torus breakdown through a global bifurcation in a glow discharge plasma, *Physical Review E*, **63**, 042702, 2001.
28. E. N. Lorenz, Computational chaos — a prelude to computational instability, *Physica D*, **35**, 299, 1989.
29. E. M. A. M. Mendes & S. A. Billings, A note on discretization of nonlinear differential equations, *Chaos*, **12** (1), 66-71, 2002.
30. E. A. Mendes & C. Letellier, Displacement in the parameter space versus spurious solution of discretization with large time step, *Journal of Physics A*, **37**, 1203-1218, 2004.
31. R. Genesio & A. Tesi, Harmonic balance methods for the analysis of chaotic dynamics in nonlinear systems, *Automatica*, **28** (3), 531-548, 1992.
32. C. Letellier & E. A. Mendes, Robust discretizations against increase of the time step for the Lorenz system, *Chaos*, in press.
33. C. Letellier, P. Dutertre & G. Gouesbet, Characterization of the Lorenz system taking into account the equivariance of the vector field, *Physical Review E*, **49** (4), 3492-3495, 1994.
34. C. Grebogi, E. Ott & J. A. Yorke, Crises, Sudden Changes in Chaotic Attractors, and Transient Chaos, *Physica D*, **7**, 181-200, 1983.
35. S. Monaco & D. Normand-Cyrot, On the sampling of a linear control system, in *Proceedings of the 24th Conference on Decision and Control*, Fort Lauderdale, pp. 1457-1482, 1985.
36. S. Monaco & D. Normand-Cyrot, A combinatorial approach to the nonlinear sampling problem, *Lecture Notes in Control and Information Sciences*, **144**, (Ed. M. Thomas & A. Wymer), Springer-Verlag, pp. 788-797, 1990.

CHAPTER 12

CONTRIBUTIONS TO THE THEORY OF NON-STANDARD FINITE DIFFERENCE METHODS AND APPLICATIONS TO SINGULAR PERTURBATION PROBLEMS

Jean M.-S. Lubuma and Kailash C. Patidar

Department of Mathematics and Applied Mathematics
University of Pretoria, Pretoria 0002 (South Africa)
jean.lubuma@up.ac.za; kailash.patidar@up.ac.za

We consider singular perturbation problems defined by first-order (systems of) ordinary differential equations, second-order ordinary differential equations, advection-reaction equations and reaction-diffusion equations. These problems, in which a small positive parameter ε is multiplied to the highest derivative, arise in various fields of science and engineering such as fluid mechanics, fluid dynamics, quantum mechanics, chemical reactor theory, etc. The main concern with such problems is the rapid growth or decay of their solutions in one or more narrow "layer region(s)". Often, the problems are dissipative or dispersive as the rapidly varying component of the solution decays exponentially (dissipates) or oscillates (disperses) from some points of discontinuity in the layer region(s) as ε tends to zero. This singular behavior of the solution makes classical numerical methods not reliable.

We provide some complements to the theory of non-standard finite difference method. We use this theory to design non-standard schemes, which replicate the above mentioned physical properties of the exact solution and which, for a class of linear problems, are ε-uniformly convergent in the sense that the parameter ε and the mesh step vary independently from one another. For a fixed ε, the schemes obtained are elementary stable or stable with respect to the monotone dependence on initial values in the case of first-order equations and advection-reaction problems; they are stable with respect to some kind of conservation laws in the case of second-order equation and they preserve the boundedness and positivity property of the solution of reaction-diffusion problems. Several numerical simulations that support the theory are provided.

1. Introduction

Ordinary and partial differential equations play a vital role in the modelling of real-life problems that occur in a variety of fields of science and technology. However, most of the models cannot be completely solved by analytic techniques. Consequently, numerical simulations are of fundamental importance in gaining some useful insights on the solutions of the differential equations.

The non-standard finite difference approach was initiated two decades ago by Mickens. An important observation, which already came out from the first papers [1,2] of this pioneer researcher on the topic was that the traditional procedures in the design of finite difference schemes had to be suitably changed if the schemes are required to have zero local truncation errors or not to contain instabilities and chaotic behavior. Subsequently, a remarkable effort was made to design exact finite difference schemes for a variety of ordinary and partial differential equations of interest in applications (see, for instance, [3] and the references therein). One of the culminating points of this continuous effort was, from the authors' point of view, the identification by Mickens of five rules for the construction of non-standard finite difference schemes as more reliable numerical methods.

The monograph [4] constitutes a self-contained and comprehensive treatment of the non-standard finite difference method. Since the publication of this book, the non-standard approach has extensively been applied to differential models originating from problems in engineering, physics, biology, chemistry, etc. In this regard, apart from the chapters in the present book, it is worthwhile mentioning the previous edited volume [5] and the special issue [6] of the Journal of Difference Equations and Applications, where a wide range of applications is presented. We also mention the paper [8] that deals with differential inclusions. Further references may be found in the survey article [7]. In all these contributions, the non-standard finite difference schemes have shown great potential in replicating the essential physical properties of the exact solutions of the involved differential models.

Despite the success of the new approach, Mickens [9] himself acknowledges the following fact: "the general rules for constructing such schemes are not precisely known at the present time. Consequently, there exists a certain level of ambiguity in the practical implementation of nonstandard procedures to the formulation of finite difference schemes for differential equations." These concerns were to some extent addressed in [10] where some mathematical foundations of the non-standard approach were provided with precise answers to the triple question below. What is a non-standard finite difference method? In which way are non-standard schemes powerful compared to the standard ones? How to construct systematically non-standard finite difference methods?

The purpose of the present chapter is threefold. Firstly, we give complements to the theory of non-standard finite difference method. In particular, the definition in [10] is extended. Secondly, we review our previous works. Finally, we focus on singularly perturbed ordinary as well as partial differential equations including advection-reaction equations and reaction-diffusion equations. In these problems, which are of practical interest, the highest derivatives are multiplied by a very small positive parameter (see e.g. [11,12,13,14]). As a result, when the parameter tends to zero the dynamics of the systems changes drastically to the extent that traditional numerical methods are not efficient. Our aim is thus to design non-standard finite difference schemes, which are suitable for the numerical treatment of singularly perturbed problems.

The rest of the chapter is organized as follows. The next section is devoted to a number of complements and extensions related to the theory of the non-standard finite difference method. In that section, an exact scheme for the Bernoulli equation is derived, along with a review of exact schemes for some other equations, which serve as motivating examples to our analysis. Section 3 deals with singular perturbation problems defined by first-order (systems of) ordinary differential equations. The design and the analysis of non-standard schemes is done by distinguishing equations with hyperbolic fixed-points from those with non-hyperbolic fixed-points as well as from linear non-autonomous equations, since the level of difficulty varies accordingly. In Section 4, we study non-standard schemes for singularly perturbed problems associated with second-order ordinary differential equations. The next two sections are concerned with singular perturbations problems for partial differential equations namely, the advection-reaction equation and the reaction-diffusion equation. Concluding remarks, discussion on our results and indications on future research directions form part of Section 7. As part of our general methodology, we have included several numerical examples throughout the different sections in order to illustrate the power of the non-standard finite difference schemes.

2. Towards a Definition of the Nonstandard Finite Difference Method

In this section, we consider an initial-value problem for a system of N first-order ordinary differential equations

$$y' \equiv \frac{dy}{dt} = f(y); \ y(0) = A. \tag{1}$$

Obviously, the setting (1) covers ordinary differential equations of arbitrary order m, restricted to $m = 2$ in this work. However, for physical reasons, from time to time it will be more suitable to consider such problems in the form

$$y'' \equiv \frac{d^2 y}{dt^2} = f(y, y') \tag{2}$$

appended with either the initial conditions

$$y(0) = A, \ y'(0) = B \tag{3}$$

or the boundary conditions

$$y(0) = A, \ y(1) = B. \tag{4}$$

In any case, we assume once and for all that the data f satisfy the necessary classical smoothness properties [15] for the problems (1), (2)-(3) and (2)-(4) to be well-posed. Depending on the context, the letters y and f representing the solutions and the data will, throughout this work, be either scalar functions or vector-functions with components $(^1y, \cdots, ^N y)$ and $(^1f, \cdots, ^N f)$, respectively.

Differential equations, as presented above, are in general pervasive in the modelling of naturally occurring phenomena. Despite the existence and uniqueness assumption made here, most of the models arising in practice cannot be completely solved by analytic techniques. Thus, numerical simulations are of fundamental importance in gaining an understanding of differential equations.

We will denote by $(y_k)_{k \geq 0}$ a sequence of approximations to the solution y at the discrete "time" $t_k := k\Delta t$ where Δt is the "time" step size:

$$y_k \approx y(t_k). \tag{5}$$

Notice that in the case when the differential equation is considered on the interval $[0, 1]$, we have $\Delta t = 1/n$ with $n > 1$ an integer.

The (standard) finite difference method is one of the oldest procedure for designing sequences of discrete solutions $(y_k)_{k \geq 0}$. As examples of finite difference schemes, which will serve as a point of departure for the approximation of the system (1), we mention the extensively used schemes

$$\frac{y_{k+1} - y_k}{\Delta t} = \theta f(y_{k+1}) + (1 - \theta)f(y_k), \tag{6}$$

$$\frac{y_{k+1} - y_k}{\Delta t} = f[\theta y_{k+1} + (1 - \theta)y_k] \tag{7}$$

known as two-stage and one-stage theta methods, respectively [16]. Here $\theta \in [0, 1]$ is given; the limit values $\theta = 0$ and $\theta = 1$ in (6)-(7) correspond to the well-known forward and backward Euler methods, respectively. Regarding the second-order equation (2), our analysis will be based on classical finite difference methods of one of the forms

$$\frac{y_{k+1} - 2y_k - y_{k-1}}{(\Delta t)^2} = f\left[\theta y_{k+1} + (1 - \theta)y_k, \frac{y_{k+1} - y_k}{\Delta t}\right] \tag{8}$$

and

$$\frac{y_{k+1} - 2y_k - y_{k-1}}{(\Delta t)^2} = f\left[\theta y_{k+1} + (1 - \theta)y_k, \frac{y_{k+1} - y_{k-1}}{2\Delta t}\right]. \tag{9}$$

For contemporary numerical analysts, the above-mentioned understanding of differential equations (1) and (2) from numerical methods, such as (6)-(9), is often limited to the study of their consistency, (zero-) stability and convergence.

The non-standard finite difference approach, to which this chapter is devoted, puts an additional emphasis on the capability of the discrete schemes to replicate significant physical properties of the solutions of the differential equations, without any restriction on the size of the mesh Δt. Such properties include among others: type of fixed points, oscillatory solution, monotonicity of solutions, conservation of energy, dissipation or dispersion of solution, etc. The precise way in which the properties are preserved is contained in the following definition [10]:

Definition 1: Let P be some property of the exact solution y of (1) or (2). A difference scheme to determine the approximate solutions $(y_k)_{k \geq 0}$ of y is called (qualitatively) stable with respect to the property P (or P-stable) if, for all step sizes Δt, the discrete solutions replicate the property P.

It should be noted that standard finite difference schemes are generally not qualitatively stable with respect to essential physical properties of solutions of problems (1) and (2). In this regard, we mention the standard finite difference scheme

$$\varepsilon\frac{y_{k+1} - 2y_k + y_{k-1}}{(\Delta t)^2} + by_k = 0 \tag{10}$$

for the approximation of the singularly perturbed boundary-value problem

$$\varepsilon y'' + by = 0, \ b > 0, \ y(0) = 0, \ y(1) = 1 \tag{11}$$

where $0 < \varepsilon \leq 1$ is a parameter. The exact solution

$$y(t) = \frac{\sin(t\sqrt{b/\varepsilon})}{\sin(\sqrt{b/\varepsilon})}, \ \varepsilon \neq b(n\pi)^{-2}, \tag{12}$$

of (11) suffers a global breakdown. More precisely, the solution becomes rapidly oscillatory or dispersive (see Figure 1, page 518) for small ε and discontinuous throughout the interval $0 \leq t \leq 1$ as $\varepsilon \to 0$. However, the discrete solution obtained from the scheme (10) fails to display these properties, as can be seen from Figure 2 (page 519). Another specific reason for mentioning the problem (11), at this stage, is that the classical scheme (10) has a more serious disadvantage. That is, the parameters ε and Δt cannot vary independently from one another [14,17]. This can be seen from Table 1 (page 519).

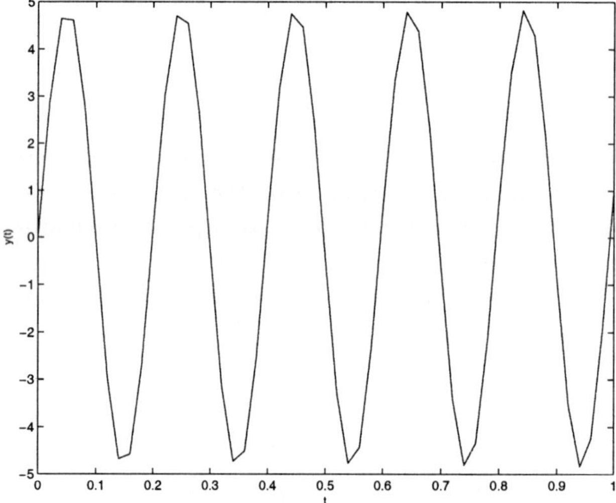

Fig. 1.   Solution $y(t)$ of (11) for $\varepsilon = 10^{-3}$, $n = 50$.

Coming back to the general setting, we state in the next definition [4,18] an ideal situation when a scheme is qualitatively stable with respect to any property.

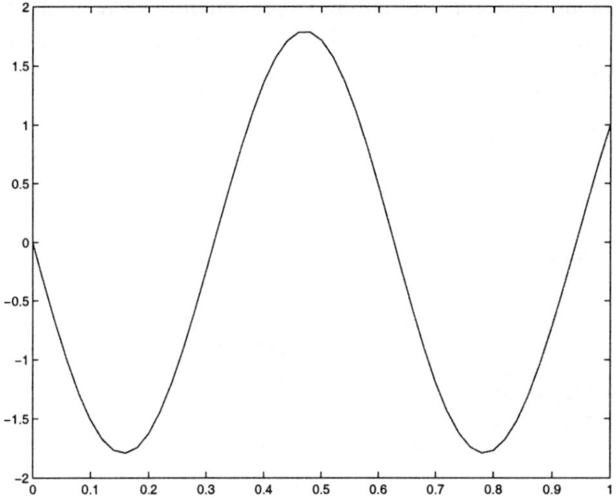

Fig. 2.   Solution of (10) for $\varepsilon = 10^{-3}$, $n = 50$.

Table 1.   Maximum errors when (10) is used to solve (11)

ε	$n = 20$	$n = 40$	$n = 80$	$n = 160$	$n = 320$
1.0	0.17E-04	0.41E-05	0.10E-05	0.26E-06	0.65E-07
10^{-2}	0.29E+00	0.81E-01	0.21E-01	0.52E-02	0.13E-02
10^{-4}	0.20E+01	0.20E+01	0.30E+01	0.29E+01	0.66E+01
10^{-6}	0.12E+01	0.12E+01	0.12E+01	0.12E+01	0.12E+01

Definition 2: A difference equation to determine the discrete solutions $(y_k)_{k \geq 0}$ of y is called an exact scheme or a dynamically consistent scheme with the differential equation (1) or (2) provided that, for any $\Delta t > 0$, the difference equation has the same general solution as the differential equation at the time t_k. In this case the approximation equation (5) becomes the identity

$$y_k = y(t_k) \tag{13}$$

when dealing with the well-posed initial-value or boundary-value problem associated with the differential equation.

Unfortunately, as mentioned above, exact schemes do not exist for most models arising in practice. Nevertheless, the relatively few existing exact schemes can be used as bench marks for the purpose of constructing qualitatively stable schemes for more complex equations. This explains why we now spend some space on designing exact schemes of a number of model equations, which are often used to test the efficiency of classical finite difference schemes.

Example 3: Our first model is the exponential equation

$$\frac{dy}{dt} = \lambda y; \ y(0) = y_0 \tag{14}$$

where $\lambda \neq 0$ is a real number. The solution of Eq. (14) is

$$y(t) \equiv E(t, y_0) = y_0 \exp(\lambda t). \tag{15}$$

The evaluation process

$$y_{k+1} = E(t_{k+1}, y_0) \tag{16}$$

may, thanks to the group property of solutions of first-order differential equations, be written in one of the equivalent forms

$$\frac{y_{k+1} - y_k}{(e^{\lambda \Delta t} - 1)/\lambda} = \lambda y_k \tag{17}$$

and

$$\frac{y_{k+1} - y_k}{(1 - e^{-\lambda \Delta t})/\lambda} = \lambda y_{k+1}, \tag{18}$$

which are exact schemes of Eq. (14).

Example 4: As a second model, we consider the vector form of Example 3, i.e.

$$y' = Ay \ ; \ y(0) = y_0, \tag{19}$$

where it is assumed that A is a diagonalizable non-singular matrix of order N. Thus, there exists a transition matrix Q such that

$$Q^{-1}AQ = \Lambda = \text{diag}(\lambda_1, \cdots, \lambda_N) \tag{20}$$

where $\lambda_1, \cdots, \lambda_N$ are the eigenvalues of A counted according to their multiplicities. Analogously to (15), the solution of (19) is

$$y(t) \equiv E(t, y_0) = \exp(tA)y_0 \tag{21}$$

where, given an arbitrary real-valued function g of real variable, a matrix function $A \mapsto g(A)$ is well-defined, via (20), by

$$g(A) := Q\text{diag}(g(\lambda_1), \cdots, g(\lambda_N))Q^{-1}. \tag{22}$$

Now considering the analogue of the evaluation (16), we end up with the algorithms

$$\left((\exp(\Delta t A) - I)A^{-1}\right)^{-1}(y_{k+1} - y_k) = Ay_k \tag{23}$$

and

$$\left((I - \exp(-\Delta t A))A^{-1}\right)^{-1}(y_{k+1} - y_k) = Ay_{k+1}, \tag{24}$$

which are exact schemes of (19). The particular case when $N = 2$ is of great interest since it provides an explicit expression, as shown in [4]. Indeed, let

$$A = \begin{bmatrix} a & b \\ c & d \end{bmatrix}.$$

Then, with

$$2\lambda_{1,2} = (a + d) \pm \sqrt{(a + d)^2 - 4(ad - bc)},$$

we have

$$Q = \begin{bmatrix} p & q \\ \frac{\lambda_1 - a}{b}p & \frac{\lambda_2 - a}{b}q \end{bmatrix} \quad \text{and} \quad Q^{-1} = \frac{b}{(\lambda_2 - \lambda_1)pq}\begin{bmatrix} \frac{\lambda_2 - a}{b}q & -q \\ -\frac{\lambda_1 - a}{b}p & p \end{bmatrix}$$

where p and q are nonzero real parameters. A simple calculation shows that

$$Q\exp(\Delta t \Lambda)Q^{-1} = \frac{e^{\lambda_2 \Delta t} - e^{\lambda_1 \Delta t}}{\lambda_2 - \lambda_1}A +$$
$$\begin{bmatrix} \frac{\lambda_2 e^{\lambda_1 \Delta t} - \lambda_1 e^{\lambda_2 \Delta t}}{\lambda_2 - \lambda_1} & 0 \\ 0 & \frac{\lambda_2 e^{\lambda_1 \Delta t} - \lambda_1 e^{\lambda_2 \Delta t}}{\lambda_2 - \lambda_1} \end{bmatrix}.$$

Setting

$$\psi \equiv \psi(\Delta t) = \frac{\lambda_2 e^{\lambda_1 \Delta t} - \lambda_1 e^{\lambda_2 \Delta t}}{\lambda_1 - \lambda_2} \tag{25}$$

and

$$\phi \equiv \phi(\Delta t) = \frac{e^{\lambda_1 \Delta t} - e^{\lambda_2 \Delta t}}{\lambda_1 - \lambda_2}, \tag{26}$$

it follows from the formula

$$y_{k+1} = Q\exp(\Delta t \Lambda)Q^{-1}y_k$$

that Eq. (23) may be written in the much simpler equivalent form:

$$\frac{y_{k+1} - \psi y_k}{\phi} = Ay_k. \tag{27}$$

A similar simple formula can be obtained from (24).

Example 5: With a small parameter $\varepsilon > 0$, we associate the constant coefficient homogeneous second-order linear equation

$$-\varepsilon y'' + ay' + by = 0, \tag{28}$$

which is actually a singular perturbation problem. Eq. (28) has the two linearly independent solutions $\exp(\lambda_1 t)$ and $\exp(\lambda_2 t)$ where the roots of the associated characteristic equation are $\lambda_{1,2} = (-a \pm \sqrt{a^2 + 4b\varepsilon})/(-2\varepsilon)$. Thus, setting $y_k = y(t_k)$, the theory of difference equations [4,19] shows that the second order linear difference equation

$$\begin{vmatrix} y_{k-1} & \exp(\lambda_1 t_{k-1}) & \exp(\lambda_2 t_{k-1}) \\ y_k & \exp(\lambda_1 t_k) & \exp(\lambda_2 t_k) \\ y_{k+1} & \exp(\lambda_1 t_{k+1}) & \exp(\lambda_2 t_{k+1}) \end{vmatrix} = 0$$

or equivalently

$$-\exp\left(-\frac{a\Delta t}{2\varepsilon}\right) y_{k+1} + 2\cosh\left(\frac{\Delta t\sqrt{a^2 + 4b\varepsilon}}{2\varepsilon}\right) y_k - \exp\left(\frac{a\Delta t}{2\varepsilon}\right) y_{k-1} = 0 \tag{29}$$

is an exact scheme of Eq. (28). Eq. (29) is equivalent to the difference scheme [4]

$$-\varepsilon \frac{y_{k+1} - 2y_k + y_{k-1}}{\frac{4}{\rho^2}\sinh^2\left(\frac{\rho\Delta t}{2}\right)} + by_k = 0, \rho = \sqrt{b/\varepsilon} \tag{30}$$

in the particular case when $a = 0$, whereas the equation becomes, after suitable manipulations, the scheme

$$-\varepsilon \frac{y_{k+1} - 2y_k + y_{k-1}}{\frac{\varepsilon\Delta t}{a}\left(\exp\left(\frac{a\Delta t}{\varepsilon}\right) - 1\right)} + a\frac{y_k - y_{k-1}}{\Delta t} = 0, \tag{31}$$

whenever $b = 0$. Notice that similar arguments show that an exact scheme of the harmonic equation in (11) is

$$\varepsilon \frac{y_{k+1} - 2y_k + y_{k-1}}{\frac{4}{\rho^2}\sin^2\left(\frac{\rho\Delta t}{2}\right)} + by_k = 0. \tag{32}$$

All the above examples model linear problems. Let us now turn to some nonlinear situations.

Example 6: We start with the problem

$$y' = \lambda y^m \; ; \; y(0) = y_0 \tag{33}$$

where $m \geq 2$ is an integer and $\lambda \neq 0$ is a real number. (The case when $m = 1$ is treated separately in Example 3.) The solution of (33) at the time $t = t_{k+1}$ is

$$y(t_{k+1}) = \frac{y(t_k)}{[1 + \lambda \Delta t(1 - m)y(t_k)^{m-1}]^{1/(m-1)}};$$

this is an exact scheme of (33), which on setting $y^k := y(t_k)$, may be re-written as in [8]:

$$\frac{y_{k+1} - y_k}{\Delta t} = \lambda \frac{(m-1)(y_{k+1})^{m-1}(y_k)^{m-1}}{\displaystyle\sum_{j=0}^{m-2} (y_{k+1})^{m-2-j}(y_k)^j}. \tag{34}$$

Example 7: We consider the logistic growth equation

$$y' = \lambda y(1 - y), \; y(0) = y_0, \quad \lambda > 0 \tag{35}$$

with its solution

$$y(t_{k+1}) = \frac{y_0}{e^{-\lambda t_{k+1}} + (1 - e^{-\lambda t_{k+1}})y_0} \tag{36}$$

at the time $t = t_{k+1}$. Setting $y_k := y(t_k)$, some few algebraic manipulations along with the group property of solutions permit us to re-write (36) in the equivalent form

$$\frac{y_{k+1} - y_k}{\frac{e^{\lambda \Delta t} - 1}{\lambda}} = \lambda y_k(1 - y_{k+1}), \tag{37}$$

which is an exact scheme of (35) obtained, for instance, in [4].

Example 8: As a generalization of (35), we consider the Bernoulli equation

$$y' = \lambda y(1 - y^{m-1}), \; y(0) = y_0, \; \lambda \neq 0 \tag{38}$$

where m is a real number and the non-linearity in the right-hand side is a typical form of reaction term that arises in chemical or physical processes modelled by advection-diffusion equations (see, for instance, [20]). We will elaborate a bit more since the finding of exact schemes for the general Bernoulli equation has not been done as such. We use the standard change of dependent variable

$$v = y^{1-m}, \tag{39}$$

assuming here and after that the expressions involving powers are meaning-ful. (In applications, $\lambda > 0$ and one is often interested in solutions satisfying the positivity and boundedness property $0 \leq y \leq 1$, cf. [21]). Eq. (38) becomes the linear equation

$$v' = (1 - m)\lambda v - \lambda(1 - m); \;\; v(0) = y_0^{1-m}, \tag{40}$$

whose solution, at the time t_{k+1}, is

$$v_{k+1} = v_k e^{(1-m)\lambda\Delta t} - \left(e^{(1-m)\lambda\Delta t} - 1\right) \tag{41}$$

where $v_k := v(t_k)$. In view of (39), we may write (41) as

$$y_{k+1}^{m-1} = \frac{y_k^{m-1}}{e^{(1-m)\lambda\Delta t} - y_k^{m-1}\left(e^{(1-m)\lambda\Delta t} - 1\right)}. \tag{42}$$

Subtract y_k^{m-1} from both sides of Eq. (42) and multiply by the common denominator of the resulting right-hand side to obtain

$$\frac{y_{k+1}^{m-1} - y_k^{m-1}}{(e^{(m-1)\lambda\Delta t} - 1)/\lambda} = \lambda y_k^{m-1}(1 - y_{k+1}^{m-1}). \tag{43}$$

Simple algebraic manipulations permit then to write (43) as the following exact scheme of (38):

$$\frac{y_{k+1} - y_k}{\frac{e^{\lambda\Delta t} - 1}{\lambda}} = \frac{\lambda y_k^{m-1}(1 - y_{k+1}^{m-1})}{\frac{y_{k+1}^{m-1} - y_k^{m-1}}{y_{k+1} - y_k} \cdot \frac{e^{\lambda\Delta t} - 1}{e^{(m-1)\lambda\Delta t} - 1}}. \tag{44}$$

Simplification can occur in the right-hand side of (44). For example, if $m \geq 2$ is an integer, then the identity

$$a^{m-1} - b^{m-1} = (a - b) \sum_{j=0}^{m-2} a^{m-j-2} b^j$$

reduces (44) to the following scheme, which is more suitable than the one obtained in [20]:

$$\frac{y_{k+1} - y_k}{\frac{e^{\lambda\Delta t} - 1}{\lambda}} = \frac{\lambda y_k^{m-1}(1 - y_{k+1}^{m-1})}{\sum_{j=0}^{m-2} y_{k+1}^{m-j-2} y_k^j} \sum_{j=0}^{m-2} e^{j\lambda\Delta t}. \tag{45}$$

The particular case when $m = 3$ is investigated in [21] where an expression similar to (45) is derived.

As mentioned earlier, exact schemes are the best ones as far as convergence (with zero truncation errors) and qualitative stability are concerned. The exact schemes in Examples 3-8 are finite difference schemes that exhibit two main features. These are: the usual denominator and numerator of the discrete derivatives are changed to more complex functions (compared for example (6) and (7) with (17), (27) and (32)); nonlinear terms are approximated in a nonlocal way (see for example (34), (37) and (45)).

For convenience, we provide in Table 2 (page 526) a list of exact schemes of ordinary and partial differential equations. It would also be interesting to investigate whether the analytical solutions in [22] can provide exact finite difference schemes.

Mickens singled out the important observation that the complex structure of the denominator and numerator of the discrete derivatives and the nonlocal form of the approximation of nonlinear terms constitute a general property of these schemes. This observation motivates the following formal definition.

Definition 9: A difference equation to determine approximate solutions (y_k) to the solution $y(t)$ of the system (1) of order $m = 1$ or of the scalar equation (2) of order $m = 2$ is called a non-standard finite difference method if at least one of the following conditions (a) and (b) is met:

(a) At least one of the derivatives $\dfrac{d^j y}{dt^j}$ of order $1 \le j \le m$ is approximated in such a way that the classical denominator $(\Delta t)^j$ and the classical difference operator $y_{k+1} - y_k$ that arises in the numerator of the discrete derivative are replaced by $\phi_j(\Delta t)$ and $y_{k+1} - \psi_j(\Delta t)y_k$ where ϕ_j and ψ_j are real-valued functions, whose values $\phi_j(\Delta t)$ and $\psi_j(\Delta t)$ can also be nonsingular matrices, in the case of first-order ($j = 1$) system of N equations. These functions satisfy the conditions

$$\phi_j(z) = z^j + O(z^{2m}) \text{ and } \psi_j(z) = 1 + O(z^{2m}) \text{ as } 0 < z \to 0, \quad (46)$$

where, in the case when $\phi_j(\Delta t)$ and $\psi_j(\Delta t)$ are matrices ($j = 1$), the notation $\dfrac{y_k}{\phi_j(\Delta t)}$ represents the vector $\phi_j(\Delta t)^{-1}y_k$, whereas the right-hand sides of the identities in (46) are to be viewed as diagonal matrices with $z^j + O(z^{2m})$ and $1 + O(z^{2m})$ as diagonal entries.

(b) Nonlinear terms that occur in $f(y)$ are approximated in a nonlocal way, i.e., by a suitable function of several points of the mesh.

Remark 10: The definition given, for the first time, in [10] is restrictive in that $\phi_j(\Delta t)$ and $\psi_j(\Delta t)$ are real numbers such that $\psi_j(\Delta t) = 1$. In the

Table 2. Exact schemes of some ODE's and PDE's

Differential Equations	Exact Finite Difference Schemes
[4]: $\frac{dy}{dt} = -\lambda y$	$\frac{y_{k+1}-y_k}{(1-e^{-\lambda \Delta t})/\lambda} = -\lambda y_k$
[4]: $\frac{d^2 y}{dt^2} + \omega^2 y = 0$	$\frac{y_{k+1}-2y_k+y_{k-1}}{\frac{4}{\omega^2}\sin^2\left(\frac{\Delta t \omega}{2}\right)} + \omega^2 y_k = 0$
[4]: $\frac{dy}{dt} = \lambda_1 y - \lambda_2 y^2$	$\frac{y_{k+1}-y_k}{(e^{\lambda_1 \Delta t}-1)/\lambda_1} = \lambda_1 y_k - \lambda_2 y_{k+1} y_k$
[4]: $2\frac{dy}{dt} + y = \frac{1}{y}$	$\frac{2(y_{k+1}-y_k)}{(1-e^{-\Delta t})} + \frac{y_k^2}{\left(\frac{y_{k+1}+y_k}{2}\right)} = \frac{1}{\left(\frac{y_{k+1}+y_k}{2}\right)}$
[4]: $\frac{d^2 y}{dt^2} = \lambda\frac{dy}{dt}$	$\frac{y_{k+1}-2y_k+y_{k-1}}{\left(\frac{e^{\lambda \Delta t}-1}{\lambda}\right)\Delta t} = \lambda\left(\frac{y_k-y_{k-1}}{\Delta t}\right)$
[4]: $u_t + u_x = u(1-u)$	$\frac{u_m^{k+1}-u_m^k}{e^{\Delta t}-1} + \frac{u_m^k-u_{m-1}^k}{e^{\Delta x}-1} = u_{m-1}^k\left(1-u_m^{k+1}\right)$ for $\Delta t = \Delta x$
[4]: $\frac{d^2 y}{dt^2} + y + \beta y^3 = 0$	$\frac{y_{k+1}-2y_k+y_{k-1}}{4\sin^2\left(\frac{\Delta t}{2}\right)} + y_k + \beta\left(\frac{\sin^2(\Delta t)}{4\sin^2\left(\frac{\Delta t}{2}\right)}\right)y_k^3 = 0$
[4]: $\frac{d^2 y}{dt^2} + y + \varepsilon y^2 = 0$	$\frac{y_{k+1}-2y_k+y_{k-1}}{4\sin^2\left(\frac{\Delta t}{2}\right)} + y_k + \varepsilon\left(\frac{\sin^2(\Delta t)}{4\sin^2\left(\frac{\Delta t}{2}\right)}\right)y_k^2 = 0$
[4]: $\frac{d^2 y}{dt^2} + y = \varepsilon(1-y^2)\frac{dy}{dt}$	$\frac{y_{k+1}-2y_k+y_{k-1}}{4\sin^2\left(\frac{\Delta t}{2}\right)} + y_k = \varepsilon\left(\frac{\sin(\Delta t)}{2\sin\left(\frac{\Delta t}{2}\right)}\right)(1-y_k^2)\left(\frac{y_k-\cos(\Delta t)y_{k-1}}{2\sin\left(\frac{\Delta t}{2}\right)}\right)$
[4]: $y_{tt} - y_{xx} = 0$	$y_m^{k+1} - 2y_m^k + y_m^{k-1} = y_{m+1}^k - 2y_m^k + y_{m-1}^k$
[4]: $\frac{d^2 y}{dt^2} + 2\varepsilon\frac{dy}{dt} + y = 0$	$\psi(\varepsilon,\Delta t) = \frac{\varepsilon e^{-\varepsilon\Delta t}}{\sqrt{1-\varepsilon^2}} + e^{-\varepsilon\Delta t}\cos\left(\sqrt{1-\varepsilon^2}\right)\Delta t, \phi(\varepsilon,\Delta t) = \frac{\varepsilon e^{-\varepsilon\Delta t}}{\sqrt{1-\varepsilon^2}}\sin\left(\sqrt{1-\varepsilon^2}\right)\Delta t ,$ $\frac{y_{k+1}-2y_k+y_{k-1}}{\phi^2} + 2\varepsilon\left(\frac{y_k-\psi y_{k-1}}{\phi}\right) + \frac{2(1-\psi)y_k+(\phi^2+\psi^2-1)y_{k-1}}{\phi^2} = 0$
[5]: $\frac{\partial c}{\partial t} + P_{n-1}(t)\frac{\partial c}{\partial x} = \lambda c(1-c)$ $P_{n-1}(t) = \sum_{i=0}^{i=n-1} a_i t^i$	$\frac{C^{k+1}(x)-C^k(\tilde{x}^k)}{(e^{\lambda\Delta t}-1)/\lambda} = \lambda C^k(\tilde{x}^k)\left(1-C^{k+1}(x)\right) ,$ $\tilde{x}^k = x - [P_n((k+1)\Delta t) - P_n(k\Delta t)],\ P_n(t) = \int_0^t P_{n-1}(\tau)d\tau.$
[5]: $\frac{\partial c}{\partial t} + P_{n-1}(t)\frac{\partial c}{\partial x} = \lambda c$	$\frac{C^{k+1}(x)-C^k(\tilde{x}^k)}{(e^{\lambda\Delta t}-1)/\lambda} = \lambda C^k(\tilde{x}^k).$
[5]: $\frac{\partial c}{\partial t} + P_{n-1}(t)\frac{\partial c}{\partial x} = \mu + \lambda c$	$\frac{C^{k+1}(x)-C^k(\tilde{x}^k)}{(e^{\lambda\Delta t}-1)/\lambda} = \mu + \lambda C^k(\tilde{x}^k).$

light of Example 4, and especially of the exact schemes (23), (24) and (27), it is natural to consider the extension proposed here.

Another interesting comparison can be made with what Mickens [4] calls "best" finite difference schemes. This pioneer author of the non-standard approach set five rules ([4, page 84]) for the construction of discrete models that have the capability to replicate the properties of the exact solution. According to Mickens, a scheme constructed by using one or more of these rules is termed a best scheme. Our definition of a non-standard finite difference scheme retains only two of Mickens' rules. These are:

Part (a) of Definition 9 This is an extension of Mickens' Rule 2 stating that denominator functions for the discrete derivatives must, in general, be expressed in terms of more complicated functions of the step sizes than those conventionally used

Part (b) of Definition 9 This is Mickens' Rule 3.

All the other rules will be expressed in the next sections in terms of Definition 1. For convenience, these rules are listed below with the Mickens' numbering:

Rule 1 The orders of the discrete derivatives must be exactly equal to the orders of the corresponding derivatives of the differential equations

Rule 4 Special solutions of differential equations should also be special discrete solutions of the finite difference models

Rule 5 The finite-difference equations should not have solutions that do not correspond exactly to solutions of the differential equations.

When Δt is small, non-standard finite difference methods behave in a similar manner as their standard counterparts. To be more precise, we have the following partial result for the second-order equations (2) and (8)-(9), a similar result for the systems (1) and (6)-(7) being established in [23].

Theorem 11: *Consider the non-standard finite difference schemes*

$$\frac{y_{k+1} - 2y_k + y_{k-1}}{\phi_2(\Delta t)} = f\left[\theta y_{k+1} + (1-\theta)y_k, \frac{y_{k+1} - y_{k-1}}{2\phi_1(\Delta t)}\right] \quad (47)$$

and

$$\frac{y_{k+1} - 2y_k + y_{k-1}}{\phi_2(\Delta t)} = \theta f\left[y_{k+1}, \frac{y_{k+1} - y_{k-1}}{2\phi_1(\Delta t)}\right] + (1-\theta)f\left[y_k, \frac{y_{k+1} - y_{k-1}}{2\phi_1(\Delta t)}\right] \quad (48)$$

where the functions ϕ_2 and ϕ_1 satisfy the condition (46). Then these schemes are of order 2 if $\theta = 0$ and of order 1 if $\theta \neq 0$ in the sense that the local truncation error τ_{k+1} is $\theta O(\Delta t) + O[(\Delta t)^2]$.

Proof. We consider only the scheme (47), the proof for the scheme (48) being similar.

Let y be the exact solution of the well-posed problem (2) and (3) or (4), which we assume to be sufficiently smooth and to have bounded derivatives. By definition

$$\tau_{k+1} = \frac{y(t_{k+1}) - 2y(t_k) + y(t_{k-1})}{\phi_2(\Delta t)} -$$
$$f\left[\theta y(t_{k+1}) + (1 - \theta)y(t_k), \ \frac{y(t_{k+1}) - y(t_{k-1})}{2\phi_1(\Delta t)}\right].$$

Using Taylor expansion of $y(t)$ around t_k, we obtain

$$\tau_{k+1} = \frac{y''(t_k)(\Delta t)^2 + \left[\sum_{i=1}^{2} y^{(4)}(\eta_i)\right](\Delta t)^4/24}{\phi_2(\Delta t)} - f\left[y(t_k) + \theta y'(\eta_3)\Delta t, \ v_k\right]$$

where

$$v_k := \frac{12y'(t_k)\Delta t + \left[\sum_{i=4}^{5} y'''(\eta_i)\right](\Delta t)^3}{12\phi_1(\Delta t)}.$$

Here, η_1, $\eta_2, \cdots$ denote some numbers between t_k and t_{k+1} or t_{k-1} and t_k. Since

$$v_k = y'(t_k) + y'(t_k)\left(\frac{\Delta t}{\phi_1} - 1\right) + \frac{\left[\sum_{i=4}^{5} y'''(\eta_i)\right](\Delta t)^3}{12\phi_1},$$

another application of Taylor expansion of f around $(y(t_k), y'(t_k))$ yields, for some (z, w) in a neighborhood of $(y(t_k), y'(t_k))$,

$$\tau_{k+1} = \frac{y''(t_k)(\Delta t)^2 + \left[\sum_{i=1}^{2} y^{(4)}(\eta_i)\right](\Delta t)^4/24}{\phi_2(\Delta t)} - f[y(t_k), y'(t_k)]$$
$$- D_1 f(z, w)\theta y'(\eta_3)\Delta t$$
$$- D_2 f(z, w)\left[y'(t_k)(\frac{\Delta t}{\phi_1} - 1) + \frac{\left[\sum_{i=4}^{5} y'''(\eta_i)\right](\Delta t)^3}{12\phi_1}\right].$$

Using the identity $f\left[y(t_k), y'(t_k)\right] = y''(t_k)$ (cf. Eq. (2)) as well as the relation (46), we obtain after the simplification

$$\tau_{k+1} = y''(t_k)\left(\frac{(\Delta t)^2}{\phi_2} - 1\right) + \frac{\left[\sum_{i=1}^{2} y^{(4)}(\eta_i)\right](\Delta t)^2}{24\phi_2}(\Delta t)^2$$
$$-D_1 f(z, w)\theta y'(\eta_3)\Delta t$$
$$-D_2 f(z, w)\left[y'(t_k)(\frac{\Delta t}{\phi_1} - 1) + \frac{\left[\sum_{i=4}^{5} y'''(\eta_i)\right](\Delta t)^3}{12\phi_1}\right]$$
$$= \theta O(\Delta t) + O[(\Delta t)^2],$$

which is the announced result. $\square$

Remark 12: It is clear from Theorem 11 that the case $\theta = 0$ will be preferred as the corresponding scheme is of order 2.

3. Singularly Perturbed First-Order Systems of ODEs

3.1. *Generalities*

The general setting of this section differs from Eq. (1) in that the derivative term is multiplied by a small positive parameter ε. Thus, we consider an initial-value problem for a system of N first-order differential equations of the form:

$$\varepsilon\frac{dy}{dt} = f(y); \quad y(0) = y_0. \tag{49}$$

For each ε the assumptions on the well-posedness of (49) and on the smoothness of the involved functions are the same as for (1). However, our emphasis will be on the singularly perturbed nature of (49) in the sense described in the following definition:

Definition 13: Let $\tilde{y}$ be a solution of the reduced problem associated with (49), i.e., $\varepsilon = 0$ in (49):

$$f(\tilde{y}) = 0. \tag{50}$$

Let y_ε denote the solution of (49) for a fixed ε. Then, the problem (49) is called singularly perturbed if

$$\lim_{\varepsilon \to 0} y_\varepsilon \neq \tilde{y}.$$

Remark 14: From (50), it follows that the solutions of the reduced problem are fixed-points (or critical points) of the differential equation in (49). Often, we will consider

$$J \equiv Jf(\widetilde{y}) = \left(\frac{\partial(f^i(\widetilde{y}))}{\partial y^j} \right)_{1 \leq i,j \leq N}, \tag{51}$$

the Jacobian of f at the fixed-point $\widetilde{y}$, as well as the following linearization of the differential equation (49):

$$\varepsilon \frac{d\delta}{dt} = J\delta; \ \delta(0) = \delta_0. \tag{52}$$

The main concern with singular perturbation problems such as (49) is that classical methods fail to provide reliable numerical results in the sense that the parameter ε and the mesh size Δt cannot vary independently. We will design numerical methods which preserve the essential physical properties of the involved differential equations. This is done in the two sub-sections below.

3.2. *Case of hyperbolic fixed points*

We recall that any constant-vector $\widetilde{y} \in \mathbb{R}^n$ satisfying (50) is called a fixed-point or critical point of the differential equation (49). We assume in this sub-section that all fixed-points $\widetilde{y}$ are hyperbolic, i.e.,

$$\mathrm{Re}\, \lambda \neq 0 \tag{53}$$

for any $\lambda \in \sigma(J)$, where $\sigma(J)$ is the spectrum of the Jacobian matrix J in (51).

We also assume that the matrix J is diagonalizable. Therefore, there exists a transition matrix Q such that

$$Q^{-1}JQ = \Lambda := \mathrm{diag}(\lambda_1, \lambda_2, \ldots, \lambda_N). \tag{54}$$

The solution of the linear equation (52) is then given by

$$\delta(t) = \exp(tJ)\delta_0, \tag{55}$$

where the exponential matrix function is defined according to the formula (22).

Our analysis is based on the following theorem due to Hartman and Grobman [16].

Theorem 15: *Let $\widetilde{y}$ be a hyperbolic fixed-point of the differential equation in (49). Then there exist a neighborhood $\mathcal{N}$ of the origin, a ball $B(\widetilde{y}, \eta) \subset$*

$\mathbb{R}^N$ *with center $\widetilde{y}$ and radius η as well as a homeomorphism $\mathcal{F} : B(\widetilde{y}, \eta) \mapsto \mathcal{N}$ such that the function $\delta(t) := \mathcal{F}(y(t))$ solves (52) if and only if $y(t)$ solves (49).*

Theorem 15 enables us to investigate the qualitative properties of a fixed-point $\widetilde{y}$ of (49) from those of the origin as fixed-point of (52). Indeed, we have the following definition:

Definition 16: A fixed-point $\widetilde{y}$ of (49) is called linearly stable if

$$\lim_{t \to \infty} \delta(t) = 0 \quad \text{or, equivalently, } \operatorname{Re} \lambda < 0 \text{ for all } \lambda \in \sigma(J). \tag{56}$$

Otherwise the fixed-point $\widetilde{y}$ is called linearly unstable.

The approximate solutions $(y_k)_{k \geq 0}$ of $y(t)$ at the time $t = t_k$ will be obtained from schemes of the general form

$$y_{k+1} = F(\Delta t; y_k). \tag{57}$$

The linearization of the scheme (57) around any of its fixed-point $\widetilde{y}$ is

$$\delta_{k+1} = JF\left(\Delta t, \widetilde{y}\right) \delta_k, \tag{58}$$

where we recall that JF stands for the Jacobian of F at $\widetilde{y}$.

Definition 17: A fixed-point $\widetilde{y}$ of the scheme (57) is called linearly stable provided that, for every Δt,

$$\lim_{k \to \infty} \delta_k = 0 \quad \text{or, equivalently, } |\lambda| < 1 \,\forall\, \lambda \in \sigma(JF).$$

Otherwise, the fixed-point is linearly unstable.

Definition 18: The scheme (57) is called elementary stable provided that the following two conditions are satisfied for every Δt:

(1) The scheme (57) has no spurious or ghost fixed-points in the sense that a point $\widetilde{y}$ is a fixed-point of (57) if and only if it is a fixed-point of (49);
(2) A fixed-point $\widetilde{y}$ has the same linear stability/instability properties for both the differential equation and the discrete scheme.

Remark 19: In Definition 18, the first item is in line with Mickens' Rule 4 and Rule 5, whereas the second one refers to the concept of P-stability (see Definition 1).

The underlying point of this sub-section is that the elimination of elementary instabilities provides more reliable schemes. In this regard, we have the first important result below.

Theorem 20: *Assume that the system of differential equations in (49) has only one hyperbolic fixed-point $\widetilde{y}$. Then the scheme*

$$\varepsilon \left[\left(\exp\left(\frac{\Delta t}{\varepsilon} J \right) - I \right) \left(\frac{J}{\varepsilon} \right)^{-1} \right]^{-1} (y_{k+1} - y_k) = f(y_k) \tag{59}$$

is elementary stable. In particular, this scheme is equivalent to

$$\varepsilon \frac{y_{k+1} - \psi y_k}{\phi} = f(y_k), \tag{60}$$

when $N = 2$, where ψ and ϕ are given as in Equations (25) and (26), but with $\varepsilon^{-1}\lambda_i$ instead of λ_i, $i = 1, 2$, only.

Proof. The Theorem is obvious since the linearization of (59) around $\widetilde{y}$ is

$$\varepsilon \left[\left(\exp\left(\frac{\Delta t}{\varepsilon} J \right) - I \right) \left(\frac{J}{\varepsilon} \right)^{-1} \right]^{-1} (\delta_{k+1} - \delta_k) = J\delta_k, \tag{61}$$

which, according to Example 4, is an exact scheme of the differential equation in (52). $\qquad\square$

The scheme (59) is obviously not applicable to the case of multiple hyperbolic fixed-points. To address this case, we will suitably modify the standard schemes (6) and (7). To this end, we first look at Example 3, Example 7 and Example 8. These examples illustrate the need for the structure of the right hand side of the differential equations to be intrinsically reflected in the discrete schemes if they are required to replicate the qualitative properties of the exact solutions. In the general framework of the system (49), its properties will be captured by a fixed nonzero number

$$q \geq \max\{|\lambda|\} \tag{62}$$

where λ traces all the eigenvalues of the Jacobian $J(f)(\widetilde{y})$ of f at all fixed-points. Moreover, in the spirit of the above-mentioned examples, we modify the denominator of the discrete derivatives in (6) and (7) through a non-negative function ϕ satisfying (46).

Our new non-standard θ-methods are

$$\varepsilon \frac{y_{k+1} - y_k}{\frac{\phi(\varepsilon^{-1}q\Delta t)}{\varepsilon^{-1}q}} = \theta f(y_{k+1}) + (1 - \theta)f(y_k) \tag{63}$$

$$\varepsilon \frac{y_{k+1} - y_k}{\frac{\phi(\varepsilon^{-1}q\Delta t)}{\varepsilon^{-1}q}} = f[\theta y_{k+1} + (1 - \theta)y_k]. \tag{64}$$

Remark 21: The choice of the number q is not so critical if the system is non-stiff. In practice, one may take $q = \max\|J(f)(\widetilde{y})\|_\infty$, where $\|\cdot\|_\infty$ is the matrix norm associated with the supremum norm on $\mathbb{R}^N$. The form of the denominator in (63) and (64) is to be used in the more realistic cases when the solution of the differential equation is not known. In the very few cases when it is known, it is recommended to use, as in Example 3, Example 7 and Example 8, the denominator which provides the exact scheme.

As mentioned in Theorem 11, the non-standard θ-methods behave in a similar manner as the standard θ-methods when Δt is small. However, the non-standard θ-methods have the following additional property that eliminates the shortcomings of the traditional methods.

Theorem 22: *Let ϕ in (46) be such that*

$$0 < \phi(z) < 1 \text{ for } z > 0 . \tag{65}$$

Setting

$$E := \cup \{\sigma(Jf(\widetilde{y})); \ \widetilde{y} \in \mathbb{R}^n, \ f(\widetilde{y}) = 0\},$$

we have the following results:

1. *For $\theta \in [0, 1/2)$, the non-standard θ-methods are elementary stable if $E \subseteq W_l$ where W_l is the wedge and half-plane*

$$W_l = \{\lambda \in \mathbb{C} | Re\lambda < 0, \arg \lambda \in [2\pi/3, \ 4\pi/3]\} \cup \{\lambda \in \mathbb{C} | Re\lambda > 0\} .$$

2. *For $\theta \in [1/2, 1]$, the non-standard θ-methods are elementary stable if $E \subseteq W_r$ where W_r is the half-plane and wedge*

$$W_r = \{\lambda \in \mathbb{C} | Re\lambda < 0\} \cup \{\lambda \in \mathbb{C} | Re\lambda > 0, \arg \lambda \in [-\pi/3, \ \pi/3]\} .$$

Proof. The proof for the two-stage scheme (63) given in [24] can be generalised to include the one-stage method as well. Here are the main ideas. Let $\widetilde{y}$ be a hyperbolic fixed-point of (49). We know that $\widetilde{y}$ is also a fixed-point of the non-standard schemes (63) and (64). The discrete analogue of the linearized equation (58) for both (63) and (64) is

$$\varepsilon \frac{\delta_{k+1} - \delta_k}{\frac{\phi(\varepsilon^{-1}q\Delta t)}{\varepsilon^{-1}q}} = \theta J\delta_{k+1} + (1 - \theta)J\delta_k. \tag{66}$$

In terms of (54) and of the change of variable $\beta_k = Q^{-1}\delta_k$, Eq. (66) is equivalent to

$$\varepsilon \frac{\beta_{k+1} - \beta_k}{\frac{\phi(\varepsilon^{-1}q\Delta t)}{\varepsilon^{-1}q}} = \theta\Lambda\beta_{k+1} + (1 - \theta)\Lambda\beta_k .$$

Thus, coordinate-wise,

$$^i\beta_k = \left[\varepsilon \frac{1 + \frac{\phi(\varepsilon^{-1}q\Delta t)}{\varepsilon^{-1}q}(1-\theta)\lambda_i}{1 - \frac{\phi(\varepsilon^{-1}q\Delta t)}{\varepsilon^{-1}q}\theta\lambda_i} \right]^k {}^i\beta_0, \text{ for } i = 1, 2 \cdots N.$$

From this relation, it is clear that for $\theta \in [0, 1/2]$ and $E \subseteq W_l$, we have $\lim_{k\to\infty} {}^i\beta_k = 0$ if and only if $Re\lambda_i < 0$. Likewise, for $\theta \in [1/2, 1]$ and $E \subseteq W_r$, we have $\lim_{k\to\infty} {}^i\beta_k = 0$ if and only if $Re\lambda_i < 0$. This proves elementary stability of the non-standard schemes (63) and (64) in view of Definition 18. □

At the end of this sub-section, we provide, for the logistic equation (35), the self-explanatory Figure 3 (page 534) where the exact solution, the discrete solutions by the standard and the non-standard schemes are plotted for different values of ε.

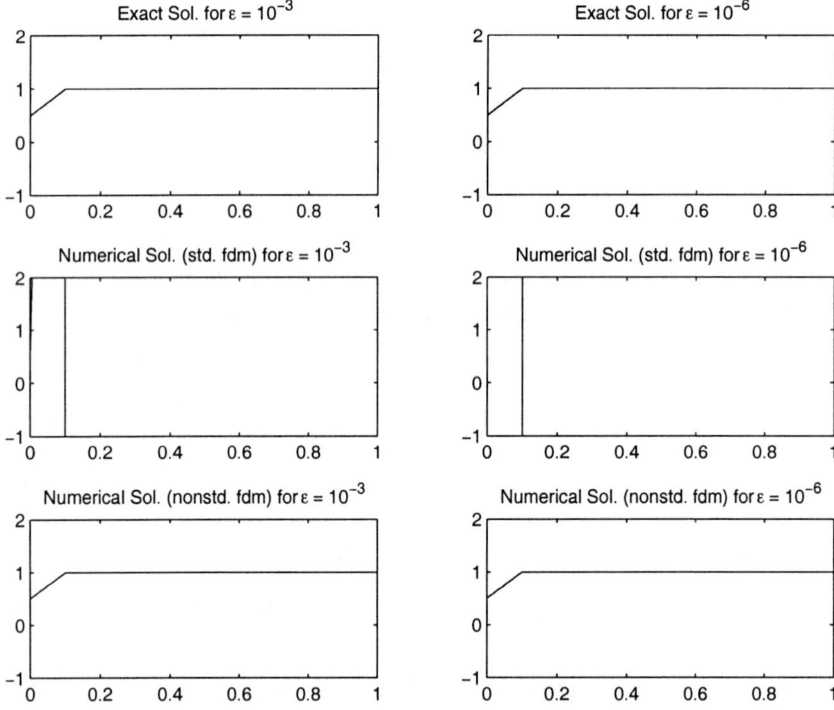

Fig. 3.　Solutions of (35) with $\lambda = 1$ and (63) with $\theta = 0$, for $t \in [0, 1]$, $y(0) = 0.5$, $\Delta t = 0.1$.

3.3. *The case of non-hyperbolic fixed-points*

In this sub-section, we consider the case when the equation (49) has at least one non-hyperbolic fixed-point. However, the analysis is restricted to the scalar equation

$$\varepsilon y' = f(y); \; y(0) = y_0, \tag{67}$$

which is (49) for $N = 1$. (The analysis for a system is still an open problem [25].)

For convenience, in the general scheme (57), we denote Δt by h. We assume that the function $F(h; y)$ in (57) has continuous derivatives with respect to both variables $h > 0$, $y \in \mathbb{R}$ and satisfies

$$F(0; y) = y \;\text{ and }\; \frac{\partial F}{\partial h}(0; y) = f(y). \tag{68}$$

We assume also that the difference scheme (57) is consistent with the differential equation (67) (see [15]). Let us note that consistency implies that (68) is satisfied when y is the solution of (67). No similar assumptions were made in the previous sub-section because of the explicit form of F related to the θ-method presented there.

Since the description of the qualitative properties of non-hyperbolic fixed-point is the source of difficulties, our point of departure will be another physical property of the differential equation (67), namely, the monotone dependence of solutions of (67) on initial values in the sense that

$$y_0 \leq z_0 \implies E(t)y_0 \leq E(t)z_0, \; t \geq 0. \tag{69}$$

The relation (69) is a consequence of the group property of the solution operator $E(t)(\cdot)$ of the differential equation (67) and of the fact that we assume that (67) has a unique solution.

Regarding the scheme (57), we assume, apart from (68), that Mickens' Rules 4 and 5 hold in the following specific way (see comments after Remark 10): for every $h > 0$, the equations

$$y = F(h, y) \;\text{ and }\; f(y) = 0 \tag{70}$$

in y have the same roots counted with their multiplicities.

On the other hand, we assume that the scheme (57) satisfies the relation:

$$\frac{\partial F}{\partial y}(h; y) \geq 0, \; \forall y \text{ and } h > 0. \tag{71}$$

By the mean-value theorem, it is easy to show that the condition (71) is necessary and sufficient for the difference scheme (57) to be stable with respect to monotone dependence on initial values, i.e.,

$$y_0 \leq z_0 \Longrightarrow E^k y_0 \leq E^k z_0, \ \forall k \geq 0 \tag{72}$$

where $k \to E^k(\cdot)$ is the discrete solution operator associated with (57).

Due to the autonomous nature of the differential equation (67), its solutions have a relatively simple structure with regard to their monotonicity. Every solution is either increasing or decreasing on the whole interval $[0, \infty)$. The increasing and decreasing solutions are separated by fixed-points.

The next theorem is proved in [26]. It provides sufficient conditions for the scheme (57) to be stable with respect to the property of monotonicity of solutions, i.e., for every $y_0 \in \mathbb{R}$ the solution (y_k) of (57) is an increasing or a decreasing sequence according as the solution $y(t)$ of equation (67) is increasing or decreasing. It also implies that stability with respect to monotone dependence on initial value coupled with stability with respect to monotonicity of solutions is a substitute of elementary stability for both hyperbolic and non-hyperbolic fixed points.

Theorem 23: *Under the conditions (70) and (71), the difference scheme (57) is elementary stable and is stable with respect to monotonicity of solutions.*

The construction of non-standard finite difference schemes fulfilling the conditions of Theorem 23 will be done by approximating the nonlinear terms in the right-hand side f of equation (67) in a nonlocal way, following Mickens' Rule 3. There are many different ways of doing this. For example

$$y^2(t_k) \approx y_k y_{k+1}, \ y_k \frac{y_{k-1} + y_{k+1}}{2}, \ y_{k-1} y_{k+1} \tag{73}$$

$$y^3(t_k) \approx y_k^2 y_{k+1}, \ y_{k-1} y_k y_{k+1}, \ \frac{y_{k-1}^2 y_{k+1}^2}{y_k}. \tag{74}$$

More generally, any linear combination of the expressions listed in (73) or (74) with the sum of the coefficients equal to 1, approximates y^2 or y^3, the error being of order $O(\Delta t)$ for sufficiently smooth $y(t)$. A typical example is, for arbitrary real numbers α and β,

$$y^2(t_k)(1 - y(t_k)) \approx \alpha y_k^2 + (1 - \alpha) y_k y_{k+1} - \beta y_k^3 - (1 - \beta) y_k^2 y_{k+1}. \tag{75}$$

In this way, the function f in equation (67) may be approximated by an expression, which contains some free parameters. These parameters are determined in such a way that the scheme satisfies a desired qualitative stability property. We now demonstrate the method with the cubic reaction term

$$f(y) = \lambda y^2(1 - y), \ \lambda > 0 \tag{76}$$

that occurs in elementary model for combustion (see [5,9]). The interest in (67) and (76) hinges on the fact that its fixed point $\tilde{y}_1 = 0$ is non-hyperbolic, whereas $\tilde{y}_2 = 1$ is a linearly stable hyperbolic fixed-point. Furthermore, the solution has interesting monotonic properties summarized in the following table:

$$
\begin{array}{lcc}
\underline{\text{Initial Condition}} & \underline{\text{Monotonicity}} & \underline{\text{Limit as } t \to \infty} \\
y_0 \in (-\infty, 0) & \text{Increasing} & 0 \\
y_0 \in (0, 1) & \text{Increasing} & 1 \\
y_0 \in (1, +\infty) & \text{Decreasing} & 1
\end{array}
\tag{77}
$$

This shows that the non-hyperbolic fixed point $\tilde{y}_1 = 0$ attracts the solutions below it and repulses the solution above it. We shall apply Theorem 23 to the design of nonstandard schemes, for (67) and (76), which produce numerical solutions with the same properties. To this end, with real parameters α and β, we consider, according to (75), the family of schemes

$$\varepsilon \frac{y_{k+1} - y_k}{\lambda \phi(h)} = \alpha(y_k)^2 + (1 - \alpha)y_k y_{k+1}$$
$$-\beta(y_k)^3 - (1 - \beta)(y_k)^2 y_{k+1} \tag{78}$$

or equivalently

$$y_{k+1} = F(h; y_k); \ \ F(h; y) = \frac{y + \varepsilon^{-1}\lambda\phi(h)\alpha y^2 - \varepsilon^{-1}\lambda\phi(h)\beta y^3}{1 + \varepsilon^{-1}\lambda\phi(h)(\alpha - 1)y + \varepsilon^{-1}\lambda\phi(h)(1 - \beta)y^2}. \tag{79}$$

We look for a set of values of the parameters α and β for which Theorem 23 applies. Condition (71) simplifies to

$$\left((\beta^2 - \beta)y^2 - 2\beta(\alpha - 1)y + \alpha^2 - \alpha\right) y^2 \varepsilon^{-2}\lambda^2\phi^2 - \tag{80}$$
$$(2\beta + 1)\varepsilon^{-1}\lambda\phi(y - \tfrac{2}{2\beta+1})^2 + \frac{\alpha^2 \varepsilon^{-1}\lambda\phi}{2\beta+1} + 1 \geq 0.$$

Simple manipulations show that (71) or (80) is met if:

$$\alpha \geq 1, \ \beta < -1/2, \tag{81}$$

and the function ϕ satisfying (46) is such that

$$0 < \phi < c \text{ where } c = -(2\beta + 1)/(\varepsilon^{-1}\lambda\alpha^2). \tag{82}$$

A possible choice for the function ϕ is $\phi(h) = c(1 - e^{-h/c})$. Furthermore, the function $F(h, y)$ can be written in the form

$$F(h; y) = y + f(y)\frac{\varepsilon^{-1}\lambda\phi}{1 + \varepsilon^{-1}\lambda\phi(\alpha - 1)y + \varepsilon^{-1}\lambda\phi(1 - \beta)y^2},$$

which under (81) yields $F(h, y) = y \iff f(y) = 0$ for every $h > 0$. Thus, under conditions (81) and (82), Theorem 23 implies elementary stability of the scheme (79) as well as its stability with respect to monotone dependence on initial values.

At the end, we have, for different values of ε, Figure 4 (page 539) regarding numerical solutions of (67) and (76) by using the standard ($\alpha = \beta = 1$) and non-standard ($\alpha = 1$, $\beta = -1$) schemes (78) for $\lambda = 1$, $h = 0.5$ and various y_0. The standard scheme is not stable with respect to monotone dependence on initial values since the discrete solution intersects each other whereas the non-standard solutions have the properties mentioned in (77).

3.4. *The case of linear non-autonomous differential equations*

In this sub-section, we consider the linear non-autonomous singularly perturbed differential equation

$$\varepsilon \frac{dy}{dt} = w(t)y + z(t); \ y(0) = y_0, \tag{83}$$

where $0 < \varepsilon \leq 1$ and $w(t)$ and $z(t)$ are smooth functions satisfying $w(t) \leq \alpha < 0$, $\forall t > 0$. This problem has been studied extensively and has, in particular, a unique solution under the above assumptions.

In the particular case when the functions $w(t)$ and $z(t)$ are constants, an exact finite difference scheme for (83) is easily obtained along the lines of Example 3. Motivated by this fact, we consider for (83) the following non-standard finite difference scheme:

$$\varepsilon \frac{y_{k+1} - y_k}{(e^{\varepsilon^{-1}w_k \Delta t} - 1)/\varepsilon^{-1}w_k} = w_k y_k + z_k, \tag{84}$$

where $w_k = w(t_k)$ and $z_k = z(t_k)$.

It is to be noted that the problem (83) is dissipative, i.e., its solution decays rapidly in the neighbourhood of the origin 0 as $\varepsilon \to 0$.

Theorem 24: *The non-standard scheme (84) is stable with respect to the dissipativity property in the sense that the scheme*

$$\varepsilon \frac{y_{k+1} - y_k}{(e^{\varepsilon^{-1}w(0)\Delta t} - 1)/\varepsilon^{-1}w(0)} = w(0)y_k,$$

is an exact scheme for the differential equation

$$\varepsilon \frac{dy}{dt} = w(0)y,$$

whose general solution is responsible for the dissipative nature of (83).

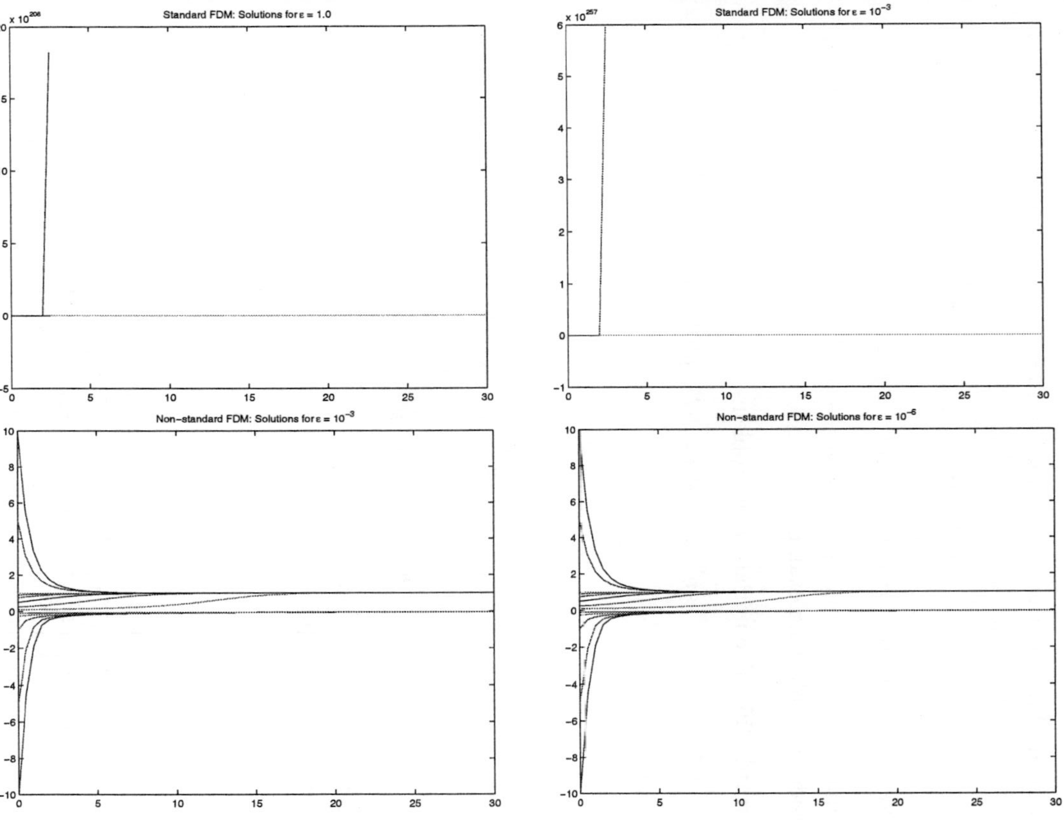

Fig. 4. Schemes (78) with $\alpha = 1$ and $\beta = 1$ (Std. FDM), $\beta = -1$ (Nonstd. FDM)

The scheme (84) is similar to the fitted scheme obtained in [17]. The following convergence result can be deduced from this reference:

Theorem 25: *The non-standard scheme (84) is ε-uniformly convergent of order one in the sense that, for $0 \leq k\Delta t \leq T$, we have*

$$\sup_{0<\varepsilon\leq 1} |y(t_k) - y_k| \leq C\Delta t,$$

where C is a constant independent of ε and Δt.

To illustrate the ε-uniform convergence result stated in Theorem 25, we consider the following example:

Example 26: The initial-value problem

$$\varepsilon y' = -y + 5(\varepsilon + t), \ y(0) = 2$$

has exact solution

$$y(t) = 2\exp(-t/\varepsilon) + 5t.$$

This leads to the results in Table 3 (page 540). We evaluate the maximum errors at all the mesh points using the formula:

$$E_{n,\varepsilon} := \max_{0\leq k\leq n} |y(t_k) - y_k|,$$

for different values of n and ε.

Further, we compute

$$E_n = \max_{0<\varepsilon\leq 1} E_{n,\varepsilon}.$$

Table 3.   Results for Example 26 (Max. Errors)

ε	$n = 10$	$n = 20$	$n = 40$	$n = 80$	$n = 160$
1.0E-2	4.50E-1	2.02E-1	8.62E-2	3.76E-2	1.72E-2
1.0E-3	4.95E-1	2.45E-1	1.20E-1	5.75E-2	2.63E-2
1.0E-4	5.00E-1	2.50E-1	1.25E-1	6.20E-2	3.08E-2
1.0E-5	5.00E-1	2.50E-1	1.25E-1	6.25E-2	3.12E-2
1.0E-6	5.00E-1	2.50E-1	1.25E-1	6.25E-2	3.13E-2
1.0E-7	5.00E-1	2.50E-1	1.25E-1	6.25E-2	3.12E-2
1.0E-8	5.00E-1	2.50E-1	1.25E-1	6.25E-2	3.12E-2
E_n	5.00E-1	2.50E-1	1.25E-1	6.25E-2	3.12E-2

4. Singularly Perturbed Second-Order Problems

4.1. *A class of nonlinear problems*

This section is devoted to scalar singular perturbation problems of the form

$$c_\varepsilon y'' + yg(y^2) = 0, \tag{85}$$

where $c_\varepsilon = \pm\varepsilon$ and we assume that the real valued function g is as smooth as needed. It is possible to investigate problem (85) by transforming it into an equivalent first order system via a standard change of dependent variables, particularly, when the problem is appended with the initial conditions (3). However, for physical reasons, we write (85) in the following equivalent form in terms of its first integral equation:

$$c_\varepsilon \frac{1}{2} \left(\frac{dy}{dt} \right)^2 + K(y^2) = \text{constant}, \tag{86}$$

where $K(t)$ is an anti-derivative of $g(t)$.

We design for (85) schemes, which replicates the property (86), which in the case of $c_\varepsilon = \varepsilon$ represents the principle of conservation of energy.

The problem (85) will be considered on the interval $[0, 1]$ with the boundary conditions

$$y(0) = \alpha_0, \ y(1) = \alpha_1, \ \alpha_0, \alpha_1 \in \mathbb{R}. \tag{87}$$

The singularly perturbed nature of the boundary value problem (85)-(87) can be defined in a way similar to Definition 13. The main concern with this problem is that its solution can be dispersive or dissipative according as $c_\varepsilon = \varepsilon$ or $c_\varepsilon = -\varepsilon$. The dispersive case was illustrated by considering the problem (11). Regarding the dissipative case, we consider the boundary value problem

$$-\varepsilon y'' + by = 0, \ b > 0, \ y(0) = 0, \ y(1) = 1 \tag{88}$$

that has the exact solution

$$y(t) = \frac{\sinh(t\sqrt{b/\varepsilon})}{\sinh(\sqrt{b/\varepsilon})}, \tag{89}$$

which suffers a local breakdown at the end point(s). More precisely, the solution decays/grows exponentially or dissipates rapidly in the neighborhood of end point(s) as $\varepsilon \to 0$ (see Figure 5, page 542).

In view of the exact schemes (30) and (32) of above-mentioned model singularly perturbed problems (28), with $a = 0$, and (11), we approximate

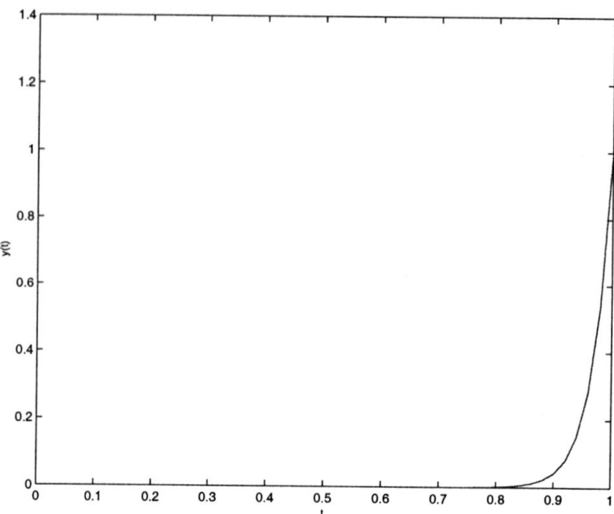

Fig. 5.   Solution $y(t)$ of (88) for $\varepsilon = 10^{-3}$.

the derivative in (85) by following Mickens' Rule 2, namely,

$$c_\varepsilon y''(t_k) \approx c_\varepsilon \frac{y(t_{k+1}) - 2y(t_k) + y(t_{k-1})}{\phi^2},$$ (90)

where, with $\rho = \sqrt{b/\varepsilon}$ and $0 < b \leq \min |g(t)|$, we have

$$\phi^2 \equiv \phi^2(\Delta t, \varepsilon) = \begin{cases} \frac{4}{\rho^2} \sin^2 \left(\frac{\rho \Delta t}{2} \right) & \text{if } c_\varepsilon = \varepsilon, \\[2mm] \frac{4}{\rho^2} \sinh^2 \left(\frac{\rho \Delta t}{2} \right) & \text{if } c_\varepsilon = -\varepsilon. \end{cases}$$ (91)

Note that ϕ^2 satisfies (46), i.e.,

$$\phi^2(\Delta t, \varepsilon) = (\Delta t)^2 + O\left((\Delta t)^4 / \varepsilon \right).$$

To approximate the term $yg(y^2)$ in (85), we consider γ a real-valued function on $\mathbb{R}^3$ that meets the consistency condition

$$\lim_{\Delta t \to 0, \; k \to \infty, \; k\Delta t = t^*} \gamma[y(t_{k-1}), y(t_k), y(t_{k+1})] = y(t^*)g[y^2(t^*)]$$ (92)

as well as the symmetry property

$$\gamma(y_{k-1}, y_k, y_{k+1}) = \gamma(y_{k+1}, y_k, y_{k-1}).$$ (93)

The above leads to the following non-standard finite difference scheme for (85):

$$c_\varepsilon \frac{y_{k+1} - 2y_k + y_{k-1}}{\phi^2} + y_k \gamma(y_{k-1}, y_k, y_{k+1}) = 0. \tag{94}$$

Theorem 27: *The scheme (94) is equivalent to the discrete analogue of (86), which reads as*

$$c_\varepsilon \left(\frac{y_{k+1} - y_k}{\phi} \right)^2 + K_{\Delta t}(y_k) = c_\varepsilon \left(\frac{y_k - y_{k-1}}{\phi} \right)^2 + K_{\Delta t}(y_{k-1}), \ \forall \, k \geq 1, \tag{95}$$

where

$$K_{\Delta t}(y_k) := \sum_{i=1}^{k} y_i (y_{i+1} - y_{i-1}) \gamma(y_{i-1}, y_i, y_{i+1}). \tag{96}$$

Proof. Simple algebraic manipulation shows that (95) is equivalent to

$$c_\varepsilon \frac{y_{k+1} - 2y_k + y_{k-1}}{\phi^2} + \frac{K_{\Delta x}(y_k) - K_{\Delta x}(y_{k-1})}{y_{k+1} - y_{k-1}} = 0. \tag{97}$$

Identification of (94) with (97) and induction on k, with the requirement $K_{\Delta t}(y_0) = 0$, yield the expression of $K_{\Delta t}(y_k)$ in (96). $\qquad\square$

Remark 28: By the mean-value theorem, it is easy to see that a possible choice for the function γ is

$$\gamma(y_{k-1}, y_k, y_{k+1}) = \frac{K(y_k y_{k+1}) - K(y_{k-1} y_k)}{y_k y_{k+1} - y_{k-1} y_k},$$

which illustrates the use of Mickens' Rule 3 on nonlocal approximation of nonlinear terms. Other choices of γ will depend on the function g under consideration.

Remark 29: In (11) and (88) where $g = b$, the scheme (94) will be an exact scheme provided that $\gamma = b$ and $\phi^2 = \frac{4}{\rho^2} \sinh^2 \left(\frac{\rho \Delta t}{2} \right)$ with $\rho = \sqrt{\frac{b}{\varepsilon}}$. The plot of the solution of the scheme for $c_\varepsilon = -\varepsilon$ is given in Figure 6 (page 544), which coincides therefore with Figure 5 (page 542). Numerical examples demonstrating that the scheme (94), for $c_\varepsilon = +\varepsilon$, is stable with respect to the property (86) are given in [8,10]. The scheme (94), for $c_\varepsilon = +\varepsilon$, is exploited in [8] for a suitable treatment of a class of differential inclusions modelling vibro-impact problems.

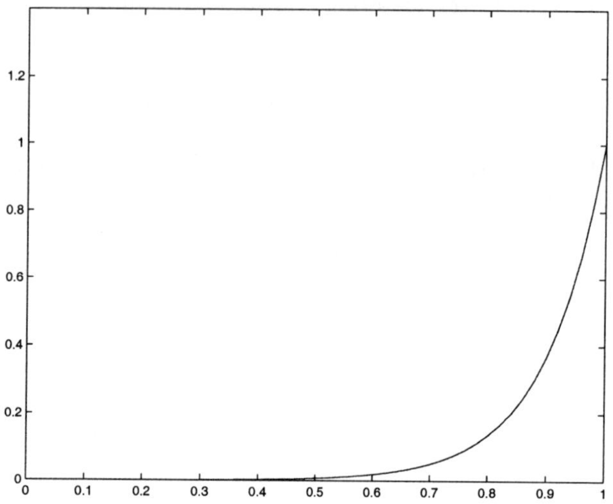

Fig. 6.   Solution of (94) for $c_\varepsilon = -\varepsilon$, $\varepsilon = 10^{-3}$, $n = 50$.

4.2. *Linear problems*

We consider the problem

$$-\varepsilon y'' + b(t)y = f(t), \quad b(t) > 0, \tag{98}$$

with the boundary conditions (87), the functions $b(t)$ and $f(t)$ being sufficiently smooth. The case of general self-adjoint singular perturbation problems is reported in [27].

Once again, due to the exact scheme of the model problems (28), we design for (98) the non-standard schemes

$$-\varepsilon \frac{y_{k+1} - 2y_k + y_{k-1}}{\phi_k^2} + b_k y_k = f_k, \tag{99}$$

where, with $b_k = b(t_k)$, $f_k = f(t_k)$ and $\rho_k = \sqrt{b_k/\varepsilon}$, we have

$$\phi_k^2 \equiv \phi_k^2(\Delta t, \varepsilon) = \frac{4}{\rho_k^2} \sinh^2\left(\frac{\rho_k \Delta t}{2}\right). \tag{100}$$

Theorem 30: *The scheme (99) is stable with respect to the dissipative property of the solution $y(t)$ of the differential equation (98) in the sense that the schemes*

$$-\varepsilon \frac{y_{k+1} - 2y_k + y_{k-1}}{\phi_0^2} + b(0)y_k = 0,$$

and

$$-\varepsilon \frac{y_{k+1} - 2y_k + y_{k-1}}{\phi_n^2} + b(1)y_k = 0,$$

are the exact schemes for the differential equations

$$-\varepsilon y'' + b(0)y = 0 \tag{101}$$

and

$$-\varepsilon y'' + b(1)y = 0, \tag{102}$$

whose general solutions are responsible for the dissipative nature of (98).

As in the case of first order equations, the scheme (99) is similar to the one obtained in [17], from where the following convergence result can be deduced.

Theorem 31: *The non-standard scheme (99) is ε-uniformly convergent of order two, i.e.,*

$$\sup_{0<\varepsilon\leq 1} |y(t_k) - y_k| \leq C\Delta t^2,$$

where C is a constant independent of ε and Δt.

To illustrate the contents of Theorem 30, we plot the singular functions $d(t) = \exp(-t\sqrt{b(0)/\varepsilon})$ (top of Figure 7, page 546) and $e(t) = \exp(-(1-t)\sqrt{b(1)/\varepsilon})$ (bottom of Figure 7) as well as the solution of the non-standard finite difference method (99) with $y_0 = d(0)$, $y_n = e(t_n)$, $b(t) = 1 + t(1-t)$ and $f_k \equiv 0$. Figure 7 illustrates that the dissipativity property is preserved. The fact that the standard scheme behaves badly in the layer regions is visualized in Figure 8, page 547.

The ε-uniform convergence result stated in Theorem 31 is illustrated in the following example:

Example 32: Consider the problem (98) with $b(t) = 1 + t(1-t)$,

$$f(t) = 1 + t(1-t) + \left[2\sqrt{\varepsilon} - t^2(1-t)\right]\exp[-(1-t)/\sqrt{\varepsilon}\,]$$
$$+ \left[2\sqrt{\varepsilon} - t(1-t)^2\right]\exp[-t/\sqrt{\varepsilon}\,]$$

$$y(0) = 0, \quad y(1) = 0.$$

Its exact solution is

$$y(t) = 1 + (t-1)\exp[-t/\sqrt{\varepsilon}\,] - t\,\exp[-(1-t)/\sqrt{\varepsilon}\,]$$

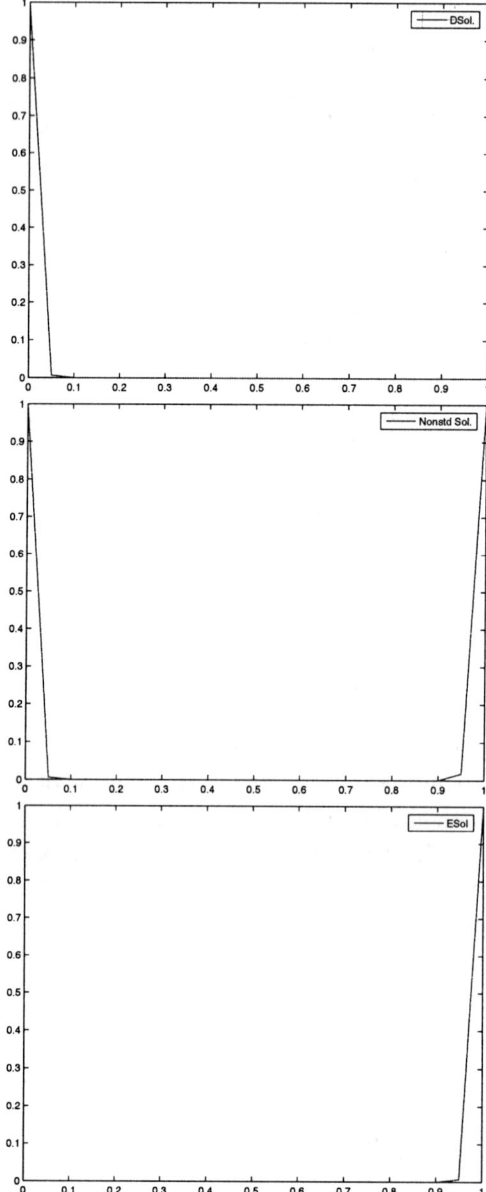

Fig. 7.   Dissipation analysis for $\varepsilon = 10^{-4}$, $n = 20$.

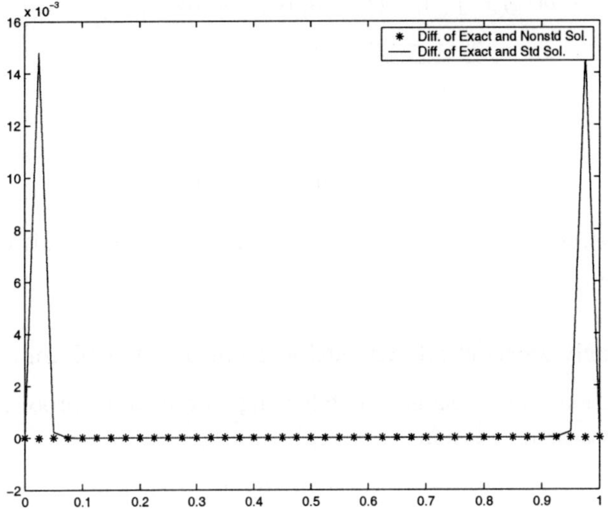

Fig. 8.   Errors of Std. & Non-std. Solutions for Ex. 32 for $\varepsilon = 10^{-5}$, $n = 40$.

In Table 4 (below), we evaluate the maximum errors at all the mesh points using the formula:

$$E_{n,\varepsilon} := \max_{0 \le k \le n} |y(t_k) - y_k|,$$

for different values of n and ε, where y_k is the approximate solution of (98).
  Further, we compute

$$E_n = \max_{0 < \varepsilon \le 1} E_{n,\varepsilon}.$$

Table 4.   Results for Example 32 (Max. Errors)

ε	$n = 4$	$n = 8$	$n = 16$	$n = 32$	$n = 64$	$n = 128$
1.0E-04	0.36E-01	0.95E-02	0.26E-02	0.86E-03	0.19E-03	0.72E-04
1.0E-05	0.36E-01	0.95E-02	0.25E-02	0.64E-03	0.25E-03	0.81E-04
1.0E-07	0.36E-01	0.95E-02	0.25E-02	0.63E-03	0.16E-03	0.40E-04
1.0E-08	0.36E-01	0.95E-02	0.25E-02	0.63E-03	0.16E-03	0.40E-04
1.0E-10	0.36E-01	0.95E-02	0.25E-02	0.63E-03	0.16E-03	0.40E-04
E_n	0.36E-01	0.95E-02	0.25E-02	0.63E-03	0.16E-03	0.40E-04

Remark 33: The non-standard finite difference scheme for the dispersive

problem $\varepsilon y'' + b(t)y = f(t), \ b(t) > 0$, is given by

$$\varepsilon \frac{y_{k+1} - 2y_k + y_{k-1}}{\phi_k^2} + b_k y_k = f_k, \tag{103}$$

where

$$\phi_k^2 = \frac{4}{\rho_k^2} \sin^2 \left(\frac{\rho_k \Delta t}{2} \right).$$

Numerical experiments show that this scheme is also ε-uniformly convergent of order two.

5. Singularly Perturbed Advection-Reaction Problems

In this section, we consider the following initial-value problem for the advection-reaction equation

$$\varepsilon \partial_t u + \partial_x u = r(u), \ u(x,0) = f(x) \tag{104}$$

where $0 < \varepsilon \leq 1$. We assume once and for all that (104) has a unique solution. In what follows, it is implicitly understood that the reaction r and the function f are as smooth as needed.

Remark 34: The problem (104) is singularly perturbed. To illustrate this fact, we consider the particular case when $r(u) = 0$ and $f(x)$ has infinite limits as x tends to $\pm\infty$. Then, when letting ε go to 0 in

$$u_\varepsilon(x,t) := f(x - \varepsilon^{-1}t),$$

the corresponding unique solution of (104), we have

$$u_0(x,t) = f(-\infty) = \infty.$$

Any function $K(t)$, which is constant with respect to x, is a solution of the reduced problem associated with (104), i.e., $\varepsilon = 0$ in the differential equation. Clearly, $K(t)$ is not equal to the infinite number $u_0(x,t)$. Notice that the problem

$$\partial_t u + \varepsilon \partial_x u = 0, \ u(x,0) = f(x), \tag{105}$$

is not singularly perturbed. Indeed, the limit of its solution $f(x - \varepsilon t)$, as $\varepsilon \to 0$, solves the corresponding reduced problem.

The following scheme, known as Lax scheme [28] in the case of a linear reaction $r(u)$, is a possible finite difference method for (104):

$$\varepsilon \frac{u_m^{k+1} - u_m^k}{\Delta t} + \frac{u_m^k - u_{m-1}^k}{\Delta x} = r(u_{m-1}^k). \tag{106}$$

Here and after, the time independent variable t is discretized by $t_k = k\Delta t$, $(k = 0, 1, 2, \cdots)$ as in Sect. 2, whereas the mesh of discrete points $x_m = m\Delta x$ $(m = \cdots, -1, 0, 1, \cdots)$ is used for the space variable x. We denote by u_m^k an approximation to the solution $u(x_m, t_k)$ at the point (x_m, t_k): $u_m^k \approx u(x_m, t_k)$.

In the linear situation, the condition $\frac{\Delta t}{\varepsilon \Delta x} \leq 1$ is essential for the standard scheme (106) to be stable in the sense of Lax-Richtmyer [29]. In the current framework of singularly perturbed problem this stability condition makes the scheme (106) not interesting in practice as the time-step size Δt has to be very small. In order to properly study the problem, we use the method of characteristics. We perform the change of variables

$$x \to x + \varepsilon^{-1}t, \ t \to t; \ U(t) := u(x + \varepsilon^{-1}t, t). \tag{107}$$

The singularly perturbed advection-reaction equation (104) becomes, for a fixed x, an initial-value problem for the first-order differential equation

$$\varepsilon \frac{dU}{dt} = r(U); U(0) = f(x), \tag{108}$$

which is the type of singularly perturbed problems analyzed in Section 3. This permits us to now extend the linear stability analysis of ordinary differential equations to the partial differential equation in (104). In fact, by a fixed-point of the said partial differential equation, we mean any zero $\widetilde{u}$ of the function r : $r(\widetilde{u}) = 0$. With $\widetilde{u}$ a hyperbolic fixed-point, i.e., a fixed-point such that

$$J \equiv r'(\widetilde{u}) \neq 0, \tag{109}$$

we associate the solution

$$\delta(x, t) = \delta_0(x - \varepsilon^{-1}t)e^{\varepsilon^{-1}Jt} \tag{110}$$

of the linearized equation

$$\varepsilon \delta_t + \delta_x = J\delta; \ \delta(x, 0) = \delta_0(x). \tag{111}$$

Then, Hartman and Grobman's theorem (see Theorem 15) shows that the solution u of (104) in a neighborhood of $\widetilde{u}$ and the solution δ of (111) in a neighborhood of 0 are such that the deviation $u - \widetilde{u}$ and δ have both, along the line passing through $(x, 0)$ and parallel to the vector $< \varepsilon^{-1}, 1 >$, the same asymptotic behaviour as $t \to \infty$. Thus, the next definition.

Definition 35: A hyperbolic fixed-point $\widetilde{u}$ is called linearly stable provided that

$$\lim_{t \to \infty} \delta(x + \varepsilon^{-1}t, t) = 0$$

or, equivalently, $J < 0$ in (110). Otherwise, the fixed-point is called linearly unstable.

In the light of (108), it is natural to consider, for (104), the time-stepping standard finite difference method

$$\varepsilon \frac{u^{k+1}(x) - u^k(\overline{x}^k)}{\Delta t} = r(u^k(\overline{x}^k)) \tag{112}$$

where

$$\overline{x}^k := x - \varepsilon^{-1} k \Delta t \tag{113}$$

is the backtrack point of x and $u^k(x)$ denotes an approximation of the solution u at the point (x, t_k).

Our aim is to modify (112) into a time-stepping method, which replicates the linear stability/instability property of fixed points of (104) described in Definition 35. Our important result in this section is in the next theorem where a new elementary stable scheme is presented.

Theorem 36: *Let ϕ be as in Theorem 22. Assume that the partial differential equation in (104) has a nonzero finite number of fixed-points $\tilde{u}$, all being hyperbolic. Put $q = \max\{|r'(\tilde{u})|; r(\tilde{u}) = 0\}$. Then, the non-standard scheme*

$$\varepsilon \frac{u^{k+1}(x_m) - u^k(\overline{x}_m^k)}{\phi(q\varepsilon^{-1}\Delta t)/q\varepsilon^{-1}} = r(u^k(\overline{x}_m^k)) \tag{114}$$

is elementary stable in the sense that for any step-size, the only fixed-points $\tilde{u}$ of the scheme are those of the partial differential equation in (104), the linear stability properties of each $\tilde{u}$ being the same for both the partial differential equation and the discrete scheme.

Proof. Let $\tilde{u}$ be a hyperbolic fixed-point of (104). Put $u^k(x_m) := \tilde{u} + \delta^k(x_m)$ in Eq. (114), where $\delta^k(x_m)$ is small enough; perform the Taylor expansion of the right-hand side of (114) around $\tilde{u}$ and retain only the linear part in $\delta^k(x_m)$ to obtain the following equation, which is the discrete analogue of the error equation (111):

$$\varepsilon \frac{\delta^{k+1}(x_m) - \delta^k(\overline{x}_m^k)}{\phi(q\varepsilon^{-1}\Delta t)/q\varepsilon^{-1}} = J\delta^k(\overline{x}_m^k).$$

By induction on k, we have

$$\delta^k(x_m) = (1 + \phi J/q)^k \delta^0(\overline{x}_m^{k^*}), \quad \text{with } k^* = k(k-1)/2.$$

Let $\widetilde{u}$ be linearly stable for the partial differential equation, i.e., $J < 0$. Then by definition of q and the condition (65), we have

$$|1 + \phi J/q| = 1 - \phi|J|/q < 1,$$

which together with the fact that $\delta^0(\overline{x}_m^{k^*})$ is bounded with respect to k imply that $\delta^k(x_m)$ tends to 0 as $k \to \infty$. This means that $\widetilde{u}$ is linearly stable for the difference schemes. Analogously, if $\widetilde{u}$ is linearly unstable for the partial differential equation, i.e., $J > 0$, then the linear instability of $\widetilde{u}$ for the difference schemes follows from

$$|1 + \phi J/q| = 1 + \phi J/q > 1. \qquad \square$$

If each backtrack-point $\overline{x}_m^k$ coincides with a space grid point, say, x_{m_k} then the scheme (114) can be exploited for numerical simulations. Under the condition $\Delta x = \varepsilon^{-1}\Delta t$ between step-sizes, this situation will arise when $\varepsilon = 1$ or ε is not so small since

$$\overline{x}_m^k = x_{m-k}.$$

The analysis of the scheme (114) done in [30] is in the above framework. However, as mentioned earlier, the condition $\Delta x = \varepsilon^{-1}\Delta t$ is not practically acceptable for singularly perturbed problems.

To circumvent these difficulties, we propose a more workable version of the scheme (114), following an idea in [20].

For any backtrack-point $\overline{x}_m^k$, there exists an integer m_k such that $x_{m_k-1} \leq \overline{x}_m^k \leq x_{m_k}$. Let Lg denote the linear Lagrange interpolation of the function g at x_{m_k-1} and x_{m_k}:

$$(Lg)(x) = \frac{x - x_{m_k}}{-\Delta x}g(x_{m_k-1}) + \frac{x - x_{m_k-1}}{\Delta x}g(x_{m_k}). \qquad (115)$$

The scheme (114) is then modified into

$$\varepsilon\frac{u^{k+1}(x_m) - Lu^k(\overline{x}_m^k)}{\phi(q\varepsilon^{-1}\Delta t)/q\varepsilon^{-1}} = r(Lu^k(\overline{x}_m^k)), \qquad (116)$$

where $Lu^k(\overline{x}_m^k)$ is visualized in Figure 9 (page 552).

Observing that the operator L is invariant under polynomials of degree ≤ 1, similar arguments to those in the proof of Theorem 36 lead to the following result:

Theorem 37: *Under the assumptions of Theorem 36, the scheme (116) is elementary stable.*

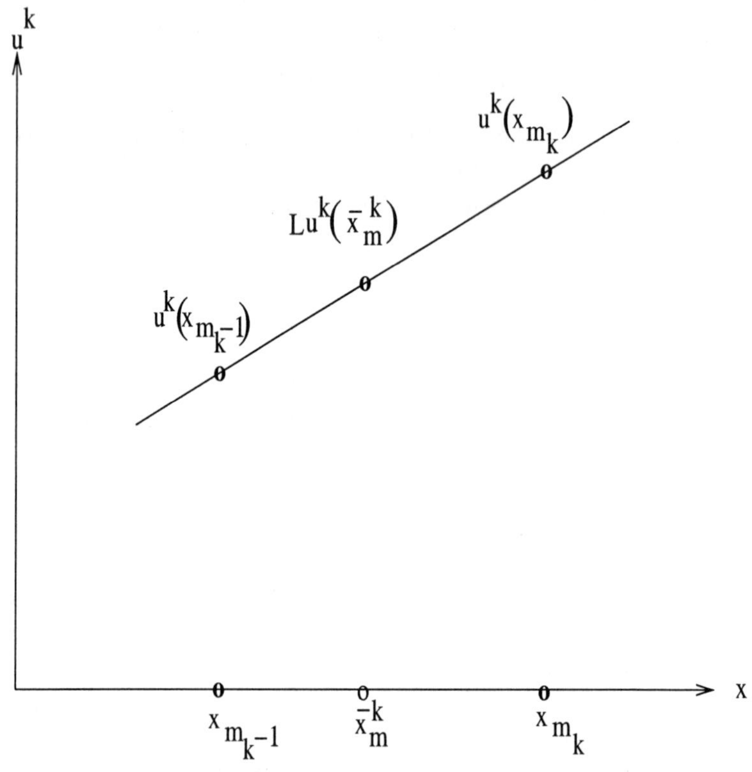

Fig. 9.   Solution u^k versus x

Remark 38: The standard finite difference schemes (112) are elementary unstable.

Remark 39: When the structure of $r(u)$ permits to design an exact scheme for (104), it is possible to construct variants of (114) and (116), which are suggested by the exact scheme. To be more specific, we take $r(u) = \lambda u(1-u)$, $\lambda > 0$, in (104). With suitable adaptations on the exact scheme of the advection-reaction equation provided in Table 2 (page 526), an exact scheme of (104) is found to be

$$\varepsilon \frac{u^{k+1}(x_m) - u^k(\bar{x}^k_m)}{(e^{\varepsilon^{-1}\lambda\Delta t} - 1)/\varepsilon^{-1}\lambda} = \lambda u^k(\bar{x}^k_m)(1 - u^{k+1}(x_m)). \tag{117}$$

Consequently, the scheme (116) can be replaced with the scheme

$$\varepsilon\frac{u^{k+1}(x_m) - Lu^k(\overline{x}_m^k)}{(e^{\varepsilon^{-1}\lambda\Delta t} - 1)/\varepsilon^{-1}\lambda} = \lambda Lu^k(\overline{x}_m^k)(1 - u^{k+1}(x_m)). \qquad (118)$$

For $\varepsilon = 1$, the scheme (118) is exploited in [20] where numerical experiments are also provided.

6. Singularly Perturbed Reaction-Diffusion Problems

We consider the problem

$$\frac{\partial u}{\partial t} = \varepsilon\frac{\partial^2 u}{\partial x^2} + r(u), \qquad (119)$$

$$u(x, 0) = f(x), \ x \in [0, 1], \qquad (120)$$

$$u(0, t) = \eta_0(t), \ u(1, t) = \eta_1(t). \qquad (121)$$

Eqs. (119)-(121) are extensively used in engineering and sciences to model a system on which reaction processes $r(u)$ lead to the diffusion in time of the quantity u. A typical example is

$$\frac{\partial u}{\partial t} = \varepsilon\frac{\partial^2 u}{\partial x^2} + \lambda u(1 - u), \ \lambda > 0, \qquad (122)$$

which, for $\varepsilon = 1$ is the so-called Fisher equation that was originally used to model mutant-gene propagation ([31,32]).

The design of exact schemes for partial differential equations being not possible in general, we decompose (119) into its space independent and stationary sub-equations:

$$\frac{du}{dt} = r(u), \ u(0) = u_0; \qquad (123)$$

$$\varepsilon\frac{d^2 u}{dx^2} + r(u) = 0, \ u(0) = \eta_0, \ u(1) = \eta_1. \qquad (124)$$

The scalar equation (123) is a particular case of Eq. (49) when $\varepsilon = 1$. Thus from the analysis of Section 3, we may approximate Eq. (123) by the non-standard scheme

$$\frac{u^{k+1} - u^k}{\frac{\phi(q\Delta t)}{q}} = r(u^k), \qquad (125)$$

assuming that all roots $\tilde{u}$ of the equation $r(u) = 0$ are such that $r'(\tilde{u}) \neq 0$.

With suitable adaptations of (94), the stationary equation (124) is approximated by the energy-preserving non-standard scheme

$$\varepsilon \frac{u_{m+1} - 2u_m + u_{m-1}}{\psi^2(\Delta x, \varepsilon)} + \gamma(u_{m-1}, u_m, u_{m+1}) = 0, \tag{126}$$

where

$$\psi^2 \equiv \psi^2(\Delta x, \varepsilon) = \frac{4}{\rho^2} \sin^2\left(\frac{\rho \Delta x}{2}\right).$$

By combining (125) and (126), we obtain the following non-standard scheme for (119):

$$\frac{u_m^{k+1} - u_m^k}{\frac{\phi(q \Delta t)}{q}} = \varepsilon \frac{u_{m+1}^k - 2u_m^k + u_{m-1}^k}{\psi^2(\Delta x, \varepsilon)} + \gamma(u_{m-1}^k, u_m^k, u_{m+1}^k). \tag{127}$$

Remark 40: In classical analysis of finite difference methods, the quantities Δt and Δx do not vary independently from each other [33]. We show that the same happens to the above non-standard scheme (127) by applying it to the linear problem

$$\frac{\partial u}{\partial t} = \varepsilon \frac{\partial^2 u}{\partial x^2} + u.$$

For this problem, the scheme (127) reads

$$\frac{u_m^{k+1} - u_m^k}{\phi(\Delta t)} = \varepsilon \frac{u_{m+1}^k - 2u_m^k + u_{m-1}^k}{[\psi(\Delta x)]^2} + u_m^k. \tag{128}$$

We use the Fourier series method [33]. The amplification factor for the scheme (128) is

$$\xi(j) = 1 - 4\varepsilon\nu \sin^2 \frac{j}{2} \Delta x + \phi(\Delta t), \ \forall j \in \mathbb{R},$$

where $\nu = \phi(\Delta t)/[\psi(\Delta x)]^2$. The scheme (128) is stable in the sense of Lax-Richtmyer if and only if the von Neumann condition $|\xi(j)| \leq 1 + K \Delta t$ is satisfied. This condition is met if

$$\nu \leq [1 + \phi(\Delta t)]/2\varepsilon, \ \text{i.e.,} \ \frac{\phi(\Delta t)}{\sin^2\left(\Delta x/2\sqrt{\varepsilon}\right)} \leq 2[1 + \phi(\Delta t)].$$

The following result states further qualitative properties of the non-standard scheme (127):

Theorem 41: *The non-standard scheme (127) is elementary stable in the limit case of the space independent variable. Furthermore, the scheme is stable with respect to the conservation of energy in the stationary case.*

To be specific in the choice of γ in (127), we consider Eq. (122) where $r(u)$ is the logistic growth reaction. We may then take

$$\gamma(u_{m-1}, u_m, u_{m+1}) = \lambda(1 - u_m)\frac{u_{m-1} + u_m + u_{m+1}}{3}. \tag{129}$$

Furthermore, since in this case the exact scheme of (123) is given in (37), we may replace the denominator in the left hand side of (127) with the one in (37). This yields the following non-standard scheme for which Theorem 41 holds:

$$\frac{u_m^{k+1} - u_m^k}{\frac{e^{\lambda \Delta t} - 1}{\lambda}} = \varepsilon \frac{u_{m+1}^k - 2u_m^k + u_{m-1}^k}{\psi^2(\Delta x, \varepsilon)}$$

$$+ \lambda(1 - u_m^k)\frac{u_{m-1}^k + u_m^k + u_{m+1}^k}{3}. \tag{130}$$

Another possible non-standard scheme, for (122), is

$$\frac{u_m^{k+1} - u_m^k}{\frac{e^{\lambda \Delta t} - 1}{\lambda}} = \varepsilon \frac{u_{m+1}^k - 2u_m^k + u_{m-1}^k}{\psi^2(\Delta x, \varepsilon)}$$

$$+ \lambda(1 - u_m^{k+1})\frac{u_{m-1}^k + u_m^k + u_{m+1}^k}{3}. \tag{131}$$

In line with Remark 40, we assume that the time and the space step sizes are related by the formula

$$\frac{\frac{e^{\lambda \Delta t} - 1}{\lambda}}{\frac{4}{\rho^2}\sin^2\left(\frac{\rho \Delta x}{2}\right)} = \frac{1}{2\varepsilon} \quad \text{or} \quad \frac{e^{\lambda \Delta t} - 1}{\sin^2\left(\frac{\sqrt{\lambda}\Delta x}{2\sqrt{\varepsilon}}\right)} = 2, \tag{132}$$

which permits to write (131) in the explicit form

$$u_m^{k+1} = \frac{\frac{1}{2}(u_{m-1}^k + u_{m+1}^k) + (e^{\lambda \Delta t} - 1)\frac{u_{m-1}^k + u_m^k + u_{m+1}^k}{3}}{1 + (e^{\lambda \Delta t} - 1)\frac{u_{m-1}^k + u_m^k + u_{m+1}^k}{3}}. \tag{133}$$

If $0 \leq u_m^k \leq 1$, it follows from (133) that $0 \leq u_m^{k+1} \leq 1$, which shows that the scheme (131) replicates the boundedness and positivity properties

$$0 \leq f \leq 1 \Rightarrow 0 \leq u \leq 1, \tag{134}$$

that the solution of (122) satisfies.

In summary, we have proved the following theorem:

Theorem 42: *Under the condition (132), the scheme (131) is stable with respect to the boundedness and positivity property (134). Furthermore, this scheme is elementary stable in the limit case of the space independent variable and it is also stable with respect to the conservation of energy in the stationary case.*

Figure 11 (page 557) illustrates Theorem 42, whereas Figure 10 (page 556) shows that the standard finite difference scheme (135) fails to preserve all the properties stated in this theorem. We use $\lambda = 1$ and the initial condition as $u(x,0) = 0.5(1 + \sin(2x))$.

The approach leading to Theorem 42 for the specific problem (122) is considered in [23] in the general setting of (119) with $\varepsilon = 1$.

Remark 43: Due to the relation (132), the scheme (131) does not depend on ε as it can be seen from (133).

Remark 44: In the classical setting of the standard finite difference scheme

$$\frac{u_m^{k+1} - u_m^k}{\Delta t} = \varepsilon \frac{u_{m+1}^k - 2u_m^k + u_{m-1}^k}{(\Delta x)^2} + \lambda u_m^k (1 - u_m^k), \qquad (135)$$

the relation (132) would be replaced with $\Delta t/(\Delta x)^2 = 1/2\varepsilon$, which is practically not acceptable as $\Delta t \gg 1$ is too large compared to $\Delta x < 1$.

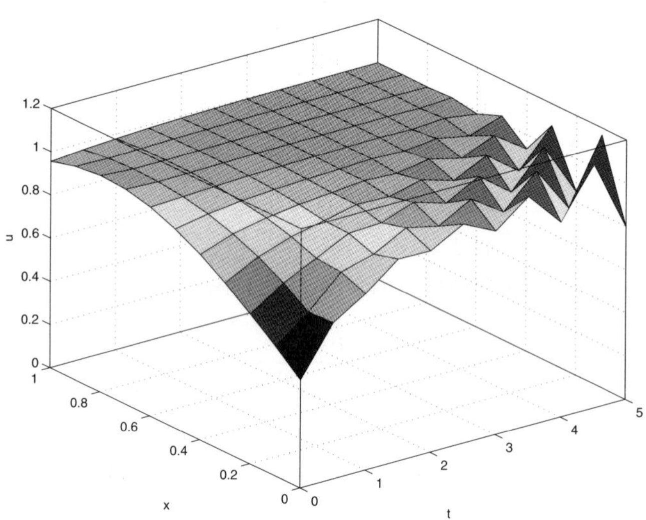

Fig. 10.   Solution of standard scheme (135) for $\varepsilon = 10^{-2}$, $\Delta x = 0.1$ and $\Delta t = 0.5$.

7. Conclusions

One of the first results obtained in this work is the construction of an exact scheme for an autonomous first-order linear system of ordinary differential equations. Motivated by this result, some theoretical aspects of the

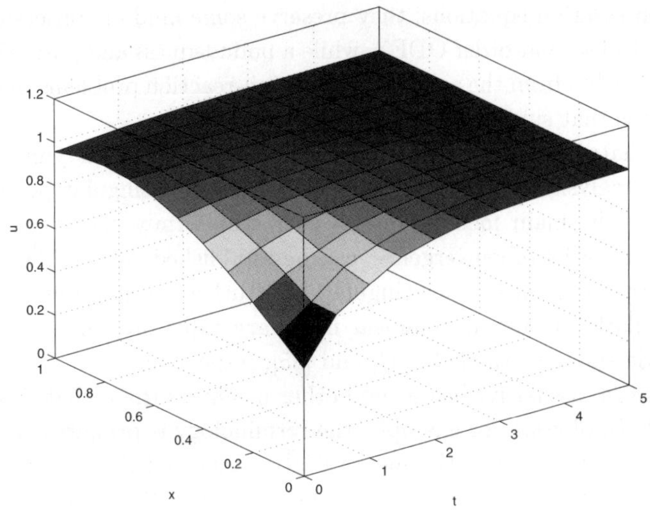

Fig. 11.   Solution of non-standard scheme (133) for $\Delta x = 0.1$ and $\Delta t = 0.5$.

non-standard finite difference method introduced in [10] are extended. In particular, taking also into account the exact schemes of second-order ordinary differential equations in [4,5], the definition of a non-standard finite difference method is revisited.

The theory is applied to singularly perturbed problems defined by ordinary or partial differential equations in which the highest derivative is multiplied by a small positive parameter ε. The solutions of these problems display the interesting physical property of growing or decaying rapidly in one or more narrow "layer region(s)". Of particular interest are the dissipative or dispersive problems, where the rapidly varying component of the solutions grows/decays exponentially (dissipates) or oscillates (disperses) from some discontinuity points as $\varepsilon \to 0$.

In this work, we use Mickens' Rule 2 on the denominator of the discrete derivatives and Rule 3 on the nonlocal approximation of nonlinear terms to design non-standard finite difference schemes, which replicate these physical properties. For a class of linear first-order and second-order ordinary differential equations, our results are in agreement with those in [14,17] regarding fitted methods. That is, the schemes are ε-uniformly convergent of order 1 and 2, respectively. Furthermore, for a fixed ε, the schemes obtained are elementary stable and stable with respect to monotone dependence on initial values in the case of first-order ordinary differential equations and advection reaction equations; they preserve some kind of conservation law in the case of second-order ODE's, while a boundedness and positivity property is inherited from the solution of diffusion-reaction problems. Numerical simulations that support the theory are provided.

As a natural follow-up of this work, we are busy studying the convergence analysis of the schemes presented here for singular perturbation problems. The main focus is on whether ε-uniformly convergent results, with precise order of convergence, can be established. We are also working on non-standard schemes for singularly perturbed turning point problems and singularly perturbed nonlinear boundary value problems treated previously via spline methods in [34] and [35], respectively.

The word dissipative has several other meanings than what is meant in this work. In physics, for example, the terminology is primarily used when there is loss of energy in the mechanical system under consideration (see [16] for some other meanings). The authors and their team are investigating the construction of non-standard finite difference schemes, which are stable with respect to other dissipative properties.

The question of constructing non-standard schemes for first-order differential equations having non-hyperbolic fixed point is resolved in this work, along the lines of [26], in the case of scalar equations. However, the problem is still open for systems [25].

Acknowledgements

The authors would like to thank Prof. R. E. Mickens for inviting them to present their work in this book. The first author is grateful to the South African National Research Foundation (NRF) for financial support within its focus area "Unlocking the Future".

References

1. R. E. Mickens, Difference equation models of differential equations having zero local truncation errors, in: I. W. Knowles and R. T. Lewis (Editors), *Differential Equations*, North-Holland, Amsterdam, 1984, 445–449.
2. R. E. Mickens, Exact solutions to difference equation models of Burgers' equation, *Numerical Methods for Partial Differential Equations* **2** (1986), 123–129.
3. R. E. Mickens, Exact solution to a finite-difference model of a nonlinear reaction-advection equation: implications for numerical analysis, *Numerical Methods for Partial Differential Equations* **5** (1989), 313–325.
4. R. E. Mickens, *Nonstandard Finite Difference Models of Differential Equations*, World Scientific, Singapore, 1994.

5. R. E. Mickens (Editor), *Applications of Nonstandard Finite Difference Schemes*, World Scientific, Singapore, 2000.

6. A. B. Gumel (Guest Editor), *Journal of Difference Equations and Applications* **9**(11-12) (2003), Special Issue dedicated to Prof R. E. Mickens on the occasion of his 60th birthday.

7. K. C. Patidar, On the use of non-standard finite difference methods, *Journal of Difference Equations and Applications*, to appear.

8. Y. Dumont and J. M.-S. Lubuma, Non-standard finite difference methods for vibro-impact problems, *Proceedings of the Royal Society of London Series A: Mathematical, Physical and Engeeniring Sciences*, to appear.

9. R. E. Mickens, Nonstandard finite difference schemes: a status report. In: Y. C. Teng, E. C. Shang, Y. H. Pao, M. H. Schultz and A. D. Pierce (Editors), *Theoretical and Computational Accoustics 97*, World Scientific, Singapore 1999, 419–428.

10. R. Anguelov and J. M-S. Lubuma, Contributions to the mathematics of the nonstandard finite difference method and applications, *Numerical Methods for Partial Differential Equations* **17** (2001), 518–543.

11. M. K. Kadalbajoo and K. C. Patidar, A survey of numerical techniques for solving singulalry perturbed ordinary differential equations, *Applied Mathematics and Computations* **130** (2002), 457–510.

12. M. K. Kadalbajoo and K. C. Patidar, Singularly perturbed problems in partial differential equations: a survey, *Applied Mathematics and Computation* **134** (2003), 371–429.

13. V. D. Liseikin, *Layer Resolving Grids and Transformation for Singular Perturbation Problems*, VSP Brill, Utrecht, 2001.

14. J.J.H. Miller, E. O'Riordan and G.I. Shishkin, *Fitted Numerical Methods for Singular Perturbation Problems*, World Scientific, Singapore, 1996.

15. J. D. Lambert, *Numerical Methods for Ordinary Differential Systems*, John Wiley & Sons, New York, 1991.

16. A. M. Stuart and A. R. Humphries, *Dynamical Systems and Numerical Analysis*, Cambridge University Press, New York, 1998.

17. E. P. Doolan, J. J. H. Miller and W. H. A. Schilders, *Uniform Numerical Methods for Problems with Initial and Boundary Layers*, Boole Press, Dublin, 1980.

18. H. Al-Kahby, F. Dannan and S. Elaydi, Non-standard discretization methods for some biological models, In: R. E. Mickens (Editor), *Applications of Nonstandard Finite Difference Schemes*, World Scientific, Singapore, 2000, 155–180.

19. R. E. Mickens, *Difference Equations: Theory and Applications*, Van Nostrand Reinhold, New York, 1990.

20. H. V. Kojouharov and B. M. Chen, Nonstandard methods for the advection-diffusion-reaction equations. In: R. E. Mickens (Editor), *Applications of Nonstandard Finite Difference Schemes*, World Scientific, Singapore, 2000, 55–108.

21. S. Rucker, Exact finite difference scheme for an advection-reaction equation, *Journal of Difference Equations and Applications* **9** (2003), 1007–1013.

22. A. D. Polynanin and V. F. Zaitsen, *Handbook of Exact Solutions for Ordinary Differential Equations*, CRC Press, New York, 1995.

23. R. Anguelov, P. Kama and J. M.-S. Lubuma, On non-standard finite difference models of reaction-diffusion equations, *Journal of Computational and Applied Mathematics* **175** (2005), 11–29.

24. J. M.-S. Lubuma and A. Roux, An improved theta method for the systems of ordinary differential equations, *Journal of Difference Equations and Applications* **9** (2003), 1023–1035.

25. R. E. Mickens, Nonstandard finite difference schemes for differential equations, *Journal of Difference Equations and Applications* **8** (2002), 823–847.

26. R. Anguelov and J. M.-S. Lubuma, Nonstandard finite difference method by nonlocal approximation, *Mathematics and Computers in Simulation* **61** (2003),465–475.

27. J. M.-S. Lubuma and K. C. Patidar, Non-standard finite difference method for self-adjoint singular perturbation problems, In: T. Simos (Editor), *Proceedings of the International Conference of Computational Methods in Science and Engineering*, (Athens, Greece, 19-23 November 2004), Lectures Series on Computer and Computational Sciences, Vol. 1, VSP International Publishers, Utrecht, 2004, 328- 331.

28. R. Dautray and J. L. Lions, *Mathematical Analysis and Numerical Methods for Science and Technology, in: Evolution Problems II*, Vol. 6, Springer-Verlag, Berlin, 1993.

29. R. D. Richtmyer and K. W. Morton, *Difference Methods for Initial-Value Problems*, Interscience, New York, 1967.

30. R. Anguelov, J. M.-S. Lubuma and S. K. Mahudu, Qualitatively stable finite difference schemes for advection-reaction equations, *Journal of Computational and Applied Mathematics* **158** (2003), 19–30.

31. J. D. Logan, *Nonlinear Differential Equations*, Wiley-Interscience, New York, 1994.

32. J. Murray, *Lectures on Nonlinear Differential Equation Models in Biology*, Clarendon Press, Oxford, 1977.

33. K. W. Morton and D. F. Mayers, *Numerical Solutions of Partial Differential Equations*, Cambridge University Press, London, 1994.

34. M. K. Kadalbajoo and K. C. Patidar, Variable mesh spline approximation method for solving singularly perturbed turning point problems having boundary layer(s), *Computers and Mathematics with Applications* **42** (2001), 1439–1453.

35. M. K. Kadalbajoo and K. C. Patidar, Numerical solution of singularly perturbed non-linear two-point boundary value problems by spline in compression, *International Journal of Computer Mathematics* **79** (2002), 271–288.

CHAPTER 13

FREQUENCY ACCURATE FINITE DIFFERENCE METHODS

A. Louise Perkins

University of Southern Mississippi Gulf Park Campus
louise.perkins@usm.edu

Peter A. Orlin

University of Southern Mississippi, Program in Scientific Computing

Farnaz Zand

University of Southern Mississippi
farnaz.zand@usm.edu

Deriving numerical approximations independently of the partial differential equation in which they appear leaves the discipline of numerical approximations at the experimental trial and error level. Comparison studies are then required to test which approximations work best for a family of equations. The resulting knowledge is not a quantification, but rather an amalgamation of a body of knowledge. Yet this is the current state-of-the-art in numerical models. Such a body of knowledge may provide intuition, but it cannot provide direct guidance.

In contrast, the approach developed here provides direct control of the numerical approximation error. While the numerical approximation error has not been eliminated, it has been reshaped to achieve desired results based directly on the physical properties for the PDE to be simulated.

Specifically, we formulate numerical approximations in terms of undefined coefficients. We then map these approximations to Frequency space where we solve for these undetermined coefficients in a physically meaningful way. The coefficients are then used directly in their respective numerical approximations. We illustrate the technique first on a spatial derivative, then in the context of an evolution equation, where we introduce a time derivative. After expanding the approximations to higher order derivatives, we illustrate the overall method on Burgers Equation.

1. Introduction

The goal of numerical simulations is to emulate salient features of natural phenomena. Traditionally, higher order accurate schemes are employed to more accurately capture the physical behavior. For more than three decades, however, we have known that higher order accurate schemes for evolution equations can be disappointing in practice. In [Hindman, 1982], while investigating the effects of errors due to coordinate transformations (i.e. numerical artifacts), they conclude that "it becomes necessary to question the very meaning of the formal order of accuracy of a method or scheme ... [to] explain the existence of extremely large errors which are formally claimed [to be small]." In this chapter we introduce a paradigm to engineer numerical approximations to more clearly distinguish between the simulated physics and the numerical artifacts. We present a method for designing derivative approximations achieving *a priori* accuracy at given frequencies. The method is then extended to allow us to design derivatives within an equation to produce known dispersion properties. We use a general, average value approximation with undetermined coefficients, together with a set of constraints that ensure convergence and consistency, to formulate a constrained optimal fitting problem both term by term, and for the entire equation. We avoid the optimal fit, however, by constructing a heuristic that produces approximations with well-defined behaviors over which we have control. We demonstrate the final approximation procedure on Burgers Equation. These constraints lead to a linear matrix formulation.

2. Average Value Spatial Difference

Let us begin by viewing Finite Difference Approximation as a resource allocation problem in the frequency domain. While the frequency domain has been widely used for the analysis of the stability of finite difference approximations to PDEs, only recently Lele [6] has shown that significant improvements in accuracy can be achieved by viewing the performance of a compact approximation in the frequency domain and adjusting the coefficients to achieve desired performance. Take a general explicit approximation to the first spatial derivative with undetermined coefficients. Define the average value spatial difference operator (on a regular grid) with these undetermined coefficients as

$$\bar{\Delta}_x(U_m^n) = \left(\frac{1}{\nu} \bar{\Delta}_x^{[+]}(U_m^n) + \frac{\nu - 1}{\nu} \bar{\Delta}_x^{[-]}(U_m^n) \right),$$ (1)

where

$$\bar{\Delta}_x^{[+]}(U_m^n) = \frac{\bar{U}_{m+}^n - U_m^n}{h}, \quad \text{and} \quad \bar{\Delta}_x^{[-]}(U_m^n) = \frac{U_m^n - \bar{U}_{m-}^n}{h}$$ (2)

and

$$\bar{U}_{m+}^n = \sum_{j=0}^{N_+} \psi_j^{[+]} U_{m+j}^n, \quad \text{and} \quad \bar{U}_{m-}^n = \sum_{j=0}^{N_-} \psi_j^{[-]} U_{m-j}^n,$$ (3)

where $\bar{U}_{m+}^n$ and $\bar{U}_{m-}^n$ are average values of $U(x,t)$ located at the right and left of the point m, and the ψ_j s are the undetermined coefficients. The ψ factors weight the approximation to the left and/or right side of the point m.

To verify convergence, consider the left and right approximations independently. Expand the right approximation in a Taylor series about the point m,

$$\bar{\Delta}_x^{[+]}(U_m^n) = \frac{1}{h} \left(\sum_{j=0}^{N_+} \psi_j^{[+]} - 1 \right) U_m^n + \left(\sum_{j=1}^{N_+} j \psi_j^{[+]} \right) \frac{\partial U_m^n}{\partial x}$$
$$+ \sum_{j=1}^{N_+} \sum_{p=2}^{\infty} \left(\frac{j^p h^{p-1}}{p!} \right) \psi_j^{[+]} \frac{\partial^p U_m^n}{\partial x^p}.$$ (4)

A similar expression can be found for the left approximation, $\bar{\Delta}_x^{[-]}(U_m^n)$.

This introduces two constraints:

$$\sum_{j=0}^{N_+} \psi_j^{[+]} = \sum_{j=0}^{N_-} \psi_j^{[-]} = 1$$ (5)

and

$$\sum_{j=1}^{N_+} j \psi_j^{[+]} = \sum_{j=1}^{N_-} j \psi_j^{[-]} = 1.$$ (6)

Eq. (5) are the convergence constraints and Eq. (6) are consistency constraints. They ensure that the approximation converges to the continuous derivative in the limit as $h \to 0$ thereby satisfying Lax's Equivalence Theorem for the approximation. Additional constraints can, of course, be added to achieve higher formal accuracy. Note, only first-order formal accuracy [2] has been achieved with these constraints. Higher orders of accuracy require a concomitant increase in the number of constraints.

We embed the convergence constraints in the approximation using the equalities

$$\psi_0^{[+]} = 1 - \sum_{j=1}^{N_+} \psi_j^{[+]} \quad \text{and} \quad \psi_0^{[-]} = 1 - \sum_{j=1}^{N_-} \psi_j^{[-]}, \tag{7}$$

and define the coefficients as

$$\psi_j^{[S]} = \psi_j^{[+]} + (\nu - 1)\psi_j^{[-]} \quad \text{and} \quad \psi_j^{[D]} = \psi_j^{[+]} - (\nu - 1)\psi_j^{[-]}, \tag{8}$$

where S and D indicate the sum and difference, respectively. Dropping the over-bar for readability, we write the Fourier transform of the result as

$$\hat{\Delta}_x = \hat{G}(\kappa_x)\hat{U}(\xi, f), \tag{9}$$

where

$$\hat{G}(\kappa_x) = -\frac{1}{\nu h} \sum_{j=1}^{N} \left\{ 2\psi_j^{[D]} \sin^2\left(\frac{j\kappa_x}{2}\right) - i\psi_j^{[S]} \sin j\kappa_x \right\}. \tag{10}$$

Here, $\hat{G}(\kappa_x)$ is the complex gain (transfer function) of our average value spatial difference approximation. In addition, we have scaled the wavenumber and temporal frequencies so that $k_x \in [-\pi, \pi]$ is the scaled wavenumber, $\kappa_t = [-\pi, \pi]$ is the scale temporal frequency, and $\xi = \frac{\kappa_x}{2}\pi h$ and $f = \kappa_t/2\pi\kappa$ are the bandwidth limited wavenumber and temporal frequencies (the maximum frequencies representable on the grid without aliasing, the Nyquist frequencies, are given by $\xi_{Nyquist} = \frac{1}{2h}$ and $f_{Nyquist} = \frac{1}{2k}$).

2.1. *Frequency Domain Formulation*

Initially we work with the spatially transformed variable, ξ, and treat the temporally transformed variable, f, as a constant. This allows us to drop the subscript on $\kappa(\kappa = \kappa_x)$, and change variables, $\hat{U}(\kappa) = \hat{U}(\xi, f)$, $\kappa \in [0, \pi]$. To examine error in the frequency domain we need the Fourier transform of a reference function (In other works we used a unit reference function, and re-fit the approximations dependent on the dominant physics [11]). For generality, we define an arbitrary reference function. This facilitates possible trade-offs between the accuracy of the approximation and desired properties.

Let $\hat{\rho}(\kappa)$ be the Fourier transform of the approximated function (the approximant) and $\sigma(\kappa)$ be a complex shaping function, which can be used to modify the properties of the approximant in transform space. Define a reference function in the form

$$\eta(\kappa) \equiv \sigma(\kappa)\hat{\rho}(\kappa), \tag{11}$$

where the Fourier transform of the approximant $\rho(\kappa)$, written as a function of κ, is

$$\hat{\rho}(\kappa) = i\frac{\kappa}{h}. \tag{12}$$

We define a shaping function as

$$\sigma(\kappa) = a(\kappa)e^{ib(\kappa)}. \tag{13}$$

Recalling Eq. (10) and defining the error gain as $G_e(\kappa_x, \kappa_t) = G(\kappa_x, \kappa_t) - \eta(\kappa_x, \kappa_t)$, our frequency space minimization problem, becomes

$$\min_{\psi_j^{[\pm]}} \int_0^{\pi} |h\hat{G}_e(\kappa)|^2 W(\kappa)d\kappa$$

$$\text{Subject to: } \sum_{j=1}^{N} j\psi_j^{[+]} = \sum_{j=1}^{N} j\psi_j^{[-]} = 1, \tag{14}$$

where [using Eq. (10) and (11)]

$$h\hat{G}_e(\kappa) = \frac{1}{\nu}\sum_{j=1}^{N}\left\{ 2\psi_j^{[D]}\sin^2\left(\frac{j\kappa}{2}\right) - i\psi_j^{[S]}\sin j\kappa \right\} + i\kappa a(\kappa)e^{ib(\kappa)}. \tag{15}$$

We note that the grid size (h) does not appear explicitly in Eq. (15); it is implicit in κ as a result of scaling onto the Nyquist interval. We include the weighting factor $W(\kappa)$ in Eq. (14) to provide an estimator for the power spectral density of $U(x,t)$.

2.2. *Frequency Domain Solutions*

We fit coefficients by setting the error gain, $\hat{G}_e(\kappa)$, to zero at selected frequencies, $\kappa_l \in (0,\pi)$ and $l = 1,\dots,N-1$. This is a heuristic choice guided by experimentation. Using Eq. (15), we obtain $N-1$ equations, one for each l,

$$\frac{1}{\nu}\sum_{j=1}^{N}\left\{ 2\psi_j^{[D]}\sin^2\left(\frac{j\kappa_l}{2}\right) - i\psi_j^{[S]}\sin j\kappa_l \right\} + i\kappa_l a(\kappa_l)e^{ib(\kappa_l)} = 0. \tag{16}$$

Adding the consistency constraint equations, which are not embedded in Eq. (16), results in N equation in N unknowns.

Write the consistency constraint equations as

$$\sum_{j=1}^{N} j\psi_j^{[S]} = \nu \ \text{ and } \ \sum_{j=1}^{N} j\psi_j^{[D]} = 2 - \nu. \tag{17}$$

Then rewrite the relations in Eq. (16) in matrix form,

$$\begin{aligned}
\Psi^{[D]} &= \nu[S_2]^{-1}K^{[D]} \\
\Psi^{[S]} &= \nu[S_1]^{-1}K^{[S]},
\end{aligned} \tag{18}$$

where

$$[S_2] = \begin{bmatrix}
2\sin^2(\frac{\kappa_l}{2}) & \cdots & 2\sin^2(\frac{N_{\kappa_1}}{2}) \\
\vdots & & \vdots \\
2\sin^2(\frac{\kappa_{N-1}}{2}) & \cdots & 2\sin^2\left(\frac{N_{\kappa_{N-1}}}{2}\right) \\
1 & \cdots & N
\end{bmatrix} \tag{19}$$

$$[S_1] = \begin{bmatrix}
\sin(\frac{\kappa_l}{2}) & \cdots & \sin(\frac{N_{\kappa_1}}{2}) \\
\vdots & & \vdots \\
\sin(\frac{\kappa_{N-1}}{2}) & \cdots & \sin\left(\frac{N_{\kappa_{N-1}}}{2}\right) \\
1 & \cdots & N
\end{bmatrix} \tag{20}$$

$$\Psi^{[D]} = \left\{\begin{matrix} \psi_1^{[D]} \\ \vdots \\ \psi_N^{[D]} \end{matrix}\right\}, \qquad \Psi^{[S]} = \left\{\begin{matrix} \psi_1^{[S]} \\ \vdots \\ \psi_N^{[S]} \end{matrix}\right\}, \tag{21}$$

$$K^{[D]} = \left\{\begin{matrix} \kappa_1 a(\kappa_1)\sin b(\kappa_1) \\ \vdots \\ \kappa_{N-1}a(\kappa_{N-1})\sin b(\kappa_{N-1}) \\ \frac{2-\nu}{\nu} \end{matrix}\right\} \tag{22}$$

and

$$K^{[S]} = \left\{\begin{matrix} \kappa_1 a(\kappa_1)\cos b(\kappa_1) \\ \vdots \\ \kappa_{N-1}a(\kappa_{N-1})\cos b(\kappa_{N-1}) \\ 1 \end{matrix}\right\} \tag{23}$$

We note that $[S_2]$ and $[S_1]$ are Vandermonde-like matrices so that their inverses exist whenever $\kappa_l \neq 0$, $\kappa_l \neq \pi$, and $\kappa_l \neq \kappa_m$.

The fitting coefficient vectors $\boldsymbol{\Psi}^{[+]}$, $(\boldsymbol{\Psi}_l^{[+]} = \psi_l^{[+]})$, and $\boldsymbol{\Psi}^{[-]}$, $(\boldsymbol{\Psi}_l^{[-]} = \psi_l^{[-]})$, $l = 1, \ldots, N$, using Eq. (8) are

$$\boldsymbol{\Psi}^{[+]} = \frac{\nu}{2}([S_1]^{-1}\mathbf{K}^{[S]} + [S_2]^{-1}\mathbf{K}^{[D]})$$

$$\boldsymbol{\Psi}^{[-]} = \frac{\nu}{2(\nu - 1)}([S]^{-1}\mathbf{K}^{[S]} - [S_2]^{-1}\mathbf{K}^{[D]}). \tag{24}$$

The fitting coefficient, ψ_0, can be found from Eq. (7),

$$\psi_0 = -\frac{\nu}{2}\mathbf{1}^T[S_2]^{-1}\mathbf{K}^{[D]}, \tag{25}$$

where $\mathbf{1}^T$ is a $1 \times N$ row vector with all elements equal to 1. Applying difference schemes with a large stencil size, such as described here, raises the question of the treatment of boundary conditions. This has been addressed by Lele [6] in connection with his compact finite approximation to the first spatial derivative. Following a similar analysis, the methods presented there can be applied to accommodate a variety of boundary conditions. Essentially, the stencil size is reduced on one side of the central point near the boundary. This is accomplished either by limiting the number of points on the left or right side of the approximation, as appropriate, or by adjusting the weighting coefficient (ν) to produce negligible coefficient weights on one side of the approximation.

We specify the spatial frequencies of the zeros, the value of ν in Eq. (16) and the reference function, Eq. (11), to solve the matrix problem given in Eq. (24) and Eq. (25). We pursue the design interactively by manually adjusting the locations of the zero error gain frequencies and ν. In so doing we are assuming that the dominant frequencies of the modeled phenomena are located in bands about the selected frequencies (spatial scales). We require that the amplitude and phase properties at the selected frequencies remain close to those of the true spatial derivative.

If we choose to focus the design on matching phase angle, which controls the accuracy of phase velocity in wave-like PDEs [15], our method will be frequency accurate across a prespecified bandwidth. For fits with 2, 3, and 4 zeros, an interactive graphical method was adequate for quickly arriving at a useful design. Additional error measures such as the magnitude and the phase angle of the integrand of the approximation $[h\hat{G}_e(\kappa)$, Eq. (15)] and the L_2 norm of the error [the first of Eq. (14)] may also be used.

Let us consider two design cases. The first design provides useful fitting accuracy over the widest spatial frequency range (an allpass design). This is the most commonly pursued design, because the power spectral density (PSD) of the solution is assumed to be uniform over the entire spatial frequency interval (cf. [6] [3] [16] [5]). The second design provides high accuracy over the low spatial frequency region and special properties over the high frequency region (a lowpass design). This design is useful in providing controlled levels of artificial viscosity at high frequencies when simulating PDEs with wave-like behavior such as the advection equation. The first design is a centered approximation, which exhibits no phase error. The second is an upwind design, which provides controlled phase error over the high spatial frequency region.

The approximation amplitude (reduced wave number) of the allpass general average derivative approximation for stencil sizes of 6, 8, and 10 points is shown in Fig. 1. There it is compared to a 10 point stencil central difference approximation and an implicit 6 point stencil optimized second-order tridiagonal scheme developed by Kim and Lee [5]. We set $\alpha(\kappa) \equiv 1$, $b(\kappa) \equiv 0$, $\nu = 2$, and estimate the PSD of U as a constant so that $W(\kappa) \equiv 1$. The highest frequency zero was placed at a frequency at which oscillations in the amplitude response due to Gibb's phenomena begin to appear. The zeros were uniformly space and the lowest zero was located at a point that reduced overall relative amplitude error of the approximation, $\epsilon_A(\kappa)$,

$$\epsilon_A(\kappa) = \frac{|h\hat{G}(\kappa)| - \kappa}{\kappa}, \tag{26}$$

over the widest range of spatial frequencies. Both the general average approximation and Kim and Lee's approximation clearly provide better frequency domain performance than higher-order central difference approximations while producing the required phase angle of $\pi/2$.

In Fig. 2, we compare the relative amplitude error of a 10 stencil point general average approximation, an optimized compact second-order tridiagonal approximation with a 6 point stencil, and a 10 point stencil central difference approximation. The zero error frequencies of the compact and general average approximations correspond to the negative cusps of their relative error curves. Both the general average and compact schemes provide better accuracy over a wider range of spatial frequencies than central difference schemes. The placement of the upper zero in the general average approximation determines the height of the relative error maxima. The placement of the lower zero, for uniformly spaced zeros, determines the relative amplitude of each of the maxima. The compact scheme achieves lower overall relative error maxima, because it has zeros in the denominator of its response, permitting the placement of numerator zeros closer to $\kappa = \pi$.

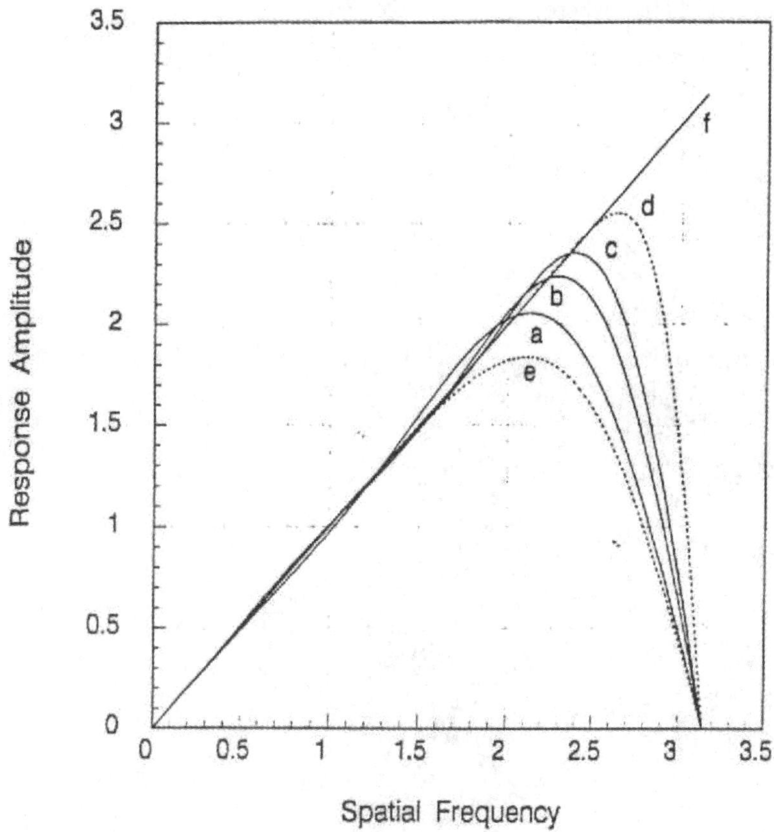

Fig. 1.   Central difference approximation response amplitude: $a = 6$ stencil point explicit general average approximation; $b = 8$ stencil point explicit general average approximation; $c = 10$ point stencil explicit general average approximation; $d = 6$ stencil point second-order, tri-diagonal implicit compact fit [5]; and $e = 10$ stencil point Taylor Series approximation.

A lowpass deviative design can be implemented by selecting a reference function that possesses a small real part of the approximation gain ($\hat{G}(\kappa)$) below a selected spatial frequency (κ_C), and an appreciable real component above this frequency. This choice is made to support use of the approximation in PDEs that have wave-like solutions [15]. For these PDEs, the positive real part of the approximation becomes a real negative exponent of e in the numerical solution. The selected shaping function is

$\sigma(\kappa) = e^{i\phi_R} f(\kappa)$, where

$$f(\kappa) = \begin{cases} 1 & \text{if } 0 \leq \kappa \leq \kappa_C \\ e^{-i(\kappa - \kappa_C)} & \text{if } \kappa_C < \kappa \leq \pi, \end{cases} \tag{27}$$

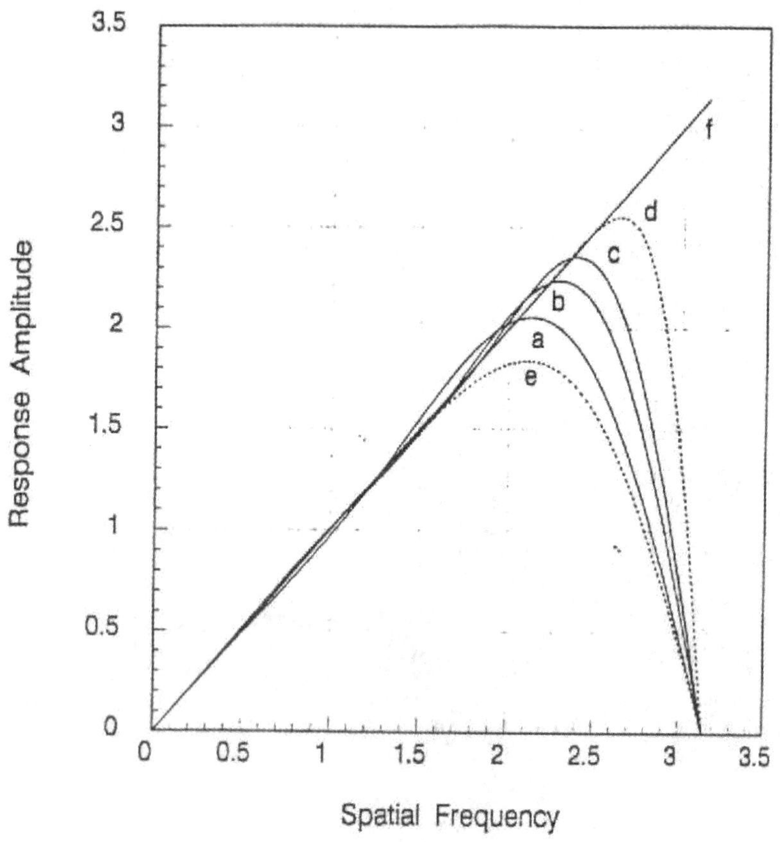

Fig. 2. Central difference relative amplitude error: Here $a = 10$ stencil point explicit general avearage approximation; $b = 6$ stencil point second-order, tridiagonal implicit compact fit [5]; and $c = 10$ stencil point Taylor Series approximation.

resulting in a reference function with $\alpha(\kappa) \equiv 1$ and $b(\kappa)$, the reference phase angle, varying linearly with κ for $\kappa > \kappa_C$. The factor ϕ_R is used to introduce a constant phase angle in the reference function. We assume that $W(\kappa)$ is unity for $\kappa \leq \kappa_C$ and negligible for $\kappa > \kappa_C$ (this design choice is in anticipation of the use of the approximation in the nonlinear Burgers equation).

The designs presented so far hold the relative phase error of the approximation within a specified tolerance. We defined the phase angle of the approximation as $\phi(\kappa) = \angle h\hat{G}(\kappa)$ so that, for the first spatial derivative, the relative phase error becomes

$$\epsilon_\phi(\kappa) = \frac{\phi(\kappa)/\pi - 0.5}{0.5}. \tag{28}$$

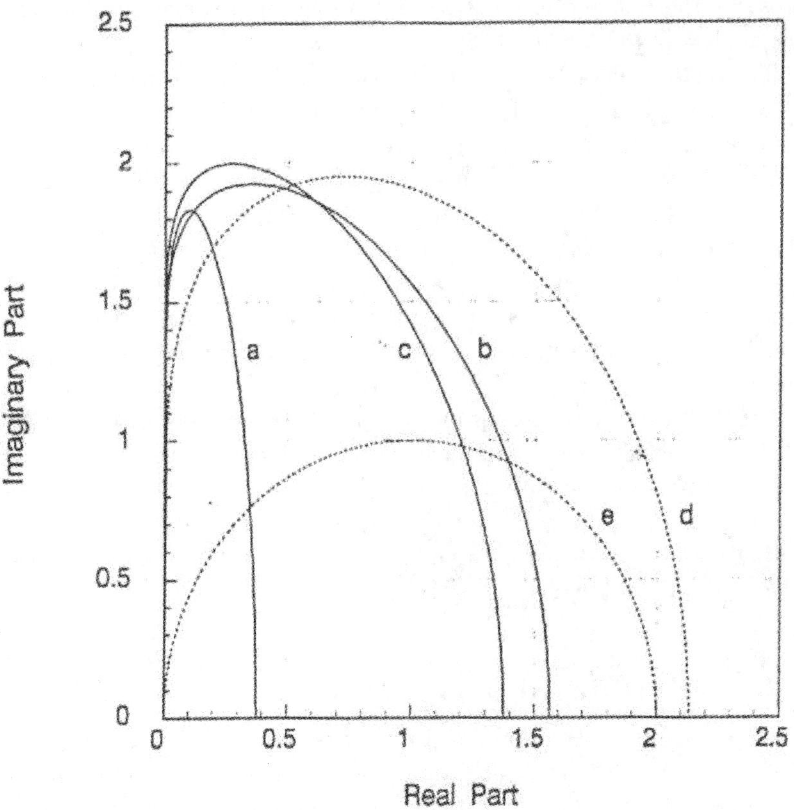

Fig. 3. Upwind difference approximation complex response. Here $a = 7$ stencil point explicit general average approximation; $b = 9$ stencil point explicit general average approximation, $c = 11$ pint stencil explicit general approximation; $d = 7$ stencil point upwind approximation [2]; and $e = 2$ stencil point upwind approximation.

The complex response of upwind general average designs for 7, 9, and 11 stencil point approximations are presented in Fig. 3, where they are compared with the 7 point stencil, sixth-order approximation developed by Collatz ([2], Appendix, Table III, un-symmetric), and a 2 stencil point, first order upwind approximation. The central difference approximation has a response that lies entirely on the positive imaginary axis. The upwind approximation designs presented here attempt to hold the approximation response near the positive imaginary axis for spatial frequencies less that κ_C and to produce a large real component for spatial frequencies greater than κ_C. This is accomplished by adjusting ν to control the location of the foot of the response curve on the real axis and by placement of the error zeros. For the 7, 9, and 11 point general average approximations, the values of ν are $\nu = [2.139, 2.505, 2.368]$.

The relative phase and amplitude errors of the approximations (ϵ_ϕ and ϵ_A) are shown in Figs. 4 and 5 for $\kappa < \pi/2$. The curves are smooth beyond this point to their limiting values, $\epsilon_\phi(\pi) = -1$ and $\epsilon_A(\pi) = 1$. The 7 and 9 stencil point approximations were designed to produce a maximum relative phase error of 0.5% for $\kappa \leq \kappa_C$. The 11 point design has a maximum relative phase error of 0.1% over the same interval. The error zeros, marked by the o symbols, were placed at the points of maximum positive relative phase error. The value of ϕ_R was set to 10^{-4} radians in each approximation to ensure that the approximations were dissipative over the entire Nyquist interval.

For PDEs with wave-like properties, the design process involves trading off the magnitude of the exponent of the real part of the solution (the dissipation) with the level of relative phase error in the approximation. Lower levels of dissipation produce smoother fits for $\kappa \leq \kappa_C$. In general, the behavior of the relative amplitude error, for $\kappa \leq \kappa_C$, mirrors that of the relative phase error.

3. Temporal Derivative

The selection of the temporal derivative approximation in a finite difference scheme is a key element in achieving high accuracy because the temporal derivative is used to propagate the solution forward in time based on values of the solution at previous time steps and on the value of the solution at nearby spatial points. For situations where the spatial extent of the grid is fixed and the intent is to achieve the maximum temporal step size, truncation error order is not always a satisfying *ab initio* estimator of simulation

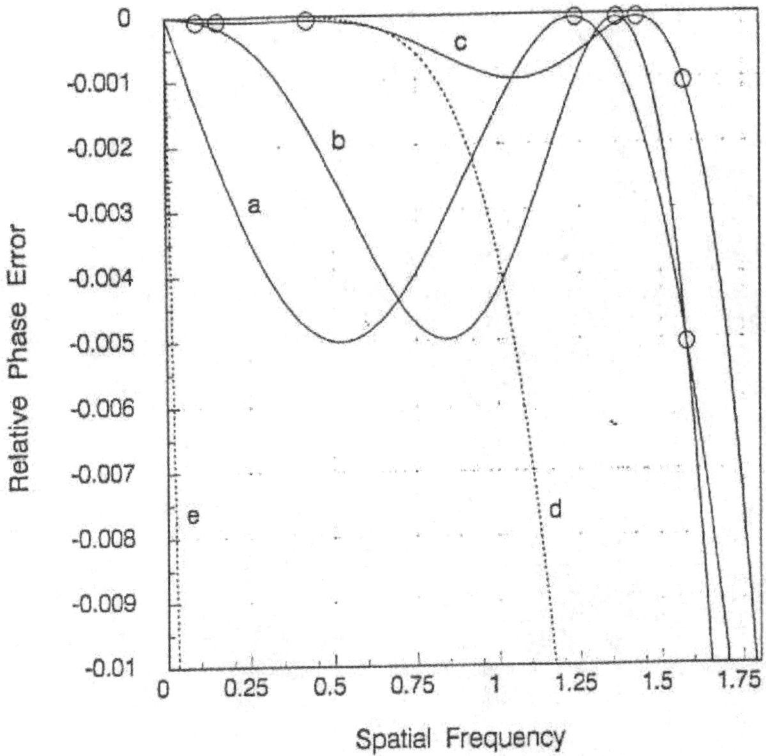

Fig. 4. Upwind difference approximation relative phase error: $a = 7$ stencil point explicit general average approximation; $b = 9$ stencil point explicit general average approximation; $c = 11$ point stencil explicit general average approximation; $d = 7$ stencil point upwind approximation [2]; and $e = 2$ stencil point upwind approximation.

error.

In pursuing the design, the focus is on the use of the approximation in partial differential equations(PDEs) with wave-like solution [15]. The dispersion relation for the continuous PDE must also be known to complete the fitting of the approximation. This approach is pursued to develop temporal approximations useful in finite difference schemes (FDSs) for full PDEs where both the spatial and temporal derivative approximations are fitted. A general explicit approximation is developed with undetermined coefficients and cast into a fitting problem in optimal form. Rather than

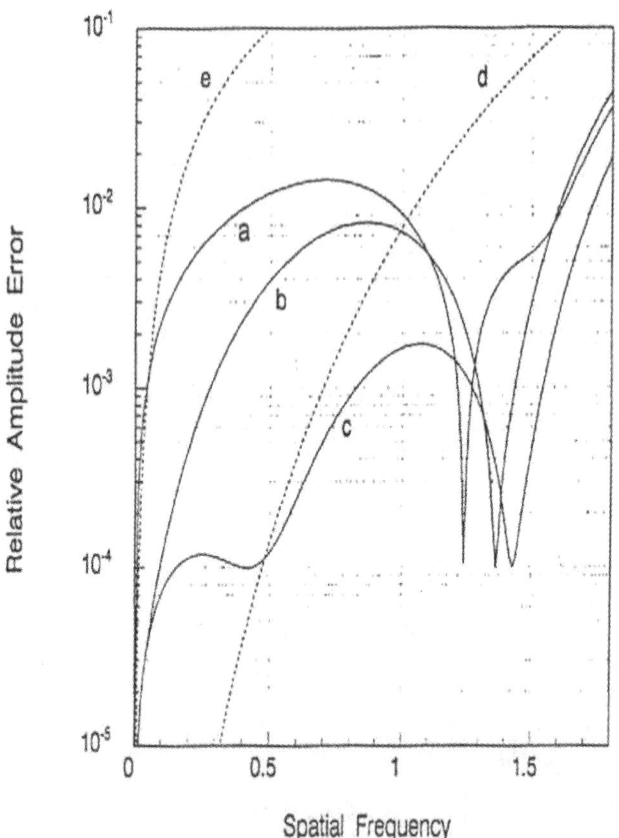

Fig. 5.   Upwind difference approximation relative phase error. Here $a = 7$ stencil point explicit general average approximation; $b = 9$ stencil point explicit general average approximation; $c = 11$ point stencil explicit general average approximation; $d = 7$ stencil point upwind approximation [2]; and $e = 2$ stencil point upwind approximation.

determine the optimal coefficients, a heuristic method based on zero error placement is used, which results in a simple linear matrix formulation, as was the case in the last section.

3.1. *Average Value Approximation*

A one-step explicit general average temporal derivative approximation is defined on a uniform spatial and temporal grid with spatial and temporal grid spacing h and k, where $U_m^n = U(mh, nk)$, analogous to the previous section,

$$\bar{\Delta}_t(U_m^n) = \frac{1}{\mu}\bar{\Delta}_t^{[+]}(U_m^n) + \frac{\mu-1}{\mu}\bar{\Delta}_x^{[-]}(U_m^n), \tag{29}$$

where

$$\bar{\Delta}_t^{[+]}(U_m^n) = \frac{1}{k}\left(U_m^{n+1} - \sum_{j=0}^{N}\theta_j^{[+]}U_{m+j}^n\right) \tag{30}$$

and

$$\bar{\Delta}_t^{[-]}(U_m^n) = \frac{1}{k}\left(U_m^{n+1} - \sum_{j=0}^{N}\theta_j^{[-]}U_{m-j}^n\right). \tag{31}$$

Here, values of $U(x,t)$ located to the right and left of the point m are subjected to weighted averaging. The weights $\theta_j^{[\pm]}$ are, again, the quantities to be determined. The factor μ is included to adjust coefficient weights about the central point m.

The general average approximation must converge as $k \to 0$. It must also satisfy the Lax Equivalence Theorem (see Ref. [12]) by converging to U_t. To insure this, convergence and consistency constraints are again placed on the values of the fitted $\phi_j^{[\pm]}$. Using the definition of the temporal derivative approximation in Eq. (29) through (31), the convergence and consistency constraints are found separately for $\bar{\Delta}_t^{[+]}(U_m^n)$ and $\bar{\Delta}_t^{[-]}(U_m^n)$. The U_m^{n+1} term is expanded in a temporal series which includes the first temporal derivative. The $U_{m\pm j}$ is expanded in a spatial series. The expansion of $\bar{\Delta}_t^{[+]}(U_m^n)$ produces

$$\bar{\Delta}_t^{[+]}(U_m^n) = \sum_{p=0}^{\infty}\frac{k^{p-1}}{p!}\left[(U_m^n)_{pt} - \alpha^p\left(\sum_{j=0}^{N}j^p\theta_j^{[+]}\right)(U_m^n)_{px}\right], \tag{32}$$

where $\alpha = h/K$ is the "grid speed" resulting from the discretization which is held constant while taking the limit. Thus, the approximation is conditionally consistent. The notation $(U_m^n)_{pt}$ is used to denote the pth temporal partial derivative of $U(x,t)$ evaluated at the point $x = mh$ and $t = nk$. A similar notation is used for the pth spatial partial derivative, $(U_m^n)_{px}$.

The first ($p = 0$) term of Eq. (32) expands to

$$\kappa^{-1}\left[U_m^n - \left(\sum_{j=0}^{N}\theta_j^{[+]}\right)U_m^n\right]. \tag{33}$$

The series will converge if this term is zero, thus resulting in the convergence constraint

$$\sum_{j=0}^{N} \theta_j^{[+]} = 1.$$

(34)

The second ($p = 1$) term of Eq. (32) expands to

$$(U_m^n)_t - \alpha \left(\sum_{j=1}^{N} j\theta_j^{[+]} \right) (U_m^n)_x.$$

(35)

The series will be consistent with the continuous temporal partial derivative if

$$\sum_{j=1}^{N} j\theta_j^{[+]} = 0.$$

(36)

A similar analysis for $\bar{\Delta}_t^{[-]}$ results in

$$\sum_{j=0}^{N} \theta_j^{[-]} = 1 \quad \text{and} \quad \sum_{j=1}^{N} j\theta_j^{[-]} = 0$$

(37)

for the convergence and consistency constraints.

Writing the convergence constraints (Eq. (34) and the first of Eq. (37)) as

$$\theta_0^{[+]} = 1 - \sum_{j=1}^{N} \theta_j^{[+]} \quad \text{and} \quad \theta_0^{[-]} = 1 - \sum_{j=1}^{N} \theta_j^{[-]},$$

(38)

defining

$$\theta_j^{[S]} = \theta_j^{[+]} + (\mu - 1)\theta_j^{[-]} \quad \text{and} \quad \theta_j^{[D]} = \theta_j^{[+]} - (\mu - 1)\theta_j^{[-]},$$

(39)

substituting Eq. (38) in Eq. (29), and taking the Fourier transform results in

$$\hat{\bar{\Delta}}_t(\kappa_x, \kappa_t) = \hat{G}(\kappa_x, \kappa_t)\hat{U}(\xi, f),$$

(40)

where

$$\hat{G}(\kappa_x, \kappa_t) = \frac{1}{k}\left[z_t - 1 + \frac{1}{\mu}\left(2\sum_{j=1}^{N} \theta_j^{[S]} \sin^2\left(\frac{jk_x}{2} \right) - i\sum_{j=1}^{N} \theta_j^{[D]} \sin j\kappa_x \right) \right],$$

(41)

subject to the constraints

$$\sum_{j=1}^{N} j\theta_j^{[S]} = \sum_{j=1}^{N} j\theta_j^{[D]} = 0, \tag{42}$$

where κ_x, $\kappa_x \in [-\pi, \pi]$ are the scaled wavenumber and temporal frequency, $\xi = \kappa_x/2\phi$ and $f = \kappa_t/2\pi k$ are the bandwidth limited wavenumber and temporal frequency, and $z_t = e^{ikt}$ and $z_t = e^{ik_x}$. The maximum.wavenumber and temporal frequency representable on the grid without aliasing, the Nyquist frequencies, are given by $\xi_{Nyquist} = 1/2h$ and $f_{Nyquist} = 1/2k$.

3.2. *Frequency Domain Formulation*

The weighting coefficients, $\phi_j^{[\pm]}$, are fitted to the temporal derivative in the frequency domain. The truncation error at the point (m, n) is

$$\hat{e}_T(x_m, t_n) = \bar{\Delta}_t(U_m^n) - U_t(x_m, t_n). \tag{43}$$

If $U(x, t)$ is linear, the Fourier transforms of each of the terms of Eq. (43) can be written symbolically as

$$\hat{\bar{\Delta}}_t = G(\kappa_x, \kappa_t)\hat{U}(\xi, f) \quad \text{and} \quad \hat{U}_t = \rho(\kappa_x, \kappa_t)]U(\xi, f). \tag{44}$$

If $U(x, t)$ is band-limited, the Fourier transform of the truncation error is

$$\hat{e}_T(\xi, f) = [G(\kappa_x, \kappa_t) - \rho(\kappa_x, \kappa_t)]U(\xi, f). \tag{45}$$

Requiring that the approximation error and its Fourier transform satisfies the Parseval identity (see Ref. [1]), the space and time domain error can be related to the wavenumber and frequency domain error by

$$hk \sum_{m=-\infty}^{\infty} \sum_{n=-\infty}^{\infty} e_T(x_m, t_n)^2 = \int_{-\frac{1}{2h}}^{\frac{1}{2h}} \int_{-\frac{1}{2\kappa}}^{\frac{1}{2\kappa}} |\hat{e}_T(\xi, f)|^2 d\xi df. \tag{46}$$

Before formally stating the fitting problem, the form of $\rho(\kappa_x, \kappa_t)$ is modified to support greater flexibility in fitting the approximation. This is done by defining the reference function as

$$\eta(\kappa_x, \kappa_t) = \zeta(\kappa_x)\rho(\kappa_t), \tag{47}$$

where $\zeta(\kappa_x)$, a shaping function is

$$\zeta(\kappa_x) \equiv a(\kappa_x)e^{ib(\kappa_x)}, \tag{48}$$

the function $a(\kappa_x)$ and $b(\kappa_x)$ are user-specified, and

$$\rho(\kappa_t) = \frac{i}{k}\kappa_t. \tag{49}$$

Recalling Eq. (41) and defining the gain error as $G_e(\kappa_x, \kappa_t) = G(\kappa_x, \kappa_t) - \eta(\kappa_x, \kappa_t)$, $G_e(\kappa_x, \kappa_t)$, it becomes

$$G_e(\kappa_x, \kappa_t) = \frac{1}{k}\left[z_t - 1 + \frac{1}{\mu}\left(2\sum_{j=1}^{N} \theta_j^{[S]} \sin^2\left(\frac{j\kappa_x}{2}\right) - i\sum_{j=1}^{N} \theta_j^{[D]} \sin j\kappa x \right) \right.$$
$$\left. - i\kappa_t a(\kappa_x)e^{ib(\kappa_x)} \right]. \tag{50}$$

3.3. *Temporal Frequency Specification*

Determining the value of G_e requires the specification of both the scaled wavenumber, κ_x, and the scales temporal frequency, κ_t. As formulated in Eq. (50), these two variables are independent so that any combination would appear to be suitable in determining the fitting coefficients of the temporal approximation. However, for linear PDEs, Fourier transforming the PDE results in an algebraic dispersion relation between the wavenumber and frequency. For a number of useful bandlimited linear PDEs with wavelike properties, the discrete wavenumber and temporal frequency are related by (see Ref. [15])

$$\kappa_t = g(\kappa_x) \quad \text{and} \quad z_t = z_x^{g(\kappa_x)/\kappa_x}. \tag{51}$$

The simplest PDE of this type we can consider is the advection equation, $U_t + cU_x = 0$, which, if the solution is suitably bandlimited, has discrete dispersion relations given by

$$\kappa_t = -\sigma\kappa_x \quad \text{and} \quad z_t = z_t^{-\sigma}, \tag{52}$$

where $\sigma = ck/h$ is the Courant number.

Using the relations $\xi = \kappa_x/2\phi$ and $f = \kappa_x/2\phi$, the first of Eq. (52) becomes $f = -c\xi$ when $\sigma \equiv 1$, the dispersion relation for the underlying continuous PDE. This is the dispersion relation preserving (DRP) criterion developed by Tam *et al.* [14] for producing a finite difference scheme which has a dispersion relation identical to the underlying PDE (Tam's definition of the DRP condition is applicable only to linear PDEs). A common practice is to develop and test approximations for linearized forms of a PDE which are then used to simulate the nonlinear PDE. This, in most instances, requires that $\sigma < 1$ which violates the DRP criterion and can result in approximations which are very accurate for the linearized PDE, but not adequately robust for the full nonlinear PDE.

Using Eq. (52), Eq. (50) becomes a function of only the scaled wavenumber. Noting this, drop the subscript on the scale wavenumber, estimate $\hat{U}(\xi, f)$ with the weighting factor $W(\kappa)$ and state the minimization problem for an explicit temporal approximation for the advection equation as

$$\min_{\theta_j^{[\pm]}} \int_0^{\pi} |k\hat{G}_e(\kappa)|^2 W(\kappa) d\kappa.$$

Subject to:

$$\sum_{j=1}^{N} j\theta_j^{[S]} = \sum_{j=1}^{N} j\theta_j^{[D]} = 0, \tag{53}$$

where

$$k\hat{G}_e(\kappa) = \sum_{j=1}^{N} \left\{ z^{-\sigma} - 1 + \frac{1}{\mu} \left(2\theta_j^{[D]} \sin^2\left(\frac{j\kappa}{2}\right) - i\theta_j^{[S]} \sin j\kappa \right) \right\}$$
$$+ i\sigma\kappa a(\kappa) e^{ib(\kappa)}. \tag{54}$$

3.4. *Frequency Domain Solutions*

The coefficients are fitted by setting the error gain, $\hat{G}_e(\kappa)$, to zero at selected frequencies, $\kappa_l \in (0, \pi)$ and $l = 1, \ldots, N - 1$. Use the following equations, one for each l,

$$2\sin^2\left(\frac{\sigma\kappa_l}{2}\right) + \frac{2}{\mu} \sum_{j=1}^{N} \theta_j^{[D]} \sin^2\left(\frac{j\kappa_l}{2}\right) - \sigma\kappa_l a(\kappa_l) \sin(b(\kappa_l))$$
$$-i\left\{ \sin\sigma\kappa_l + \frac{1}{\mu} \sum_{j=1}^{N} \theta_j^{[S]} \sin j\kappa_l - \sigma\kappa_l a(\kappa_l) \cos(b(\kappa_l)) \right\} = 0, \tag{55}$$

and add the consistency constraint equations, Eq. (42), to again produce N equations in N unknowns.

Separate Eq. (55) into its real and imaginary parts. Cast the two sets of equations into matrix form and solve for the differences and sums of the fitting coefficients,

$$\Theta^{[D]} = \frac{\mu}{2}[S_2]^{-1}\mathbf{K}_\theta^{[D]}$$
$$\Theta^{[S]} = \mu[S_1]^{-1}\mathbf{K}_\theta^{[S]}, \tag{56}$$

where

$$[S_2] = \begin{bmatrix} \sin^2(\frac{\kappa_l}{2}) & \cdots & \sin^2(\frac{N_{\kappa_1}}{2}) \\ \vdots & & \vdots \\ \sin^2(\frac{\kappa_{N-1}}{2}) & \cdots & \sin^2\left(\frac{N_{\kappa_{N-1}}}{2}\right) \\ 1 & \cdots & N \end{bmatrix},$$
(57)

$$[S_1] = \begin{bmatrix} \sin(\frac{\kappa_l}{2}) & \cdots & \sin(\frac{N_{\kappa_1}}{2}) \\ \vdots & & \vdots \\ \sin(\frac{\kappa_{N-1}}{2}) & \cdots & \sin\left(\frac{N_{\kappa_{N-1}}}{2}\right) \\ 1 & \cdots & N \end{bmatrix},$$
(58)

$$\Theta^{[D]} = \left\{ \begin{array}{c} \theta_1^{[D]} \\ \vdots \\ \theta_N^{[D]} \end{array} \right\}, \qquad \Theta^{[S]} = \left\{ \begin{array}{c} \theta_1^{[S]} \\ \vdots \\ \theta_N^{[S]} \end{array} \right\},$$
(59)

$$K_\theta^{[D]} = \left\{ \begin{array}{c} \sigma\kappa_1 a \sin b(\kappa_1) + 2\sin^2\left(\frac{\sigma\kappa_1}{2}\right) \\ \vdots \\ \sigma\kappa_{N-1} a(\kappa_{N-1})\sigma \sin b(\kappa_{N-1}) + 2\sin^2\left(\frac{\sigma\kappa_{N-1}}{2}\right) \\ 0 \end{array} \right\},$$
(60)

and

$$K_\theta^{[S]} = \left\{ \begin{array}{c} \sigma\kappa_1 a(\kappa_1)\cos b(\kappa_1) - \sin \sigma\kappa_1 \\ \vdots \\ \sigma\kappa_{N-1} a(\kappa_{N-1})\cos b(\kappa_{N-1}) - \sin \sigma\kappa_{N-1} \\ 0 \end{array} \right\}.$$
(61)

Note that $[S_2]$ and $[S_1]$ are again Vandermonde-like matrices. Their inverses exist whenever $\kappa_l \neq 0$, $\kappa_l \neq \pi$ and $\kappa_l \neq \kappa_m$.

The fitting coefficients vectors $\Theta^{[+]}$, $((\Theta^{[+]})_l = \theta_l^{[+]})$ and $\Theta^{[-]}$, $((\Theta^{[-]})_l = \theta_l^{[-]})$, $l = 1, \ldots, N$, using Eq. (39), are

$$\Theta^{[+]} = \frac{\mu}{2}\left([S_1]^{-1}\mathbf{K}_\theta^{[S]} + \frac{1}{2}[S_2]^{-1}\mathbf{K}_\theta^{[D]} \right)$$

$$\Theta^{[-]} = \frac{\mu}{2(\mu-1)}\left([S_1]^{-1}\mathbf{K}_\theta^{[S]} + \frac{1}{2}[S_2]^{-1}\mathbf{K}_\theta^{[D]} \right).$$
(62)

The center coefficient is calculated from Eq. (38) and (39) as

$$\theta_0 = 1 - \frac{1}{\mu} \sum_{j=1}^{N} \theta_j^{[S]}. \tag{63}$$

3.5. *Fitting Example*

To illustrate the design process, an approximation is fitted which provides accurate phase and amplitude approximations over the lower half of the Nyquist interval, $\kappa \leq \pi/2$, and controlled error over the upper half, $\kappa > \pi/2$. This is referred to as a lowpass design since the advection equation is used for the underlying PDE and the error is placed in the upper portion of the Nyquist interval in order to provide dissipation when used with an appropriate spatial approximation. As such, the design is a single time level high order Euler approximation. Example fits for 7, 9 and 11 stencil point approximations are developed to provide phase and amplitude accurate approximations for scaled wavenumber in the lower half of the spatial Nyquist interval. Because the advection equation was chosen to particularize the approximation, a large negative value of the imaginary part of the frequency domain response in the upper spatial Nyquist interval is required to produce numerical dissipation.

The fractional amplitude and phase error of the approximation are defined as

$$\epsilon_A(\kappa) = \frac{|\kappa \hat{G}(\kappa)| - \kappa}{\kappa} \quad \text{and} \quad \epsilon_\phi(\kappa) = -\frac{\angle(k\hat{G}(\kappa)) + \pi/2}{\pi\,2}, \tag{64}$$

where the value of the amplitude and phase of the continuous derivative are κ and $-\pi/2$, respectively.

The fractional phase error for 7, 9 and 11 point fits are shown in Fig. 6. The fractional phase error for the low frequency portion of the Nyquist interval are shown in Fig. 7. The approximations were designed to provide a maximum 1% phase error for the 7 stencil point fit, a 0.5% maximum phase error for the 9 point fit, and 0.1% maximum phase error for the 11 point fit. The attempt during the design was to extend the region of low phase error to the highest wavenumber consistent with the maximum fractional phase error. The value of $b(\kappa)$ for each fit was set to a small negative value to insure that the phase angle over the fitted interval was less than $-\pi/2$.

Two first derivative approximations are shown for comparison with the fitted approximation, a two point Euler and a two point Lax approximation.

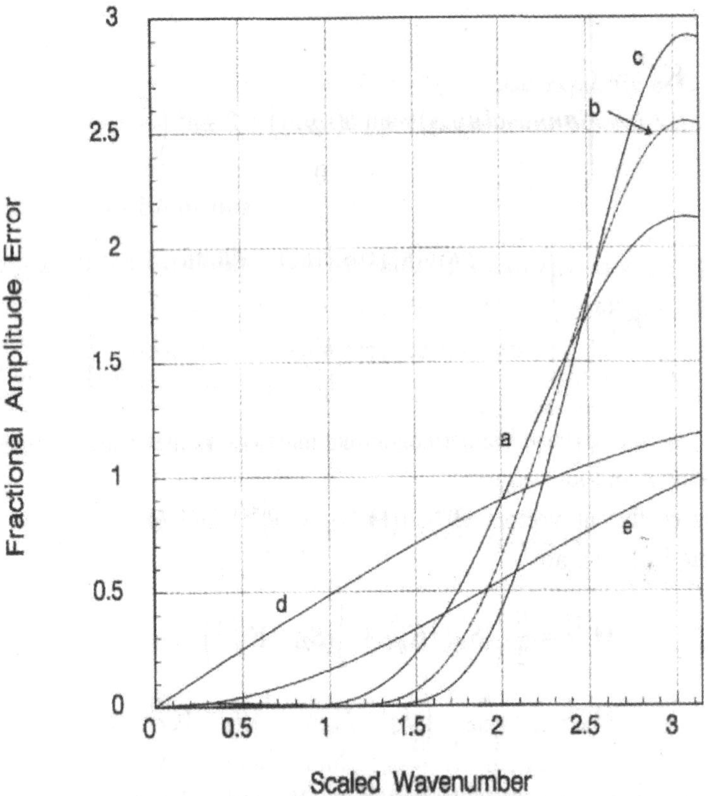

Fig. 6. Nyquist interval fractional amplitude error: $a = 7$ stencil point explicit general average approximation, $b = 9$ stencil point, explicit general average approximation, $c = 11$ stencil point explicit general average approximation, $d = 2$ stencil point Euler approximation, and $e = 2$ stencil point Lax approximation.

The Euler approximation

The fractional amplitude error for the fitted approximations compared with the Euler and Lax approximations are shown in Figs. 8 and 9. Both the Euler and Lax approximations exhibit appreciable fractional amplitude error over the lower portion of the Nyquist interval. The fitted approximations exhibit error on the order of less than 1% over the lower portion of the Nyquist interval. This performance is achieved at the expense of increased amplitude error in the upper Nyquist interval. However, since the object of the design was to produce a lowpass approximation suitable for

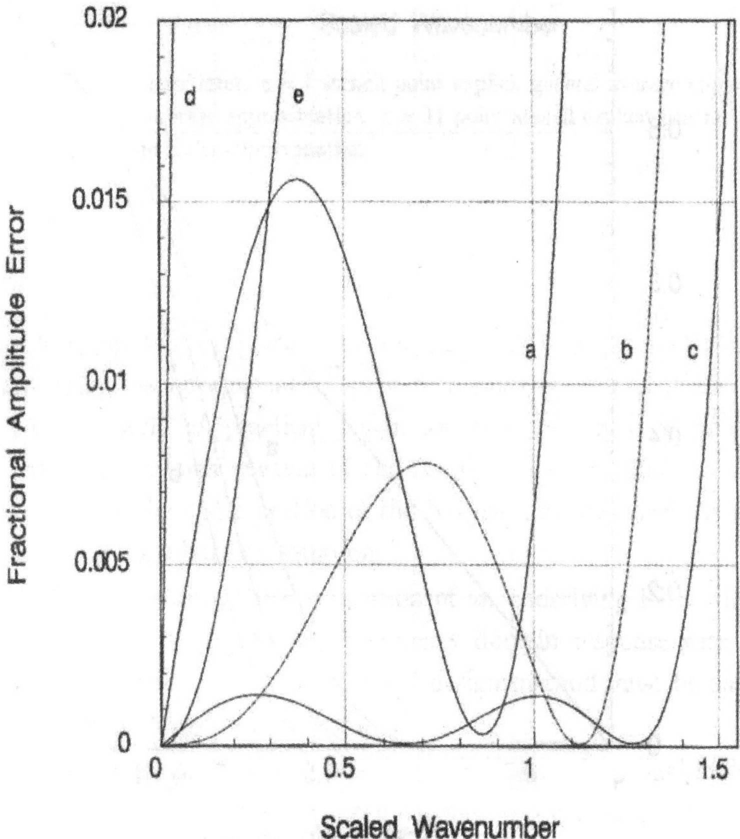

Fig. 7. Passband fractional amplitude error. Here $a = 7$ stencil point explicit general average approximation, $b = 9$ stencil point explicit general average approximation, $c = 11$ point stencil explicit general avearage approximation, $d = 2$ stencil point Euler approximation, and $e = 2$ stencil point Lax approximation.

simulating the advection equation with high accuracy in the lower half of the Nyquist interval and high dissipation in the upper half, the increased amplitude error occurs at values of greatest dissipation. Hence, this design succeeds by moving both amplitude and phase error to the upper portion of the Nyquist interval and then dissipating the error.

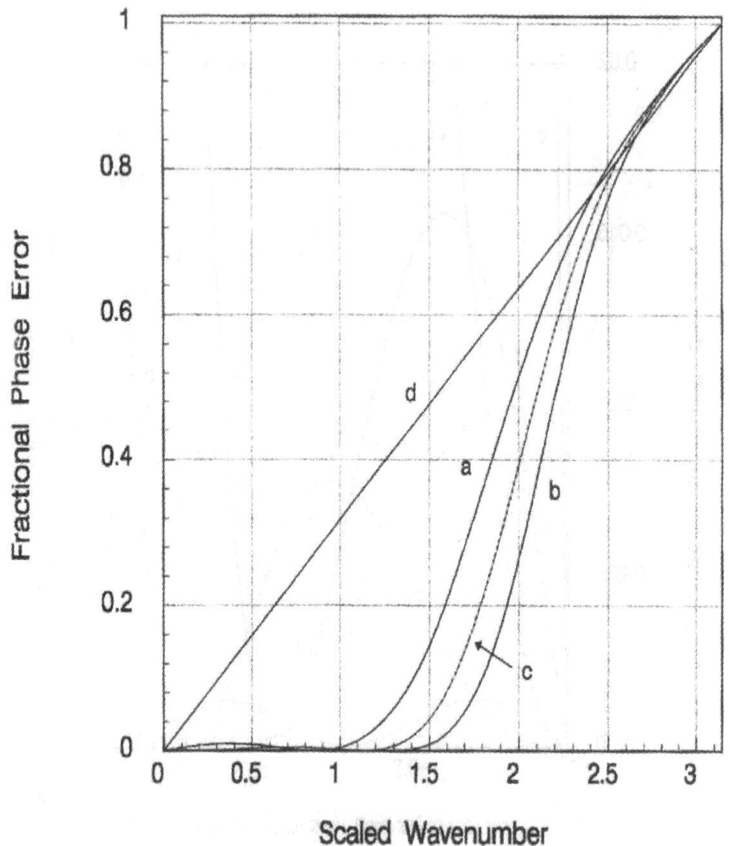

Fig. 8. Nyquist interval fractional phase error. Here $a = 7$ stencil point explicit general average approximation, $b = 9$ stencil point explicit general average approximation, $c = 11$ point stencil explicit general average approximation, $d = 2$ stencil point Euler approximation, and $e = 2$ stencil point Lax approximation.

4. Higher Order Spatial Derivatives

A design method for a general approximation to an rth order spatial derivative with undetermined coefficients is now developed using explicit rather than embedded constraints and eliminating as arbitrary weighting factor used in the previous sections. To accomplish this we constrain the design space so that the approximation converges and satisfies the Lax Equivalence Theorem. A general, frequency accurate, r^{th} derivative approximation can

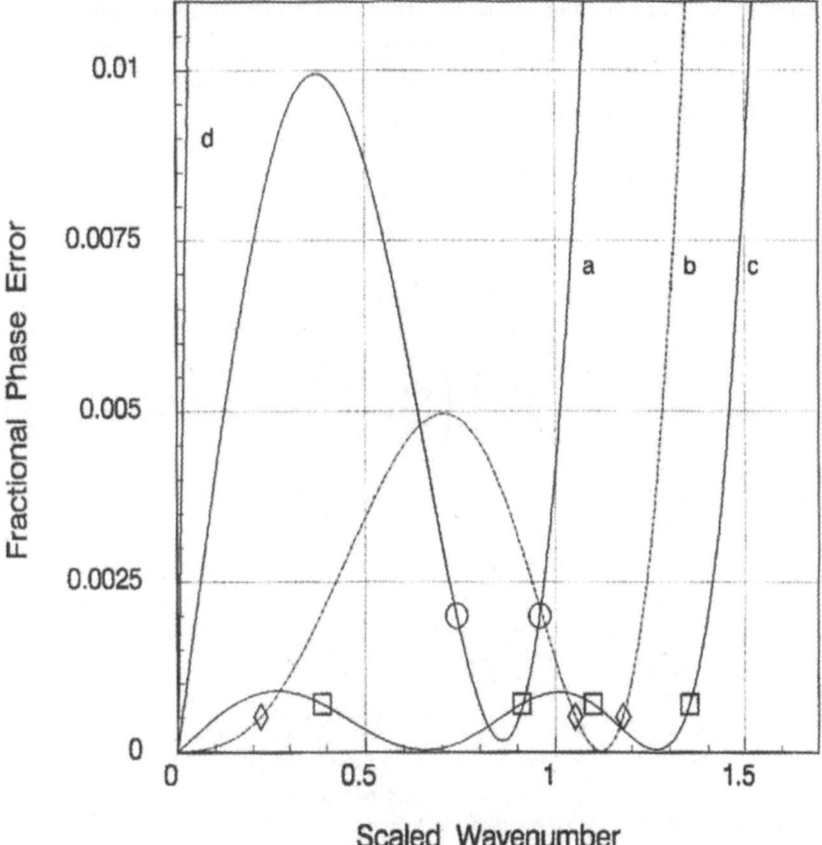

Fig. 9.   Passband fractional phase error. Here $a = 7$ stencil point explicit general average approximation, $b = 9$ stencil point explicit general average approximation, $c = 11$ point stencil explicit general average approximation, and $d = 2$ stencil point Euler approximation.

be written as

$$\bar{\Delta}_{rx} = \bar{\Delta}_{rx}^{[+]} + \bar{\Delta}_{rx}^{[-]}. \tag{65}$$

Here the individual right and left approximations about an arbitrary point m are

$$\bar{\Delta}_{rx}^{[+]} = \frac{1}{h^r} \sum_{j=0}^{N_+} \psi_j^{[+]} U_{m+j}^n \quad \text{and} \quad \bar{\Delta}_{rx}^{[-]} = \frac{1}{h^r} \sum_{j=0}^{N_-} \psi_j^{[-]} U_{m-j}^n, \tag{66}$$

where the $\psi_j^{[\pm]}$ are coefficients to be fitted (the undetermined coefficients). To examine the properties of this approximation, consider the right and left approximations independently. Expand the right approximation ($\bar{\Delta}_{rx}^{[+]}$, the first of Eq. (66), in a Taylor series about the point m,

$$
\bar{\Delta}_{rx}^{[+]} = \sum_{p=0}^{r-1} \frac{1}{p!} \left(\sum_{j=0}^{N_+} j^p \psi_j^{[+]} \right) (U_m^n)_{px} h^{p-r}
$$
$$
+ \frac{1}{r!} \left(\sum_{j=0}^{N_+} j^r \psi_j^{[+]} \right) (U_m^n)_{rx} \tag{67}
$$
$$
+ \sum_{p=r+1}^{\infty} \frac{1}{p!} \left(\sum_{j=0}^{N_+} j^p \psi_j^{[+]} \right) (U_m^n)_{px} h^{p-r}.
$$

A similar expression can be found for the left approximation, $\bar{\Delta}_{rx}^{[-]}$, in terms of the $\bar{\psi}_j^{[-]}$ coefficients.

This expansion requires two sets of constraints. The first set of constraints, derived from the first term of Eq. (67), ensures that the sums converge as $h \to 0$. Combining the constraints for the $\psi^{[+]}$ and $\psi^{[-]}$ coefficients and setting the coefficients to zero produce

$$
\sum_{j=0}^{N_+} j^p \psi_j^{[+]} + \sum_{j=0}^{N_-} (-j)^p \psi_j^{[-]} = 0 \qquad \text{for } p = 0, \dots, r-1. \tag{68}
$$

Applying the Lax consistency condition to the second term of Eq. (67) and its counterpart for the $\bar{\Delta}_{rx}^{[-]}$, the coefficients of the rth derivative term must converge to 1, as $h \to 0$, which produces the consistency constraint

$$
\sum_{j=1}^{N_+} j^r \psi_j^{[+]} + \sum_{j=1}^{N_-} (-j)^r \psi_j^{[-]} = r!. \tag{69}
$$

The constraints, as formulated in Eqs. (68) and (69), could be combined to solve for the unique Taylor-series-based approximation. Instead, introducing additional stencil points($\psi_j^{[+]}$ and $\psi_j^{[-]}$) beyond that required for a unique solution to the constraint equations results in additional degrees of freedom, which can be manipulated to reduce the approximation error on a fixed size grid. For example, sequentially setting the coefficients of the third term in the right-hand side of Eq. (67) to zero (and the corresponding terms for the left approximation) and adding these conditions as constraints to the fitting problem, result in uniquely determined approximations of the

r^{th} spatial derivative with higher formal order of truncation error as was done by Collatz [2]. But in doing so, the additional degrees of freedom, which allow the formulation of "performance requirements" to allocate the error, are sacrificed. Hence, the approach here is to distribute the potential reduction in error resulting from the larger stencil size based on selective accuracy criteria applied in the frequency domain.

The $\psi_0^{[\pm]}$ coefficients overlap and appear only in the $p = 0$ constraint of Eq. (68). They are uniquely determined by the remaining coefficients. Letting

$$\psi_0 = \psi_0^{[+]} + \psi_0^{[-]}, \tag{70}$$

and rewriting the $p = 0$ constraint as

$$\sum_{j=1}^{N_+} \psi_j^{[+]} + \sum_{j=1}^{N_-} \psi_j^{[-]} = -\psi_0, \tag{71}$$

the remainder of the convergence constraints for $p > 0$ becomes

$$\sum_{j=1}^{N_+} j^p \psi_j^{[+]} + \sum_{j=1}^{N_-} (-j)^p \psi_j^{[-]} = 0 \qquad \text{for } p = 1, \ldots, r - 1. \tag{72}$$

Notationally, the expression for the rth spatial derivative approximation given in Eqs. (65) and (66) becomes

$$\bar{\Delta}_{rx} = \frac{1}{h^r} \left(\sum_{j=1}^{N_+} \psi_j^{[+]} U_{m+j}^n + \psi_0 U_m^n + \sum_{j=1}^{N_-} \psi_j^{[-]} U_{m-j}^n \right), \tag{73}$$

where $\psi_j^{[\pm]}$ are the coefficients to the fitted subject to the constraints of Eq. (71) and (72).

The fitting parameter ψ_0 controls the weighting of the approximation about the center point. Changing ψ_0 adjusts the relative phase shift (lead or lag) of the fitted approximation. This property is illustrated below in as simple fitting example.

4.1. *Frequency Domain Formulation*

Transforming Eq. (73) to the frequency domain produces

$$\hat{\Delta}_{rx} = \frac{1}{h^r} \left(\sum_{j=1}^{N_+} \psi_j^{[+]} z_x^j + \psi_0 + \sum_{j=1}^{N_-} \psi_j^{[-]} \psi_j^{[-]} z_x^{-j} \right) \hat{U}(\xi, \omega), \tag{74}$$

where $z_x = e^{ikx}$, $\kappa_x \in [-\pi, \pi]$, and ξ and ω are the spatial and temporal frequencies. Assume that $\hat{U}(\xi, \omega)$ is bandlimited so that ξ and ω are always less than the Nyquist frequencies of the spatial and temporal sampling intervals. The spatial frequency, ξ, is allowed to vary, while the temporal frequency, ω, is held constant. This permits dropping the subscript on $\kappa_x (\kappa = \kappa_x)$ and a change in variables, $\hat{U}(\xi, \omega) \to \hat{U}(\kappa)$.

To control error in the frequency domain, the Fourier transform of a reference function for the rth derivative is required. An arbitrary reference function is defined, which facilitates possible trade-offs between the accuracy of the approximation and other desired properties. Define the reference function in the form

$$\eta(\kappa) = a(\kappa)e^{ib(\kappa)} \left(i\frac{\kappa}{h}\right)^r \hat{U}(\kappa) \tag{75}$$

where $\alpha(\kappa)$ and $b(\kappa)$ are arbitrary "tuning" functions, which can be used to stretch and rotate the transform of the continuous derivative. An attempt to exactly fit to the continuous derivative requires $\alpha(\kappa) = 1$ and $b(\kappa) = 0$. Other choices of these functions, particularly $b(\kappa)$ functions, which are linear in κ, are useful in adjusting fitted phase angle error over specified ranges of spatial frequencies. The complex fitting error can now be written as

$$\epsilon(\kappa) = \hat{\hat{\Delta}}_{rx}(\kappa) - \eta(\kappa). \tag{76}$$

Using Parseval's identity, the problem of minimizing the integrated error over the Nyquist interval becomes

$$\min_{\psi_j^{[\pm]}} \int_0^\pi |h^r \epsilon(\kappa)|^2 W^2(\kappa) d\kappa, \tag{77}$$

subject to the constraints of Eq. (71) and (72).

The weighting factor, $W(\kappa)$, is included in Eq. (77) to provide an estimator for the power spectral density of $U(x, t)$.

4.2. *Heuristic Formulation*

Rather than pursue the optimal solution of Eq. (77), we again introduce a heuristic discrete problem which will be formulated where the fitting error is defined at M fitting frequencies, κ_l, where $l = 1, \ldots, M$. Noting that $z = cos(\kappa)$, using Eqs. (74) and (75) and separating real and imaginary parts, $h^r \epsilon(\kappa_l) = 0$ becomes

$$\sum_{j=1}^{N_+} \psi_j^{[+]} \cos(j\kappa_l) + \sum_{j=1}^{N_-} \psi_j^{[-]} \cos(j\kappa_l) = \kappa_l^r a(\kappa_l) \cos\left(b(\kappa_l) + \frac{r\pi}{2}\right) - \psi_0 \quad (78)$$

and

$$\sum_{j=1}^{N_+} \psi_j^{[+]} \sin(j\kappa_l) + \sum_{j=1}^{N_-} \psi_j^{[-]} \sin(j\kappa_l) = \kappa_l^r a(\kappa_l) \sin\left(b(\kappa_l) + \frac{r\pi}{2}\right). \quad (79)$$

Using the discrete fitting frequencies, the problem can be cast in matrix form as

$$[M]\boldsymbol{\Psi} = \mathbf{R}, \quad (80)$$

where

$$[M] = \begin{bmatrix} \cos(\kappa_1) & \cdots & \cos(N+\kappa_1) & \cos(\kappa_1) & \cdots & \cos(N-\kappa_1) \\ \sin(\kappa_1) & \cdots & \sin(N+\kappa_1) & -\sin(\kappa_1) & \cdots & -\sin(N-\kappa_1) \\ \vdots & & \vdots & \vdots & & \vdots \\ \cos(\kappa_M) & \cdots & \cos(N+\kappa_M) & \cos(\kappa_M) & \cdots & \cos(N-\kappa_M) \\ \sin(\kappa_M) & \cdots & \sin(N+\kappa_M) & -\sin(\kappa_M) & \cdots & -\sin(N-\kappa_M) \\ 1 & \cdots & 1 & 1 & \cdots & 1 \\ 1 & \cdots & N_+ & -1 & \cdots & -N_- \\ \vdots & & \vdots & \vdots & & \vdots \\ 1 & \cdots & N_+^{r-1} & (-1)^{r-1} & \cdots & (-N_-)^{r-1} \\ 1 & \cdots & N_+^{r} & (-1)^{r} & \cdots & (-N_-)^{r} \end{bmatrix} \quad (81)$$

$$\boldsymbol{\Psi} = \left\{ \begin{array}{c} \psi_1^{[+]} \\ \vdots \\ \psi_{N_+}^{[+]} \\ \psi_1^{[-]} \\ \vdots \\ \psi_{N_-}^{[-]} \end{array} \right\} \quad (82)$$

$$\mathbf{R} = \left\{ \begin{array}{c} \kappa_1^r a(\kappa_1)\cos(b(\kappa_1) + \frac{r\pi}{2}) - \psi_0 \\ \kappa_1^r a(\kappa_1)\sin(b(\kappa_1) + \frac{r\pi}{2}) \\ \vdots \\ \kappa_M^r a(\kappa_M)\cos(b(\kappa_M) + \frac{r\pi}{2}) - \psi_0 \\ \kappa_M^r a(\kappa_M)\sin(b(\kappa_M) + \frac{r\pi}{2}) \\ -\psi_0 \\ 0 \\ \vdots \\ 0 \\ r! \end{array} \right\}. \tag{83}$$

The matrix $[M]$ is once again a Vandermonde-like matrix, which is invertible when $0 < \kappa_l < \pi$ and $\kappa_l \neq \kappa_m$. However, finding the inverse of the matrix can become computationally difficult due to ill-conditioning.

The dimensions of the matrix are $(2M + r + 1) \times (N_+ + N_-)$. Hence, for r odd, $(N_+ + N_-)$ must be even. For r even, $(N_+ + N_-)$ must be odd.

A simple example illustrates the solution of the matrix problem for the case $M = 0$, the fully constrained condition. Letting $r = 1$ for a first spatial derivative approximation and choosing $N_+ = 1$ and $N_- = 1$, $[M]$ and $\mathbf{R}$ become

$$[M] = \begin{bmatrix} 1 & 1 \\ 1 & -1 \end{bmatrix} \quad \text{and} \quad \mathbf{R} = \left\{ \begin{array}{c} -\psi_0 \\ 1 \end{array} \right\} \tag{84}$$

so that

$$\Psi = \frac{1}{2} \left\{ \begin{array}{c} -\psi_0 + 1 \\ -\psi_0 - 1 \end{array} \right\}. \tag{85}$$

For $\psi_0 = 0$, the approximation stencil is $(1/2h)\{-1, 0, 1\}$, the first central difference. for $\psi_0 = -1$ the approximation stencil becomes $(1/h)\{0, -1, 1\}$, the first upwind difference for $U_t + cU_x = 0$. For $\psi_0 = 1$, the approximation stencil becomes $(1/h)\{-1, 1, 0\}$, the first upwind difference for $U_t - cU_x = 0$. the choice of the value of ψ_0 weights the approximation on either side of the central point and changes the phase angle and magnitude of its complex response. Similar expressions can be produced for the higher derivative approximations, including those with additional row constraints corresponding to the elimination of higher orders of truncation error. These include most of the commonly used derivative approximations [2] in addition to custom approximation designed for specific applications.

4.3. *Approximation Fitting*

To illustrate the application of the fitting procedure, the first and second spatial derivatives are fitted for use in the linear advection-diffusion equation. To do this, the continuous equation,

$$U_t + cU_x - bU_{xx} = 0, \tag{86}$$

is assumed to be bandlimited to the temporal and spatial Nyquist interval. Transforming Eq. (86) to the spatial frequency domain, its response function, $Q(\kappa)$, becomes

$$Q(\kappa_x) = -\kappa_x(\sigma_2 \kappa_x + i\sigma_1), \tag{87}$$

the Nyquist-scaled continuous frequency equation. Here, $\sigma_1 = c\kappa/h$ (The Courant-Friedrichs-Levy number), $\sigma_2 = b\kappa/h^2$ (the grid Reynolds number). The relations, $\omega = \kappa_t/K$ and $\xi = \kappa_x/h$, and the bandlimited assumption, $\kappa_t, \kappa_x \in [-\pi, \pi]$, have been used to scale the continuous characteristic equation to the Nyquist interval of the approximation.

The first and second spatial derivative approximation, $\bar{\Delta}_x$ and $\bar{\Delta}_{2x}$, are fitted individually using the error defined in Eq. (76). The responses of the fitted approximations are combined using

$$\zeta_T(\kappa) = -\sigma_1 \bar{\Delta}_x(\kappa) + \sigma_2 \bar{\Delta}_{2x}(\kappa), \tag{88}$$

which when combined with Eq. (87), yields the total fitted error equation

$$\epsilon_T(\kappa) = Q(\kappa) - \zeta_T(\kappa), \tag{89}$$

where the subscript on κ_x has been dropped; the error is a function of only spatial frequency.

Rather than attempting to fit directly to ϵ_T, performance parameters based on the dissipation and phase or group velocity of the combined derivative approximation are used. To this end, define the scaled spatial dissipation factor and the scaled phase velocity of the continuous equation as

$$\delta_c = -\sigma_2 \kappa^2 \quad \text{and} \quad v_c = -\sigma_1. \tag{90}$$

The scaled dissipation and phase velocity factors for the combined approximation are

$$\delta_a = Re(Q(\kappa)) \quad \text{and} \quad v_a = \frac{Im(Q(\kappa))}{\kappa}. \tag{91}$$

Using Eq. (90) and (91), the relative errors for dissipation and phase velocity are

$$R_\delta = \frac{\delta_a}{\delta_c} = -\frac{Re(Q(\kappa))}{\sigma_2 \kappa^2} \quad \text{and} \quad R_v = \frac{v_a}{v_c} = -\frac{Im(Q(\kappa))}{\sigma_1 \kappa}. \tag{92}$$

The fitted frequency accurate approximation for the second spatial derivative uses 3 positive coefficients and 2 negative coefficients resulting in one fitting zero (one degree of freedom). The first derivative approximation uses 3 positive and 3 negative coefficients resulting in 2 fitting zeros (2 degrees of freedom). Compare this with both the central difference first and second spatial derivative approximations with 1 positive and 1 negative coefficient and symmetric approximations with 3 positive and 3 negative coefficients due to Collatz [2] (Appendix, Table III),

$$y' = \frac{1}{60h}(-y_{-3} + 9y_{-2} - 45y_{-1} + 45y_1 - 9y_2 + y_3) \tag{93}$$

and

$$y'' = \frac{1}{180h^2}(2y_{-3} - 27y_{-2} + 270y_{-1} - 490y_0 + 270y_1 - 27y_2 + 2y_3) \tag{94}$$

are used. Both Collatz approximations are formally 6th order accurate.

The design objective here is to produce combined first and second derivative approximations, which produce low phase velocity error (waveform distortion) in the lower half of the Nyquist interval ($\kappa \leq \pi/2$) and acceptable dissipation error over the same interval. Essentially, the design objective is to illustrate that the same number of grid points as used in the Collatz approximations can be allocated to attain better performance. For illustrative purposes, the design is performed at $\sigma_1 = 0.5$ and $\sigma_2 = 1.0$.

The results of the first sample design are shown in Figs. 10–13. In Fig. 10, the phase velocity factor excess ($|R_v - 1|$) for the Frequency Accurate(FA), Collatz(CZ), and Central Difference(CD) is shown as a function of spatial frequency. The FA approximation achieves less than 6% phase velocity error over the lower half of the Nyquist interval, while the CZ and CD approximations produce 7% and 37% worst-case phase velocity error, respectively. The ability to use the FA approximation to design to phase velocity requirements means that the sharp increase in phase velocity error of the CZ and CD approximation at low spatial frequencies can be traded for relatively unform error across a wider frequency interval. In this case, the performance of the CZ and CD approximations at $\frac{1}{2}$ and $\frac{1}{20}$ of the Nyquist interval, respectively, is extended to $\frac{1}{2}$ of the Nyquist interval. the advantages of smaller phase velocity error at the expense of dissipation error are discussed in [10].

Fig. 11 shows the dissipation factor ratio excess ($|R_\delta - 1|$) for the three approximations. The achievement of low-phase velocity error for the FA approximation results in a loss of accuracy in modeling dissipation. While the

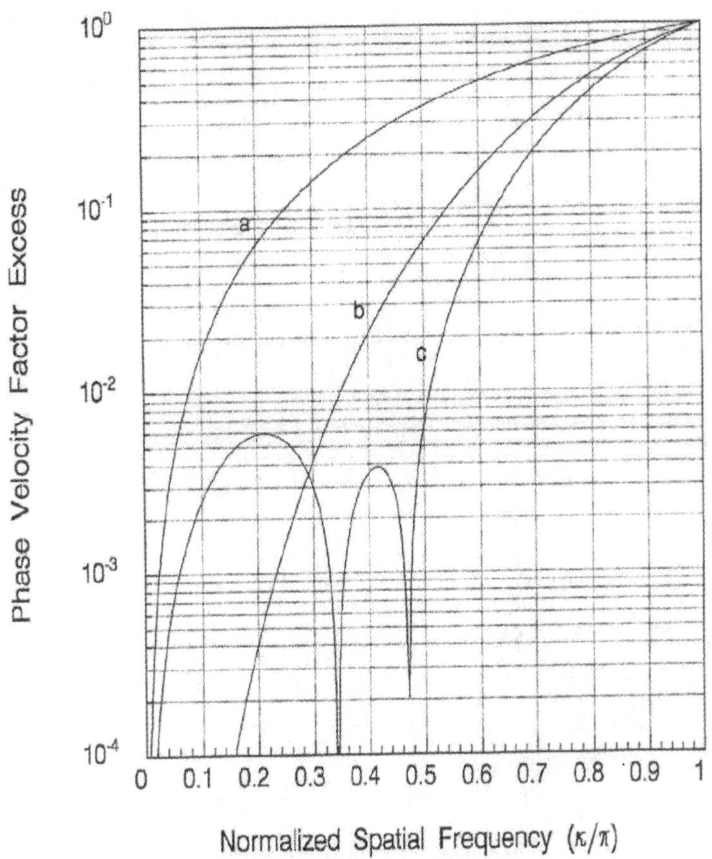

Fig. 10. Phase Velocity Factor Excess, $|R_v - 1|$. The fractional error in the approximation phase velocity ffactor relative to the scaled, bandlimited continuous equation. Here $a = 3$ point stencil central difference approximation; $b = 7$ point stencil Collatz approximation, and $c = 7$ point stencil frequency accurate approximation.

FA approximation exhibits the same worst-case dissipation error as the CZ approximation over the lower half of the Nyquist interval, it exhibits higher error at almost all other frequencies. However, it nearly matches the performance of the CZ approximation over the high half of the Nyquist interval and avoids the extreme error at the highest frequencies. The CD approximation exhibits better dissipation performance than the FA approximation only for the lower $\frac{1}{4}$ of the Nyquist interval.

The reason for this performance is illustrated in Figs. 12 and 13. In Fig. 12, the complex response for the scaled continuous frequency equation

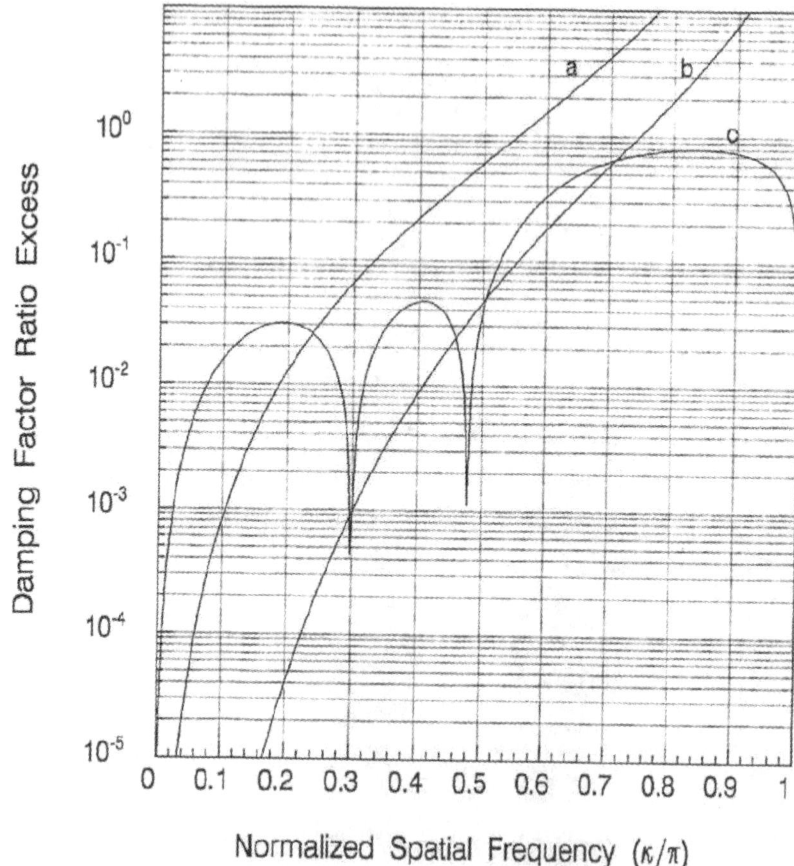

Fig. 11.   Damping factor ratio excess $|R_\delta - 1|$. The fractional error in the approximation dissipation factor relative to the scaled, bandlimited continuous equation. Here $a = 3$ point stencil central difference approximation; $b = 7$ point stencil Collatz approximation; and $c = 7$ point stencil frequency accurate approximation.

is a section of a parabola whose real and imaginary parts go from zero to negative infinity. The approximations, however, are sections of arcs that begin and end on the real axis. All three approximations match the shape of the continuous curve at low spatial frequencies. However, as the spatial frequency increases, the complex response of the approximations departs from the continuous curve. through careful choice of the fitting parameters, it is possible to hold the complex response of the FA approximation near that of the scaled continuous equation over a relatively long range of spatial frequencies. It is also possible to place the high frequency of the FA approximation at a point on the real axis that corresponds to the highest frequency real component of the scaled continuous equation. Since the real part of the complex response corresponds to dissipation, a near-zero dissipation error is possible at the upper end of the Nyquist interval.

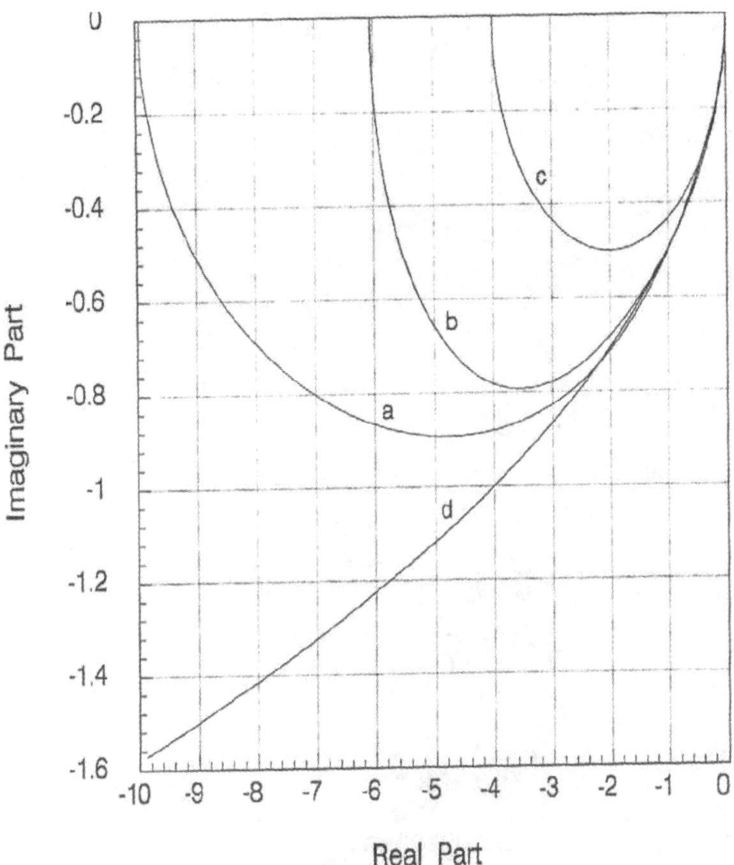

Fig. 12.   Complex Response of the approximations and the scaled, bandlimited continuous equation, $Q(\kappa)$. Here $a = 7$ stencil point frequency accurate approximation; $b = 7$ point stencil Collatz approximation; $c = 3$ stencil point central difference approximation; and d is the scaled, bandlimited continuous equation.

The response amplitude is the magnitude of the radius vector from the origin to the complex response curve as a function of spatial frequency. Fig. 13 shows the response amplitude for each of the approximations and the continuous equation. The response amplitude of the FA approximation exceeds that of the continuous equation over most of the Nyquist interval and is very near the correct value at the upper Nyquist frequency. The CZ and CD approximations produce response amplitudes with much larger deviations. The deviation of the approximation response amplitude from the continuous equation results in both phase velocity and dissipation error.

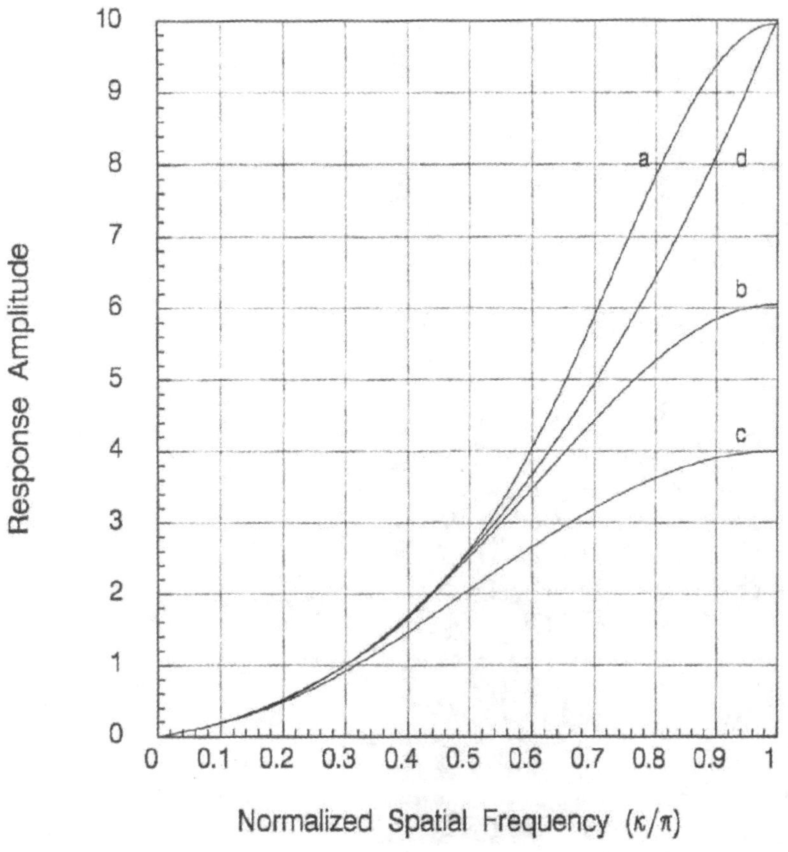

Fig. 13.   Response amplitude. Here $a = 7$ stencil point frequency accurate approximation; $b = 7$ point stencil Collatz approximation; $c = 3$ stencil point central difference approximation; and d is the scaled bandlimited continuous equation.

5. A Frequency Accurate Finite Difference Scheme for Burgers Equation

We now approximate a nonlinear problem, Burgers equation, using an explicit finite difference scheme(FDS) that combines the techniques presented in the previous three sections. The fit minimizes error at *a priori* spatial frequencies for a linearized form of the equation.

Our FDS for the inviscid Burgers equations is developed by transforming general linear approximations with undetermined coefficients for the temporal and spatial derivatives to the frequency domain. There, the fitting error is defined through the dispersion relation and the coefficients are determined which minimize the error at selected spatial frequencies. The intent is to produce a linear fit which minimizes phase and group velocity error for the FDS at Courant-Friedrichs-Levy(CFL) numbers less than one, to improve stability when used in the nonlinear Burgers equation. Rather than attempting to directly fit both temporal and spatial approximations, the entire FDS is fitted using a specified spatial derivative approximation, and subsequently determining a compensating temporal derivative approximation.

5.1. *Dispersion Relations*

Using a nonlinear equation in flux formulation,

$$U_t + f(U)_x = 0, \tag{95}$$

choose $f(U)$ as

$$f(U) = cU + \frac{a}{2}U^2, \tag{96}$$

where c and a are constants.

Linearize Eq. (95) by assuming that $a|\bar{U}|/2 \ll c$, where the overbar indicates a maximum. The linearized equation, $U_t + cU_x = 0$, has the dispersion relation $f = -c\xi$ where f and ξ are the temporal frequency and the wavenumber. If the equation is discretized on a uniform grid so that $U_m^n = U(mh, nk)$, then the wavenumber and temporal frequency Nyquist limits are $\xi = 1/2h$ and $f = 1/2k$. Now, require that the solution be bandlimited, so that $\xi = \kappa_x/2\phi$ and $f = \kappa_t/2\pi\kappa$; where $\kappa_x, \kappa_t \in [-\pi, \pi]$ are the scaled wavenumber and frequency. With these definitions, the discretized dispersion relation for the advection equation becomes

$$\kappa_t = -\sigma\kappa_x, \tag{97}$$

where $\sigma = c/\alpha$ is the CFL number and $\alpha = h/k$, the "grid speed," is a factor resulting from the discretization.

Defining the variables $z_t = e^{ikt}$ and $z_x = e^{ik_x}$ the scaled discrete dispersion relation becomes

$$z_t = z_x^{-\sigma}. \tag{98}$$

When $\sigma = 1$, the approximation satisfies the DRP condition since Eq. (97) and (98) can be transformed to $f = -c\xi$ through the definitions of the wavenumber and temporal frequency.

A linear dispersion relation for a single time level FDS is now derived for the linearized form of Eq. (95), which can be written in operator form as

$$\Delta_t(U_m^n) + c\Delta_x(U_m^n) = 0, \tag{99}$$

where

$$\Delta_t(U_m^n) = \frac{1}{k}(U_m^{n+1} - \mathsf{T}(U_m^n)) \quad \text{and} \quad \Delta_x(U_m n) = \frac{1}{h}\mathsf{L}(U_m^n). \tag{100}$$

Here, $\mathsf{T}(U_m^n)$ and $\mathsf{L}(U_m^n)$ are fitted linear temporal and spatial derivative operators. Using Eq. (100), rearranging Eq. (99), and taking the Fourier transform produces the dispersion relation for the approximation,

$$z_t = \hat{\mathsf{T}}(z_x) - \sigma\hat{\mathsf{L}}(z_x). \tag{101}$$

Fitting the coefficients of the FDS such that Eq. (101) is satisfied results in a FDS which has the DRP property when $\sigma = 1$. However, it is not always possible to obtain a FDS that DRP. Hence, we consider a modified equation by defining the complex function $Q(\kappa) = g(\kappa_x)z_t$, where $g(\kappa_x)$ is a user specified function. Substituting Eq. (98) and dropping the subscript x on κ_x and z_x, the fitting equation becomes

$$Q(\kappa) - \hat{\mathsf{T}}(z) + \sigma\hat{\mathsf{L}}(z) = \epsilon(\kappa), \tag{102}$$

where $\epsilon(\kappa)$ is the fitting error.

Relating the spatial and temporal error to the frequency domain through Parseval's theorem [1], the general fitting problem, stated as minimization problem, is

$$\min_{\hat{\mathsf{T}},\hat{\mathsf{L}}} \int_{-\pi}^{\pi} |\epsilon(\kappa)|^2 W(\kappa)d\kappa, \tag{103}$$

subject to constraints on the coefficients that insure convergence and consistency. The function $W(\kappa)$ is included to weight the minimization for the problem dynamics. Rather than pursuing the minimization, a heuristic method which sets $\epsilon(\kappa) = 0$ at selected spatial frequencies κ_l will again be used.

The coefficients produced from Eq. (103) must result in a stable FDS. To develop the conditions for stability, rewrite the fitting error as

$$\epsilon = \epsilon_0(\kappa) z_t \tag{104}$$

and define the response function of the FDS as

$$h(\kappa) = g(\kappa) - \epsilon_0(\kappa), \tag{105}$$

so that, using Eq. (101),

$$h(\kappa) z_t = \hat{\mathsf{T}}(z) - \sigma \hat{\mathsf{L}}(z). \tag{106}$$

The FDS is stable, in the sense of von Neumann, if $|h(\kappa)| \le 1$. Following the heuristic approach, the coefficients of $\hat{\mathsf{T}}$ and $\hat{\mathsf{L}}$ are fitted at selected frequencies where $\epsilon_0 = 0$. At these frequencies $h(\kappa) = g(\kappa)$. At other frequencies, the fitting error must be such that $|h(\kappa)| \le 1$ for a stable scheme. Here, explicit design requirements on $|h(\kappa)|$ are included as part of the fitting process to insure stability.

5.2. *Frequency Domain Fitting*

The average value spatial difference operator on a regular grid with undetermined coefficients is defined as in Eq. (95), and is repeated here

$$\Delta_x(U_m^n) = \frac{1}{\nu} \Delta_x^{[+]}(U_m^n) + \frac{\nu - 1}{\nu} \Delta_x^{[-]}(U_m^n), \tag{107}$$

where

$$\Delta_x^{[+]}(U_m^n) = \frac{1}{h} \left(\sum_{j=0}^{N} \psi_j^{[+]} U_{m+j}^n - U_m^n \right) \tag{108}$$

and

$$\Delta_x^{[-]}(U_m^n) = \frac{1}{h} \left(U_m^n - \sum_{j=0}^{N} \psi_j^{[-]} U_{m-j}^n \right). \tag{109}$$

A two-time level, explicit general average temporal derivative approximation is defined as

$$\Delta_t(U_m^n) = \frac{1}{\mu} \Delta_t^{[+]}(U_m^n) + \frac{\mu - 1}{\mu} \Delta_t^{[-]}(U_m^n), \tag{110}$$

where

$$\Delta_t^{[+]}(U_m^n) = \frac{1}{k}\left(U_m^{n+1} - \sum_{j=0}^{N}\theta_j^{[+]}U_{m+j}^n\right) \tag{111}$$

and

$$\Delta_t^{[-]}(U_m^n) = \frac{1}{k}\left(U_m^{n+1} - \sum_{j=0}^{N}\theta_j^{[-]}U_{m-j}^n\right). \tag{112}$$

The factors ν and μ weight the approximating to the left or right of the point m.

To insure that each approximation converges and satisfies the Lax consistency condition, the fitted coefficients are constrained so that

$$\sum_{j=0}^{N}\psi_j^{[+]} = \sum_{j=0}^{N}\psi_j^{[-]} = \sum_{j=0}^{N}\theta_j^{[+]} = \sum_{j=0}^{N}\theta_j^{[-]} = 1, \tag{113}$$

$$\sum_{j=1}^{N}j\psi_j^{[+]} = \sum_{j=1}^{N}j\psi_j^{[-]} = 1 \tag{114}$$

and

$$\sum_{j=1}^{N}j\theta_j^{[+]} = \sum_{j=1}^{N}j\theta_j^{[-]} = 0. \tag{115}$$

The operator form of the approximations used in the previous section are found from Eq. (107) through (112) if the fitted coefficients are known. However, the object here is to determine the coefficients. To do this we combine the undetermined coefficients $\theta_j^{[\pm]}$ and $\psi_j^{[\pm]}$, with the fitting coefficients for the FDS, $V_j^{[S/D]}$, so that

$$V_j^{[S]} = \frac{1}{\mu}\theta_j^{[S]} - \frac{\sigma}{\nu}\psi_j^{[D]} \quad \text{and} \quad V_j^{[D]} = \frac{1}{\mu}\theta_j^{[D]} - \frac{\sigma}{\nu}\psi_j^{[S]}, \tag{116}$$

where

$$\theta_j^{[S]} = \theta_j^{[+]} + (\mu - 1)\theta_j^{[-]}, \qquad \theta_j^{[D]} = \theta_j^{[+]} - (\mu - 1)\theta_j^{[-]}, \tag{117}$$

$$\psi_j^{[S]} = \psi_j^{[+]} + (\nu - 1)\psi_j^{[-]} \quad \text{and} \quad \psi_j^{[D]} = \psi_j^{[+]} - (\nu - 1)\psi_j^{[-]}. \tag{118}$$

Substituting the expressions for the $V_j^{[S/D]}$ into Eq. (107) through (112) and Eq. (99), and Fourier transforming the result produces the frequency equation for fitting the FDS,

$$Q(\kappa) - 1 + 2\sum_{j=1}^{N} V_j^{[S]} \sin^2\left(\frac{j\kappa}{2}\right) - i\sum_{j=1}^{N} V_j^{[D]} \sin(j\kappa) = 0, \qquad (119)$$

where Eq. (113) have been solved for $\theta_0^{[\pm]}$ and $\psi_0^{[\pm]}$ and substituted in Eq. (119) to imbed the convergence constraints.

Using Eqs. (114)–(116), the consistency constraints are rewritten as

$$\sum_{j=1}^{N} jV_j^{[S]} = \sigma\left(\frac{\nu-2}{\nu}\right) \quad \text{and} \quad \sum_{j=1}^{N} jV_j^{[D]} = -\sigma. \qquad (120)$$

Specify $N-1$ fitting frequencies, separate Eq. (119) into its real and imaginary parts, and cast them in matrix form as

$$[S_{2v}]\mathbf{V}^{[S]} = -\frac{1}{2}\mathbf{G}_{V_R}$$
$$[S_{1v}]\mathbf{V}^{[D]} = \mathbf{G}_{V_I}, \qquad (121)$$

where

$$[S_{2v}] = \begin{bmatrix} \sin^2(\frac{\kappa_1}{2}) & \cdots & \sin^2(\frac{N_{\kappa_1}}{2}) \\ \vdots & & \vdots \\ \sin^2(\frac{\kappa_{N-1}}{2}) & \cdots & \sin^2\left(\frac{N_{\kappa_{N-1}}}{2}\right) \\ 1 & \cdots & N \end{bmatrix}, \qquad (122)$$

$$[S_{1v}] = \begin{bmatrix} \sin(\frac{\kappa_1}{2}) & \cdots & \sin(\frac{N_{\kappa_1}}{2}) \\ \vdots & & \vdots \\ \sin(\frac{\kappa_{N-1}}{2}) & \cdots & \sin\left(\frac{N_{\kappa_{N-1}}}{2}\right) \\ 1 & \cdots & N \end{bmatrix}, \qquad (123)$$

$$\mathbf{V}^{[D]} = \left\{ \begin{array}{c} V_1^{[D]} \\ \vdots \\ V_N^{[D]} \end{array} \right\}, \mathbf{V}^{[S]} = \left\{ \begin{array}{c} V_1^{[S]} \\ \vdots \\ V_N^{[S]} \end{array} \right\}, \qquad (124)$$

$$\mathbf{G}_{V_R} = \left\{ \begin{array}{c} \text{Re}(Q(\kappa_1) - 1) \\ \vdots \\ \text{Re}(Q(\kappa_{N-1}) - 1) \\ -\frac{2\sigma(\nu-2)}{\nu} \end{array} \right\} \quad \text{and} \quad \mathbf{G}_{V_I} = \left\{ \begin{array}{c} \text{Im}(Q(\kappa_1)) \\ \vdots \\ \text{Im}(Q(\kappa_{N-1})) \\ -\sigma \end{array} \right\}. \qquad (125)$$

The last rows of the matrices and right-hand sides are the consistency constraints, Eq. (120).

Cast the relation between the coefficients of the FDS and the coefficients of the temporal and spatial derivative approximations, Eq. (116), in vector form and solve for the vectors of temporal derivative coefficients so that

$$\boldsymbol{\Theta}^{[S]} = \mu\left(\mathbf{V}^{[S]} + \frac{\sigma}{\nu}\boldsymbol{\Psi}^{[D]}\right) \quad \text{and} \quad \boldsymbol{\Theta}^{[D]} = \mu\left(\mathbf{V}^{[D]} + \frac{\sigma}{\nu}\boldsymbol{\Psi}^{[S]}\right). \tag{126}$$

Now, pose the fitting problem to determine the coefficients of the temporal approximation given the desired properties of the FDS and a spatial approximation with known coefficients. In this case, the vectors $\boldsymbol{\Psi}^{[S/D]}$ in Eq. (126) are treated as constants. Alternatively, both the coefficients of the FDS and the spatial approximation can be fitted in tandem. If this course is chosen, the spatial derivative approximation can be cast in matrix form and written as

$$\boldsymbol{\Psi}^{[D]} = \frac{\nu}{2}[S_{2\Psi}]^{-1}\mathbf{G}_{\Psi_R} \quad \text{and} \quad \boldsymbol{\Psi}^{[S]} = \nu[S_{1\Psi}]^{-1}\mathbf{G}_{\Psi_I}, \tag{127}$$

where the matrices have the same form as Eq. (122), and the vectors are

$$\boldsymbol{\Psi}^{[D]} = \left\{ \begin{matrix} \psi_1^{[D]} \\ \vdots \\ \psi_N^{[D]} \end{matrix} \right\}, \quad \boldsymbol{\Psi}^{[S]} = \left\{ \begin{matrix} \psi_1^{[S]} \\ \vdots \\ \psi_N^{[S]} \end{matrix} \right\}, \tag{128}$$

$$\mathbf{G}_{\Psi_R} = \left\{ \begin{matrix} \mathrm{Re}(\eta(\kappa_1)) \\ \vdots \\ \mathrm{Im}(\eta(\kappa_{N-1})) \\ \frac{2-\nu}{\nu} \end{matrix} \right\} \quad \text{and} \quad \mathbf{G}_{\eta_I} = \left\{ \begin{matrix} \mathrm{Im}(\eta(\kappa_1)) \\ \vdots \\ \mathrm{Im}(\eta(\kappa_{N-1})) \\ 1 \end{matrix} \right\}, \tag{129}$$

with $\eta(\kappa) = i\zeta(\kappa)\kappa$ and $\zeta(\kappa)$ a user-defined shaping function. Combining the matrix expressions for the $V^{[S/D]}$ and $\zeta^{[S/D]}$ in Eq. (126) produces the $\Theta^{[S/D]}$ fitting equations

$$\boldsymbol{\Theta}^{[S]} = -\frac{\mu}{2}([S_{2V}]^{-1}\mathbf{G}_{V_R} - \sigma[S_{2\Psi}]^{-1}\mathbf{G}_{\Psi_R}),$$
$$\boldsymbol{\Theta}^{[D]} = \mu([S_{1V}]^{-1}\mathbf{G}_{V_I} - \sigma[S_{1\Psi}]^{-1}\mathbf{G}_{\Psi_I}). \tag{130}$$

The θ coefficients are found using Eq. (117). The θ_0 coefficient is found from Eqs. (113) and (117) as

$$\theta_0 = 1 - \frac{1}{\mu}\sum_{j=0}^{N}\theta_j^{[S]}. \tag{131}$$

The ψ coefficients are found using Eq. (127) and (118). The ψ_0 coefficient is

$$\psi_0 = -\frac{1}{2} \sum_{j=1}^{N} \psi_j^{[D]}. \tag{132}$$

Equations (130) are solved using $2(N-1)$ distinct fitting frequencies for the FDS and the spatial approximation. The independent shaping functions $g(\kappa)$ and $\eta(\kappa)$ as well as the free parameters μ and ν are available to adjust the fit. The fitting frequencies for the V and ψ coefficients are independent.

5.3. *Nonlinear Stabilization*

To illustrate the design method, define the test FDS

$$r_m^{n+1} = \mathsf{T}(r_m^n) - \sigma(p + qr_m^n)\mathsf{L}(r_m^n), \tag{133}$$

where $p \in [0, 1]$ is a parameter controlling the level of the linear term and $q \in [0, 1]$ is a parameter controlling the level of the nonlinear term. $\sigma \in (0, 1]$ is the CFL number.

The response level of the FDS in the low frequency region contributes to the growth of an instability when $q \neq 0$. To control the instability a spatial digital filter is applied to the grid, as needed [11].

Given an approximate solution, r_m^{n+1}, found from Eq. (133), the resulting filtered solution is

$$(r_m^{n+1})_{\text{filt}} = \sum_{j=0}^{3} f_j r_{m-j+s}^{n+1}, \tag{134}$$

where s is a shift factor.

The filter coefficients are determined here using a 3 zero digital filter [4] based on the polynomial

$$P(z) = (z + 1)(z - e^{ia\pi}(z - e^{-ia\pi})), \tag{135}$$

which has a zero at $\kappa = \pi$ and two zeros at $\kappa = \pm a\pi$, $a \in (0, 1)$.

The Lax Equivalence Theorem requires that when the filter, Eq. (134), is expanded in a spatial Taylor series about r_m^{n+1}, $(r_m^{n+1})_{\text{filt}} \to r_m^{n+1}$ as $h \to 0$. This is achieved if $\sum_{j=0}^{3} f_j = 1$, implying unity response at $\kappa = 0$.

The spatial filter is similar to a Shapiro filter (see Ref. [13]) with the exception that it does not provide a zero phase shift for all spatial frequencies. To compensate for this property, phase shift the filter using the shift factor s in Eq. (134). For this filter, a phase shift (lag) of -2 minimizes the phase shift with as small positive residual. A phase shift of -1 minimizes phase shift with a small negative residual of equal magnitude. Hence, alternate use of the phase shifts on invocations of the filter results in residual phase shifts which cancel, on average.

The spatial filter is invoked when the ratio of the energy in the high Nyquist interval, $\pi/2 < \kappa \le \pi$, is greater than a desired fraction of the energy in the low Nyquist interval, $0 \le \kappa \le \pi/2$. The energy ratio is determined using a matched pair of 3 zero high and low pass filters and Parseval's identity to estimate the energy in each band. Denoting the filter coefficients as $f_j^{[H]}$ and $f_j^{[L]}$ for the high and low pass filters, respectively, the energy estimates for a periodic grid with spatial grid points $m \in [0, M]$ are

$$E_H = \sum_{m=0}^{M} \left(\sum_{j=0}^{3} f_j^{[H]} r_{m-j}^{n+1} \right)^2 \quad \text{and} \quad E_L = \sum_{m=0}^{M} \left(\sum_{j=0}^{3} f_j^{[L]} r_{m-j}^{n+1} \right)^2 \tag{136}$$

and the estimated energy ratio is $\rho_E = E_H/E_L$.

The computational sequence propagates the solution using Eq. (133) and the high/low energy ratio is then calculated using Eq. (136). If ρ_E exceeds a predetermined value, the filter, Eq. (134), is invoked with the appropriate shift factor. This scheme is pursued to allow the instability to grow to a point where it will influence the trajectory of the solution but not become so large as to destroy it.

5.4. *Numerical Example*

The coefficients of the FDS are fitted heuristically by specifying fitting frequencies and the desired response. The design was pursued graphically using a previously fitted spatial derivative approximation, which provides high accuracy at low spatial frequencies and high dissipation at high spatial frequencies. Seven stencil point spatial and temporal derivative approximations were used. The fitting frequencies and the reference function were manually adjusted to produce the FDS coefficients and satisfy the stability requirement $|h(\kappa)| \le 0$. Specify the reference function, $g(\kappa)$, indirectly in terms of functions of its norm and phase angle. Define the influence distance, λ, as

$$\lambda = -\frac{1}{\ln |g(\kappa)|}, \tag{137}$$

the distance, in the linear advection equation, that a wave propagates before being dissipated to $1/e$ of its starting value (the e-folding scale). The

Table 1. The coefficients of the temporal and spatial dervative operators. j is the offset index relative to the central point of the stencil.

j	$\mathsf{T}(x_j)$ Coefficients	$\mathsf{L}(x_j)$ Coefficients
3	$-2.554506029755566 \times 10^{-2}$	$-5.109012077846825 \times 10^{-2}$
2	$1.277555127484613 \times 10^{-1}$	$2.232162645786716 \times 10^{-1}$
1	$-1.788758446042557 \times 10^{-1}$	$-1.404347763557288 \times 10^{-1}$
0	1.181575293419925	1.178128436285669
-1	$-2.380298195192641 \times 10^{-1}$	1.661081793865152
-2	$1.613299352388024 \times 10^{-1}$	$-5.399767860736688 \times 10^{-1}$
-3	$-2.821001698611358 \times 10^{-2}$	$8.871479593866080 \times 10^{-2}$

influence distance provides a measure of dissipation of the fitted FDS. It is dimensionless with putative units of "grid squares."

Define the phase velocity factor, f_C, in terms of $\gamma = \angle g$, the phase of $g(\kappa)$, so that $c(\kappa) = f_C c_0$ where

$$f_C = \frac{\gamma}{\sigma \kappa}, \tag{138}$$

and c_0 is the phase velocity.

Similarly, define the group velocity factor, f_V, $V_G(\kappa) = f_V c_0$ as

$$f_V = -\frac{d\gamma}{\sigma d\kappa}. \tag{139}$$

In a phase accurate approximation, both f_C and f_V are very nearly unity.

Noting that $g + \mathrm{Re}(g) + i\mathrm{Im}(g)$ and using Eq. (137) and (138), the real and imaginary parts of the reference equation for the FDS, in terms of its influence distance and phase velocity factor, are

$$\mathrm{Re}(g) = \pm \frac{e^{-\frac{1}{\lambda}}}{1 + \tan^2(f_C \sigma \kappa)} \quad \text{and} \quad \mathrm{Im}(g) = \pm \frac{e^{-\frac{1}{\lambda}} \tan^2(f_C \sigma \kappa)}{1 + \tan^2(f_C \sigma \kappa)}, \tag{140}$$

where the signs are determined by the quadrant in the complex plane.

Require that $f_C \approx 1$ over the Nyquist interval and that $\lambda > 10$ for $\kappa \leq \pi/2$. To insure high dissipation at high spatial frequencies, further require that λ drop off rapidly for $\kappa > \pi/2$. Select two points to specify g, at $\kappa = 0.65\pi$ where $\lambda = 475$ and $f_C = 1$, and at $\kappa = 0.88\pi$ where $\lambda = 0.721$ and $f_C = 0.5$. Set $\lambda = \inf$ and $f_C = 1$ at $\kappa = 0$ and $\lambda = 0$ and $f_C = 0$ at $\kappa = \pi$. Then, linearly interpolate between these set point values when fitting the V coefficients of the FDS. The fitted values of the coefficients of T and L are shown in Table 1. The L coefficients were fitted at $\kappa = 0.11\pi$, $\kappa = 0.2108\pi$ and $\nu = 6.547614$. The FDS (the V coefficients) were fitted at $\kappa = 0.22\pi$, $\kappa = 0.3\pi$, $\mu = 2$ and $\sigma = 0.5$.

Table 2.   The coefficients of the stabilizing and high and low pass filters. j is the offset index relative to the point at which the filter is being applied.

j	Stabilizing filter	low pass filter	High pass filter
0	0.1250027758673328	0.1381966011250105	0.1381966011250105
1	0.3749972241326672	0.3618033988749895	-0.3618033988749895
2	0.3749972241326672	0.3618033988749895	0.3618033988749895
3	0.1250027758673328	0.1381966011250105	-0.1381966011250105

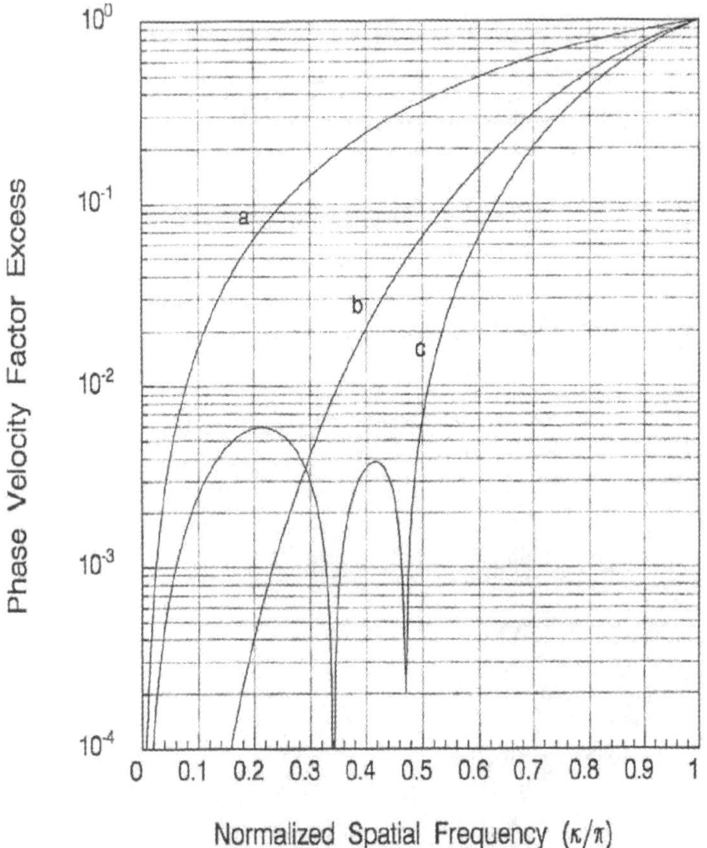

Fig. 14.   The response amplitude of the T and L operators. a = The L operator, b = the response of the continous spatial derivative, c = The T operator. The points of zero error in the L operator are indicated by the symbol o.

The response amplitude of the $\hat{T}$ and $\hat{L}$ operators is shown in Fig. 14. The ideal response of the spatial derivative is shown for reference. The spatial operator maintains high accuracy over the lower frequency portion of the Nyquist interval. The deviation form the ideal response at higher frequencies is necessary to insure phase accuracy while providing dissipation. The location of the points of zero fitting error are marked by the symbol o. Note that the fitting of the spatial operator results in an additional zero near 0.6π. The temporal operator resulting form the design requirements of the FDS compensates from the error in the spatial operator.

The influence distance properties of the fitted FDS are shown in Fig. 15. Also shown in the figure are the influence distance properties of the Euler/upwind FDS and the filtered form of the fitted FDS. The fitted FDS exhibits large influence distances over the lower portion of the Nyquist interval. The location of the fitting frequencies are marked by the symbol o. The attempt during the design was to place the higher frequency zero as far to the right as possible, and then to place the lower frequency zero at a point which resulted in a smooth response over the Nyquist interval. The result is a fit which maintains large influence distances over the lower half of the Nyquist interval and a rapid decrease to small values over the upper half of the interval.

The phase velocity factor, f_c, of the fitted FDS is unity over the entire Nyquist interval within an error of $\pm10^{-6}$. The group velocity factor, f_V, is unity, within an error of $\pm10^{-3}$. Hence, the design can be considered to be phase accurate.

The response amplitude of the digital filters for stabilizing the FDS and determining the energy ratio are shown in Fig. 16. The response amplitude of a Shapiro filter with a shaping factor of 0.5 is shown for reference. The polynomial form of the Shapiro filter is $P(z) = (z+1)^2$ resulting in two zeros at $\kappa = \pi$. The polynomial form of the stabilizing filter, given in Eq. (135), with $a = 0.997\pi$ results in three zeros clustered near $\kappa = \pi$. The high and low pass digital filters used to determine the energy ratio also have the same polynomial form given in Eq. (135) with $a = 0.2\pi$ and $a = 0.8\pi$, respectively. The values of the filter coefficients are given in Table 2.

For the linear case, $p = 1$ and $q = 0$, in Eq. (133) the accuracy of the simulation is controlled by the properties of the influence distance and the stabilizing filter. If the initial conditions are such that the stabilizing filter is not invoked, then the PSD of the evolved waveform essentially matches the shape of the influence distance curve as would be anticipated. If the stabilizing filter is invoked, then the PSD of the evolved wave is that of the

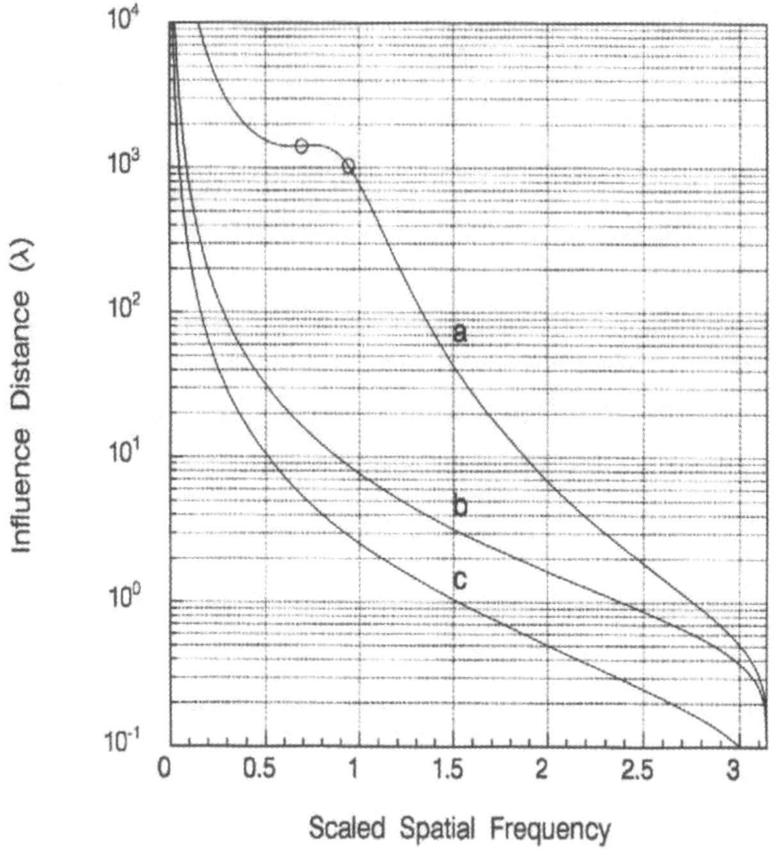

Fig. 15.   Fitted FDS influence distance. a =The general average FDS, b =The Euler upwind FDS, c =The general average FDS after application of the stabilizing filter. The points of zero error in the FDS are indicated by the symbol o.

response function of the stabilizing filter.

Intermediate settings of p and q with p, $q \in [0,1]$, produce waves with varying degrees of nonlinearity propagating to the right. The nonlinear performance of the FDS is illustrated for $p = 0$ and $q = 1$, the "one hump" wave, which has an analytic asymptotic solution [15]. The one hump wave begins as an arbitrary waveform which is greater than zero over a contiguous portion of the grid and zero everywhere else. The area of the initial wave form is an invariant. That is, given an initial value, $r(x,0) \geq 0$ for $a \leq x \leq b$ and $r(x,0) = 0$ elsewhere, and $A = \int_a^b r\,dx$, the long-term evolved waveform,

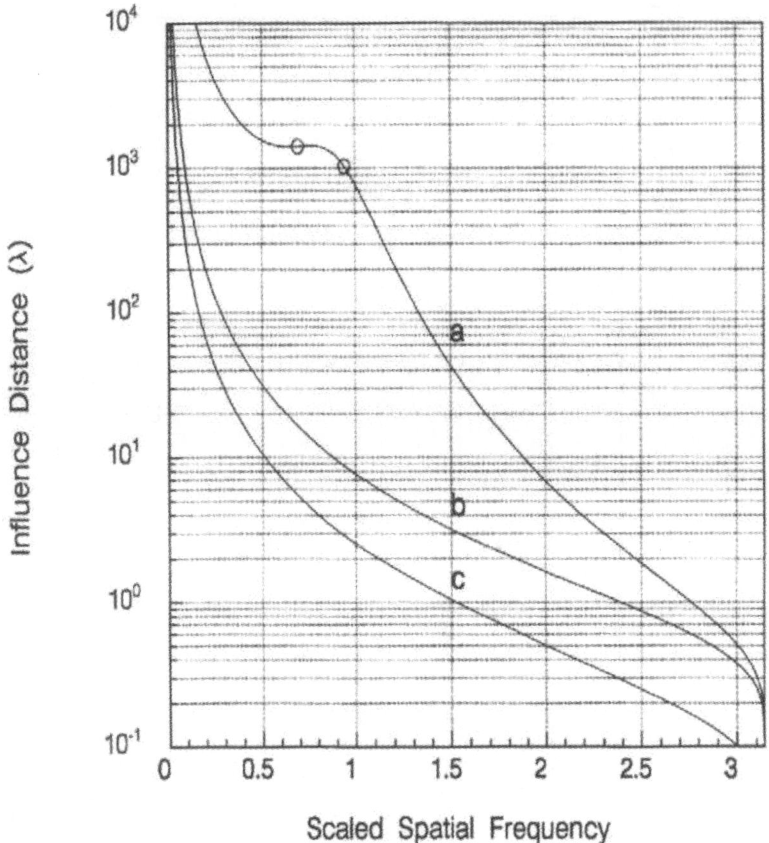

Fig. 16.　The responses of the digital filters. a = The high pass energy estimating filter,, b = The low pass energy estimating filter, c = The stabilizing filter (the response amplitude is the same regardless of the shift), d = The Shapiro filter.

if

$$r = \frac{x}{t}\frac{G(v)}{v}, \tag{141}$$

where

$$G(v) = \begin{cases} 1 & \text{if } v \in (0,1) \\ 0 & \text{otherwise} \end{cases} \tag{142}$$

and $v = x/\sqrt{2At}$, the area remains constant.

Selecting as an initial value a triangular wave with a shape similar to the fully evolved waveform, the initial value and the FDS evolve to the wave-

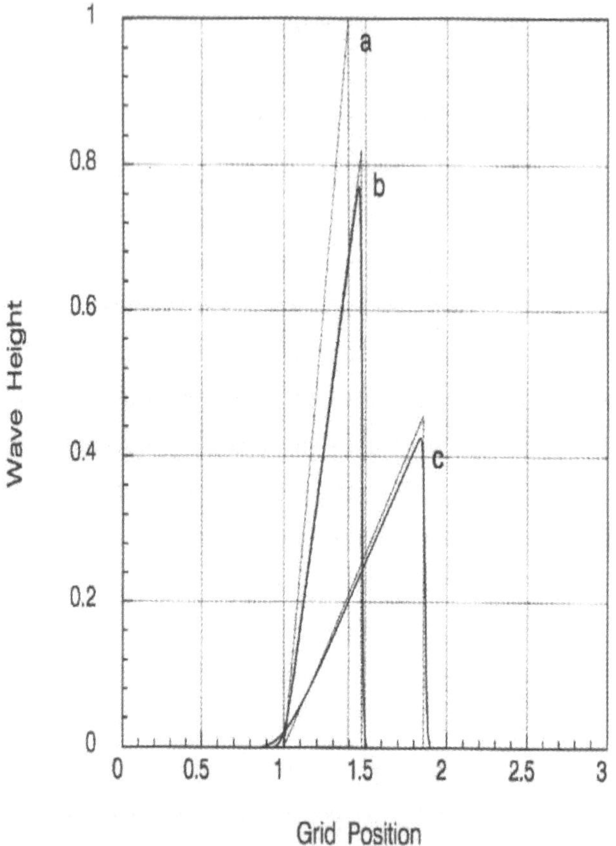

Fig. 17.　The evolved one hump wave solution to Burgers' equation. The general average FDS is shown with the solid line. The analytic asymptotic solution is shown with the dashed line. Here a is the Initial value, b is the evolution at one hundred time steps, and c is the evolution at 800 time steps.

forms shown in Fig. 17 at 100 and 800 time steps. Numerical experiments with other waveforms satisfying the initial conditions resulted in evolution to a similar shape.

The area of the curve, A, which is invariant in the analytic solution, varies by less than 0.3% in the simulation–the bulk of the variation occurring on the first time step where the stabilizing filter initially shapes the wave. The variation of A after this initial adjustment is approximately two orders of magnitude less.

The energy ratio for invoking the stabilizing filter was selected to be

$\rho_E = 7 \times 10^{-6}$. The tradeoff in selecting this value is between the slope of the face of the shock and the curvature at the peak and foot of the simulated wave versus the introduction of appreciable levels of spurious oscillation. Smaller values of ρ_E produce shallower slopes and more highly curved feet and peaks. Larger values of ρ_E produce steeper slopes and sharper feet and peaks at the expense of spurious oscillation. Hence, ρ_E plays a role similar to viscosity in the simulation in addition to controlling stability.

5.5. *Conclusion*

Rather than minimize the overall error, a heuristic method was pursued for each approximation. The heuristic was selected to insure that the fitting error vanished at *a priori* selected frequencies. We first used the method to build a spatial derivative approximation in isolation. We continued engineering additional physical approximations, culminating with Burgers Equation, with good success. Even more important, we built each approximation in a controlled way.

In our final equation, the linearized Burgers equation was used to fit the spatial and temporal coefficients in such a way as to allow monitoring of departures from the DRP condition. The heuristic placed the higher frequency zero as far to the right of the Nyquist interval as possible, and then placed lower frequency zeros so as to produce and overall smooth response over the Nyquist interval. To apply the resulting method to the nonlinear Burgers equation, a spatial digital filter was invoked to eliminate unstable growth in the low frequency region when the high/low energy ratio exceeds a threshold value.

The method provided for alternate designs based on the desired performance of the FDS. In the examples shown, accurate simulation of the phase and group velocity in the linearized equation was the primary design criterion. The method provides a means to simulate PDEs with strong non-linearities with a computationally efficient scheme, trading formal order of accuracy for an engineering approach. Concurrently, we decreased storage requirements and increased runtime efficiency.

The method demonstrates that, in general, deriving numerical approximations independently of the partial differential equation leaves the approximation at the experimental trial and error level. Comparison studies are then required to test which approximations work best for a family of equations. The resulting knowledge is not a quantification, but rather an amalgamation of a body of knowledge. Yet this is the current state-of-the-art in numerical models. Such a body of knowledge may provide intuition, but it cannot provide direct guidance. In contrast, the approach developed here provides direct control of the numerical approximation error. While the numerical approximation error has not been eliminated, it has been re-shaped to achieve desired results based directly on the physical properties for the PDE to be simulated.

To perform this "reshaping," the nature of the numerical approxima-tion error is parameterized to allow matching it to the physical effects one is trying to simulate, in order to achieve an overall simulation where the physical effects are well modeled despite the presence of numerical error. As the numerical approximation is at least as complicated as the PDE–whose solution may not be known–this is usually a difficult problem. Indeed, even for equations whose solution is known, the direct method proposed here for the viscid Burgers equation must be re-adopted to achieve an new bal-ance between the physical equation and numerical approximation. For each equation guidance comes from knowledge of the physical response within the approximation space used to fit the coefficients. Current research is de-ricted toward the viscid Burgers and Kortweg-deVries equations, both of which present unique modeling challenges.

One trade-off between the use of an implicit compact scheme, such as Kim and Lee's, and our explicit general difference approximation is the opportunity to use a reduced number of stencil points to achieve comparable levels of accuracy. Another is that implicit schemes usually require the inversion of a matrix. The final determination may well lie in the choice of derivative approximations for the remaining terms of the PDE. For example, with the advection equation, the choice of the temporal integration scheme introduces errors that may reduce the accuracy of the spatial derivative approximation ([5] [7] [16]).

6. Historical Notes

Lele [6] used as a starting point an implicit Pade approximation to the first spatial derivative originally developed by Hermite and described by Collatz [2]. Haras and Ta'asan [3] extended Lele's work to include the defi-nition of an optimal formulation of the fitting problem, which includes both the error resulting from the spatial approximation and the error resulting from the choice of a temporal integration scheme. They present an exam-ple fit using the advection equation. Lele's scheme has recently attracted attention in the aerodynamics community, where Yu *et al* [16] used Lele's coefficients to analyze performance with various order Runge-kutta schemes in the advection equation. Kim and Lee [5] present and analysis using co-efficients selected to maximize spatial frequency resolution f r various wave forms and Runge-kutta temporal integration schemes. Tam *et al* [14]. use a frequency domain formulation to develop a dispersion relation preserving (DRP) scheme for fitting implicit approximations to both the temporal and spatial derivatives of the advection equation. They use this scheme to examine error in simulating the linear Euler equation with discontinuous initial data. In so doing, they also examine the effects of the selective inclusion of artificial damping.

Using the fitted spatial approximation in a partial differential equation requires both temporal and spatial approximations working in tandem. For example, for the linear advection equation, $u_t + cu_x = 0$, this implies developing a first temporal derivative approximation capable of being fitted in tandem with the spatial derivative approximation along with development of appropriate convergence and consistency constraints. And integrated approach to fitting temporal and spatial approximations in tandem was first describe by Tam *et al* [14]. Tam notes that an important aspect of an accurate FDS approximation is the fidelity with which the dispersion relation of the Partial Differential Equation(PDE) is matched by the FDS. If the two are identical over the entire Nyquist interval of the Fourier transform of the approximation, then the approximation is said to be DRP. an example of a DRP FDS can be obtained by using a forward Euler temporal derivative approximation and a first-order upwind spatial derivative approximation in the FDS for the advection equation. The dispersion relations of the continuous PDE and the resulting FDS are the same in their respective frequency domains if a Courant-Friedrichs-Levy number of one is used, so that the FDS is DRP. Indeed, and essential aspect of the DRP condition is that the combination of temporal and spatial derivative approximation reduces to a discretized form of the dispersion relation of the PDE in the spatial frequency domain. Hence, the notion of a DRP condition for an isolated spatial derivative, as put forth in the Technical Note, is puzzling.

The definition of a temporal derivative suitable for use in constructing a fully fitted FDS is described in [9]. The fitting of FDS using frequency accurate approximations for both the temporal and spatial derivatives is described in [8]. In the latter paper, the emphasis is on fitting both the spatial and temporal derivatives to the response of the PDE in the frequency domain. This is primarily accomplished by fitting the temporal derivative approximation to produce the required response while maintaining a stable solution over the Nyquist interval.

Acknowledgements

This work was sponsored, in part, by the Naval Research Laboratory.

References

1. R. N. Bracewell. *The Fourier Transform and Its Applications*. McGrawHill, New York, 1986.
2. L. Collatz. *The Numerical Treatment of Differential Equations*. Springer-Verlag, New York, 1966.
3. Z. Haras and S. Ta'asan. Finite difference schemes for long-time integration. *J. Comp. Phys*, 114:265–279, 1994.
4. L. B. Jackson. *Digital Filters and Signal Processing*. Kluwer, Boston, 1986.
5. J. W. Kim and D. J. Lee. Optimized compact finite difference schemes with maximum resolution. *AIAA J.*, 34:887–893, 1996.
6. S. K. Lele. Compact finite difference schemes with spectral-like resolution. *J. Comp. Phys*, 103:14–42, 1992.
7. P. Orlin. *Frequency Accurate Finite Difference Approximations*. PhD thesis, University of Southern Mississippi, University of Southern Mississippi, Hattiesburg, MS, May 1996.
8. P. Orlin and A. L. Perkins. A frequency finite accurate difference scheme for burgers equation. *J. Comp Acoustics*, 6:321–335, 1998.
9. P. Orlin, A. L. Perkins, and G. Heburn. A frequency accurate temporal derivative finite difference approximation. *J. Comp Acoustics*, 5:371–382, 1997.
10. A. L. Perkins, P. Orlin, and L. F. Smedstead. A numerical methodology for reducing total simulation error. In *Conf Proc 4th International Conf on Numerical Methods and Applications*, Sophia, Bulgaria, 1998. Bulgarian Academy of Sciences.
11. Alice Qiao. PhD thesis, University of Southern Mississippi, Hattiesburg, MS.
12. R. D. Richmeyer and K. W. Morton. *Difference Methods for Initial-Value Problems*. Interscience Publishers, J. Wiley and Sons, New York, 1967.
13. F. G. Shuman. Numerical method in weather prediction ii smoothing and filtering. *Monthly Weather Review*, 85:357–361, 1957.
14. C. K. Tam, J. C. Webb, and Z. Dong. A study of the short wave components in computational acoustics. *J. Comput. Acoustics*, 1:1–30, 1993.
15. G. B. Whitham. *Linear and Nonlinear Waves*. Wiley & Sons, New York, 1973.
16. S. T. Yu, K. C. Hseih, and Y. L. Tsai. Simulating waves in flows by runge-kutta and compact difference schemes. *AIAA J.*, 33:421–429, 1995.

CHAPTER 14

NONSTANDARD DISCRETIZATION METHODS ON LOTKA-VOLTERRA DIFFERENTIAL EQUATIONS

Lih-Ing W. Roeger

Department of Mathematics and Statistics
Texas Tech University
lih-ing.roeger@ttu.edu

We apply some nonstandard discretization methods on the 2-dimensional and 3-dimensional Lotka-Volterra systems and compare the local stability of the equilibria. For 2-dimensional systems, we will present a class of symplectic nonstandard methods and show that some of the numerical methods preserve the local stability and the positivity of the solutions. One of the symplectic method, Kahan's method, will be applied to three-dimensional May-Leonard and Lotka-Volterra competitive systems. Kahan's method will preserve not only the local stability but also Hopf bifurcations.

1. Introduction

Lotka-Volterra model is a differential equation system used to model the interaction between two or more species in ecology system. Let x_i denote the densities of species i, the model is as the following

$$x_i' = x_i(r_i + \sum_{j=1}^{n} a_{ij}x_j), \quad i = 1, 2, \ldots, n.$$

Differential equation models are suitable for populations in which generations are overlapped. Insects, for instance, have distinct generations. In these species, difference equations are more appropriate than differential equations. One way to produce difference equation is to discretize a differential equation. Numerical methods had been applied to differential equations to produce difference equations for epidemic models [2,8] and for Lotka-Volterra systems [9,10,11,12,15,16,20,21,22,23]. A resulting difference

615

equation is said to be *dynamically consistent* with its continuous counterpart if they both exhibit the same qualitative behavior, such as stability, bifurcation, and chaos [10].

In this chapter, we will focus on the 2-dimensional and 3-dimensional Lotka-Volterra systems and on nonstandard discretization methods, especially the methods by Kahan and Mickens. Noncanonical symplectic methods will be discussed in Section 2. We will present a class of nonstandard numerical methods that are symplectic for the 2-dimensional Lotka-Volterra models. Kahan's method, Mickens's method, and some other nonstandard methods will also be discussed. Some nonstandard methods can be easily applied to three-dimensional systems. We will show the special case of 3-dimensional Lotka-Volterra competition models, namely the May-Leonard competition models, in Section 3. Finally, in Section 4, we apply three nonstandard methods on the more general 3-dimensional Lotka-Volterra competitive systems and show that Kahan's method will preserve not only the local stability but periodic solutions as well.

2. Two-Dimensional Lotka-Volterra Systems

2.1. *Predator-prey System*

The Lotka-Volterra model

$$x' = x - xy,$$
$$y' = -y + xy, \tag{1}$$

is a normalized mathematical model for predator-prey interaction. For positive initial conditions, $x(0) = x_0 > 0$ and $y(0) = y_0 > 0$, all solutions, except for the fixed-point at $(1, 1)$ are periodic [17].

Sanz-Serna [24] showed that an unconventional numerical method by W. Kahan

$$\frac{X - x}{h} = \frac{X + x}{2} - \frac{Xy + xY}{2}, \tag{2}$$
$$\frac{Y - y}{h} = -\frac{Y + y}{2} + \frac{Xy + xY}{2},$$

where

$$\begin{aligned}
X &= x(t + h) = x(k + 1), \\
Y &= y(t + h) = y(k + 1), \\
x &= \quad x(t) = \quad x(k), \\
y &= \quad y(t) = \quad y(k), \ k = 0, 1, 2, \ldots,
\end{aligned} \tag{3}$$

produces numerical solutions that do not spiral because it preserves the weighted area $(dx \wedge dy)/(xy)$ and therefore the transformation (2) is symplectic with respect to a noncanonical symplectic structure. Gander et al. [5] applied symplectic Euler method for the predator-prey system (1) and obtain the following that also preserves $(dx \wedge dy)/(xy)$.

$$\frac{X - x}{h} = x - xy, \qquad (4)$$
$$\frac{Y - y}{h} = -y + Xy.$$

Mickens [15] applied a different nonstandard numerical method and obtained the following discrete system

$$\frac{X - x}{\phi(h)} = (2x - X) - Xy, \qquad (5)$$
$$\frac{Y - y}{\phi(h)} = -Y + (2Xy - XY),$$

where $\phi(h) = h + O(h^2)$. This system is equivalent to

$$X = \frac{x(1 + 2\phi(h))}{1 + \phi(h) + \phi(h)y},$$
$$Y = \frac{y(1 + 2\phi(h)X)}{1 + \phi(h) + \phi(h)X}.$$

Therefore, it preserves the positivity of the solutions. Mickens's method (5) is probably the first nonstandard method that produced limit periodic cycles and solutions that are positive invariant for the predator-prey model (1).

Because the right-hand side of system (1), f and g, exhibit a symmetry $f(y, x) = -g(x, y)$ and $g(y, x) = -f(x, y)$, Mounim et al. [16] suggested in the numerical method that the right-hand side of each equation should transform in the opposite of the other under the simultaneous interchanges $x \leftrightarrow y$ and $k \leftrightarrow k+1$. Use the notation X and Y in (3), the above procedure means $x \leftrightarrow Y$ and $y \leftrightarrow X$. Their two methods are

$$\frac{X - x}{h} = x - (2Xy - xy), \qquad (6)$$
$$\frac{Y - y}{h} = -Y + (2Xy - XY),$$

and

$$\frac{X - x}{h} = (2x - X) - Xy, \tag{7}$$

$$\frac{Y - y}{h} = -(2Y - y) + Xy.$$

The two models (6) and (7) also ensure the positivity of the solutions. It is easy to verify that the positive cone $R_+^2 = \{(x, y)|x \geq 0, y \geq 0\}$ is invariant under these two methods by solving X and Y explicitly in terms of x and y. These two methods also give periodic cycles. However, the two methods, (6) and (7), approximate the exact solution of (1) better than the symplectic Euler's method (4) and Mickens's method (5). See Figure 1.

The above three methods, Mickens's method (5), and Mounim's two methods (6) and (7) are symplectic with respect to the same noncanonical symplectic structure as Kahan's method. In fact, the transformations defined by (5), (6), and (7) also preserve the weighted area $(dx \wedge dy)/(xy)$. This can be shown by following the proof in [24]. I will show the example of (6). Differentiate (6) and rearrange we can get

$$(1 - 2hy)\, dX = (1 + h - hy)\, dx + h(2X - x)\, dy,$$

$$-h(2y - Y)\, dX + (1 - h + hX)\, dY = (1 + 2hX)\, dy.$$

Take the wedge product of these equations to obtain

$$(1 - 2hy)(1 - h + hX)\, dX \wedge dY = (1 + h - hy)(1 + 2hX)\, dx \wedge dy. \tag{8}$$

On the other hand, we can rewrite the equations (6) as

$$(1 - 2hy)\, X = (1 + h - hy)\, x,$$

$$(1 - h + hX)\, Y = (1 + 2hX)\, y.$$

Multiplication of these two equations leads to

$$(1 - 2hy)(1 - h + hX)\, XY = (1 + h - hy)(1 + 2hX)\, xy.$$

This result along with (8) implies $(dX \wedge dY)/(XY) = (dx \wedge dy)/(xy)$. We can show that (5) and (7) are also symplectic in a similar way.

Following the rule that the right-hand side of each equation transforms in the opposite of the other under the simultaneous interchanges $x \leftrightarrow Y$ and $y \leftrightarrow X$, we are able to construct a class of noncanonical symplectic discretization methods [23]. For the following Lotka-Volterra system

$$x' = Ax + Bxy,$$

$$y' = Cy + Dxy, \tag{9}$$

we have the following nonstandard discretization method:

$$\frac{X - x}{h} = A(\alpha_1 x + \alpha_2 X) + B(\beta_1 xy + \beta_2 Xy + \beta_3 xY + \beta_4 XY),$$

$$\frac{Y - y}{h} = C(\alpha_1 Y + \alpha_2 y) + D(\beta_1 XY + \beta_2 Xy + \beta_3 xY + \beta_4 xy), \quad (10)$$

where $\alpha_1 + \alpha_2 = 1$ and $\beta_1 + \beta_2 + \beta_3 + \beta_4 = 1$. The method (10) will be denoted as method $(\alpha_1 x + \alpha_2 X, \beta_1 xy + \beta_2 Xy + \beta_3 xY + \beta_4 XY)$. Then if β_i's satisfy the following condition,

$$\beta_1 \beta_4 = 0, \quad \beta_2 \beta_4 = 0 \quad \text{and} \quad \beta_1 \beta_3 = 0, \quad (11)$$

the above method (10) is sympletic with a noncanonical structure.

Theorem 1: [23] *If A, B, C, D, α_i, and β_i are real constants, and β_i satisfy (11). Then the transformation defined by (10) preserves the weighted area $(dx \wedge dy)/(xy)$.*

Proof. Without loss of generality, let $h = 1$. Following the proof in [24], our goal is to show that we can find two functions $F(x, y, X, Y)$ and $G(x, y, X, Y)$ such that

$$F(x, y, X, Y) \, dX \wedge dY = G(x, y, X, Y) \, dx \wedge dy,$$

$$F(x, y, X, Y) \, XY = G(x, y, X, Y) \, xy. \quad (12)$$

Then $(dX \wedge dY)/(XY) = (dx \wedge dy)/(xy)$ follows.

Differentiate the two equations in (10) and rearrange to get

$$\begin{aligned}
(1 - A\alpha_2)dX &- B(\beta_2 y + \beta_4 Y)dX - B(\beta_3 x + \beta_4 X)dY = (1 + A\alpha_1)dx \\
&+ B(\beta_1 y + \beta_3 Y)dx + B(\beta_1 x + \beta_2 X)dy, \\
&- D(\beta_1 Y + \beta_2 y)dX + (1 - C\alpha_1)dY \\
&- D(\beta_1 X + \beta_3 x)dY = D(\beta_3 Y + \beta_4 y)dx \\
&+ (1 + C\alpha_2)dy + D(\beta_2 X + \beta_4 x)dy.
\end{aligned}$$

Take the wedge product of these equations to obtain

$$\begin{aligned}
[(1 &- A\alpha_2)(1 - C\alpha_1) - D(1 - A\alpha_2)(\beta_1 X + \beta_3 x) \\
&- B(1 - C\alpha_1)(\beta_2 y + \beta_4 Y) \\
&+ BD(\beta_1 \beta_2 Xy - \beta_2 \beta_4 Xy - \beta_1 \beta_3 xY + \beta_3 \beta_4 xY)] \, dX \wedge dY \\
= [(1 &+ A\alpha_1)(1 + C\alpha_2) + D(1 + A\alpha_1)(\beta_2 X + \beta_4 x) \\
&+ B(1 + C\alpha_2)(\beta_1 y + \beta_3 Y) \\
&+ BD(\beta_1 \beta_2 Xy - \beta_2 \beta_4 Xy - \beta_1 \beta_3 xY + \beta_3 \beta_4 xY)] \, dx \wedge dy.
\end{aligned}$$

On the other hand, we rewrite the equations (10) as

$$(1 - A\alpha_2)X - B(\beta_2 y + \beta_4 Y)X = (1 + A\alpha_1)x + B(\beta_1 y + \beta_3 Y)x$$
$$(1 - C\alpha_1)Y - D(\beta_1 X + \beta_3 x)Y = (1 + C\alpha_2)y + D(\beta_2 X + \beta_4 x)y$$

Multiplication and rearrange of these equations leads to

$$\begin{aligned}
[(1 - A\alpha_2)(1 - C\alpha_1) &- D(1 - A\alpha_2)(\beta_1 X + \beta_3 x) \\
&- B(1 - C\alpha_1)(\beta_2 y + \beta_4 Y) \\
&+ BD(\beta_2\beta_3 xy + \beta_1\beta_2 Xy + \beta_3\beta_4 xY + \beta_1\beta_4 XY)]\ XY \\
= [(1 + A\alpha_1)(1 + C\alpha_2) &+ D(1 + A\alpha_1)(\beta_2 X + \beta_4 x) \\
&+ B(1 + C\alpha_2)(\beta_1 y + \beta_3 Y) \\
&+ BD(\beta_1\beta_4 xy + \beta_1\beta_2 Xy + \beta_3\beta_4 xY + \beta_2\beta_3 XY)]\ xy.
\end{aligned}$$

In the above two equations, both sides can cancel out $BD\beta_2\beta_3 xyXY$. Then under the condition (11), we have

$$\begin{aligned}
F(x, y, X, Y) = {} & (1 - A\alpha_2)(1 - C\alpha_1) - D(1 - A\alpha_2)(\beta_1 X + \beta_3 x) \\
& -B(1 - C\alpha_1)(\beta_2 y + \beta_4 Y) + BD(\beta_1\beta_2 Xy + \beta_3\beta_4 xY), \\
G(x, y, X, Y) = {} & (1 + A\alpha_1)(1 + C\alpha_2) + D(1 + A\alpha_1)(\beta_2 X + \beta_4 x) \\
& +B(1 + C\alpha_2)(\beta_1 y + \beta_3 Y) + BD(\beta_1\beta_2 Xy + \beta_3\beta_4 xY).
\end{aligned}$$

The theorem is proved. $\square$

The theorem says that the parameters α_i's have no contribution to the symplectic methods. The condition on β_i's, (11), says that at most two nonzero β_i's can be in the equation. Hence, there are 7 possible cases for β_i's: $(\beta_1, \beta_2, \beta_3, \beta_4) = (1, 0, 0, 0)$, $(0, 1, 0, 0)$, $(0, 0, 1, 0)$, $(0, 0, 0, 1)$, $(\beta, 1 - \beta, 0, 0)$, $(0, \beta, 1 - \beta, 0)$, and $(0, 0, \beta, 1 - \beta)$. Only xy and Xy, Xy and xY, xY and XY can appear together to produce symplectic methods for the system (9).

The nonstandard method (10) generalizes W. Kahan's method and Mounim and de Dormale's methods. As a result, the parameters β_i's chosen for Kahan's method (2) and Mounim's methods (6) and (7), are $(\beta_1, \beta_2, \beta_3, \beta_4) = (0, 1/2, 1/2, 0)$, $(-1, 2, 0, 0)$, and $(0, 1, 0, 0)$. They are methods $((X + x)/2, (xY + Xy)/2)$, $(x, 2Xy - xy)$, and $(2x - X, Xy)$ respectively. All three parameter sets satisfy the condition (11) for β_i's. Therefore, they all are symplectic with the same noncanonical structure. Although the symplectic Euler's method (4) and Mickens's method (5) do not satisfy the condition for β, (11), they also have the same symplectic structure.

In Table 1, we give some examples of symplectic methods. The first row (column) shows 5 (22) choices for the x (xy) term in the first equation of (9).

There are total 65 symplectic methods out of 110 possible combinations. For the Lotka-Volterra predator-prey model (1), the positive cone R_+^2 is invariant under 10 of the symplectic methods. Note that the non-positivity preserving schemes should not be considered as discrete-time models for the Lotka-Volterra system. In Figure 1, we show some discretization methods on the predator-prey model (1). The exact solutions are shown as a solid line and the numerical methods are shown in circle curves. The step-size is chosen to be the same among all of the methods, $h = 0.1$, except for the symplectic Euler's method (4) in which $h = 0.05$. Figure 1(f) demonstrates one of the methods, method (x, Xy) found using rule (10). Method (x, Xy) is the simplest symplectic methods in Table 1 that preserve the positivity of solutions and the weighted area $(dx \wedge dy)/(xy)$.

Table 1. Some symplectic numerical methods for the Lotka-Volterra model (9) according to (10). Entries in the first row are some choices for x, and in the first column some choices for xy in the first equation of (10). The mark "o" indicates the combination is symplectic, "x" not symplectic. The mark "+" means the positive cone R_+^2 is invariant for the predator-prey model (1) under this method. "M_1" and "M_2" are the two methods by Mounim et al. "K" is the method by Kahan. Sixty-five methods are symplectic with respect to the noncanonical structure.

	x	X	$2x - X$	$2X - x$	$(x + X)/2$
1. xy	o	o	o	o	o
2. Xy	o, +	o	o, +, M_2	o	o
3. xY	o	o	o	o	o
4. XY	o, +	o	o, +	o	o
5. $2xy - Xy$	o	o	o	o	o
6. $2xy - xY$	x	x	x	x	x
7. $2xy - XY$	x	x	x	x	x
8. $2Xy - xy$	o, +, M_1	o	o, +	o	o
9. $2Xy - xY$	o, +	o	o, +	o	o
10. $2Xy - XY$	x	x	x	x	x
11. $2xY - xy$	x	x	x	x	x
12. $2xY - Xy$	o	o	o	o	o
13. $2xY - XY$	o	o	o	o	o
14. $2XY - xy$	x	x	x	x	x
15. $2XY - Xy$	x	x	x	x	x
16. $2XY - xY$	o, +	o	o, +	o	o
17. $(xy + Xy)/2$	o	o	o	o	o
18. $(xy + xY)/2$	x	x	x	x	x
19. $(xy + XY)/2$	x	x	x	x	x
20. $(Xy + xY)/2$	o	o	o	o	o, K
21. $(Xy + XY)/2$	x	x	x	x	x
22. $(xY + XY)/2$	o	o	o	o	o

The method (10) is symplectic for any 2-dimensional Lotka-Volterra system of the type (9). For systems that have x^2 or y^2 terms in the equation, the method is not symplectic with the same noncanonical structure. For example, apply the method (x, xy) to the following system

$$x' = -x + 2xy + x^2,$$
$$y' = y - 2xy - y^2, \tag{13}$$

with x^2 (y^2) being discretized by xX (yY). Then the discrete system becomes

$$\frac{X - x}{h} = -x + 2xy + xX,$$
$$\frac{Y - y}{h} = Y - 2XY - yY. \tag{14}$$

Differentiate the system and apply wedge product we obtain

$$\frac{dX \wedge dY}{XY} = \left(\frac{1 - hY}{1 - hx}\right)\frac{dx \wedge dy}{xy}.$$

and if we apply method (X, xy) to the system (13) we obtain

$$\frac{dX \wedge dY}{XY} = \left(\frac{1 + h - hY}{1 + h - hx}\right)\frac{dx \wedge dy}{xy}.$$

The transformations by the two methods (x, xy) and (X, xy) are not symplectic. We suspect that symplectic method is only for two dimensional Lotka-Volterra models without x^2 and y^2 terms.

The rule of simultaneous interchanging $x \leftrightarrow Y$ and $y \leftrightarrow X$, is difficult to apply to higher dimension Lotka-Volterra systems. But there is one exception, Kahan's method. Because Kahan's method take the average of the nonlocal terms, it works for all autonomous differential equations $x' = f(x)$, where $f(x)$ is at most quadratic [19]. We will show in the next subsection that Kahan's method preserves the stability criteria for 2-dimensional Lotka-Volterra competitive system.

2.2. *Competitive System*

The following is a 2-dimensional Lotka-Volterra competitive system

$$x' = x(r_1 - a_{11}x - a_{12}y),$$
$$y' = y(r_2 - a_{21}x - a_{22}y), \tag{15}$$

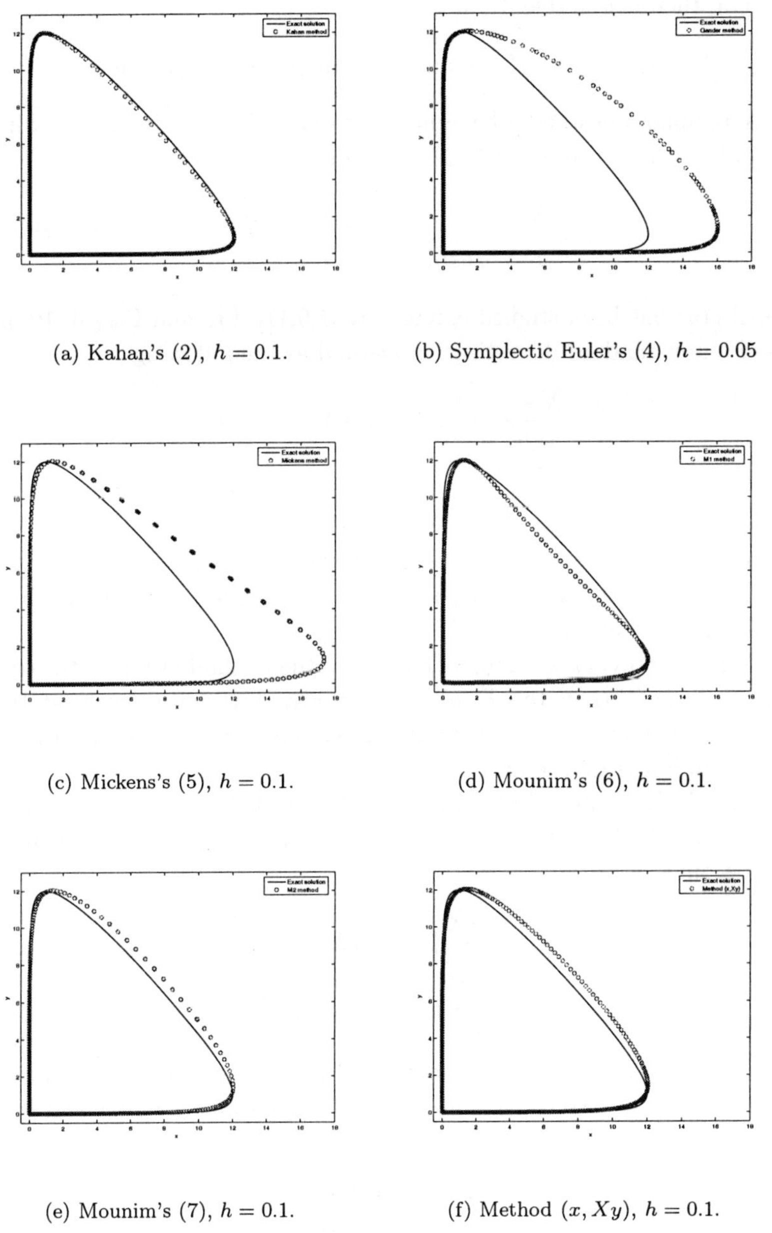

(a) Kahan's (2), $h = 0.1$.

(b) Symplectic Euler's (4), $h = 0.05$

(c) Mickens's (5), $h = 0.1$.

(d) Mounim's (6), $h = 0.1$.

(e) Mounim's (7), $h = 0.1$.

(f) Method (x, Xy), $h = 0.1$.

Fig. 1. Some numerical solutions for the predator-prey model (1). All of the methods in the figures produce limit cycles.

where, all parameters are nonnegative. If there exists a positive equilibrium (x^*, y^*), then it is stable if

$$\tilde{T} = a_{11}x^* + a_{22}y^* > 0 \text{ and } \tilde{D} = (a_{11}a_{22} - a_{12}a_{21})x^*y^* > 0.$$

A popular nonstandard method is to consider equation. (15) with a piecewise constant arguments [1] to obtain

$$X = x\exp(r_1 - a_{11}x - a_{12}y),$$
$$Y = y\exp(r_2 - a_{21}x - a_{22}y). \tag{16}$$

Model (16) has been studied extensively [7,9,11]. Liu and Elaydi [10] proposed a nonstandard discretization method as the following

$$\frac{X - x}{\phi(h)} = r_1 x - a_{11}xX - a_{12}Xy,$$
$$\frac{Y - y}{\phi(h)} = r_2 y - a_{21}xY - a_{22}yY.$$

Their discretized system possesses dynamical consistency with the continuous model (15). Other methods applied to the competitive system can also be found in the literature [1,10].

From the previous section, we know that the method (10) plus the transformation of x^2 (or y^2) to xX (yY) will not give a symplectic method. However, they may preserve local stability of system (15). We will use Kahan's method to illustrate the idea.

Using Kahan's method $((x+X)/2, (xY+Xy)/2)$ and replace x^2 and y^2 by xX and yY, the difference equation system from the continuous model (15) becomes

$$\frac{X - x}{h} = r_1\left(\frac{x + X}{2}\right) - a_{11}xX - a_{12}\left(\frac{Xy + xY}{2}\right),$$
$$\frac{Y - y}{h} = r_2\left(\frac{y + Y}{2}\right) - a_{21}\left(\frac{Xy + xY}{2}\right) - a_{22}yY, \tag{17}$$

where the step size h satisfies $0 < h < 1$. Since the system (17) is linear in X and Y, we can express X and Y explicitly as a function of x and y. We have

$$X = x\,\frac{f_1(x, y)}{g(x, y)},$$
$$Y = y\,\frac{f_2(x, y)}{g(x, y)},$$

where

$$g(x,y) = 2h^2[a_{11}a_{21}x^2 + 2a_{11}a_{22}xy + a_{12}a_{22}y^2] + hx[a_{21}(2 - hr_1)$$
$$+ 2a_{11}(2 - hr_2)] + hy[a_{12}(2 - hr_2) + 2a_{22}(2 - hr_1)]$$
$$+ (2 - hr_1)(2 - hr_2),$$
$$f_1(x,y) = a_{21}hx(2 + hr_1) + hy[2a_{22}(2 + hr_1) - a_{12}(2 + hr_2)]$$
$$+ (2 - hr_2)(2 + hr_1),$$
$$f_2(x,y) = a_{12}hy(2 + hr_2) + hx[2a_{11}(2 + hr_2) - a_{21}(2 + hr_1)]$$
$$+ (2 - hr_1)(2 + hr_2).$$

From the above equations, we can see that by choosing step-size h sufficiently small such that $2 - hr_1 \geq 0$, $2 - hr_2 \geq 0$, $2a_{22}(2 + hr_1) - a_{12}(2 + hr_2) \geq 0$, and $2a_{11}(2 + hr_2) - a_{21}(2 + hr_1) \geq 0$, then the positive cone R_+^2 is invariant.

The discrete model (17) and the continuous model (15) share the same equilibria. For the discrete model (17), the Jacobian matrix evaluated at positive equilibrium (x^*, y^*) is

$$J = I - \frac{2h}{\tilde{\Phi}(h)} \begin{pmatrix} 2a_{11}x^* + h\tilde{D} & 2a_{12}x^* \\ 2a_{21}y^* & 2a_{22}y^* + h\tilde{D} \end{pmatrix},$$

where

$$\tilde{\Phi}(h) = h^2\tilde{D} + 2h\tilde{T} + 4.$$

The characteristic equation associated with the Jacobian matrix is $p(\lambda)$. By the Jury conditions [17], (x^*, y^*) is stable if

$$p(1) = 4h^2\tilde{D}/\tilde{\Phi}(h) > 0,$$
$$p(-1) = 16/\tilde{\Phi}(h) > 0,$$
$$|p(0)| = |1 - 4h\tilde{T}/\tilde{\Phi}(h)| < 1.$$

If $\tilde{T} > 0$ and $\tilde{D} > 0$ then $p(1) > 0$, $p(-1) > 0$, and $|p(0)| < 1$ follow immediately. It is not difficult to show that if $p(1) > 0$, $p(-1) > 0$, and $|p(0)| < 1$ then $\tilde{T} > 0$ and $\tilde{D} > 0$. Therefore, we have the following theorem.

Theorem 2: [20] *Assume that the 2-dimensional Lotka-Volterra system (15) has a positive equilibrium (x^*, y^*). Then the stability criteria of the equilibrium (x^*, y^*) are the same for both the continuous system (15) and the discrete system (17).*

Theorem 2 says that Kahan's discretization method preserves the local dynamics of the 2-dimensional Lotka-Volterra competitive system. This is not surprising, since it was shown [19] that Kahan's method will preserve the local stability of any autonomous system $x' = f(x)$, where $f(x)$ is at most quadratic.

We conjecture that all symplectic methods in Table 1 will also preserve local stability for the competition system (15) and the following predator-pray system

$$x' = x(r_1 - a_{11}x - a_{12}y), \tag{18}$$
$$y' = y(r_2 + a_{21}x - a_{22}y).$$

However, none of the symplectic methods in Table 1 (with x^2 transformed to xX and y^2 transformed to yY) preserve the positivity of the solutions for the competitive system (15) and the predator-prey system (18) unless the step-size h is sufficiently small.

3. The May-Leonard Competitive System

The following is a 3-dimensional May-Leonard competitive system [3,13].

$$x_1' = x_1(1 - x_1 - \alpha_1 x_2 - \beta_1 x_3),$$
$$x_2' = x_2(1 - \beta_2 x_1 - x_2 - \alpha_2 x_3), \tag{19}$$
$$x_3' = x_3(1 - \alpha_3 x_1 - \beta_3 x_2 - x_3).$$

Chi et al. [3] studied the above system (19) under the assumption

$$0 < \alpha_i < 1 < \beta_i. \tag{20}$$

Let $A_i = 1 - \alpha_i$ and $B_i = \beta_i - 1$. They showed that the three-species equilibrium P_0 is globally asymptotically stable if $B_1 B_2 B_3 < A_1 A_2 A_3$, unstable if $B_1 B_2 B_3 > A_1 A_2 A_3$, and if $B_1 B_2 B_3 = A_1 A_2 A_3$, a degenerate Hopf bifurcation occurs and there exists a family of neutrally stable periodic solutions. In the case $\alpha_i = \alpha$ and $\beta_i = \beta$, P_0 is globally asymptotically stable if $\alpha + \beta < 2$, unstable if $\alpha + \beta > 2$, and a Hopf bifurcation occurs when $\alpha + \beta = 2$.

Since May-Leonard system is well understood, it is an excellent example for testing the nonstandard methods. In this section, we will apply two of them to this model. The first one is similar to the 2-dimensional system (16) by piecewise constant argument and extend it to 3-dimensional system to obtain

$$X_1 = x_1 \exp\left[r(1 - x_1 - \alpha_1 x_2 - \beta_1 x_3)\right],$$
$$X_2 = x_2 \exp\left[r(1 - \beta_2 x_1 - x_2 - \alpha_2 x_3)\right], \tag{21}$$
$$X_3 = x_3 \exp\left[r(1 - \alpha_3 x_1 - \beta_3 x_2 - x_3)\right].$$

We need to restrict the parameter r to be

$$0 < r \leq 2.$$

The second method we will consider is Kahan's method. Apply Kahan's method to the May-Leonard model (19) and rearrange we have the following

$$X_1 = \frac{(2 + h - \alpha_1 h X_2 - \beta_1 h X_3)\, x_1}{2 - h + 2hx_1 + \alpha_1 hx_2 + \beta_1 hx_3},$$
$$X_2 = \frac{(2 + h - \beta_2 h X_1 - \alpha_2 h X_3)\, x_2}{2 - h + 2hx_2 + \beta_2 hx_2 + \alpha_2 hx_3}, \tag{22}$$
$$X_3 = \frac{(2 + h - \alpha_3 h X_1 - \beta_3 h X_2)\, x_3}{2 - h + 2hx_3 + \alpha_3 hx_1 + \beta_3 hx_2}.$$

There are other types of discrete-time competition models. For example, Liu and Elaydi [10] used Mickens's nonstandard discretization method and derived a 2-dimensional Lotka-Volterra competitive model that is dynamically consistent with the continuous model. If we apply their nonstandard discretization method to the continuous M-L model, the discrete M-L model takes the form

$$X_1 = \frac{(1 + \phi(h))\, x_1}{1 + \phi(h)\, (x_1 + \alpha_1 x_2 + \beta_1 x_3)},$$
$$X_2 = \frac{(1 + \phi(h))\, x_2}{1 + \phi(h)\, (\beta_2 x_1 + x_2 + \alpha_2 x_3)}, \tag{23}$$
$$X_3 = \frac{(1 + \phi(h))\, x_3}{1 + \phi(h)\, (\alpha_3 x_1 + \beta_3 x_2 + x_3)},$$

where $\phi_i(h) = h + o(h)$ and $0 < \phi_i(h) < 1$. The local dynamics of the discrete M-L model (23) is similar to that of the discrete M-L model (21). See [21]. The two discrete-time M-L models, (21) and (23) preserve the positivity of solutions but not the model by Kahan's method (22).

3.1. *May-Leonard Model of Exponential Form*

3.1.1. *Symmetric May-Leonard competition model*

The discrete symmetric May-Leonard model (21) is the model under the condition $\alpha_i = \alpha$ and $\beta_i = \beta$ for $i = 1, 2, 3$ and $0 < \alpha < 1 < \beta$. This is a discrete analog of the symmetric May-Leonard model (19). We show that the local dynamics between the two models are similar, but not the same. We only consider the local dynamics at the interior equilibrium $P_0 = (1,1,1)/(1+\alpha+\beta)$.

To consider the stability of the equilibrium P_0, let $A = 1 - \alpha > 0$ and $B = \beta - 1 > 0$. And let

$$\mu_e = \frac{(B-A)(B-A+3)}{A^2 + AB + B^2} + r. \tag{24}$$

We have the following theorem.

Theorem 3: [21] *For system* (21), *assume* $\alpha_i = \alpha = 1 - A$ *and* $\beta_i = \beta = B + 1$ *for all* $i = 1,2,3$. *Let* μ_e *be defined as in* (24). *Then the interior equilibrium* P_0 *is locally asymptotically stable if* $\mu_e < 0$, *and unstable if* $\mu_e > 0$. *And as* $r \to 0^+$, *the stability criterion approaches that of the continuous M-L model, i.e.* P_0 *is stable if* $\alpha+\beta < 2$ *and unstable if* $\alpha+\beta > 2$.

Proof. The interior equilibrium is $P_0 = (p,p,p) = (1,1,1)/(1+\alpha+\beta)$. The Jacobian matrix evaluated at P_0 is:

$$J_{P_0} = I - rp \begin{pmatrix} 1 & \alpha & \beta \\ \beta & 1 & \alpha \\ \alpha & \beta & 1 \end{pmatrix}.$$

One of the eigenvalues of J_{P_0} is $1-r$. The other two eigenvalues are complex conjugates. They are

$$\lambda, \bar{\lambda} = \frac{1}{2}[2 + rp(\alpha + \beta - 2)] \pm i \frac{\sqrt{3}}{2} rp(\beta - \alpha).$$

Then $|\lambda| < 1$ if

$$\frac{1}{4}[2 + rp(\alpha + \beta - 2)]^2 + \frac{3}{4}[rp(\beta - \alpha)]^2 < 1.$$

This leads to $(\alpha + \beta - 2) + rp(\alpha^2 - \alpha - \alpha\beta - \beta + 1 + \beta^2) < 0$. Substitute $A = 1 - \alpha$ and $B = \beta - 1$ into the inequality. We obtain

$$(B - A) + rp(A^2 + AB + B^2) < 0.$$

Since $p = 1/(1 + \alpha + \beta) = 1/(B - A + 3) > 0$, we can divide the above inequality by $p(A^2 + AB + B^2)$ and get a new inequality

$$\frac{(B-A)(B-A+3)}{A^2 + AB + B^2} + r < 0. \tag{25}$$

We conclude that P_0 is stable if $\mu_e < 0$.

Since $B - A + 3 = \alpha + \beta + 1$ and $A^2 + AB + B^2$ are both greater than zero, as $r \to 0^+$, the inequality (25) is equivalent to $B - A < 0$ or $\alpha + \beta < 2$ which is the stability criterion for the symmetric M-L model (19). $\qquad\square$

Similar to the continuous system, a discrete Hopf bifurcation (Neimark-Sacker bifurcation) may occur at P_0 near $\mu_e = 0$. Near P_0 and $\mu_e = 0$, the 3-dimensional system (21) can be transformed to a 2-dimensional system on its center manifold. Then to consider Hopf bifurcation, we can restrict our system to the following 2-dimensional system in w_1-w_2 space,

$$\begin{pmatrix} W_1 \\ W_2 \end{pmatrix} = \begin{pmatrix} \mathrm{Re}(\lambda) & -\mathrm{Im}(\lambda) \\ \mathrm{Im}(\lambda) & \mathrm{Re}(\lambda) \end{pmatrix} \begin{pmatrix} w_1 \\ w_2 \end{pmatrix} + \text{h.o.t.} \tag{26}$$

The dynamics near P_0 of the discrete M-L model (21) is the same as the dynamics near the origin of the two-dimensional system (26). Therefore, we can apply the Hopf bifurcation theorem for maps [6].

Theorem 4: [21] *A Neimark-Sacker bifurcation occurs at P_0 in the symmetric M-L model (21) when $\mu_e = 0$.*

Proof. There are four conditions to verify (i)–(iv) in [6]. Let μ_e be defined as in (24) and F be defined by the two-dimensional system (26), such that $F : \mathbb{R} \times \mathbb{R}^2 \to \mathbb{R}^2; (\mu_e, \mathbf{w}) \to F(\mu_e, \mathbf{w})$. Then, condition (i) holds: $F(\mu_e, 0) = 0$ for μ_e near 0.

Let

$$\lambda(\mu_e) = \frac{1}{2}[2 + rp(\alpha + \beta - 2)] + \mathrm{i}\,\frac{\sqrt{3}}{2} rp(\beta - \alpha),$$

then $\lambda(\mu_e)$ and $\bar\lambda(\mu_e)$ are eigenvalues of J_{P_0}. Under the assumption $\beta > \alpha$, the imaginary part of $\lambda(\mu_e)$ is not zero. And we know that $\mu_e = 0$ if and only if $|\lambda(\mu_e)| = 1$. Therefore, condition (ii) holds: $DF(\mu_e, 0)$ has two non-real eigenvalues $\lambda(\mu_e)$ and $\bar\lambda(\mu_e)$ for μ_e near 0, with $|\lambda(0)| = 1$.

When $\mu_e = 0$ we have $B < A$. After a tedious calculation, we obtain

$$|\lambda(\mu_e)|^2 = \frac{A^2 + AB + B^2}{(B - A + 3)^2}\,\mu_e^2 + \frac{A - B}{B - A + 3}\,\mu_e + 1.$$

Then the derivative of $|\lambda(\mu_e)|$ evaluated at $\mu_e = 0$ is

$$\frac{d}{d\mu_e}|\lambda(\mu_e)|_0 = \frac{A - B}{2(B - A + 3)} > 0$$

since $B - A + 3 = \alpha + \beta + 1 > 0$ and $B < A$. Therefore, the condition (iii) holds: $\dfrac{d}{d\mu_e}|\lambda(\mu_e)| > 0$ at $\mu_e = 0$.

Substitute $r = -f(A, B)$ into $\lambda(\mu_e)$ so that $\mu_e = 0$. Then we obtain

$$\lambda(0) = \frac{(A+B)^2 + 2AB}{2(A^2 + AB + B^2)} + \frac{\sqrt{3}\,(A-B)(A+B)}{2(A^2 + AB + B^2)}\,\mathrm{i}.$$

Since $A > B > 0$, the real part of $\lambda(0)$ is positive and the imaginary part of $\lambda(0)$ is never zero. The eigenvalues $\lambda(0)$ and $\bar{\lambda}(0)$ cannot equal ± 1, $\pm i$, or $(-1 \pm \sqrt{3}\,i)/2$, which are the possible roots of $x^k = 1$, for $k = 1, 2, 3, 4$. Therefore condition (iv) $\lambda^k(0) \neq 1$ for $k = 1, 2, 3, 4$ holds.

By the Neimark-Sacker bifurcation theorem([6], Theorem 15.31, page 474), conditions (i), (ii), (iii), and (iv) are sufficient conditions for a bifurcation to occur at $\mu_e = 0$ in system (26). Therefore, at P_0 in system (21), there is a discrete Hopf bifurcation. $\square$

3.1.2. *Asymmetric May-Leonard competition model*

For the asymmetric discrete M-L model (21), the local dynamics are very similar to the dynamics of the symmetric case. The interior equilibrium is $P_0 = (p_1, p_2, p_3)$ where p_1, p_2, and p_3 satisfy

$$\begin{aligned}
p_1 + \alpha_1 p_2 + \beta_1 p_3 &= 1, \\
\beta_2 p_1 + p_2 + \alpha_2 p_3 &= 1, \\
\alpha_3 p_1 + \beta_3 p_2 + p_3 &= 1.
\end{aligned} \tag{27}$$

The Jacobian matrix evaluated at the interior equilibrium P_0 has the form:

$$J_{P_0} = I - \mathrm{diag}(r p_i)\,\tilde{\mathcal{M}},$$

where

$$\mathrm{diag}(r p_i) = \begin{pmatrix} r p_1 & 0 & 0 \\ 0 & r p_2 & 0 \\ 0 & 0 & r p_3 \end{pmatrix} \text{ and } \tilde{\mathcal{M}} = \begin{pmatrix} 1 & \alpha_1 & \beta_1 \\ \beta_2 & 1 & \alpha_2 \\ \alpha_3 & \beta_3 & 1 \end{pmatrix}.$$

Let $A_i = 1 - \alpha_i$ and $B_i = \beta_i - 1$. Because the parameters α_i and β_i satisfy $0 < \alpha_i < 1 < \beta_i$, we have $0 < A_i < 1$ and $0 < B_i$. Let Δ be the determinant of the matrix $\tilde{\mathcal{M}}$.

$$\begin{aligned}
\Delta = {}& B_1 B_2 B_2 + B_1 B_2 + B_2 B_3 + B_3 B_1 + A_1 B_2 + A_2 B_3 + A_3 B_1 \\
& + A_1 A_2 + A_2 A_3 + A_3 A_1 (1 - A_2).
\end{aligned} \tag{28}$$

Let

$$\begin{aligned}
\Delta_1 &= A_1 A_2 + A_2 B_3 + B_3 B_1, \\
\Delta_2 &= A_2 A_3 + A_3 B_1 + B_1 B_2, \\
\Delta_3 &= A_3 A_1 + A_1 B_2 + B_2 B_3.
\end{aligned} \tag{29}$$

Then $\Delta > 0$ and $\Delta_i > 0$ for $i = 1, 2, 3$ and we have $P_0 = (\Delta_1/\Delta, \Delta_2/\Delta, \Delta_3/\Delta)$ by solving the linear system (27) for P_0 using Cramer's Rule.

Let

$$\mu_e = \frac{\Delta \left(B_1 B_2 B_3 - A_1 A_2 A_3\right)}{\Delta_1 \Delta_2 \Delta_3} + r. \tag{30}$$

Then we have the following result similar to the symmetric model.

Theorem 5: [21] *For system* (21), *assume* $\alpha_i = 1 - A_i$ *and* $\beta_i = B_i + 1$. *Then the interior equilibrium* P_0 *is locally asymptotically stable if* $\mu_e < 0$, *and unstable if* $\mu_e > 0$. *As* $r \to 0^+$, *the stability criterion approaches that of the asymmetric continuous M-L model* (19), $B_1 B_2 B_3 < A_1 A_2 A_3$.

Proof.   Similar to the proof of Theorem 3, to determine the stability of P_0, we need to consider the eigenvalues of the Jacobian matrix J_{P_0}. For the discrete asymmetric system (21), the characteristic polynomial for the Jacobian matrix evaluated at $P_0 = (p_1, p_2, p_3)$ has the form $(x - 1 + r)(x^2 + a_1 x + a_2) = 0$ because $1 - r$ is an eigenvalue that corresponding to a one-dimensional stable manifold. The coefficients a_1 and a_2 are

$$a_1 = r(p_1 + p_2 + p_3 - 1) - 2,$$

$$a_2 = 1 - r(p_1 + p_2 + p_3 - 1) + \frac{r^2}{\Delta^2}\Delta_1 \Delta_2 \Delta_3.$$

The discriminant of $x^2 + a_1 x + a_2 = 0$ is

$$a_1^2 - 4a_2 = -r^2 \left(\frac{4}{\Delta^2}\Delta_1 \Delta_2 \Delta_3 - (p_1 + p_2 + p_3 - 1)^2\right)$$

$$= -\frac{r^2}{\Delta^2}\left(4\Delta_1 \Delta_2 \Delta_3 - (A_1 A_2 A_3 - B_1 B_2 B_3)^2\right).$$

The discriminant is less than zero because all of the parameters A_i's and B_i's are positive and both negative terms $-A_1^2 A_2^2 A_3^2$ and $-B_1^2 B_2^2 B_3^2$ can be canceled out by expanding $4\Delta_1\Delta_1\Delta_3$. Therefore, the two eigenvalues satisfying $x^2 + a_1 x + a_2 = 0$ are complex numbers. And the Jury condition can be reduced to

$$|\lambda| < 1 \iff a_2 < 1.$$

This leads to

$$\mu_e = \frac{\Delta(B_1 B_2 B_3 - A_1 A_2 A_3)}{\Delta_1 \Delta_2 \Delta_3} + r < 0. \tag{31}$$

Since Δ and Δ_i's are all positive, if we let $r \to 0^+$, the inequality (31) becomes $B_1 B_2 B_3 - A_1 A_2 A_3 < 0$ which is the stability criterion for the asymmetric continuous M-L model (19). $\qquad\square$

We have shown that both eigenvalues λ and $\bar{\lambda}$ in Theorem 5 are complex numbers. The condition $|\lambda| = 1$ implies $\mu_e = 0$. At $\mu_e = 0$ we have $B_1 B_2 B_3 < A_1 A_2 A_3$. After a lengthy calculation we obtain

$$|\lambda(\mu_e)|^2 = 1 + \frac{A_1 A_2 A_3 - B_1 B_2 B_3}{\Delta} \mu_e + \frac{\Delta_1 \Delta_2 \Delta_3}{\Delta^2} \mu_e^2.$$

Then the derivative of $\lambda(\mu_e)$ evaluated at $\mu_e = 0$ is

$$\frac{d}{d\mu_e}|\lambda(\mu_e)|_0 = \frac{A_1 A_2 A_3 - B_1 B_2 B_3}{2\Delta} > 0.$$

$\frac{d}{d\mu_e}|\lambda(\mu_e)|_0$ is positive because if $\mu_e = 0$, then $B_1 B_2 B_3 - A_1 A_2 A_3 < 0$ and Δ is positive. Hence we have proved the following theorem.

Theorem 6: [21] *A Hopf bifurcation occurs at P_0 in the asymmetric M-L model (21) when $\mu_e = 0$.*

The two theorems in this section tell us that the discrete-time model (21) is far from being dynamically consistent with the continuous M-L model (19).

3.2. *Kahan's Method on May-Leonard Model*

3.2.1. *Symmetric May-Leonard competition model*

For the discrete symmetric M-L model (22) when $\alpha_i = \alpha$ and $\beta_i = \beta$ for all $i = 1, 2, 3$, let $A = 1 - \alpha$ and $B = \beta - 1$. Let $0 < h < 1$. The Jacobian matrix evaluated at the interior equilibrium P_0 is a circulant matrix [18] and has the following form

$$J_{P_0} = I + \frac{2h}{(2+h)\Phi_1(h)} \begin{pmatrix} a & b & c \\ c & a & b \\ b & c & a \end{pmatrix},$$

where

$$a = h^2(A^2 + AB + B^2) + 4h(A - B + AB) + 4(B - A + 3),$$
$$b = 2h(A + 2B + B^2) - 4(1 - A)(B - A + 3),$$
$$c = -2h(2A + B - A^2) - 4(1 + B)(B - A + 3),$$

and

$$\Phi_1(h) = h^2(A^2 + AB + B^2) + 2h(A - B)(B - A + 3) + 4(B - A + 3)^2$$
$$= h^2(A^2 + AB + B^2) + 2(B - A + 3)\big[(B - A)(2 - h) + 6\big] > 0.$$

$\Phi_1(h)$ is positive because $B - A + 3 = \alpha + \beta + 1 > 0$ and $2 - h > 0$ and either $A \geq B$ or $A \leq B$ will make one of the two equations in $\Phi_1(h)$ to be positive.

We have the following theorem that says the local stability condition is exact the same for the continuous M-L model (19) and the discrete M-L model (22).

Theorem 7: [20] *The interior equilibrium P_0 of the discrete symmetric M-L model (22) is locally asymptotically stable if $B < A$, and unstable if $B > A$. The stability criterion is the same as that of the continuous M-L model (19).*

Proof. The eigenvalues of the Jacobian matrix are $(2 - h)/(2 + h) < 1$ and two complex conjugate numbers

$$\lambda, \bar{\lambda} = \frac{1}{\Phi_1(h)}\big[-h^2(A^2 + AB + B^2) + 4(B - A + 3)^2$$
$$\pm\, \mathrm{i}\, 2\sqrt{3}\, h(A + B)(B - A + 3)\big]. \tag{32}$$

Then we have

$$|\lambda| < 1 \iff 4h(B - A)(B - A + 3)\Phi_1(h) < 0 \iff B - A < 0, \tag{33}$$

since $h > 0$, $B - A + 3 > 0$, and $\Phi(h) > 0$. The local stability of the discrete M-L model (22) is the same as the continuous M-L model (19). $\quad\square$

In the previous proof, we see that $|\lambda| = 1$ if and only if $B - A = 0$. Therefore, if we let $\lambda(\mu) = \lambda$ denote the eigenvalue given in (32) and $\mu = B - A$, then $|\lambda(\mu)| = 1$ if and only if $\mu = 0$. Rewrite λ in (32) as

$$\lambda(\mu) = \frac{1}{\Phi_1(h)}\big[-h^2(A^2 + AB + B^2) + 4(B - A + 3)^2 + \mathrm{i}\, 2\sqrt{3}\, h(A + B)(B - A + 3)\big].$$

The derivative of $|\lambda(\mu)|$ evaluated at $\mu = B - A = 0$ is

$$\frac{d|\lambda(\mu)|}{d\mu}\bigg|_{\mu=0} = \frac{2h}{12 + h^2 A^2} > 0.$$

Hence we have proved the following theorem.

Theorem 8: [20] *A Hopf bifurcation occurs at P_0 in the symmetric M-L system (22) when $B = A$.*

The Hopf bifurcation condition is exactly the same as the one for the continuous M-L model.

3.2.2. *Asymmetric May-Leonard Competition Model*

In this section, we will study the asymmetric discrete M-L model (22), where the only restrictions on α_i and β_i are given by (20).

Let $A_i = 1 - \alpha_i$ and $B_i = \beta_i - 1$, for $i = 1, 2, 3$. Let Δ, Δ_1, Δ_2, and Δ_3 are defined as (28) and (29) in the previous section. The asymmetric discrete M-L system (22) has the same interior equilibrium $P_0 = (p_1, p_2, p_3)$ as the continuous M-L model (19). The Jacobian matrix evaluated at $P_0 = (p_1, p_2, p_3)$ by a computer algebra system is

$$J_{P_0} = I + \frac{2h}{(2+h)\Phi_2(h)}\mathcal{M},$$

where

$$\Phi_2(h) = \Delta_1\Delta_2\Delta_3 h^2 + 2\Delta(A_1 A_2 A_3 - B_1 B_2 B_3)h + 4\Delta^2.$$

The elements of the matrix $\mathcal{M}$ are as follows.

$$m_{11} = 4\Delta(\Delta - \Delta_1) - 2h\Delta_2\Delta_3(\Delta_1 - A_2 B_3 - A_2 + B_3) - \Phi_2(h),$$
$$m_{12} = 2h\Delta_1\Delta_3(A_1 + B_1 + B_3 + B_1 B_3) - 4\Delta\Delta_1(1 - A_1),$$
$$m_{13} = -2h\Delta_1\Delta_2(B_1 + A_1 + A_2 - A_1 A_2) + 4\Delta\Delta_1(1 + B_1),$$
$$m_{21} = -2h\Delta_2\Delta_3(B_2 + A_2 + A_3 - A_2 A_3) + 4\Delta\Delta_2(1 + B_2),$$
$$m_{22} = 4\Delta(\Delta - \Delta_2) - 2h\Delta_1\Delta_3(\Delta_2 - A_3 B_1 - A_3 + B_1) - \Phi_2(h),$$
$$m_{23} = 2h\Delta_2\Delta_1(A_2 + B_1 + B_2 + B_1 B_2) - 4\Delta\Delta_2(1 - A_2),$$
$$m_{31} = 2h\Delta_3\Delta_2(A_3 + B_2 + B_3 + B_2 B_3) - 4\Delta\Delta_3(1 - A_3),$$
$$m_{32} = -2h\Delta_3\Delta_1(B_3 + A_1 + A_3 - A_1 A_3) + 4\Delta\Delta_3(1 + B_3),$$
$$m_{33} = 4\Delta(\Delta - \Delta_3) - 2h\Delta_1\Delta_2(\Delta_3 - A_1 B_2 - A_1 + B_2) - \Phi_2(h).$$

The function $\Phi_2(h)$ can be shown to be positive. The matrix $\mathcal{M}$ has an eigenvalue $-\Phi_2(h)$ with an associated eigenvector $P_0^T = (p_1, p_2, p_3)^T$. This can be done by a computer algebra system.

We can also show that $(2 - h)/(2 + h)$ is an eigenvalue of the Jacobian matrix. The characteristic polynomial of $\mathcal{M}$ is $(x + \Phi_2(h))(x^2 + a_1 x + a_2) = 0$, where

$$a_1 = 2(2 + h)\left[h\Delta_1\Delta_2\Delta_3 + (A_1 A_2 A_3 - B_1 B_2 B_3)\Delta\right],$$
$$a_2 = (2 + h)^2\Delta_1\Delta_2\Delta_3\,\Phi_2(h). \tag{34}$$

The discriminant of the quadratic equation $x^2 + a_1 x + a_2 = 0$ is

$$a_1^2 - 4a_2 = -4(2+h)^2 \Delta^2 \left[4\Delta_1 \Delta_2 \Delta_3 - (A_1 A_2 A_3 - B_1 B_2 B_3)^2 \right] < 0.$$

So $\mathcal{M}$ has two complex eigenvalues and so does $J_{P_0} = I + \frac{2h}{(2+h)\Phi_2(h)}\mathcal{M}$. The complex conjugates eigenvalues of $\mathcal{M}$ are

$$
\begin{aligned}
z, \bar{z} &= -a_1/2 \pm \mathrm{i}\,\sqrt{4a_2 - a_1^2}\,/2, \\
&= -(2+h)\left[h\Delta_1 \Delta_2 \Delta_3 + (A_1 A_2 A_3 - B_1 B_2 B_3)\Delta \right] \\
&\qquad \pm \mathrm{i}\,(2+h)\Delta\sqrt{4\Delta_1 \Delta_2 \Delta_3 - (A_1 A_2 A_3 - B_1 B_2 B_3)^2}. \quad (35)
\end{aligned}
$$

If $z = -a_1/2 + \mathrm{i}\,\sqrt{4a_2 - a_1^2}\,/2$ is a complex number and $k > 0$, then we have

$$
\begin{aligned}
|1 + kz| < 1 &\iff (1 - a_1 k/2)^2 + (k\sqrt{4a_2 - a_1^2}\,/2)^2 < 1 \\
&\iff k(a_2 k - a_1) < 0 \\
&\iff a_2 k - a_1 < 0.
\end{aligned}
$$

That is $|1 + kz| < 1$ if and only if $a_2 k - a_1 < 0$. We can use the results to prove the following theorem.

Theorem 9: [20] *Assume $\alpha_i = 1 - A_i$ and $\beta_i = B_i - 1$ for $i = 1, 2, 3$. Then the equilibrium $P_0 = (p_1, p_2, p_3)$ of the discrete asymmetric M-L model (22) is locally asymptotically stable if $B_1 B_2 B_3 < A_1 A_2 A_3$, and unstable if $B_1 B_2 B_3 > A_1 A_2 A_3$. The stability criterion is the same as that of the continuous M-L model (19).*

Proof. Let $k = \frac{2h}{(2+h)\Phi_2(h)}$, then $J_{P_0} = I + k\mathcal{M}$. We know that J_{P_0} has one eigenvalue $(2-h)/(2+h) < 1$ and two complex conjugate eigenvalues $1 + kz$ and $1 + k\bar{z}$, where z and $\bar{z}$ are as (35). We know that $|1 + kz| < 1$ if and only if $a_2 k - a_1 < 0$, where a_1 and a_2 are as (34). The condition $a_2 k - a_1 < 0$ can be simplified to

$$2(2+h)(B_1 B_2 B_3 - A_1 A_2 A_3)\Delta < 0.$$

Since $h > 0$ and $\Delta > 0$, the above inequality is the same as

$$B_1 B_2 B_3 - A_1 A_2 A_3 < 0.$$

Therefore P_0 is stable if $B_1 B_2 B_3 < A_1 A_2 A_3$. $\qquad\square$

We have shown that $J_{P_0} = I + k\mathcal{M}$ has two complex conjugate eigenvalues $1 + kz$ and $1 + k\bar{z}$. Let $\mu = B_1 B_2 B_3 - A_1 A_2 A_3$ and $\lambda(\mu) = 1 + kz$. Then we have $|\lambda(\mu)| = 1$ if and only if $\mu = 0$. Using similar arguments as for the symmetric case, when $\mu = 0$, the asymmetric case (22) has a 2-dimensional center manifold at P_0. When $\mu = B_1 B_2 B_3 - A_1 A_2 A_3$, we obtain

$$|\lambda(\mu)|^2 = 1 + \frac{4h\Delta(B_1 B_2 B_3 - A_1 A_2 A_3)}{\Delta_1 \Delta_2 \Delta_3 h^2 - 2h\Delta(B_1 B_2 B_3 - A_1 A_2 A_3) + 4\Delta^2}.$$

We substitute $B_3 = (\mu + A_1 A_2 A_3)/B_1 B_2$ into the above equation for $|\lambda(\mu)|^2$, then find its derivative with respect to μ. The derivative of $|\lambda(\mu)|$ evaluated at $\mu = 0$ is

$$\left. \frac{d|\lambda(\mu)|}{d\mu} \right|_{\mu=0} = \frac{2h(A_1 A_2 + B_1 B_2 + A_1 B_2)B_1 B_2}{h^2 \Delta_2^2 A_1^2 A_2 B_2 + 4\Delta_2(A_1 A_2 + B_1 B_2 + A_1 B_2)^2} > 0.$$

We have proved the following theorem.

Theorem 10: [20] *A Hopf bifurcation occurs at P_0 in the asymmetric M-L model (22) when $B_1 B_2 B_3 = A_1 A_2 A_3$.*

Like the symmetric discrete M-L model, the asymmetric discrete M-L model (22) has exactly the same conditions as the continuous M-L model for the Hopf bifurcation to occur. The model by Kahan's method (22) preserves Hopf bifurcations.

3.3. *Numerical Examples*

Figures 2, 3, and 4 show the numerical results when $\alpha = 0.4$ and β is changing from 1.3 to 1.8. For Kahan's model (22), the step size is chosen to be $h = 0.1$. For the discrete-time model (21), $r = 0.6$. All of the solutions in the figures have the same initial values $(0.1, 0.2, 0.3)$ except for Figures 2(d) and 3(d), where we used one more initial value $(0.1, 0.2, 0.8)$.

Figures 2 shows the solutions for the continuous May-Leonard model (19). When $\alpha + \beta < 2$, the interior equilibrium is stable. When $\alpha + \beta = 2$, degenerate Hopf bifurcation occurs. In Figure 2(d), when $\alpha + \beta = 2$ and a Hopf bifurcation occurs, we show two neutrally stable periodic solutions.

Figure 3 shows the solutions for the discrete-time model (22). The figures look exactly like the continuous May-Leonard system in Figure 19). But we have chosen the step size quite large $h = 0.1$. Hopf bifurcation for the model also occurs at $\alpha + \beta = 2$, Figure 3(d).

Figure 4 shows the solutions for the discrete-time model (21). Hopf bifurcation is non-degenerate. Figures 4(b) and (c) show limit cycles when $\beta = 1.4$ and $\beta = 1.5$.

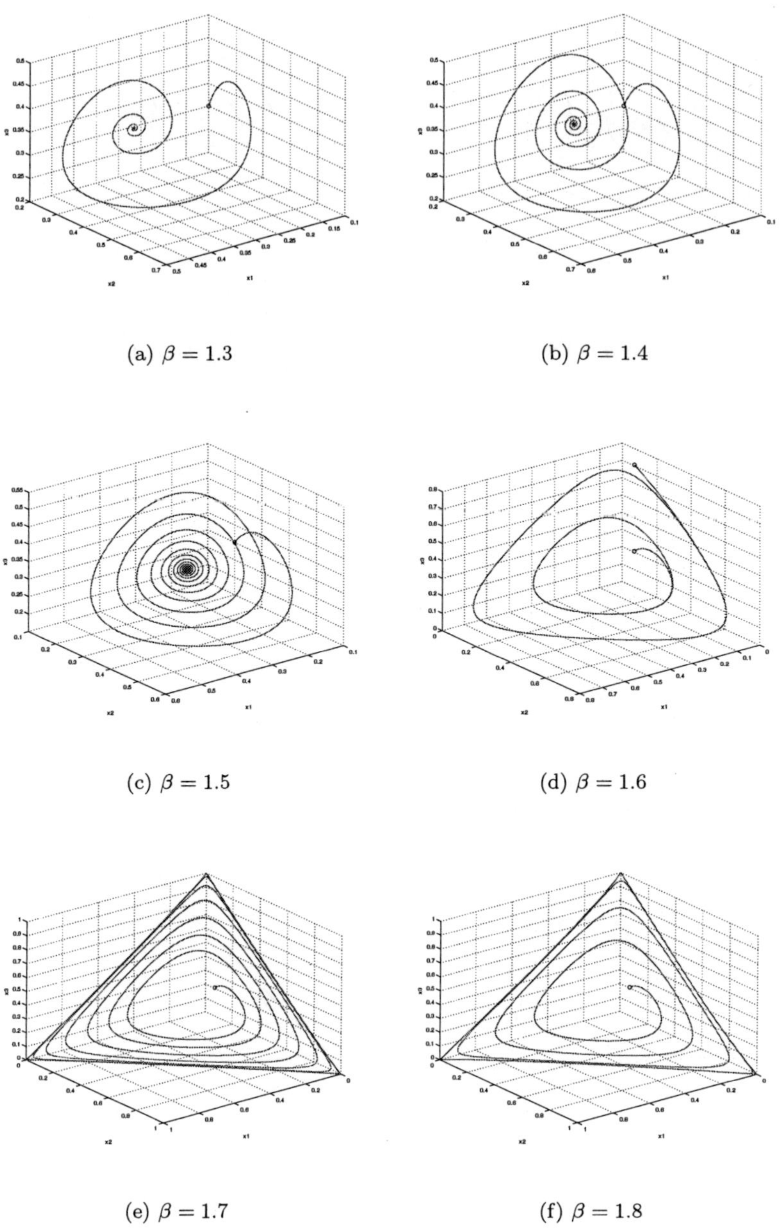

(a) $\beta = 1.3$

(b) $\beta = 1.4$

(c) $\beta = 1.5$

(d) $\beta = 1.6$

(e) $\beta = 1.7$

(f) $\beta = 1.8$

Fig. 2. The continuous M-L model (19), $\alpha = 0.4$.

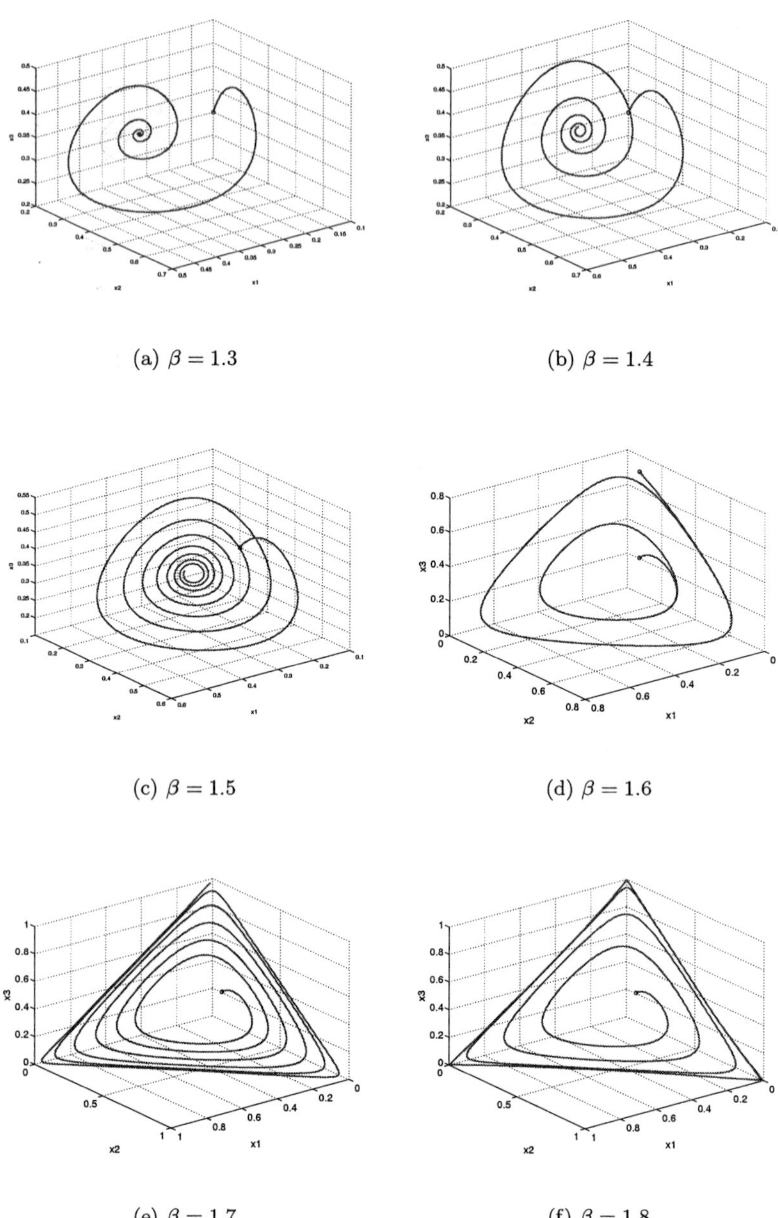

Fig. 3. The discrete M-L model by Kahan's method (22), $\alpha = 0.4$ and $h = 0.1$.

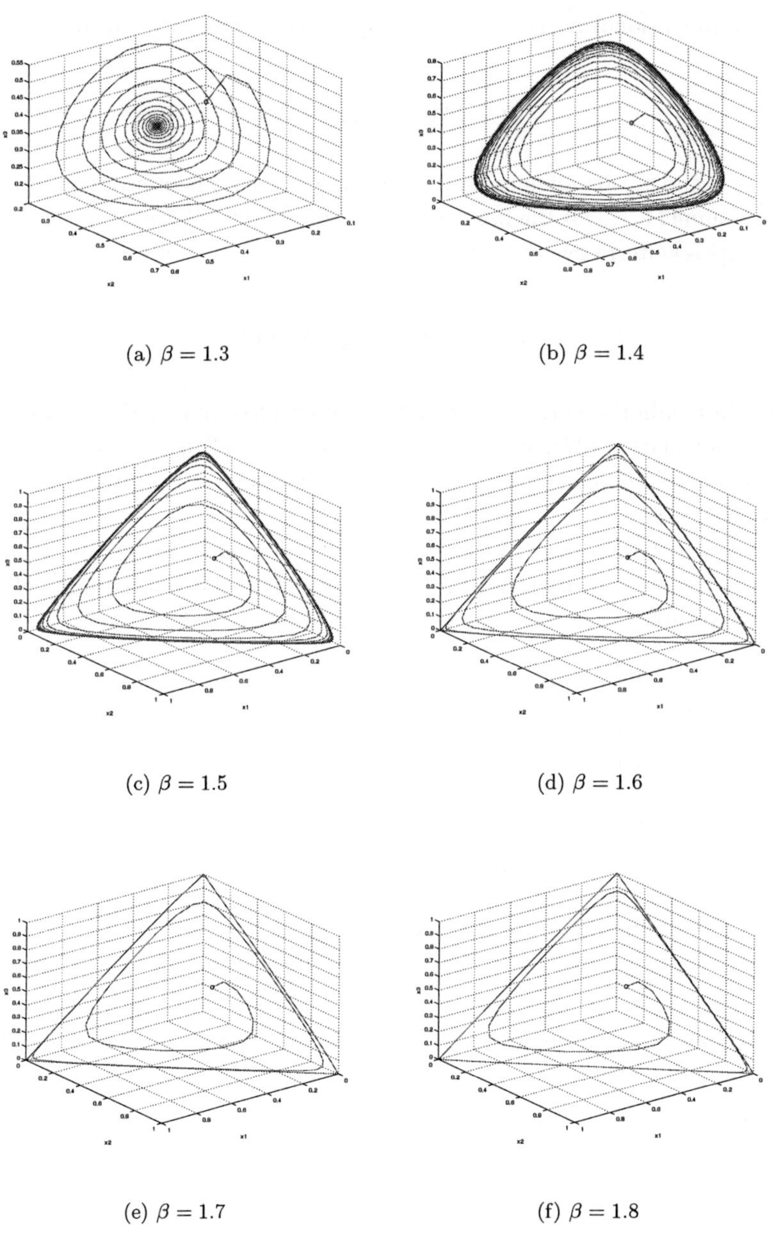

(a) $\beta = 1.3$

(b) $\beta = 1.4$

(c) $\beta = 1.5$

(d) $\beta = 1.6$

(e) $\beta = 1.7$

(f) $\beta = 1.8$

Fig. 4. The discrete M-L model (21), $\alpha = 0.4$ and $r = 0.6$.

4. Three-Dimensional Lotka-Volterra Competitive System

The 3-dimensional Lotka-Volterra competitive system is given by

$$u_i' = u_i\left(r_i - \sum_{j=1}^{3} a_{ij} u_j\right), \tag{36}$$

where r_i and a_{ij} are all nonnegative. Assume there is an interior equilibrium and assume $(1,1,1)$ is the interior equilibrium of (36). Set $x_i = u_i - 1$. Then the Lotka-Volterra system (36) becomes

$$x_i' = -(1 + x_i)\left(\sum_{j=1}^{3} a_{ij} x_j\right). \tag{37}$$

Now the equilibrium $(1,1,1)$ of equation (36) corresponds to the equilibrium $\mathbf{x} = 0$ of equation (37). We will discretize system (37) using three different methods and compare the local dynamics and Hopf bifurcations near the equilibrium $\mathbf{x} = 0$.

The Jacobian matrix of the continuous L-V model (37) at the origin is the matrix $-\mathcal{A} = -(a_{ij})$. Let the trace, the sum of the determinants of the principal submatrices, and the determinant of $\mathcal{A}$ be denoted as

$$T = a_{11} + a_{22} + a_{33},$$
$$M = a_{11}a_{22} - a_{12}a_{21} + a_{11}a_{33} - a_{13}a_{31} + a_{22}a_{33} - a_{23}a_{32}, \tag{38}$$
$$D = \det(\mathcal{A}).$$

Then the equilibrium $\mathbf{x} = 0$ is stable if

$$T > 0, D > 0, \text{ and } D < TM,$$

and a Hopf bifurcation occurs at $\mathbf{x} = 0$ if

$$T > 0, D > 0, \text{ and } D = TM.$$

Applying Kahan's method to model (37), we have the analogous discrete L-V competition model:

$$X_i = \frac{2x_i - h\sum_{j=1}^{3} a_{ij}x_j - h(1 + x_i)\sum_{j\neq i} a_{ij}X_j}{2 + ha_{ii}(1 + x_i) + h\sum_{j=1}^{3} a_{ij}x_j}, i = 1, 2, 3, \tag{39}$$

where $X_i = x_i(k + 1) = x_i(t + h)$ and $x_i = x_i(k) = x_i(t)$.

The second method to obtain a discrete analog of model (37) is to consider a variation with piecewise constant arguments for certain terms on the right side of the system

$$\frac{dx_i(t)}{dt} = -\left(1 + x_i(t)\right)\left(\sum_{j=1}^{3} a_{ij}x_j(\lceil t \rceil)\right), \ 0 \le k \le t < k+1, \ i = 1,2,3,$$

where $\lceil t \rceil$ denotes the greatest integer in t. Integrating both sides of equations on $[k, k+1]$ and letting $t \to k+1$, then we obtain the following discrete model

$$X_i = \left(1 + x_i\right) \exp\left(-\sum_{j=1}^{3} a_{ij}x_j \right) - 1, \ i = 1,2,3.\dots \tag{40}$$

The third discrete L-V model is derived from model (37) by a method proposed by Liu and Elaydi [10] using Mickens's method [14] as the following

$$\frac{X_i - x_i}{\phi_i(h)} = -\left(1 + X_i\right)\left(\sum_{j=1}^{3} a_{ij}x_j \right), \ i = 1,2,3,$$

where $\phi_i(h) = h + o(h)$ and $0 < \phi_i(h) < 1$ for $i = 1,2,3$. To simplify our analysis, we assume $\phi_i(h) = \phi(h)$ for all $i = 1,2,3$. Then the above model can be written as

$$X_i = \left(x_i - \phi(h) \sum_{j=1}^{3} a_{ij}x_j \right) \Big/ \left(1 + \phi(h) \sum_{j=1}^{3} a_{ij}x_j \right), \ i = 1,2,3. \tag{41}$$

The four systems (37), (39), (40), and (41), have the same equilibria. We will analyze the local behaviors near the equilibrium $\mathbf{x} = 0$ for all of the discrete L-V models and compare them with the continuous L-V model. Note that the Lotka-Volterra model (36) is positively invariant, meaning that given positive initial values, the solution will remain positive. This is an important issue due to the biological background of the model. The two discrete models, (40) and (41), are also positively invariant. However, Kahan's scheme does not preserve this property unless h is sufficiently small.

4.1. *Kahan Competitive Model*

The Kahan competitive model (39) has the same equilibria as the L-V model (37). To understand the local stability of the equilibrium $\mathbf{x} = 0$ for model (39), we evaluate the Jacobian matrix at $\mathbf{x} = 0$. If all of the eigenvalues of the Jacobian matrix satisfy $|\lambda| < 1$, then $\mathbf{x} = 0$ is stable.

If $p(\lambda) = \lambda^3 + a_1\lambda^2 + a_2\lambda + a_3$ is the characteristic equation for a matrix of order 3. Then Jury condition [4] says that all of the roots of $p(\lambda)$ satisfy $|\lambda| < 1$ if

$$p(1) > 0, \ p(-1) < 0, \ |a_3| < 1, \text{ and } |1 - a_3^2| > |a_2 - a_3 a_1|.$$

The following theorem says that Kahan's method preserves local stability for the Lotka-Volterra competitive system (37). The stability criteria of the equilibrium $\mathbf{x} = 0$ are the same for (37) and (39).

Theorem 11: [22] *For the Kahan competitive model* (39), *regardless of the step size* h $(0 < h < 1)$, *the equilibrium* $\mathbf{x} = 0$ *is stable if* $T > 0$, $D > 0$, *and* $D < TM$.

Proof. For model (39), the Jacobian matrix evaluated at $\mathbf{x} = 0$ is

$$J = I - \frac{2h}{\Phi(h)}(4\mathcal{A} + 2h\mathcal{B} + Dh^2 I),$$

where

$$\Phi(h) = 8 + 4Th + 2Mh^2 + Dh^3, \tag{42}$$

$\mathcal{A} = (a_{ij})$, T, M, and D are defined as in (38) and matrix

$$\mathcal{B} = \begin{pmatrix} c_{22} + c_{33} & c_{21} & c_{31} \\ c_{12} & c_{33} + c_{11} & c_{32} \\ c_{13} & c_{23} & c_{11} + c_{22} \end{pmatrix},$$

c_{ij}'s are the minors obtained from the determinant of the sub-matrix of $\mathcal{A}$ by removing all elements of the ith row and jth column of $\mathcal{A}$.

The characteristic equation of the matrix J is $p(\lambda) = \lambda^3 + a_1\lambda^2 + a_2\lambda + a_3$, where

$$\begin{aligned}
a_1 &= (-24 - 4Th + 2Mh^2 + 3Dh^3)/\Phi(h), \\
a_2 &= (24 - 4Th - 2Mh^2 + 3Dh^3)/\Phi(h), \\
a_3 &= (-8 + 4Th - 2Mh^2 + Dh^3)/\Phi(h).
\end{aligned} \tag{43}$$

Applying the Jury conditions, the equilibrium $\mathbf{x} = 0$ is stable if

$$\begin{aligned}
&8Dh^3/\Phi(h) > 0, \\
&-64/\Phi(h) < 0, \\
&-8h(4 + Mh^2)(4T + Dh^2)/\Phi(h)^2 < 0, \\
&1024h^4(TM - D)(16T + 8Dh^2 + DMh^4)/\Phi(h)^4 > 0.
\end{aligned} \tag{44}$$

We would like to show that the four inequalities in (44) are satisfied if and only if $T > 0$, $D > 0$, and $D < TM$. If $T > 0$, $D > 0$, and $D < TM$, then $M > 0$ and all of the inequalities in (44) follow immediately since $0 < h < 1$.

Suppose all of the four inequalities in (44) are satisfied. Then following the second inequality we have $\Phi(h) > 0$. Then from the first inequality we get $D > 0$. Since T is the trace of $\mathcal{A}$ and all of the elements of $\mathcal{A}$ are nonnegative, $T \geq 0$. Following the third inequality, we obtain $(4 + Mh^2)(4T + Dh^2) > 0$. And since $T \geq 0$, $4T + Dh^2 > 0$. This implies $4 + Mh^2 > 0$.

Following $4T + Dh^2 > 0$, one of the factors in the fourth inequality will be positive; $16T + 8Dh^2 + DMh^4 = 16T + 4Dh^2 + Dh^2(4T + Dh^2) > 0$. So we get $D < TM$. Then, of course, we should have $T > 0$. $\square$

From Theorem 11, we see that if $T > 0$ and $D > 0$ then the stability of the equilibrium $\mathbf{x} = 0$ of the Kahan model (39) depends on $D - TM$. If $D < TM$, then $\mathbf{x} = 0$ is stable, and if $D > TM$, unstable. At $D = TM$, a Hopf bifurcation may occur.

To show a Hopf bifurcation occurs, we need to restrict the system to its center manifold and find its normal form [25] on the center manifold. We are interested in the existence of a Hopf bifurcation; therefore, only the linear part of the model is needed.

Let $\mu = D - TM$. If the characteristic equation $p(\lambda) = 0$ has one real and two complex conjugate roots such that the complex roots satisfy $|\lambda| = 1$, then the real root is $-a_3$ and the complex roots are

$$\lambda, \bar{\lambda} = \frac{8 + 4Th - 2Mh^2 - Dh^3}{\Phi(h)} + \mathrm{i}\,\frac{4h\sqrt{4M - 2TMh + 6Dh + TDh^2}}{\Phi(h)},$$

where $\Phi(h)$ is as in (42). It is easy to show that $|\lambda| = 1$ if and only if $\mu = 0$. When $\mu = 0$, the three roots are

$$\lambda_1 = \frac{2 - Th}{2 + Th}, \quad \lambda, \bar{\lambda} = \frac{4 - Mh^2}{4 + Mh^2} \pm \mathrm{i}\,\frac{4h\sqrt{M}}{4 + Mh^2}.$$

Since the coefficients of the characteristic equation $p(\lambda)$, a_1, a_2, and a_3, are all continuous functions of μ, the roots of $p(\lambda) = 0$ are also continuous functions of μ. Therefore, there exists a neighborhood of $\mu = 0$ such that $p(\lambda) = 0$ has one real root λ_1, $|\lambda_1| < 1$, and two complex conjugate roots, λ and $\bar{\lambda}$. On its center manifold near $\mathbf{x} = 0$ and $\mu = 0$ we should have a 2-dimensional system that is similar to (26)

$$\begin{pmatrix} W_1 \\ W_2 \end{pmatrix} = \begin{pmatrix} \mathrm{Re}(\lambda) & -\mathrm{Im}(\lambda) \\ \mathrm{Im}(\lambda) & \mathrm{Re}(\lambda) \end{pmatrix} \begin{pmatrix} w_1 \\ w_2 \end{pmatrix} + \text{h.o.t.} \tag{45}$$

The local dynamics near $\mathbf{x} = 0$ at $\mu = 0$ of the Kahan competitive model (39) are the same as the dynamics near the origin of the 2-dimensional system (45). On the center manifold we have the following theorem.

Theorem 12: [22] *For the Kahan competitive model* (39), *assume* $T > 0$ *and* $D > 0$. *Then a Hopf bifurcation occurs at* $\mathbf{x} = 0$ *if* $D = TM$. *The Hopf bifurcation criterion is the same as that of the continuous model* (37).

Proof. According to the Poincaré-Andronov-Hopf Bifurcation for maps [6, p. 474, Theorem 15.31], there are four conditions to verify (i)–(iv). Let $\mu = D - TM$ and F be defined by the 2-dimensional system (45), such that $F : \mathbb{R} \times \mathbb{R}^2 \to \mathbb{R}^2$; $(\mu, \mathbf{w}) \to F(\mu, \mathbf{w})$. Then, condition (i) holds: $F(\mu, 0) = 0$ for μ near 0.

In the system (45), λ is a continuous function of μ near $\mu = 0$. It is not difficult to show that $|\lambda(\mu)| = 1$ if and only if $\mu = 0$. At $\mu = 0$, we have $M = D/T > 0$ and

$$\lambda(0) = \frac{4 - Mh^2}{4 + Mh^2} + \mathrm{i}\,\frac{4h\sqrt{M}}{4 + Mh^2}.$$

Therefore, in the neighborhood of $\mu = 0$ ($M \approx D/T$), the imaginary part of λ is not zero. Since $0 < h < 1$, we can choose h small enough, such that the real part of $\lambda(0)$ is positive. Then the eigenvalues $\lambda(0)$ and $\bar{\lambda}(0)$ cannot be equal to ± 1, $\pm \mathrm{i}$, or $(-1 \pm \sqrt{3}\,\mathrm{i})/2$, which are the possible roots of $x^k = 1$, for $k = 1, 2, 3, 4$. Therefore, conditions (ii) and (iv) hold: (ii) $DF(\mu, 0)$ has two non-real eigenvalues $\lambda(\mu)$ and $\bar{\lambda}(\mu)$ for μ near 0, with $|\lambda(0)| = 1$, (iv) $\lambda^k(0) \neq 1$ for $k = 1, 2, 3, 4$.

Next we show that condition (iii) $d|\lambda(\mu)|/d\mu > 0$ is also satisfied. Substitute $D = \mu + TM$ into $p(\lambda)$. At $\mu = 0$, the two complex roots are $\alpha_0 \pm \mathrm{i}\,\beta_0$, where

$$\alpha_0 = \frac{4 - Mh^2}{4 + Mh^2}, \quad \beta_0 = \frac{4h\sqrt{M}}{4 + Mh^2}.$$

In the neighborhood of $\mu = 0$, the two complex conjugate roots $\alpha \pm \mathrm{i}\,\beta$ are

$$\alpha = \alpha_0 + \alpha_1 \mu + O(\mu^2), \ \beta = \beta_0 + \beta_1 \mu + O(\mu^2),$$

where α_1 and β_1 can be found to be

$$\alpha_1 = \frac{2h(4 - 4Th - Mh^2)}{(4 + Mh^2)^2(T^2 + M)},$$

$$\beta_1 = \frac{2h(4T + 4Mh - TMh^2)}{(4 + Mh^2)^2(T^2 + M)M^{1/2}}.$$

Then

$$\frac{d|\lambda(\mu)|}{d\mu}\bigg|_{\mu=0} = \alpha_0\alpha_1 + \beta_0\beta_1 = \frac{2h}{(4+Mh^2)(T^2+M)} > 0.$$

Therefore, we conclude the condition (iii) holds: $d|\lambda(\mu)|/d\mu > 0$ at $\mu = 0$. Hence at $\mathbf{x} = 0$ in model (39), there is a Hopf bifurcation. $\square$ Theorem 12 says that Kahan's discretization method preserves Hopf bifurcations for 3-dimensional competitive Lotka-Volterra model. Therefore, it also preserves periodic solutions.

4.2. *The Discrete Model of the Exponential Type*

For the discrete competitive L-V model of the exponential type (40). The Jacobian matrix evaluated at the equilibrium $\mathbf{x} = 0$ is

$$J = I - \mathcal{A}.$$

Its characteristic equation is $p(\lambda) = \lambda^3 + a_1\lambda^2 + a_2\lambda + a_3$, where

$$a_1 = -3 + T, \ a_2 = 3 - 2T + M, \ \text{and} \ a_3 = -1 + T - M + D. \tag{46}$$

T, M, and D are defined as (38). We also have a similar theorem considering the stability of the fixed point $\mathbf{x} = 0$ for model (40). But this theorem shows that the discrete model of exponential type (40) is not dynamically consistent with the continuous L-V model (37). Their stability criteria are quite different.

Theorem 13: [22] *For the discrete Lotka-Volterra model of exponential type (40), the equilibrium $\mathbf{x} = 0$ is locally asymptotically stable if (i) $D > 0$, (ii) $8 - 4T + 2M - D > 0$, (iii) $0 < T - M + D < 2$, and (iv) $D < (T - M + D)(M - D)$.*

Proof.   Proof of this theorem is similar to Theorem 11. Using the Jury conditions, the four inequalities for $\mathbf{x} = 0$ to be stable are

$$D > 0, \ -8 + 4T - 2M + D < 0, \ 0 < T - M + D < 2, \ \text{and}$$

$$\left[(T - M + D)(M - D) - D\right]\left[(T - M + D)(4 - 2T + M - D) + D\right] > 0.$$

Then if the first three inequalities are satisfied, condition (iv) can be deduced from the fourth inequality. $\square$

Using similar arguments we had for the Kahan competitive model (39), when $\mu = D - (T - M + D)(M - D) = 0$, the discrete model (40) has a two-dimensional center manifold at $\mathbf{x} = 0$. A Hopf bifurcation occurs on the center manifold near $\mathbf{x} = 0$ at $\mu = 0$ for the discrete model (40).

Theorem 14: **[22]** *For the discrete Lotka-Volterra model of the exponential type (40), assume $D > 0$, $8 - 4T + 2M - D > 0$, and $0 < T - M + D < 2$. Then a Hopf bifurcation occurs at $\mathbf{x} = 0$ if $D = (T - M + D)(M - D)$.*

Proof.   The proof is similar to Theorem 12. We will only show that $d|\lambda(\mu)|/d\mu \neq 0$ at $\mu = 0$. Let

$$\mu = D - (T - M + D)(M - D).$$

Substitute $T = M - D + (D - \mu)/(M - D)$ into the characteristic equation $p(\lambda) = \lambda^3 + a_1\lambda^2 + a_2\lambda + a_3 = 0$, where a_1, a_2, and a_3 are as in (46). Then, at $\mu = 0$, the two complex roots are $\alpha_0 \pm i\,\beta_0$, where

$$\alpha_0 = \frac{2 - M + D}{2}, \quad \beta_0 = \frac{\sqrt{(M - D)(4 - M + D)}}{2}.$$

In the neighborhood of $\mu = 0$, the two complex conjugate roots $\alpha \pm i\beta$ are

$$\alpha = \alpha_0 + \alpha_1\mu + O(\mu^2), \quad \beta = \beta_0 + \beta_1\mu + O(\mu^2),$$

where α_1 and β_1 can be found to be

$$\alpha_1 = \frac{(M - 2D)(M - D)}{2K_0},$$

$$\beta_1 = \frac{(M - D)(M^2 - 3MD + 2D + 2D^2)}{2K_0[(M - D)(4 - M + D)]^{1/2}},$$

and $K_0 = D^2 + (M - 2D)(M - D)^2$. Then

$$\left.\frac{d|\lambda(\mu)|}{d\mu}\right|_{\mu=0} = \alpha_0\alpha_1 + \beta_0\beta_1 = \frac{(M - D)^2}{2K_0}.$$

The above equation is zero only if $M = D$. But if $M = D$, then at $\mu = 0$ we obtain $D = 0$, a contradiction. Therefore $d|\lambda(\mu)|/d\mu \neq 0$ at $\mu = 0$. $\square$

4.3. *The Discrete Model of the Fractional Type*

For the discrete L-V model of the fractional type (41), we denote ϕ as $\phi(h)$. The Jacobian matrix evaluated at the equilibrium $\mathbf{x} = 0$ is

$$J = I - \phi\mathcal{A}.$$

Its characteristic equation is $p(\lambda) = \lambda^3 + a_1\lambda^2 + a_2\lambda + a_3$, where

$$a_1 = -3 + T\phi, \quad a_2 = 3 - 2T\phi + M\phi^2, \quad a_3 = -1 + T\phi - M\phi^2 + D\phi^3, \quad (47)$$

where T, M, and D are also defined as in (38). The following theorem is similar to Theorems 11 and 13.

Theorem 15: [22] *For the discrete Lotka-Volterra model of the fractional type* (41), *the equilibrium* $\mathbf{x} = 0$ *is locally asymptotically stable if (i)* $D > 0$, *(ii)* $8 - 4T\phi + 2M\phi^2 - D\phi^3 > 0$, *(iii)* $0 < (T - M\phi + D\phi^2)\phi < 2$, *and (iv)* $D < (T - M\phi + D\phi^2)(M - D\phi)$.

Proof. The proof of Theorem 15 is similar to Theorem 11. Using the Jury conditions, the equilibrium $\mathbf{x} = 0$ is stable if $D\phi^3 > 0$, $8 - 4T\phi + 2M\phi^2 - D\phi^3 > 0$, $|1 - T\phi + M\phi^2 - D\phi^3| < 1$, and

$$[(T - M\phi + D\phi^2)(M - D\phi) - D]$$
$$\times [(T - M\phi + D\phi^2)(4 - 2T\phi + M\phi^2 - D\phi^3) + D\phi^2]\phi^4 > 0.$$

Then if the first three inequalities are satisfied, condition (iv) can be deduced from the fourth inequality.

The stability criteria are different from that of the continuous model (37). As $\phi \to 0^+$ or $\phi \to 1^-$, the stability criteria approach that of the Kahan model (39) and the discrete model of the exponential type (40) respectively. $\qquad\square$

Using similar arguments we had for Kahan competitive model (39), when $D - (T - M\phi + D\phi^2)(M - D\phi) = 0$, the discrete model (41) has a two-dimensional center manifold at $\mathbf{x} = 0$. The following theorem says a Hopf bifurcation occurs at the origin for the discrete model (41).

Theorem 16: [22] *For the discrete Lotka-Volterra model of the fractional type* (41), *assume* $D > 0$, $8 - 4T\phi + 2M\phi^2 - D\phi^3 > 0$, *and* $0 < (T - M\phi + D\phi^2)\phi < 2$. *Then a Hopf bifurcation occurs at* $\mathbf{x} = 0$ *if*

$$D = (T - M\phi + D\phi^2)(M - D\phi).$$

Proof. The proof is similar to proofs of Theorems 12 and 14. We will only that $d|\lambda(\mu)|/d\mu \neq 0$ at $\mu = 0$. Let

$$\mu = D - (T - M\phi + D\phi^2)(M - D\phi). \qquad (48)$$

Substitute $T = M\phi - D\phi^2 + (D - \mu)/(M - D\phi)$ into the characteristic equation $p(\lambda) = 0$. Then, at $\mu = 0$, the two complex roots are $\alpha_0 \pm i\,\beta_0$, where

$$\alpha_0 = \frac{2 - M\phi^2 + D\phi^3}{2}, \beta_0 = \frac{\phi\sqrt{(M - D\phi)(4 - M\phi^2 + D\phi^3)}}{2}.$$

In the neighborhood of $\mu = 0$, the two complex conjugate roots $\alpha \pm i\beta$ are

$$\alpha = \alpha_0 + \alpha_1\mu + O(\mu^2), \quad \beta = \beta_0 + \beta_1\mu + O(\mu^2),$$

where α_1 and β_1 can be found to be

$$\alpha_1 = \frac{(M - 2D\phi)(M - D\phi)\phi}{2K_1},$$

$$\beta_1 = \frac{(M - D\phi)(2D + M^2\phi - 3MD\phi^2 + 2D^2\phi^3)\phi}{2K_1[(M - D\phi)(4 - M\phi^2 + D\phi^3)]^{1/2}},$$

and

$$K_1 = D^2 + (M - 2D\phi)(M - D\phi)^2.$$

Then

$$\left.\frac{d|\lambda(\mu)|}{d\mu}\right|_{\mu=0} = \alpha_0\alpha_1 + \beta_0\beta_1 = \frac{(M - D\phi)^2\phi}{2K_1}.$$

The above equation is zero only if $M = D\phi$. But if $M = D\phi$, then at $\mu = 0$ we obtain $D = 0$, a contradiction. Therefore $d|\lambda(\mu)|/d\mu \neq 0$ at $\mu = 0$.

As $\phi \to 0^+$, the Hopf bifurcation criterion μ in (48) approaches $D - TM$ and as $\phi \to 1^-$, μ approaches $D - (T - M + D)(M - D)$. The Hopf bifurcation criteria approach that of the Kahan model (39) and the model of the exponential type (40) respectively.   □

We see that the discrete model of the fractional type (41) is a model between Kahan's discrete model (39) and the discrete model of the exponential type (40).

5. Conclusions and Open Problems

Kahan's method is symplectic with a noncanonical structure. It preserves the local stabilities and Hopf bifurcations of 2-dimensional and 3-dimensional Lotka-Volterra competitive systems.

For 3-dimensional Lotka-Volterra system, the two nonstandard methods (40) and (41) preserve the positivity of solutions but do not preserve local dynamics.

For 2-dimensional Lotka-Volterra systems in the form (9), we have generalized W. Kahan's method and produced a class of symplectic numerical methods that give solutions that do not spiral. Some open problems concerning the symplectic methods are

(1) Besides Kahan's method, which symplectic methods can be applied to three-dimensional systems?

(2) How should one proceed if the Lotka-Volterra system is modified to include x^2 and/or y^2 terms?
(3) Do all symplectic methods preserve local stability and Hopf bifurcations?
(4) The symplectic method generated by interchanging $x \leftrightarrow Y$ and $y \leftrightarrow X$ does not give the Euler symplectic method and Mickens's symplectic method. How can we find the most general class that are symplectic?
(5) How do we construct symplectic methods for the Leslie predator-prey model?

$$x' = x(r_1 - a_{11}x - a_{12}y),$$
$$y' = y(r_2 - a_{22}\frac{y}{x}).$$

Acknowledgments. The author would like to thank Ronald Mickens for inviting her to contribute to this volume, and for his constant support.

References

1. H. Al-Kahby, F. Dannan, and S. Elaydi. Non-standard discretization methods for some biological models. In R. Mickens, editor, *Applications of Nonstandard Finite Difference Schemes*, chapter 4, pages 155–180. World Scientific, New Jersey, 2000.
2. L. J. S. Allen. Some discrete-time SI, SIR, and SIS epidemic models. *Mathematical Biosciences*, 124: 83–105, 1994.
3. C.-W. Chi, S.-B. Hsu, and L.-I. Wu. On the asymmetric May-Leonard model of three competing species. *SIAM J. Appl. Math.*, 58(1): 211–226, 1998.
4. L. Edelstein-Keshet. *Mathematical Models in Biology*. McGraw-Hill, 1988.
5. M. J. Gander and R. Meyer-Spasche. An introduction to numerical integrators preserving physical properties. In R. Mickens, editor, *Applications of Nonstandard Finite Difference Schemes*, chapter 5, pages 181–243. World Scientific, New Jersey, 2000.
6. J. Hale and H. Koçak, *Dynamics and Bifurcations*, Springer-Verlag, New York, 1991.
7. J. Hofbauer, V. Hutson, and W. Jansen. Coexistence for systems governed by difference equations of Lotka-Volterra type. *J. Math. Biol.*, 25:553–570, 1987.
8. S. Jang and S. N. Elaydi. Difference equations from discretization of a continuous epidemic model with immigration of infectives. *Canadian Applied Mathematics Quarterly*, 11(2), 2005.
9. H. Jiang and T. D. Rogers. The discrete dynamics of symmetric competition in the plane. *J. Math. Biol.*, 25:573–596, 1987.
10. P. Liu and S. N. Elaydi. Discrete competitive and cooperative models of Lotka-Volterra type. *J. Comp. Anal. Appl.*, 3 (2001) 53–73.

11. Z. Lu and W. Wang. Permanence and global attractivity for Lotka-Volterra difference systems. *J. Math. biol.*, 39:269–282, 1999.

12. S. M. Moghadas, M. E. Alexander, and B. D. Corbett. A non-standard numerical scheme for a generalized Gause-type predator-prey model. *Physica D*, 188: 134–151, 2004.

13. R. M. May and W. J. Leonard. Nonlinear aspects of competition between three species. *SIAM J. Appl. Math.*, 29 (1975) 243–253.

14. R. E. Mickens. *Nonstandard Finite Difference Models of Differential Equations.* World Scientific, Singapore, 1994.

15. R. E. Mickens. A nonstandard finite-difference scheme for the Lotka-Volterra system. *Applied Numerical Mathematics*, 45: 309–314, 2003.

16. A. S. Mounim and B. de Dormale. A note on Mickens's finite-difference scheme for the Lotka-Volterra system. *Applied Numerical Mathematics*, 51: 341–344, 2004.

17. J. D. Murray, *Mathematical Biology*, Springer-Verlag, New York, 1993.

18. J. Ortega, *Matrix Theory*, Plenum Press, New York, 1987.

19. L.-I. W. Roeger. Local stability of Euler's and Kahan's methods. *J. Diff. Equ. Appl.*, 10:601–614, 2004.

20. L.-I. W. Roeger. Discrete May-Leonard competition models III. *J. Diff. Equ. Appl.*, 10:773–790, 2004.

21. L.-I. W. Roeger. Discrete May-Leonard competition models II. *Discrete and Continuous Dynamical Systems Series B*, 5(3):841–860, 2005.

22. L.-I. W. Roeger. A nonstandard discretization method for Lotka-Volterra models that preserves periodic solutions. To appear in JDEA.

23. L.-I. W. Roeger. A class of nonstandard symplectic discretization methods for Lotka-Volterra system. submitted.

24. J. M. Sanz-Serna. An unconventional symplectic integrator of W. Kahan. *Appl. Numerical Math.*, 16:245–250, 1994.

25. S. Wiggins. *Introduction to Applied Nonlinear Dynamical Systems and Chaos*, volume 2 of *Texts in Applied Mathematics*. Springer-Verlag, New York, 1990.